U0300999

“十三五”
国家重点图书

Series on Advanced Electronic Packaging Technology and Key Materials

先进电子封装技术与关键材料丛书

汪正平（C.P. Wong） 刘胜（Sheng Liu） 朱文辉（Wenhui Zhu） 主编

Advanced Polyimide Materials

Synthesis, Characterization and Applications

先进聚酰亚胺材料

——合成、表征及应用

杨士勇（Shi-yong Yang） 主编

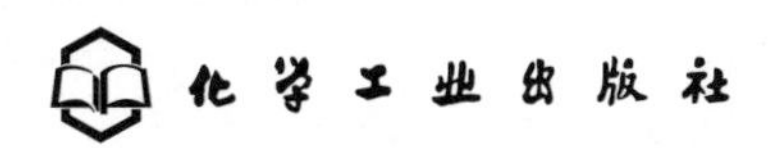

·北京·

内 容 简 介

先进聚酰亚胺材料，由于具有优异的耐热性、综合力学性能、电学性能和化学稳定性等，在航空、航天、微电子、平面显示、电气绝缘等高技术领域具有广泛的应用前景。

本书系统介绍了先进聚酰亚胺材料的合成、性能和应用。包括先进聚酰亚胺薄膜、先进聚酰亚胺纤维、碳纤维复合材料用树脂基体、超级工程塑料和泡沫材料、微电子用聚酰亚胺材料，还介绍了气体分离膜、质子交换膜、可溶性和低 k 值聚酰亚胺材料。

本书可供化学工业、高分子材料、航空航天、微电子制造与封装、平面显示等领域的相关研究人员、工程师、研究生以及高年级的本科生参考使用。

图书在版编目（CIP）数据

先进聚酰亚胺材料：合成、表征及应用＝Advanced Polyimide Materials：Synthesis，Characterization and Applications：英文/杨士勇主编. —北京：化学工业出版社，2020.12
（先进电子封装技术与关键材料丛书/汪正平，刘胜，朱文辉主编）
ISBN 978-7-122-33498-5

Ⅰ.①先… Ⅱ.①杨… Ⅲ.①聚酰亚胺-英文 Ⅳ.①TQ323.7

中国版本图书馆 CIP 数据核字（2018）第 294554 号

责任编辑：吴　刚　　　装帧设计：关　飞
责任校对：宋　玮

出版发行：化学工业出版社(北京市东城区青年湖南街 13 号　邮政编码 100011)
印　　装：中煤（北京）印务有限公司
710mm×1000mm　1/16　印张 29¼　字数 1026 千字　2020 年 12 月北京第 1 版第 1 次印刷

购书咨询：010-64518888　　　售后服务：010-64518899
网　　址：http://www.cip.com.cn
凡购买本书，如有缺损质量问题，本社销售中心负责调换。

定　　价：298.00 元

Preface to the Series

The technical level and development scale of the integrated circuit (IC) industry is one of the important indicators to measure a country's industrial competitiveness and comprehensive national strength, and is the source of modern economic development. The application of IC has already become routine in various industries, such as military satellites, radar, civilian automotive electronics, smart equipment, and consumer electronics,ect. Atpresent, the IC industry has formed three major industrial chains of design, manufacturing and packaging testing, which have become the indispensable pillar in the IC industry.

IC packaging is an indispensable process in the IC industry, which is the bridge from chip to device and device to system. It is a key fundamental manufacturing part of the IC industry and a competitive commanding height for the core device manufacturing of the IC industry.

With the rapid development of IC technology, higher and higher requirements for miniaturization, multi-function, high reliability and low cost of electronic products are put forward. Facing this situation, the electronic packaging materials and technologies are undergoing rapid development, promoting lots of advanced packaging materials. Advanced electronic packaging materials and technologies are the core of IC packaging.

In order to promote the development of China's advanced electronic packaging industry and meet the urgent needs of researchers ranged from teaching and scientific study to engineering developing in the field of electronic packaging, the editorial committee has invited famous specialists to write the *series on Advanced Electronic Packaging Technology and Key Materials* in recent years (English version). The Seriesincludes: "*Advanced Polyimide Materials*", "*LED Packaging Technologies*", "*Freeform Optics for LED Packages and Applications*", "*Thermal Management of IC/LED Packages and Applications*", "*Modeling, Analysis, Design, and Tests for Electronics Packaging beyond Moore*","*Ultrasonic Technologies in Advanced Electronic Packaging*", "*TSV(through-silicon via technology) Package*" *etc.*

This series of books systematically describes the advanced electronic packaging from three aspects: advanced packaging materials, advanced packaging technologies and advanced packaging simulation design methods. This series covers the most advanced packaging materials such as polyimide materials and packaging technologies such as freeform optical technology, ultrasonic technology, TSV(through-silicon via technology) packaging, and advanced packaging simulation design methods such as heat management, reliability design and multi-physics analysis and applications. In addition, this series also makes a planning outlook and forecast for the development trend of advanced electronic packaging.

This series of books is of great worth for workers engaged in scientific research, production and application in electronic packaging and related industries, and also has great reference significance to teachers and students of related majors in higher education institutions.

We believe that the publication of this series of books will play a positive role in promoting the development of China's IC industry and advanced electronic packaging industry.

Finally, we would like to express our sincere gratitude to our colleagues who have worked hard in the preparation of this book. We also express our heartfelt thanks to those who participated in organizing the publication of this Series!

C.P. Wong
IEEE Fellow
Member of Academy of Engineering of the USA
Member of Chinese Academy of Engineering
Former Bell Labs Fellow
Dean of Engineering,The Chinese University of Hong Kong
Regents' Professor, Georgia Institute of Technology, Atlanta, GA 30332,USA

Sheng Liu, Ph.D.
IEEE Fellow, ASME Fellow
ChangJiang Scholar Professor
Dean, School of Power and Mechanical Engineering,
Executive Director,Institute of Technological Sciences,
Wuhan University,
Wuhan, Hubei, China

Wenhui Zhu, Ph.D.
National Invited Professor
College of Mechanical and Electrical Engineering,
Central South University
Changsha, Hunan, China

Contents

Preface

Aromatic polyimides, an important class of heteroaromatic polymers, have many desirable characteristics, including excellent thermal stability in the range of −250°C to 550°C, high mechanical strength and toughness, high electric insulating properties, low thermal dimensional expansion, low dielectric constants and dissipation factors at a wide range of frequencies, as well as high radiation and wear resistance, etc. Additionally, aromatic polyimides can be processed for many different materials, such as films, fibers, carbon fiber composites, engineering plastics, foams, porous membranes, coatings, and varnishes, etc. Based on their excellent combined properties and versatile processability, aromatic polyimide materials have found extensive applications in most high-tech fields such as electric insulating, microelectronics and optoelectronics, aerospace and aviation industries.

Although the first report on the synthesis of aromatic polyimide was published in 1908, the first commercial polyimide was introduced by Dupont in the late 1960s when a successful synthetic route to high-molecular-weight polyimides was developed. However, even today the most common methods for synthesis of polyimides are still not completely understood. This is because the polymerization reaction of aromatic dianhydrides with diamines is dramatically affected by various reaction conditions, such as solvent, moisture, impurity, temperature, etc. Even the addition mode of monomers can affect the molecular weights of the formed polyimides.

There are already several books on polyimides available to readers. They are *Polyimides* by D. Wilson, H.D. Stezenberger, and P.M. Hergenrother in 1990, *Polyimides: Fundamentals and Applications* by Malay K. Ghosh and K.L. Mittal in 1996, *Polyimides: Chemistry, Relationship Between Structure and Properties and Materials* (in Chinese) by Meng-Xian Ding in 2006, *Polyimides: Monomer Synthesis, Polymerization and Materials Preparation* (in Chinese) by Meng-Xian Ding in 2011. However, all of these books describe the basic chemistry, synthesis, characterization, and fundamentals for applications of aromatic polyimides, and there is no book focusing on a systematic description of advanced polyimide materials for high-tech applications. The authors thought that this might be due to the highly proprietary nature of advanced polyimide materials. In addition, there are no books dedicated to materials processing and testing. In recent years, with the rapid development of many high-tech industries all over the world, there is an urgent need for both academic and engineering knowledge for advanced polyimide materials. The authors feel obligated to explore these subjects and contribute to this community by sharing their recent findings so as to promote the healthy development of high technologies.

It is now appropriate to present the state-of-the-art knowledge and research in this very active field. We hope this book is able to provide useful knowledge for our colleagues and to facilitate research and development in the field of advanced polymer materials. To facilitate the exchange of original research results and reviews on the design, synthesis, characterization, and applications of polyimide materials, we have written this book entitled *Advanced Polyimide Materials: Synthesis, Characterization, and Applications* to address many important aspects of the polyimide materials fields. This book aims to embrace important interdisciplinary topics in fundamental and applied research of advanced polyimide materials. Thus, important and interesting topics of research frontiers for a wide range of scientific and engineering areas are presented. We believe that this is a good reference book for readers interested in the design, synthesis, and applications of polyimide materials.

I express our sincere appreciation to Prof. Qing-Hua Zhang and Prof. De-Zhen Wu, who contributed to Chapter 2, Advanced Polyimide Fibers; Prof. Qing-Hua Lu and Dr. Feng Zhang, who contributed to Chapter 5, Polyimides for Electronic Applications; Dr. Xiao-Hua Ma, who contributed to Chapter 6, Polyimide Gas Separation Membranes; Assoc. Prof. Jian-Hua Fang, who contributed to Chapter 7, Polyimide Proton Exchange Membranes; Prof. Yi Zhang and Prof. Wei Huang, who contributed to Chapter 8, Soluble and Low-*k* Polyimide Materials. I wish to thank Prof. Jiang Zhao, Prof. Chen-Yang Liu, and Prof. Ji-Zheng Wang at my institute for valuable advice and continuous encouragement.

I am very grateful to my colleagues, Assoc. Prof. Ai-Jun Hu, Assoc. Prof. Mian Ji, Assoc. Prof. Hai-Xia Yang, Prof. Jin-Gang Liu, Dr. Li-Li Yuan, who have contributed a large amount to this book on the research into the advanced polyimide materials at my institute. I am very grateful to my students, Zhen-He Wang, Fu-Lin Liu, Dian-Rui Zhou, and Jin-Yi Zhang who have contributed to the figures, tables, photos, and references in the book. I am also very grateful to the Senior Engineer Wei-Dong Zhao, Chao Cui, and Dr. Lin-Ying Pan at the Research Institute of Aerospace Materials and Processing in Beijing and the Senior Engineer Guang-Qiang Fang at the Institute of Aerospace System Engineering in Shanghai. Finally, it is my great pleasure to thank Mrs Gang Wu, who has contributed so generously with illustrations and helpful suggestions to improve the book.

This book is supported by China Sci-Tech projects including 973 Program (2014CB643600). I look forward to receiving any comments, criticisms, and suggestions from the readership, which would be of benefit to the book and the authors. Finally, I would like to express my gratitude to my family for their support of my work.

This edition is co-published with Elsevier Inc. In accordance with Elsevier's edition, this book follows the typeset of Elsevier, including, but not limited to, fonts, size, subscript, superscript, normal or italic letters, as a courtesy. Meanwhile, the appendix provides the conversion of SI and CGS units for reference.

Shi-Yong Yang
Institute of Chemistry, Chinese Academy of Sciences (ICCAS), Beijing, China

List of Contributors

Jie Dong, Donghua University, Shanghai, China

Jian-Hua Fang, Shanghai Jiao Tong University, Shanghai, China

Ai-Jun Hu, Institute of Chemistry, Chinese Academy of Sciences, Beijing, China

Wei Huang, Shanghai Jiao Tong University, Shanghai, China

Mian Ji, Institute of Chemistry, Chinese Academy of Sciences, Beijing, China

Qing-Hua Lu, Shanghai Jiao Tong University, Shanghai, China

Xiao-Hua Ma, Institute of Chemistry, Chinese Academy of Sciences, Beijing, China

De-Zhen Wu, Beijing University of Chemical Technology, Beijing, China

Hai-Xia Yang, Institute of Chemistry, Chinese Academy of Sciences, Beijing, China

Shi-Yong Yang, Institute of Chemistry, Chinese Academy of Sciences, Beijing, China

Li-Li Yuan, Institute of Chemistry, Chinese Academy of Sciences, Beijing, China

Qing-Hua Zhang, Donghua University, Shanghai, China

Yi Zhang, Sun Yat-sen University, Guangzhou, China

Feng Zheng, Shanghai Jiao Tong University, Shanghai, China

About the Editor

Shi-Yong Yang is a professor at the Institute of Chemistry, Chinese Academy of Sciences (ICCAS), Beijing, China. He obtained his PhD from Nankai University, Tianjin, China, in 1988, his MS degree from ICCAS in 1985, and his BS degree from Lanzhou University, Lanzhou, China in 1982. From 1988 to 1992, he was an assistant professor at Shanghai Institute of Organic Chemistry, Chinese Academy of Sciences, Shanghai, China. After spending more than 4 years (March 1992–May 1996) as a postdoc at The University of New York at Buffalo and The University of Chicago, he joined the ICCAS as associate professor in 1996 and professor in 1998 and was the Director of the Laboratory of Advanced Polymer Materials, ICCAS, from December 2002 to May 2017. His research interests have been focused on the development of advanced polyimide materials, including the polyimide films for electrical insulating and microelectronic packaging applications, thermoset polyimide matrix resins of high-temperature carbon fiber composites, polyimide coatings for microelectronic manufacturing and packaging applications, polyimide super engineering plastics, and polyimide foams for aerospace and aviation applications, etc. He has filed and owned more than 100 patents in China, published more than 200 papers, and given more than 50 keynote speeches and invited talks.

Chapter 1

Advanced Polyimide Films

Shi-Yong Yang and Li-Li Yuan
Institute of Chemistry, Chinese Academy of Sciences, Beijing, China

1.1 Introduction

Due to the excellent combination of thermal, mechanical, and electrical properties, aromatic polyimide films have been extensively used in many high-tech fields such as electrical insulation, electronic packaging, and aerospace industries. The typical applications include the tapes for winding of magnet wires, the base films for flexible print circuits (FPCs), tape automated bonding (TAB) carrier tape, magnetic recording tape, and thermal-controlled shielding film for satellites, etc.

Polyimide film, first developed and commercialized in the late of 1960s by Dupont [1], was prepared by casting a solution of polyimide precursor-polyamic acid (PAA) on to the surface of the substrate followed by thermal-baking to give a self-supporting gelled film. The gelled film was then converted by thermal baking into polyimide film (Fig. 1.1). The conversion can be accomplished thermally by baking at temperatures in excess of 300 °C (thermal imidization method) or chemically [2], by using of a mixture of an anhydride and an aromatic base, such as acetic anhydride and pyridine (chemical imidization method). Moreover, the chemical conversion also needs a final heating treatment to remove the organic volatiles and to ensure the complete imidization.

The PAA solution was usually prepared by the polycondensation of aromatic dianhydrides and aromatic diamines in dipolar aprotic solvent at low temperatures (Fig. 1.2), in which the important aromatic dianhydrides include PMDA, BPDA, BTDA, ODPA, etc., and aromatic diamines include ODA, PDA, BAPP, etc. The solvents include dimethylacetamide (DMAc), dimethylformamide (DMF), and *N*-methyl-2- pyrrolidone (NMP), etc.

The Dupont Kapton films were produced derived from PMDA and ODA or/and PDA using DMAc as solvent, while Ube Upilex films were based on BPDA and ODA or PDA. Ube Industries have developed an alternative casting method to produce BPDA-based polyimide films [3-5]. Because the BPDA-based polyimide was first disclosed in US patent filing by Dupont [6], Ube had to find an alternative solvent such as phenolic instead of DMAc to cast the PAA-(BPDA/PDA) solution for the production of polyimide films. It was found that the PAA derived from BPDA and ODA is soluble in *p*-chlorophenol, which can be cast from this solvent on to a support surface, and then heated to >300 °C to complete the thermal imidization and drive out all of the high-boiling solvent. The PI-(BPDA/ODA) film was commercialized as Upilex R. The PI-(BPDA/PDA) film, another Ube product (Upilex S), was also produced in a mixed solvent containing *p*-chlorophenol as the main solvent using a similar procedure to that of Kapton films.

In 1968, it was found that stretching orientations have obvious effects on the mechanical properties of polyimide films. At that time, the researchers at Dupont developed the chemical imidization method to produce polyimide films. A chemical conversion agent (a mixture of acetic anhydride and β-picoline) was first added to the chilled PAA solution in DMAc, which was then cast on to a heated drum to give a gelled film which was a partly imidized polymer containing large quantities of solvent. After being

Advanced Polyimide Materials. DOI: https://doi.org/10.1016/B978-0-12-812640-0.00001-9

peeled off from the drum surface, the self-supported gelled film was then one-way or two-way stretched and simultaneously imidized under heating to produce a fully imidized polymer (polyimide) film. Experimental results indicated that the orientation of polyimide films produced by mono-or bi-axial stretching exhibited improved tensile strength and tensile modulus and significantly reduced water uptake and dissipation factor [7]. For instance, after being mono-axially oriented, the BPDA-based PAA gelled film was converted thermally to the polyimide film. The resulting polyimide film was brittle transverse to the draw direction, but major changes of properties in the draw direction were observed. At draw ratios of 1.75×, the maximum mechanical properties were achieved with tensile strength of 140 MPa and tensile modulus of 7.0 GPa, respectively. It was also found that polyimide films derived from PMDA and *m*-phenylenediamine (MPD), *para*-phenylenediamine (PDA) or benzidine showed very poor hydrolytic stability, however the bridged diamines such as ODA give polyimide films with very good hydrolytic stability. Hence, the chemical structures of polyimide films have obvious influences on the mechanical, thermal, and chemical properties.

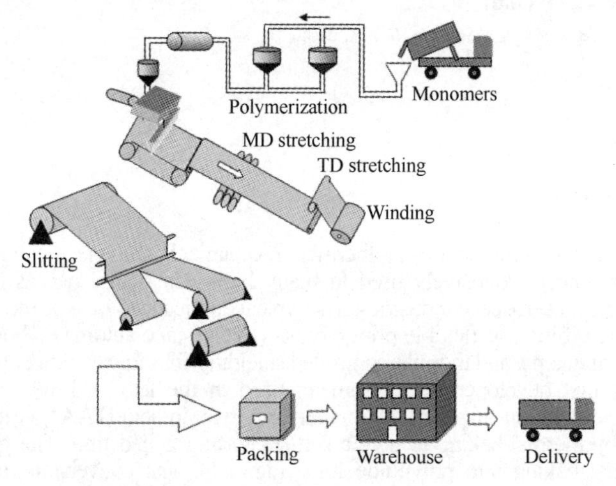

FIGURE 1.1 Production process of polyimide films

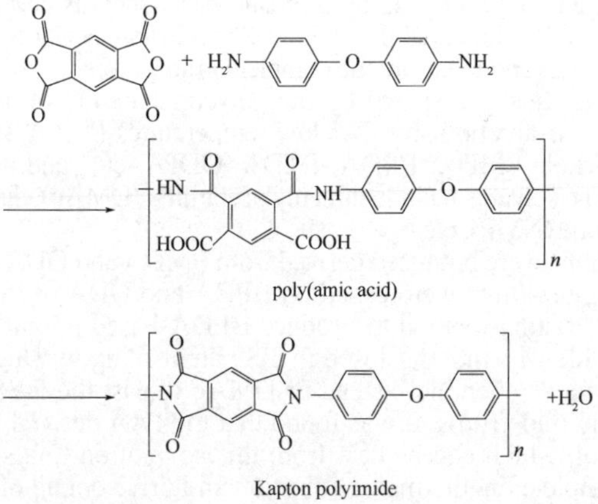

FIGURE 1.2 Chemistry of poly(amic acid)and polyimides

For many years, Kapton H films have been accepted as the standard in commercial polyimide films. The Upilex S films derived BPDA and PDA exhibited higher tensile strength and modulus as well as very low coefficient of thermal expansion (*CTE*). However, the elongation at breakage of Upilex S was reduced about 50% compared with Kapton H. In addition, Upilex R films derived BPDA and ODA exhibited comparable combined properties with Kapton H films.

In recent years, polyimide films have been extensively used in microelectronic manufacturing and packaging industries, especially as the metal interconnect board substrates for FPC, TAB, chip on film (COF), chip scale packaging, ball grid array, etc. With the miniaturization and thin filming in electrical and electronic parts, line thinning of electric circuits is rapidly progressing. Dimensional changes of polyimide films during long-term servicing may cause accidents such as disconnection and short-circuit in the thinner lined circuit structures. Hence, highly accurate dimensional stability and high elastic modulus as well as low moisture absorption are required for polyimide films used for microelectronic applications.

In this chapter, the chemistry, preparation, and applications of the advanced polyimide films will be discussed.

1.2 Chemistry of Polyimide Films

Because of the extended rigid planar aromatic and heteroaromatic structures, aromatic polyimides are usually infusible and insoluble. Hence, polyimide films are generally prepared by casting of polyimide precursor-PAA resin solution on the support surface to give self-standing gelled films, which are then imidized by thermally baking at <180 °C, as shown in Fig. 1.2. This process made it possible to bring the first significant commercial polyimide film product (Kapton H) to market, and it is still the method of choice in the majority of polyimide film production.

When an aromatic diamine and an aromatic dianhydride are mixed in dipolar aprotic solvent such as DMAc, DMF, or NMP, PAA resin is rapidly formed at ambient temperatures. The reaction mechanism involves the nucleophilic attack of the amino group on the carbonyl of the anhydride group, followed by the opening of the anhydride ring to form amic acid group, as shown in Fig. 1.3. This process is an equilibrium reaction and is irreversible because a high-molecular-weight PAA resin is readily produced. The reaction is much faster than the reverse reaction by several orders of magnitude. If there is not a large difference in reaction rate, the high-molecular-weight PAA is not produced. Hence, it is a key issue to examine the driving forces that favor the forward reaction over the reverse reaction. In addition, the acylation reaction of amines is an exothermic reaction and its equilibrium is favored at low temperatures. The forward reaction in dipolar solvents is a second-order reaction and the reverse reaction is first-order. The equilibrium is favored at high monomer concentrations to form higher-molecular-weight PAA resins.

The conversion of PAA to polyimide can be completed either by thermal imidization or by chemical imidization methods, which will be discussed in the following sections.

1.2.1 Thermal Imidization

The most commonly used pathway for conversion of cast PAA gelled films to the corresponding polyimide films is the thermal imidization. This method is suitable for production of continuous polyimide films thermally cured at elevated temperatures to allow the diffusion of by-product (water) and organic solvents without forming bristles and voids in the final polyimide films. In this method, the cast PAA film on the surface of a circular seamless steel support tape is dried and heated gradually to about 180 °C, and then cooled and peeled off to give a self-standing gelled film which is mono- or biaxially stretched and imidized by gradually heating from about 200 °C to 350 °C to give the fully polyimide films. It was noted that too rapid heating might cause the formation of bubbles and voids in the final polyimide film products. After the PAA solution in solvents such as DMAc or DMF is cast on the support surface and dried at low temperature (<180 °C) to a nontacky state, the resulted gelled film still contains a substantial amount of the solvent, typically up to 25% (*w*) depending on the drying conditions. In the subsequent heating, imidization reaction takes place not in a true solid state but rather in a very concentrated viscous solution, at least during the initial and intermediate stages of thermal imidization.

The residual solvent remaining in the gelled films plays an important role in the production of polyimide films. The thermal imidization proceeds faster in the presence of dipolar amide solvents such as DMAc or DMF, in which a specific solvation existed, allowing the favorable conformation of the amic acid group to cyclize [8]. The amide solvent also has a plasticizing effect, resulting in the increases in mobility of the reactive acidic groups due to its basicity and acceptability of protons. Because the protons of the carboxylic group are strongly hydrogen-bonded to the carbonyl group of the amide solvent, the cyclization of the *o*-carboxyamide group results in dihydrogen bonding and release of the solvent molecule along with water of condensation.

It should be mentioned that the thermal imidization process of PAA to polyimide is very complicated, and it is very difficult to describe with a simple kinetic expression. There are several interrelated elementary reactions and dynamically changing physical properties such as diffusion rate, chain mobility, solvation, and acidity, etc. The thermal imidization proceeds rapidly at the initial stage and tapers off to a plateau, which is a typical diffusion-limited kinetic process. With the imidization degree (ID) increaseing, the T_g or stiffness of the polyimide backbone chain increases. When the T_g approaches the reaction temperature, the imidization rate slows down rapidly. At a higher temperature, a higher ID is achieved.

At the initial stage, the imidization proceeded rapidly due to the five-member ring closure of amic acid in the favorable conformation (1), then the imidization rate slowed in the later stage due to the unfavorable conformation (2) (Fig. 1.3), in which there was a rearrangement of conformation (2) to conformation (1) before ring closure. Such a conformational rearrangement requires rotational motion of the adjoining polymer chain and strongly bound solvent molecules. The effect of the conformation of amic acid on the imidization rate is also consistent with the observation [8] that the thermal cyclization of model compounds *N*-substituted phthamamic acids was strongly influenced by the steric effect imposed by *N*-substituents. The vacuum thermolysis of neat *N*-substituted phthamamic acids afforded the corresponding imides and anhydrides. It was concluded that conformation required for imide formation is apparently influenced by the steric effect of the *N*-substituents.

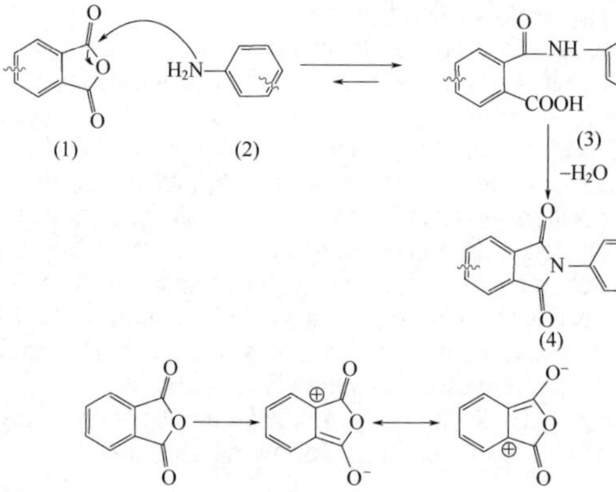

FIGURE 1.3 Reaction mechanism of polyimide formation.

The polycondensation of dianhydrides with diamines to produce poly(amic acid) is an exothermic reaction. On heating, the equilibrium shifts toward the left (Fig. 1.4). The reversion should result in a higher level of anhydride and amino groups, and a lower molecular weight. There are several factors that play important roles in the synthesis of PAA resins, such as the reactivity of monomers, solvents, moisture, monomer adding mode and molecular-weight distribution, etc.

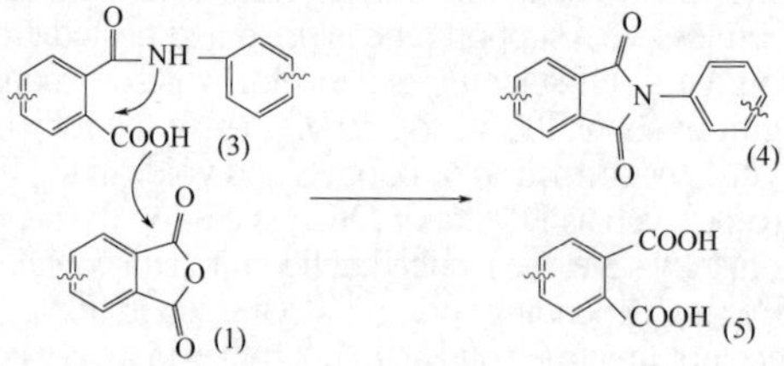

FIGURE 1.4 Mechanism of thermal ring closure of amic acid to imide.

1.2.1.1 Reactivity of Monomers

The mechanism of polycondensation between dianhydride and diamine in amide solvent to form PAA resin is a nucleophilic substitution reaction at one of the carbonyl carbon atoms of dianhydride with diamine. Hence, the reaction rate is primarily governed by the electrophilicity of the carbonyl groups of the dianhydride and the nucleophilicity of the amino nitrogen atom of the diamine. The phthalic anhydride group is a strong electron acceptor and a relatively strong Lewis acid, in which the two carbonyl (C=O) groups are situated at the *ortho*-position to each other, and their strong electron-withdrawing effects activate each other towards nucleophilic reaction. The effect is particularly enhanced by the preferred carbonyl conformation locked in the coplanar aromatic ring (Fig. 1.5).

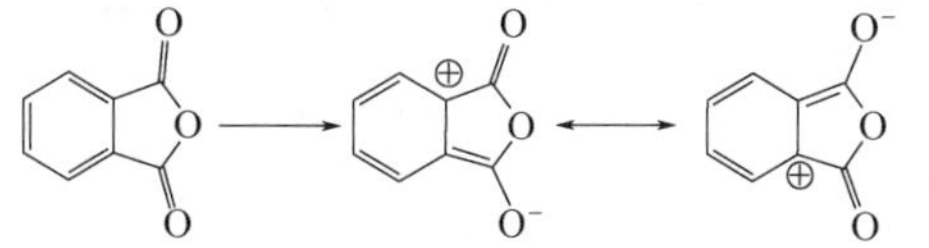

FIGURE 1.5 Resonance effect on phthalic anhydride group

PMDA is the most reactive dianhydride monomer, because the four carbonyl (C=O) groups attached to one benzene ring in a coplanar conformation, showing the strongest tendency to accept an electron. The electrophilicity of carbonyl carbons of each dianhydride can be measured in terms of the electron affinity, a tendency of the molecule to accept an electron. Table 1.1 compares the electron affinity (E_a) for various aromatic dianhydrides by polarographic measurement. PMDA exhibits E_a of 1.90, higher than DSDA (1.57) and BTDA (1.55). BPDA and ODPA show E_a of 1.38 and 1.30, respectively. The ether-bridged dianhydride (BPADA, and EDA) possess E_a in the low level of 1.10-1.19.

TABLE 1.1 Electron Affinity(E_a)of Aromatic Dianhydrides

Aromatic Dianhydrides	Abbreviation	E_a(eV)
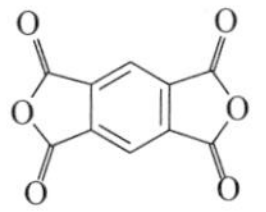	PMDA(Pyromellitic dianhydride)	1.90
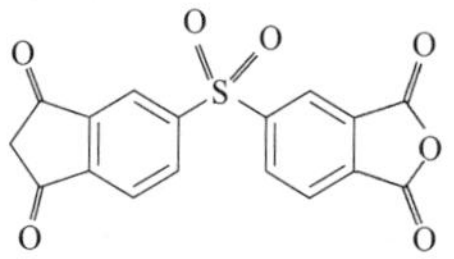	DSDA(3,3′4,4′-Diphenylsulfonetetracarboxylic dianhydride)	1.57
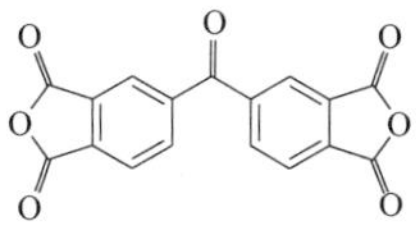	BTDA(3,3′,4,4′-benzophenonetet acarboxylic dianhydride)	1.55
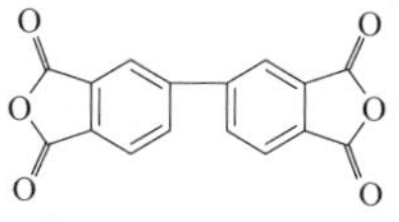	BPDA(3,3′,4,4′-biphenyltetracarboxylic dianhydride)	1.38
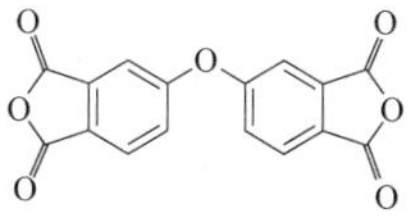	ODPA(4,4′-Oxydiphthalic anhydride)	1.30
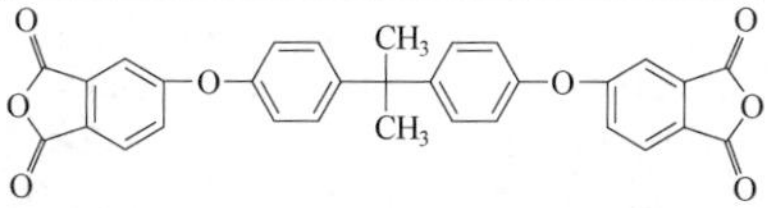	BPADA 4,4′-(4,4′-isopropylidenediphenoxyl)bis(phthalic anhydride)	1.12
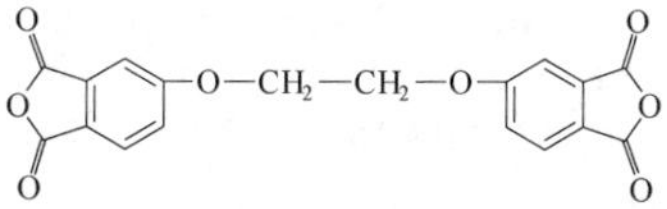	EDA 5,5′-(ethane-1,2-diylbis(oxy))bis (isobenzo furan-1,3-dione)	1.10

Due to the strong electron-accepting property of dianhydrides and high electron density of amino groups of diamines, charge transfer interactions and electrostatic interactions between dianhydrides and diamines have obvious effects on PAA formation. The acylation of a diamine with a dianhydride is preceded by a charge transfer interaction in which diamine is a donor and dianhydride is an acceptor. The reaction mixture shows a strong color of charge transfer complex at the beginning, and the color gradually fades as the acylation proceeds. For the ether-bridged bisphthalic anhydrides, the ether-bridge group strongly reduces E_a of the dianhydride. Compared with BPDA, which lacks a bridge group, the electron-withdrawing bridge groups such as SO_2 and C=O increase the E_a values substantially. On the other hand, the electron-donating bridge groups such as —O significantly reduce E_a values. The dianhydrides with low E_a are not affected by the atmospheric moisture [9] while PMDA and BTDA must be handled under strictly moisture-free conditions at all times.

Unlike the E_a value of dianhydride, the ability of diamine to give off an electron, the ionization potential (I), does not seem to correlate well in a simple manner. The reactivity of diamines instead correlates well with its basicity (pK_a) in a Hammett's relation. The reaction rates (k) of various diamines toward PMDA are in relation to their pK_a (Table 1.2). The chemical structures of aromatic diamines have significant effects on the rate of acylation reaction, moreso than those of aromatic dianhydride. The rate constants differ by four orders of magnitude between aromatic amines with electron-donating substituents and those with electron-withdrawing ones.

TABLE1.2 Basicity(pK_a) of Aromatic Diamines and Their Reactivity Toward PMDA

Diamines	Abbreviation	pK_a	lg k
H_2N–C6H4–NH_2	PPD *p*-Phenylenediamine	6.08	2.12
H_2N–C6H4–O–C6H4–NH_2	4,4′-ODA 4,4′-Oxydianiline	5.20	0.78
H_2N, NH_2 (meta)	MPD *m*-Phenylenediamine	4.80	0
H_2N–C6H4–C6H4–NH_2	Benzidine	4.60	0.37
H_2N–C6H4–C(O)–C6H4–NH_2	4,4′-Diaminobenzophenone	3.10	−2.15

1.2.1.2 Effect of Solvents

The most commonly used solvents are dipolar aprotic amide solvents such as DMF, DMAc, NMP. One of the important properties of the solvent selected is its basicity (Lewis base). The starting materials are weakly basic aromatic diamine monomers and dianhydrides, and the reaction product PAA is a strong aprotic acid. The *ortho*-amic acid is a relatively strong carboxylic acid because of the electron-withdrawing effect of the *ortho*-amide group and the stabilization by internal hydrogen bonding of dissociated carboxylate with amide hydrogen. Hence, the strong acid-base interaction between the amic acid and the amide solvent is a major source of *exo*-thermicity of the reaction, and one of the most important driving forces. The rate of PAA formation is usually faster in more basic and more polar solvents.

If solvent contains an amount of water as impurity, some anhydride groups are hydrolyzed to *ortho*-dicarboxylic groups which are unreactive end groups. As a result, the equivalent amount of amino groups will also be left in an unreactive state if the stoichiometric amounts of dianhydride and diamine monomers are used in the beginning. If an additional dianhydride equivalent to the unreacted amino group is added to the above solution, it reacts with the amino end groups, resulting in the increase in molecular weight of PAA. Although the *ortho*-dicarboxylic acid group is inactive at ambient temperature, it will thermally dehydrate to the anhydride and reacts with amine at high temperatures, thus giving an off-stoichiometric situation with the dianhydride in excess. Hence, the molecular weight of the polyimide product is primarily governed by the stoichiometric relation of the monomers.

1.2.1.3 Effect of Moisture

PAAs are known to undergo hydrolytic degradation even at ambient temperature. This is attributed to the presence of a small amount of anhydride groups existing in an equilibrium concentration, which plays an important role in the hydrolytic degradation of PAA. The anhydride group in PAA solution can be hydrolyzed to form *ortho*-dicarboxylic group in the presence of water. The reaction is driven by the enhanced nucleophilicity of water in a dipolar aprotic solvent and by strong acid-base interaction of the product with the dipolar solvent. The *ortho*-dicarboxylic groups are stable as one of the end groups of PAA and do not revert to the anhydride. After the anhydride groups are completely consumed, more anhydride groups are produced to reestablish the reaction equilibrium. It was found that water has a significant effect on the molecular weight of PAAs during polymerization [10]. The common source of water is in the solvents and the monomers as an impurity, and water can also be produced by the imidization of amic acid groups. Although the rate of the imidization and the formation of water is relatively slow at ambient temperatures, it is still significant enough to cause a gradual decrease in molecular weight over a long period of time. Hence, if long-term storage is necessary, PAA resin solution should be kept refrigerated.

1.2.1.4 Monomer Adding Mode

Dianhydride and dianmine monomer adding mode is an important factor for the preparation of PAA. If 25% (mol) of additional dianhydride (PMDA) or diamine (ODA) is added to the high molecular weight PAA solution, obvious decreases in the solution viscosity at 35 °C are measured, demonstrating that the addition of additional monomer could cause reequilibrium of the main equilibrium reaction. Addition of 25% (mol) PMDA resulted in a rapid initial decrease in its viscosity, but within 30 h the rate of the decrease tapered off at a higher level, and at 1000 h the viscosity was only slightly lower than that of the original sample after the equivalent aging period.

When a diamine-excess solution was compensated with the equivalent amount of additional dianhydride (PMDA), the viscosity of PAA solution was rapidly recovered to its original level. On the other hand, addition of diamine (ODA) to the dianhydride-excess solution resulted in a rapid decrease in viscosity to a much lower level. The diamine-excess condition can be corrected to restore the ultimate molecular weight, and the dianhydride-excess condition results in a permanent decrease in the molecular weight of PAA. Hence, the reactive dianhydride (PMDA) as a strong dehydrating agent is not stable in the presence of amic acid groups in solution, which can dehydrate the amic acid to the imide by hydrolyzing itself to the *ortho*-dicarboxylic acid; and the *ortho*-dicarboxylic acid group is unreactive in the presence of PAA, acting as a chain end group to control the molecular weight of the resulting polymers. Hence, the preferred mode of monomer addition during PAA preparation is to add the dianhydride to the diamine solution. In this manner, the reacting solution does not contain an excess of dianhydride at any time.

1.2.1.5 Effect of Molecular Weight Distribution

Although PAA solution should be kept at low temperature to maintain its viscosity without apparently change for further film processing, the long-term storage always results in its viscosity decreasing to some extent. The more rapid and significant decrease in viscosity occurs in the initial aging period. This cannot be explained by the effect of water or other inadvertently introduced contaminants such as impurities in the monomers (Fig. 1.6).

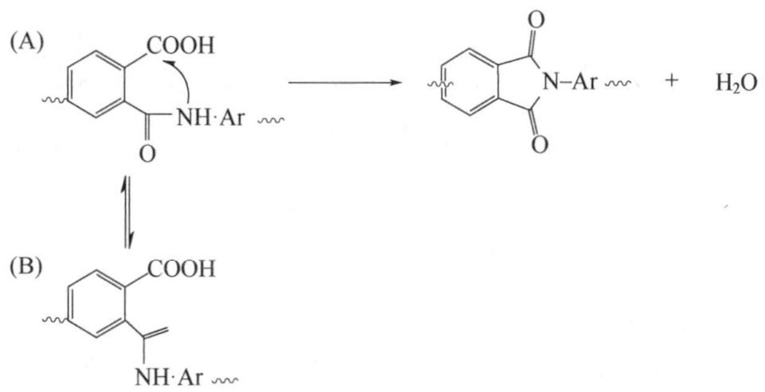

FIGURE 1.6 Effect of excess anhydride on PAA formation.

The change in the molecular weight distribution of the initially produced poly(amic acid) was found to be the major cause [11]. When the insoluble dianhydride such as PMDA is added to the diamine (ODA) solution in the form of solid or slurry, the polycondensation reaction proceeds in a heterogeneous manner, giving a condition of interfacial polymerization between solid dianhydride and the solution of diamine. Because the reaction rate of PMDA and ODA is very fast and viscosity of PAA is very high, the interfacial polymerization is governed by the diffusion rates of two monomers approaching from the opposite sides of the interface. At the interface where the exact stoichiometry is satisfied, a high-molecular-weight PAA was obtained. After the polycondensation reaction is completed, the polymer solution contains some unusually high-molecular-weight fractions, which contribute more heavily to the weight average molecular weight (M_w) than to the number average molecular weight (M_n). During the polycondensation, the M_w is initially high and then decreases markedly in the initial rapid drop of the solution viscosity. However, M_n remains essentially constant at the same period. Hence, the polydispersity (M_w/M_n) is initially high and then decreases toward a value of 2, which is the most probable molecular weight distribution. For most polyimides, the viscosity average molecular weight is much closer to M_w than to M_n [12]. Hence, the solution viscosity decreases with M_w decreasing while M_n remains relatively unchanged.

1.2.2 Chemical Imidization

The cyclodehydration of PAAs to polyimides can be readily achieved using chemical dehydration at ambient temperature. The commonly used reagents are the mixtures of acid anhydrides and tertiary amines in dipolar aprotic solvents [2,13-16]. The dehydration agents include acetic anhydride, propionic anhydride, *n*-butyric anhydride, benzoic anhydride, etc, and the amine catalysts are pyridine, methylpyridines, lutidine, trialkylamines, etc.

High-molecular-weight polyimides can be obtained using trialkylamines with high pK_a (>10.65) as catalyst. However, low basic tertiary amines usually result in lower-molecular-weight polyimides. The highest-molecular-weight polyimides can be obtained using heteroaromatic amines such as pyridine, 2-methylpyridine, and isoquinoline ($5.23 < pK_a < 5.68$) as catalysts. The chemical imidization reaction seems simple, but the mechanism is quite complex. The polycondensation products are very different depending on the type of dehydrating agents, monomer components, reaction temperature, and other factors.

The use of *N, N*-dicyclohexylcarbodiimide (DCC) produces essentially quantitative conversion of amic acids to isoimides instead of imides [17]. The combination of trifleoroacetic anhydride-triethylamine and ethylchloroformate-triethylamine also afforded high yields of isoimides [17]. It was revealed that isoimides and imides are formed via a mixed-anhydride intermediate, which is formed by acylation of the carboxylic group of amic acid (Fig. 1.7).

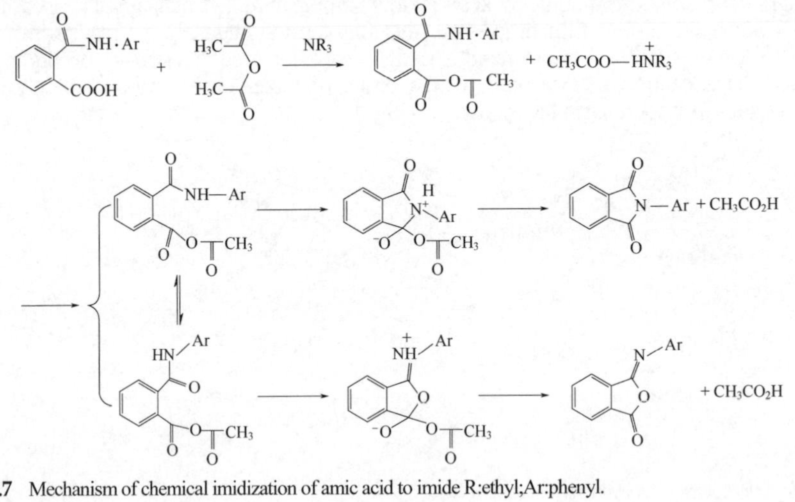

FIGURE 1.7 Mechanism of chemical imidization of amic acid to imide R:ethyl;Ar:phenyl.

The imidization process involves the simultaneous PAA imidization, solvent diffusion, and the development of ordering aggregation. There was complexation between solvent NMP and PAA. The imidization kinetics were investigated by isolation of the processes of decomplexation between NMP and PAA and imidization. The solvent content in the partially imidized film is crucial in determining the ordering degree and texture of the film. The ordering degree can be improved at higher heating rate and thicker film in the imidization. Different morphology in polyimide films can be obtained using thermal or chemical imidization [18]. The surface of the polyimide films by thermal imidization showed microdomains of an almost spherical shape, and their size and packing was dependent on the heating temperature, while it exhibited net structure by chemical imidization. The chemical structures of PAA resins can influence the combined properties of the final polyimide films; the control of ID of PAA resins in solution is especially important. After the mixture of acetic anhydride and pyridine were added into PAA solution, the imidization occurred immediately to yield a copolymer, i.e., poly(amic acid-imide) (PAAI), whose solubility is directly relative to ID. There is a critical point for ID, below which PAAI solution is homogenous and above which phase separation will occur immediately. Hence, homogeneous copolymer PAAI solutions with different IDs can be obtained before the critical point of phase separation in the imidization process; and the control of ID can provide a pathway to prepare polyimide films with the required morphology and properties.

In the imidization process, the solution viscosity was temporarily increased at the initial reaction stage due to the interchain mixed-anhydride formation, which was then gradually decreased back to the normal level with the cyclization reaction. Polyimide was formed by intramolecular nucleophilic substitution at the anhydride carbonyl by the amide nitrogen atom (Figs. 1.7, 1.8 and 1.9), while isoimides were formed as a result of substitution by the amide oxygen (Figs. 1.8 and 1.9).

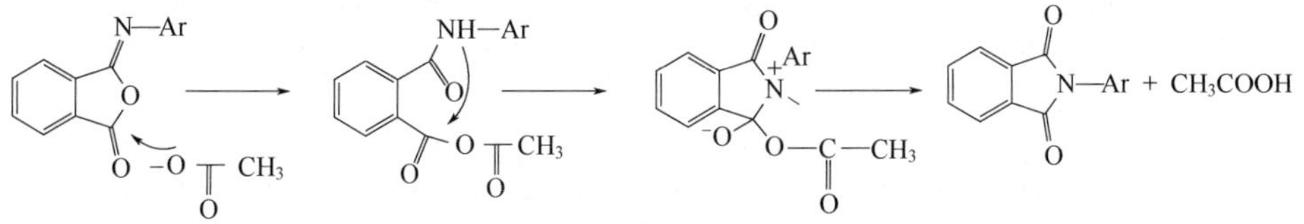

FIGURE 1.8 Mechanism of rearrangement of isoimide to imide.

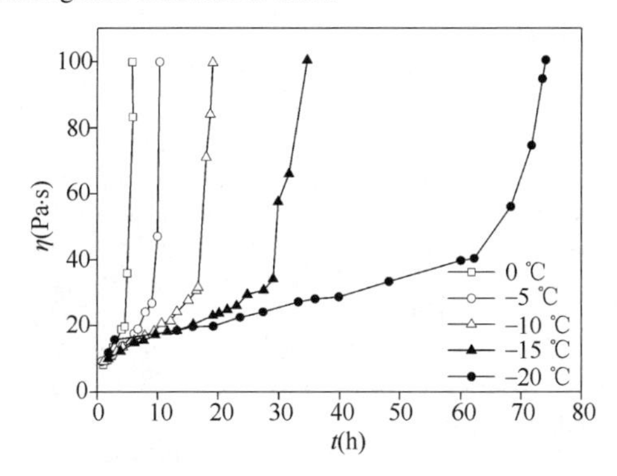

FIGURE 1.9 Effect of temperature on the solution viscosities of PAA-(PMDA/ODA) with different reaction times.

The cyclization of a model compound, *N*-phenylphthalamic acids, with acetic anhydride proceeds smoothly at room temperature in DMAc in the presence of a tertiary amine, in which the amine acts as a catalyst as well as an acid acceptor. The stoichiometric amount of amines in the PAA solution could still lead to the completion of imidization reaction, but only at a lower rate. Strong base, such as trimethylamine (pK_a 10.6), as a catalyst produced exclusively normal imides, however, a mixture of imide and isoimide was formed if lower basic pyridine (pK_a 5.2) was employed as a catalyst. The rate of formation of isoimide was faster than that of imide in the initial stage and the formation of imide overtook at the later stage. Hence, the isoimide should not be considered as the intermediate to the polyimide. After completion of the cyclization, the isoimide slowly rearranged to imide. The rearrangement reaction was found to be efficiently catalyzed by acetate ion [17], as shown in Fig. 1.8.

In the presence of stronger amines such as trimethylamine, a high concentration of acetate ions is formed, resulting in the concurrent transformation of the isoimide to the imide during the cyclization. In contrast, trifluoroacetic anhydride or *N,N*-dicyclohexylcarbodiimide as dehydrating agent affords exclusively isoimides for both model compounds and PAAs. In general, isoimides exhibit intense, bright-yellow, or yellowish-orange color derived from a strong absorption at around 350nm-400 nm in the UV-visible spectrum.

Polyimide films with different levels of anisotropy can be obtained by varying the level of imidization with catalysts prior to drying or by relieving the in-plane stress by unconstraining the films during cure. Bi-axially stretched polyimide films by thermal imidization method exhibit a higher degree of molecular orientation than films cured in an unconstrained state [19,20]. Since the degree of molecular orientation is the result of restricting the in-plane shrinkage of the polyimide film produced during the process of thermal curing, this technique can be used to prepare films with varying degrees of molecular orientation. Chemical imidization can also be used to prepare polyimide films with varying degrees of molecular orientation. The birefringence in PMDA-ODA films prepared by chemical imidization was about a factor of 3 higher than the that of films prepared by thermal imidization.

The varied degree of molecular orientations in PMDA-ODA films have been prepared both by varying the constraint on the films during the thermal curing and by varying the level of chemical imidization prior to cure. Polyimide films with a high degree of molecular orientation can be prepared by partially chemically imidizing PAA film prior to cure. Excess molar amounts of acetic anhydride and β-picoline were added to the PAA-(PMDA/ODA) solution in DMAc. The PAA solution was cast onto glass plates, followed by heating to 100 °C/30 min, then to 250 °C/30 min to give partly imidized films. The lifted film was fixed in a frame, and was then heated to 400 °C for 5 min to remove the last traces of solvent and complete the imidization.

It was found that the PAA films exhibit a large optical anisotropy, indicating preferential alignment of the long axis of the molecule in the plane of the film. Imidization increased the birefringence of the film by a factor of 2.5 and reduced the film thickness. The only parameter that affected the anisotropy of the films was the method of imidization. Chemical imidization was found to increase the birefringence by a factor of 3, indicating of a higher degree of molecular orientation parallel to the film surface. This effect was not observed in thicker (>25 μm) films using X-ray diffraction where the orientation function was independent of the method of imidization.

The chemical imidization pathways of polyimide (MPDA-ODA) and (BPDA-PDA) films were systematically investigated [21]. PAA resin solution was first mixed with dehydrating agents (the mixture of acetic anhydride and tertiary amine as the catalyst), and was then cast on the support surface to give a partly imidized PAA gel film. The gel film was then thermally converted to fully imidized polyimide films at high temperatures. The gel point was dependent on ID despite the temperature and the molar ratio of catalyst to acetic acid. Experimental results indicated that the ID was about 35% for PMDA/ODA and about 22% for BPDA/PPD. The effect of catalyst on imidization was in the order trimethylamine > 3-methylpyridine > pyridine > isoquinoline > 2-methylpyridine. The stretching of the films greatly reduced the coefficient of linear thermal expansion (*CTE*) either in the longitudinal or transversal direction. Compared to the film from PAA, the partly imidized film had greater stretching ratio, so that the unaxially stretched polyimide film from partly imidized PAA had higher tensile strength and tensile modulus, but lower elongation in the stretching direction.

1.2.2.1 Effect of Temperature on Viscosity and Gelation Time

The ID was calculated as the gelation of PAA resin occurring at a given temperature of ≤0 °C. Table 1.3 shows the gelation times at different temperatures and the ID of PAA-(PDMA/ODA) and PAA-(BPDA/PPD). Obviously, the gelation time depended on the temperatures. The higher the temperature, the shorter the gelation times. At 0 °C, PAA-(PMDA-ODA) exhibits a gelation time of 4 h, shorter than 8 h at –5 °C and 17 h at –10 °C, respectively. However, the ID value is almost constant no matter what temperature the gelation occurs at. PAA-(PMDA/ODA) shows ID of about 33.1%-37.2%, higher than PAA-(BPDA/PDA) (ID=21.2%-23.2%), which might be attributed to the backbone rigidity of the PAA resins. The rigid PAA-(BPDA/PDA) resin has low solubility, causing the gelation to occur at a lower level of imidization.

TABLE 1.3 Dependence of Imidization Degree and Gelation Time of PAAs on Reaction Temperatures

Temperature(°C)	PAA-(PMDA/ODA)		PAA-(BPDA/PDA)	
	Gelation Time(h)	Degree of Imidization(%)	Gelation Time(h)	Degree of Imidization(%)
0	4	36.7	3.5	21.3
−5	8	37.2	6.5	23.2
−10	17	33.1	13	23.0
−15	29	34.4	24	22.6
−20	63.5	34.7	49	21.2

The effects of temperature on viscosity of PAA were measured using a 10% (w) PAA solution with an initial viscosity of 14.3 Pa·s, which contains dehydrating agents with mole ratios of n[PAA (calculated as repeating units)]: n(acetic anhydride): n(pyridine)=1:2.5:1.5. Fig. 1.9 shows the effect of temperatures on the plot of PAA solution viscosity versus reaction time. At lower temperature of −20 °C, the gelation time can extend to longer than 63.5 h for PAA-(PMDA/ODA), and 49 h for PAA-(BPDA/PDA). At 0 °C, the gelation time is only 4 h for PAA-(PMDA/ODA), and 3.5 h for PAA-(BPDA/PDA), respectively. Increasing the temperature from −20 °C to 0 °C reduced the gelation time from 63.5 h to 4 h for PAA-(PMDA/ODA), and 49 h to 3.5 h for PAA-(BPDA/PDA), respectively.

1.2.2.2 Effect of Molar Ratio of Pyridine to Acetic Anhydride on PAA Viscosity

PAA-(PMDA/ODA) solution (10% of solid concentration) in DMAc has a viscosity of 25.3 Pa·s at −15 °C, the effect of molar ratio of pyridine to acetic anhydride on viscosity is shown in Fig. 1.10. As the molar ratio of pyridine/acetic anhydride increased from 1.5:2.5 to 4.0:2.5, the gelation time decreased from 29 h to 18 h, indicating that the amount of the base catalyst can obviously accelerate the imidization of poly(amic acid) into polyimide.

1.2.2.3 Effect of Catalysts on PAA Imidization

The catalysts play an important role in the PAA imidization. PAA-(PMDA/ODA) solution (10% (w)) in DMAc with a viscosity of 14.3 Pa·s at −15 °C and a ratio of catalyst to acetic anhydride of 1.5:2.5 per mole of PAA was used to investigate the effect of catalyst on PAA imidization. It was found that there was an imidization reaction order: trimethylamine > 3-methylpyridine > pyridine > isoquinoline > 2-methylpyridine (Fig. 1.11). The gelation time for the three former catalysts was 4 h, 24.5 h, 29 h, and 47.5 h, respectively. However, 2-methylpyridine had a very low catalytic effect on imidization, the very weak imidization of PAA in solution could be observed after storage for several days at room temperature. This might be attributed to the steric effect of the methyl group located at the α-position of pyridine ring. Isoquinoline has a gelation time of 47.5 h at −15 °C, which can accelerate the imidization reaction at higher temperature and might be the good candidate for industrial catalyst for polyimide film production line.

1.2.2.4 Effect of Initial Viscosity on Gelation Time

PAA-(PMDA/ODA) solution (10% (w)) in DMAc with a ratio of catalyst to acetic anhydride of 1.5:2.5 per mole of PAA was used to investigate the effect of initial viscosities on gelation time.

The three PAAs with different initial viscosities of 25.3 Pa·s (PAA_1), 20 Pa·s (PAA_2) and 12 Pa·s (PAA_3) at –15 °C was selected. Fig. 1.12 compares the dependence of gelation time on the initial viscosities. Apparently, the gelation time is independent of the initial viscosity. All the PAA solutions with different initial viscosities exhibited almost the same gelation time of 29 h at –15 °C.

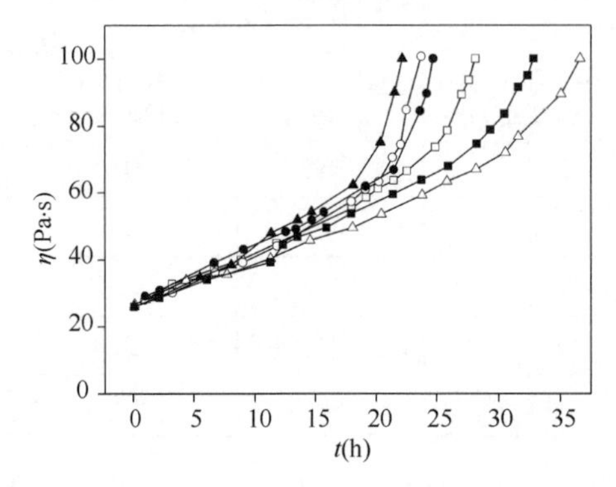

FIGURE 1.10 Effect of the molar ratios of pyridine to acetic anhydride on viscosities of PAA-(PMDA/ODA) at–15 °C n(Pyridine):n(acetic anhydride): △ 1.5:2.5;■ 2:3; □ 2.5:2.5 ●3:2.5; ○3.5:2.5; ▲ 4:2.5.

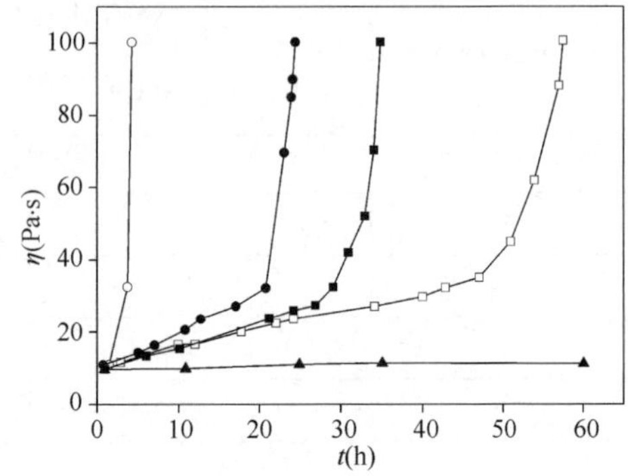

FIGURE 1.11 Effects of catalysts on the imidization of PAA-(PMDA/ODA) ○ Triethylamine;● 3-methylpyridine; ■ pyridine; □ isoquinoline; ▲ 2-methylpyridine.

1.2.2.5 Properties of the Chemically Imidized Films

Table 1.4 compares the effect of stretching ratios on *CTE* of the fully imidized films. The stretched films exhibit greatly reduced *CTE*s both in the longitudinal direction (*M*) and in the transversal direction (*T*), although the *M* direction yields lower *CTE* than *T* direction. For instance, the unstretched PI-(PMDA/ODA) film has a *CTE* of 35.1×10^{-6} °C^{-1}, much higher than the stretched films (8.5×10^{-6} °C^{-1} at *M* direction and 26.4×10^{-6} °C^{-1} at *T* direction with 1.20 stretching ratio). This might be attributed to the high degree of molecular orientation caused by the partial chemically imidized PAA films prior to curing and the stretch drawing during the imidization. The molecular orientation can enhance the polyimide's dimensional stability and its mechanical properties.

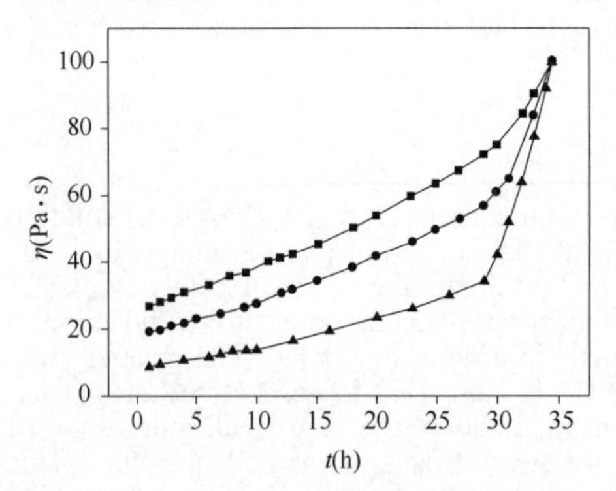

FIGURE 1.12 Dependence of gelation time on the initial viscosity of PAA solution ■ PAA 1:25.3 Pa·s ● PAA2:20Pa·s; ▲ PAA 3:12 Pa·s.

Meanwhile, PI-(BPDA/PPD) films exhibit lower *CTE* than PI-(PMDA/ODA) films under the same conditions, attributed to the fact that the former has a more rigid molecular chain structure of PI-(BPDA/PDA) than the latter. Table 1.5 compares the mechanical properties of stretched polyimide films. Higher tensile strength and modulus have been obtained in the *M* direction. The films produced by chemical imidization can be stretched at higher stretching ratios, yielding higher strength and modulus than that by thermal imidization.

Overall, the chemical imidization can produce aromatic polyimide films with lower *CTE*, higher strength and modulus, either in *M* direction or *T* direction, than thermal imidization. The gelation time of partly imidized PAA resin was dependent on the ID despite the temperature and the molar ratio of catalyst to acetic acid, it was about 35% for PAA-(PMDA-ODA) and about 22% for PAA-(BPDA/PPD). PAA-(BPDA/PPD) can produce fully imidized film PI-(BPDA/PDA) with higher tensile strength (up to 613 MPa), tensile module (up to 9.3 GPa) and lower *CTE* than PAA-(PMDA/ODA).

TABLE 1.4 ***CTE*** **Values of the Stretched Polyimide Films(50 °C–250 °C)**

Polyimide	Stretching Ratio	Direction*	$CTE(10^{-6}℃^{-1})$
PMDA/ODA	Unstretched		35.1
	1.20	*M*	8.5
		T	26.4
	1.13	*M*	14.9
		T	28.9
BPDA/PPD	Unstretched		23
	1.27	*M*	10
		T	15
	1.08	*M*	13
		T	19

**M,longitudinal direction;T,transversal direction.*

TABLE 1.5 Mechanical Properties of Axially Stretched Films

Polyimide	Stretching Ratio	direction	Tensile Strength(MPa)	Tensile Modulus(GPa)	Elongation(%)
BPDA/PPD[a]	1.27	*M*	613	9.3	12
		T	271	3.8	40
	1.08	*M*	392	4.9	26
		T	245	3.9	33
BPDA/PPD[b]	1.15	*M*	455	6.1	20
		T	260	3.8	35
Upilex-S[c]			547	9.3	47
PMDA/ODA[a]	1.2	*M*	236	1.7	39
		T	132	1.4	97
	1.13	*M*	223	1.6	59
		T	183	1.3	77
PMDA/ODA[b]	1.10	*M*	211	1.5	63
		T	130	1.4	75
Kapton HN[c]			231	2.5	72

[a]*From partly imidized PAA.*
[b]*From PAA.*
[c]*The commercial products technical data.*

1.3 Thermal Curing of Polyimide Films

The production of aromatic polyimide films is usually performed by a two-step process. The first step involves the synthesis of PAA in aprotic polar solvents, such as DMF, DMAc, or NMP. In the second step, the conversion of PAA by either thermal or chemical imidization produces polyimides. In a typical process, PAA solution was cast on the support surface followed by thermal baking to give a self-standing gelled film. The cast PAA film is imidized to the corresponding insoluble polyimide films by either heating at temperatures in excess of 300 °C (thermal imidization) or through the addition of chemical dehydrating agents (chemical imidization) followed by heating at elevated temperature.

1.3.1 Thermal Imidization Process

During the thermal process, important factors that affect the ultimate properties of polyimide films include the solvent evaporation, polymer chain orientation, chemical conversion, water loss (resulting weight loss in the film), thickness reduction, and change in chain mobility (leading to the development of birefringence). It is very difficult to predict the effect of each factor on the final properties of the film, because all of these factors are interrelated and occur simultaneously. Therefore, a large number of studies have been devoted to investigating the influence under the initial conditions only, such as the chemical structure of the molecules, the thickness of cast films, and the solvents used in the reaction. However, the real-time changes in the backbone structures and the mechanical properties during the imidization process are still unclear.

The thermal imidization process of PAA solution derived from PMDA and ODA in NMP to form polyimide films is divided into three stages: the solvent evaporation stage, imidization stage, and annealing stage. The solvent evaporation stage (the first stage) occurs at low-temperature, the mechanical properties and T_g of polymer films were increased with removal of the solvent. The evaporations of hydrogen-bonded solvents on PAA and the dehydration during the imidization reaction occur above 150 °C (the second stage). During the imidization stage, the interplay between the solvent evaporation and the imidization was the key factor that determined the enhancement of the mechanical properties and T_g. The ID approached 94% when the temperature was increased to 250 °C. A "complete" imidization (the third stage) was achieved by annealing between 350 °C and 400 °C. The enhancement in the mechanical properties of final PI films may result from the increased T_g and the crystallized structure formed during the annealing stage.

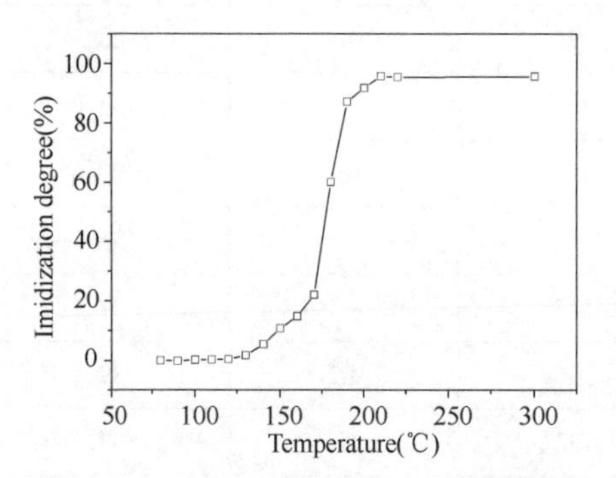

FIGURE 1.13 Imidization degree of PI-(PMDA/ODA) films as a function of the temperature from 70 °C to 300 °C.

Fig. 1.13 shows the ID of PI-(PMDA/ODA) films as a function of the temperature from 70 °C to 300 °C. The ID value started to evidently increase as the temperature increased from 150 °C and reached a constant value at approximately 220 °C. There was no apparent change in the ID above 220 °C because the uncertainty of the FT-IR method was at least 5%, which hindered the accurate determination of the ID at the late stage. According to the curing temperature, the imidization process can be divided into three stages: (1) $T < 150$ °C: solvent evaporation stage; (2) $T = 150$° C-250 °C: imidization stage; and (3) $T > 250$ °C: annealing stage.

1.3.1.1 Solvent Evaporation Stage (T<150 °C)

In this stage, the imidization reaction hardly occurs before the temperature reaches 150 °C. The PAA films thermally treated at temperatures of lower than 100 °C are easily dissolved in NMP. The solvent-PAA interactions were established by Brekner and Feger [22,23]. The solvent molecules in PAA films exist in two forms: the free solvent molecules and the molecules that are hydrogen-bonded to the PAA molecules. First, the evaporation of the free solvent molecules occurs at a relatively low temperature to form a PAA film with 4:1 complexes between the solvent and the repeating unit of the PAA, as shown in Fig. 1.14. Then, the decomplexation of the two NMP molecules H-bonded to the amide moiety occurs firstly because they have lower decomplexation energy than the two additional NMP molecules attached to the carboxylic acid moiety. As the temperature increases further, the two additional NMP molecules H-bonded to the carboxylic acid decompose. Experimental results showed that the solvent contents of PAA films thermally treated at 80 °C/0 min and 80 °C/10 min. were between 4:1 and 2:1, indicating the presence of solvent molecules H-bonded to both the amide moiety and the carboxylic acid moiety. The PAA films thermally treated at 100 °C/10 min (PAA-100-10) showed a solvent content close to 2:1, implying that the solvent molecules H-bonded to the amide moiety were almost removed. Hence, the dissociation of H-bonds between NMP and the amide moiety of PAA occurs between 95 °C and 110 °C, and the dissociation temperature range of H-bonds attached to the carboxylic acid appears at approximately 165 °C, implying a difference in the decomplexation energy of these two types of H-bonds. The temperature range of imidization spans from 150 °C to 350 °C, and the maximum reaction rate occurs at approximately 205 °C. The imidization reaction has a broad tail on the high temperature side, and the symmetrical Lorentzian function cannot accurately describe this reaction.

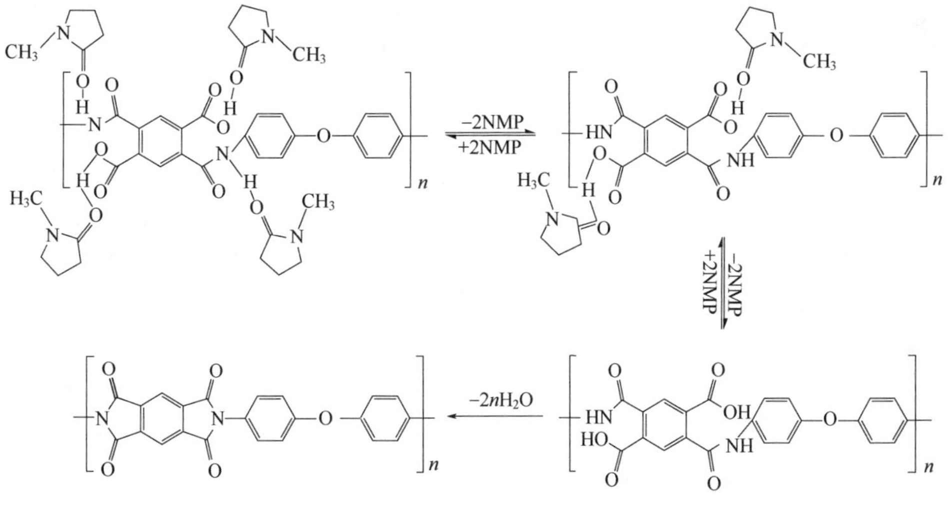

−2NMP
+2NMP
−2NMP
+2NMP
$-2nH_2O$

FIGURE 1.14 Schematic illustration of the evaporation of NMP from PAA solution.

1.3.1.2 Imidization Stage (T=150 °C-250 °C)

For an accurate determination of the ID at this stage, NMP molecules are first removed by washing the films with water for 24 h and drying under vacuum at 50 °C for 24 h because NMP is soluble in water. The TGA curves of PAA-100-10 (ID=0) without NMP and the "complete" imidized film PI-400-10 (ID=100%) are used as two references. The weight loss of PAA-100-10 at 400 °C is approximately 8.8%, indicating that NMP molecules have been completely removed (Fig. 1.15). Experimental results indicate that the solvent molecules attached to—COOH have been completely removed once treated above 200 °C for 30 min. The imidization peak is at 270 °C and the residual NMP molecules slightly accelerate the imidization due to the plasticization effect [24]. Therefore, the interplay between the solvent evaporation and the imidization reaction is the key factor that determines the improvement of the mechanical properties and T_g. Because the polyimide film still has an imidization peak near 400 °C in dynamic thermogravimetric ananlysis (DTGA), an annealing procedure is required in the fabrication of a fully imidized polyimide film. The changes in T_g with ID are an important issue during the thermal imidization. The interplay between T_g and ID is very complex. T_g increases as ID increases, and a higher T_g will prevent further imidization because the chain diffusion slows down below T_g.

DMA was used to determine the T_g [25,26]. The temperature, at which the maximum rate of turndown of the storage modulus (E') occurs or the temperature at the loss modulus (E'') and tan δ peaks, can be considered to be T_g. As the temperature increases, E' exhibits two drops, in which the first drop corresponds to the T_g of the samples, which is very close to the treatment temperature. Both the evaporation of the solvent and the imidization reaction can result in an increase of the storage modulus, thus, the drop in the E' curves can be an indicator of the glass transition. The second drop that occurs at 370 °C may correspond to the T_g after the imidization is completed. It is supported by the result for PAI-250-30 (ID=94%), and there is only one drop on the E' curve, which occurs similarly at 370 °C.

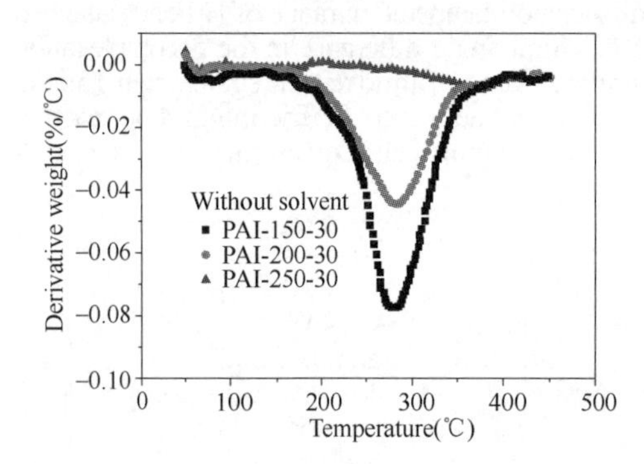

FIGURE 1.15 Dependence of derivative weights on temperature.

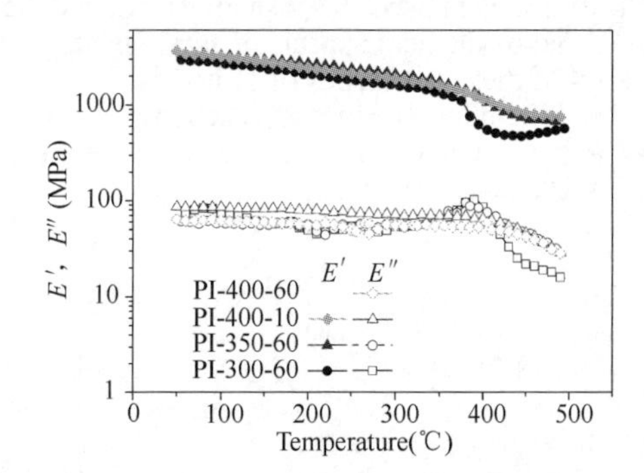

FIGURE 1.16 DMA curves of the polyimide films at a frequency of 1 Hz and a heating rate of 5 °C/min.

Lastly, PAI-150-30 (ID=19%) shows a very low solubility (~4%) in NMP, even after the sample is immersed in the solvent for 10 days. PAI-200-30 and PAI-250-30 are insoluble in NMP, indicating that the interchain imidization reaction forms a cross-linked percolation network at ID ~ 19% [27]. However, the form of intermolecular links and the relative proportions depend on the reaction conditions and the type of imide molecules used [22,23].

1.3.1.3 Annealing Stage (T>250 °C)

The annealing procedure is required for the fabrication of polyimide films because the imidization reaction has a broad tail on the high-temperature side. Because PI-400-10 is considered a complete imidization film, ID values of the other samples are calculated from the weight loss at 400 °C as 97% and 99% for PI-300-60 and PI-350-60, respectively. This reflects the clear temperature dependence of the imidization reaction. Fig. 1.16 shows the DMA results of the samples. E' in the glass region is app roximately 2.9 GPa for PI-400-10. The T_g obtained from the tanδ curves gradually increases from 395 °C for PI-300-60 and from 440 °C for PI-400-10 and PI-400-60. Therefore, the elevated annealing temperature especially above T_g increases the T_g and the storage modulus, and the annealing treatment is critical to the final properties of the polyimide film. It should be noted that PI-400-10 and PI-400-60 does not display an E'' peak and their E' decreases gradually during the transition region. These features may relate to the broad order-disorder transition.

Fig. 1.17 compares the mechanical properties of the polyimide films at different stages. Generally, the mechanical properties of the samples gradually increase as the thermal treatment temperature increases. At the solvent evaporation stage, the tensile strength and the Young's modulus of PAA films increase from 38 MPa and 1.6 GPa to 68 MPa and 1.8 GPa, respectively. The improvement in the mechanical properties is mainly due to the removal of the solvent. At the imidization stage, the tensile strength and the tensile modulus of the PAI films increase to 100 MPa and 2.4 GPa (2.4 GPa for E' from the DMA results). As the thermal temperature and the ID increase, the in-plane orientation of the molecular chain and T_g can be further improved [28]. Finally, the effect of the annealing temperature on the mechanical properties is obvious, although there is only a slight increase in ID. The tensile strength and the Young's modulus for the PI samples increase from 120 MPa and 2.5 GPa (2.5 GPa from the DMA results) for PI-300-60 to 150 MPa and 2.8 GPa (2.9 GPa from the DMA results) for PI-400-10 without losing the elongation at breakage. However, the PI-400-60 film exhibits the highest Young's modulus (3.0 GPa), but it loses the elongation at breakage. The mechanical properties of the films are closely related with the T_g (the right x-axis), and the improved mechanical properties may be the result of the ordered structure of the molecular chains, which are induced by annealing. Therefore, annealing near the T_g of the samples induces crystallization in polyimide films, and such a prominent annealing effect has been reported in the literature [29,30]. Conclusively, the imidization is completed by annealing at 400 °C and the crystallization degree increases as the annealing time increases. When annealing is conducted at 400 °C, the weight of the film decreases rapidly during the initial heating step due to the secondary imidization, and it continues to fall slowly as time elapses. This indicates that a long annealing time at 400 °C can cause the decomposition of the film, which may damage the properties of PI films. Thus, the optimized treatment condition may be between 350 °C/60 min and 400 °C/10 min.

Fig. 1.18 summarizes the changes in ID, tensile properties, and T_g of PAA films at three different imidization stages. In the solvent evaporation stage, the imidization reaction does not occur, and the effect of solvent evaporation determines the changes in mechanical and thermal properties that occur in the samples. The mechanical properties and T_g of the PAA films have been improved gradually due to the evaporation of solvent. Two types of solvents, which are attached to —NH or —COOH, can be distinguished by TGA and DTG. In the imidization stage, the ID of the samples can also be accurately monitored by TGA. The changes in the stiffness and the inplane orientation of the chains results from the imidization reaction. The tensile strength (100 MPa) and the Young's modulus (2.4 GPa) of the PAI-250-30 sample (ID=94%, T_g=393 °C) are similar to the values obtained for the final PI films. The annealing at 350 °C-400 °C can further improve the properties of PI films due to the "complete" imidization and the development of the crystallized structures.

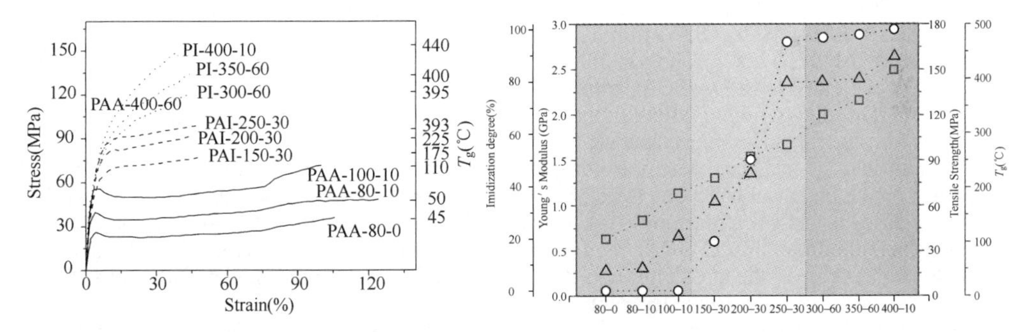

FIGURE 1.17 Tensile strain-stress curves of the films with different imidization degree.

FIGURE 1.18 Changes of imidization degree,tensile properties and T_g of PAA films at three different imidization stages.

Overall, the imidization process include three stages. During thermal imidization, there are obvious influences of curing process conditions on the physical and chemical evolutions of PAA-(PMDA/ODA) in NMP. An optimized condition (100 °C/10 min, 250 °C/30 min and 400 °C/10 min) has been recommended to fabricate high-quality polyimide films by a discontinuous process. A TGA method is developed here to accurately determine the ID; moreover, it can quantitatively distinguish the locations of solvent evaporation and the imidization reaction. These methods can be used to investigate the thermal imidization process for other polyimide systems with different chain structures and/or different solvents.

TABLE 1.6 Properties of Polyimide Films as Function of Final Curing Temperatures of 250 °C-450 °C

Cure Temperature(°C)	Δn	$n_{average}$	$\alpha_{in\text{-}plane}(\times 10^{-6}\ °C^{-1})$	Stress(δ)(MPa)	Modulus(E)(GPa)
PI-(BPDA/PDA):T_g=340 °C					
250	0.1826	1.7641	3±1		7.1±1
300	0.1960	1.7714	3±1	2±1	7.2±1
350	0.1979	1.7637	3±1	2±1	7.3±1
400	0.2186	1.7771	5±1	10±1	8.2±1
PI-(PMDA/ODA):T_g=420 °C					
250	0.0710	1.6857	26±1	16±1	2.2±1
300	0.0716	1.6870	26±1	16±1	2.3±1
350	0.0708	1.6869	26±1	17±1	2.3±1
400	0.0676	1.6914	32±1	18±1	2.4±1
450	0.0899	1.7010	36±1	25±1	2.5±1

1.3.2 Influence of Curing Temperatures on Film's Properties

Table 1.6 compares the film properties of two different polyimides as a function of final curing temperatures ranging from 250 °C to 450 °C. At 350 °C, the stiff PI-(BPDA/PDA) film has a high birefringence ($n_{average}$=1.7637, Δn=0.1979), low *CTE* ($\alpha_{in\text{-}plane}$=(3±1)×10^{-6} °C^{-1}), and high modulus (E=7.3±1 GPa), indicating that the polymer has high in-plane chain axis orientation. In contrast, the relatively flexible PI-(PMDA/ODA) film has a low birefringence ($n_{average}$=1.6869, Δn=0.0708), higher *CTE* ($\alpha_{in\text{-}plane}$=(26±1)×10^{-6} °C^{-1}), and low modulus (E=2.3±1 GPa), consistent with low in-plane chain axis orientation. The PI-(BPDA/PDA) film shows a stress level (δ) of 2±1 MPa, much lower than PI-(PMDA/ODA) film (δ=17±1 MPa), probably due to the increase in the *CTE* mismatch between the polyimide film and the underlying substrate.

The influence of final curing temperature on morphology and properties of polyimide films are dependent on the chemical backbone structures of the polyimides. PI-(BPDA/PDA) film (T_g=340 °C) shows increases in optical anisotropy (Δn), which is gradual from 0.1826-0.2186 with rising of the curing temperature from 250 °C to 400 °C. The PI-(PMDA/ODA) film remains relatively constant with curing temperature up to the glass transition temperature, and then increases dramatically from 400 °C to 450 °C above the glass transition temperature (T_g=420 °C).

Changes in the average refractive index are generally due to changes in density. Polyimide film increases its density with an increase in crystallinity. The average refractive index of PI-(BPDA/PDA) film increases only slightly with curing temperature up to 400 °C, while PI-(PMDA/ODA) film remains relatively constant with curing temperature up to 350 °C, and then increases significantly at higher curing temperatures. The crystallinity in PMDA/ODA film increases when cured above the glass transition temperature.

The stress in PI-(BPDA/PDA) film is low at a curing temperature of <350 °C, and then rises sharply at a higher curing temperature (400 °C) (Table 1.6). Similar results are observed in PI-(PMDA/ODA) film, where the stress increases only slightly at curing temperatures of lower than its T_g. The stress level rises significantly at curing temperatures of higher than T_g. This is attributed to the semicrystalline nature of the polyimide films. The semicrystalline polyimides can sustain stress above their glass transition temperatures because the modulus remains relatively high. In addition, shrinkage forces resulting from density increases contribute to increases in the in-plane stresses.

Fig. 1.19 compares the ID as a function of curing temperature for two different polyimide films (PI-(PMDA/ODA) and PI-(BTDA/ODA-MPD). At lower temperatures, the ID varies significantly with the chemical structures of the polyimide films. At 250 °C, all the polyimides show ID of over 95%. The ID increases gradually from 95% to 100% by heating at 250 °C-350 °C. The ID of PI-(PMDA/ODA) material appears to decrease by about 4% from its maximum at 350 °C to 450 °C, probably attributable to the development of crystallinity. It was found that the imide group in PI-(BPDA/PDA) preferentially aligns parallel to the plane of the film [31].

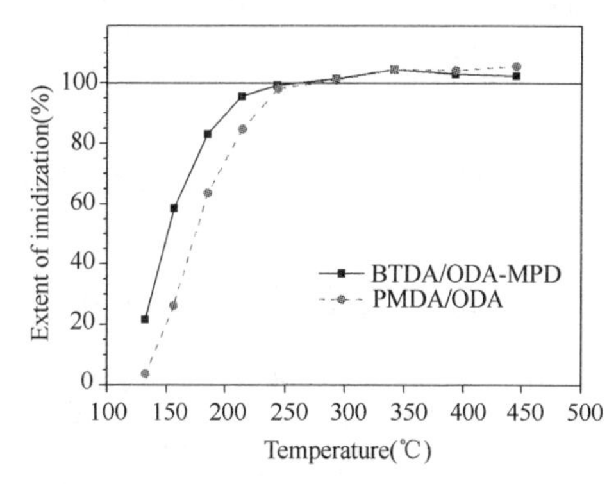

FIGURE 1.19 Imidization degree as a function of curing temperatures for three different polyimide films.

PI-(PMDA/ODA) film cured at 400 °C shows a broad glass transition at approximately 420 °C by DMA, compared with PI-(BPDA/PDA) with a broad and slightly more prominent glass transition observed at 340 °C. The broad glass transitions are attributed to the semicrystalline morphology, and the small magnitude is due to both the polyimide backbone stiffness and the presence of crystallinity. The T_g of PI-(PMDA/ODA) film was measured at about 420 °C, which is not affected by the changes in final curing temperatures. The magnitude of glass transition is also not affected by curing temperatures of up to 350 °C. However, the magnitude decreases slightly at the curing temperature of 400 °C and reduces sharply at 450 °C. This is attributed to the increase in crystallinity in PI-(PMDA/ODA) films cured near or above the glass transition temperature.

The (004) peak in X-ray diffraction patterns of PI-(BPDA/PDA) films cured at 400 °C is very prominent in transmission and absent in reflection. This is due to the preferential alignment of the chain axis in the plane of the film. The (002) diffraction peak of PI-(PMDA/ODA) film is prominent in transmission yet barely visible in reflection, due to the preferential alignment of the chain axis in the plane of the film. The reflection and transmission scans show some evidence of crystalline order. However, the crystalline peaks are not very prominent and are overshadowed by the more prominent broad amorphous peak. The (002) peak of PI-(PMDA/ODA) film in transmission increases in magnitude with changes in the curing temperatures. Other diffraction peaks are sharper and more prominent in the film cured at 450 °C, evidence of higher crystalline order. This behavior is consistent with the behavior observed by DMA.

PI-(PMDA/ODA) and PI-(BPDA/PDA) films are stiff and anisotropic with lower in-plane *CTEs* and stresses. These polyimides crystallize when cured above their glass transition temperatures. Due to the relatively high glass transition temperature (420 °C), PI-(PMDA/ODA) does not have appreciable crystallinity when cured using a standard heating rate of 2 °C/min up to temperatures as high as 400 °C. The birefringence and the average refractive index in PI-(PMDA/ODA) film remain relatively constant up to 400 °C. However, the birefringence decreases slightly at 400 °C, then increases significantly at 450 °C. PI-(BPDA/PDA) film is slightly more rigid and linear than PI-(PMDA/ODA). The stiffness and linearity of PI-(BPDA/PDA) lead to a higher degree of anisotropy. The birefringence increases gradually with curing temperature up to the region of the glass transition (350 °C). At 400 °C, the birefringence index and the average refractive index increase substantially. These increases are attributed to increases in crystalline order and in density.

The molecular anisotropy in polyimide films has a dramatic effect on both in-plane and out-plane *CTE*s. Because the coefficient of volumetric expansion (*CVE*) is constant, unaffected by orientation [32], the out-of-plane *CTE* increases while the in-plane *CTE* decreases with increasing in-plane chain axis alignment. The *CVE* can be measured from specific volume vs. temperature measurements. PI-(BPDA/PDA) film shows only a gradual increase in the slope of the specific volume data in the temperature range of the glass transition by DMA (340 °C), consistent with the less prominent and relatively broad glass transition observed by DMA. PI-(BPDA/PDA) film shows an in-plane *CTE* of 6×10^{-6} °C^{-1} and an out-of-plane *CTE* of 144×10^{-6} °C^{-1}, compared with PI-(PMDA/ODA) film which has an in-plane *CTE* of 33×10^{-6} °C^{-1} and an out-of-plane *CTE* of 126×10^{-6} °C^{-1}, respectively. The anisotropy nature of the *CTE*s is an important functional property which is extremely sensitive to molecular orientation. PI-(BPDA/PDA) film exhibits obvious anisotropy, has an out-of-plane *CTE* of 25 times larger than the in-plane *CTE*. This sensitivity of *CTE* to molecular orientation must be taken into account in designing polyimide films for microelectronic applications.

1.4 Structures and Properties of Polyimide Films

1.4.1 Advanced Polyimide Films

Advanced polyimide films have great combined properties, such as high mechanical strength and modulus, low thermal dimensional expansion, high adherence to metal and oxide material surfaces (copper, aluminum) and acceptable prices in market, etc., suitable for applications in manufacturing of high-density electronic packaging substrates. The common chemical structures of the advanced polyimide films were produced by the chemical imidization method derived from the aromatic dianhydrides (PMDA, or/and BPDA) and aromatic diamines (ODA, or/and PDA), etc. The representative successful commercial products in the current market include Kapton EN, Upilex S, Apical HP, etc.

The thermal and mechanical properties of advanced polyimide films can be adjusted by appropriately selecting the ratios of the aromatic dianhydrides such as PMDA and BPDA and diamines such as ODA and PDA in the formation of PAAs [33]. The PI-(BPDA/PDA) film has the highest modulus, the highest T_g, and the lowest *CTE* value among the polyimide films derived from BPDA, PDA, and ODA. These best properties are attributed to the rigid chemical structure of BPDA and PDA due to their nonflexible linkages in the molecular backbone structures. On the other hand, ODA has a flexible ether linkage due to the possibility of bending and rotation of the molecular chain. As the mole ratio of PDA/PDA+ODA in the copolyimide PI-(BPDA/PDA-ODA) films increases, the elastic modulus increases accordinglys from 3.8×10^{-6} to 7.8 GPa (Fig. 1.20) and T_g increases from 290 °C to 420 °C (Fig. 1.21), while *CTE* decreases from 36×10^{-6} to15×10^{-6} °C^{-1}. It was found that even if a small proportion of ODA is incorporated with the polyimide backbone, the structure is disturbed and the obvious drop in T_g occurs, indicating that a simple additivity of component could cause obvious changes in T_g and that the effect of component would be sensible for rigidity.

In addition, the homopolyimide PI-(PMDA/PDA) film was expected to show the higher modulus and T_g than the homopolyimide PI-(BPDA/PDA) systems due to the greater rigidity of PMDA compared to BPDA. Unfortunately, the homopolyimide PI-(PMDA/PDA) film prepared is too brittle to measure the mechanical properties. Even the copolyimide PI-(PMDA/PDA-ODA (50/50)) film also showed a very low elongation (19%) and higher *CTE* (33×10^{-6} °C^{-1}). The copolyimide films PI-(BPDA- PMDA/PDA) have large elongation at breakage, high modulus, as well as low *CTE*, when the mole ratio of BPDA/MPDA is greater than 50% (mol). On the other hand, the copolyimide films PI-(BPDA-PMDA/ODA) showed much higher *CTE*s and elongation at breakages than that of the corresponding PI-(BPDA-PMDA/ PDA) film.

Compared with the three-component copolyimide films derived from BPDA, MPDA and PDA, the four-component copolyimide films derived from BPDA-PMDA and PDA-ODA showed complex relationships between the thermal and mechanical properties and the polymer backbone structures. Only 10%—30% (mol) of ODA was allowed to incorporate in the copolyimide backbone for balancing of *CTE* and elongation at breakage of the films. High ODA loadings resulted in the *CTE* increasing significantly. The copolyimide films derived from BPDA-PMDA (50/50) and PDA-ODA showed almost linear relationships between the *CTE* and the PDA loadings (Fig. 1.22) and between the elastic modulus and the BPDA loadings (Fig. 1.23). The copolyimide film was prepared derived from BPDA-PMDA (50/50) and PDA- ODA (70/30), which showed the best combination of properties with *CTE* of 19×10^{-6} °C^{-1}, elongation at breakage of 60%, modulus of 5.0 GPa, and T_g of 360 °C, successively.

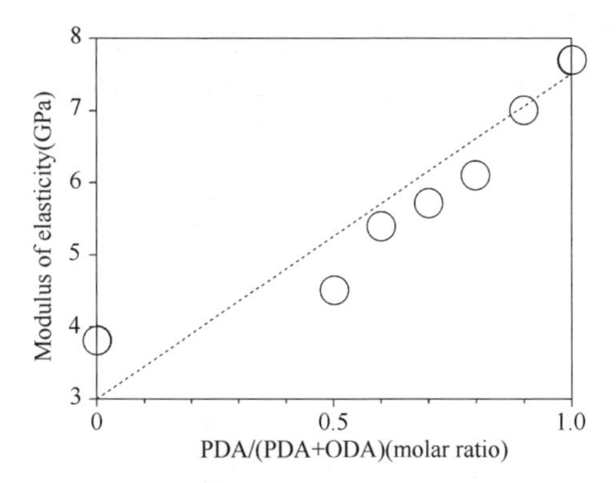

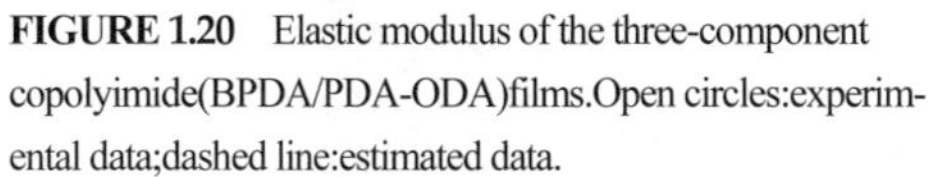
FIGURE 1.20 Elastic modulus of the three-component copolyimide(BPDA/PDA-ODA)films.Open circles:experimental data;dashed line:estimated data.

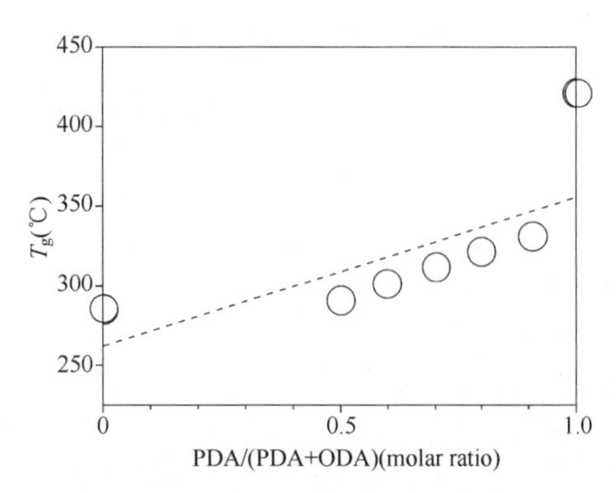

FIGURE 1.21 T_g of the three-component copolyimide (BPDA/PDA-ODA)films.Open circles:experimental data;dashed line:estimated data.

Table 1.7 compares the properties of the three-component and four-component copolyimide films. A trade-off correction was clearly observed between the mechanical properties and the thermal properties, and the balanced properties of the polyimide films can be controlled, to some extent, by adjusting the polyimide backbone structures.

Because the apparent viscosity of PAA resin is mutually restricted by its solid concentration and/or polymer molecular weight, the PAA resin solutions derived from rigid aromatic dianhydrides and diamines always have too high solution viscosities for casting films on the support surface. Hence, PAA resins with relatively high concentrations ($\geqslant$15%) and low solution viscosity in the range of 10 mPa·s-30×10^4 mPa·s at room temperature are desired for the manufacture of polyimide film with high modulus and low thermal expansion.

In order to determine the influence of PAA molecular weights on their solution-cast processing performance, a series of molecular-weight-controlled polyimide precursors (PI-PEPA) with designed calculated M_w were prepared from the reaction of BPDA and PDA using PEPA as a chain-extendable and end-capping reagent at concentrations of 20% in DMAc, as depicted in Fig. 1.24. The molecular weights were controlled by termination main-chains with the calculated molar fractions of PEPA to yield PAAs with calculated M_w of 5×10^3 g·mol^{-1}to 50×10^3 g·mol^{-1}. An analogous series of the unreactive phthalic-end-capped PAA resins (PAA-PA) with equal controlled M_w of 5×10^3 g·mol^{-1}-50×10^3 g·mol^{-1} was also made for comparison.

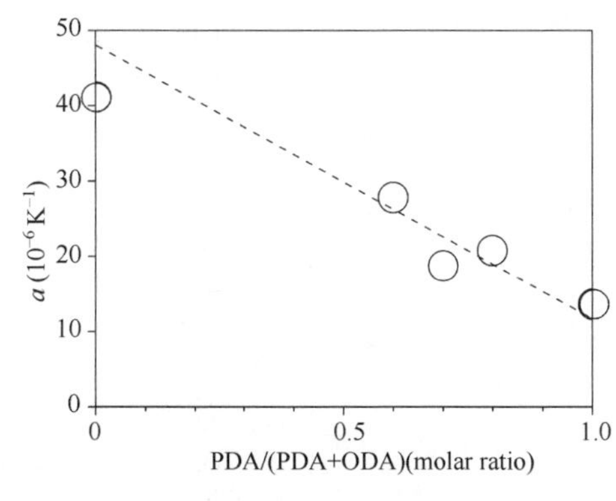

FIGURE 1.22 *CTE* of the four-component copolyimide(BPDA-PMDA/PDA-ODA)films. Open circles:experimental data;dashed line:estimated data.

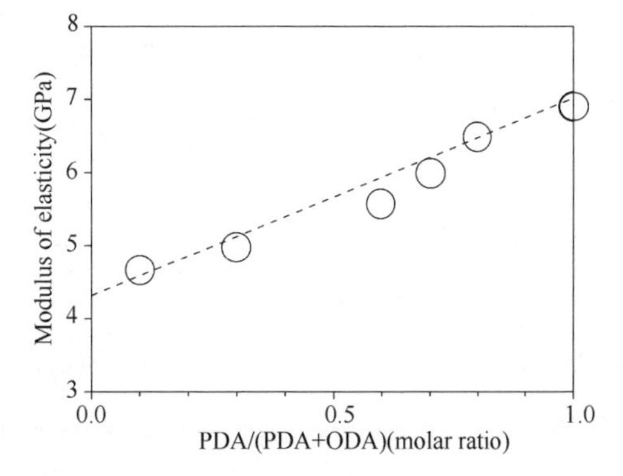

FIGURE 1.23 Elastic modulus of the four-component copolyimide(BPDA-PMDA/PDA-ODA(90/10))films.Open circles:experimental data;dashed line:estimated data.

TABLE1.7 Properties of Three-and Four-Component Copolyimide Films

	Three-component PI Film(BPDA/MPDA (50/50)-PDA)	Four-component PI Film(BPDA/PMDA(50/50)-PDA/ODA(70/30))
T_g(°C)	369	360
$CTE\times10^{-6}$ °C^{-1}(50–200 °C)	14	19
Tensile modulus(GPa)	6.4	5.0
Elongation at breakage(%)	38	60

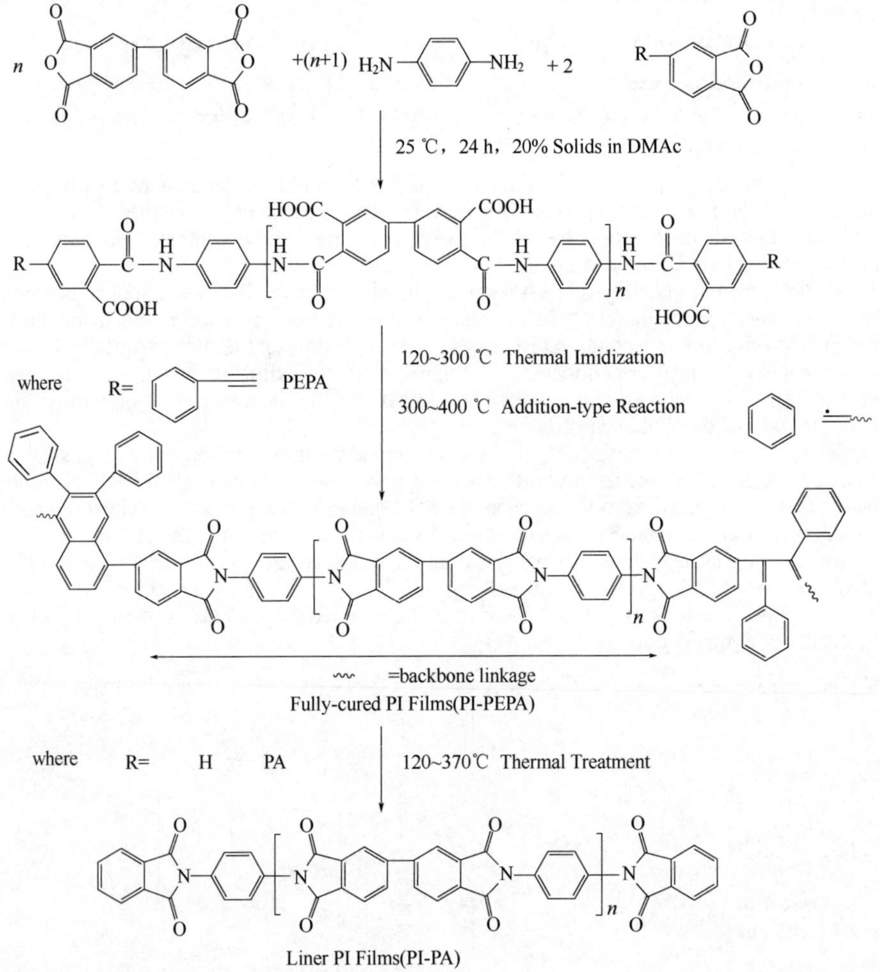

n + (*n*+1) H_2N–C₆H₄–NH_2 + 2

25 ℃, 24 h, 20% Solids in DMAc

where R= PEPA

120~300 ℃ Thermal Imidization

300~400 ℃ Addition-type Reaction

=backbone linkage

Fully-cured PI Films(PI-PEPA)

where R= H PA

120~370℃ Thermal Treatment

Liner PI Films(PI-PA)

FIGURE 1.24 Synthesis of molecular-weight controlled polyimides.

1.4.1.1 Film-forming Ability of Molecular-weight-controlled PAAs

The suitable apparent viscosity of PAA resin with high concentration is an important solution-cast processing parameter for the large-scale production of high-quality polyimide films. Fig. 1.25 shows the dependence of solution viscosity on the concentration for the molecular-weight-controlled PAAs with calculated M_w of 25×10^3 g·mol^{-1} and the molecular-weight-uncontrolled PAAs. A systematic rising trend in the apparent viscosities is observed with increasing concentration, and two regions with quite different slopes suggest critical concentrations (c^* and $c^{*\prime}$) at about 10% for PAAs (BPDA-PDA) without molecular weight limitation and 20% for molecular-weight-controlled PAAs (PAA-PEPA-25 and PAA-PA-25), respectively. It is noteworthy that the apparent viscosity of PAA resin varies as the fifth power of the concentration as the concentration exceeds 10%, and reaches over 9000 Pa·s with the concentration in excess of 20%.

In comparison, apparent viscosities of PAA-PEPA-25 and PAA-PA-25 are no more than 82 Pa·s with the concentration below 20%, after which an abrupt growth is noticed by a smaller exponential rate, proportional to the cube of PI precursor concentration. The observed huge disparity in the concentration dependence of solution viscosity indicates that molecular weight control is a promising pathway to readily adjust PAA resin to the desired practical film solutioncast processing conditions [3,4].

Fig. 1.26 describes the molecular weight dependence of apparent viscosity for molecular-weight-controlled PAA resins with concentration of 20% in DMAc at 25 °C. The apparent viscosities of both PAA-PEPA and PAA-PA increase gradually to 120 Pa·s when PAA calculated $M_w \leqslant 30\times10^3$ g·mol^{-1}, after which the seemingly exponential growths to 1160 Pa·s are displayed, rising in decreasing order: PAA-PA > PAA-PEPA. Obviously, the critical calculated M_w for molecular-weight-controlled PAAs imply that the overlapping of PAA main-chains and complexation-mediated solubilization between amide-acid chains and solvent become more prominent as the calculated M_w are over 30×10^3 g·mol^{-1}. [5] The interactions and physical-type entanglements between polymer chains, primarily driven by hydrogen bonds and charge-transfer complexes (CTC), are more likely to play critical roles in concentrated solutions and govern their apparent viscosities, leading to the observed phenomenon that the molecular weights studied of molecular-weight-controlled precursors strongly affect the apparent viscosity behavior. Moreover, PAA-PEPA series might possess weaker internal friction upon shear deformation at the same controlled calculated M_w than PAA-PA ones. Therefore, introduction of PEPA end-capping into BPDA-PDA main-chain could be more effective to improve the solution-cast processability.

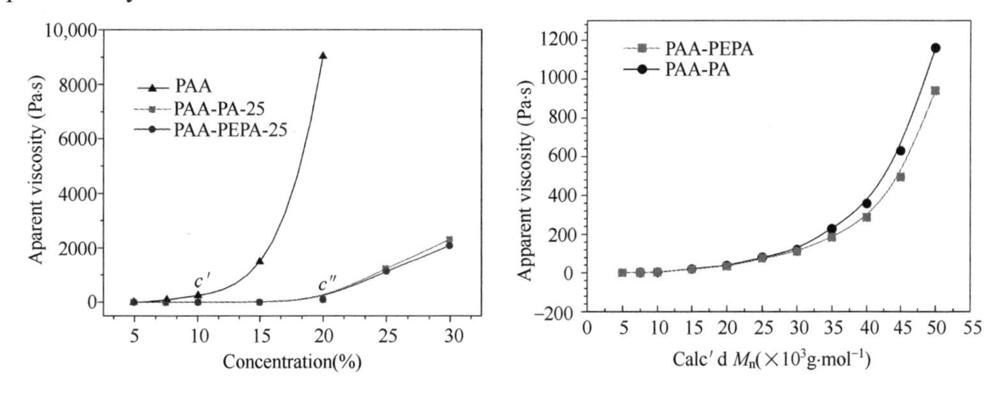

FIGURE 1.25 Dependence of apparent viscosity on concentrations of the molecular-weight-controlled PAAs with calculated M_w of 25×10^3 g·mol^{-1} and the molecular-weight-uncontrolled PAAs.

FIGURE 1.26 Molecular weight dependence of apparent viscosity for the initial molecular-weight-controlled PAA with concentration of 20%.

1.4.1.2 Mechanical Properties of Fully Cured Polyimide Films

The fully cured PI films were fabricated by classical two-step polymerization method in which the aforementioned molecular-weight-controlled PAA-PEPA resins with appropriate viscosities were thermally activated cyclized and successively stepwise-cured finally at 400 °C/1 h. For comparative purposes, the aforesaid PAA-PA controls end-capped with unreactive phthalic groups were thermally imidized up to 370 °C/1 h to yield typical linear PI films specified as PI-PA. All the films obtained were smooth, creasable, and yellow to orange in color.

Mechanical property comparatively recorded for PI-PEPA and PI-PA as a function of PAA calculated M_w is plotted in Fig. 1.27 and representatively summarized in Table 1.8. The tensile strength of PI-PEPA-5 to PI-PEPA-20 with PAA-PEPA calculated $M_w \leqslant 20\times10^3$ g·mol^{-1}, as illustrated in Fig. 1.27A, increases impressively by 38%, reaching 231.7 MPa. And then it appears to level off at 237.6 MPa when calculated M_w is above 20×10^3 g·mol^{-1}. This suggests a critical calculated M_w of PAA-PEPA at around 20×10^3 g·mol^{-1}, above which very small rises in the molecular weights do not affect the strength remarkably. In contrast, each film in PI-PA series has much lower strength at the identical calculated M_w value. For instance, PAA-PAs with calculated $M_w < 15\times10^3$ g·mol^{-1} were not transformed to the free-standing PI films, because weak molecular interactions cannot offset internal stresses of shrinkage in the imidization process. But PAA-PEPA-5 was successfully converted to PI-PEPA-5 with the strength stabilized at 168.5 MPa, presumably by virtue of chain-extension effect occurring in the phenylethynyl moiety [34,35]. In Fig. 1.27B, the elasticity modulus of PI-PEPA decreases slightly from 7.7 to 7.3 GPa as the theoretical PAA-PEPA M_w grows to 20×10^3 g·mol^{-1}, probably due to the reduced chain crosslinking densities; it becomes comparable for both PI versions with a further rise in PAA calculated M_n. The elongations at break (Fig. 1.27C) see a steady increase in the whole PAA calculated M_w of 382% and 325% for PI-PEPA and PI-PA, respectively. And PI-PEPA samples show generally higher elongations to failure; e.g., PI-PEPA-25 exhibits elongation of 8.5%, 39% higher than PI-PA-25 (6.1%).

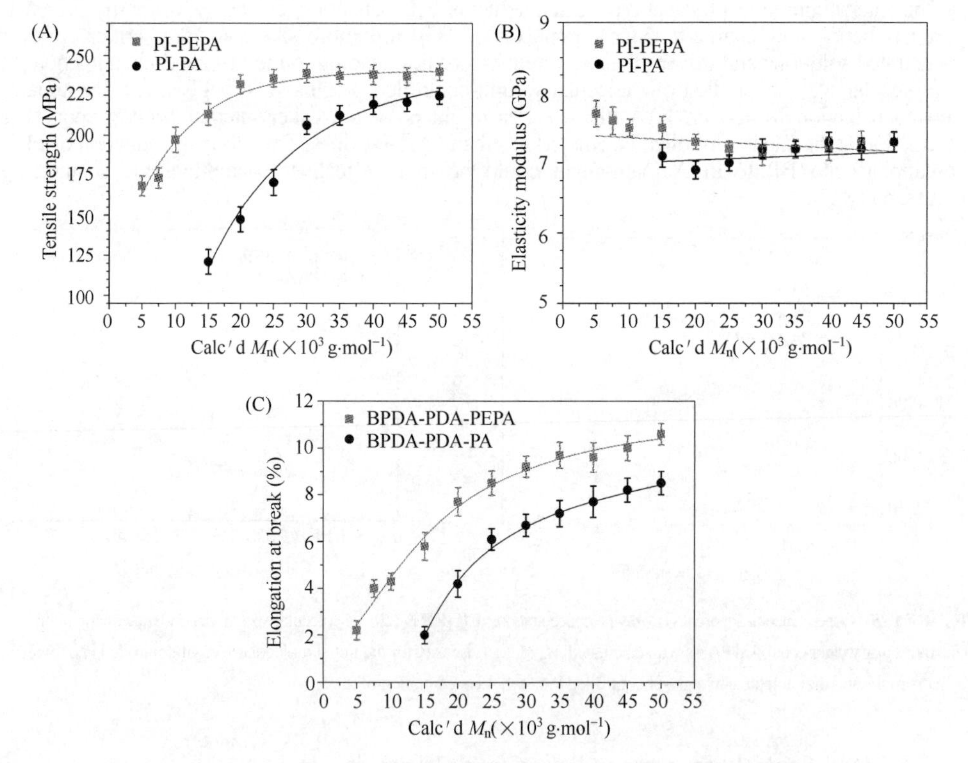

FIGURE 1.27 Film mechanical property as a function of PAA calculated M_w(A)tensile strength;(B)elasticity modulus;(C)elongation at breakage.

The mechanical property variation is principally related to the molecular weight of PAA. The fully cured PI films obtained from reactive phenylethynyl-functionalized PAAs with calculated M_w>20×10^3 g·mol^{-1}

surprisingly demonstrate much higher tensile strengths and elongations to deformation as compared to the linear polyimide controls with unreactive phthalic end-groups. Generally, the physical property is sensitive to the state of molecular interactions such as crosslinking density, the degree of crystallinity, and molecular stacking. Hence, the conspicuous enhancement in the tensile property of the fully cured PI film is assumed to be favored by the evolution in the aggregation structure effectuated by the thermal-curing reaction of phenylethynyl end-groups at elevated curing temperatures.

TABLE 1.8 Typical Mechanical Properties of the Fully Cured Polyimide Films Versus Linear Films[a]

PI Films	Tensile Strength σ_t(MPa)	Elasticity Modulus E(GPa)
PI-PEPA-5	168.5	7.7
PI-PEPA-15	213.0	7.5
PI-PEPA-20	231.7	7.3
PI-PEPA-25	235.6	7.2
PI-PEPA-50	239.5	7.3
PI-PA-15	120.9	7.1
PI-PA-20	147.4	6.9
PI-PA-25	170.6	7.0
PI-PA-50	224.5	7.3

[a] *σ_t, tensile strength; E, elasticity modulus; δ, elongation at break; s, standard deviation.*

1.4.1.3 Effect of Thermal Curing on Aggregation State of Polyimide Films

The aggregation structure of polyimide film is intensely affected by the variables that control the thermal-curing process, e.g., temperature and time, which in turn would directly play a critical role in dictating mechanical and thermal properties [36,37]. It was expected that thermal transitions and concomitant morphology changes caused by the thermally activated, free-radical-predominant curing of phenylethynyl functional group under various processing conditions should be indicative of the curing behavior and mechanical response of PI films. PAA-PEPA-25 was selected to experience each curing procedure compared with PAA-PA-25.

Fig. 1.28 depicts the impact of thermal-curing processes on the crystallization behavior and morphological features of PI-PEPA-25 and PI-PA-25 films characterized by WAXD. Broad amorphous halos are visible for both series of films heated at lower temperatures (300 °C). And yet the relative intensities of three distinguishable diffraction peaks superimposed in the range of 18°-26° increase drastically, and each peak's half-width decreases with rising in the final curing temperature from 370 °C to 400 °C, which reveals a rapid enhancement in the regularity of intramolecular arrangements in polyimide films [38]. Interestingly, the crystalline peaks in PI-PEPA-25 ultimately thermally processed at 450 °C transform to obtuse and less-structured patterns, indicating a reduction in chain ordering, while that in PI-PA-25 counterpart becomes sharper. The semicrystallized state disparity could be ascribed to the fact that the presence of the phenyl-ended moiety in the chain-extension structure marginally widens the extended molecular chains of layer structure in eclipsed crystals, thereby generating less-dense chain stacking and slight depression in the T_m [39,40].

The effect of curing temperatures on the ratios of crystalline regions estimated from the diffractogram of the resulting PI film is compared in Fig. 1.28C. As the final heating temperature is ramped from 350 °C to 425 °C, the degrees of crystallinity in PI-PEPA-25 are 7.22%, 14.68%, 24.98%, and 27.26%, respectively, appearing to increase linearly. And the remarkable discrepancies of 2.17%, 6.95%, 3.89%, and 4.79% are observed as well, implying that the curing reaction occurring in the phenylethynyl moiety strengthens the ability to crystallize. According to Lambert et al. [41] the negligible proportion of the bulky crosslinks (cyclotrimer and polyene) significantly improves the rate of secondary nucleation in a way which causes the growth rate to increase drastically with decreasing crosslinking density, and the chain-extension moiety can be incorporated into the crystals. The cyclodimer as strong intermolecular bond localized at chain ends evolves gradually and constrains conformational freedom in the main-chain with elevated temperatures. It is reasonable to assume that the suppression of chain mobility due to the chain-extension chemistry is favorable to developing sparsely packed chains, inducing the ordered phase more easily than the rearrangement in intrinsically extended segments in the linear PI-PA-25 control [42].

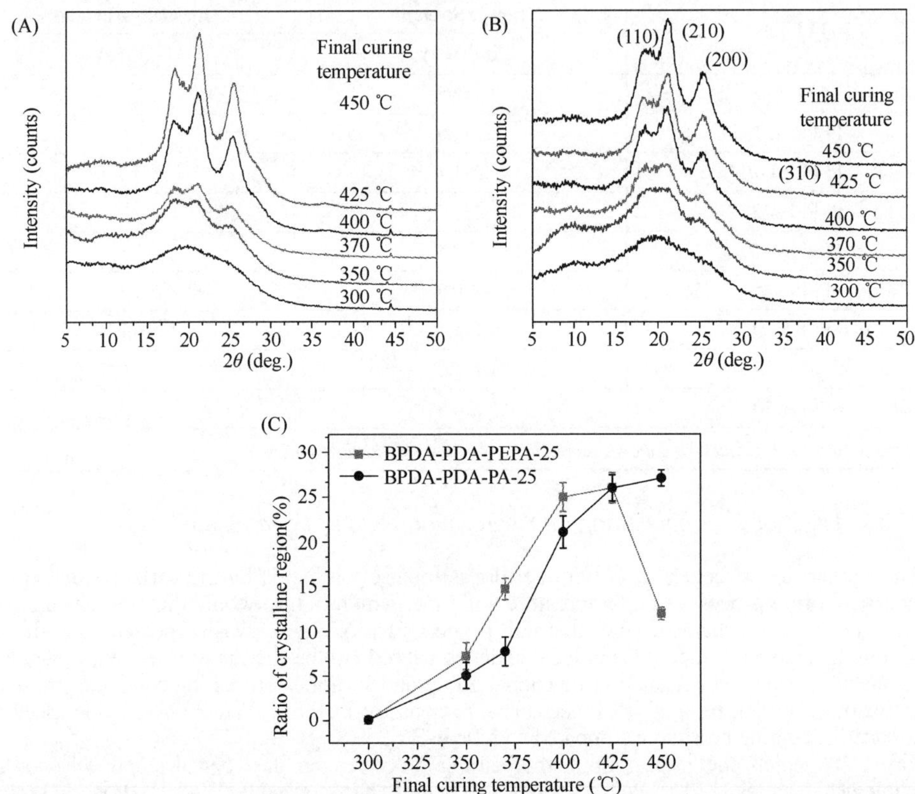

FIGURE 1.28 XRD patterns of PI-PEPA-25 films(A)and PI-PA-25 films(B) cured at different temperatures,and the effect of final curing temperature on the ratio of crystalline region(C)

1.4.1.4 Effect of Thermal Curing on Molecular Orientation and CTE of PI Films

The effects of curing procedures on molecular orientation and thermal dimension stability of polyimide films prepared from PAA-PEPA-25 and PAA-PA-25 are contrastively explored in Fig. 1.29. Both films with BPDA-PDA based backbone architectures show large persistence lengths in the chain direction and low chain flexibility, and the high birefringence (Δn) and low *CTE* consistent with a high level of in-plane chain orientation are normally manifested.

In Fig. 1.29A, the uncured polyimide film bearing the reactive phenylethynyl end-cap has lower birefringence and much higher *CTE* than the linear control with the unreactive phthalic end-group, declaring a great influence of phenylethynyl moiety on chain configuration by hindrance to rotation. The birefringence associated with the optical anisotropy increases by 33.7% for PI-PEPA-25, as compared to 15.9% for PI-PA-25 control, until the PAA films are ultimately thermally treated at 370 °C.

In fact, the super-molecular structure is governed by the competition in orientation-relaxation in the polymer [43]. The rapid increase in the extent of molecular alignment parallel to the thermally cured film plane could be entailed not only by the fast upward trend in crystalline order as testified by WXRD, but also by the formation and retention of regular packing in molecular chains because of charge-transfer interactions over several consecutive repeat units in the amorphous regions [6]. At the same time, relaxation originated from random thermal motions could be effectively prevented by crystalline domain and short-range order region especially when the final curing temperature in CP1-CP3 is well below the T_g. But PI-PEPA-25 finally heated from 425 °C to 450 °C shows reduced birefringence and conversely PI-PA-25 counterpart exhibits slightly increased Δn. This tendency is similar to that in the degree of crystallinity, since the in-plane orientation engendered by short-range order is likely to be fairly countered by segmental relaxation above the T_g [42].

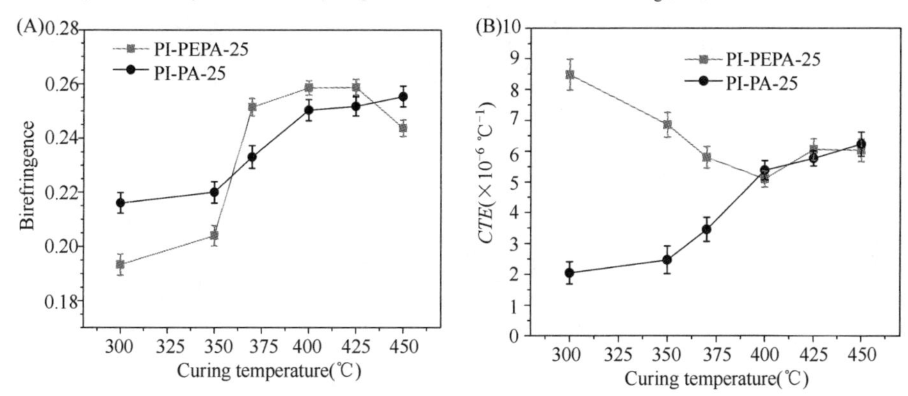

FIGURE 1.29 (A)Birefringence and (B) *CTE* of PI Films thermally treated at various final curing temperatures.

As expected in Fig. 1.29B, the *CTE* of the thermally cured material derived from PAA-PEPA-25 decreases dramatically from 8.5×10^{-6} to 5.1×10^{-6} °C^{-1}, while that of PI-PA-25 is on the rise when the final curing temperature is set from 350 °C to 400 °C. This apparent disparity is associated with the different reasons for alterations in the birefringence and the *CTE*: the growth in Δn depends on the increase in a net rise in the preferential alignment of the imide rings in the in-plane direction; the increase in the *CTE* is contingent on the upward level of in-plane orientational relaxation. Nonetheless, the *CTE* data increase by around 18.4% for both PI films upon ultimately conditioning at 400 -450 °C.

1.4.1.5 Effect of Thermal Curing on Mechanical Properties of Polyimide Films

In order to further apprehend the thermal-curing effect of phenylethynyl moieties in the relatively high-molecular-weight PI film, the influence of the curing procedure on the mechanical properties of PI film is evaluated in Fig. 1.30. There is an optimal curing procedure (CP3) with the ultimate temperature as high as 400 °C for phenylethynyl-end-capped PAA resin (e.g., PAA-PEPA-25) transforming to the fully cured polyimide film (PI-PEPA-25). Tensile strength and elongation at break increase by 55% and 83% with advancement of the final curing temperature to 400 °C, reaching the highest figures (247.1 MPa and 8.6%). And then they drop substantially to 186.7 MPa and 3.4%, respectively, showing that the failure mode has turned to brittle fracture. A similar trend is visible for the linear PI-PA-25, and the largest strength and elongation of the material heating up to 370 °C are 25% and 26% lower than PI-PEPA-25. By increasing the curing parameters studied, the alterations in the crosslinking structures triggered by thermal-curing reaction, densification, and then thermo-oxidative degradation drives the elasticity modulus of both PI films to rise monotonously as anticipated.

On the basis of the deduction from WXRD and thermomechanical analyses, the highly disordered crystals in intramolecular chains develop faster with evolution in chain-extension structures by the curing reaction of phenylethynyl moieties until the final curing temperature proceeds to the critical point (400 °C). The plastic deformation of aromatic PIs can be interpreted in terms of local order consisting of bundles of parallel-packed main-chains [7,8]. Consequently, relatively more microcrystals and short-range ordered domains with stronger molecular interactions could reinforce and toughen PI film. However, once the final curing step exceeds the critical point, the disorientation and degradation of molecular chains occur in the amorphous phase and crystalline area, holding the key to disrupting polymer chains and breaking ordered fractions.

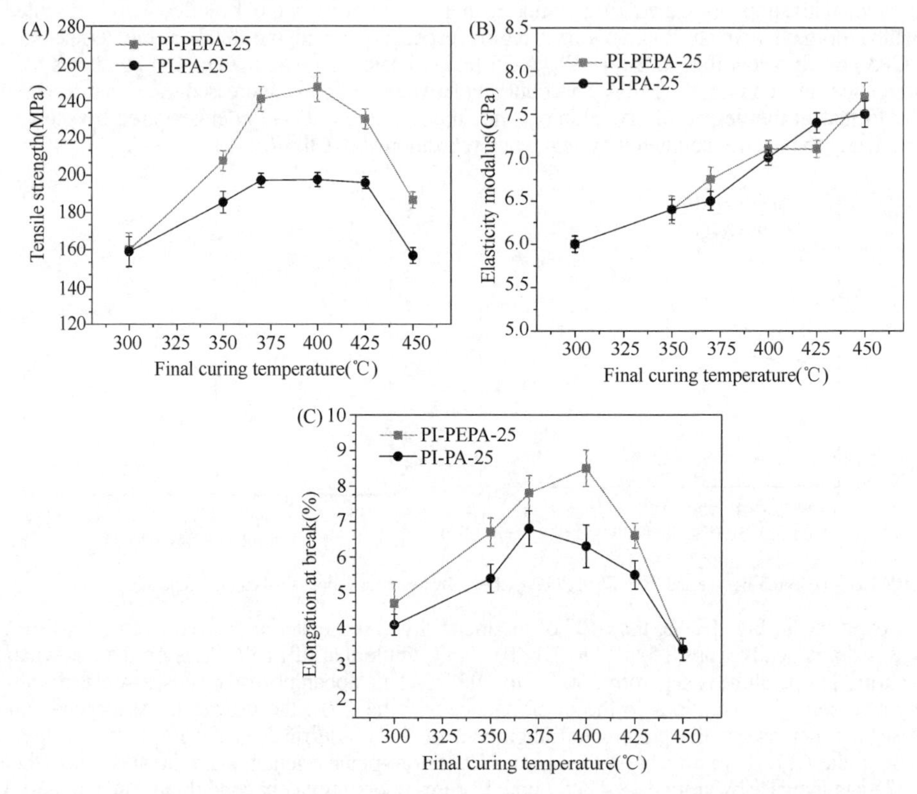

FIGURE 1.30 Effect of curing temperature on mechanical property of PI films(A)tensile strength;(B)elasticity modulus;(C)elongation at breakage.

Overall, the molecular-weight-controlled PAA resins end capped by reactive phenylethynyl groups with calculated $M_w > 20\times10^3$ g·mol^{-1} exhibit moderate viscosities and high solid concentrations, showing enhanced wetting/spreading ability to form continuous PAA films. After thermal curing at 400 °C/1 h, the PAAs are successfully converted to the fully cured polyimide films which display better mechanical properties and thermal resistances than the high-molecular-weight, unreactive phthalic end-capped control, suitable for high-temperature microelectronic packaging applications. This property improvement may be put down to the thermal-curing effect of phenylethynyl end-groups on promoting ordering/packing of chain segments, resulting into an increment in the molecular interactions to reinforce and toughen PI films.

1.4.2 Low-*CTE* Polyimide Films

Aromatic polyimide films usually have much higher linear *CTE*s in the film plane direction ($(40-80)\times10^{-6}$ °C^{-1}) than those of metal substrates (e.g., 17×10^{-6} °C^{-1} for copper foil). When polyimide films were formed on a metal substrate via thermal imidization after the solution casting of PAA solution, the polyimide/metal laminates undergo thermal stress arising from the *CTE* mismatch during the cooling process from cure temperature to room temperature. Consequently, serious problems are caused such as curling, cracking, and detaching of polyimide films. Considerable efforts have been made to attempt to decrease the film's *CTE*s. Systematic researches on the structure-*CTE* relationship in various aromatic polyimide systems revealed that low-*CTE* polyimide films ($<20\times10^{-6}$ °C^{-1}) have unexceptionally linear/stiff polyimide chain backbones. A typical low-*CTE* polyimide film is the Upilex S based on BPDA and PDA, which exhibits *CTE* values in the range of 5×10^{-6} to 20×10^{-6} °C^{-1} depending on the film thickness and processing conditions. The low *CTE* characteristics are closely related to a thermal imidization-induced in-plane chain orientation phenomenon.

Recently, demands in electronic applications have required polyimide films with both *CTE* and coefficient of humidity expansion (*CHE*), apart from the excellent mechanical properties. For PFC applications, adhesive-free polyimide film/Cu laminates (FFL, flexible copper clad laminates) are usually fabricated by direct coating of PAA solution onto copper foil. In this case, the *CTE*s of polyimide films must be precisely controlled to avoid the serious problems mentioned above. Another important requirement for polyimide films is the dimensional stability against water absorption or to decrease water absorption itself. In general, due to the highly polarizable imide groups in the backbone structures, polyimide films absorb water more easily in air than polyester films. For instance, commercially available Kapton H film (PI-(PMDA/ODA) has a moisture uptake of 2.5%-2.8%(*w*), much higher than polyester films (0.4%-0.6%). It is known that the lower water absorption characteristics of polyester films are attributed to the ester linkages in their backbone structures. Hence, polyimide films with ester-linkages in their backbone structures have been considered to be useful for reducing water absorption.

1.4.2.1 Ester-Bridged Low-CTE Polyimide Films

Hasegawa et al. have systematically investigated the structure-property relationship of polyimide films with ester-linkages in the polymer backbones (poly(ester-imide)s, PEsI) [44]. An ester-containing dianhydride monomer (Fig. 1.31), bis(trimellitic acid anhydride)phenyl ester (TAHQ), was used to react with various aromatic diamine monomers with stiff/linear structures, such as PDA, *trans*-1,4-cyclohexanediamine (CHDA), 2,2-bis(trifluoromethyl)benzidine (TFMB), 4-aminophenyl-4-aminophenyl -benzoate (APAB), and 4,4-daminobenzanilide (DABA) to give PEI precursors (PEsAA) solution, which was then cast on substrate to form PEsAA films and thermally imidized at high temperatures to give PEsI films.

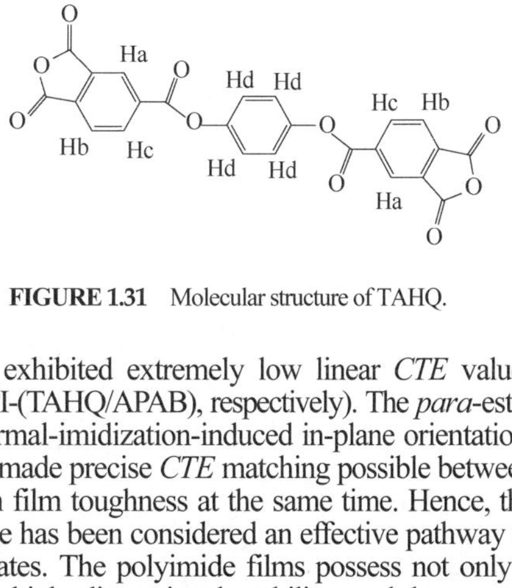

FIGURE 1.31 Molecular structure of TAHQ.

Experimental results indicated that the PEsI films exhibited extremely low linear *CTE* values (3.2×10^{-6} °C^{-1} for PI-(TAHQ/PDA) and 3.3×10^{-6} °C^{-1} for PI-(TAHQ/APAB), respectively). The *para*-ester linkages behave as a rod-like segment, favorable for thermal-imidization-induced in-plane orientation. Copolymerization with flexible 4,4′-oxydianiline (ODA) made precise *CTE* matching possible between PEsI/copper substrate, with a significant improvement in film toughness at the same time. Hence, the introduction of ester linkages into the polyimide backbone has been considered an effective pathway to produce polyimide films suitable for use as PFC substrates. The polyimide films possess not only a copper-level low *CTE* but also high film toughness, high dimensional stability, and low water absorption.

TAHQ exhibits good polymerizability with aromatic diamines. The reduced viscosities (η_{red}) of the PEsAAs obtained by polycondensation of TAHQ and diamines in DMAc or NMP were measured in the range of 1.10-5.19 $dL\cdot g^{-1}$, indicating the high molecular weights of the PEsAAs. Hence, TAHQ has sufficiently high reactivity.

A series of TAHQ-derived homo-PEsI films, PI-(TAHQ-PDA), PI-(TAHQ-CHDA), PI-(TAHQ-TFMB), and PI-(TAHQ-ODA), were prepared. The combinations of three diamines (PDA, CHDA, and TFMB) with TAHQ are expected to give low *CTE* owing to their rigid/linear chain structures favorable for the in-plane orientation. And ODA was used as a typical flexible structure of diamine, expected to be effective as a comonomer for the improvement of film toughness. PI-(TAHQ-PDA) film is a high-quality, yellow-colored clear film. No cracks were observed on a 180 °C folding test, showing good flexibility. No distinct glass transition temperature up to 450 °C was observed, displaying an excellent dimensional stability. This may be attributed to suppressed molecular motions (internal rotation) in the amorphous regions and a semicrystalline morphology as mentioned later. PI-(TAHQ/PDA) film exhibited an extremely low *CTE* (3.2×10^{-6} °C^{-1}), comparable to that of silicon wafer and a considerably high tensile modulus (*E*) of 8.86 GPa and strength (224 MPa). A high birefringence (Δn=0.219) was measured, corresponding to high in-plane orientation in the films. The elongation at breakage (E_b) representing toughness was not sufficiently high (E_b=5.4%) and the water absorption was 1.6%. In comparison, PI-(TAHQ/TFMB) also showed similar excellent properties such as no glass transition or high T_g behavior, high tensile modulus and strength, and low water absorption. However, the *CTE* was increased to 31.5×10^{-6} °C^{-1}. In contrast, PI-(TAHQ/ODA) film showed common properties with the flexible-chain PI films, i.e., high *CTE*=51.2×10^{-6} °C^{-1} and low *E*=2.9 GPa. However, the T_g was unexpectedly high compared with other ODA-derived films. Fig. 1.32 depicts the backbone structures of three ODA-based films. PI-(TAHQ/ODA) film shows a T_g of 320 °C, much higher than that of the corresponding ether-linked PI-(HQDA/ODA) film (T_g=245 °C) and that of PI-(BPDA/ODA) film (T_g=245 °C) [30], respectively, indicating that the internal rotations are suppressed in the ester-linked polymer backbone structures. The PI-(TAHQ/ODA) film has very high toughness (E_b=67.2%) in contrast to PI-(TAHQ/PDA). Regarding water absorption properties, PI-(TAHQ/PDA) film showed obviously lower water absorption (1.6%) than the conventional Kapton H film (2.5%) due to the substitution of imide groups by ester groups in the polymer backbone structures. It is notable that the lower water absorption of the PI-(TAHQ/ODA) (0.6%) is comparable to that of fluorinated PI-(TAHQ/TFMB) film (0.7%). These results give a hint that copolymerization using flexible ODA can provide the balanced properties attaining low *CTE*, high T_g, low water absorption, and high toughness.

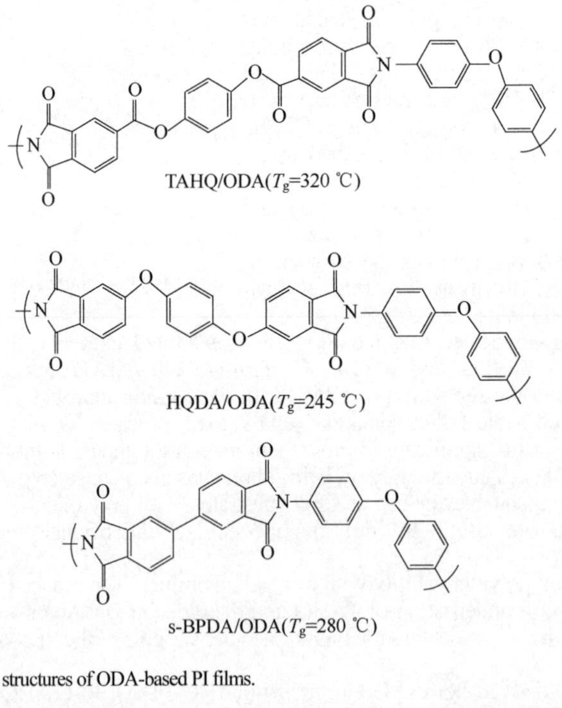

TAHQ/ODA(T_g=320 ℃)

HQDA/ODA(T_g=245 ℃)

s-BPDA/ODA(T_g=280 ℃)

FIGURE 1.32 Molecular structures of ODA-based PI films.

Polyimide films with stiff main chains such as PI-(BPDA/PDA) and PI-(PMDA/ODA) tend to have complex morphologies in contrast to common semicrystalline polymers such as PET, which are composed of a simple crystalline/amorphous two-phase morphology. This is due to insufficient molecular mobility, and thereby the chain rearrangement for crystallization is disturbed. PI-(TAHQ/PDA) film displayed a much stronger and sharper pattern peaking at 2θ=21.6 degrees than PI-(BPDA/ODA) film, suggesting a semicrystalline structure. The crystallinity in the PI-(TAHQ/PDA) film probably contributes to the unclear T_g behavior observed in the E' curve. The *CTE* may also be influenced by its crystalline morphology. It was concluded that *CTE* is primarily governed by the extent of in-plane orientation rather than crystallinity. Principally, crystallinity should also control the extent of water absorption (Table 1.9).

In addition, an ester-containing aromatic diamine, APAB, was also used to react with the stiff/linear aromatic dianhydrides including PMDA, BPDA, and TAHQ to prepare the PEsI films. The reduced viscosities of the APAB-derived PEsI films were measured in the range of 1.09dL·g^{-1}-2.81 dL·g^{-1}, indicating a sufficiently high polymerization reactivity of APAB. The cast films were all highly flexible.

TABLE 1.9 Properties of TAHQ-derived PEsI Films

PI Films	T_g(°C)	*CTE*(×10^{-6}°C^{-1})	Δn	T_5 in N_2(°C)	W_a(%)	Tensile Strength (MPa)	Tensile Modulus (GPa)	Elongation (%)
TAHQ/PDA	ND	3.2	0.219	480.7	1.6	224	8.7	5.4
TAHQ/FMB	ND	31.5	0.135	486.5	0.7	204	5.8	27.6
TAHQ/ODA	320	51.2	0.101	462.2	0.6	236	2.9	67.2

TABLE 1.10 Properties of APAB-derived PEsI Films

PI Films	T_g(°C)	*CTE*(×10^{-6}°C^{-1})	Δn	T_5 in N_2(°C)	W_a(%)	Tensile Strength (MPa)	Tensile Modulus (GPa)	Elongation (%)
PMDA-APAB	ND	2.0	0.172	530.9	1.6	270	7.7	6.4
BPDA-APAB	ND	3.4	0.183	534.4	0.7	250	7.6	6.0
TAHQ-APAB	ND	3.3	0.199	470.6	0.7	224	7.1	10.6

Table 1.10 summarizes the properties of APAB-derived PEsI films. It is noteworthy that both PI-(PMDA/APAB) and PI-(BPDA/APAB) films showed extremely low *CTE* values comparable with that of silicon wafer in accordance with considerably high birefringence. Thus, the results of the TAHQ- and APAB-derived PEsI films revealed that the *para*-aromatic ester linkages play a great role in the imidization-induced in-plane orientation, resulting in low *CTE* values. All of the APAB-derived PEsI films showed no appreciable glass transitions detected by DMA. High tensile modulus and high strength were also acquired in addition to high thermal stability.

PI-(TAHQ/APAB) film exhibits similar properties to PI-(PMDA/APAB) and PI-(BPDA/APAB) films. The film toughness was somewhat improved by the combination of TAHQ and APAB. PI-(TAHQ/APAB) film also has a low water absorption. A sharp and strong reflection peaking at 22.0 degrees was observed in WAXD pattern, implying the presence of a semicrystalline morphology. As mentioned above, the chain linearity/rigidity is the most important factor for imidization-induced in-plane orientation (low *CTE* generation). This situation is seen in the comparison of the tensile modulus (E) and *CTE* between PI-(PMDA/4,4-ODA) film (E=3.0 GPa, *CTE*=41×10^{-6} °C^{-1}) and PI-(PMDA/3,4-ODA) (E=5.0 GPa, *CTE*=33×10^{-6} °C^{-1}) [38].

Fig. 1.33 compares the chain vectors (arrows) and possible conformations of several polyimide films. The chain segment of the former is largely bent at the ether linkage. The chain vector in the latter is also bent at the ether linkage but the linearity of the vector is somewhat recovered by taking a crank-shaft-like conformation. The higher modulus (higher degree of in-plane orientation) of PI-(PMDA/3,4-ODA) can be interpreted in terms of its possible conformation with a higher chain linearity. The *para*-aromatic ester unit can also be regarded as a similar crank-shaft-like linkage.

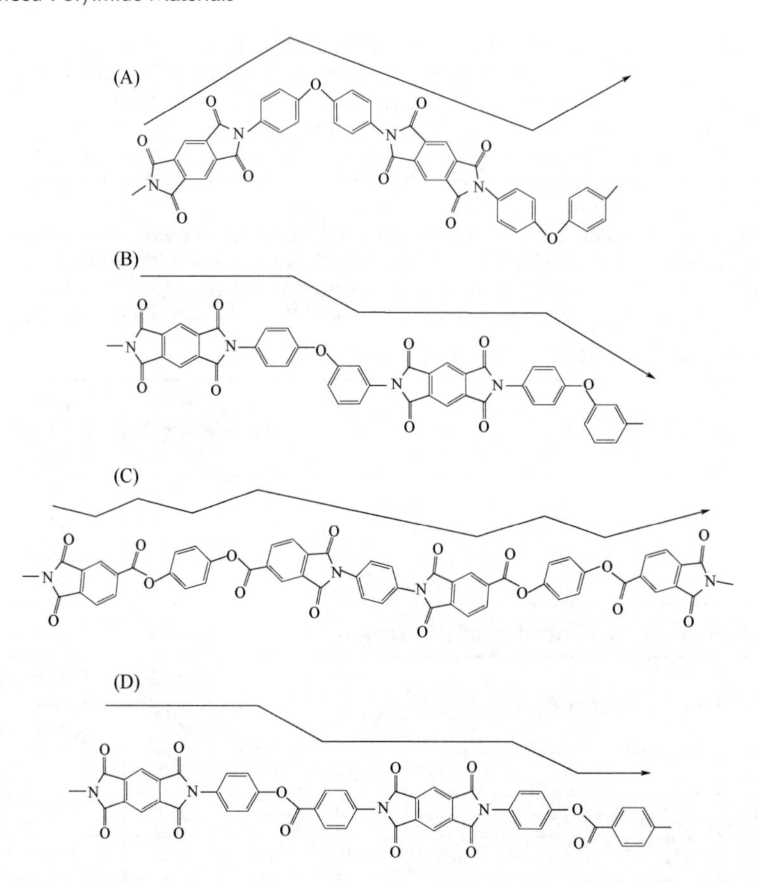

FIGURE 1.33 Comparison of the chain vectors (arrows)and possible conformations(A)PMDA/4,4'-ODA,(B)PMDA/3,4'-ODA, (C)TAHQ/PDA,and (D)PMDA/APAB.

In order to gain PEsI films with the target balanced properties, ODA was used to copolymerize with the stiff/linear aromatic dianhydrides and diamines, giving three coployimide films, including PI-(TAHQ/PDA-ODA (70/30)), PI-(PMDA/APAB-ODA (60/40)), and PI-(TAHQ/APAB-ODA (70/30). Table 1.11 compares the properties of the three copolyimids films.

The PI-(TAHQ/PDA-ODA(70/30)) film shows good combined properties, including a very low *CTE* (11.7×10^{-6} °C^{-1}), a relatively low water absorption (1.0%), and a high toughness (E_b=35.0%). In comparison, PI-(PMDA/APAB-ODA (60/40)) film also exhibits a very low *CTE* (11.9×10^{-6} °C^{-1}) and sufficient toughness (E_b=25.5%), but a high water absorption (2.5%), probably due to the presence of the PMDA/ODA sequence. PI-(TAHQ/APAB-ODA(70/30)) film also shows a great combination of thermal, mechanical, and water-resistant properties, including a low water absorption (0.7%), a low *CTE* (14.8×10^{-6} °C^{-1}), and a high toughness (E_b=35.8%). Obviously, the introduction of the flexible part into PI-(TAHQ/APAB) film is an effective pathway to accomplish the target combined properties.

Furthermore, the effect of substituents on the properties of the PEsI films with ester-linkages in the polymer backbones was systematically investigated [45]. The film properties of a series of PEsI films were evaluated for applications of high-temperature base film in FPC boards. Three ester-containing tetracarboxylic dianhydrides, including TAHQ, methyl-substituted TAHQ (M-TAHQ), and methoxy-substituted TAHQ (MeO-TAHQ), were employed to react with aromatic diamines such as PDA to give PEsAAs, which were then coated on glass and thermally imidized to PEsI films. The ester-containing monomers (TAHQ series and APAB series) were all highly reactive and led to PEsAAs possessing high inherent viscosities ranging from 1.09dL·g^{-1} to 9.33 dL·g^{-1}. The incorporation of methyl- and methoxy-substituents into rigid TAHQ-based PEsI films caused no significant decrease in T_g, but allowed molecular motions above the T_gs. The substituents especially the methoxy group contributed to a significant decrease in water absorption without sacrificing other target properties. For practical FPC application, a flexible diamine (ODA) was copolymerized into the highly esterified rigid PEsI films. One of the PEsI copolymer films exhibited excellent combined properties: a low *CTE* (17.8×10^{-6} K^{-1}) completely consistent with that of copper foil as a conductive layer, considerably low water absorption (0.47 %(*w*)), a high T_g (363 °C), and improved toughness (E_b>40%).

The ester-containing aromatic diamines, i.e., methyl substituted bis(4-aminophenyl) terephthalate (M-BPTP) and substituted 4-aminophenyl -4′-aminobenzoate (M-APAB) were used to prepare a series of the PEsI films (Fig. 1.34) [46].

TABLE1.11 Properties of the Copolyimide Films

Copolyimide Films	T_g(°C)	*CTE* ($\times10^{-6}$°C^{-1})	Δn	T_5 in N_2 (°C)	W_a (%)	Tensile Strength (MPa)	Tensile Modulus (GPa)	Elong-ation(%)
TAHQ/PDA-ODA (70/30)	ND	11.7	0.163	503.2	1.0	250	6.0	35.0
PMDA/APAB-ODA (60/40)	382	11.9	0.148	561.0	2.5	192	3.9	25.5
TAHQ/APAB-ODA (70/30)	395	14.8	0.170	486.8	0.7	295	6.3	35.8

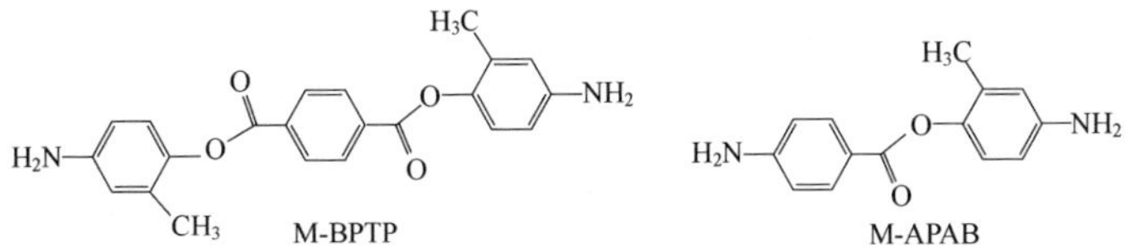

FIGURE 1.34 Molecular structures of M-BPTP and M-APAB.

The PEsI films consisting of stiff/linear backbone structures and possessing lower *CTE* values than that of copper foil (17.7×10^{-6} °C^{-1}) were obtained by a simple copolymerization approach using adequate amounts of flexible monomers to achieve the *CTE* matching in flexible copper-clad laminates (FCCLs) and drastic improvement of film toughness at the same time. Table 1.12 compares the combined properties of the PEsI films modified with flexible diamine. The incorporation of the ether-linkages by using 4,4′-ODA into PI-(TAHQ/BPTP) film improved the film toughness from E_b=4.5% to 12.7%, which was not so drastic an enhancement. The properties of PI-(TAHQ/M-BPTP) film were dramatically improved by copolymerization with 4,4′-ODA as indicated from a significantly enhanced E_b (50.7%) and a considerably reduced Wa (0.35%) as well as a controlled *CTE* (10.0×10^{-6} °C^{-1}), although the concomitant decrease in the tensile modulus (from 7.74 GPa to 6.3 GPa) was not drastic.

1.4.2.2 Fluorinated Low-CTE Polyimide Films

An effective method of reducing the moisture uptake of low thermal expandable polyimide films by attaching CF_3 groups to the polymer backbone has been developed. A novel fluorinated ester-bridged aromatic diamine, bis(2-trifluoromethyl-4- aminophenyl)terephthalate (CF_3-BPTP) was employed to prepare a series of fluorinated ester-bridged polyimide films with controlled ester-bridged segments and fluorine contents in the polymer backbone (Fig. 1.35). The PI films were prepared by copolymerization of BPDA as aromatic dianhydride monomer and the aromatic diamine monomer mixture consisting of PDA and different amounts of CF_3-BPTP. Experimental results indicated that the film's water uptakes (W_a) were reduced with increasing of the CF_3 groups loadings in the ester-bridged polyimide backbones while keeping the films with low enough *CTE*s. By controlling CF_3 group loadings, polyimide films with desirable combinations of thermal, mechanical, and dielectric properties for application in high-density and thinner FPCs have been obtained. Thus, polyimide films with *CTE* of ≤20×10^{-6} °C^{-1} at 50-200 °C, glass transition temperature (T_g) of ≥310 °C, Young's modulus of ≥6.0 GPa, W_a of ≤1.2%, dielectric constant (ε) of 3.4 have been obtained. The two-layer flexible copper clad laminate (2L-FCCL) prepared by coating the polyimide precursor resin solution on the surface of copper foil followed by being thermally imidized at elevated temperature did not cause apparent curling due to their closed *CTE* values.

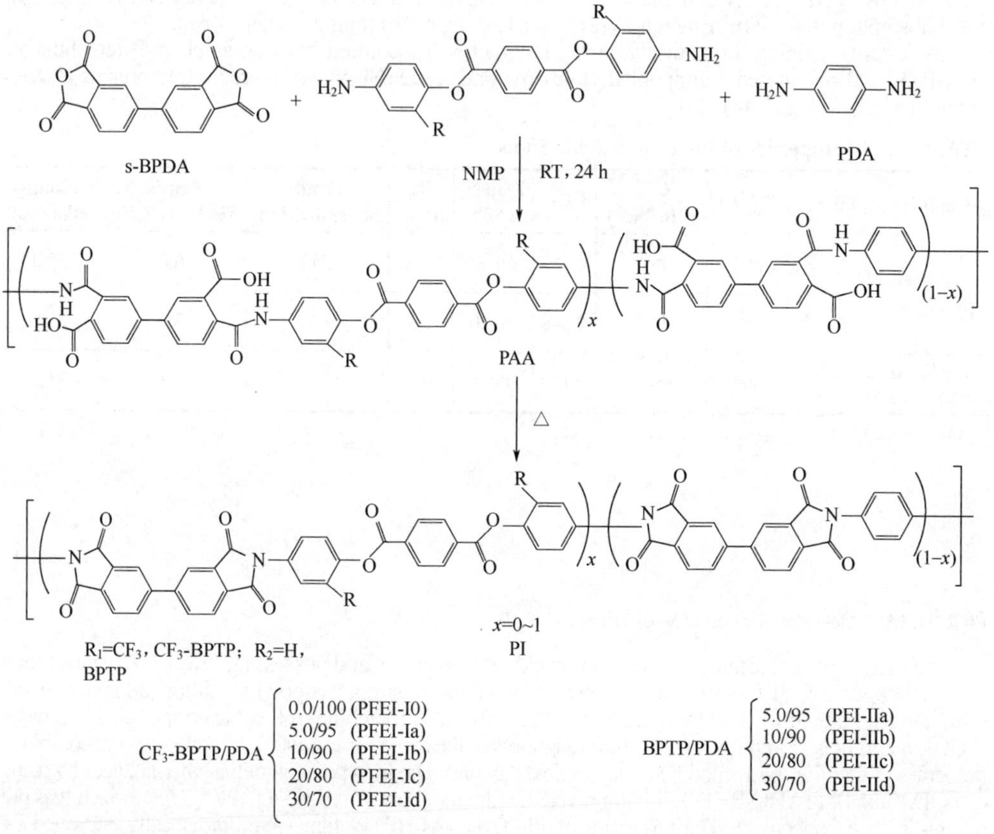

s-BPDA + H_2N–…–NH_2 + PDA

NMP RT, 24 h

PAA

△

x=0~1

PI

$R_1=CF_3$, CF_3-BPTP; R_2=H, BPTP

CF_3-BPTP/PDA: 0.0/100 (PFEI-I0); 5.0/95 (PFEI-Ia); 10/90 (PFEI-Ib); 20/80 (PFEI-Ic); 30/70 (PFEI-Id)

BPTP/PDA: 5.0/95 (PEI-IIa); 10/90 (PEI-IIb); 20/80 (PEI-IIc); 30/70 (PEI-IId)

FIGURE 1.35 Synthesis of the fluorinated ester-bridged polyimide films.

1.4.2.2.1 Film Forming Ability

The effect of the ester-bridged diamine concentration on the PAA solution viscosities is shown in Fig. 1.36. The PAA-(BPDA/PDA) solution has very high viscosity, resulting in difficulty in casting films. As the fluorinated ester-bridged aromatic diamine (CF_3-BPTP) is inserted in the polymer backbone structure of PAA-(BPDA/PDA) resin, the solution viscosity can be decreased significantly. As the mole ratio of CF_3-BPTP/PDA was increased from 5% to 30%, the produced PAA resin solution viscosity decreased from 38.18 Pa·s to 2.17 Pa·s. In comparison, the unfluorinated esterbridged aromatic diamine (BPTP) only yielded the PAA resin solution viscosity descended from 86.70 Pa·s to 41.52 Pa·s under the same conditions, demonstrating that the fluorinated ester-bridged diamine was more effective than the corresponding fluorine-free one in reducing of the PAA resin solution viscosity, thus improving the film-casting property. This phenomenon might be interpreted by the presence of bulky CF_3 substituent in the polyimide backbone, which inhibited the chain packing and weakened the intermolecular interactions of the rigid polymer backbones.

TABLE 1.12 Properties of PEsI Films Modified With Flexible Diamines

Dianhydride	Rigid Diamines %(mol)	Flexible Diamine %(mol)	η_{red} (dL·g^{-1})	T_g (°C)	*CTE* (×10^{-6}°C^{-1})	Δn	W_a (%)	Tensile Strength (MPa)	Tensile Modulus (GPa)	Elongation at Breakage (%)
TAHQ	BPTP(70)	ODA(30)	14.4	ND	4.7	0.134	0.77	193	5.9	12.7
	M-BPTB(70)	ODA(30)	17.8	410	10.0	0.168	0.35	248	6.4	50.7
	M-APAB(70)	ODA(30)	1.77	384	17.1	0.155	0.76	267	5.2	46.0
M-TAHQ	BPTP(70)	ODA(30)	16.5	420	12.1	0.184	0.55	220	6.2	14.6
	M-BPTB(70)	ODA(30)	15.3	390	28.4	0.151	0.41	191	5.2	33.6

1.4.2.2.2 Film Mechanical Properties

The effect of the ester-bridged segment concentrations in polyimide on film mechanical properties was shown in Fig. 1.37. The ester-bridged polyimide films showed higher Young's modulus in the range of 6.0 GPa—6.9 GPa, comparable to that of the commercial PI-(BPDA/PDA) film (Upilex S, 6.0 GPa) at the same film thickness and treatment procedure. This could be ascribed to the following two reasons: (1) the polyimide backbone composed of a *para*-linked aromatic ester-bridged segment and a rigid BPDA-PDA skeleton has a high level of in-plane orientation; (2) the strong dipole-dipole interaction between the imide carbonyl groups and ester linkages in the polymer backbones exhibited strong inter-molecular interactions, resulting in a physical cross-linking effect in the polyimide morphology. Besides, with increasing loading of the ester-bridged segments in polymer backbone, the film's modulus gradually reduced from 6.9 GPa (5 %(mol) CF_3-BPTP) to 6.7 GPa (10 %(mol) CF_3-BPTP), and to 5.9 GPa (30 %(mol) CF_3-BPTP). Moreover, the tensile strengths of the ester-bridged polyimide films declined linearly from 185 MPa to 140 MPa with an increase in the concentration of the ester-bridged segments in polyimide backbone.

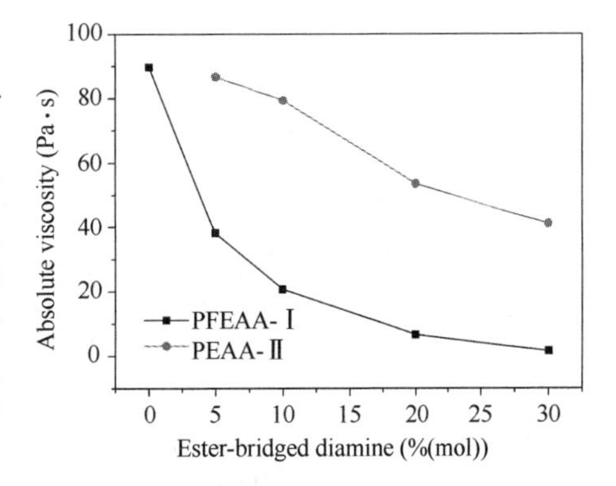

FIGURE 1.36 Dependence of PAA resin absolute viscosity on the ester-bridged diamine concentrations.

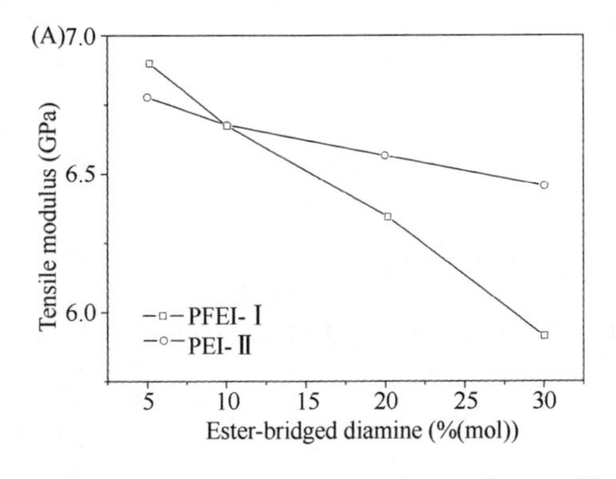

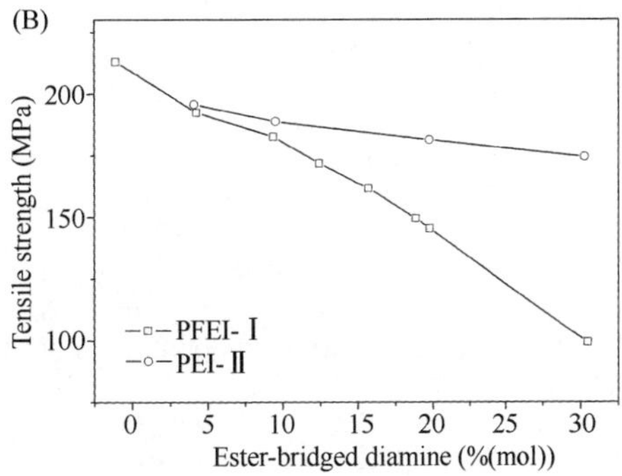

FIGURE 1.37 Effect of ester-bridged segment loading on polyimide film's tensile modulus(A)and tensile strength(B).

In contrast, PEI-II series films derived from BPTP exhibited higher Young's modulus and higher tensile strengths than PFEI-I series films at the same concentration of ester-bridged segments, indicating that the insertion of CF_3 groups in the ester-bridged polyimide backbone resulted in the decline of both film's modulus and strength probably due to its weaker concentration reactivity.

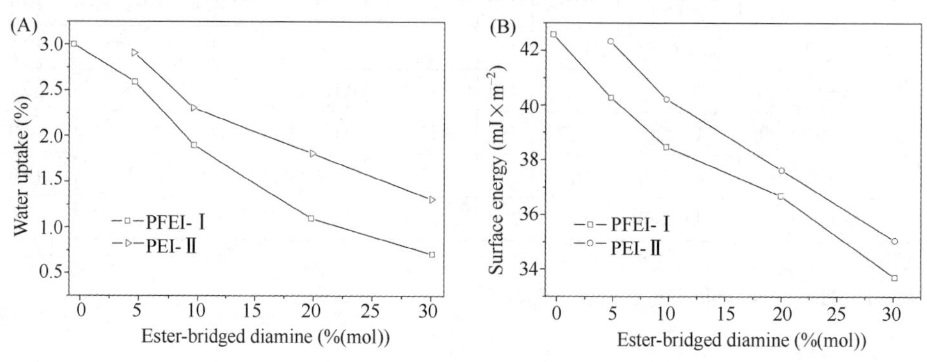

FIGURE 1.38 Effect of ester-bridged segment loadings on polyimide film's water uptakes(A)and surface energy(B).

1.4.2.2.3 Water Uptake

The water uptake of aromatic polyimide film for high-density FPC application is a key issue. The lower the water uptake, the lower the humidity expansion of FPC. Fig. 1.38 shows the dependence of the ester-bridged segment loading in the polyimide backbone on the film water uptakes. The PFEI-I series films exhibited water uptakes decreased from 2.6% for PFEI-I_a (5%(mol) of CF_3-BPTP) to 0.7% for PFEI-I_d (30%(mol) of CF_3-BPTP) with increasing of the ester-bridged segment loadings. This could be interpreted by the polymer structural factors: one was the decreased imide contents in the ester-bridged polyimides; another reason was the existence of the hydrophobic trifluoromethyl substituent in the polyimide backbone, which reduced the surface energy and intermolecular cohesive force, resulting in inhibition of the moisture adsorption. Moreover, the fluorinated ester-bridged polyimide films (PFEI series films) showed lower water uptakes than the corresponding fluorine-free ones (PEI-II series films).

1.4.2.2.4 Thermal Dimension Stabilities

The in-plane *CTE* of the fluorinated ester-bridged polyimide films were measured by TMA analysis, as shown in Fig. 1.39. The polyimide films showed small dimensional increases until the scanned temperature reached to the glass transition temperature, then the dimensional increases changed obviously (Fig. 1.39A). The *CTE* values increased linearly with the increasing of the ester-bridged segment loadings in the polyimide backbones (Fig. 1.39B). The PFEI-I_0 had the lowest *CTE* of 3.3×10^{-6} °C^{-1} due to its most linear/rigid backbone structure, compared to 8.3×10^{-6} °C^{-1} of PFEI-I_a (5%(mol) of CF_3-BPTP), and 18.3×10^{-6} °C^{-1} of PFEI-I_d (30%(mol) of CF_3-BPTP), respectively, probably attributed to the big free volume and looser chain packing induced by CF_3 groups. On the other hand, the incorporation of fluorine-free BPTP did not cause drastic dimensional increase like the fluorinated CF_3-BPTP, whose *CTE* values varied from 3.7×10^{-6} to 6.3×10^{-6} °C^{-1} with increasing of the ester-bridged segments. Hence, the rod-like *para*-aromatic ester-bridged segment behaved like the rigid poly(BPDA/PDA) skeleton, which was also an indispensable structure factor for the in-plane orientation, resulting in lower thermal expansion. Meanwhile, although appropriate insertion of CF_3 substituent into polyimide backbone led to a slight increase in *CTE* value, it was beneficial for adjusting the *CTE* value precisely close to that of copper foil for FPC application. In the fluorinated ester-bridged polyimide films, PFEI-I_d film possessed much closer *CTE* (18.3×10^{-6} °C^{-1}) to that of copper foil (17.8×10^{-6} °C^{-1}) and low water uptake of 0.7%(*w*), thus making it a desirable candidate for FPC's substrate film.

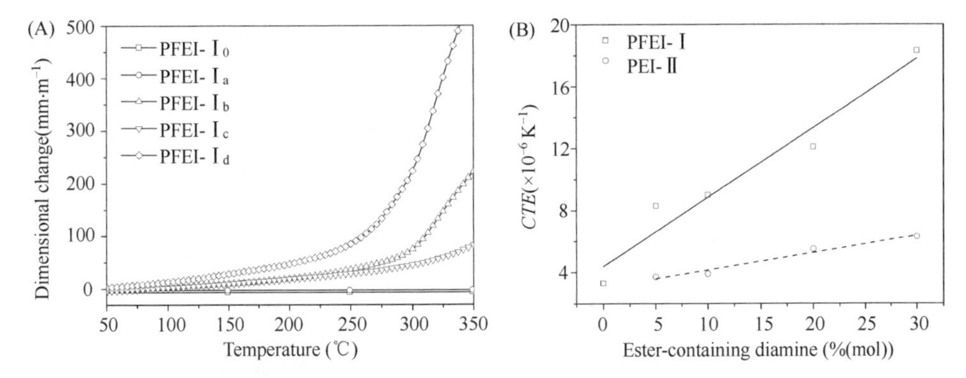

FIGURE 1.39 Dimensional stabilities of the PFEI-I series films(A)and the dependence of *CTE* values on the ester-bridged segment loadings(B).

1.4.3 Transparent Polyimide Films

In recent years, aromatic polyimide films have been widely used in microelectronics and aerospace industries, however the colors from yellow to brown have limited their applications in optical and display devices. The coloration of the common aromatic polyimide films is primarily attributed to the intermolecular CTC formed due to the molecular aggregation between polymer chains in the solid states. With the rapid development of optoelectronic engineering, optical films with high thermal resistance are highly desired due to the increasing demands for high reliability, high integration, and high signal transmission speed in optoelectronic devices [47,48]. For instance, in the fabrication of flexible active matrix organic light emitting display devices (AMOLEDs), the processing temperature on the flexible polymer film substrates might be higher than 300 °C [49-51]. Most of the common polymer optical films, such as polyethylene terephthalate (PET) or polyethylene naphthalate (PEN) lose their optical and mechanical properties at such a high processing temperature. Thus, colorless and optically transparent polymer optical films with high-temperature resistance have attracted a great deal of attention from both the academic and engineering fields. This driving force has greatly promoted the development of optical polymer films with outstanding thermal stability.

Generally, optical polymer films can be divided into three types according to their servicing temperatures or glass transition temperatures (T_g), including conventional optical films ($T_g < 100$ °C), common high-temperature optical films (100 °C$\leqslant T_g < 200$ °C), and high temperature optical films ($T_g \geqslant$ 200 °C), as shown in Fig. 1.40. The typical chemical structures for the optical polymer films and their T_g values are illustrated in Fig. 1.41. Although conventional polymer optical films possess excellent optical transparency, the lower T_g (PET, T_g=78 °C, PEN, T_g=123 °C) limit their practical applications in advanced optoelectronic industries. Hence, high-temperature polymers such as polyamideimide (PAI), polyetherimide (PEI), and polyimide (PI), with high optical transparency have been considered as one of the hot topics in the advanced polymer optical film field in recent years.

For development of colorless transparent polyimide (CTPI) films, one of the most challenging issues is to balance the thermal properties, the optical transparent properties, and the mechanical properties. Fig. 1.42 shows the molecular structure design for CTPI films, including favorable designs and unfavorable ones. The favorable designs that could improve both the high-temperature stability and optical transparency include introduction of alicyclic substituents, such as cyclobutane, cyclohexane, cardo groups (diphenylcyclohexane, adamantane, etc.); highly electronegative groups, such as trifluoromethyl (—CF_3); and asymmetrical or twisted rigid substituents, such as asymmetrically substituted biphenyl or diphenylether moieties. These functional groups or structure segments have been widely adopted to develop novel CTPI films, endowing CTPI films with both excellent thermal stability ($T_g \geqslant$300 °C) and good optical transmittance (>85% in the visible light region). Simultaneously, these substituents or structure segments could prohibit the formation of CTC induced by the electron-donating and electron-withdrawing moieties in CTPI films.

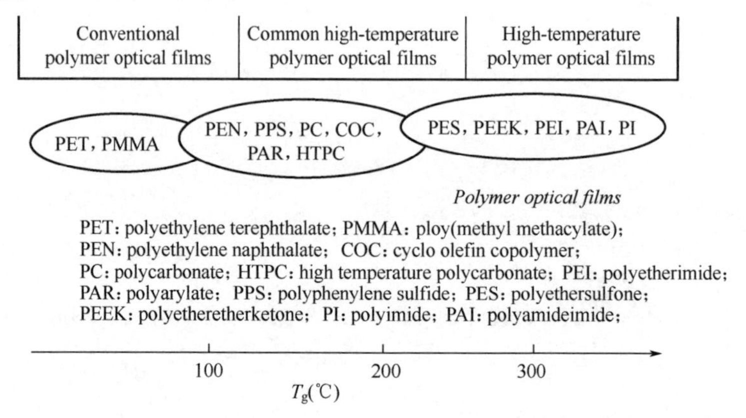

FIGURE 1.40 Classification of optical polymer films.

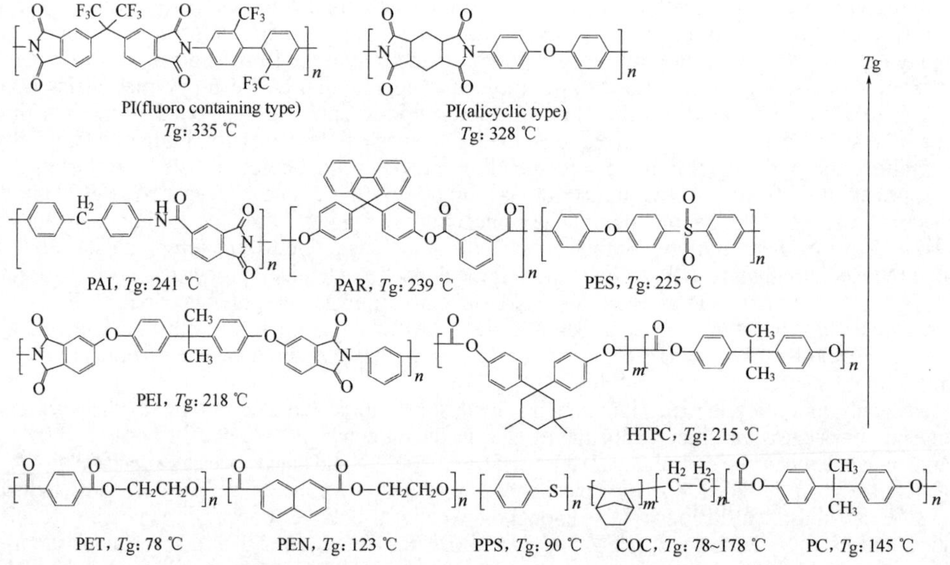

PI(fluoro containing type) Tg: 335 ℃

PI(alicyclic type) Tg: 328 ℃

PAI, Tg: 241 ℃ PAR, Tg: 239 ℃ PES, Tg: 225 ℃

PEI, Tg: 218 ℃ HTPC, Tg: 215 ℃

PET, Tg: 78 ℃ PEN, Tg: 123 ℃ PPS, Tg: 90 ℃ COC, Tg: 78~178 ℃ PC, Tg: 145 ℃

FIGURE 1.41 Typical chemical structures and T_gs of the optical polymer films.

Monomer design and synthesis is a key issue in the preparation of CTPI films because most of the properties of CTPI films, including optical transparency, mechanical strength, *CTE*, etc., are dependent on the molecular structures of the starting monomers, including dianhydrides and diamines. Currently, the development of CTPI films has been hampered by the high cost of the limited commercially available monomers. Table 1.13 summarizes some typical commercially available monomers for CTPI films [52-60]. For these monomers, including dianhydrides and diamines, the alicyclic moiety or highly electronegative —CF_3 or bulky —SO_2— moieties in their structures are beneficial to reducing or inhabiting the intra- and/or intermolecular CTC formation in the derived CTPI films, thus improving their optical transparency. Meanwhile, the high chemical bond energies and bulky molecular volumes for these groups endow the derived CTPI films with good thermal stability. It was found that the dianhydride monomers have more effects on the properties of CTPI films than the diamines. Up until now, fluorine-containing or alicyclic dianhydrides have been intensively investigated to develop novel

CTPI films. However, the high cost and synthetic difficulties have seriously hindered the rapid development and commercialization of CTPI films.

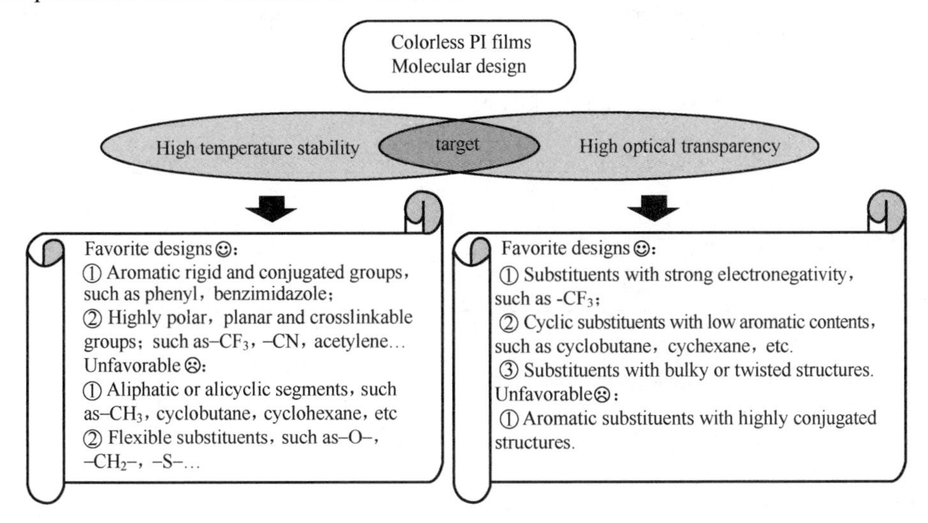

FIGURE 1.42 Molecular structure design for colorless transparent polyimide films.

The reactivity and purity of monomers also have obvious influence on the physical and chemical properties of CTPI resins, including molecular weights and their distribution, inherent viscosities, solubility in organic solvents, appearance, color, and so on. These features have great effects on the combined properties of the final optical polymer films, including their color, optical transparency, mechanical strength, thermal stability, and dielectric properties. Besides, the film-forming technologies including casting procedure, the uniaxial or biaxial stretching process, high-temperature curing program, and even the final winding and rewinding process also have apparent influence on the combined properties of CTPI films. Hence, the manufacture of CTPI films is a multidisciplinary technology.

CTPI resin synthesis is another key feature in the development of high-quality CTPI films. Polycondensation is the stepwise reactions between bifunctional or polyfunctional components with the liberation of small molecules, such as water, alcohol, hydrogen halide, etc. Generally, the polycondensation reaction is carried out in fully dried polar solvents with high boiling points, such as dimethyl sulfoxide (DMSO), DMF, DMAc, NMP, g-butyrrolactone (GBL), and meta-cresol in order to achieve homogeneous reaction and obtain high-molecular-weight CTPI resins. In some cases, molecular weights of the CTPI resins have to be controlled in order to achieve good processability. A stoichiometric imbalance in the monomers (excess of one of the reactants over another) or addition of mono-functional reagent is usually used to control the molecular weight of the obtained polymers.

TABLE1.13 Typical Monomers for CTPI Synthesis

Monomers	Chemical Structure	Usage
1S,2R,4S,5R-cyclohexane tetracarboxylic dianhydride(H''-PMDA)		Colorless PI film
1R,2S,4S,5R-cyclohexane tetracarboxylic dianhydride(HPMDA)		Colorless PI film

TABLE1.13 (Continued)

Monomers	Chemical Structure	Usage
3,3′,4,4′-Bicyclohexyl tetracarboxylic dianhydride(HBPDA)	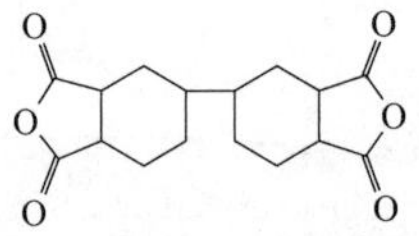	Colorless PI film
1,2,3,4-Cyclobutane tetracarboxylic dianhydride(CBDA)	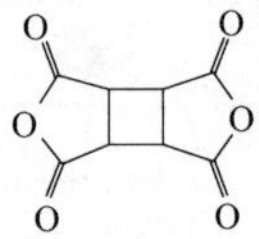	Colorless PI film
2,3,5-Tricarboxycyclopentylacetic dianhydride(TCA-AH)	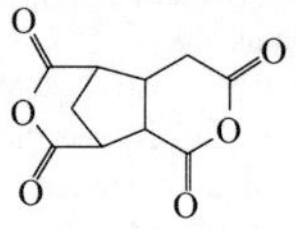	Colorless PI film
3,4-Dicarboxy-1,2,3,4-tetrahydro-1-naphthalene succinic dianhydride(TDA)	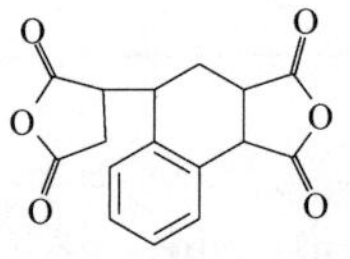	Colorless PI film
3,4-Dicarboxy-1,2,3,4-tetrahydro-6-fluoro-1-naphthalene succinic dianhydride(FTDA)	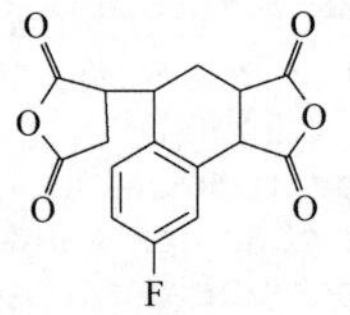 F	Colorless PI film
3,4-Dicarboxy-1,2,3,4-tetrahydro-6-chloro-methyl-1-naphthalene succinic dianhydride (CMTDA)	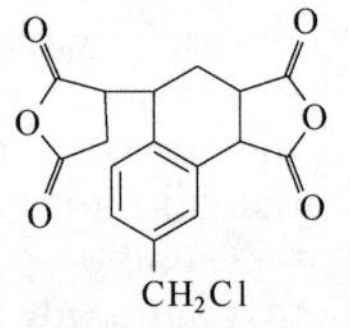 CH_2Cl	Colorless PI film
2,2′-Bis(3,4-diacarboxyphenyl)hexafluoro-propane dianhydride(6FDA)	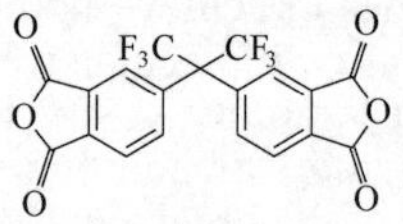 F_3C CF_3	Colorless PI film
2,2′-Bis(Trifluoromethyl)-4,4′-diaminobiphenyl(TFDB)	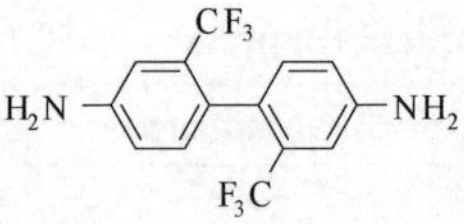 CF_3 H_2N NH_2 F_3C	Colorless PI film
1,4-Diaminocyclohexane	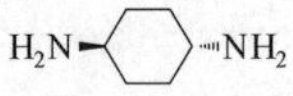 H_2N NH_2	Colorless PI film
1,2,4-Cyclohexanetricarboxylic anhydride(HTA)	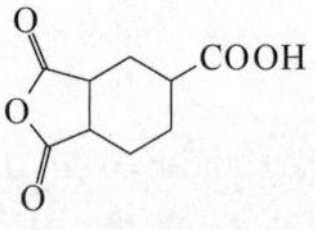 COOH	Colorless PAI film

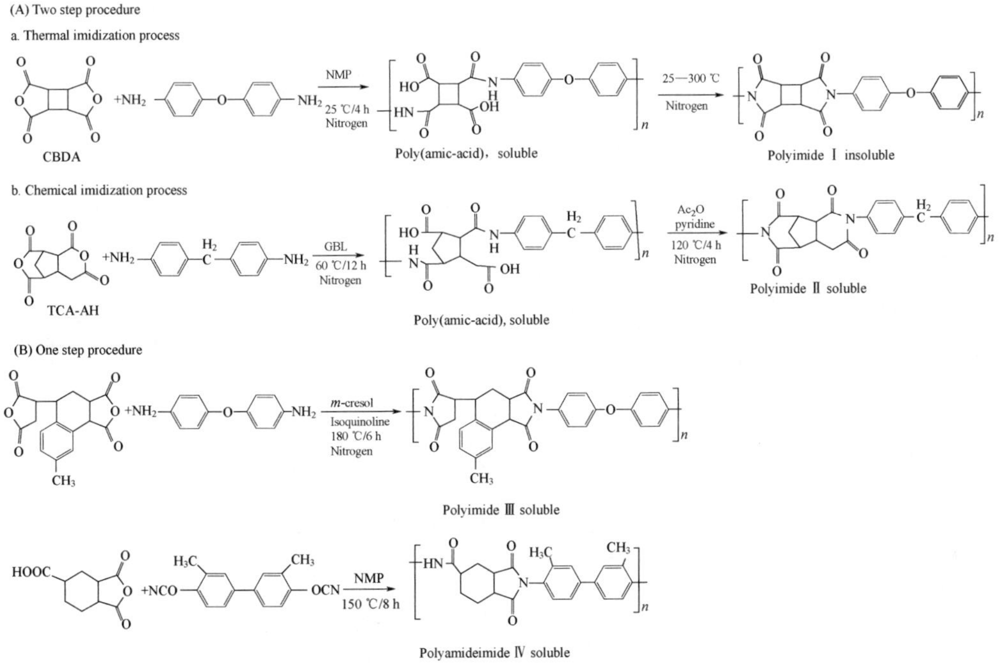

(A) Two step procedure

a. Thermal imidization process

CBDA

Poly(amic-acid), soluble

Polyimide I insoluble

b. Chemical imidization process

TCA-AH

Poly(amic-acid), soluble

Polyimide II soluble

(B) One step procedure

Polyimide III soluble

Polyamideimide IV soluble

FIGURE 1.43 Synthesis of transparent polyimide resins.

The synthesis of CTPI resins can usually be performed via two procedures, including a two-step synthesis via PAA precursor, followed by imidization reaction either at elevated temperatures up to 300 °C or by chemical dehydration reaction with acetic anhydride under the catalysis of pyridine or other organic alkalis (Fig. 1.43A); and a one-step high-temperature polycondensation procedure to produce polyimide directly (Fig. 1.43B). The first reaction consists of the polymerization of an aromatic diamine and a dianhydride at room temperature in a polar aprotic solvent (DMAc or NMP). The resulting PAA is soluble in the reaction medium. Then, the viscous PAA solution is cast into a thin layer on glass or stainless steel substrates and heated from room temperature to high temperatures up to 300 °C in inert atmosphere, resulting in the evaporation of solvents and ring closure with intramolecular elimination of water to facilitate the imidization reaction [61]. The obtained CTPI resins are usually not soluble in organic solvents. Alternatively, if the final CTPI resins are soluble in organic solvents, the PAA precursors can be imidized via a chemical dehydration procedure [62,63]. Similarly, organo-soluble polyimide resins can also be synthesized via a high-temperature polycondensation procedure [64]. In this procedure, as shown in Fig. 1.43B, the PAA precursor is not separated from the reaction medium, which is dehydrated in situ to form the CTPI resins. The solid preimidized CTPI resins are usually soluble in polar solvents and their solution can be cast into films at a relatively low temperature (evaporation of solvents). This is quite beneficial for producing colorless films.

Overall, CTPI resins can be synthesized via various polycondensation procedures. Several critical factors, including the reactivity, purity, and stoichiometry of the starting monomers, and polymerization temperature and time will determine the final physical (molecular weight and its distribution, density, etc.) and chemical (solubility, solvent resistance, etc.) characteristics of the resulting CTPI resins. These features will directly exert an influence on the properties of the final CTPI films, such as optical, mechanical, thermal, and dielectric properties.

The fluorinated polyimides derived from 6FDA (2,2-bis(3,4-dicarboxyphenyl) hexafluoropropanedianhydride) and aromatic diamines are well known to be very transparent [65-69]. And the polyimides derived from an alicyclic diamine, DCHM (4,4′-diaminodicyclohexylmethane) and aromatic dianhydrides, also showed good transparency and thermal stability (Fig. 1.44). Hence, semialicycli polyimides have been investigated systematically to seek the colorless transparent polyimide films.

The ultraviolet-visible absorption spectra and fluorescence spectra of PI (6FDA/DCHM) and PI (6FDA/PDA) films were investigated, as well as their thermal properties [70]. Both films showed high transparency in the visible region (Fig. 1.45), there are no absorption bands above 370 nm. The bulky group —$C(CF_3)_2$— prevents molecular packing in the polyimide films, resulting in weak intermolecular interaction between the diamine moiety as an electronic donor, and a di-imide moiety as an electronic acceptor. It is noteworthy that the absorption spectrum of PI (6FDA/DCHM) film exhibits a very small platform region from 260nm to 300 nm, and PI (6FDA/PDA) film shows a shoulder absorption at 260 nm, similar to that of Kapton H films. This is attributed to the intramolecular charge transfer. The very weak intramolecular charge transfer of PI (6FDA/DCHM) film, compared with PI (6FDA/PDA), can be explained by the weak electron-donating property of DCHM [69]. The absorption intensity for PI (6FDA/DCHM) above 300 nm is weaker than that for PI (6FDA/PDA), implying that intermolecular charge transfer in PI (6FDA/DCHM) is very weak even if it was formed, compared with that for PI (6FDA/PDA). It can be concluded that the introduction of 6FDA into polyimide chains would weaken the intermolecular charge transfer due to the steric hindrance [48]. On the other hand, the weak electron-donating property of DCHM would reduce not only intermolecular charge transfer but also intramolecular charge transfer. Thus, by introducing 6FDA and DCHM into polyimide chain, a very transparent polyimide film was obtained. Both films have T_gs of about 250 °C, indicating that thermal stability of polyimides was not reduced by introducing the alicyclic groups.

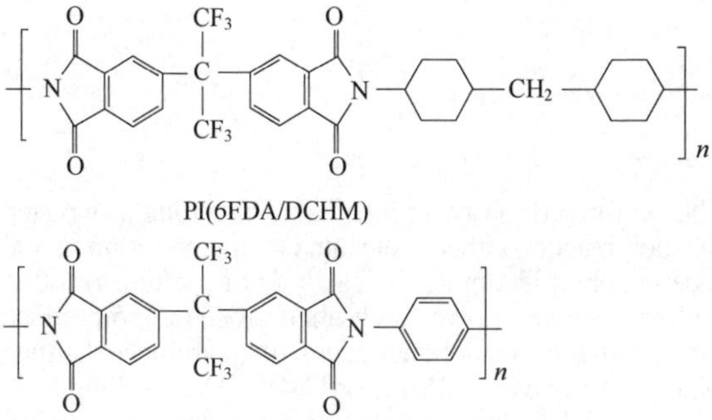

FIGURE 1.44 The chemical structures of the transparent polyimide films.

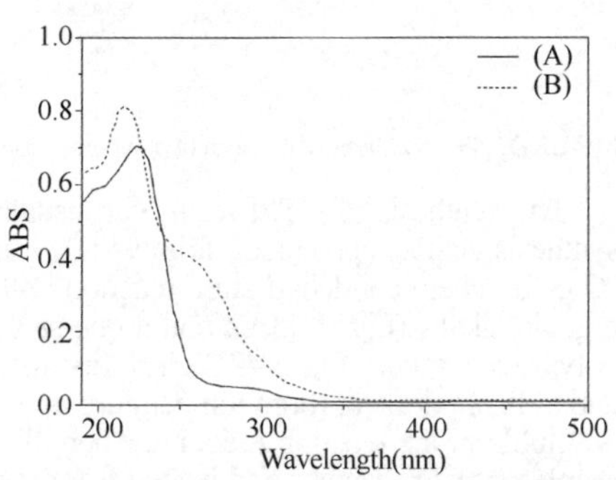

FIGURE 1.45 The UV absorption spectra of PI-(6FDA/DCHM)(A)and PI-(6FDA/PDA)(B)thin film of 0.6μm.

A series of transparent polyimides were prepared derived from 2,3,5-tricarboxycyclopentylaceticacid dianhydride (TCAAH) with different aromatic diamines, including an aromatic diamine, 4,4'-diaminodiphenylmethane (DPM), a fluorinated diamine, 2,2-bis(4-diaminodiphenyl)hexafluoropropane (6FdA), and an alicyclic diamine, DCHM (Fig. 1.46). All of the polyimide films showed high transparency in the visible region, no absorption tail was found above 330 nm (Fig. 1.47). The absorption shoulders at 248 nm of PI (TCAAH/DPM) and PI (TCAAH/6FdA) are due to the absorption of phenyl groups in the diamine moieties. The T_gs of these polyimides were measured in the region of 251 °C-336 °C, and the decomposition temperatures were measured in the region of 427 °C-457 °C.

A series of transparent polyimide films were prepared derived from 1,2,3,4-cyclobutanetetracarboxylic dianhydride (CBDA) with different diamines, including an aromatic diamine (MDA), a fluorinated diamine (6FdA), and an alicyclic diamine (DCHM) (Fig. 1.48). All of the films showed high transparency in the visible region, i.e., no absorption tail was found above 375 nm (Fig. 1.49). In particular, PI (CBDA/DCHM) film showed the highest transparent property with almost no absorption above 310 nm. Absorption shoulders at 235 nm of PI (CBDA/DPM) and PI (CBDA/6FdA) are due to the absorption of phenylene groups in the diamine moieties. The glass transition temperatures of the polyimides were measured at 277 °C-360 °C and the decomposition temperatures at around 450 °C under deaerated conditions.

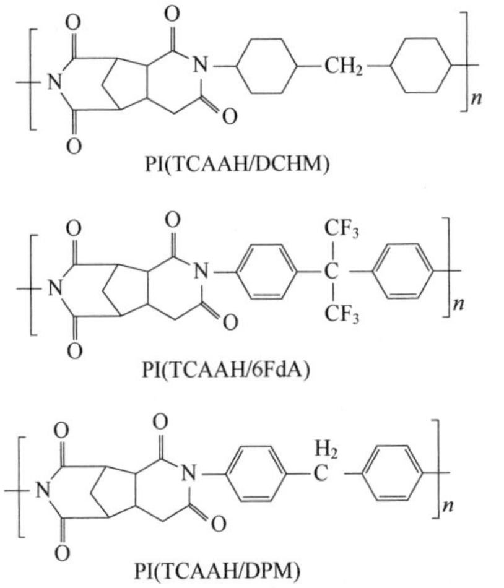

PI(TCAAH/DCHM)

PI(TCAAH/6FdA)

PI(TCAAH/DPM)

FIGURE 1.46 Chemical structures of transparent polyimides.

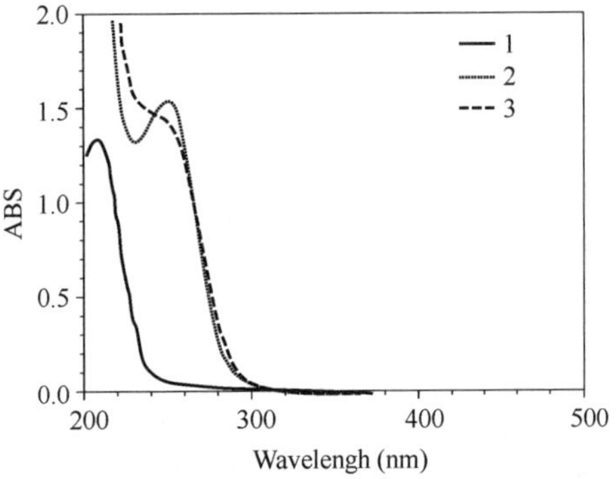

FIGURE 1.47 UV absorption spectra of transparent polyimide films(0.3 μm) (1)PI-(TCAAH/DCHM); (2)PI-(TCAAH/6FdA); (3)PI-(TCAAH/DPM).

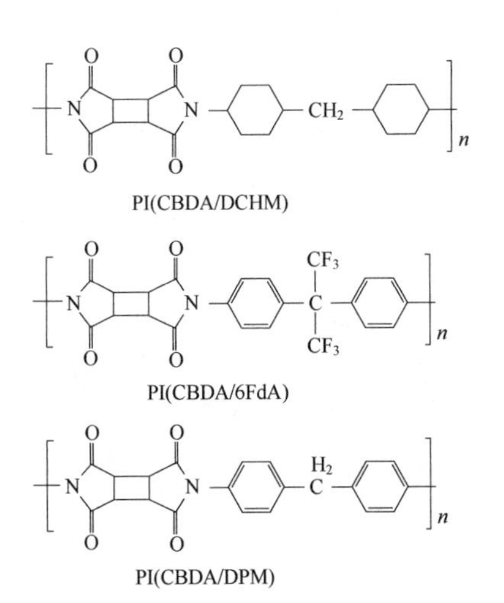

FIGURE 1.48 Chemical structures of the transparent polyimides.

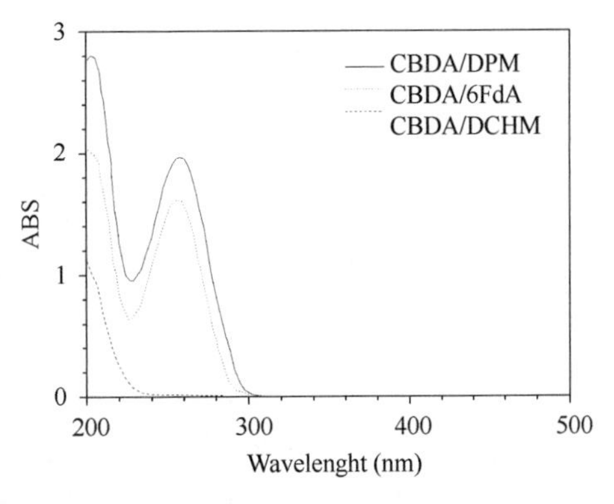

FIGURE 1.49 UV absorption spectra of the CTPI films(ca.0.3 μm).

Ha et al. have prepared a series of novel CTPI films, including fully aliphthatic polyimides either from adamantane-based diamines [71-73] or from an aliphatic dianhydride [74], highly transparent and hydrophobic fluorinated polyimides derived from perfluorodecylthio substituted diamine monomers [75], colorless PIs with chlorine side groups [76], fluorine-containing colorless PI-montmorillonites nanocomposite films [77], colorless PI-organoclay nanocomposite films [78], highly optically transparent PIs containing carbazole moieties [79], polynorbornene-chlorinated PI copolymers films [80], and optically transparent PI/graphene nanocomposite films [81]. Various applications for the developed colorless PI films as flexible substrates for organic electroluminescent devices [82] and as hole-transporting layers for hybrid organic LEDs [83] have also been investigated.

A highly transparent and colorless semiaromatic polyimide film has been prepared derived from alicyclic dianhydride and aromatic diamine at ICCAS (Fig. 1.50). A novel *meta*-substituted aromatic diamine containing trifluoromethyl and sulfonyl groups in the backbone, 2,2′-bis[4-(3-amino-5-trifluoromethylphenoxy) phenyl]sulfone (*m*-6FBAPS), was synthesized, which was then used to react with 1,2,4,5-cyclohexanetetracarboxylic dianhydride (CHDA) to yield a semiaromatic polyimide film. In comparison, a series of other semiaromatic polyimides were also prepared from CHDA and various aromatic diamines.

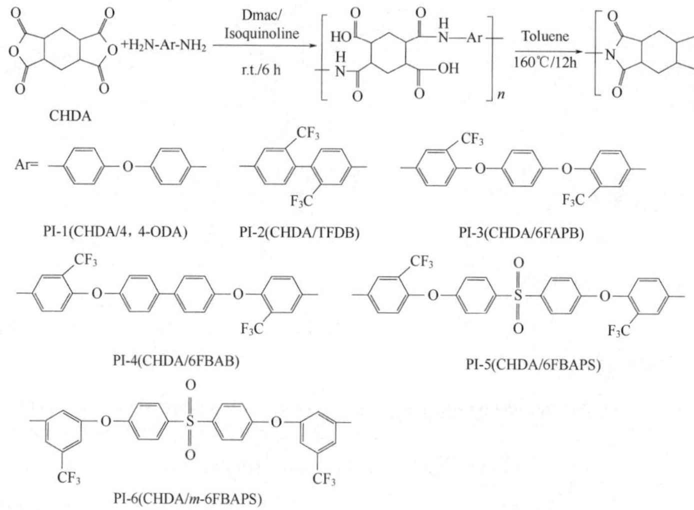

CHDA + H_2N-Ar-NH_2 → (Dmac/Isoquinoline, r.t./6 h) → (Toluene, 160℃/12h)

Ar=

PI-1(CHDA/4，4-ODA) PI-2(CHDA/TFDB) PI-3(CHDA/6FAPB)

PI-4(CHDA/6FBAB) PI-5(CHDA/6FBAPS)

PI-6(CHDA/*m*-6FBAPS)

FIGURE 1.50 Synthesis of semiaromatic polyimides.

The optical transparency of the semiaromatic polyimide films was evaluated by their physical appearance, UV-vis spectra and color intensities. These films were transparent and essentially colorless even as the film thickness was as high as 70 mm, being significantly different from the deep-yellow or brown color of traditional aromatic polyimide films (Fig. 1.51). The improved optical transparency is attributed to the effect of the alicyclic moieties in the polymer structure. The optical transparency obtained from the UV-vis spectra, including cutoff wavelength (absorption edge) and transmittance at different wavelength, are summarized in Table 1.14. All of the semiaromatic polyimide films showed excellent optical transparency with UV cutoff wavelengths of less than 314 nm and transmittance at 450 nm of higher than 91%. In comparison, the typical aromatic polyimide film derived from PMDA and 4,4′-ODA (ref-PIb) showed a deep-yellow color with transmittance at 450 nm of 2%.

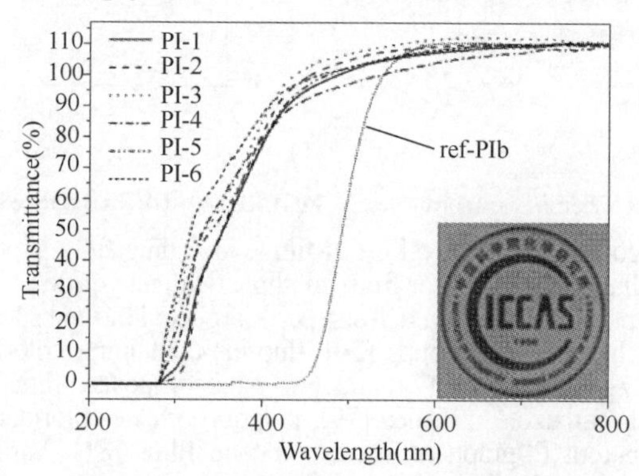

FIGURE 1.51 UV-vis spectra of semiaromatic polyimide films(film thickness:25 mm). The insert is the photograph of ref-PIb film.

The extremely transparent and entirely colorless films were obtained from the polymers incorporated with bulky electron-withdrawing trifluoromethyl- and sulfonyl- groups in diamine moieties, such as PI-2, PI-5, and PI-6, in which PI-6 showed the best transparency with T_{400} of 90% and T_{450}, of 94%, respectively. Their excellent optical properties are attributed to the distorted molecular conformation combined with the weakened electron-accepting and electron-donating properties of dianhydride and diamines, which significantly restrained the formation of inter-/intra-molecular charge transfer interactions.

TABLE1.14 Optical Transparency of Semiaromatic Polyimide Films

PIs	λ_0(nm)	T_{400}(%)	T_{450}(%)	T_{500}(%)
PI-1	293	84	91	94
PI-2	292	85	93	95
PI-3	302	88	92	95
PI-4	314	87	93	95
PI-5	298	88	93	96
PI-6	297	90	94	97
ref-PIb	444	0	2	58

λ_0, UV cutoff wavelength;T_{400}(%),T_{450}(%),T_{500}(%),transmittance at 400,450,and 500 nm, respectively; ref-PIb, aromatic polyimide film derived from PMDA and 4,4′-ODA.

Furthermore, the method of preparing transparent and conductive indium tin oxide (ITO)/polyimide films by high-temperature radio-frequency magnetron sputtering has been developed at ICCAS. A highly transparent and thermally stable polyimide substrate (Fig. 1.52) was prepared and used for the fabrication of ITO/PI films via radio-frequency magnetron sputtering at an elevated substrate temperature. The effect of the deposition conditions, i.e., the oxygen flow rate, substrate temperature, sputtering power, and working pressure, on the optical and electrical properties of the ITO/PI films were investigated from the microstructural aspects. The results indicate that the optical and electrical properties of ITO were sensitive to the oxygen. Moreover, it was beneficial to the improvement of the ITO conductivity through the adoption of a high substrate temperature and sputtering power and a low working pressure in the deposition process. A two-step deposition method was developed in which a thick bulk ITO layer was overlapped by deposition on a thin seed ITO layer with a dense surface to prepare the highly transparent and conductive ITO/PI films.

The optical properties of the ITO-m/PI and ITO-b/PI films were evaluated by their physical appearances, transmittance spectra, and color intensities. Fig. 1.53 shows the film transmittance spectra and photographs. Both films were highly transparent and entirely colorless, showing average transmittances in the range 400nm-800 nm over 81% and values of b^* less than 3 (Table 1.15). ITO-*b*/PI exhibited a very low resistivity of 9.03×10^{-4} Ω·cm, two orders of magnitude lower than that of ITO-*m*/PI. The difference in the conductivity could be explained by the surface morphology perspective. The ITO/PI film after annealing at 240 °C gave a transmittance of 83% and a sheet resistance of 19.7 $\Omega\cdot m^{-2}$.

Mitsubishi Gas Chemical Company reported a method for producing colorless and transparent PI films by a solution casting procedure. The bi-axially stretched colorless PI films exhibit excellent optical transparency, heat resistance, and reduced dimensional changes [84].

The films were produced with the soluble CTPI resin as the starting materials, which were derived from CBDA and aromatic diamines by a one-step high- temperature polycondensation route. The CTPI film was bi-axially stretched in the machine direction by 1.01 times and in the transverse direction by 1.03 times at 250 °C for 11 min under a stream of nitrogen. Then, the CTPI film was dried by blowing nitrogen containing 1000×10^{-6}oxygen at a flow rate of 3.3 $m\cdot s^{-1}$ at 280 °C for 45 min. The obtained CTPI film had a thickness of 200 μm, a total light transmittance of 89.8%, a yellow index of 1.9, and a haze of 0.74%. The solvent residual ratio in the film was 0.5% by weight. Due to their unique optical and thermal properties, the colorless CPPI films might find extensive applications in optoelectronic applications, such as transparent conductive film, transparent substrates for flexible display, flexible solar cells, and flexible printing circuit board (FPCB). Similar procedures were also reported [85].

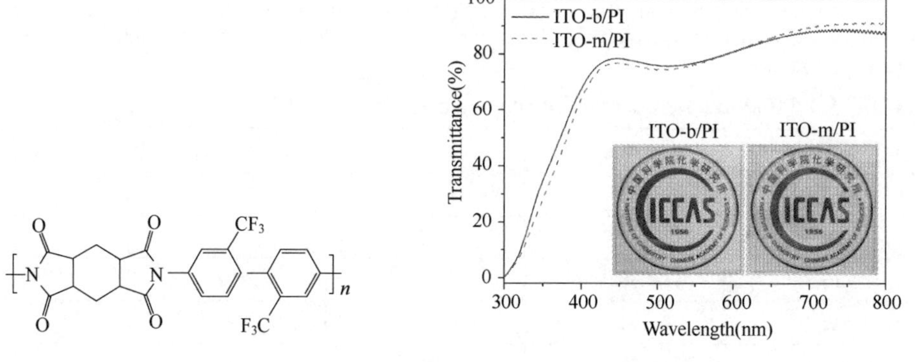

FIGURE 1.52 Chemical structure of the transparent polyimide film.

FIGURE 1.53 Transmittance spectra of the ITO/PI films. The insets show the photographs of (A)ITO-m/PI and(B)ITO-b/PI films.

TABLE1.15 Optical and Electrical Properties of the As–Deposited ITO–m/PI and/To–/PI Films and Annealed ITO–m/PI Film

Sample	T_{av} (%)[a]	b^*	Carrier Concentration(com^{-3})	Carrier Mobility(cm^2·V·s^{-1})	Resistivity (Ω·cm^{-1})	Sheet Resistance (Ω·m^{-2})
ITO-m/PI	81.9	2.27	1.17×10^{19}	11.71	4.56×10^{-2}	2.3×10^{3}
ITO-b/PI	81.6	2.74	2.26×10^{20}	30.62	9.03×10^{-4}	45.2
Annealed ITO-/PI	83.0	2.19	4.70×10^{21}	3.31	3.94×10^{-4}	19.7

[a] *Average transmittance of the ITO/PI films in the range 400-800nm.*

Additionally, DuPont Company [86] and Kolon Industries [87] have reported the methods of low-color polyimide films derived from the copolymers of fluoro-containing dianhydride, 2,2-bis(3,4-dicarboxyphenyl)-hexafluoropropane dianhydride (6FDA), BPDA, and fluoro-containing diamine, 2,2′-bis(trifluoro-methyl) benzidine (TFMB). The copolymers were prepared via PAA precursors, followed by chemical imidization of the PAAs to afford the gel-like PAA films or soluble CTPI resins. Then, the CTPI films were produced from these intermediates at high temperatures up to 300 °C. Flexible and tough CTPI films with low color and high transparency were obtained.

NASA (National Aeronautics and Space Administration, USA) Langley research center have investigated the molecularly oriented transparent polyimide films for space applications [88]. In large space structures with designed lifetimes of 10-30 years, there exists a need for high temperature (200 °C-300 °C) stable, flexible polymer films that have high optical transparency in the visible light region. For this purpose, a colorless and transparent polyimide film, LaRC-CP1, derived from 6FDA and fluoro-containing diamine, 2,2-bis[4-(4-aminophenoxy)phenyl] hexafluoropropane (BDAF), has been developed by NASA. This film is prepared from soluble polyimide resin. The LaRC-CP1 film was mono-axially stretched by 1.5, 1.75 and 2 times the original length of the film. The tensile properties of the film increased with increased stretching ratio. The tensile strength of 2.0× stretched film increased from 93.0MPa to 145.4 MPa after stretching treatment; and the elongations increased from 16% to 65%. After stretching treatment, the dimensional stability, stiffness, elongation, and strength of the films were greatly enhanced, which are crucial for the applications in space environments.

1.4.4 Atomic Oxygen-resistant Polyimide Films

Atomic oxygen (AO) and vacuum ultraviolet (VUV) can create hazardous conditions to degrade polymer materials used in the exterior surfaces of a spacecraft in low Earth orbit (LEO). Usually, the exterior surfaces of spacecraft are covered with multilayer thermal insulation (MLI), in which the outer surface of MLI consists of bare polyimide films, such as Kapton H, Upilex R, etc. The AO- and VUV-induced erosion of these polymer films always results in the deterioration of MLI properties. In order to protect polyimide films from AO bombardment, optically transparent inorganic thin films have been used as overlay protective coatings. The inorganic thin films used includes ITO, tin dioxide (SnO_2), Al_2O_3, silicon dioxide(SiO_2) or silicon dioxide mixed with PTFE Teflon, etc. [89]. However, there are three major problems in the current hard coating technology. One is pinholes in the protective coatings produced in the coating process. It was found that even small pinholes can cause serious undercutting in the polymer layers [90]. Hence, the number of pinholes or defects in the inorganic coating must be controlled to a very low level, enabling the polymer films to be survived. This is a big technological challenge for current thin-film processes. Another problem is the cracks produced at areas where cutting, punching, or a sharp bend is performed on the inorganic coated polymer films. The cracks in size are larger than pinholes, which are usually the major sources for MLI damage. The last problem is collision with space debris or micrometeoroids in orbit, which may result in openings in the inorganic coatings where AO can react with the polymer underneath. A hard coating with micrometer thickness can be easily penetrated by these high-energy particles.

The protective inorganic coatings are always used in the actual space systems such as the International Space Station; some of the coatings are quite durable when used properly but there are still some problems that occur due to their lack of self-healing at defect sites. Hence, protective coatings with self-healing capability is desired for the space crafts in LEO.

Kapton polyimide film is extensively used in solar arrays, spacecraft thermal blankets, and space inflatable structures. Upon exposure to AO in LEO, the film is severely eroded. An effective approach to prevent this erosion is to incorporate polydedral oligomeric silsesquioxane (POSS) into the polyimide backbone structures by copolymerizing the monomer mixture of POSS-containing diamine and aromatic diamine with aromatic dianhydride in a polar solvent to give high-molecular-weight PAA solution, which is then cast on a support surface to give a PAA film [91]. After removal of some part of the solvent by evaporation and partial chemical or thermal imidization, a self-standing gel film is peeled off from the support surface, followed by imidization to produce a POSS-Kapton polyimide film (Fig. 1.54). The POSS is a unique family of nanoscale inorganic/organic hybrid cage-like structures that contain silicon/oxygen framework ($RSiO_{1.5}$), where R is a hydrocarbon group used to tailor compatibility with the polymer and chemical stability. Thus, a series of POSS-Kapton films with different POSS-diamine monomer loadings of 0%, 5%, 10%, 20%, and 30%, corresponding to Si_8O_{11} cage loadings of 0%, 1.75%, 3.5%, 7.0%, and 8.8%, were prepared, including main-chain (MC) and side-chain (SC) types of POSS-kapton films.

POSS-kapton films were exposed to AO in the space environment on the Materials International Space Station Experiment (MISSE) platform. The AO flux and fluence varied from experiment to experiment, depending on exposure duration, timing of the experiment within the solar cycle, and orientation of MISSE sample tray. The relative velocity of the AO with respect to ram surface is approximately 7.4 $km \cdot s^{-1}$, corresponding to oxygen atoms striking the surface with an average velocity of 4.5 eV. The MISSE-1 flight experiment was retrieved after 3.9 years. The samples were exposed to the ram and therefore all components of the LEO environment, including AO and VUV light, with the O-atom fluence being around 8×10^{21} $atoms \cdot cm^{-2}$, similar to the fluence on the companion MISSE-2 where the fluence was accurately measured to be 8.43×10^{21} $atoms \cdot cm^{-2}$.

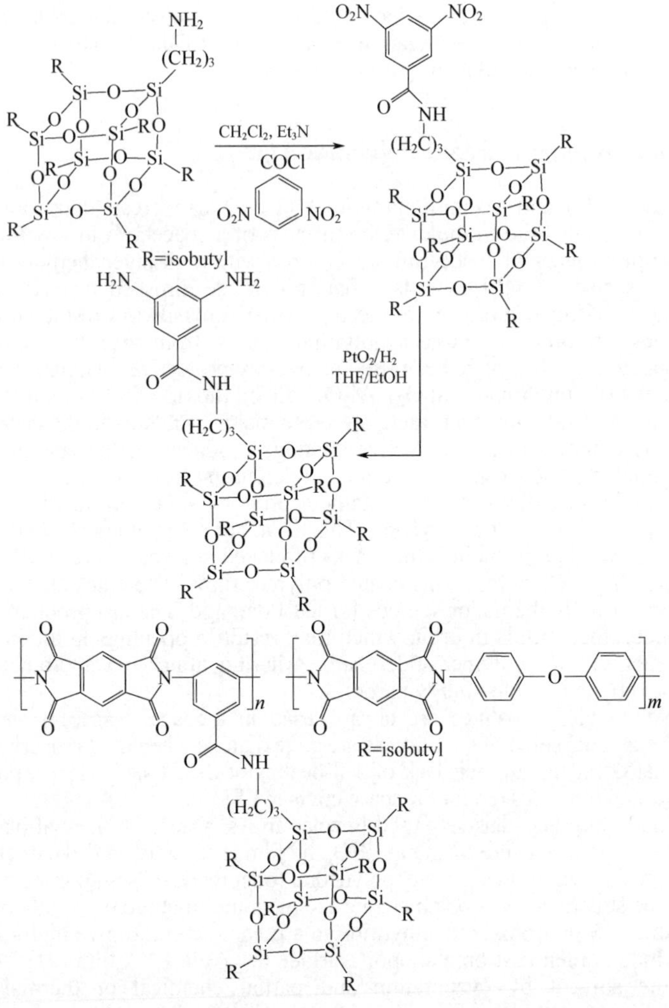

FIGURE 1.54 The chemical structure of MC and SC POSS-Kapton films.

The SC-POSS-kapton polyimide film was exposed to AO in the laboratory at room temperature, with a total fluence of about 2.7×10^{20} atoms·cm^{-2}. Table 1.16 shows the laboratory AO erosion data of SC-Kapton film [92]. The experimental results indicate dramatic decreases in the surface erosion with increasing contents of SC POSS. Both SC POSS-Kapton and MC POSS-Kapton films exhibit comparable atomic-oxygen erosion yields. More specifically, in an atomic-oxygen exposure with a total fluence of 3.53×10^{20} O-atoms·cm^{-2}, 7.0 %(w) Si_8O_{12} SC POSS-Kapton film showed an erosion yield that was 3.3% that of Kapton H film. It was found that MC POSS-Kapton and SC POSS-Kapton films showed the comparable AO erosion yields for a given %(w) POSS cage.

The effect of temperatures on the atomic-oxygen erosion yield of POSS-Kapton films with different Si_8O_{12} loadings of 0%, 3.5%, and 7.0% were investigated. The samples were exposed to the hyperthermal O-atom beam at different temperatures of 25%, 100%, 150%, 220%, and 300 °C, respectively. Each sample set, corresponding to a particular temperature, was exposed to 50,000 pulses of the hyperthermal O-atom beam. Table 1.17 shows the laboratory AO erosion depths of MC-Kapton films at different temperatures. The erosion of the 0%(w) MC POSS-Kapton film control showed the strongest temperature dependence, with the erosion depth increasing by a factor of about 3.6 from 25 °C to 300 °C. And the 3.5 and 7.00%(*w*) Si_8O_{12} MC POSS-Kapton films showed less temperature dependence in their erosion. The erosion depths of these samples increased by factors of 2.2 and 2.4, respectively, with the increase in temperature from 25 °C to 300 °C.

TABLE1.16 Laboratory Atomic Oxygen Erosion Data of SC-Kapton Film

%(*w*) Si_8O_{12} in SC-Kapton Film	Kapton-equivalent Fluence($\times10^{20}$ O-atoms·cm^{-2})	Erosion Depth(μm)	% Erosion of Kapton H Reference
1.75	2.71	1.99±0.01	24.5
3.5	2.66	1.29±0.05	16.2
7.0	2.68	0.39±0.04	4.90
8.8	2.68	0.13±0.02	1.64
10.5	2.71	0.25±0.03	3.06
12.3	2.71	0.11±0.03	1.39
14.0	2.71	undetectable	~0

TABLE1.17 The Laboratory Atomic Oxygen Erosion Depths of MC-Kapton Films at Different Temperatures

Test Temperatures(°C)	Erosion Depths of MC-Kapton Films With Different %(*w*) of Si_8O_{12} POSS Loadings(μm)			
	0.0	3.5	7.0	Kapton H 23 °C Reference
300	10.37±0.47	1.24±0.17	0.67±0.16	3.14±0.13
220	7.47±0.37	0.94±0.21	0.78±0.08	3.36±0.20
150	5.36±0.23	1.02±0.11	0.41±0.07	3.59±0.11
100	4.09±0.38	0.82±0.07	0.43±0.06	3.55±0.11
25	3.17±0.24	0.63±0.08	0.30±0.08	3.50±0.12

A Kapton H reference material, held at 23 °C,accompanied each exposure.

The MC POSS-Kapton films with 0%(*w*), 1.75%(*w*), and 3.5%(*w*) were flown on MISSE-1 (ISS in 3.9 years). Various images of the samples were taken throughout the flight, and it was found that the 0%(*w*) MC POSS-Kapton film control sample was completely eroded in less than 4 months. On the basis of the estimated O-atom fluence of 8×10^{21} atoms·cm^{-2} and an erosion yield of 3.00×10^{-24} cm^3 per atom for Kapton H film, the 0%(*w*) MC POSS-Kapton film should have eroded a minimum of 240 μm. Experimental results indicated that the 1.75%(*w*) and 3.5%(*w*) MC POSS-Kapton films were eroded by 5.8μm and 2.1 μm, respectively.

Miyazaki et al. investigated the tolerance of polysiloxane-block polyimide film against AO [93,94]. The commercially available silicon containing polyimide film (BSF30), a polysiloxane- block- polyimide is selected for investigation (Fig. 1.55). An AO beam was irradiated on the polysiloxane- block-polyimide film at the Combined Space Effects Test Facility of JAXA in Tsukuba, Japan. To investigate the AO tolerance, mass change measurement, cross-sectional transmission electron microscopic (TEM) observation, and X-ray photoelectron spectroscopic (XPS) analysis were performed.

Experimental results indicated that the mass loss of polysiloxane-block polyimide is only 1% or less than that of Kapton H. Cross-sectional TEM observation and XPS analysis revealed that the

AO-protective SiO_2 layer is self-organized by AO irradiation. Furthermore, the self-organized SiO_2 layer is intentionally damaged to investigate reorganization of a new layer on it. AO irradiation of the damaged surface revealed that the new layer is built with a 500nm-deep eroded region. The "self-healing" ability of polysiloxane-block- polyimide was observed, suggesting that polysiloxane-block-polyimide film has high potential to provide many advantages of a space-use material, especially for LEO spacecraft.

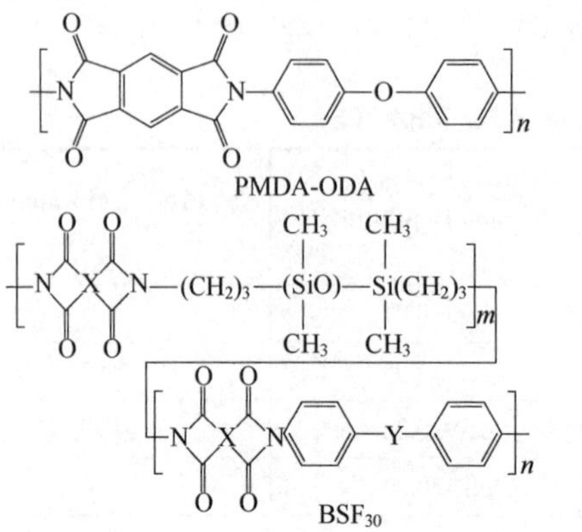

FIGURE 1.55 Molecular structures of PMDA-ODA polyimide (SP-510) and Si-containing polyimide (BSF30).

A series of AO-resistant and transparent phosphine-containing polyimide coatings have been developed. A *meta*-substituted aromatic diamine, [3,5-bis(3-aminophenoxy) phenyl]diphenylphosphine oxide (m-BADPO) was first synthesized by the Williamson reaction of 3,5-difluorophenyldiphenyl- phosphine oxide (DFPPO) and meta-aminophenol. The diamine was then polymerized with several commercially available aromatic dianhydrides to afford a series of aromatic polyimides (PI-1-PI-4) (Fig. 1.56). The *meta*-substituted molecular skeleton and the pendant bulky phenylphosphine oxide (PPO) group endowed the polyimides with desired properties for potential applications in space environments.

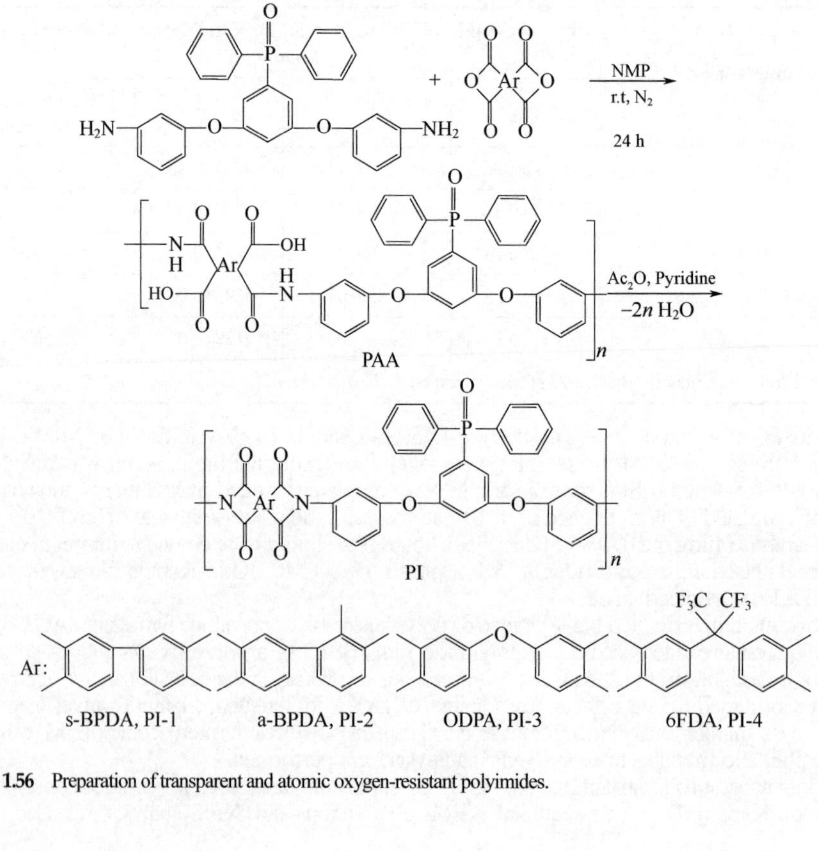

FIGURE 1.56 Preparation of transparent and atomic oxygen-resistant polyimides.

It was found that the solubility of the polyimides was enhanced due to the synergic effects of meta structure and the bulky PPO groups, making it possible to fabricate the polyimide films via solution procedure. The films exhibited flexible and tough natures with light color and high transparency in the visible light region (Fig. 1.57). Transmittance up to 87% at 400 nm was achieved in the films. PI-4 showed a cutoff wavelength at 319 nm, which was 7 nm lower than that of polyimide from p-BADPO and 6FDA (PI-4) at the same thickness. The transmittance of the PI-4 film at 400 nm (85%) was also superior to its *para*-linked analog (PI-4: 82%). Yellow index (YI) is usually adopted as a criterion evaluating the color of a polymer film. This value describes the color changes of a film sample from clear or white toward yellow. Lower YI value usually indicates a weak coloration for a polymer film. The YI values (b^* values) of the m-BADPO-PI films (thickness: ~40 m) ranged from 6.4 to 13.4 dependent on the aromatic dianhydrides used. The YI values of 7.2 for PI-3 and 6.4 for PI-4 were apparently lower than those of their p-BADPO analogs (PI-3: 9.7 and PI-4: 8.0). The good transparency and low coloration of PI-3 and PI-4 films, on the one hand, could be attributed to its loose molecular packing induced by the bulky pendant PPO group in the diamine moiety and the flexible ether or 6F linkage in the aromatic dianhydride. On the other hand, the irregular structures caused by their *meta*-substituted molecular backbone were also advantageous to reducing the formation of charge transfer complexes (CTCs). Influence of the steric hindrance and inductive effects of the PPO groups and the meta structures on the optical properties of the polyimide films could also be reflected from the refractive index (n) values of the films. Generally, polymers containing substituents with low molar refractions or bulky molecular volumes often exhibit low n values according to the Lorentz-Lorenz equation. For instance, the Kapton type PI-(PMDA/ODA) has a n value of 1.6478 at 1320 nm, whereas PI-(6FDA/ODA) with a much looser molecular packing exhibits a lower n value of 1.5565. Obviously, the pendant bulky PPO substituents and the *meta*-substituted structure are expected to decrease the refractive indices of the polyimide films.

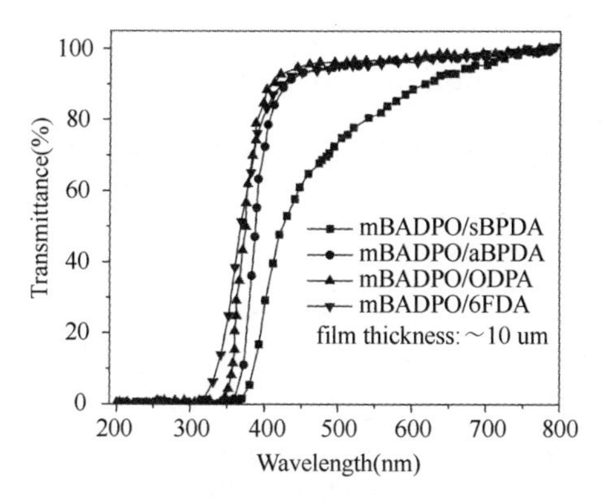

FIGURE 1.57 UV-visible spectra of the transparent and atomic oxygen-resistant polyimide films.

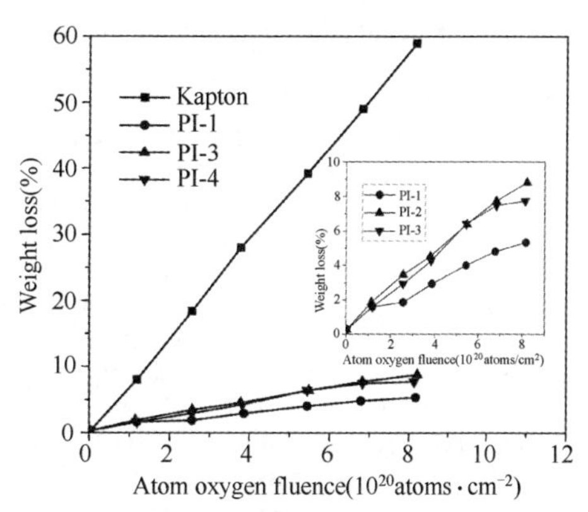

FIGURE 1.58 Mass loss versus atomic oxygen fluence for the transparent and atomic oxygen-resistant polyimide films.

The polyimide films, PI-1, PI-3, PI-4 together with the Kapton standard (20×20× 0.05 mm) were exposed to AO at the ground-based simulation facility. The erosion rate of the Kapton was set to be 3.0×10^{-24} $cm^3\cdot atom^{-1}$ as a reference. The samples were exposed to AO with a total fluence of 8.13×10^{20} $atoms\cdot cm^{-2}$. Fig. 1.58 depicts the mass loss versus AO fluence for the polyimide films and Kapton reference. It can be observed that the polyimide samples containing the PPO group exhibited a weight loss of 5.48%(w)-8.76%(w) while that of Kapton H was 56.76%(w). In addition, the PPO-containing polyimide films exhibited a nonlinear weight loss rate while Kapton H exhibited a linear one. The erosion yields of the PPO-containing polyimide films were much lower than that of Kapton H. PI-3 shows the lowest erosion yield of 6.59×10^{-25} $cm^3\cdot atom^{-1}$ while PI-4 exhibits the highest one. Obviously, the PPO groups in the present polyimides decrease their erosion yields in AO environments.

The surface composition of the AO-exposed PI films was examined by XPS. The experimental results indicated that a layer of phosphorus oxide at the surface of PI-1 film was formed. This inert layer might inhibit the further erosion of the underlying polyimide film. Fig. 1.59 compares SEM images of Kapton H and PI-1 samples after AO exposure. Both of the surfaces of PI-1 and Kapton films exhibited a carpet-like appearance.

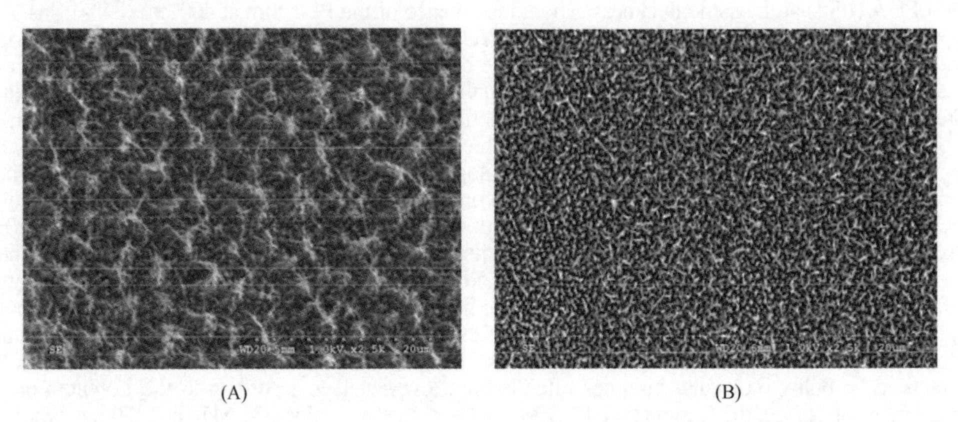

FIGURE 1.59 SEM images of Kapton H (A) and PI-1 (B) exposed to atomic oxygen (8.13×10^{20} atoms/cm).

1.5 Surface Modification of Polyimide Films

Aromatic polyimide films have been extensively used in many high-tech fields such as electrical insulation, microelectronics, military aircraft and spacecraft, etc., due to their attractive thermal, mechanical, and electrical properties. These important applications usually require the polyimide films to have a modified surface to improve its adhesion to other materials such as metals, oxides, or other polymers. Plasma treatment is the common accepted technology for this purpose. Plasma are collections of charged particles, most commonly occurring in the gaseous state, with stars, fluorescent lights, and neon signs among the more familiar examples. Plasma technology has been applied in the microelectronics industry for stripping of organic etching masks and removal of organic surface contaminants since the late 1960s, of which oxygen plasma has been used to simulate the degradation of polyimide films used as a protective material on satellite systems and space stations by AO encountered at orbital velocities in LEO [95,96].

The components of gaseous plasma that can interact with polymer surfaces are electrons, reactive-neutral atoms and molecules, photons including those in the highly energetic VUV region, and energetic positive ions. The region between a plasma and an adjacent solid surface is known as a space charge sheath. There is a voltage drop across this sheath that accelerates positive ions toward the solid surface. These ions can have a major effect on chemical reactions occurring on the substrate surfaces and can impact enough energy to a solid surface to eject particles (sputtering). Whereas the goal of plasma etching is removal of materials, plasma surface modification is intended only to alter the nature of the film surface without changing the bulk properties. It is often desirable to improve upon the surface properties (wettability and adhesion to subsequently deposited or laminated materials) of polyimide films. Plasma technology has been used to enhance the wettability of many polymer film surfaces [97-99]. Depending on the plasma system configuration and selection of gases used for treatment, material removal and modification can occur simultaneously, proceeding by chemical reactions, by physical means, or by a combination of these two mechanisms.

As mentioned above, polyimide films are often used as dielectric materials in the fabrication of thin-film electronic packages such as TAB and FPCs. FPCs are now in common use as low-cost IC chip carriers. One strategy for reducing cost is to fabricate these carriers in a roll format. Polyimide films are commonly the preferred dielectric for this application because of their high thermal stability and good mechanical properties. Kapton H and Upilex R and S films are the commonly used polyimide films, of which Upilex S film has low moisture regain and a different thermal coefficient of expansion than Kapton H.

The plasma reactor configurations have obvious effects on the surface modification of polyimide films. Oxygen microwave (remote) and DC plasmas are employed to treat the polyimide film surface. Compared with the untreated polyimide films, microwave plasma treatment reduced practical adhesion levels for Kapton H and Uplix S, while DC-glow treatment produced enhanced adhesion about triple the values of the untreated films. Analysis of polymer surface and characterization of the plasma environment are required to understand the dependence of the resulting adhesion on the plasma system configuration. There are many factors that can affect the adhesion strength including thickness of the film being peeled, width of the peeled line, peel rate, ductility of the metals, and peel technique.

To illustrate the effects of modification by the reactive-neutral oxygen atoms, the surface chemistry of Kapton H, Upilex S, and Upilex R films treated downstream from oxygen microwave plasmas has been extensively investigated. Deionized (DI) water contact angles on Kapton H films were measured as a function of time of treatment downstream from oxygen plasmas at 30, 60, and 120 W (O atom concentrations increase in that order). For short treatments (less than 15 s), no discernible dependence on O-atom concentration was observed, indicating that initially a reaction with a relatively high rate constant occurs, that reduces the DI water contact angle from 72 degrees to about 45 degrees. With longer treatments, the effect of variation in O-atom concentrations becomes apparent, indicating the occurrence of at least one other reaction with a lower rate constant. Eventually, at each power, the surface comes to the same steady-state contact angle. In each case, after samples were subjected to a DI water rinse, subsequently measured contact angles increased to roughly 48 degrees, similar to the value obtained following the initial, fast reaction.

Upilex S film shows similar surface modification with Kapton H. Although the surface compositions are similar, 1.0% of silicon is detected in the commercially available Upilex S film. The silicon incorporation could be intentional (e.g., as part of an additive to reduce frictional drag) or unintentional (as a contaminant, e.g., from silicon rubber rollers used in the curing process). The degree of surface modification depended not only on treatment time, but also on the number density of oxygen atoms in the plasma [100]. Several gas additives, including nitrogen, can be used to increase the AO concentrations in the plasma.

Ion beam treatment can also be used to promote polyimide film adhesion. Several ion beam techniques can be used, such as surface treatment of polyimide film prior to deposition, simultaneous beam irradiation and metal deposition, and treatment following deposition. Ion beam treatment is not a form of plasma modification, but bombardment by energetic ion beams can be similar to that occurring at the surface residing in the plasma. There are several examples of improved adhesion for metal films vapor deposition onto polyimide film surface pretreated by irradiation with argon (and oxygen) ion beams. The degree of enhancement in etching or modification due to ion bombardment depends on the dose of ions incident onto the surface, such as the ion density, energy, and duration of exposure. The energy of ions bombarding the surface in plasma systems typically does not exceed a few hundred eV. The ion penetration depth at this energy is of the order of a few tenths of a nanometer, but the ions can extend several nanometers into the polyimide film due to the effect of the energy transfer through the solid. The surface properties of polyimides subjected to interactions with energetic ions are altered with respect to chemical composition and polymer backbone structures.

Energetic ions at 200 eV were used to modify the surface of both Kapton H and Upilex S films prior to sequential metallization with chromium and copper. With argon ions, Kapton H film gave 90 degrees peel strengths that were about twice the values measured for untreated polyimide films. On the other hand, very little adhesion improvement was obtained for Upilex S film with the same treatment. Kapton H film treated with argon ions and subsequently by oxygen ions, or using only oxygen ions showed 90 degrees peel values enhanced by a factor of 2 relative to the untreated film. The latter treatments resulted in a factor of four increase in peel values for Upilex S film. Hence, chemical structure of the polyimide films plays an important role in the surface modification and resulting improvements in adhesion.

A low-energy (500 eV) Ar^+ ion beam was used to modify the surface of Kapton H film. Adhesion of thin gold film overlayers to the treated Kapton H film increased by an order of magnitude. Adhesion of copper to the treated Kapton H film was increased by a factor of 5. Table 1.18 summarizes the improvements in surface adhesion of polyimide films to the multilayer of vacuum-deposited metals plus plated copper using various plasma and beam techniques. The surface-treating technique by reactive neutrals + ions + electrons + photons give a surface-actived Kapton H film using O_2 DC glow which has 90 degrees peel strength to the multilayer of Cr (20 nm) plus Cu (>10 μm) of 490-657 $N\cdot m^{-1}$, and Upilex S film using O_2 sputter etching with 90degrees peel strength to the multilayer of Ta (50nm) plus Cu (>8 μm) of 784-882 $N\cdot m^{-1}$. In addition, the surface treatment by ions + electrons + photons also gives high adhesion surfaces both for Kapton H and Upilex S films. The 90 degrees peel strength of the Kapton H film treated using Ar^+ sputter etching to the multilayer of E-beam evaporated Cr (20nm) + plated Cu (>8 μm) was 882$N\cdot m^{-1}$, comparable to that of the treated Upilex S film. Experimental results showed that the ions surface treatment yielded Upilex S film with better adhesion to copper layer than Kapton H film, implying that the chemical structure of polyimide films have obvious effects on the polymer surface modifications. In comparison, the reactive neutrals treating technique does not show apparent adhesion enhancement for Kapton H film.

1.6 Applications of Polyimide Films

Aromatic polyimide films exhibit a unique combination of thermal, mechanical, electrical, dielectric, and properties, making it ideal for a variety of applications in many high-tech industries. Polyimide films can maintain their excellent combined properties over a wide temperature range, and exhibit excellent chemical resistance without any dissolution by most common organic solvents. Additionally, polyimide films do not melt or burn due to its highest UL-94 flammability rating: V-0. Polyimide films can be used both in electrical insulation applications, including wire and cable tapes, formed coil insulation, motor slot liners, magnet wire insulation, transformer and capacitor insulation, magnetic and pressure-sensitive tapes and tubing, and in microelectronic packaging applications, including FPC, TAB, and COF, etc. However, each special use requires the polyimide film to have its desirable combined properties.

1.6.1 Electric Insulating Applications

The traditional applications of polyimide films are the motor and magnet wire industry due to their unique properties. The electrical and mechanical strengths of polyimide films enable thinner insulation designs and conserving space for conductors, yielding more power without increasing motor size. Polyimide films can provide exceptional overload protection and long motor life, even in the most harsh servicing environments and conditions. Polyimide films have superior chemical resistance to most organic solvents, hydrocarbons, lubricants, resins, and varnishes, and UL 94-V0 flammability rating, and will not melt, ignite, or propagate flame.

The typical motor applications include magnet wire, turn-to-turn, strand, coil, slot liner, and ground insulation. Currently, there are several types of polyimide films which can be used in motors, generators, and transformers, including: (1) Kapton HN film which exhibit exceptional and unique balance of physical, chemical, and electrical properties over a wide temperature range, particularly at high temperatures. (2) Kapton CR film which is developed specifically to withstand the damaging effects of partial discharge. The corona produced by the partial discharge can cause the eventual breakdown of an insulation material or system. Kapton CR film has a corona resistance or voltage endurance that is significantly higher than standard Kapton HN film, and also provides great higher thermal conductivity, allowing better dissipation of heat in motors and other electrical equipment. (3) Kapton WR film which can be exposed continuously to hot water, combating the effect of water on insulation systems where hydrolytic stability is important.

TABLE 1.18 The Surface Adhesion of Polyimide Films Treated by Various Plasma and Beam Techniques to the Vacuum–Deposited Metals Plus Copper Layer

Films	Surface Treating Method	Deposited Metals	Peel Strength of As-deposited Metal Films 90° Peel ($N \cdot m^{-1}$)		
			Untreated	Treated	Reference
1. Reactive neutrals + ions + electrons + photons					
Kapton H		20 nm Cr-sputtered			
	O_2 DC glow	+300 nm Cu-sputtered +>10 μm Cu-plated	196	490—657	50
	O_2 RF Plasma	100 nm Fe- E-beam evaporated	<50	784	73
	O_2 RIBE	Cu	20	686	74
Upilex S	O_2 sputter etch	50 nm Ta-sputtered	20	784	
	O_2 + Ar sequential sputter etching	+8 μm Cu plated		882	75
2. Ions + electrons + photons					
Kapton H	Ar+ sputter etch	20 nm Cr E-beam evaporated +8 μm Cu plated	49	882	77
Upilex S	Ar+ sputter etch	50 nm Ta-sputtered +8 μm Cu	49	833	76
3. Ions only					
Kapton H	Ar+ ion beam (200 eV)	20-50 nm Cr E-beam evaporated/sputtered +8 μm Cu	235-274	519—568	76
Upilex	Ar+ ion beam (200 eV)	20-50 nm Cr E-beam evaporated (or sputtered) +8 μm Cu	49	833	76
4. Reactive neutrals only					
Kapton H	O_2 microwave plasma downstream	20 nm Cr-sputtered +300 nm Cu-sputtered +>10 μm Cu-plated	225	157	50

The polyimide films used in electrical insulating areas are usually coated with Teflon FEP or Teflon PFA polyfluorocarbon resins, which were melt-extrudable copolymers of poly(tetrafluoro-ethylene). The heat-sealable Kapton films are always used as primary insulation on magnet wires. The thermally stable coatings confer on polyimide films heat sealability to conductor metals and act as high-temperature adhesives. The film tapes are used to wrap conductors and are then heat-sealed at high temperature. The heat-sealable Kapton films include FN, FCR, FWN, FWR, etc. Tables1.19, 1.20, and 1.21 summarize typical polyimide films for the electrical insulation industry.

TABLE 1.19 Properties of Nonheat-sealable Kapton Polyimide Films

Properties	Product Designation				Testing Method
	100HN	200HN	100CR	200CR	
Minimum MD ultimate tensile strength (MPa)	179	186	179	117	ASTM D882
Minimum MD ultimate elongation (%)	55	55	55	35	ASTM D882
Minimum dielectric strength ($kV \cdot mm^{-1}$)	244	197	236	177	ASTM D149
Typical volume resistivity ($\Omega \cdot cm^{-1}$)	10^{17}	10^{17}	10^{17}	10^{17}	ASTM D257
Typical dielectric constant, at 1 kHz	3.4	3.4	3.4	4.0	ASTM D150
Typical dissipation factor, at 1 kHz	0.002	0.002	0.003	0.003	ASTM D882

TABLE 1.20 Properties of Heat-sealable Kapton Polyimide Films

Properties	Product Designation				Testing Method
	150FN019	150FCR019	150FWN019	150FWR019	
Minimum MD ultimate elongation (%)	60	55	65	45	ASTM D882
Minimum dielectric strength ($kV \cdot mm^{-1}$)	169	142	173	173	ASTM D149
Minimum heat seal peel strength, Teflon to Kapton ($N \cdot cm^{-1}$)	16.2	17.4	17.4	18.7	ASTM D5213

TABLE 1.21 Comparison on Properties of PMDA-based and BPDA-based Polyimide Films

	Kapton H (PMDA/ODA)	Upilex R (BPDA/ODA)	Upilex S25 (BPDA/PDA)	Apical NP (PMDA/PDA-ODA)
T_g (°C)	373	285	420	—
Tensile strength (MPa)	231	430	520	302
Tensile modulus (GPa)	2.8	3.9	9.1	4.0
Elongation at breakage (%)	86	160	42	80
CTE(10^{-6} °C^{-1}) (50-200 °C)	31	—	12	16
Water uptake (%)	2.8	1.4	1.4	2.3

1.6.2 Electronic and Optoelectronic Applications

1.6.2.1 Electronic Applications

Polyimide films used for electronic applications such as FPC, TAB, and COF must have low *CTE*, low coefficients of humility expansion, high elastic modulus, and low moisture uptakes. For instance, TAB tape is a polyimide film substrate on the surface of which is fabricated very fine metal circuits, and the substrate has openings or "windows" for mounting integrated circuit (IC) chips. Sprocket holes for precisely feeding the TAB tape are produced near both edges of the tape. IC chips are embedded in the windows on the TAB tape and bonded to the metal interconnects on the tape surface. TAB tape is used to automate and simplify the process of mounting IC chips, improving manufacturing productivity and enhancing the electrical characteristics of electronic equipment containing mounted IC chips.

There are two constructions for TAB tapes, one is the three-layer construction which was composed of a polyimide substrate film on the surface of which an electrically conductive metal foil has been laminated with an intervening layer of polyester, acrylic, epoxy, or polyimide-based adhesives. Another is the two-layer construction composed of a polyimide substrate film on the surface of which a conductive metal layer has been directly laminated without an intervening layer of adhesive.

The polyimide substrate film in TAB tapes are thus required to be excellent in thermal resistance, ensuring that the substrate film is able to withstand the high-temperature operations in the soldering of IC chips bonded to the metal interconnections on TAB tape and when the TAB tape carried with IC chips is bonded to a printed circuit for wiring electronic equipment. However, the heat was incurred in the process of laminating polyimide film with metal foil or a metal layer. Then, the chemical etching of the metal foil or metal layer to form metal interconnections may elicit different degrees of dimensional change in the polyimide film and metal, sometimes causing considerable deformation of the TAB tape. Such deformation can greatly hinder or even render impossible subsequent operations. Hence, making the thermal expansion coefficient of polyimide film close to that of the metal so as to reduce deformation of the TAB tape becomes a key issue in the production of polyimide films. Moreover, reducing dimensional change attributed to the tensile and compressive forces in TAB tape is important for achieving finer-pitch metal interconnects, reducing strain on the metal interconnects and reducing strain on the mounted IC chips. Hence, polyimide film used as the substrate must have a higher elastic modulus.

In order to satisfy the special requirements in microelectronic packaging and electrical insulating applications, polyimide films with different backbone structures have been developed. The most successful commercial polyimide films are the PMDA-based Kapton H films by Dupont and the BPDA-based Upilex R and S films by Ube. Upilex R, which was prepared using the same diamine as Kapton H, has a significantly lower T_g (285 °C versus about 385 °C), but is comparable to Kapton H in mechanical and electrical properties both at ambient and elevated temperatures. Moreover, Upilex R shows lower H_2O uptake, lower 250 °C shrinkage and excellent hydrolytic resistance, particularly to aqueous NaOH solution, and has found extensive application in traction motors, which require mechanical and thermal durability above 300 °C.

Upilex S film is significantly different from Kapton H. It is much stiffer, has substantially higher tensile strength, significantly lower high-temperature shrinkage, and lower thermal and hygroscopic coefficients of expansion. It has much lower permeability to gas and to water vapor. Upilex S film shows excellent hydrolytic durability, which is superior to Kapton H. It is particularly in circuit applications that Upilex S appears to offer significant potential competition to Kapton H.

Kapton H film has played a major role in aerospace wire and cable, traction motors, magnet wire, transformers, capacitors, and many other areas. Upilex S has lower elongation at breakage than Kapton H; this brittleness limits its utility as electrical insulating tapes for wire and cable as well as motor and generator applications. Table 1.21 compares the mechanical properties of PMDA-based and BPDA-based polyimide films.

Ube and Kaneka have developed several commercialized polyimide films for electronic applications (Table 1.22). The Apical HP films by Kenaka exhibit a great combination of mechanical and thermal properties with heat shrinkage of 0.06% at 200 °C/2 h, *CTE* of 11×10^{-6} $°C^{-1}$, elastic modulus of 5.6GPa-6.0 GPa, elongation at breakage of 42%-47%, and moisture uptake of 1.2%.

1.6.2.2 Optoelectronic Applications

CTPI films have been intensively investigated for potential applications in flexible optoelectronic devices, such as flexible light-emitting diodes (F-LED), flexible solar cells or photovoltaic cells (F-PV), flexible thin-film transistors (F-TFT), flexible printing circuit boards (FPCs), and so on. The optoelectronic applications require that the CTPI films have excellent combined properties which are not available from the commodity polyimide films. To satisfy these requirements, engineers have screened and tested many types of high-performance polymer films, such as PET, PEN, PEI, etc. Up to now, CTPI films have been considered as one of the most competitive candidates as the flexible plastic substrates for flexible optoelectronic devices, including FPC, flexible display (TFT-LCDs or AMOLEDs, etc.), touch panel, electronic paper, and thin photovoltaic cells. The flexible plastic substrates with both optical transparency and high temperature resistance have great potential applications in these areas due to their superior flexibility, lightness, cost-effectiveness, and processability compared with the fragile and expensive glass analogs.

TABLE 1.22 Comparison of Properties of Polyimide Films for Electronic Applications

	Apical NP 25/ 75 μm	Apical HP 25/ 50 μm	Upilex S 25/75 μm	Kapton EN 25/ 50 μm
Heat shrinkage (%) 200 °C/2 h	0.06/0.07	0.06/0.06	0.1/0.01	—
CTE (10^{-6} °C^{-1}) (50-200 °C)	16/16	11/11	12/20	32/32
Tensile strength (MPa)	302/285	346/335	520/360	310/310
Tensile modulus (GPa)	4.0/3.8	6.0/5.6	9.1/6.9	4.8/5.2
Elongation at breakage (%)	80/102	42/47	42/50	55/55
Moisture uptake (%)	2.3/2.7	1.2/1.2	1.4/0.8	2.2/2.2
CHE (10^{-6} °C^{-1})	13/13	7.0/7.0	--/--	—

This developing trend provides great opportunities for the CTPI optical films. With the structural support and optical signal transmission pathway and medium, flexible substrates will play very important roles in advanced optoelectronic display devices. The characteristics and functionalities of flexible substrates are the important factors that affect the quality of flexible devices. Currently, there are mainly three types of substrates for flexible displays: thin glass, transparent plastic (polymer), and metal foil. The transparent polyimide films not only have good optical transmittance comparable to the thin glass, but also possess good flexibility and toughness similar to metal foils. Thus, CTPI films are ideal candidates for flexible display. Use of flexible plastic substrate is considered to be one of the promising technologic breakthroughs in optical displays due to its attractive features, such as thinness, light weight, and good flexibility. The development of flexible substrates is experiencing a roadmap from plane (current) to bended (2015) then to rollable (2018) finally to foldable (2020) in the coming years [101]. The radius of curvature for highly transparent flexible substrates might reach <3 mm at 2020. At that time, transparent polymer film substrates might be the best candidates to meet the demands in device flexibility.

In order to achieve a practical application for transparent polymer film substrates in flexible display, several issues have to be addressed. First, the thermal stability of the transparent substrate should meet the application demands. For instance, thin film transistors (TFTs) are currently fabricated on the common optical polymer films or sheets, which are produced at low temperature due to the low thermal stability of used plastic substrates (< 250 °C). To date, there are four types of producing technologies for TFT fabrications in AMOLED, including amorphous silicon (a-Si) TFTs, low-temperature polysilicon (LTPS) TFTs, oxide TFTs, and organic TFTs (OTFTs) [66]. The LTPS TFTs exhibit the highest field-effect mobility and stable electrical performance. However, the procedure requires a high process temperature of about 500 °C during silicon crystallization. a-Si TFTs process has been widely used to produce AMOLED devices, which show uniform electrical characteristics over large areas, reasonable field-effect mobility, low-temperature process (<300 °C), and low cost compared to the other techniques. Hence, colorless and transparent polymer substrates with good thermal resistance above 300 °C are highly desired in advanced flexible display engineering.

Second, the CTPI film substrate should have low water vapor transmission rate (WVTR) and oxygen transmission rate (OTR) features. When a polymer film substrate is used for the flexible OLED application, the WVTR and OTR feature of the film matrix become critical because most high-performance semiconductor organic compounds built on the substrate show degraded performance when exposed to environmental moisture [48]. WVTR and OTR of the currently flexible substrates are severely limited to be below 10^{-4} $cm^3 \cdot m^{-2}$ per day and 10^{-6} $g \cdot m^{-2}$ per day, respectively, for AMOLED and organic solar cells [102]. Hence, polymer film substrates cannot effectively protect the water and oxygen permeants. In general, CTPI films have WVTR values of 10^0-10^2 $g \cdot m^{-2}$ per day depending on the aggregation structures of their molecular chains.

Third, the polymer film substrate should have comparable *CTE* values to the inorganic or metal components in display devices. The *CTE* value of polymer film substrate is quite important for its application as flexible substrate, especially when it is used with other heterogeneous materials, such as metal, glass, or ceramic. The CTPI film substrates usually have *CTE* values of higher than 30×10^{-6} °C^{-1}; However, the inorganic components, such as SiNx gas barrier layer, have *CTE* values of lower than 20×10^{-6} °C^{-1}. The unmatched *CTE* values between the polymer films with other materials are thought to be one of the most important reasons for delamination, cracking, and other failures in the devices [73]. Hybrid with inorganic additives, such as silica, titania via sol-gel route has been attempted to reduce the *CTE* values of CTPI films [103,68].

Recently, ITRI (Industrial Technology Research Institute, Taiwan) have developed a unique flexible-universal-plane (FlexUP) solution for flexible display applications [104]. This new technique relies on two key innovations: flexible substrate and a debonding layer (DBL). As for the flexible substrate, ITRI developed a CTPI substrate, which contains a high content (>60%(w)) of inorganic silica particles in the PI matrix. The CTPI substrate exhibits good optical transmittance (90%), high T_g (>300 °C), low *CTE* (28×10^{-6} °C^{-1}), and good chemical resistance. In addition, the CTPI substrate with additional barrier treatment shows a WVTR value less than 4×10^{-5} g·m^{-2} per day. Moreover, this barrier property suffered only to a minor drop, to 8×10^{-5} g·m^{-2} per day, after the flexible panel had been bent 1000 times at a radius of 5 cm. A 6-inch flexible color AMOLED display device was successfully fabricated using this substrate. Using this CTPI substrate, the flexible touch panel was successfully prepared.

A 7-inch flexible VGA transmissive-type active matrix TFT-LCD display with a-Si TFT was successfully fabricated on CTPI film substrate by ITRI [105]. The CTPI film substrate has the features of high T_g (>350 °C) and high light transmittance (>90%), which ensure the successful fabrication of 200 °C a-Si:H TFT in the flexible device. The flexible panel showed resolution of 640×RGB×480, pixel pitch of 75×225 mm, and brightness of 100 nit. This technique is fully a-Si TFT backplane compatible, making it very attractive for applications in high-performance flexible displays. Similarly, a-Si TFTs depo sited on clear plastic substrates (from DuPont) at 250 °C-280 °C was reported [106]. The free-standing clear plastic substrate has a T_g value of higher than 315 °C and a *CTE* value of lower than 10×10^{-6} °C^{-1}. The maximum process temperature of 280 °C is close to the temperature used in industrial a-Si TFT production on glass substrates (300 °C-350 °C).

Toshiba Corp. Japan have successfully fabricated a flexible 10.2 in. WUXGA (1920×1200) bottom-emission AMOLED display device driven by amorphous indium gallium zinc oxide (IGZO) TFTs on a CTPI film substrate [107]. First, a transparent PI film was formed on a glass substrate and then a barrier layer was deposited to prevent the permeation of water. Then, the gate insulator, IGZO thin film, source-drain metal, and passivation layer were successively deposited to afford the IGZO TFT. Second, the flexible AMOLED panel was fabricated using the IGZO TFT, color filter, white OLED, and encapsulation layer. Finally, the OLED panel was debonded from the glass substrate to afford the final AMOLED panel. The threshold voltage shifts of amorphous IGZO TFTs on the PI substrates under bias-temperature stress have been successfully reduced to less than 0.03 V, which is equivalent to those on glass substrates. ITRI also reported high-performance flexible amorphous IGZO TFTs on CTPI-based nanocomposites substrates [79]. Similarly, a high-heat-resistant PC film with the T_g of 240 °C and optical transmittance higher than 90% in the visible light region has been reported by General Electric [108]. A transparent, high barrier, and high heat substrate for organic electronics was successfully prepared based on the CTPI film.

Over the years, the flexible printing circuit boards (FPC) have been the largest market for high-temperature polymer films, such as PI, polyamideimide, and polyetherimide films. The flexible nature of FPCs allows their convenient use in compact electronic equipment such as portable computers, digital cameras, watches, and panel boards. Generally, the traditional FPC is mainly prepared from FCCLs. FCCLs consist of a layer of PI film bonded to copper foil. Depending on the intended use of the laminate, copper may be applied to one (single-sided) or both sides (double-sided) of the PI film. PI film almost dominates the portion of the FCCL market in which heat resistance is needed to withstand the soldering temperatures. Recently, with the development of flexible displays, necessity for a transparent film substrate in place of glass substrate is increasing. Correspondingly, a transparent film substrate for FCCLs is increasingly desired. However, most of the all-aromatic PI films currently used in FCCLs show colors from yellow to deep brown, and thus cannot be used in transparent FCCLs.

Very recently, there has been vigorous activity in developing and commercializing transparent FPC products. This is mainly driven by the urgent needs for such products for mobile communication optoelectronics. Various polymeric optical films, including PEN, PAI, and PI films have been used as the substrates together with the transparent conductive films (usually indium-tin oxide (In_2O_3 + SnO_2)

(ITO) film) in these new products. Toyobo Corp., Japan recently patented a colorless and transparent FCCL and the derived FPCB based on a PAI film. The PAI film was synthesized from 1,2,4-cyclohexanetricarboxylic anhydride (HTA) and aromatic diisocyanate monomers and the curing procedure was 200 °C/1 h, 250 °C/1 h, and 300 °C/30 min under nitrogen. The film exhibited good thermal stability with T_g of 300 °C, light transmittance of 89%, good tensile properties with tensile strength of 140 MPa, elongation at break of 30%, tensile modulus of 3.9 GPa, and low *CTE* of 33×10^{-6} °C^{-1}. The single-side FCCL from the PAI film and copper coil showed good soldering resistance, high bonding strength (10.6 N·cm^{-1}), and good dimensional stability under the condition of 150 °C for 30 min. In addition, the FCCL showed good optical transparency with a transmittance of 75% at the wavelength of 500 nm. It can be anticipated that polyimide optical films will play an increasingly important role in the future development of transparent FPCs.

Thin film solar cells or flexible photovoltaics (PV) have been intensively investigated in energy industries due to their potential ability to reduce the cost per watt of solar energy and improve lifetime performance of solar modules [109]. Conventional thin-film solar cells are usually manufactured on transparent conducting oxide-coated 3 to 5mm-thick soda-lime glass substrates and offer no weight advantage or shape adaptability for curved surfaces. Fabricating thin film solar cells on flexible polyimide substrates seems to offer several advantages in practical applications, such as weight saving, cost saving and easy fabrication. The polyimide substrates for thin film solar cell fabrications should be optically transparent and withstand the high processing temperatures. For instance, the current cadmium telluride (CdTe) cell fabrication techniques, the processing temperatures are in the range of 450 -500 °C. Most of the transparent polymers will degrade at such high temperatures. Undoubtedly, the lack of a transparent polymer film which is stable at the high processing temperature of solar cells is one of the biggest obstacles for the application of polymer substrates in flexible solar cells. Wholly aromatic polyimide films, such as Kapton and Upilex films can withstand high temperatures of around 450 °C. However, they exhibit deep colors and strongly absorb visible light. CdTe solar cells on such polyimide substrates will yield only low current due to large optical absorption [83]. Hence, the development of CTPI films with good high-temperature stability makes it possible to produce highly efficient solar cells. One of the most promising reports on the successful applications of CTPI films in flexible solar cells fabrication might be the work carried out in Swiss Federal Laboratories for Materials Science and Technology (Empa) [84]. As part of the Empa's continuous work on developing high-efficiency thin-film solar cells aiming at enhancing their performance and simplifying the fabrication processes, they utilized colorless PI film (developed by DuPont) as the flexible substrate for CdTe thin-film PV modulus in 2011. A conversion efficiency of 13.8% using the new substrates was achieved, which was the new record among this type of solar cells at that time.

Overall, CTPIs represent a class of new materials with both high technological content and high additional value. Excellent comprehensive properties make them good candidates for advanced optoelectronic devices. It can be anticipated that, with the ever-increasing demands of optoelectronic fabrication, CTPI films will attract more attention from both academia and industry. The demand will continue to grow for displays of smart phones, tablet PCs, and other types of mobile electronic devices. Furthermore, these displays will be continuously improved in terms of visibility, flexibility, durability, and light weight. Currently, CTPI optical films are facing great development opportunities. However, several obstacles still exist at present, that should be overcome for the wide applications of CTPI films in advanced high-tech fields. First, very limited commercially available CTPI film products greatly increase their cost, which leads to a very limited application only in high-end optoelectronic products. Low-cost CPIs are highly desired for their wide applications. Second, the combined properties of current CTPI films should be further enhanced, such as their optical transmittances at elevated temperatures, and gas barrier properties. Third, the manufacturing technology for CTPI films should be optimized in order to increase their uniformity, colorlessness, and dimensional stability at high temperatures.

1.6.3 Aerospace Applications

1.6.3.1 Multilayer Thermal Blanket

Aromatic polyimide films, especially the Kapton H film derived from PMDA and ODA were extensively used in aerospace until the 1970s. However, there is no standard definition for high-performance polymers because the requirements and environments for different applications vary significantly. Multilayer insulation, the so-called multilayer thermal blanket, is a typical example of aromatic polyimide films for space applications. Spacecraft themselves and each of the instruments within are generally covered with various metalized polymer films for controlling the temperature of the instruments in the spacecraft. Surface temperature of a metalized polymer film is a function of the ratio of solar absorptivity (α) to low-temperature emissivity (ε) of the film. At equilibrium, a low α/ε ratio provides a low surface temperature (high values provide a high surface temperature). Therefore, by choosing the proper film and the appropriate metal, it is possible to specify a thermal control surface within a wide range of α and ε values. A passive (nonactive) thermal control system such as MLI significantly helps to maintain spacecraft systems and components at specified temperature limits. Kapton polyimide films and Teflon fluoropolymer films have long been accepted as space-stable insulating materials. The outer layer uses 50 μm-thick aluminized Kapton film with 11 layers of double aluminized thin Kapton films. The ASCA spacecraft was a powerful X-ray observatory satellite of ISAS in Japan launched in 1993. Because the maximum allowable fluctuation of the focal point is 0.5 mm, the surface of 3.4 m-long, high-precision Extendible Optical Bench (made of carbon fiber reinforced plastics, CFRP) was covered with a Upilex-R polyimide film MLI. Upilex R film, derived from BPDA and ODA, possesses a higher optical transparency of the film compared with Kapton H derived from PMDA and ODA. This is an advantage for the system due to the low absorptivity. A new MLI covering over 90% of the outer layer covering of the spacecraft ASCA is composed of two outer layers of aluminized 25 μm Upilex R and five layers of double-aluminized 12 μm polyester films with a separator net. Until 1987, Kapton MLI was the only MLI for the high-temperature areas of spacecraft and ASCA is the first satellite completely covered by an another polyimide MLI. Colorless polyimide films were first reported by NASA for varying the molecular structure and reducing the electronic interaction between chromophoric centers for space applications. Because of the requirement for radiation resistivity in space, the highly transparent polyimide films were prepared derived from aromatic dianhydride and special diamine monomers [89].

1.6.3.2 Flexible and/or Rigid Extendable Structures

A flexible solar array is an attractive example of combining an extendable advanced composite with aromatic polyimide films. Japan's first spacecraft SFU deployed the two large flexible solar arrays in low earth orbit (LEO). Due to the high T_g and outstanding mechanical properties, even in very-low-temperature environments, aromatic polyimide films are the most successful, widely used polymeric material in space. Until 10 years ago, a spacecraft was usually equipped with a rigid type power generator (solar paddle). However, as a spacecraft becomes larger, it requires much more electric power. It is thought that a flexible solar array is the most attractive way for power generation in space. In the SFU spacecraft and the array configuration with the extendable mast, each deployed solar array is 2.4 m wide and 9.7 m long. The array is composed of two assembly boards and the mast canister. The extendable/retractable mast is a continual coilable mast consisting of three GFRP spring rods (longerons) and radial spacers. The main source of its spring force is generated by the bending strain energy of the GFRP longerons. The radial spacers were made of molded Upilex R, because of its excellent space environmental stability without creep behavior originating from the high intermolecular interaction of benzoimide rings. Therefore, no mechanical backlash exists (since there are no pin-joint hinges), resulting in high dimensional stability. Each array blanket consists of 48 hinged low *CTE* Upilex S film, whose dimensions are 202 mm wide and 2400 mm long. About 27,000 solar cells are mounted on two array blankets and generate 3.0 kW of power. One-hundred-micrometer-thick silicon cells with 100 μm cover glass are adhered by S-691-RTV silicon type adhesive on the polyimide panels.

1.6.3.3 Large Deployable Antenna

Development of a large deployable antenna on Muses-B is another example of advanced technology in space. The Muses-B antenna launched in 1997 was used aboard the satellite for Space-VLBI (very long baseline interferometry). A 10 m-diameter parabolic antenna with a mesh surface was deployed with steps extending the six extendable masts in LEO. This incredibly complex system consists of 6000 fine cables of high modulus Kevlar 149 aramid covered by a Conex aramid net. In order to achieve high surface accuracy, a cable must precisely keep its length without creep under a tension field in space. The high-strength Kevlar 149 cable exhibits very little elongation when stressed and has negative *CTE* over a wide range of temperatures. Because the lengths and tension force of the cable were critical, the tension force of each extendable mast was strictly controlled by the tensioners on the top of the masts. It was the first application of high-performance organic fibers for a large deployable parabolic antenna surface in space.

1.7 Summary

Advanced polyimide films have been extensively used in electrical insulation, microelectronic and opto-electric display manufacturing and packaging, aerospace and aviation industries, etc., due to their excellent combined properties including thermal dimensional stability, high mechanical strength, modulus, and toughness, high electrical insulating performance, and low dielectric constant and dissipation factor, etc. There are two pathways in polyimide film production: one is the thermal imidization, and the other is chemical imidization. The polyimide film-forming processes are very complicated chemical reactions and physical changes. The precise control of the chemical reactions and physical changes are the key issues in the production of high-quality advanced polyimide films. There are obvious effects of chemical structures on the combined properties of the polyimide films. The surface modifications of polyimide films by plasma technology are also important for electronic applications. Many high-performance and functional polyimide films have been developed in recent years, including the low-thermal-expansion and high-modulus polyimide films, corona-resistant polyimide films, etc.

REFERENCES

[1] C.E. Sroog, A.L. Endrey, S.V. Abramo, C.E. Berr, W.M. Edwards, K.L. Olivier, Aromatic polypyromellitimides from aromatic polyamic acids, J. Polym. Sci. Part A: General Papers 3 (4PA) (1965) 1373.

[2] E.A. Laszlo. Process for preparing polyimides by treating polyamide-acids with lower fatty monocarboxylic acid anhydrides. USA. US3179630, 1965.

[3] Y. Sasaki, H. Inoue, I. Sasaki, H. Itaya, M. Kashima, H. Itatani. Stable, homogeneous polyimide soln., esp. For films-prepd. By reacting 3,3′-4,4′-di:Phenyl ether. In phenol or halo-phenol. US4290936-A, 1981.

[4] Y. Sasaki, H. Inoue, Y. Negi, K. Sakai. Continuous prepn. Of polyimide film from polyimide soln. Dope|by extruding dope of specified viscosity through i-die, supporting film on endless moving surface and evaporating solvent. US4473523-A, 1984.

[5] H. Itaya, T. Inaike, S. Yamamoto, H. Itatani. Polyimide resin moulding compsn. -obtd. By dissolving aromatic polyimide in solvent mixt. Of naphthol and/or resorcin and phenol and/or cresol. US4568715-A, 1986.

[6] M.E. Walter. Polyamide-acids, compositions thereof, and process for their preparation. US3179614, 1965.

[7] M. Kochi, T. Uruji, T. Iizuka, I. Mita, R. Yokota, High-modulus and high-strength polybiphenyltetracarboximide films, J.Polym. Sci. Part CPolym. Lett. 25 (11) (1987) 441-446.

[8] J.W. Verbicky, L. Williams, Thermolysis of n-alkyl-substituted phthalamic acids-steric inhibition of imide formation, J. Org. Chem. 46 (1) (1981) 175-177.

[9] T. Takekoshi, J.E. Kochanowski, J.S. Manello, M.J. Webber, Polyetherimides. 1. Preparation of dianhydrides containing aromatic ether groups, J. Polym. Sci. Part A:Polym. Chem. 23 (6) (1985) 1759-1769.

[10] M. Bessonov, Polyimides--Thermally Stable Polymers, Consultants Bureau, New York, 1987.

[11] C.C. Walker, High-performance size exclusion chromatography of polyamic acid, J. Polym. Sci. Part A:Polym. Chem. 26 (6) (1988) 1649-1657.

[12] P.J. Flory, Principles of Polymer Chemistry, Cornell University Press, Ithaca, NY, 1953.

[13] L.E. Andrew. Aromatic polyimides from meta-phenylene diamine and para-phenylene diamine. USA. US3179633, 1965.

[14] E.A. Laszlo. Aromatic polyimide particles from polycyclic diamines. US3179631, 1965.

[15] R.H. William. Process for preparing polyimides by treating polyamide-acids with aromatic monocarboxylic acid anhydrides: Google Patents, 1965.

[16] R.J. Angelo. Treatment of aromatic polyamide-acids with carbodiimides: Google Patents, 1966.
[17] R.J. Cotter, C.K. Sauers, J.M. Whelan, Synthesis of n-substituted isomaleimides, J. Org. Chem. 26 (1) (1961) 10.
[18] R.L. Kaas, Auto-catalysis and equilibrium in polyimide synthesis, J. Polym. Sci. Part A:Polym. Chem. 19 (9) (1981) 2255-2267.
[19] T.P. Russell, H. Gugger, J.D. Swalen, Inplane orientation of polyimide, J. Polym. Sci. Part B-Polym. Phys. 21 (9) (1983) 1745-1756.
[20] S. Numata, T. Miwa, Thermal-expansion coefficients and moduli of uniaxially stretched polyimide films with rigid and flexible molecular chains, Polymer 30 (6) (1989) 1170-1174.
[21] P.-c Ma, Y. Hou, Partly imidized polyamic acid and its uniaxial stretched polyimide films, Chem. Res. Chinese Universities 29 (2) (2013) 396-400.
[22] M.J. Brekner, C. Feger, Curing studies of a polyimide precursor, J. Polym. Sci. Part A:Polym. Chem. 25 (7) (1987) 2005-2020.
[23] M.J. Brekner, C. Feger, Curing studies of a polyimide precursor. 2. Polyamic acid, J. Polym. Sci. Part B-Polym. Chem. 25 (9) (1987) 2479-2491.
[24] T.C.J. Hsu, Z.L. Liu, Solvent effect on the curing of polyimide resins, J. Appl. Polym. Sci. 46 (10) (1992) 1821-1833.
[25] H.-T. Chiu, J.-O. Cheng, Thermal imidization behavior of aromatic polyimides by rigid-body pendulum rheometer, J. Appl. Polym. Sci. 108 (6) (2008) 3973-3981.
[26] I. Sava, S. Chisca, M. Bruma, G. Lisa, Effect of thermal curing on the properties of thin films based on benzophenonetetracarboxylic dianhydride and 4,4'-diamino-3,3'- dimethyldiphenylmethane, J. Thermal Analysis Calorimetry 104 (3) (2011) 1135-1143.
[27] R.W. Snyder, B. Thomson, B. Bartges, D. Czerniawski, P.C. Painter, Ftir studies of polyimides - thermal curing, Macromolecules 22 (11) (1989) 4166-4172.
[28] E. Unsal, M. Cakmak, Real-time characterization of physical changes in polyimide film formation: From casting to imidization, Macromolecules 46 (21) (2013) 8616-8627.
[29] S. Isoda, H. Shimada, M. Kochi, H. Kambe, Molecular aggregation of solid aromatic polymers. 1. Small-angle x-ray-scattering from aromatic polyimide film, J. Polym. Sci. Part B-Polym. Phys. 19 (9) (1981) 1293-1312.
[30] M. Hasegawa, N. Sensui, Y. Shindo, R. Yokota, Structure and properties of novel asymmetric biphenyl type polyimides. Homo- and copolymers and blends, Macromolecules 32 (2) (1999) 387-396.
[31] D. Boese, H. Lee, D.Y. Yoon, J.D. Swalen, J.F. Rabolt, Chain orientation and anisotropies in optical and dielectric-properties in thin-films of stiff polyimides, J. Polym. Sci. Part B-Polym. Phys. 30 (12) (1992) 1321-1327.
[32] M.T. Pottiger, J.C. Coburn, J.R. Edman, The effect of orientation on thermal-expansion behavior in polyimide films, J. Polym. Sci. Part BPolym. Phys. 32 (5) (1994) 825-837.
[33] H. Inoue, Y. Sasaki, T. Ocawa, Properties of copolyimides prepared from different tetracarboxylic dianhydrides and diamines, J. Appl. Polym. Sci. 62 (13) (1996) 2303-2310.
[34] P.M. Hergenrother, J.G. Smith, Chemistry and properties of imide oligomers end-capped with phenylethynylphthalic anhydrides, Polymer 35 (22) (1994) 4857-4864.
[35] C.C. Roberts, T.M. Apple, G.E. Wnek, Curing chemistry of phenylethynyl-terminated imide oligomers: Synthesis of C-13-labeled oligomers and solid-state NMR studies, J. Polym. Sci. Part A:Polym. Chem. 38 (19) (2000) 3486-3497.
[36] T.A. Bullions, M.P. Stoykovich, J.E. McGrath, A.C. Loos, Monitoring the reaction progress of a high-performance phenylethynyl-terminated poly(etherlmide). Part ii: Advancement of glass transition temperature, Polym. Compos. 23 (4) (2002) 479-494.
[37] K. Kim, T. Yoo, J. Kim, H. Ha, H. Han, Effects of dianhydrides on the thermal behavior of linear and crosslinked polyimides, J. Appl. Polym. Sci. 132 (2015) 6.
[38] Y. Jung, Y. Yang, S. Lee, S. Byun, H. Jeon, M.D. Cho, Characterization of fluorinated polyimide morphology by transition mechanical analysis, Polymer 59 (2015) 200-206.
[39] T.W. Poon, B.D. Silverman, R.F. Saraf, A.R. Rossi, P.S. Ho, Simulated crystalline-structures of aromatic polyimides, Phys. Rev. B 46 (18) (1992) 11456-11462.
[40] W.S. Lambert, P.J. Phillips, J.S. Lin, Small-angle x-ray-scattering studies of crystallization in cross-linked linear polyethylene, Polymer 35 (9) (1994) 1809-1818.
[41] M. Kochi, R. Yokota, T. Iizuka, I. Mita, Improving tensile mechanical-properties of aromatic polyimides by thermal imidization after cold drawing of poly (amic acids), J. Polym. Sci. Part B-Polym. Phys. 28 (13) (1990) 2463-2472.
[42] H. Chung, J. Lee, W. Jang, Y. Shul, H. Han, Stress behaviors and thermal properties of polyimide thin films depending on the different curing process, J. Polym. Sci. Part B-Polym. Phys. 38 (22) (2000) 2879-2890.
[43] S.P. Ma, T. Sasaki, K. Sakurai, T. Takahashi, Morphology of solution-cast thin-films of wholly aromatic thermoplastic polyimides with various molecular-weights, Polymer 35 (26) (1994) 5618-5625.

[44] M. Hasegawa, K. Koseki, Poly(ester imide)s possessing low coefficient of thermal expansion and low water absorption, High Perform. Polym. 18 (5) (2006) 697-717.

[45] M. Hasegawa, Y. Tsujimura, K. Koseki, T. Miyazaki, Poly(ester imide)s possessing low *CTE* and low water absorption (ii). Effect of substituents, Polym. J. 40 (1) (2008) 56-67.

[46] M. Hasegawa, Y. Sakamoto, Y. Tanaka, Y. Kobayashi, Poly(ester imide)s possessing low coefficients of thermal expansion (*CTE*) and low water absorption (iii). Use of bis(4-aminophenyl)terephthalate and effect of substituents, Eur. Polym. J. 46 (7) (2010) 1510-1524.

[47] M.-C. Choi, Y. Kim, C.-S. Ha, Polymers for flexible displays: From material selection to device applications, Prog. Polym. Sci. 33 (6) (2008) 581-630.

[48] S. Logothetidis, Flexible organic electronic devices: Materials, process and applications, Mater. Sci. Eng. B-Adv. Funct. Solid-State Mater. 152 (1-3) (2008) 96-104.

[49] J.-J. Huang, Y.-P. Chen, S.-Y. Lien, K.-W. Weng, C.-H. Chao, High mechanical and electrical reliability of bottom-gate microcrystalline silicon thin film transistors on polyimide substrate, Curr. Appl. Phys. 11 (1) (2011) S266-S270.

[50] S. Nakano, N. Saito, K. Miura, T. Sakano, T. Ueda, K. Sugi, et al., Highly reliable a-IGZO TFTs on a plastic substrate for flexible amoled displays, J. Soc. Inf. Display 20 (9) (2012) 493-498.

[51] H. Yamaguchi, T. Ueda, K. Miura, N. Saito, S. Nakano, T. Sakano, et al. 74.2 l: Late-News paper: 11.7-inch flexible amoled display driven by a-IGZO TFTs on plastic substrate. SID Symposium Digest of Technical Papers. Wiley Online Library, 2012, pp. 1002-1005.

[52] Q. Jin, T. Yamashita, K. Horie, R. Yokota, I. Mita, Polyimides with alicyclic diamines. 1. Syntheses and thermal-properties, J. Polym. Sci. Part A:Polym. Chem. 31 (9) (1993) 2345-2351.

[53] M. Nishikawa, Y. Matsuki, N. Bessho, Y. Iimura, S. Kobayashi, Characteristics of polyimide liquid crystal alignment films for active matrix LCD use, J. Photopolymer Sci. Technol. 8 (2) (1995) 233-240.

[54] Y. Tsuda, K. Etou, N. Hiyoshi, M. Nishikawa, Y. Matsuki, N. Bessho, Soluble copolyimides based on 2,3,5-tricarboxycyclopentyl acetic dianhydride and conventional aromatic tetracarboxylic dianhydrides, Polym. J. 30 (3) (1998) 222-228.

[55] H. Suzuki, T. Abe, K. Takaishi, M. Narita, F. Hamada, The synthesis and x-ray structure of 1,2,3,4-cyclobutane tetracarboxylic dianhydride and the preparation of a new type of polyimide showing excellent transparency and heat resistance, J. Polym. Sci. Part A:Polym. Chem. 38 (1) (2000) 108-116.

[56] T.-L. Li, S.L.-C. Hsu, Preparation and properties of a high temperature, flexible and colorless ito coated polyimide substrate, Eur. Polym. J. 43 (8) (2007) 3368-3373.

[57] Y.-z Guo, H.-w Song, L. Zhai, J.-g Liu, S. Yang, Synthesis and characterization of novel semi-alicyclic polyimides from methyl-substituted tetralin dianhydride and aromatic diamines, Polym. J. 44 (7) (2012) 718-723.

[58] M. Hasegawa, K. Kasamatsu, K. Koseki, Colorless poly(ester imide)s derived from hydrogenated trimellitic anhydride, Eur. Polym. J. 48 (3) (2012) 483-498.

[59] M. Hasegawa, D. Hirano, M. Fujii, M. Haga, E. Takezawa, S. Yamaguchi, et al., Solution-processable colorless polyimides derived from hydrogenated pyromellitic dianhydride with controlled steric structure, J. Polym. Sci. Part A:Polym. Chem. 51 (3) (2013) 575-592.

[60] M. Hasegawa, M. Fujii, J. Ishii, S. Yamaguchi, E. Takezawa, T. Kagayama, et al., Colorless polyimides derived from 1s,2s,4r,5r-cyclohexanetetra carboxylic dianhydride, self-orientation behavior during solution casting, and their optoelectronic applications, Polymer 55 (18) (2014) 4693-4708.

[61] M. Hasegawa, M. Horiuchi, K. Kumakura, J. Koyama, Colorless polyimides with low coefficient of thermal expansion derived from alkylsubstituted cyclobutanetetracarboxylic dianhydrides, Polym. Int. 63 (3) (2014) 486-500.

[62] S.-H. Hsiao, H.-M. Wang, W.-J. Chen, T.-M. Lee, C.-M. Leu, Synthesis and properties of novel triptycene-based polyimides, J. Polym. Sci. Part A:Polym. Chem. 49 (14) (2011) 3109-3120.

[63] T. Matsumoto, E. Ishiguro, S. Komatsu, Low temperature film-fabrication of hardly soluble alicyclic polyimides with high tg by a combined chemical and thermal imidization method, J. Photopolymer Sci. Technol. 27 (2) (2014) 167-171.

[64] S.D. Kim, S.Y. Kim, I.S. Chung, Soluble and transparent polyimides from unsymmetrical diamine containing two trifluoromethyl groups, J. Polym. Sci. Part A:Polym. Chem. 51 (20) (2013) 4413-4422.

[65] Y. Oishi, K. Itoya, M. Kakimoto, Y. Imai, Preparation and properties of molecular composite films of block copolyimides based on rigid rod and semi-flexible segments, Polym. J. 21 (10) (1989) 771-780.

[66] M.K. Kolel-Veetil, H.W. Beckham, T.M. Keller, Dependence of thermal properties on the copolymer sequence in diacetylene-containing polycarboranylenesiloxanes, Chem. Mater. 16 (16) (2004) 3162-3167.

[67] E.F. Palermo, A.J. McNeil, Impact of copolymer sequence on solid-state properties for random, gradient and block copolymers containing thiophene and selenophene, Macromolecules 45 (15) (2012) 5948-5955.

[68] C.H. Choi, B.H. Sohn, J.-H. Chang, Colorless and transparent polyimide nanocomposites: comparison of the properties of

homo- and copolymers, J. Ind. Eng. Chem. 19 (5) (2013) 1593-1599.

[69] K.-i Fukukawa, M. Okazaki, Y. Sakata, T. Urakami, W. Yamashita, S. Tamai, Synthesis and properties of multi-block semi-alicyclic polyimides for thermally stable transparent and low *CTE* film, Polymer 54 (3) (2013) 1053-1063.

[70] Q. Li, K. Hone, R. Yokota, Absorption and fluorescence spectra and thermal properties of novel transparent polyimides, J. Photopolymer Sci. Technol. 10 (1) (1997) 49-54.

[71] A.S. Mathews, I. Kim, C.-S. Ha, Fully aliphatic polyimides from adamantane-based diamines for enhanced thermal stability, solubility, transparency, and low dielectric constant, J. Appl. Polym. Sci. 102 (4) (2006) 3316-3326.

[72] A.S. Mathews, I. Kim, C.-S. Ha, Synthesis, characterization, and properties of fully aliphatic polyimides and their derivatives for microelectronics and optoelectronics applications, Macromol. Res. 15 (2) (2007) 114-128.

[73] A.S. Mathews, I. Kim, C.-S. Ha, Fully aliphatic polyimides - influence of adamantane and siloxane moieties, Macromol. Symp. 249 (2007) 344-349.

[74] P.K. Tapaswi, M.-C. Choi, Y.S. Jung, H.J. Cho, D.J. Seo, C.-S. Ha, Synthesis and characterization of fully aliphatic polyimides from an aliphatic dianhydride with piperazine spacer for enhanced solubility, transparency, and low dielectric constant, J. Polym. Sci. Part A:Polym. Chem. 52 (16) (2014) 2316-2328.

[75] P.K. Tapaswi, M.-C. Choi, S. Nagappan, C.-S. Ha, Synthesis and characterization of highly transparent and hydrophobic fluorinated polyimides derived from perfluorodecylthio substituted diamine monomers, J. Polym. Sci. Part A:Polym. Chem. 53 (3) (2015) 479-488.

[76] M.-C. Choi, J. Wakita, C.-S. Ha, S. Ando, Highly transparent and refractive polyimides with controlled molecular structure by chlorine side groups, Macromolecules 42 (14) (2009) 5112-5120.

[77] J.-H. Park, J.-H. Kim, J.-W. Park, J.-H. Chang, C.-S. Ha, Preparation and properties of fluorine-containing colorless polyimide nanocomposite films with organo-modified montmorillonites for potential flexible substrate, J. Nanosci. Nanotechnol. 8 (4) (2008) 1700-1706.

[78] J.-H. Kim, M.-C. Choi, H. Kim, Y. Kim, J.-H. Chang, M. Han, et al., Colorless polyimide/organoclay nanocomposite substrates for flexible organic light-emitting devices, J. Nanosci. Nanotechnol. 10 (1) (2010) 388-396.

[79] A.S. Mathews, D. Kim, Y. Kim, I. Kim, C.-S. Ha, Synthesis and characterization of soluble polyimides functionalized with carbazole moieties, J. Polym. Sci. Part A:Polym. Chem. 46 (24) (2008) 8117-8130.

[80] M.-C. Choi, J.-C. Hwang, C. Kim, S. Ando, C.-S. Ha, New colorless substrates based on polynorbornene-chlorinated polyimide copolymers and their application for flexible displays, J. Polym. Sci. Part A:Polym. Chem. 48 (8) (2010) 1806-1814.

[81] G.Y. Kim, M.-C. Choi, D. Lee, C.-S. Ha, 2d-aligned graphene sheets in transparent polyimide/graphene nanocomposite films based on noncovalent interactions between poly(amic acid) and graphene carboxylic acid, Macromol. Mater. Eng. 297 (4) (2012) 303-311.

[82] H. Lim, C.M. Bae, Y.K. Kim, C.H. Park, W.J. Cho, C.S. Ha, Preparation and characterization of ito-coated colorless polyimide substrates, Synth. Metals 135 (1-3) (2003) 49-50.

[83] M.-C. Choi, J.-C. Hwang, C. Kim, Y. Kim, C.-S. Ha, Synthesis of poly(n-9-ethylcarbazole-exo-norbornene-5,6-dicarboximide) for holetransporting layer in hybrid organic light-emitting devices, J. Polym. Sci. Part A:Polym. Chem. 48 (22) (2010) 5189-5197.

[84] J. Oishi, S. Hiramatsu, S. Kihara, H. Sotaro, H. Kihara. Manufacture of resin film for printed wiring board, involves forming polyamic acid or organic solvent containing polyimide on substrate, spraying gas, evaporating organic solvent, and peeling self-supportive film from substrate. US8357322.

[85] H. Oguro, S. Kihara, T. Bito, H. Kihara, T. Mifuji. Production of solvent-soluble polyimide comprises polycondensing tetracarboxylic acid component with diamine component in solvent in presence of tertiary amine. US7078477.

[86] C.D. Simone, B.C. Auman, P.F. Carcia, R.A. Wessel. Polyimide film used for electronic display, comprises perfluoro-imide moiety obtained by contacting dianhydride component and diamine component. US7550194.

[87] H.M. Cho, Y.H. Jeong, H.J. Park. Powder used for manufacture of polyimide film, contains imidized compound of polyamic acid obtained by polymerizing diamine component and acid dianhydride component, and has preset imidization rate and absolute molecular weight. US8846852.

[88] C.C. Fay, D.M. Stoakley, A.K. St Clair, Molecularly oriented films for space applications, High Perform. Polym. 11 (1) (1999) 145-156.

[89] R. Yokota, Recent trends and space applications of polyimides, J. Photopolymer Sci. Technol. 12 (2) (1999) 209-216.

[90] D. Dooling, M. Finckenor. Material selection guidelines to limit atomic oxygen effects on spacecraft surfaces, 1999.

[91] T.K. Minton, M.E. Wright, S.J. Tomczak, S.A. Marquez, L. Shen, A.L. Brunsvold, et al., Atomic oxygen effects on poss polyimides in low earth orbit, ACS Appl. Mater. Interfaces 4 (2) (2012) 492-502.

[92] S.J. Tomczak, M.E. Wright, A.J. Guenthner, B.J. Pettys, A.L. Brunsvold, C. Knight, et al. Space survivability of main-chain and side-chain poss-kapton polyimides. In: Kleiman JI (ed). Protection of materials and structures from space environment, edn, vol. 1087, 2009, p. 505.

[93] E. Miyazaki, M. Tagawa, K. Yokota, R. Yokota, Y. Kimoto, J. Ishizawa, Investigation into tolerance of polysiloxane-block-polyimide film against atomic oxygen, Acta Astronautica 66 (5-6) (2010) 922-928.

[94] K. Yokota, S. Abe, M. Tagawa, M. Iwata, E. Miyazaki, J.-I. Ishizawa, et al., Degradation property of commercially available si-containing polyimide in simulated atomic oxygen environments for low earth orbit, High Performance Polym. 22 (2) (2010) 237-251.

[95] M.A. Golub, T. Wydeven, Reactions of atomic oxygen ($O(^3P)$) with various polymer-films, Polym. Degrad. Stabil. 22 (4) (1988) 325-338.

[96] R.A. Synowki, J.S. Hale, J.A. Woollam, Low earth simulation and materials characterization, J. Spacecraft. Rockets 30 (1) (1993) 116-119.

[97] R.H. Hansen, J.V. Pascale, T. De Benedi(ctis), P.M. Rentzepis, Effect of atomic oxygen on polymers, J. Polym. Sci. Part A: General Papers 3 (6PA) (1965). 2205-2214.

[98] M. Strobel, S. Corn, C.S. Lyons, G.A. Korba, Surface modification of polypropylene withcf4, cf3h, cf3cl, and cf3br plasmas, J. Polym. Sci. Part A:Polym. Chem. 23 (4) (1985) 1125-1135.

[99] A.D. Katnani, A. Knoll, M.A. Mycek, Effects of environment and heat-treatment on an oxygen plasma-treated polyimide surface and its adhesion to a chromium overcoat, J. Adhesion Sci. Technol. 3 (6) (1989) 441-453.

[100] L.J. Matienzo, F.D. Egitto, Polymer oxidation downstream from oxygen microwave plasmas, Polym. Degrad. Stabil. 35 (2) (1992) 181-192.

[101] J. Chen, C.T. Liu, Technology advances in flexible displays and substrates, IEEE Access 1 (2013) 150-158.

[102] J.-S. Park, H. Chae, H.K. Chung, S.I. Lee, Thin film encapsulation for flexible AM-OLED: A review, Semicond. Sci. Technol. 26 (2011) 3.

[103] Y.-W. Wang, W.-C. Chen, Synthesis, properties, and anti-reflective applications of new colorless polyimide-inorganic hybrid optical materials, Compos. Sci. Technol. 70 (5) (2010) 769-775.

[104] J. Chen, J.-C. Ho, Frontlinetechnology: A flexible universal plane for displays Inf. Display 27 (2) (2011) 6.

[105] Y.-H. Yeh, C.-C. Cheng, K.-Y. Ho, P.-C. Chen, M.H. Lee, J.-J. Huang, et al. 7-inch color VGA flexible TFT LCD on colorless polyimide substrate with 200 °C a-Si: H TFTs. In. 2007 SID international symposium, digest of technical papers, vol xxxviii, books i and ii, edn, Vol. 38, 2007, pp. 1677-1679.

[106] K. Long, A.Z. Kattamis, I.C. Cheng, H. Gleskova, S. Wagner, J.C. Sturm, Stability of amorphous-silicon TFTs deposited on clear plastic substrates at 250 °C to 280 °C, IEEE Elect. Dev. Lett. 27 (2) (2006) 111-113.

[107] K. Miura, T. Ueda, N. Saito, S. Nakano, T. Sakano, K. Sugi, et al. Flexible amoled display driven by amorphous InGaZnO TFTs. Proceedings of 2013 Twentieth International Workshop on Active-Matrix Flatpanel Displays and Devices (AM-FPD 13): TFT Technologies and FPD Materials, 2013, pp. 29-32.

[108] M. Yan, T.W. Kim, A.G. Erlat, M. Pellow, D.F. Foust, H. Liu, et al., A transparent, high barrier, and high heat substrate for organic electronics, Proc. IEEE 93 (8) (2005) 1468-1477.

[109] L.M. Fraas, L.D. Partain, Solar Cells and Their Applications, Vol. 236, John Wiley & Sons, New York, 2010.

Chapter 2

Advanced Polyimide Fibers

Qing-Hua Zhang[1], Jie Dong[1] and De-Zhen Wu[2]
[1]Donghua University, Shanghai, China, [2]Beijing University of Chemical Technology, Beijing, China

2.1 Introduction

Polyimide (PI) fibers refer to a type of polymeric materials containing imide rings in macromolecular chains. The highly conjugated structure has endowed it with outstanding mechanical properties, relatively high thermal stability, good solvent resistance, and excellent light stability, which give it greater advantages compared with some other high-tech polymeric fibers when employed in severe conditions, possessing a wide application prospect in areas of aerospace and environmental conservation [1-4].

In the 1960s, the cutting-edge lab at Dupont and the relative research institution from the former Soviet Union began researching PI fibers [5]. However, no quick popularization and applications of PI fibers were obtained due to the immaturity of PI synthesis and spinning technology, as well as the high cost of the fiber production. Afterwards, Rhône Poulenc France developed a type of poly(amide-imide) fiber based on *m*-aromatic polyamide, and Kermel France used the brand name Kermel for commercial development [6]. Nowadays, the Kermel-Tech poly(amide-imide) fiber is designed and developed for the specific requirements on temperature and chemical reactions. The long-term working temperature of this aromatic fiber reaches as high as 220 °C, and the highest acceptable temperature is close to 240 °C. The glass transition temperature (T_g) is 340 °C, at present, and it has been widely used in the high-temperature filtration area for energy production and various manufacturing industries as it can maintain excellent mechanical proper H_2O ties at extremely high working temperatures.

In the 1980s, Lenzing AG Austria launched a novel type of PI fiber with the brand name of P84, which was produced from toluene diisocyanate, diphenylmethane-diisocyanate, and 3,3′,4,4′-benzophenonetetracarboxylic dianhydride (BTDA) [7]. The P84 PI fibers possess an irregular foliaceous cross-section, increasing the surface area by 80% compared with the normal round cross-section. To date, the P84 PI fiber has been extensively used in high-temperature filtration areas [8].

In recent years, the production technology of PI fiber in China has been developing rapidly. In the 1960s, a small amount of PI fibers was first produced in the Shanghai Synthetic Fiber Research Institute. The PI fibers were used for the radiation protection cladding of cables, radiation-resistant strings, etc. However, large-scale production of PI fibers was not achieved. Since the 2000s, based on the unique overall performance of PI fiber and requirements in some special areas, many scientists in Chinese universities and research institutions have begun to investigate PI fibers again. Currently, there are several companies in China focusing on the production technology of advanced PI fibers, including Changchun Polyimide Materials Co., Ltd, Jiangsu Aoshen Hi-tech Materials Co., Ltd, and Jiangsu Shino New Materials & Technology Co., Ltd, etc. A series of commercially available PI fibers including the thermal-resistant PI fibers and high-performance PI fibers have been produced using

Advanced Polyimide Materials. DOI: https://doi.org/10.1016/B978-0-12-812640-0.00002-0

different production techniques. Fig. 2.1 shows photographs of typical commercial PI fibers, which can be used in the areas of environmental protection, aerospace, sophisticated weapons, and personal protection, etc.

2.2 Synthesis of Spinning Resin Solutions

PIs have various chemical structures, which can be synthesized using multi-approaches, and this adjustability is unique among polymeric materials. At present, there are two main routes for synthesizing PIs: one approach is to synthesize PI from monomers containing imide rings, and the other approach is to prepare the precursor poly(amic acid) (PAA) at first, continuing with thermal or chemical cyclization to form imide rings. The second synthetic route is widely used now due to the advantages that dianhydride and diamine monomers can derive from a wealth of sources and a simple preparation process. This process is the most common route for synthesizing PIs, having a great potential for industrial development. Based on the difference of reaction processes and mechanisms, the synthesis of PIs is usually divided into "one-step" and "two-step" methods, respectively, which will be illustrated in detail, combined with specific examples in the following sections.

FIGURE 2.1 Relative product photos of Superlon polyimide fibers (from left to right: short-cut polyimide fibers, filaments, and color yarns).

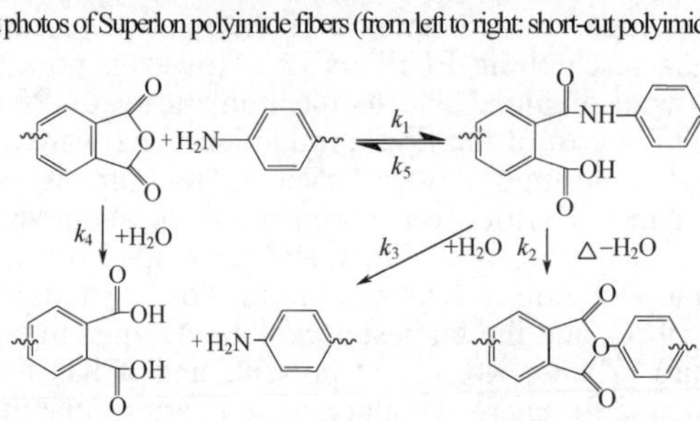

FIGURE 2.2 The main reaction of the "two-step" method to produce polyimide fibers.

2.2.1 "Two-Step" Polymerization Method

2.2.1.1 Synthesis of Poly(Amic Acid) Solution

In a typical "two-step" method, as shown in Fig. 2.2, PAA is firstly obtained by polycondensation of equimolar dianhydrides and diamines in aprotic polar solvents, such as *N,N*-dimethyl formamide (DMF), *N,N*-dimethylacetamide (DMAc), *N*-methyl-2-pyrrolidone (NMP) and other ap rotic polar solvents, and then converted to PI through a chemical or thermal imidization process [9].

The reaction of synthesizing PAA in aprotic polar solvents by diamine and dianhydrides is reversible [10,11], and the forward reaction is regarded as forming charge transfer complex among dianhydrides and diamines. The equilibrium constant at room temperature is as high as 10^5 L·mol^{-1} [12], therefore, it is easy to obtain PAA with high molecular weights. The electron affinity of dianhydride monomers and the alkalinity of diamines are the key factors that affect reaction rates. In general, dianhydride monomers containing electron-withdrawing groups, including C=O, O=S=O, and $-CF_3$, are beneficial in improving the acylating abilities of dianhydrides. However, it is very difficult to obtain PAA with high molecular weight through a polycondensation at low temperature when diamines containing electron-withdrawing groups, especially these groups located in the para- or ortho-positions of the amino groups. The chemical structures and electron affinities (E_a) of some common dianhydrides are listed in Table 1.1 in Chapter 1, Advanced Polyimide Films, and the chemical structures of aromatic diamines, their alkalinities (pK_a), and acylation rates constantly reacting with pyromellitic dianhydride (PMDA) ($\lg k$) are listed in Table 1.2 in Chapter 1, Advanced Polyimide Films. In addition to the monomer structures, factors that affect PAA synthesis also include the following:

1. *Reaction temperature*. The synthesis of PAA belongs to an exothermic reaction, and increasing the reaction temperature is beneficial for the reverse reaction, reducing the relative molecular weights of PAA. Therefore, temperature of the polycondensation is always controlled to −5 °C to 20 °C. As for those monomers with low reactivities, normally, increasing temperature is in favor of forward reaction, in fact, the "high-temperature one-step" method is often used to prepare high-molecular PIs when the reactivity of diamines or dianhydrides is relatively low.

2. *Concentration of reaction monomer*. In addition to the reaction temperature, the concentration of monomers also plays an important role in the molecular weights of PAA. The forward reaction of preparing PAA is a bimolecular reaction, yet the reverse reaction is an unimolecular reaction; as a result, increasing temperature is beneficial to the forward reaction; however, high concentration usually leads to high viscosity, resulting in uneven mass and heat transfer, which hinders the chain propagation. For preparing high-quality spinning solutions, the concentration of monomers is usually adjusted in the range of 10%-25%(w).

3. *Molar ratio of dianhydride and diamine*. In theory, when the molar ratio of dianhydride and diamine is close to 1:1, the molecular weight and inherent viscosity of synthesized PAA are highest, in fact, dianhydride monomer is sensitive to trace moisture, and easily deliquesces to form carboxylic groups thus decreasing the reactivity, so it is best to control the molar ratio at (1-1.02):1 of dianhydride and diamine monomers.

4. *Solvent*. The aprotic polar solvents are frequently used for PAA synthesis, such as DMAc, DMF, dimethyl sulfoxide (DMSO), NMP, and so on. The dissolving capacity varies from one solvent to another. In view of the requirements of environmental protection and safety, DMAc and NMP are the most widely used at present. NMP is inert and environmentally friendly, and does not associate with PAA, showing a better dissolving capacity, in favor of preparing high-molecular-mass PAA resin.

5. *Other factors*. In addition to the above-mentioned factors, monomer purity, the feeding method, and moisture content will also affect the PAA molecular weights. In practice, various factors should be comprehensively considered to prepare high-molecular PAA, providing a good base for producing high-performance PI.

2.2.1.2 Imidization Reaction

The imidization of PAA is an important issue in preparing PI through the "two-step" method, it has a great significance for improving the processing and the properties of polymers, so that it has drawn the wide attention of many researchers to study the imidization process of PAA. The research in imidization process mainly involves the detection of cyclization degree, building cyclization kinetics equations and cyclization mechanism. Spectroscopy is the most common method used to detect imidization degree, in which the infrared spectrum has been widely used because of its simple operation, no damage to samples, and online monitoring [13-15].

Normally, asymmetric carbonyl stretching (1780 cm^{-1}), imide deformation (725 cm^{-1}) and C-N stretching mode (1380 cm^{-1}) in FTIR are always used to differentiate PI from PAA. Pryde et al. [15] discussed the difference in imidization degree calculated from these three peaks. In their conclusions, the intensities of characteristic peaks at 1780 cm^{-1} and 725 cm^{-1} were too weak, considerable errors were easily made when calculating the imidization degree. However, the peak at 1380 cm^{-1} was quite strong, slightly affected by the other chemical groups around, which was more suitable for calculating imidization degree quantitatively. Recently, the ratio of *A*1380/*A*1500 (*A* refers to the intensity) has been the common choice to detect imidization degree [16].

Many researches confirm that there exists a temperature-time effect in the imidization process of PAA, that is, two stages can be observed during the thermal imidization: as illustrated in Fig. 2.3, the initial fast stage and subsequent slow stage, and the cyclization slows down or even stops to a certain extent at a certain temperature. The reaction is accelerated immediately with rising temperature, but it slows down after a certain time, until the temperature increases again or complete imidization is carried out; this is the so-called "kinetic stop" [17]. One explanation is that the glass transition temperature of the precursor polymer increases as cyclization goes on, leading to a decrease in chain mobility, which slows down the imidization reaction. Another explanation attributes the "kinetic stop" to the decrease in supporting effect of the residual solvent. There exists a complex compound between PAA and solvent, as for DMAc. In Brekner and Feger's research [18], the complexation ratio may be 1 : 4 and 1 : 2 for PAA/DMAc, and the latter is more stable, which cannot be removed through the vacuum method at room temperature. This complex dissociation can only occur with elevating temperature. Shibayev et al. [18] believed that the complexed solvent could reduce the energy barrier of cyclization, thus accelerating imidization. However, the complex dissociation removes solvent with increasing temperature, lowering the imidization rate.

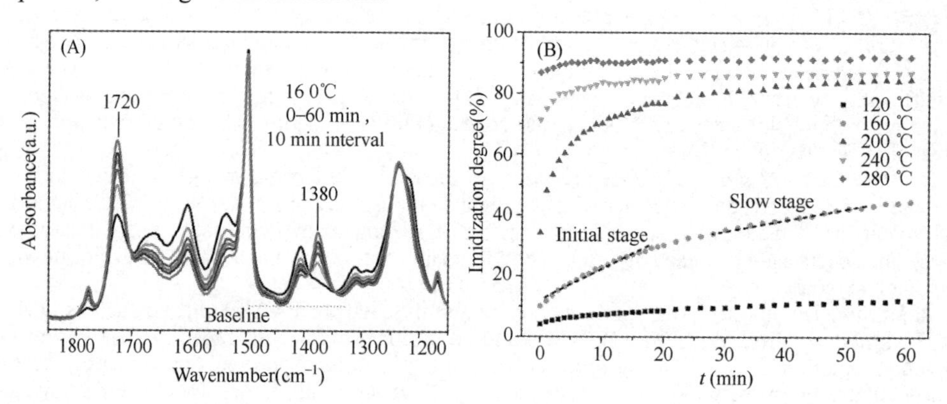

FIGURE 2.3 (A) FTIR spectrum of poly(amic acid) during thermal imidization. (B) Time-dependence of imidization degree.

2.2.2 "One-Step" Polymerization Method

Different from the "two-step" method, in the "one-step" polycondensation, PI is directly synthesized by equimolar dianhydride and diamine monomers in high boiling solvents. At a high temperature, the intermediate PAA is cyclized spontaneously to PIs, and the generated water is removed along with nitrogen flow in order to obtain high-molecular-weight PIs.

In the 1980s, Kaneda et al. [19] from Kyoto University reported that PI solutions were prepared in the "one-step" polycondensation reaction of the mixed dianhydride from 3,3′,4,4′-biphenyltetra-carboxylic dianhydride (BPDA) and PMDA with various aromatic diamines, such as 3,3′-dimethyl-4,4′-diaminodiphenyl or 3,4′-oxydianiline (3,4′-ODA), in *p*-chlorophenol and *m*-cresol at 180 °C. The obtained inherent viscosity varies from 3.2 $dL\cdot g^{-1}$ to 5.2 $dL\cdot g^{-1}$. Besides, they discussed the influence of concentration, reaction time and the amount of carboxylic acid added on the molecular mass of polymers. It should be noted that they found that *p*-hydroxyl phenyl acid showed a remarkable catalytic effect in this system. Stephen Z. D. Cheng and coworkers [20] from Akron University prepared a series of PIs via the "one-step" polycondensation, and they found that there existed an obvious phase transition in PI/*m*-cresol system by means of wide angle X-ray diffraction (2D WAXD), polarized light microscope, and differential scanning calorimetry. At room temperature, the crystallosolvate I state formed initially when the concentration was higher than 40%(*w*), and the system showed an anisotropic feature in the range of 45%-95%(*w*), in which crystallosolvate I evolved into crystallosolvate II. The original aggregation structure and the evolution mechanism of the solution have been reported in their research, laying a solid foundation for preparing high-performance PI fibers.

However, phenolic solvents, such as *p*-chlorophenol and *m*-cresol were generally used in the traditional "one-step" polycondensation, whose strong irritant smells and high toxicities restricted their wide applications. On the other hand, this method was only fit for organo-soluble PIs, however, these high-molecular-mass insoluble PIs could not be obtained as precipitation formed during the high-temperature reaction. Therefore, this method is limited by the selection of monomers and solvents. To overcome these problems in the "one-step" polycondensation, great efforts have been made in recent years, mainly reflecting two aspects: designing new monomers and selection of new friendly organic solvents. Sakagrchi et al. [21] from Japan Toyobo successfully synthesized high-molecular-weight poly(benzoxazole-imide) in polyphosphoric acid (PPA) with a yield of 92% for the first time. The reaction temperature was controlled to 160 °C-200 °C; meanwhile, the influence of P_2O_5 content, reaction temperature, and solid content on the inherent viscosities of the resulting PIs were investigated in detail. Inspired by the above method, Chen et al. [22] prepared a series of PIs containing benzoxazole and benzimidazole units through the "one-step" polymerization (as shown in Fig. 2.4), and they studied the structures and properties of the obtained high-molecular-weight PI in detail. It should be noted that the tensile strength and modulus of the dry-jet wet spun fibers reached as high as 3.12GPa and 220 GPa, respectively. Compared with phenol solvents, the PPA is more environmental friendly, less toxic, and PI with a special structure could form a liquid-crystal structure in PPA, in favor of preparing high-performance materials.

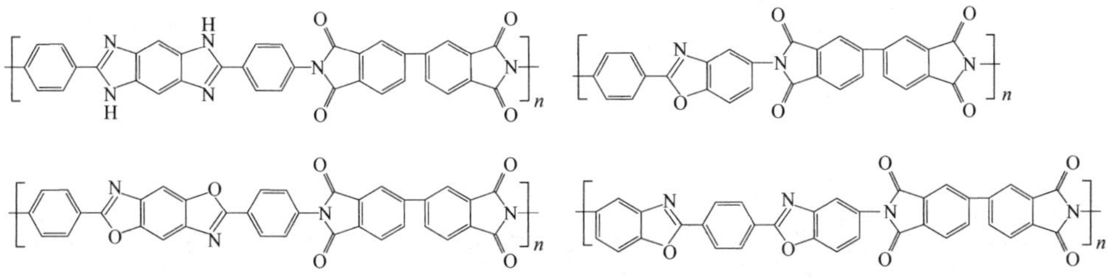

FIGURE 2.4 Chemical structure of polyimide containing heterocyclic units synthesized in PAA via "one-step" method.

Zhang and coworkers [23] have also tried to synthesize high-molecular-weight PIs in PPA, and the obtained PIs showed better thermal stability and higher thermal decomposition activation energy compared with PI synthesized from the traditional "two-step" polycondensation. Hasanain and Wang [24] synthesized more than 10 kinds of PIs with different structures via the high-temperature "one-step" polymerization in salicylic acid, and the obvious feature of this polymerization pathway was that the reaction time reduced sharply, to less than 2 h in general. Meanwhile, the solid solvent can be recycled and reused. In recent years, efforts have focused on preparing PIs in "green solvents." Tsuda et al. [25] reported that a soluble PI with an inherent viscosity of 0.5 $dL \cdot g^{-1}$ was synthesized through the high-temperature "one-step" method in the imidazole hexafluorophosphate ionic liquid, and no catalyst was added in this system, in which ionic liquids acted as catalyst to a certain degree.

Compared with the solvent system, the macromolecular structure has a greater influence on the properties of soluble PIs prepared by the "one-step" polymerization. In general, the introduction of ether linkage, trifluoromethyl, bulky side groups, and asymmetrical units into macromolecular chains resulted in great benefits for improving solubilities of PIs, as the symmetry and regularity of macromolecules were affected or even damaged, increasing the free volume, as well as decreasing interaction of molecular chains. For example, Chung et al. [26] reported high-molecular-weight PIs synthesized by a series of commercial dianhydride and diamine monomers, containing both a benzimidazole ring and trifluoromethyl side groups using NMP as the solvent at 190 °C via the "one-step" method, and these samples showed good solubilities in DMAc, DMF, DMSO, and other solvents. As reported by Zhang and coworkers [27], a series of high molecular-weight PIs were prepared in NMP at 190 °C for 12 h by introducing asymmetrical heterocyclic benzimidazole ring and trifluoromethyl groups into the backbones, and these organo-soluble PIs exhibit the number-average molecular mass (M_n) of (3.1-4.1) × 10^4. Besides, the gel-sol transition was observed in PI/NMP system. As depicted in Fig. 2.5, as the concentration of PI increased to 13%(w), the system showed an obvious banded texture, exhibiting an anisotropic gel behavior. Increasing the temperature to 65 °C, the system transformed into an isotropy solution. It has been proved that this gel-sol transition was mainly caused by the inner ordered domains, and relative research provided vital reference for producing PI fibers with high strength and high modulus [28].

In conclusion, the synthesized PI from the "one-step" method at a high temperature has higher molecular weights and narrower molecular weight distributions. It also solved the problems of unstable storage process as well as avoiding complicated cyclization process in the "two-step" method, having more latent values for development from the point view of polymer processing. However, compared with the traditional "two-step" method, "one-step" polycondensation relied more on the solvent system, molecular structure design and good interaction between macromolecules and solvents. In addition, the selection of monomers for the high-temperature "one-step" method is also expensive and complicated, which are the main factors that have restricted its large-scale application.

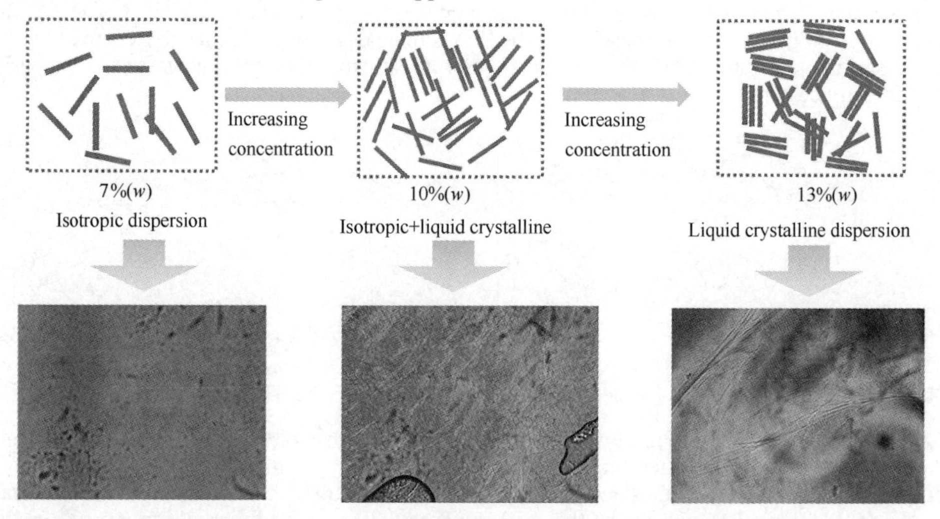

FIGURE 2.5 Polarized optical micrographs (POM) of polyimide/NMP systems with different concentrations at 25 °C.

2.3 Preparation of Polyimide Fibers

At present, three methods are generally employed to fabricate PI fibers including melt-spinning, electro-spinning, and solution spinning. The solution spinning includes dry-spinning, wet-spinning, dry-jet wet-spinning, and liquid crystalline spinning. In the following parts, introduction of the above-mentioned methods is expounded, combined with specific processing characteristics and applications.

2.3.1 Wet-Spinning Method

Wet-spinning is one of the most widely used methods for the fabrication of PI fibers. A diagram of the wet-spinning process is shown in Fig. 2.6; the spinning solution is extruded into the coagulation bath through the spinneret, and then the solvent in the spinning solution and the non-solvent in the coagulation bath interdiffuse, which results in the phase separation. In this situation, polymer dense phase and dilute phase appear in solution, and filament turns into solid fiber gradually along with phase separation. The wet-spinning process can be divided into "one-step" and "two-step" methods according to the differences in the prepared spinning solutions. In the "two-step" wet-spinning process, PAA solution is spun into precursor fibers firstly, and then PI fibers can be obtained through the chemical or thermal imidization. In the early research, the "two-step" method was generally thought to be disadvantageous for preparing PI fibers with high strength and modulus due to the volatilization of water that existed in as-spun fibers always leads to microporous defects in PI fibers during the complicated thermal imidization process, which weakens mechanical properties of the final fibers. However, with the development of synthesis and spinning technologies, the above statement is considered to be immature. In the "one-step" wet-spinning method, as-spun PI fibers can be directly obtained from the spinning solution of soluble PI. Therefore, microporous defects are avoided without volatilization of water existing in as-spun fibers, and it is in favor of aggregation structure controlling, which is beneficial for the fabrication of PI fibers with high strength and modulus.

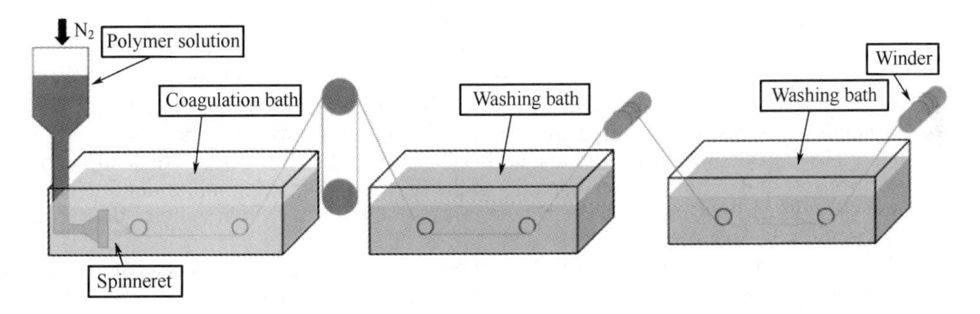

FIGURE 2.6 Diagram of wet-spinning process.

2.3.1.1 "Two-Step" Wet-Spinning Method

In the "two-step" wet-spinning process, defects generated in the fiber-forming process and complicated post-process due to volatilization of water existing in as-spun fibers are the main pivotal problems, which obstruct the thermal stretching process and finally weaken the final performance of fibers, so fabrication of as-spun fibers with a dense structure is very important. For this purpose, solving the problem of the formation of defects in as-spun fibers becomes a key point in improving the performance of PI fibers and an extensive discussion on controlling the micromorphology of as-spun fibers has been held at home and abroad. Goel et al. [29,30] synthesized a type of PAA from 4,4′-diaminodiphenylmethane (MDA) and PMDA in DMF and then PAA fibers were produced from the mixed coagulation bath of DMF and H_2O. After chemical imidization treatment ($V_{aceticanhydride}$: $V_{pyridine}$=1 : 1), thermal stretching process was carried out at 300 °C. The tensile strength of the obtained PI fibers only reached about 0.05 GPa because of the existence of many micropores in fibers and the thermal stretching process made the pores even larger according to the SEM images, which proves the importance of the morphology adjustment in the fabrication process of as-spun fibers. Dorogy and Clair [31] discussed the effects of coagulation bath conditions on the fiber-forming process based on the BTDA/ODA/DABP (3,3′-diamino-benzophenone) system in detail. They analyzed the pores in as-spun fibers produced in the mixed coagulation bath of H_2O and ethanol, H_2O and ethylene glycol, H_2O and DMAc. The cross-section of the resulting fibers was the most dense when the coagulation bath was a mixture of 80% ethanol/H_2O solution or 75% ethylene glycol/H_2O solution, but many microporous defects were generated when using the other coagulant.

Park and Farris [32] prepared a type of PI fiber with a dense structure by the wet-spinning method from the partial imidization PAA solution and thermal imidization process. The tensile strength and modulus of final fibers were 0.4 GPa and 5.2 GPa, respectively. However, compared with the pure PAA spinning solution with no imidization, the partially imidized PAA spinning solution has lower diffusion velocity, which is helpful for forming denser structures in the resulting fibers and avoiding the formation of skin-core structure. Meanwhile, this study provides a new method for the high draft in the dry-jet wet-spinning process, because PAA fibers cannot bear the pressure from the high draft. Zhang and coworkers [33] analyzed the dual diffusion process of the PAA solution in H_2O, ethanol, and ethylene glycol in detail, based on available theoretical and experimental ternary phase diagrams to optimize the fabrication process of PI fibers by the "two-step" wet-spinning method. As shown in Fig. 2.7, solidification abilities of the three kinds of solvents were H_2O>ethylene glycol>ethanol. It is noteworthy that the interaction parameter of the ethanol/PAA system was only 0.28 which is extremely small. This result illustrated that the solidification ability of ethanol was relatively weak compared with the other two solvents, and H_2O showed the strongest solidification ability. In the three non-solvents, the polymer contents at the dividing point were all under 5%(*w*), thus the forming process of PAA tended to be by the nucleation growth, which is helpful to produce as-spun fibers with dense structures.

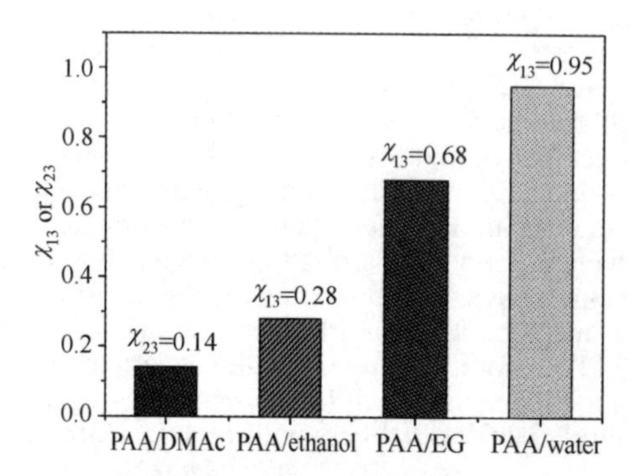

FIGURE 2.7 Interaction parameter of PAA/DMAc and coagulant/PAA system at 27 °C.

In recent years, with the development of synthesis and spinning technologies, the integrated performance of PI fibers fabricated by the "two-step" wet-spinning method was improved constantly. Inspired by the heterocyclic Aramid fiber named Armos from the Soviet Union Tver Chemical Fiber Freedom Company, introducing heterocyclic moieties into PI main chains via heterocyclic diamine monomers and fabrication of high-performance PI fibers by the "two-step" wet-spinning method has become a new research hotspot, and achievements are prominent. Liu et al. [34] carried on copolymerization from 2-(4-aminophenyl)-5-aminobenzimidazole (PABZ)/ PMDA/ ODA and fabricated high-performance PI fibers by the wet-spinning method. As shown in Table 2.1, the monomer ratio of PABZ and ODA had a critical influence on the mechanical properties of PI fibers. When the monomer ratio was PABZ/ODA=7/3, tensile strength and modulus of PI fibers were enhanced about 2.5 times and 26 times, reaching 1.53 GPa and 220 GPa, respectively. According to the test results of FTIR and DMA, strong intermolecular hydrogen bonding existed in the fibers owing to the introduction of PABZ, which was much helpful for the improvement of fibers' mechanical properties.

TABLE2.1 Mechanical Properties of PMDA/ODA/PABZ PI Fibers

ODA:PABZ Molar Ratios	Inherent Viscosity ($dL \cdot g^{-1}$)	Strength (GPa)	Modulus (GPa)	Elongation (%)
10:0	3.56	0.61	8.5	9.0
7:3	2.73	0.92	56.6	6.6
5:5	2.35	1.26	130.9	5.8
3:7	1.89	1.53	220	3.2

A series of heterocyclic PI fibers were prepared from BPDA/PPD/BIA (BIA: 2-(4-aminophenyl)-5-aminobenzimidazole) and BPDA/PPD/PRM (PRM: 2,5-bis(4-aminophenyl)-pyrimidine), and many deep studies on some critical point were carried out, such as fiber's structure, mechanical properties, thermal stabilities, and so on.

In recent years [35-40], a series of PI and co-PI fibers have been produced through the integrated continuous two-step wet-spinning method, and the chemical structures are shown in Fig. 2.8. In their studies, the BPDA/PDA homo-PI fibers possessed tensile strength, initial modulus, and elongation of about 1.0 GPa, 50 GPa, and 1.1%, respectively, attributing to the rigid chain structure. However, the stiffness of the polymer chains also results in poor spinnability and low mechanical properties of the resulting fibers. Therefore, with the aim of improving the mechanical properties and processibility of the PI fibers, the most efficient approach is modifying the chemical structures of PIs by incorporating other moieties into the BPDA/PDA backbones. By optimizing the chemical structure and controlling the spinning conditions, the co-PI fibers exhibited tensile strength higher than 3.0 GPa and modulus higher than 130 GPa.

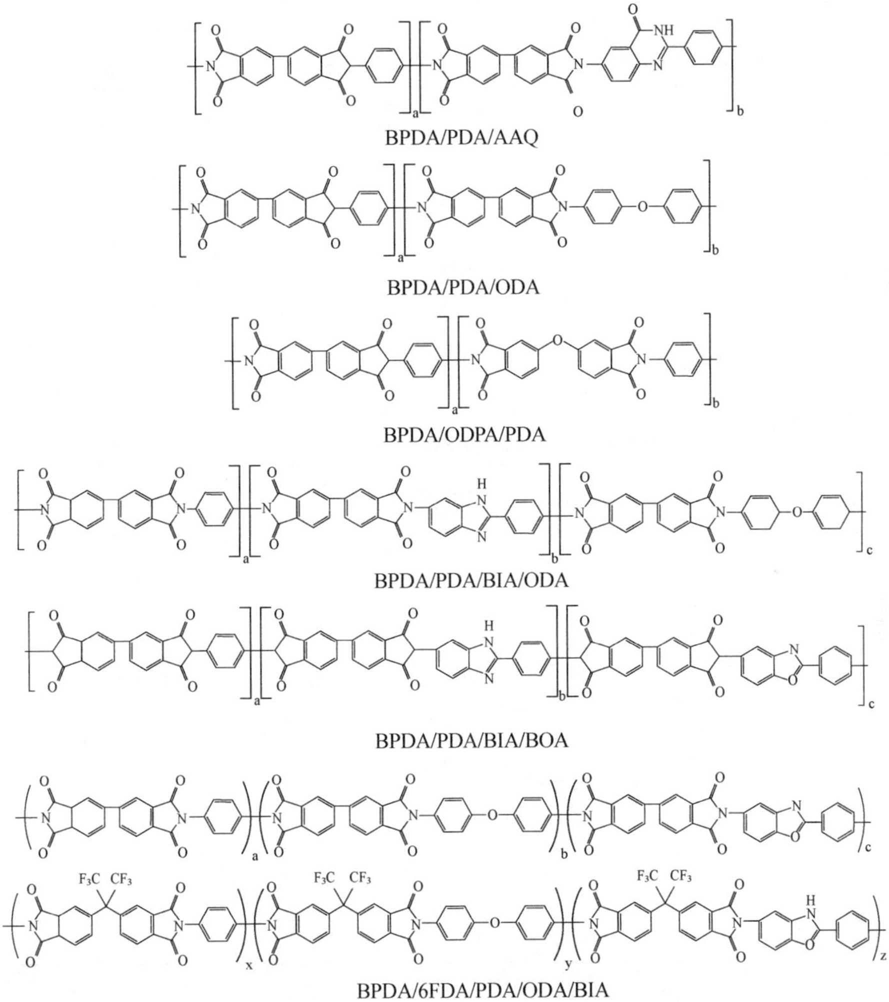

BPDA/PDA/AAQ

BPDA/PDA/ODA

BPDA/ODPA/PDA

BPDA/PDA/BIA/ODA

BPDA/PDA/BIA/BOA

BPDA/6FDA/PDA/ODA/BIA

FIGURE 2.8 Typical chemical structures of copolyimide fibers prepared in Wu's laboratory.

A new heterocyclic diamine monomer with high reactivity and rigid-rod structure named 2-(4-aminophenyl)-6-amino-4(3H)-quinazolinone (AAQ) has been synthesized, and was introduced into the main chain of PI to fabricate fibers with high tensile strength and modulus (Table 2.2). The results indicated that there were strong intermolecular hydrogen bonds in fibers since the introduction of AAQ monomer. In addition, orientation degrees of molecular chain and crystalline structures in fibers were much improved on account of the rigid-rod structure of the AAQ monomer, which is of much significance for improving fibers' mechanical properties. They also introduced 4,4-oxydiphthalic anhydride (ODPA) and ODA containing ether groups into the BPDA/PDA polymer backbone, and discovered that the introduced ether groups were helpful for the reduction in the microvoids' size in the fibers. By combining the advantages of BIA and ODA units, a series of BPDA/PDA/BIA/ODA co-PI fibers have been prepared. The mechanical properties of the co-PI fibers were improved due to the reduction in the microvoids and strengthened intermolecular interactions, respectively.

In conclusion, PI fibers with high strength and modulus can be fabricated by the introduction of heterocyclic units into PI macromolecular chains through the "two-step" wet-spinning method. Some common heterocyclic diamine monomers and their fibers' performance are listed in Table 2.3.

TABLE2.2 Mechanical Properties of Polyimide Fibers From BPDA/AAQ/PDA [39]

AAQ:PDA	Hydrogen-Bond Degree (%)	Strength (GPa)	Modulus (GPa)	Elongation (%)	T_g (°C)
0	19.69	1.2	64.4	2.1	338.3
1:9	41.47	1.3	67.2	2.3	341.3
3:7	41.99	2.7	104.8	3.1	363.4
5:5	42.46	2.8	115.2	3.1	382.4
7:3	48.73	2.7	113.1	2.7	400.1
9:1	53.58	1.9	104.3	2.0	400.7

TABLE2.3 Common Heterocyclic Diamine Monomers and Their Fibers' Performance

Diamine	Chemical Structure of Diamines	Tendency (GPa)	Modulus (GPa)	Elongation (%)	Reference
BIA	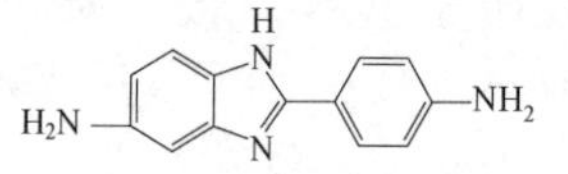 H_2N–(benzimidazole, H N, N)–(phenyl)–NH_2	3.2	114.6	3.4	[41]
BOA	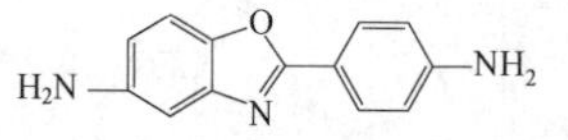 H_2N–(benzoxazole, O, N)–(phenyl)–NH_2	2.6	91.8	3.0	[41]
AAQ	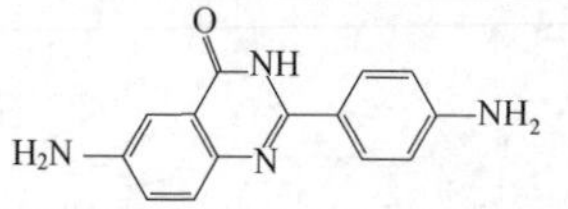 H_2N–(quinazolinone, O, NH, N)–(phenyl)–NH_2	2.8	115.2	3.1	[39]
PRM	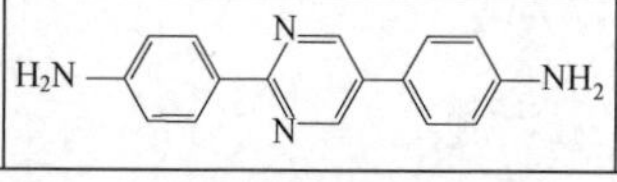 H_2N–(phenyl)–(pyrimidine, N, N)–(phenyl)–NH_2	3.0	130	*	[42]

Corresponding value has not been found.

2.3.1.2 "One-Step" Wet-Spinning Method

The appearance of soluble PIs promotes the development of fabrication technologies based on the "one-step" spinning method for PI fibers. In traditional "one-step" wet-spinning or dry-jet wet-spinning processes, phenolic solvents (*m*-cresol, *p*-chlorophenol, *m*-chlorophenol) were widely used, and alcohol (methanol, ethanol, and ethylene glycol) or mixture of alcohol/water were generally used as coagulants. PI fibers with high tensile strength and modulus can be obtained by high draft of as-spun PI fibers. In an early research, Kaneda et al. [43] from Kyoto University synthesized a series of soluble PIs in p-chlorophenol solvent based on BPDA and various diamine monomers, and the PI fiber with a tensile strength of 3.1 GPa and modulus of 128 GPa were prepared by "one-step" wet-spinning in the coagulation bath of ethanol/H_2O mixture. Moreover, compared with Kevlar 49 from DuPont, the above fibers have lower water absorption and stronger acid resistance. Cheng's group [44] from Akron University also prepared PI fibers in methanol/H_2O coagulation bath from soluble PI/*m*-cresol solution with a content of 12%-15%(*w*) synthesized from BPDA and 2,2′-bistrifluoromethyl-4,4′- diaminobiphenyl. After a high draft of 10 times at 380 °C, the tensile strength and modulus reached 3.2 GPa and 130 GPa, respectively. Furthermore, the modulus of obtained fibers reduced only 5% after treatment at 400 °C for 5 h, which illustrated the outstanding high-temperature strength retention of the prepared PI fibers. As mentioned earlier, the "one-step" method is beneficial to prepare PI fibers with high tensile strength and modulus. However, phenolic solvents are usually hypertoxic and difficult to thoroughly remove from the as-spun fibers. Therefore, if PIs with special structures and high organo-solubilities are synthesized, the "one-step" fabrication process of PI fibers is more promising. As reported by Zhang and coworkers [27], a series of co-PIs were synthesized from BTDA, 2,2′-bis(trifluoromethyl)-4, 4′-diaminobiphenyl (TFMB) and BIA with different diamine ratios via the "one-step" polymerization at 190 °C in the NMP solvent, and the inherent viscosity and number-average molecular weight of the obtained PI/NMP solution were 1.83-2.32 $dL \cdot g^{-1}$ and 31,300-41,000, respectively. And then, a series of high-performance PI fibers whose tensile strength, modulus, and elongation at break were 1.37 GPa-2.13 GPa, 29.9 GPa-101.9 GPa, and 4.57%-2.14%, respectively, were fabricated by a "one-step" wet-spinning and thermal stretching process. These works provide important references for preparing high-performance PI fibers by the "one-step" wet-spinning method in common and environmentally friendly organic solvents. Chen et al. [4] also have successfully synthesized a type of heterocyclic PI with an inherent viscosity of 6.98 $dL \cdot g^{-1}$ by "one-step" method at high temperature in PPA, and the synthetic route is shown in Figs. 2.9. In the meantime, a series of PI fibers with high strength and modulus were prepared by dry-jet wet-spinning and their tensile strength were in the range of 2.45 GPa-3.12 GPa. Compared with phenolic solvents, PPA is more environmental friendly with low toxicity, and PI can form liquid-crystal state in PPA, which is in favor of obtaining high-performance fibers.

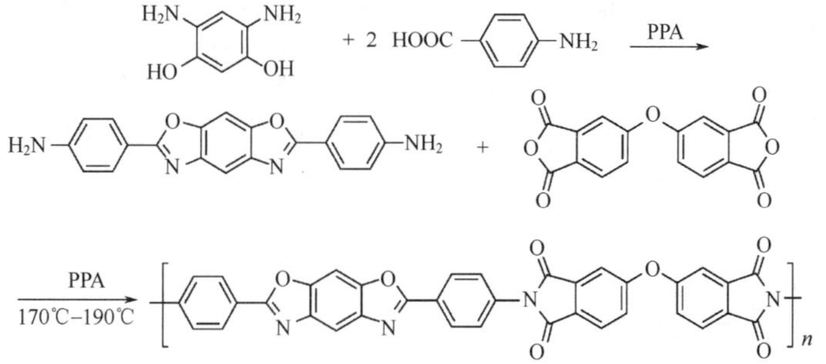

FIGURE 2.9 Synthesis route of poly(imide-co-benzoxazole) by "one-step." Polymerization at high temperature in PPA.

2.3.2 Dry-Spinning Method

Fibers prepared by the dry-spinning process are based on a specific fiber-forming principle, which is different from the melt-spinning or wet-spinning method. In the melt-spinning process, polymer melts are gradually solidified along the spinning line with the decreased temperature. PI fibers are difficult to fabricate by the melt-spinning method due to the rigid backbones and strong intermolecular interactions of PI chains, which result in the unmelt behavior of PIs even under a high temperature. In the wet-spinning, the coagulation of as-spun fibers is a dual diffusion process, namely, the solvent in the solution diffuses into the coagulation bath and the non-solvent diffuses into the fibers, which is a relatively complex process. For the dry-spinning method, fibers are quickly solidified due to the rapid evaporation of solvent in the spinning solution and the spinning process is easily affected by the temperature and volume of air in the column. So far, several commercial fibers, including polyurethane and cellulose acetate fibers, have been successfully prepared by the dry-spinning process. Compared with the wet-spinning, advantages of the dry-spinning process are as follows: (1) higher spinning rates, more environmentally friendly, and more convenient for the solvent recovery; (2) the spinning solution in the hot column of the dry-spinning process will solidify rapidly by an unidirectional diffusion of the solvent, which is beneficial for the formation of denser inner structures of the resulting fibers, as shown in Fig. 2.10. However, the prerequisite of polymeric solutions for the dry-spinning is much harsher, for example, the apparent viscosity of the solution should be fit for the spinning conditions, the rheological properties should be good enough, and its higher requirements on equipment compared with the wet-spinning process.

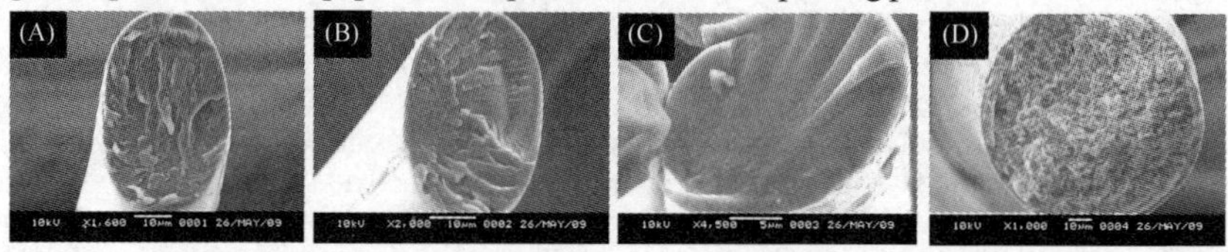

FIGURE 2.10 (A) SEM pictures of the cross-section for the as-spun PAA fiber by dry-spinning, (B) thermally imidized PI fiber, (C) drawn PI fiber with Dr=2.5, and (D) cross-section of the as-spun PAA fiber with the same backbone by the wet-spinning method.

Early studies about PI fibers prepared via the dry-spinning process started in the 1960s. Investigators in DuPont were aware of the outstanding combination of properties of PI fibers, and then they tried to produce high-strength and high-modulus PI fibers via dry-spinning, and they applied for corresponding patents. Samuel and Edgar [45] successfully synthesized a PAA spinning solution with the inherent viscosity of 1.3 $dL \cdot g^{-1}$, and they attempted to prepare PI fibers using this solution in a spinning column at 265 °C with a spinning rate of 160 $m \cdot min^{-1}$. The final tensile strength and modulus of the PI fiber were around 3.5 $g \cdot d^{-1}$ and 50 $g \cdot d^{-1}$, respectively. About 20 years later, researchers in Dupont tried to prepare co-PI fibers via the dry-spinning process again by using 3,4′-ODA and p-PDA as the diamines, and PMDA as the dianhydride. The tensile strength and modulus of the final co-PI fibers reached 15.6 $g \cdot d^{-1}$ and 534 $g \cdot d^{-1}$ after treating the precursor co-PAA fibers under 150, 200, and 300 °C for 20, 20, and 30 min, respectively [46]. Fig. 2.11 depicts the schematic diagram of preparing PI fibers by dry-spinning process. Here, a brief description is given in the following to introduce some characteristics of this spinning method.

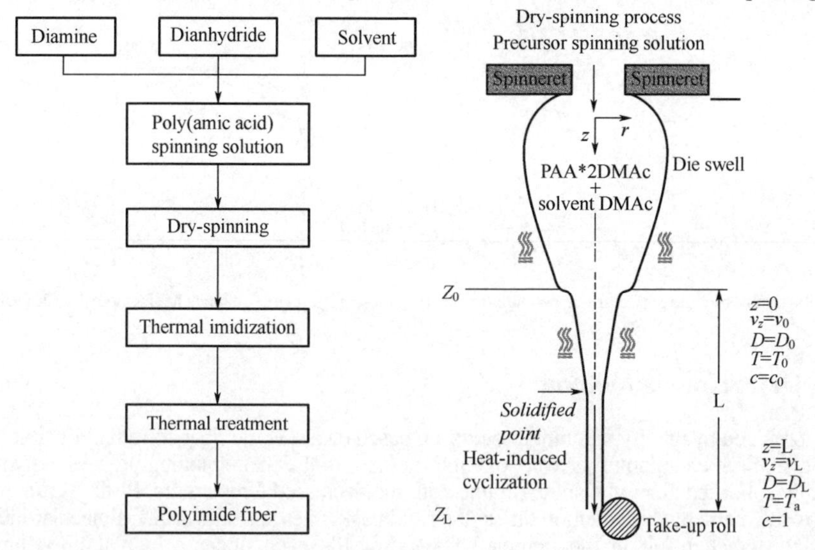

FIGURE 2.11 Schematic diagram of preparing PI fibers via dry-spinning.

In general, the solvent in the spinning solution will evaporate rapidly, which is attributed to the great difference of temperature between the hot air in the column and the spinning solution. That is, the solvent will undergo a flash process resulting in the solidification of polymer solutions. In the beginning, the spinning solution contains large quantities of solvents (around 80%-90%(w)), and the diffusion rate of the solvent in the filament is mainly controlled by the heat and mass transfer rate. Solidified filament forms in the one-third part of the spinning line, and the evaporation rate of solvent decreases to around zero. In fact, the strong hydrogen bonding interaction can form between PAA and the solvent (like DMF, DMAc), and the residual solvent is difficult to fully remove due to the fact that the filament stays for a short time in the spinning column. As shown in Fig. 2.12, the residual solvent content in the as-spun fiber is closely related to the spinning speed, which dramatically increases with increasing spinning speed, and the final content is mainly in the range of 15%-30%(w), this residual solvent plays an important role in the following thermal imidization and hot-drawing process of the fibers [47].

There is a big difference in the dry-spinning process between PI fibers and other polymeric fibers: the precursor PAA will be partially thermally imidized to PI in the high-temperature spinning column. As Fig. 2.13 shows, the imidization process for the PAA in the spinning column can be divided into two different stages along the spinning line: in the first stage, the imidization degree changes little due to the relatively low temperature of the filament because the rapid evaporation of the solvent will take away a great deal of heat; in the late stage, the imidization degree increases quickly due to the raised temperature of the solidified filament. The partial imidization of the PAA fiber shows a great significance for its storage stability and the following thermal treatment. Meanwhile, the imidization degree shows a higher value under low spinning speeds due to the relatively long imidization time.

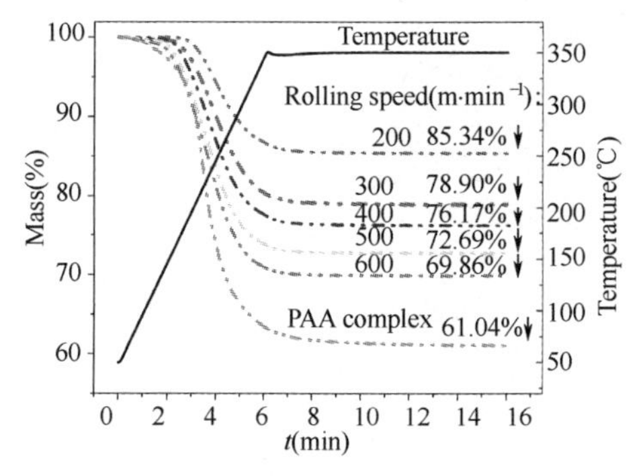

FIGURE 2.12 The content of the residual solvent in as-spun fibers as a function of the spinning speed.

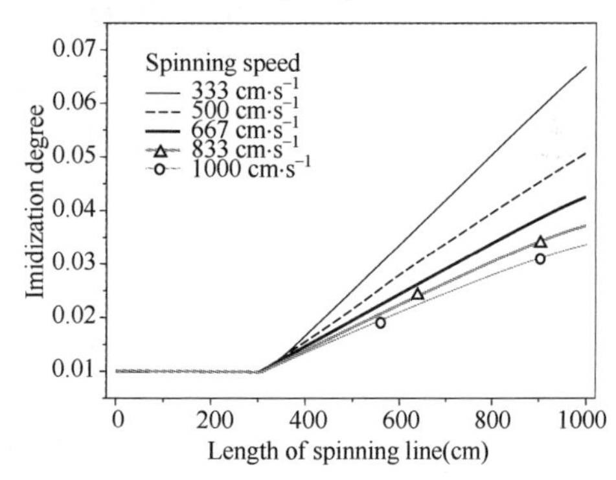

FIGURE 2.13 The imidization degree of as-spun fibers as a function of spinning line length and spinning speed.

Zhang and coworkers have made a great efforts on the preparation of PI fibers by dry-spinning method [48-50]. In order to investigate the partial imidization reaction of PAA in the spinning column, a dry-spinning model for the PI fibers was established. Interestingly, by building corresponding spinning equations, a credible model for spinning profiled PI fibers was successfully obtained. Herein, a brief introduction was given.

A trilobal cross-section of the spinneret hole and surface tension on the fiber with trilobal cross-section model were given as designed in Fig. 2.14, and the streamline diagram and velocity vector of the spinning solution in the designed spinneret were successfully simulated by the software of Polyflow. Near the spinneret, the filament contains a high amount of free solvent, and flash evaporation takes place at the high temperature of the air flow in the spinning column. The mass fraction of the DMAc thus decreases rapidly and the filament solidifies at z=150 cm, as shown in Fig. 2.15. At the same time and as a result of flash evaporation near the spinneret, the temperature of the filament decreases rapidly to a minimum at z=85 cm. It then quickly reaches the temperature of the hot air (253 °C) at z=290 cm.

The formation of profiled PI fibers in the high-temperature column can be controlled by the spinning parameters (temperature, drawing ratio, hot air velocity). In the dry-spinning process model, the air velocity is changed from 20 $cm \cdot s^{-1}$ to 100 $cm \cdot s^{-1}$, as shown in Fig. 2.16. At the spin line z<120 cm, the profile degree drops due to the solvent evaporation and the viscoelastic behavior of the filament; the air velocity has no influence on the profile degree. At z>120 cm, a high air velocity results in a relatively high value of the profile degree. It must be attributed to the increased air velocity that results in higher rate of solvent evaporation. The filament shape is thus "freezed" at short times.

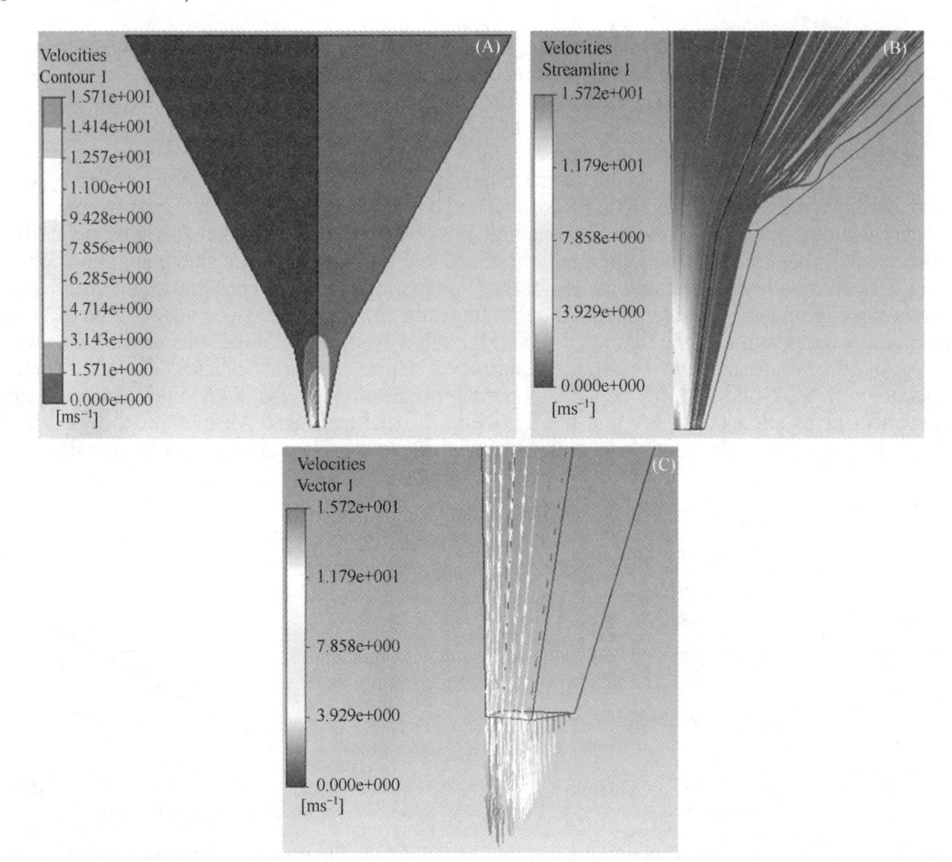

FIGURE 2.14 (A) The designed spinneret with a trilobal cross-section; (B) the streamline diagram; and (C) velocity vector of the spinning solution in the designed spinneret.

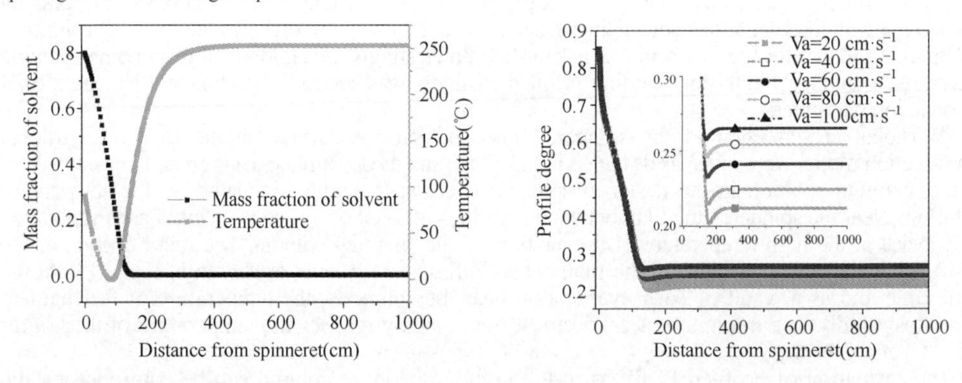

FIGURE 2.15 Variation of solvent mass fraction and temperature along the spin line.

FIGURE 2.16 Effect of air velocity on profile degree.

The impact of air temperature in the spinning column on the profile degree is shown in Fig. 2.17. The final profile degree decreases for lower temperature; the maximum profile degree is declined at T=253 °C. As we know, the higher temperature leads to solvent evaporating more rapidly and the short solidification time of the filament. Besides, the high air temperature increases imidization reaction degree of PAA, causing a larger viscosity of spinning dope before solidification. Thus, the cross-section deformation under surface tension at high temperature reduces, and the final profile degree is larger compared to one at lower temperatures. Based on the above results, bringing the solidification point close to the spinneret helps to obtain (within certain limits) a larger final profile degree.

By keeping the mass flow constant but increasing the take-up velocity, a larger draw ratio (DR) can be achieved. Fig. 2.18 shows the dependence of the profile degree on the DR from 1 to 3. A clear trend emerges: the profile degree increases with the DR. On the one hand, a high DR results in thinner filaments with a higher specific surface area, which accelerates solvent evaporation and solidification. On the other hand, a high DR increases the velocity of the filament and decreases residence time in the column, which goes against solvent evaporation. The combination of these two opposing effects results in a lager profile degree with the DR increase. The final value reaches in our simulation 0.24 at DR=1 and 0.40 at DR=3.

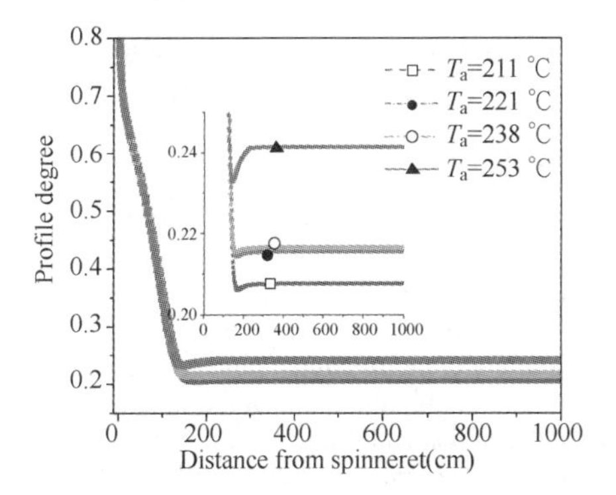

FIGURE 2.17 Effect of temperature in spinning column on profile degree.

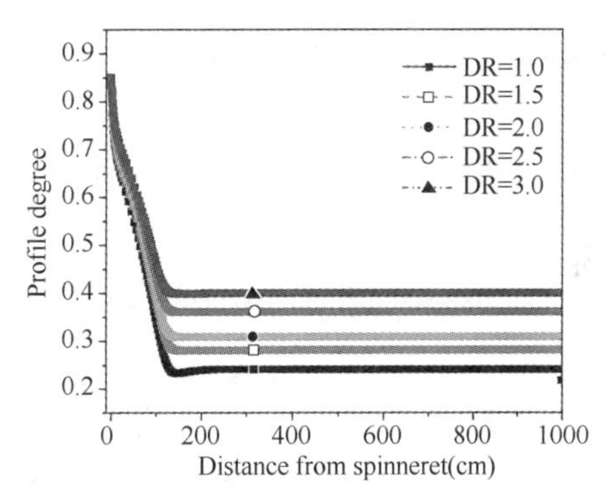

FIGURE 2.18 Effect of draw ratio on profile degree.

In summary, in the dry-spinning process, precursor PAA fibers are firstly solidified as the evaporation of the solvent and then partial imidization reaction takes place, which greatly enhances the stability of the PAA fibers. Therefore, the phase separation process and imidization reaction of the PAAs in the high-temperature column have a great effect on the microstructure, morphologies, and properties of the final PI fibers.

2.3.3 Other–Spinning Methods

2.3.3.1 Liquid-Crystal Spinning Process

In the liquid-crystal spinning process, the spinning solution will be subject to the strong shearing and stretching effects when extruded from the spinneret. The nematic liquid-crystal microregions will orient along the fibers' axis, and the molecular chains will further orient in the air gap and will be fixed when the solution comes into the coagulation bath. Therefore, fibers spun via the liquid-crystal spinning process always show high strength and high modulus even without thermally stretching. Few studies about PI fibers fabricated by the liquid-crystal spinning have been reported due to the fact that soluble PAAs are difficult to form liquid-crystal phase. Neuber et al. [51] successfully synthesized a poly(amic ethyl ester) (PAE) solution with an inherent viscosity of 5.89 $g \cdot cm^{-3}$. This PAE/NMP solution can form lyotropic liquid-crystal phase when the solid content reached 40%(w) at 80 °C. Corresponding PI fibers were successfully prepared via the liquid-crystal spinning process using acetone as the coagulant. The final tensile strength and modulus of the fibers were 0.7 GPa and 68.0 GPa, respectively.

2.3.3.2 Melt-Spinning Process

Melt-spinning is a common spinning method to fabricate polymer fibers. However, PI fibers were rarely prepared by this spinning method due to low solubility and infusibility properties of most PIs because of their rigid backbones. However, more and more poly(ether imide)s have been developed in recent years by introducing flexible ether or methylene units into polymer chains that endow these PIs with good spinnability by the melt-spinning process. Researchers in Teijin Chemicals Ltd. tried to produce poly(ether imide) fibers via the melt-spinning process at temperatures of 345 °C-475 °C. However, the mechanical properties of the final fibers were unsatisfactory. Irwin [52] also tried to fabricate PI fibers via the melt-spinning process under 300 °C-400 °C, and rolling speed reached 300 $m \cdot min^{-1}$-500 $m \cdot min^{-1}$. The tensile strength of as-spun fibers was just 0.59 GPa, and the final strength and modulus of fibers after thermal treatment reached 1.55 GPa and 48 GPa, respectively. Based on the above analysis, it's easy to conclude that the melt-spinning temperature of PI fibers is always too high, yet the prerequisite of this spinning method is that PIs can be melted at certain temperatures. These limitations mean that the melt-spinning process is not suitable for preparing PI fibers on a large-scale.

2.4 Structure and Properties of Polyimide Fibers

2.4.1 Aggregation Structure of Polyimide Fibers

PI fibers possess several distinguished characteristics, such as high modulus, high strength, excellent thermal dimensional stability and chemical resistance, which are not only the result of the rigid-rod polymer backbones but also of the highly oriented polymer chains and ordered lateral packing. PI materials are always regarded as the semi-crystalline polymers. However, crystallinity and orientation degree of PI fibers would increase clearly, and typical crystalline regions would form in the fibers after the thermal stretching process, which are the main factors for preparing high-strength and high-modulus PI fibers. As reported by Wakita et al. [53], common PI materials are amorphous, which means that most PI cannot form 3D crystalline structures. Saraf [54] thought PIs tend to form "liquid-crystalline-like ordered regions" (LC-like). The packing of PI chains was relatively disordered in a liquid-crystalline-like ordered structure due to variation in conformation around the chain axis and the distribution of interchain distances. In the amorphous region, molecular chains are in a random arrangement, and charge transfer complexation occurs between electron-donating diamine moieties and electron-accepting dianhydride moieties, which easily form "mixed layer packing". Wu and coworkers [55] has proposed a hypothetical structure model of the PI fibers as pictured in Fig. 2.19. The co-PI fibers are composed of fibrils and exhibit skin-core structure due to the different solidification rates between the external and inner structure. The packing compactness of the fibrils near the surface is higher than that in the core, resulting in different abilities of deformations between the skin to the core in the fibers at high DRs. For the ordered structure region, the molecules tend to align in a distinct series of layers, while there are only liquid-like short-range order within their layers, which is named a "smectic a liquid crystal-like" ordered layer structure. The ordered chain packing along the *c*-axis and the short-range ordered lateral packing in the *ab*-plane are formed.

Crystalline structures and lattice parameters of PI fibers have been investigated in detail [44]. A novel PI fiber has been spun based on BPDA and DMB via the dry-jet wet-spinning. The fibers were stretched and annealed at elevated temperatures above 400 °C to achieve excellent mechanical properties. As for seven-times drawn fibers, the BPDA-DMB molecular packs into a triclinic unit cell with dimensions of a=2.10(2) nm, b=1.523(8) nm, c=4.12(7) nm, α=61.2°, β=50.7°, γ=78.9°, and the number of chain repeating units per unit cell (Z) is 16. After annealing at elevated temperatures, the unit cell of fibers changed a little, that is a=2.048(6) nm, b=1.529(5) nm, c=4.00(2) nm, α=62.1°, β=52.2°, γ=79.6°. After heat-drawing and annealing treatment, both the crystallinity and crystal orientation increase. Obata et al. [56] prepared a wholly aromatic PI fiber with a rigid-rod chain based on PMDA and benzidine (BZ) and analyzed the molecular conformation and crystal structure of fibers using WAXD measurement and computer simulation. The unit cell parameters were a=0.857 nm, b= 0.551 nm, c=1.678 nm, and the crystalline structure belonged to an orthorhombic crystal system.

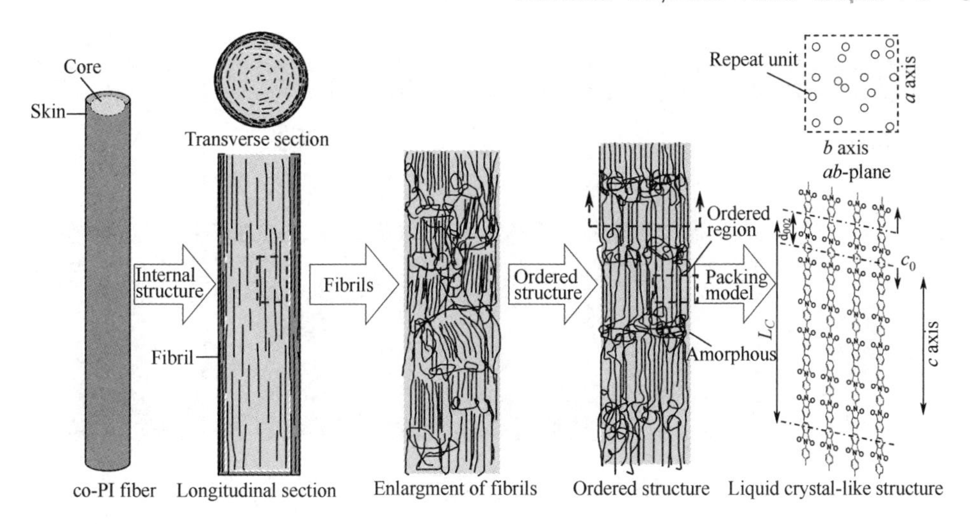

FIGURE 2.19 The sketch of a hypothetical structure model of the copolyimide fibers.

Evolution of the aggregation structure of co-PI fibers containing flexible and rigid diamines in the heat-drawing process has been investigated in recent years. Zhang's group prepared a series of co-PI fibers by introducing a new rigid diamine 2,6-(4,4′-diaminodiphenyl) benzo[1,2-d:5,4-d′] bisoxazole (PBOA) into the relative flexible homo-PI backbones of BPDA and ODA. In Fig. 2.10, all of the as-spun fibers exhibited obscure equatorial arcs, named "amorphous halos," indicating relatively low ordering degrees of the lateral stack of the polymer chains in the as-spun fibers. However, by thermal treatment, significant changes appear in the fibers' diffraction patterns. All drawn PBOI fibers had similar lattices with the BPDA-ODA oriented PI thin films showing orthogonal geometry. Several [001] diffractions appear on the meridian (drawing direction) and obscure [*hkl*] diffractions can be observed in the quadrants, reflecting at least with the lattice parameters of $\alpha=\beta=90°$. Furthermore, the appearance of the [*hk*0] planes on the equator can be explained by assuming $\gamma=90°$. Accordingly, it can be assigned that the crystal structure of these PBOI fibers is orthorhombic symmetry. The distinct diffraction arcs of (001) lattice planes along the meridian represent the rigid-rod polymer chains having a very high degree of preferred orientation aligned with the fiber axis. Meanwhile, equatorial reflections become much sharper and several clear diffraction points can be observed indicating the formation of a well-defined intermolecular lateral packing order, which means that the regularity of the molecular stacking has been substantially improved and crystalline regions have formed in the thermal drawn PBOI fibers.

Meanwhile, by means of small-angle X-ray scattering (SAXS), Yin et al. [57] intensively investigated the morphology evolution of the PI fibers. The micromorphology of PI fibers changes dramatically by introducing a relatively rigid diamine, PBOA, into the BPDA-ODA backbones. As shown in Fig. 2.20, the intense and elongated streaks near the beam stop along the meridian direction reflecting the oriented needle-like structures of microvoids which aligned parallel to the fiber axis. After the thermal stretching, for all resulting fibers, the scattering pattern exhibits a sharp and elongated equatorial streak superimposed with an obvious meridional intensity distribution, suggesting the formation of periodic lamellar structures in the drawn PBOI fibers.

Additionally, this meridional intensity becomes stronger with increasing PBOA concentrations. 1D SAXS scattering profiles of PI fibers along the stretching direction are shown in Fig. 2.21B; a scattering maximum is consistently seen in all fibers, which can be attributed to the long period repeat from a crystalline and amorphous two-phase stacking morphology. The scattering maximum gradually moves to a lower q value with increasing PBOA content. The long period L obtained from the correlation function ranges from 19.3 Å to 22.0 Å, and increasing PBOA content will result in higher long periods. In addition, it is seen that the calculated crystal thickness 1C and amorphous layer thickness lA slightly increase with increasing PBOA content, which are in the range of 8.2 Å-9.3 Å and 11.1 Å-12.8 Å, respectively, based on the correlation function as shown in Fig. 2.21C.

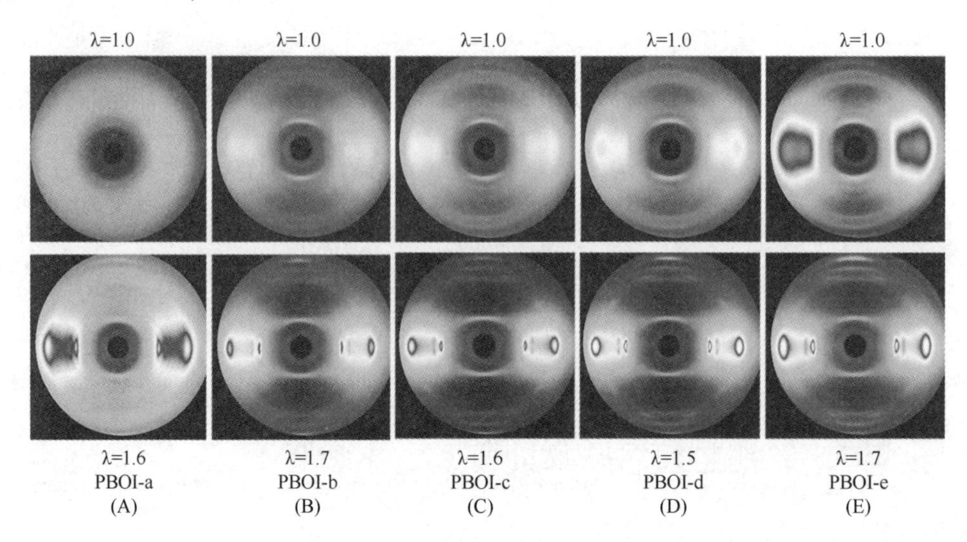

FIGURE 2.20 2D WAXD patterns of BPDA/ODA/PBOA copolyimide fibers (molar ratio: (A) ODA/PBOA=9/1; (B) 8/2; (C) 7/3; (D) 6/4; (E) 5/5).

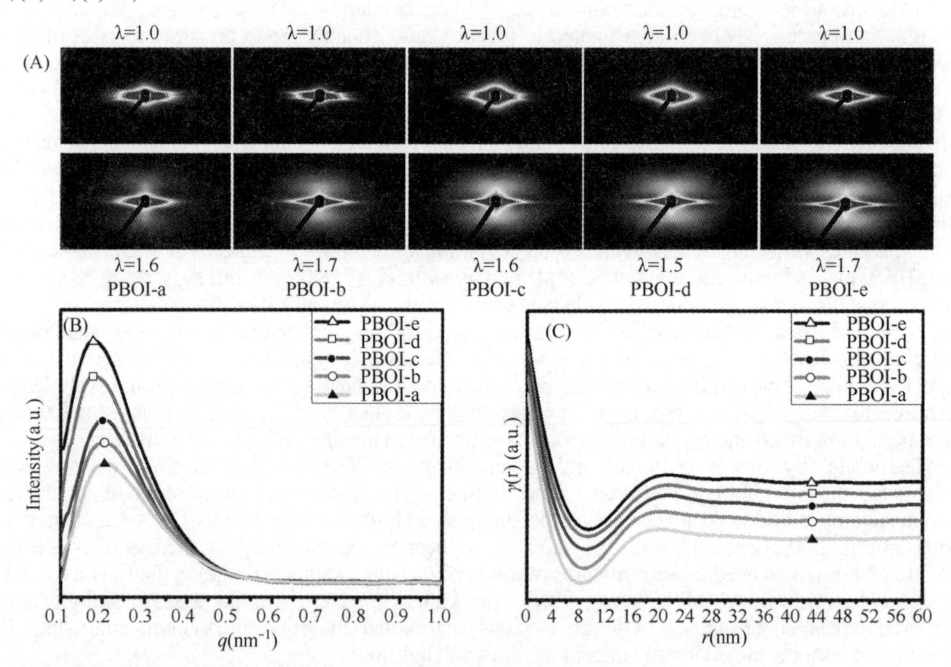

FIGURE 2.21 (A) 2D SAXS patterns of the BPDA/ODA/PBOA copolyimide fibers; (B) 1D SAXS scattering intensity distribution of PBOI fibers along stretching direction; (C) corresponding correction function diagrams calculated from Fig. 2.21B.

2.4.2 Chemical Structure–Property Relationship

As is known that subtle variations in the structures of the dianhydride and diamine components have a tremendous effect on the properties of the final PI fibers. Different chemical composition directly leads to various performances of the PI fibers. Herein, we focused on the relationship between chemical structure and properties.

2.4.2.1 Typical Homo-Polyimide Fibers

Among all the preparation systems, the PMDA/ODA system is considered as the simplest one due to the high reactivity of the two monomers when reacted with other dianhydrides/diamines. The PMDA/ODA polymer backbone consists of rigid PMDA units and flexible ODA units, and thus leading to the contradiction in the performance of the resultant PI fibers. The rigid polymer chains gave rise to the excellent thermal-oxidative stability with 10% weight loss temperature (T_{d10}) over 540 °C in air atmosphere and glass transition temperature of 410 °C. On the other hand, the flexible ODA units contorted the chain symmetry and thus made it harder for the chain to pack into the crystal lattice, resulting in a poor degree of lateral molecular arrangement. The fibers derived from PMDA/ODA system exhibited the tensile strength, initial modulus and elongation up to 0.4 GPa, 5.2 GPa, and 11.1%, respectively. The PI fibers derived from BPDA and p-PDA are preferred due to the stiff and linear chain structure, strong intermolecular association, and high molecular orientation. The fibers prepared in BPDA/PDA system possessed the tensile strength, initial modulus, and elongation of 1.07 GPa, 50.39 GPa, and 1.12%, respectively, with the T_g around 340 °C [35]. However, the stiffness polymer chains resulted in the poor processability of the fibers simultaneously. The BPDA/ODA PI fibers exhibited the tensile strength and initial modulus of 0.8 GPa and 30 GPa with the drawing ratio of 5.5. The fibrils and microvoids could be identified from the fractured cross-sectional morphologies, which had affected the mechanical properties of the PI fibers. The fibers also exhibited excellent thermal stabilities with the T_{d5} around 400 °C in air atmosphere, a negative value of linear coefficient of thermal expansion below 400 °C, and T_g ranged from 276 °C to 297 °C with the frequency increased from 0.1 Hz to 100 Hz.

2.4.2.2 Copolyimide Fibers with Hydrogen Bonding Interactions

The heterocyclic diamine, AAQ or BIA, is a versatile monomer with an asymmetry structure feature, which has strong inter-or intramolecular hydrogen bonding formation capability that strengthens the mechanical properties of resulting fibers. The hydrogen bonding interactions are usually investigated by deconvoluting the Fourier transform infrared (FTIR) spectra in a wave number range of 1700cm^{-1}-1750 cm^{-1}. For instance, the percentage of the hydrogen bonding associated imides in the BPDA/PDA/AAQ co-PI fibers increases with increasing AAQ content and the AAQ moiety has a positive effect on preventing the side reactions during the thermal cycloimidization process [39]. The hydrogen bonding effectively enhanced the mechanical properties of the co-PI fibers, reaching the optimum tensile strength of 2.8 GPa and modulus of 115 GPa, at the AAQ/PDA molar ratio of 5/5. The 5% weight loss temperatures (T_{d5}) of the BPDA/PDA/AAQ fibers under nitrogen were in the range of 599 °C-604 °C, implying that hydrogen bonding interactions also play a positive role in improving the thermal stability of the PI fibers. Moreover, the fibers exhibited gradually enhanced glass transition temperatures (T_g) from 340 °C to 400 °C with increasing AAQ content. The BPDA/PDA/AAQ co-PI fibers possessed a highly ordered structure along the fiber direction with the optimum orientation degree of 0.86. Moreover, the ordered chain repeat length was slightly decreased with the incorporation of quinazolinone moiety in the co-PI fibers, ascribing to the relatively flexible structure of the AAQ unit.

Dong et al. [58] investigated the effects of TFMB/BIA molar ratio on the structures and properties of BTDA/TFMB/BIA fibers. The heterocyclic diamine BIA provided a strong intermolecular hydrogen bonding interaction that strengthened the mechanical properties of the PI fibers, while the diamine TFMB possessed a rigid nonplanar structure with two bulky trifluoromethyl groups, resulting in great benefits for improving solubility of the PIs. Consequently, the final PI fibers exhibited the tensile strength of 2.25 GPa and modulus of 102 GPa with a drawing ratio of 3.0 when the molar ratio of TFMA/BIA was 50/50. The fibers exhibited excellent thermal stabilities in the range of 531 °C-550 °C in nitrogen and 528 °C-544 °C in air for a 5% weight loss. Meanwhile, the improved thermal stability with more BIA contents is related to the increased intermolecular interactions and crystal structures. And the corresponding T_gs of the fibers displayed a clear increasing trend with increasing BIA content, reaching a value of 360 °C. Similarly, the PI fibers derived from BPDA, 4-amino-*N*- (4-aminophenyl) benzamide (DABA) and BIA possessed the tensile strength and initial modulus up to 1.96 GPa and 108.3GPa, respectively, attributed to the well-defined ordered and dense structure. The chemical structure and molecular packing significantly affected the thermal stability of fibers, resulting in gradual increased T_gs from 350 °C to 380 °C with increasing DABA content [59].

It is pointed out that those hydrogen bonding interactions would act as physical cross-linking points and lead to the benzimidazole/imide "mixed layer" packing in which the imide ring of one chain and the (amine) phenyl ring of another chain adopt a coplanar "sandwich" conformation. The crystallinity of the fibers enhanced from 1.8% to 24.5% when the annealing temperatures increased from 390 °C to 400 °C, which was manifested as a sudden crystallization process. The increased crystallinity of the fibers led to the improvement in mechanical properties simultaneously, and the tensile strength increased from 1.31 GPa to 1.68 GPa in the same temperature range.

2.4.2.3 Copolyimide Fibers Containing Ether Units

The incorporation of hydrogen bonding interactions greatly enhanced the mechanical properties of PI fibers. Nevertheless, with the increased rigid units, it is difficult for the polymer chains to rearrange themselves to reduce the size of microvoids and for fiber spinning. On the contrary, the introduction of flexible units into the rigid polymer backbone can provide opportunities to reduce the size of microvoids because of the increased flexibility of polymer chains. For instance, the fibers containing ODPA moieties exhibited the tensile strength and initial modulus in the range of 1.07GPa-1.58 GPa and 50.39 GPa-67.75 GPa, respectively. The optimum mechanical properties of the PI fibers containing ODA contents were obtained with the tensile strength of 2.53 GPa at a PDA/ODA molar ratio of 5/5. The BPDA/PDA/ODA fibers exhibited thermal stabilities of up to 563 °C in nitrogen and 536 °C in air for a 5% weight loss and T_gs above 270 °C The average radius, length, misorientation, and internal surface roughness of microvoids were found to decrease with increased ODPA or ODA contents, suggesting the gradually formed homogenous structures in the fibers and accounting for the improved mechanical properties. However, the values of T_{d5} both in nitrogen and air reduced with increasing ODPA or ODA content, which is attributed to the increased C-O bonds in the polymer chains because the C-O bonds are much more easily broken than C-C bonds. And the values of T_g and tan delta exhibited a clear decreased and increased tendency, respectively, due to the increased flexibility of the polymer chain.

The effect of the BIA and ODA moieties on the properties of the PI fibers was systematically investigated. The results showed that both the introduction of BIA and ODA could improve the mechanical properties of PI fibers, but the improvement induced by BIA moieties was larger than that of ODA moieties. For instance, the fibers with the PDA/BIA/ODA molar ratio of 6/1/3 possessed the tensile strength and initial modulus of 2.19 GPa and 60.85 GPa, while the value increased to 2.72 GPa and 94.33 GPa when the molar ratio of p-PDA/BIA/ODA was 6/3/1, respectively. Due to the different mechanisms in improving the mechanical properties of PI fibers, the fibers with more ODA moieties exhibited the more disrupted lateral molecular packing in the fibers and reduced thermal-oxidative stabilities. However, the case was on the contrary for the BIA moiety. The fibers exhibited excellent thermal and thermal-oxidative stability, with a T_{d5} of 578 °C in nitrogen and 572 °C in air.

2.4.2.4 Copolyimide Fibers Containing Benzoxazole Units

The heterocyclic benzoxadazole moiety has an analogous chemical structure to benzimidazole and also shows unique properties. Inspired by the fact that Zylon fiber (PBO) containing benzoxazole units possesses very high tensile strength of 5.8 GPa, modulus of 270 GPa, excellent thermal stability as well as good environmental resistance, the introduction of benzoxazole units is expected to enhance the mechanical and thermal properties, as well as decrease the water absorption of the PI fibers. Huang et al. introduced 2-(4-aminophenyl)-5-aminobenzoxazole (BOA) into the BPDA/*p*- PDA polymer backbone [60]. The mobility of the polymer chains was improved with the incorporated third monomer and favored large drawing during the thermal treatment process and reduced the influence of the voids. Therefore, the mechanical properties were subtly improved, reaching the tensile strength of 1.0 GPa and modulus of 92 GPa. However, further addition of the third monomer reduced the stacking density of the polymer chains oppositely, which was not conducive to the improvement in the mechanical properties of PI fibers.

Yin et al. [61] investigated the effects of different molar ratios of BIA/BOA on the performances of the BPDA/BIA/BOA PI fibers. The optimal mechanical properties of co-PI fibers were obtained when the molar ratio of BIA/BOA was 7/3, with a tensile strength and modulus of 1.74 GPa and 74.4 GPa, respectively. SAXS results illustrated that addition of BOA accounts for the disappearance of the periodic lamellar structure for homo-BPDA/BIA PI fibers. Co-PI fibers with the molar ratio of BIA/BOA=9/1 and 7/3 have the smallest radius of the microvoids, suggesting the most homogenous structure. The prepared co-PI fibers showed thermal stability with T_gs ranging from 367 °C to 302 °C, and T_{d5} in N_2 is 540 °C-530 °C. The corresponding T_gs of the fibers displayed a clear decreasing trend with increasing BOA content. Additionally, the hydrophobic behavior of the composite fibers is greatly enhanced by incorporating BOA content so that the fibers can be employed in the hot and humid environment. Sun et al. investigated the effects of BIA/BOA molar ratios on the properties of BPDA/PDA/BIA/BOA co-PI fibers [62]. Due to the presence of hydrogen bonding interaction and high orientation in the fibers, the mechanical properties of the BPDA/PDA/BIA/BOA fibers were improved when both BOA and BIA were incorporated into the polymer chains. The optimum tensile strength was 2.26 GPa with the BIA/BOA molar ratio of 3/1, whereas the optimum initial modulus was 145.0 GPa with the BIA/BOA molar ratio of 1/3. Moreover, the co-PI fibers exhibited excellent thermal properties with 5% weight loss temperature ranging from 563 °C to 570 °C in nitrogen and glass transition temperature ranging from 308 °C to 321 °C.

2.4.2.5 Copolyimide Fibers Containing Fluorinated Groups

The reduced dielectric permittivity of PI fibers is of significance when occupied in microelectronic industries. There is thus no doubt that the introduction of fluorine groups is one of the most efficient approaches to reducing the dielectric permittivity of the PI fibers due to the low molar polarization ratio of the F atom. The 4,4'-(hexafluoroisopropylidene)diphthalic anhydride (6FDA)/2,2-bis(4-(4-aminophenoxy(-phenyl)-hexa-fluoropropane (4BDAF) fibers had a maximum tensile strength and modulus of 200 MPa and 6 GPa, respectively. A dielectric constant of 2.5 was measured over a frequency range between 1.0 MHz and 1.8 Hz. Also, the BPDA/PDA/TFMB fibers prepared by Huang et al. [41] showed enhanced mechanical properties with increased TFMB contents. It is stated that the mobility of polymer chains could be improved, thereby a larger drawing ratio could be obtained in the heat-treatment stage. Chang et al. [63] have investigated the effects of 6FDA on the properties of the BPDA/6FDA/PDA/ODA/BIA co-PI fibers. The increased 6FDA moieties in the system resulted in unexpected great changes in the properties of the resultant PI fibers. Regarding mechanical performances of the PI fibers, the tensile strength and initial modulus of the fibers decreased from 2.56 GPa to 0.13 GPa and 91.55 GPa to 2.99 GPa, respectively. The PI fibers decomposed at lower temperature with the increased 6FDA contents, which was attributed to the decreased stacking density and increased free volume of the polymer chains induced by the introduction of $-CF_3$ pendant group. The dielectric permittivity was found to decrease from 3.46 to 2.78 in the frequency of 10 MHz with the increased 6FDA contents, associated with the increased free volume of the polymer chains and macrovoids structures in the fibers.

2.4.3 Properties of Polyimide Fibers

2.4.3.1 Mechanical Properties

Tensile strength and modulus of PI fibers mainly depend on the chemical structure, molecular weight, orientation degree, and crystallinity, as well as the dispersion of defects of the fibers. At present, the tensile strength and modulus of heat-resistant PI fibers are in the range of 0.5 GPa-1.0 GPa and 10 GPa-40 GPa, respectively. However, for the high-strength and high-modulus PI fibers, the tensile strength is over 2.5 GPa and the modulus is as high as 90 GPa. Fig. 2.22 compares the tensile properties between PI fibers containing pyridine units and several commercial polymeric fibers [64]. Apparently, PI fibers show higher specific strength and specific modulus than the well-known Kevlar 29, Twaron fibers and so on, and the specific strength is even higher than Kevlar 49, which indicates a wide application prospect in the area of composite materials.

2.4.3.2 Thermal Stability

For the wholly aromatic PI fibers, the initial decomposition temperatures are generally over 500 °C, and a specific PI fiber synthesized from BPDA/p-PDA shows an initial decomposition temperature of over 600 °C, which is one of polymers with the best thermal stability properties.

2.4.3.3 Chemical Resistance

PI fibers show excellent chemical resistance in an acidic environment. Fig. 2.23 shows the change of tensile strengths of three different PI fibers (PI-1, PI-2, and commercial P84 fiber) and the Poly(phenylenesulfide) (PPS) fiber after treatment in a 10%(vol) HCl aqueous solution at different times. It is obvious that the strength retention of PI-1 and PI-2 fibers are as high as 30%-40%, which is much better than the P84 fiber of Evonik. Unfortunately, all PI fibers show poor resistance in an alkaline environment, in which the polymer chains will degrade rapidly. However, this characteristic also brings some benefits for PI: for example, the diamine and dianhydride monomers of PI fibers can be recycled in a base solution.

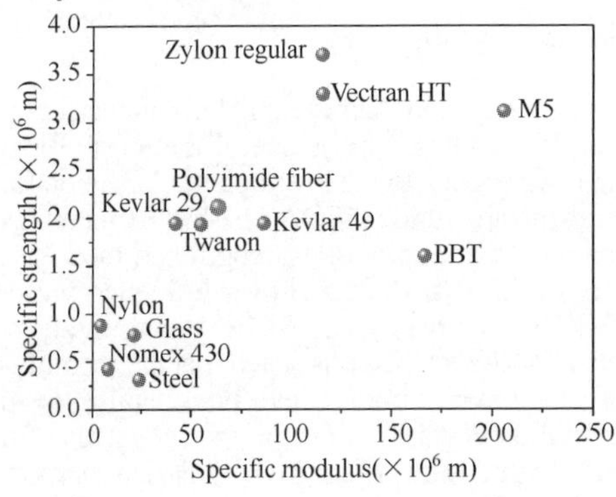

FIGURE 2.22 Comparison of specific strength and modulus of PI fibers containing pyridine units with other polymeric fibers.

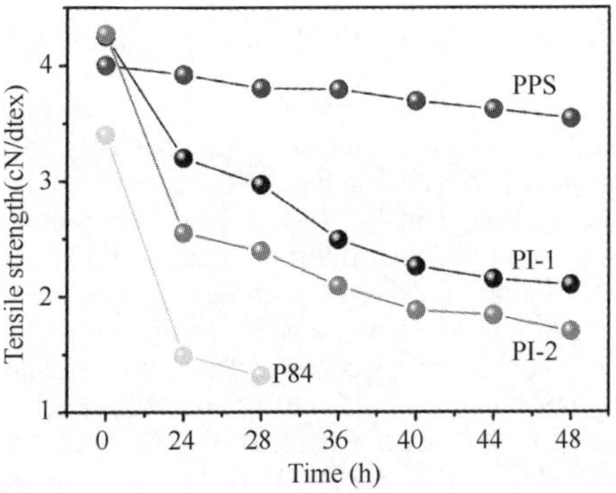

FIGURE 2.23 Tensile strength of PI fibers, P84 fiber and PPS fiber as a function treating time in a 10 %(vol) HCl aqueous solution (PI-1 was the thermal treated fiber provided by Jiangsu Aoshen Hi-tech Materials Co., Ltd, and PI-2 fiber was the as-spun fiber).

2.4.3.4 Flame-Retardant Properties

PI fibers have been regarded as one of the most excellent flame-retardant materials and exhibit outstanding self-extinguishing performance, which can satisfy the flame-retardant requirements in most areas. Meanwhile, the fire-resistant properties of PI fibers can be affected by the chemical structures; for example, the limiting oxygen index (LOI) for PMDA/ODA PI is around 37%, while, for BPDA-p-PDA, the LOI reaches 66%, satisfying the application of PI fibers in some special environments.

2.4.3.5 Other Properties

Most PI fibers possess excellent radiation resistance, and the strength retention can reach 90% after being irradiated under a 1×10^8 Gy fast electron irradiation. Meanwhile, PI fibers have superior dielectric properties and the dielectric constant of the PMDA-ODA PI fiber is around 3.4. By introducing fluorine units or bulky pendant groups, the dielectric constant can be as low as 2.8-3.0, and dielectric loss reaches around 10^{-3}, making PI fibers ideal reinforcements for wave-transparent composites.

2.5 Applications of Polyimide Fibers

2.5.1 Production of Polyimide Fibers

Currently, there are several companies focusing on the development of advanced PI fibers in China, including Jiangsu Aoshen Hi-tech Materials Co., Ltd, Changchun Polyimide Materials Co., Ltd, and Jiangsu Shino New Materials & Technology Co., Ltd. Of these, Jiangsu Aoshen Hi-tech Materials Co., Ltd and Donghua University jointly developed a new technology to prepare PI fiber through dry-spinning, building the first dry-spinning PI fiber production line with annual output of 1000 tons. The products, branded Suplon PI fibers, include filaments, staple fibers, chopped fibers, colorful yarn, and so on. Changchun Polyimide Materials Co., Ltd cooperated with Changchun Institute of Applied Chemistry Chinese Academy of Sciences, and mainly used the wet-spinning technology to produce PI fibers. A series of Yilun products are commercially available, including PI filaments, staple fibers, chopped fibers, and so on. Additionally, Jiangsu Shino New Materials Technology Co., Ltd. have been focusing on the high-performance PI fibers supported by Beijing University of Chemical Technology. A series of Shilon products have been manufactured via an integrated continuous two-step wet-spinning method, as shown in Fig. 2.24. The tensile strength and modulus of the high-performance PI fibers are measured up to 3.5 GPa and 140 GPa, respectively.

FIGURE 2.24 PI fibers with high strength and high modulus.

2.5.2 Application of Polyimide Fibers

PI fiber has been considered as one of the most important high-performance fibers, with the development of PI synthesis technology and fiber spinning technology, the process of industrialization for PI fibers with outstanding mechanical properties, good thermal stability, and radiation resistance gradually accelerated, playing a more and more important role in the fields of high-temperature filtration, special protection, aerospace, national defense, new building materials, environmentally friendly and fireproof applications, and so on. Fig. 2.25 depicts several PI fiber products, including a house-bag duster for high temperature, high-temperature roller, high-temperature- resistant fabric, etc.

1. *High-temperature PI filter bags*. PI fibers are applied in environment-protection areas mainly as high-temperature filtration materials. The total annual sales of house-bag dusters are more than 30 billion yuan, requiring around 1×10^8 m^2 filter materials. At present, high-performance filter materials are insufficient and expensive, and product qualities are relatively low with short service life and low dedusting efficiency, which cannot meet the increasingly stringent environmental requirements. House-bag dusters produced with PI fibers can effectively catch dust grains, and PI fibers with special cross-sections can make dust gather on the surface of filtration materials, which can avoid pore block, improving the filter efficiency. In addition, PI fibers are the best high-temperature filtration materials due to their outstanding chemical resistance, acid resistance, and excellent thermal stability, and they are widely employed in the fields of thermal power generation, cement production, metal smelting, and so on, playing an important role in protecting the environmental pollution arising from PM 2.5 and PM 10.

2. *Special protection*. PI fibers show very low thermal conductivities. For example, the coefficient of thermal conductivity is around 0.03 $W \cdot m^{-1} \cdot K^{-1}$ at 300 °C. The LOI of PI fibers is in the range of 35%-50%, and PI can be self-extinguishing, incombustible, non-melting, and produces no smoke when on fire, which can significantly improve the safety and fighting ability of fire-fighters when used in the area of fire control. Meanwhile, these fibers have excellent fire-resistant properties and high thermal stabilities, thus, PI fibers can be utilized to fabricate heat-resistant workwear, fiber blankets, and flame-retardant textiles that can be used by firemen and metallurgical workers.

3. *National security*. Compared with other polymeric fibers, the PI fiber exhibits excellent thermal stability, radiation resistance, low water absorption, low density, and so on, making it a good candidate for application in the fields of power, national defense, and aerospace.

4. *High-temperature insulting papers and structural materials*. Papers made from PI fibers show a better combination performance than the Aramid papers, and they can be used as insulating materials in power generators and dry type transformers. Honeycomb-structure composite materials fabricated by PI fibers can be used to prepare radar jammers, cabin, and aviation lightweight plates.

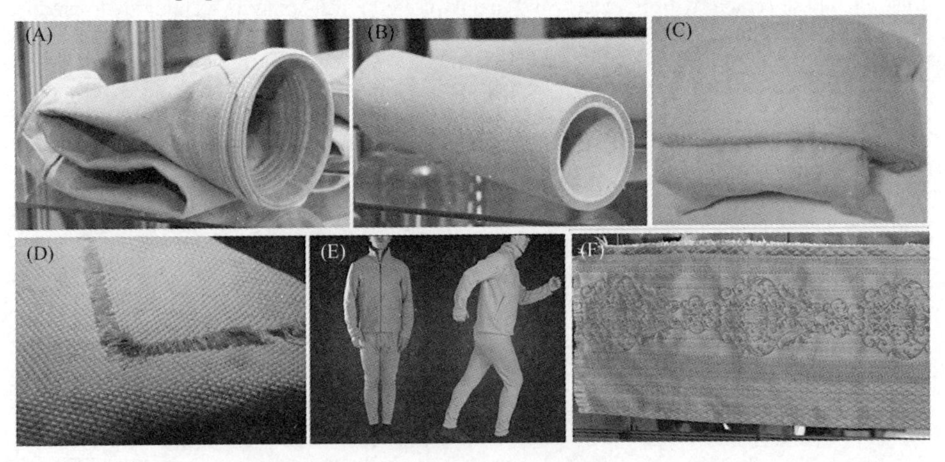

FIGURE 2.25 Typical products of polyimide fibers prepared by Jiangsu Aoshen Hi-tech Materials Co., Ltd. (A) High-temperature filtration bag; (B) high-temperature roller; (C) high-temperature-resistant batting; (D) polyimide clothing; (E) polyimide fleece jackets; (F) flame-retardant carpet.

FIGURE 2.26 PI-fiber-reinforced composite materials.

5. *PI-fiber-reinforced composites*. The PI fibers with high-strength-high-modulus are used as reinforcements in the preparation of composite materials. The PI-fiber-reinforced epoxy resin composite exhibits a bending strength and modulus higher than 650 MPa and 43 GPa, respectively, promising it to be one of the preferred materials used in supersonic aircraft, aerospace equipment, nuclear facilities, high-temperature electrical insulation, bullet-proof vests, and other high-tech

industries. Fig. 2.26 shows the high-strength-high-modulus fiber-reinforced composite materials.

6. *PI fiber ropes*. The PI fiber ropes can be obtained by merging and knitting PI fibers under constant tension. The PI fiber ropes can be applied in space vehicles, nuclear industry fire-proofing materials due to their outstanding mechanical properties, lightweight, low creep, high radiation resistance, and good wear resistance.

2.6 Summary

As a new kind of high-performance polymeric fiber material, PI fiber has attracted more and more attention due to its excellent combined properties. Currently, PI fibers are mainly used as high thermally resistant materials. With the rapid development in monomer synthesis and fiber-spinning technology, higher and higher strength and modulus PI fibers will be developed and produced on a large scale, enabling them to be used to fabricate advanced composites and other high-tech fields. The development of high-strength and high-modulus PI fibers will lead a new round of high technology revolution in the high-performance polymer fiber field.

REFERENCES

[1] D.J. Liaw, K.L. Wang, Y.C. Huang, K.R. Lee, J.Y. Lai, C.S. Ha, Advanced polyimide materials: syntheses, physical properties and applications, Prog. Polym. Sci. 37 (7) (2012) 907-974.

[2] R. Irwin, W. Sweeny, Polyimide fibers, J. Polym. Sci. Polym. Symp. 19 (1967) 41-48. Wiley Online Library.

[3] Q.H. Zhang, M. Dai, M.X. Ding, D.J. Chen, L.X. Gao, Mechanical properties of BPDA-ODA polyimide fibers, Eur. Polym. J. 40 (11) (2004) 2487-2493.

[4] M.E. Walter, Aromatic polyimides and the process for preparing them, Google Patents, 1965.

[5] V. Gabara, High-performance fibers, Ullmann's Encyclopedia of Industrial Chemistry, 1989.

[6] H.B. Xiang, Z. Huang, L.Q. Liu, L. Chen, J. Zhu, Z.M. Hu, Structure and properties of polyimide (BTDA-TDI/MDI co-polyimide) fibers obtained by wet-spinning, Macromol. Res. 19 (7) (2011) 645-653.

[7] G. Bhat, R. Schwanke, Thermal properties of a polyimide fiber, J. Therm. Anal. Calorim. 49 (1) (1997) 399-405.

[8] V. Svetlichnyi, K. Kalnin'sh, V. Kudryavtsev, M. Koton, Charge transfer complexes of aromatic dianhydrides, Doklady. Akademii. Nauk. SSSR (Engl. Transl.) 237 (1977) 612-615.

[9] V. Zubkov, M. Koton, V. Kudryavtsev, V. Svetlichnyi, Quantum chemical-analysis of reactivity of aromatic diamines in acylation by phthalicanhydride, Zhurnal. Organ. Icheskol. Khimii. 17 (8) (1981) 1682-1688.

[10] C.P. Yang, S.H. Hsiao, Effects of various factors on the formation of high molecular weight polyamic acid, J. Appl. Polym. Sci. 30 (7) (1985) 2883-2905.

[11] L. Frost, I. Kesse, Spontaneous degradation of aromatic polypromellitamic acids, J. Appl. Polym. Sci. 8 (3) (1964) 1039-1051.

[12] N.C. Stoffel, E.J. Kramer, W. Volksen, T.P. Russell, Solvent and isomer effects on the imidization of pyromellitic dianhydride-oxydianilinebased poly(amic ethyl ester)s, Polymer 34 (21) (1993) 4524-4530.

[13] Y. Seo, S.M. Lee, D.Y. Kim, K.U. Kim, Kinetic study of the imidization of a poly (ester amic acid) by FT-Raman spectroscopy, Macromolecules 30 (13) (1997) 3747-3753.

[14] Y. Xu, Q. Zhang, Two-dimensional fourier transform infrared (FT-IR) correlation spectroscopy study of the imidization reaction from polyamic acid to polyimide, Appl. Spectrosc. 68 (6) (2014) 657-662.

[15] C. Pryde, IR studies of polyimides. I. Effects of chemical and physical changes during cure, J. Polym. Sci. Pol. Chem. 27 (2) (1989) 711-724.

[16] J. Dong, Y. Xu, Q. Xia, C. Yin, Q. Zhang, Investigation on cyclization process of co-polyimides containing 2-(4-aminophenyl)-5-aminobenzimidazole units, High Perform. Polym. 26 (5) (2014) 517-525.

[17] M.J. Brekner, C. Feger, Curing studies of a polyimide precursor, J. Polym. Sci. Pol. Chem. 25 (7) (1987) 2005-2020.

[18] L. Shibayev, S. Dauengauer, N. Stepanov, L. Chetkina, N. Magomedova, V. Bel'skii, Effect of hydrogen bonds on the solid phase cyclodehydration of polyamic acids, Polym. Sci. USSR 29 (4) (1987) 875-881.

[19] T. Kaneda, T. Katsura, K. Nakagawa, H. Makino, M. Horio, High-strength-high-modulus polyimide fibers I. One-step synthesis of spinnable polyimides, J. Appl. Polym. Sci. 32 (1) (1986) 3133-3149.

[20] J.Y. Park, D. Kim, F.W. Harris, S.Z. Cheng, Phase structure, morphology and phase boundary diagram in an aromatic polyimide (BPDAPFMB)/m-cresol system, Polym. Int. 37 (3) (1995) 207-214.

[21] Y. Sakaguchi, Y. Kato, Synthesis of polyimide and poly (imide-benzoxazole) in polyphosphoric acid, J. Polym. Sci. Pol. Chem. 31 (4) (1993) 1029-1033.

[22] X. Chen, Z. Li, F. Liu, Q. Sun, J. Li, Synthesis and properties of poly (imide-benzoxazole) fibers from 4,40-oxydiphthalic dianhydride in polyphosphoric acid, Eur. Polym. J. 64 (2015) 108-117.

[23] L. Jin, Q. Zhang, Y. Xu, Q. Xia, D. Chen, Homogenous one-pot synthesis of polyimides in polyphosphoric acid, Eur. Polym. J. 45 (10) (2009) 2805-2811.

[24] F. Hasanain, Z.Y. Wang, New one-step synthesis of polyimides in salicylic acid, Polymer 49 (4) (2008) 831-835.

[25] Y. Tsuda, T. Yoshida, T. Kakoi, Synthesis of soluble polyimides based on alicyclic dianhydride in ionic liquids, Polym. J. 38 (1) (2006) 88-90.

[26] I.S. Chung, C.E. Park, M. Ree, S.Y. Kim, Soluble polyimides containing benzimidazole rings for interlevel dielectrics, Chem. Mater. 13 (9) (2001) 2801-2806.

[27] J. Dong, C. Yin, W. Luo, Q. Zhang, Synthesis of organ-soluble copolyimides by one-step polymerization and fabrication of high performance fibers, J. Mater. Sci. 48 (21) (2013) 7594-7602.

[28] J. Dong, C. Yin, Y. Zhang, Q. Zhang, Gel-sol transition for soluble polyimide solution, Polym. Sci. Pol. Phys. 52 (6) (2014) 450-459.

[29] R. Goel, A. Hepworth, B. Deopura, I. Varma, D. Varma, Polyimide fibers: structure and morphology, J. Appl. Polym. Sci. 23 (12) (1979) 3541-3552.

[30] R. Goel, I. Varma, D. Varma, Preparation and properties of polyimide fibers, J. Appl. Polym. Sci. 24 (4) (1979) 1061-1072.

[31] W.E. Dorogy, A.K. Clair, Wet spinning of solid polyamic acid fibers, J. Appl. Polym. Sci. 43 (3) (1991) 501-519.

[32] S.K. Park, R.J. Farris, Dry-jet wet spinning of aromatic polyamic acid fiber using chemical imidization, Polymer 42 (26) (2001) 10087-10093.

[33] C. Yin, J. Dong, Z. Li, Z. Zhang, Q. Zhang, Ternary phase diagram and fiber morphology for nonsolvent/DMAc/polyamic acid systems, Polym. Bull. 72 (5) (2015) 1039-1054.

[34] X. Liu, G. Gao, L. Dong, G. Ye, Y. Gu, Correlation between hydrogen-bonding interaction and mechanical properties of polyimide fibers, Polym. Adv. Technol. 20 (4) (2009) 362-366.

[35] H. Niu, S. Qi, E. Han, G. Tian, X. Wang, D. Wu, Fabrication of high-performance copolyimide fibers from 3, 3′, 4, 4′-biphenyltetracarboxylic dianhydride, p-phenylenediamine and 2-(4-aminophenyl)-6-amino-4 (3H)-quinazolinone, Mater. Lett. 89 (2012) 63-65.

[36] S. Mu, Z. Wu, S. Qi, D. Wu, W. Yang, Preparation of electrically conductive polyimide/silver composite fibers *via* in-situ surface treatment, Mater. Lett. 64 (15) (2010) 1668-1671.

[37] J. Chang, H. Niu, M. Zhang, Q. Ge, Y. Li, D. Wu, Structures and properties of polyimide fibers containing ether units, J. Mater. Sci. 50 (11) (2015) 4104-4114.

[38] J. Chang, H. Niu, M. He, M. Sun, D. Wu, Structure-property relationship of polyimide fibers containing ether groups, J. Appl. Polym. Sci. 132 (34) (2015). Available from: https://doi.org/10.1002/app.42474.

[39] H. Niu, M. Huang, S. Qi, E. Han, G. Tian, X. Wang, High-performance copolyimide fibers containing quinazolinone moiety: preparation, structure and properties, Polymer 54 (6) (2013) 1700-1708.

[40] M. Zhang, H. Niu, J. Chang, Q. Ge, L. Cao, D. Wu, High-performance fibers based on copolyimides containing benzimidazole and ether moieties: molecular packing, morphology, hydrogen-bonding interactions and properties, Polym. Eng. Sci. 55 (11) (2015) 2615-2625.

[41] L. Luo, Y. Wang, J. Zhang, J. Huang, Y. Feng, C. Peng, The effect of asymmetric heterocyclic units on the microstructure and the improvement of mechanical properties of three rigid-rod co-PI fibers, Macromol. Mater. Eng. 301 (7) (2016) 853-863.

[42] T. Sukhanova, Y.G. Baklagina, V. Kudryavtsev, T. Maricheva, F. Lednický , Morphology, deformation and failure behaviour of homo-and copolyimide fibres: 1. Fibres from 4,4′-oxybis (phthalic anhydride)(DPhO) and p-phenylenediamine (PPh) or/and 2,5-bis (4-aminophenyl)-pyrimidine (2, 5-PRM), Polymer 40 (23) (1999) 6265-6276.

[43] T. Kaneda, T. Katsura, K. Nakagawa, H. Makino, M. Horio, High-strength-high-modulus polyimide fibers II. Spinning and properties of fibers, J. Appl. Polym. Sci. 32 (1) (1986) 3151-3176.

[44] M. Eashoo, Z. Wu, A. Zhang, D. Shen, C. Tse, F.W. Harris, High performance aromatic polyimide fibers, 3. A polyimide synthesized from 3,3′,4,4′-biphenyltetracarboxylic dianhydride and 2, 2′-dimethyl-4, 4′-diaminobiphenyl, Macromol. Chem. Phys. 195 (6) (1994) 2207-2225.

[45] I.R. Samuel, S.C. Edgar, Formation of polypyromellitimide filaments, US Patent: 3415782, 1968.

[46] R.S. Irwin, Filament of polyimide from pyromellitic acid dianhydride and 3,4′-oxydianiline, US Patent: 4640972, 1987.

[47] Y. Xu, S. Wang, Z. Li, Q. Xu, Q. Zhang, Polyimide fibers prepared by dry-spinning process: imidization degree and mechanical properties, J. Mater. Sci. 48 (22) (2013) 7863-7868.

[48] G. Deng, S. Wang, X. Zhao, Q. Zhang, Simulation of polyimide fibers with trilobal cross section produced by dry-pinning technology, Polym. Eng. Sci. 55 (9) (2015) 2148-2155.

[49] G. Deng, Q. Xia, Y. Xu, Q. Zhang, Simulation of dry-spinning process of polyimide fibers, J. Appl. Polym. Sci. 113 (5) (2009) 3059-3067.
[50] S. Wang, J. Dong, Z. Li, Y. Xu, W. Tan, X. Zhao, Polyimide fibers prepared by a dry-spinning process: enhanced mechanical properties of fibers containing biphenyl units, J. Appl. Polym. Sci. 133 (31) (2016). Available from: https://doi.org/10.1002/app.43727.
[51] C. Neuber, H.W. Schmidt, R. Giesa, Polyimide fibers obtained by spinning lyotropic solutions of rigid-rod aromatic poly(amic ethyl ester)s, Macromol. Mater. Eng. 291 (11) (2006) 1315-1326.
[52] R.S. Irwin, Polyimide-esters and filaments, US Patent: 4383105, 1983.
[53] J. Wakita, S. Jin, T.J. Shin, M. Ree, S. Ando, Analysis of molecular aggregation structures of fully aromatic and semialiphatic polyimide films with synchrotron grazing incidence wide-angle X-ray scattering, Macromolecules 43 (4) (2010) 1930-1941.
[54] R.F. Saraf, Effect of processing and thickness on the structure in a polyimide film of poly (pyromellitic dianhydride-oxydianiline), Polym. Eng. Sci. 37 (7) (1997) 1195-1209.
[55] M. Zhang, H. Niu, Z. Lin, S. Qi, J. Chang, Q. Ge, et al., Preparation of high performance copolyimide fibers via increasing draw ratios, Macromol. Mater. Eng. 300 (11) (2015) 1096-1107.
[56] Y. Obata, K. Okuyama, S. Kurihara, Y. Kitano, T. Jinda, X-ray structure analysis of a wholly aromatic polyimide with a rigid-rod chain, Macromolecules 28 (5) (1995) 1547-1551.
[57] C. Yin, J. Dong, W. Tan, J. Lin, D. Chen, Q. Zhang, Strain-induced crystallization of polyimide fibers containing 2-(4-aminophenyl)-5-aminobenzimidazole moiety, Polymer 75 (2015) 178-186.
[58] J. Dong, C. Yin, Z. Zhang, X. Wang, H. Li, Q. Zhang, Hydrogen-bonding interactions and molecular packing in polyimide fibers containing benzimidazole units, Macromol. Mater. Eng. 299 (10) (2014) 1170-1179.
[59] C. Yin, J. Dong, Z. Zhang, Q. Zhang, Structure and properties of polyimide fibers containing benzimidazole and amide units, J. Polym. Sci. Polym. Phys. 53 (2014) 183-191.
[60] S. Huang, Z. Gao, X. Ma, X. Qiu, L. Gao, The properties, morphology and structure of BPDA/PPD/BOA polyimide fibers, e-Polymers 12 (1) (2012) 990-1002.
[61] C. Yin, J. Dong, D. Zhang, J. Lin, Q. Zhang, Enhanced mechanical and hydrophobic properties of polyimide fibers containing benzimidazole and benzoxazole units, Euro. Polym. J. 67 (2015) 88-98.
[62] M. Sun, J. Chang, G. Tian, H. Niu, D. Wu, Preparation of high-performance polyimide fibers containing benzimidazole and benzoxazole units, J. Mater. Sci. 51 (6) (2016) 2830-2840.
[63] J. Chang, W. Liu, M. Zhang, L. Cao, Q. Ge, H. Niu, et al., Structures and properties of polyimide fibers containing fluorine groups, RSC Adv. 5 (87) (2015) 71425-71432.
[64] F. Gan, J. Dong, W. Tan, D. Zhang, X. Zhao, Q.H. Zhang, Fabrication and characterization of co-polyimide fibers containing pyrimidine units, J. Mater. Sci. 52 (16) (2017) 9895-9906.

FURTHER READING

M.X. Ding, Polyimides: Chemisty, Relationship Between Structure and Properties and Materials. Science Press, Beijing , 2006, pp. 37-38.

W. Volksen, P. Cotts, D. Yoon, Molecular weight dependence of mechanical properties of poly (p, p'oxydiphenylene pyromellitimide) films. J. Polym. Sci. Pol. Phys. 25 (12) (1987), 2487-2495.

Chapter 3

Polyimide Matrices for Carbon Fiber Composites

Shi-Yong Yang and Mian Ji
Institute of Chemistry, Chinese Academy of Sciences, Beijing, China

3.1 Introduction

In the past 30 years, advanced carbon fiber composites have been extensively used in many high-tech fields such as aerospace and aviation industries. The major driving force is to reduce the structural weight due to the composite's high stiffness/weight and high strength/weight ratios, as well as the potential in lowering part cost. For instance, 1 kg of weight saved in a Boeing 747 is worth about US $450, while a similar weight saved on the space shuttle is worth more than US $30,000 [1].

Carbon fiber composites consist of carbon fiber as reinforcing fibers and thermoset or thermoplastic polymer resin as matrices. The important resin matrices currently include thermoset epoxy (Ep)-based resins, thermoset bismaleimide (BMI)-based resins, and thermoset polyimide (PI)-based resins. The carbon composites are made from woven fabrics or parallel arrays of unidirectional fibers (tapes). Such fabrics and tapes are usually preimpregnated with the matrix resins by either hot-melt or solution-coating processes.

In the hot-melt process (dry method), the matrix resin is rolled into a thin film of desired thickness to provide the correct fiber/resin ratio. The reinforcement fiber and resin film are then joined and passed between heated rollers to saturate the bundles of fibers, producing a prepreg. In solvent coating (wet method), fibers or woven fabric are passed through baths containing matrix resin solution in a low-boiling-point solvent. The preimpregnated fabric is then passed through heated ovens to remove most of the solvent. The prepreg is rolled at the end of the line with a release paper interleaf.

For fabrication of carbon fiber composites, the individual prepreg plies were laid up to the designed orientation. In this process, the physical characteristics of the prepreg such as tack and drape, etc., are important. Additionally, the shelf life (out-time) of the prepreg must be long enough to ensure the lay-up of large complex parts before resin viscosity becomes too high. The prepreg stack can be thermally cured either in a hot press or in an autoclave. In the former method, the matching tool-steel male and female dies are closed to form a cavity of the shape of the component. The prepreg stack is placed in the lower mold section, which was then covered by the upper one. The two halves of the mold are brought together in a press. Usually, the prepreg stack is "staged" in a vacuum oven or at lower temperature to remove residual solvent or volatiles produced in the chemical reactions prior to placing in the mold.

In the autoclave method, the lay-up is consolidated against an open die or mold, then a vacuum bag is formed over the lay-up and mold surface, allowing simultaneous application of both pressure and vacuum. The thermal curing process of the composite component is usually proprietary and complex. Hence, precise control of the applied temperature and pressure are very important. All of the fibers must be "wetted out" by the melt resin, and the laminate must be consolidated before cross-linking occurs. Otherwise, high porous laminate with poor mechanical properties will be produced.

Advanced Polyimide Materials. DOI: https://doi.org/10.1016/B978-0-12-812640-0.00003-2

The epoxy-based carbon composites (C_f/Ep) have dominated the aerospace and aviation industries due to their easy process parameter control for large composite component manufacturing and low production costs. However, the long-term servicing temperature of C_f/Ep composites is limited to lower than 180 °C. Currently, there are many practical applications in aerospace and aviation industries requiring the long-term use of temperatures in excess of 180 °C, such as military aircraft, missiles, and aero-engines. Hence, BMI resin-based carbon fiber composites (C_f/BMI), due to its long-term servicing temperature of 230 °C (450°F), has offered a "half-way house" in temperature performance between epoxies and polyimides. Moreover, Cf/BMI composite components can be fabricated using epoxy-like conditions without evolution of void-producing volatiles. Thermoset polyimides-based carbon composites (C_f/PI) exhibit long-term servicing temperatures of over 300 °C (600°F), having the ability to be fabricated by hot-press, autoclave, resin transfer molding (RTM), etc. The major applications are in aero-engines and military aircraft. Fig. 3.1 compares the long-term servicing performance of C_f/Ep, C_f/BMI and C_f/PI composites.

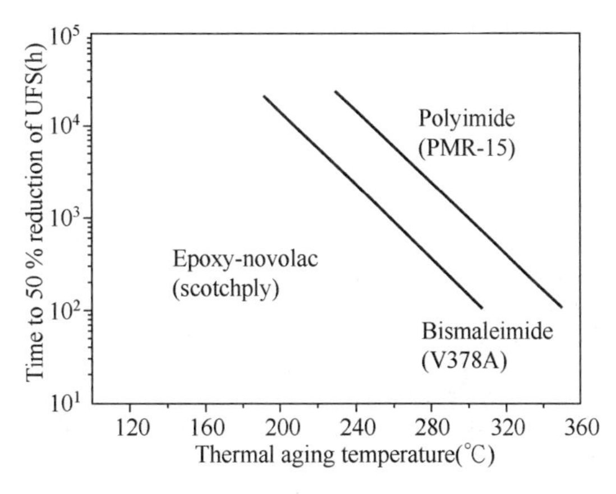

FIGURE 3.1 Comparison on the long-term servicing life of three carbon fiber composites at different temperatures.

However, production of high-quality composite components for high-temperature applications is very difficult due to the poor processing properties of the matrix resins. In order to obtain high-quality, low-void-content components or parts, the matrix resins must have appropriate processing characteristics, including the melt resin viscosity and stability as well as the processing temperature windows. Hence, a large amount of research work has been performed on improving the melt processing properties of thermoset polyimide matrix resins.

Thermoset polyimide resins are usually reactive endcapped prepolyimide resins with controlled molecular weights, which can be prepared from the condensation of the aromatic dianhydrides (or its diester derivates) and aromatic diamines in the presence of reactive endcapping agents. The reactive endcappers are usually mono-anhydride compounds with a thermally sensitive functional group susceptible to polymerization, copolymerization, or cross-linking. Examples of the reactive endcapping agents include maleic anhydride (MA), 5-norbornene-2,3-dicarboxylic acid (NA), 4-phenylethynyl- phthalic anhydride (PEPA).

MA is used to produce the building blocks of BMI resins, BMI. A typical BMI building block is 4,4-bismaleimidodiphenylmethane (Fig. 3.2) produced from the reaction of an aromatic diamine and MA. Thermal curing occurs through the maleimide double bonds without the evolution of organic volatiles at temperatures of 150-250 °C, producing a highly cross-linked and brittle product. Monomeric maleimides are usually employed for the BMI resin formulation in order to improve the resin's processing characteristics. Detailed descriptions of the BMI resins can be found in the Ref. [2].

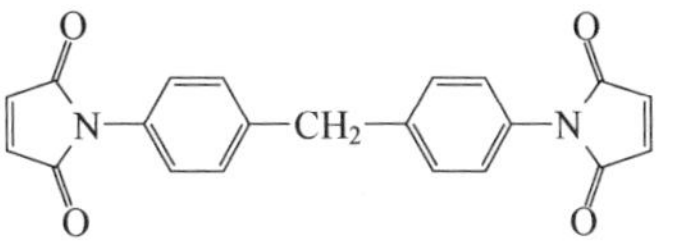

FIGURE 3.2 Chemical structure of a representative BMI building block used for advanced C_f/BMI composite.

In the following sections, NA and PEPA-endcapped thermoset polyimides will be described, especially in the synthesis, characterization and processing properties and composite properties.

3.2 NA-endcapped Thermoset Matrix Resins

NA-endcapped prepolyimide resins are the leading resin systems for long-term servicing temperature of over 316 °C (600°F), the most successful matrix resin is PMR-15. Since its initial development, large sums of R&D funds have been committed to its successful commercialization. This section will describe its chemistry, properties, applications, and drawbacks.

3.2.1 Chemistry

NA-endcapped prepolyimide resins are prepared by the polymerization of monomeric reactants (PMR) approach originally developed by NASA Lewis Research Center [3-6]. The PMR resins are produced by the reaction of aromatic tetracarboxylic acid diester and aromatic diamine using norbornene dicarboxylic acid monoester (NE) as the reactive endcapping cross-linker (Fig. 3.3). The solvents used are the low-boiling-point aliphatic alcohols, such as methanol, ethanol, isopropyl alcohol, etc., which can be easily removed at relative low temperature. The impregnating properties of the resin solutions on the carbon fiber surface can be adjusted by the calculated molecular weight (Calc'd M_w). Hence, the PMR resin is a low-viscosity and high-solid-content solution, highly desirable for impregnating carbon fibers to produce high-quality carbon fiber prepreg. Additionally, this route prevents premature formation of polyamic acids at ambient temperature, allowing the production of resins with long storage life. Thus, PMR-15 is prepared derived from 3,3′,4,4′-benzophenonedicarboxylic dimethyl ester (BTDE), 4,4-diamino-diphenylmethane (MDA) and NE using methanol as solvent, where BTDE and NE are first prepared by the esterification of BTDA and NA in methanol, respectively. And KH-304 resin, developed by Institute of Chemistry, Chinese Academy of Sciences in the late of 1970s, using ethyl alcohol instead of methyl alcohol in PMR-15.

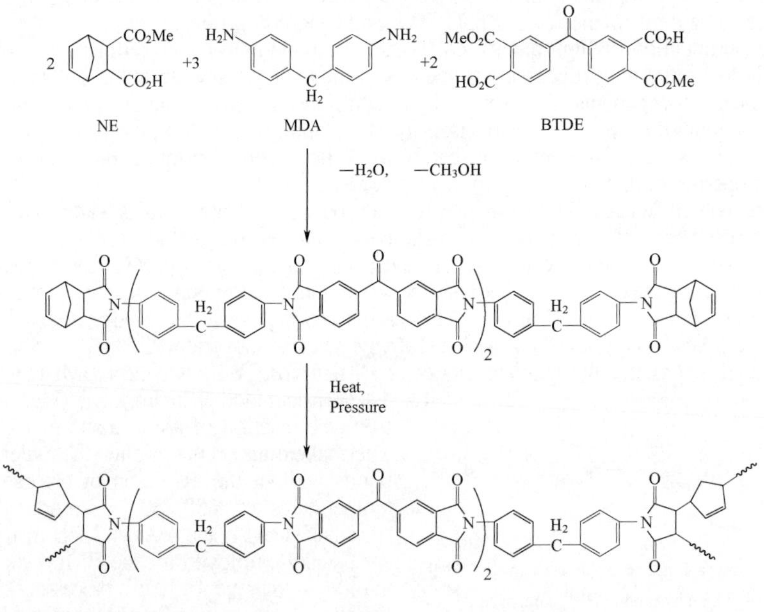

2 NE (CO_2Me, CO_2H) +3 MDA (H_2N, NH_2, CH_2) +2 BTDE (MeO_2C, HO_2C, CO_2H, CO_2Me)

$-H_2O$, $-CH_3OH$

Heat, Pressure

FIGURE 3.3 Chemistry of PMR-15 matrix resins.

PMR is a unique approach to temperature-resistant cross-linked polyimides because the prepolyimide is synthesized in situ during processing of the resin. The resin undergoes a condensation reaction to form the prepolyimide, which, after completion of the condensation reaction, polymerizes to a temperature resistant three-dimensional cross-link network.

This approach eliminated the need for prepolymer synthesis and overcame many of the shortcomings associated with the use of addition-type amide-acid prepolymers. The method involves the use of acid esters rather than anhydrides, thereby preventing the formation of prepolymers at room temperature (RT). The di-alkylester of aromatic tetracarboxylic acid, an aromatic diamine, and the mono-alkylester of 5-norbornene-2, 3-dicarboxylic acid are dissolved in methanol or ethanol and the low-viscosity solution is used to impregnate fibers or fabric to provide a prepreg by removing most of the solvent, which contains the monomeric reactants in almost unchanged form. On subsequent heating to 150 °C-200 °C, the monomers undergo in situ condensation to form the reactive norbornene (NA)-endcapped low-molecular-weight imide prepolymer. The final cure (cross-linking) can be performed at temperatures of 280-320 °C via the reverse Diels-Alder/cross-linking polymerization. The final addition cure reaction is very complex and is still not fully understood. However, the conventional view is that a reverse Diel-Alder reaction of the norbornene endcaps takes place, producing a maleimide end group and cyclopentadiene molecule. Then the maleimide endcap group in situ reacts with the cyclopentadiene molecule to produce a cross-linked polymer with the evolution of only a small amount of organic volatiles.

The aromatic diamines and the aromatic tetracarboxylic diahydrides can be selected widely. However, it was found that the best overall balance of melt processing characteristics, thermo-oxidative stability, and physical properties in laminates is provided by the monomer combination of dimethylester of BTDE, 4,4′-methylenedianiline (MDA) and 5-norbornene-2,3-dicarboxylic acid half ester (NE), in a molar ratio of 2.087:3.087:2. This molar ratio corresponds to a Calc'd M_w of 1500 for the imidized prepolymer.

The temperature profile during the imidization (polycondensation of monomeric reactants) determines the molecular weight distribution of the prepolyimide. The bisnadimide formation was favored at low and ambient temperatures due to the distinct reaction rates of the aromatic diamine with NE and BTDE. The cure reaction of bisnadimides has been extensively studied [7-9]. Cyclopentadiene is released during the thermal cure and can simultaneously react with nadimide to form a 1:1 adduct which subsequently can be involved in addition reactions with maleimide or nadimide. And the reactivity of the nadimide is strongly influenced by the chemical structure of the *N*-substitute. Calorimetric and thermogravimetric analysis [10] indicated that these isomers behaved differently in air than in nitrogen, suggesting different mechanisms of cure depending on the atmosphere. The data are consistent with the proposed reverse Diels-Alder mechanism leading to a loss in cyclopentadiene in nitrogen, and a more direct chain extension without weight loss in air.

3.2.2 Structures and Properties

3.2.2.1 NA-endcapped Polyimides for 316 °C/ 600 °F Applications

In the development of PMR-15, many experiments were designed to optimize both the constituents and the molecular weights. The best overall balance of processing characteristics, composite thermomechanical and physical properties with 316 °C (600°F) thermo-oxidative stability is provided by a monomer combination consisting of BTDE, MDA, and NE in a molar ratio of 2.087:3.087:2. This molar ratio corresponds to a Calc'd M_w of 1500.

However, there are two problems for PMR-15, one is the low toughness of the cured neat resins due to the thermoset feature, which can produce microcracks in the C_f/PMR-15 composite laminates. Early research work indicated that thermal curing of C_f/PMR-15 composite laminate produced microcracks due to the low toughness of the cured thermoset polyimide resins, leading to a reduction in mechanical performance. Fig. 3.4 shows an optical micrograph of a thermal cycled laminate. The residual strength after 1500 cycles (–55 °C/30 min to 232 °C/30 min) is about 40% of initial values for both compressive and interlaminar shear strengths.

FIGURE 3.4 Microcracks of the thermally cycled C_f/PMR-15 laminate.

Another problem is the use of MDA, which is a suspected carcinogen; although MDA has been used in industries for many years, such as an epoxy curing agent, the US authorities have restricted its use in many applications from 2001. Hence, the replacement of MDA by less toxic aromatic diamines while still retaining the overall processing properties became an extremely difficult task.

Both the monomeric constituents and the cross-linking density of PMR-15 type resins were changed, the primary aim was to improve the thermal stability of the carbon fiber composites by replacement of MDA. Earlier studies indicated that no other aromatic diamines tested provided the same combination of processing and thermal properties as MDA except diaminotriphenylmethane (DATPM). The success of DATPM in the replacement of MDA in PMR-15 formulation was thought to be the synergistic combination of two cross-linking sites.

Jigajinni et al. carried out a study on the structure-property relationship in PMR polyimide resins in order to develop MDA-free PMR-15 analogs with novel diaminoarene building blocks [11]. The evaluation of PMR-15 prepolymers and resins derived from diaminoanthraquinones [12] and α,ω-di(3-aminophenyl) alkanes [13] were first reported. Then diamines in triazole, quinoxaline, pyridopyrazine, and pyrazinopyrazine have been used to prepare a series of PMR-15-type polyimide resins. It was found that several resins exhibited higher thermal and thermos oxidative stability than PMR-15 itself, and most of the resins displayed promising physical characteristics.

In 1991, LaRC-RP46 was developed by NASA Langley Research Center [14,15]. It is an MDA-free, low-toxicity PMR polyimide resin for 300 °C long-term servicing applications. Besides low toxicity, this material shows the same or improved characteristics as PMR-15, such as low raw material costs, good processability, and reproducibility, high fracture toughness, and high temperature endurance. LaRC-RP46T is a derivative of LaRC-RP46 in order to further improve the processability, which was prepared by in situ polymerization of three monomer reactants: monoisopropyl ester of 5-norbornene-2,3-dicarboxylic acid (NE); 3,4′-oxydianiline (3,4′-ODA); and diisopropyl ester of BTDE using methanol as solvent (Fig. 3.5).

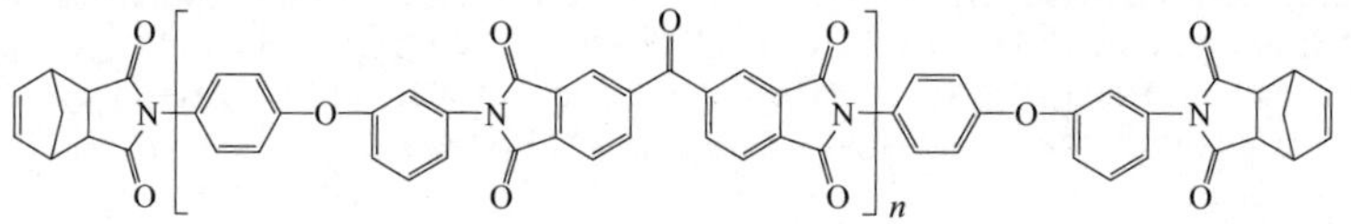

FIGURE 3.5 Chemical structure of LaRC-RP46T resin.

By varying the molecular weight between the cross-link sites (Calc'd M_w) from 1500 to 21,000 g·mol^{-1}, a series of resins with different cross-link densities were produced [16]. Cross-ply composite laminates were compression molded from the LaRC-RP-46T/IM-7 graphite fibers prepreg and were cut into a series of triangular test specimens. The test specimens were then subjected to 1500 thermal cycles (from 0 °C/10 min to 232 °C/10 min) in order to investigate the damages associated with thermal cycling. The extent of the microcrack damage was determined by measuring the number of transverse microcracks and by noting changes in the density and moisture absorption of each specimen.

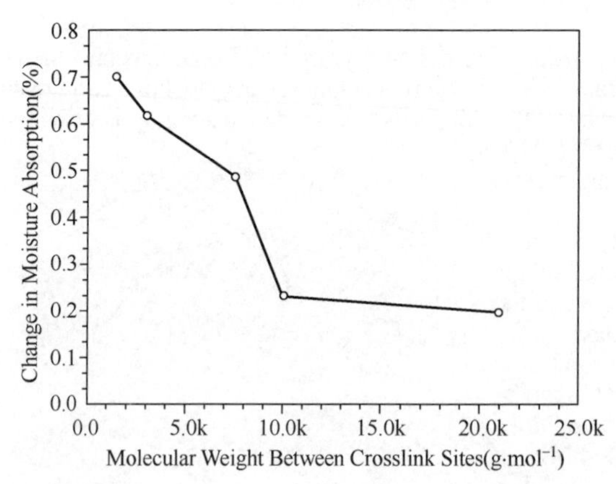

FIGURE 3.6 Changes in moisture absorption versus Calc'd M_w.

Fig. 3.6 shows the changes in moisture absorption versus Calc'd M_w. As the Calc'd M_w increases from 1500 to 10,000 g·mol^{-1}, the moisture absorption decreases from 0.7% to 0.2%, then a plateau appears. Similarly, the microcrack density is also decreased from 70 microcracks·cm^{-2} to 20 microcracks·cm^{-2} with increasing of the Calc'd M_w from 1500 to 7500 g·mol^{-1}, then a plateau is reached at Calc'd M_w of >7500 g·mol^{-1}. Clearly, there is a closed relationship between microcrack densities and moisture absorptions. As moisture absorption is lower than 0.45%, the microcrack density decreases to the lowest level. The experimental results indicate that LaRC-46 shows a better microcracking resistance than commercially available PMR-15 when subjected to the same thermal cycling environments. The microcrack density decreases with increasing the Calc'd M_w. When the Calc'd M_w approaches 10,000 or more, the microcrack density reaches a low plateau.

Xiao et al. prepared a PMR-type thermosetting polyimide with improved impact toughness and excellent thermal and thermos-oxidative stability [17]. It was recognized that improvement in impact toughness might reduce the formation of microcracks in polymer composite laminates. The improved PMR resins were designed by changing the molar ratios of the flexible ether-bridged segment and rigid phenyl group in PMR polyimide backbone using the molecular architectonic technique. A series of improved PMR resins were prepared by in situ polymerization of four monomer reactants: diethyl ester of 3,3′,4,4′-oxydiphthalic acid (ODPE), *p*-phenylenediamine (*p*-PDA), and MDA, monoethyl ester of 5-norbornene-2,3-dicarboxylic acid (NE) using ethyl alcohol as solvent (Fig. 3.7), in which *p*-PDA content in the diamine mixture was designed as 0 (KHPI-1), 33.3% (KHPI-2), 50.0% (KHPI-3), and 66.6%(KHPI-4). It was found that the processability of the B-staged resins could be controlled and improved by adjusting the molar ratio of the rigid phenyl group to the flexible diphenylmethyl segments in the resin. KHPI-2 and KHPI-3 could be easily processed to afford thermoset polyimides with good qualities. The thermally cured KHPI-3 resin showed a glass transition temperatures (T_g) value of as high as 394.6 °C (DMA), giving an Izod nonnotched

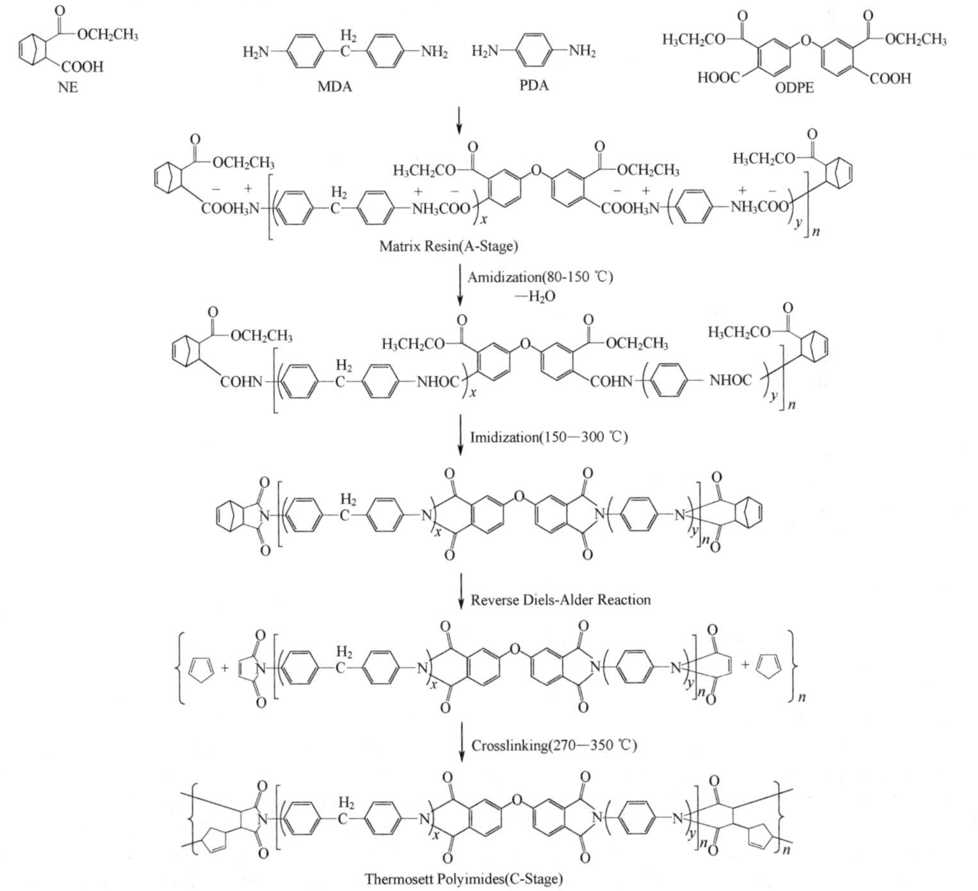

FIGURE 3.7 Chemical and thermal process for KHPI polyimides.

impact strength of 22.8 kg·cm·cm^{-2} and tensile strength of 77.6 MPa, respectively, much higher than that of PMR-15 (impact strength: 4.0 kg·cm·cm^{-2}, tensile strength: 56.8 MPa). Fig. 3.8 compares the thermal and thermo-oxidative stability of the thermoset polyimides at 320 and 350 °C, respectively. In nitrogen, the thermoset polyimides exhibit weight losses of 3.2%-5.6 %(w) after isothermal aging for more than 500 h at 320 °C and 5.8%-6.4 %(w) for 430 h at 350 °C, respectively. In flowing air, the weight losses were measured in the range of 5.4%-8.4 %(w) after 400 h isothermal aging at 320 °C and 12%-18 %(w) after 140 h at 350 °C, respectively. The KHPI-3 exhibits better thermal and thermo-oxidative stability than others.

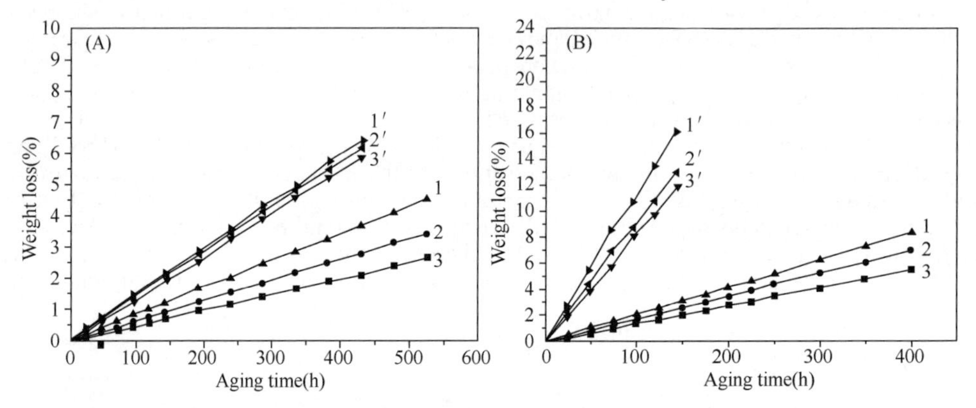

FIGURE 3.8 Thermal and thermo-oxidative stability of the thermoset polyimides at (A) 320 °C and (B) 350 °C with (A) isothermal aging in nitrogen and (B) isothermal ageing in airflow. (1) KHPI-1 at 320 °C; (2) KHPI-2 at 320 °C; (3) KHPI-3 at 320 °C; (1′) KHPI-1 at 350 °C; (2′) KHPI-2 at 350 °C; (3′) KHPI-3 at 350 °C.

3.2.2.2 NA-endcapped Polyimides for 371 °C/700°F Applications

Based on the great success of PMR-15 used for 316 °C applications, the requirement for carbon fiber composites having the ability to resist thermal and thermos-oxidative degradation at higher temperatures (370-400 °C) has driven the intensive research in this area over the past decades because of the many applications in gas turbine engines, missiles, and high-speed civil transportations [18]. Hence, the second generation of PMR-polyimides (PMR-II) have been developed based on diethyl ester of hexafluoroisopropylidene-bis(phthalic acid) (6FDE), NE and *p*-PDA (Fig. 3.9). The PMR-II-50 resin, a reprensatative example of PMR-II, has successfully been used in the manufacturing of advanced aircraft engines [19]. However, PMR-II polyimide composites usually exhibit some shortcomings compared with PMR-15, including relatively poor processability and low retention of mechanical properties at the high servicing temperatures. One of the major concerns for the improvement of mechanical properties at high temperature was the increase in glass transition temperature of the materials and the improvement in proccessability. Hence, dramatic increase in the service temperature without sacrificing processability became a great technical challenge.

Hao et al. [20] prepared a series of thermoset polyimides with high glass transition temperatures and high storage modulus retention by introducing carbonyl groups into the PMR-II polyimide backbone using a molecular architectonic technique to increase the interaction between the polymer chains and to improve polymer morphology while retaining the processability of the material (Fig. 3.10). The thermoset polyimides, prepared by incorporating carbonyl groups into the polymer backbone of PMR-II, exhibit good features as a matrix resin for carbon-fiber composites. The cured polyimide neat resins containing a 1:1 mole ratio of the carbonyl group [—C(O)—] to the hexafluoroisopropyl group [—$C(CF_3)_2$—] in the polymer backbone, show very high glass transition temperature (440 °C) and an excellent thermal decomposition temperature (576 °C). The onset temperature of the storage modulus curve in DMA was measured at about 400 °C. The cured neat resins exhibit outstanding thermal and thermo-oxidative stability at elevated temperature (371 °C) determined by isothermally aging the sample in nitrogen or a flowing air atmosphere (Fig. 3.11). In nitrogen, the original weight losses were only 1.3%-1.8 %(w). after isothermal aging of 550 h. In flowing air, weight losses were measured of 5%-8 %(w) after 220 h of isothermal aging and 8%-12 %(w) after 350 h, indicating excellent thermal stability and thermo-oxidative stability.

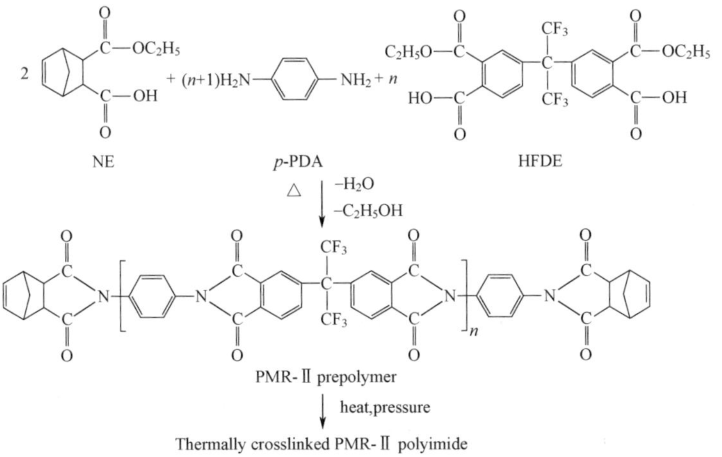

FIGURE 3.9 Chemistry of PMR-II polyimide resins.

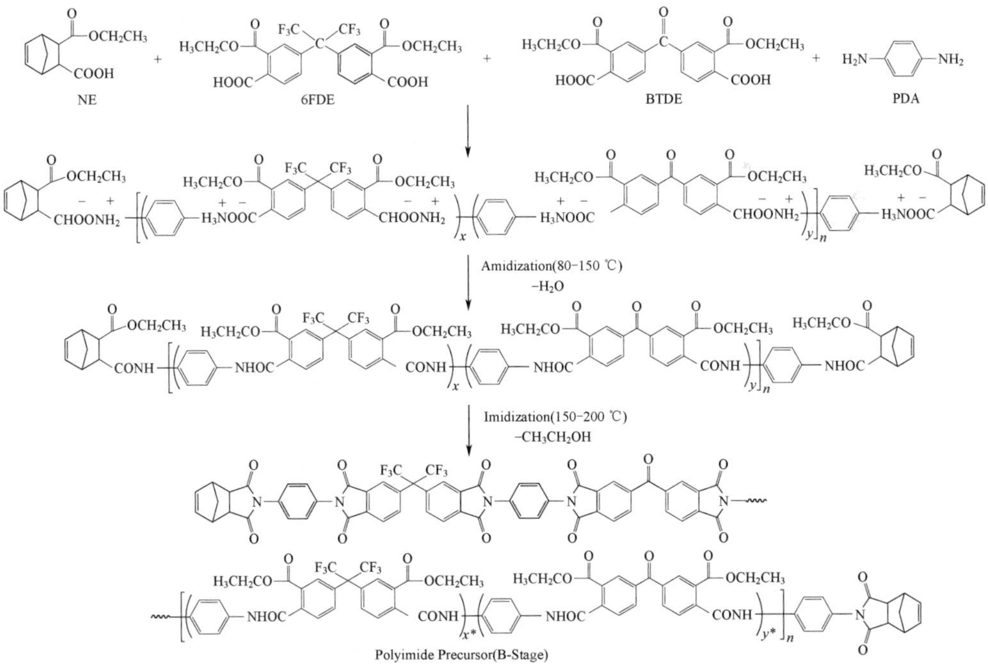

FIGURE 3.10 Chemistry of the improved PMR-II-50 resin.

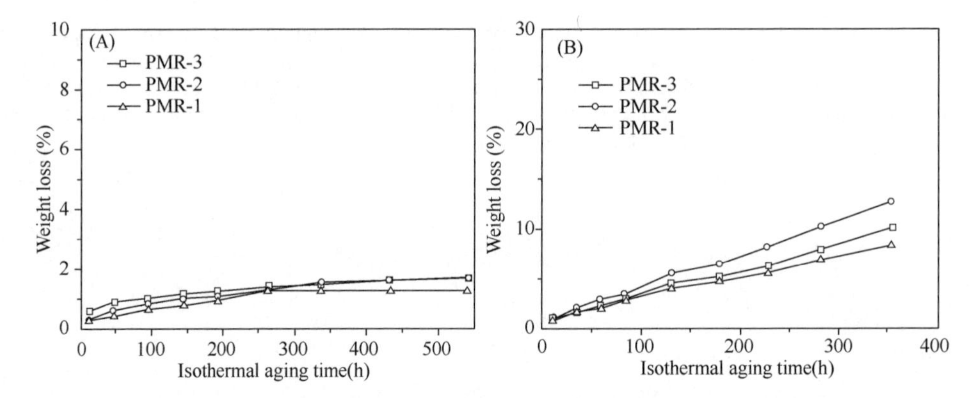

FIGURE 3.11 The thermal and thermo-oxidative stabilities of the improved PMR-II resins.

The carbon fiber composites also showed a high onset temperature of the storage modulus (420 °C) and high storage modulus retention (>85%) at 400 °C, exhibiting excellent mechanical strength retention at 371 °C, compared to that of C_f/PMR-II-50 composite. In order to explore the matrix resins for high-temperature applications, fluorinated PMR thermosetting polyimides have been prepared based on diethyl ester of 4,4-(2,2,2-trifluoro-1-phenylethylidene)diphthalic anhydride (3FDE), similar to 6FDE in chemical structure except that one of the trifluoromethyl groups (CF_3—) in 6FDE was replaced by a stable aromatic phenyl group [21]. Thus, a series of fluoro-PMR-type resin solutions (50%(*w*)) with different Calc'd M_w were prepared by mixing the monomer solutions of NE, 3FDE and *p*-PDA in anhydrous ethyl alcohol at RT, including 3FPMR-15 (Calc'd M_w=1500, *n*=3), 3FPMR-30 (Calc'd M_w=3000, *n*=5), and 3FPMR-50 (Calc'd M_w=5000, *n*=9) (Fig. 3.12).

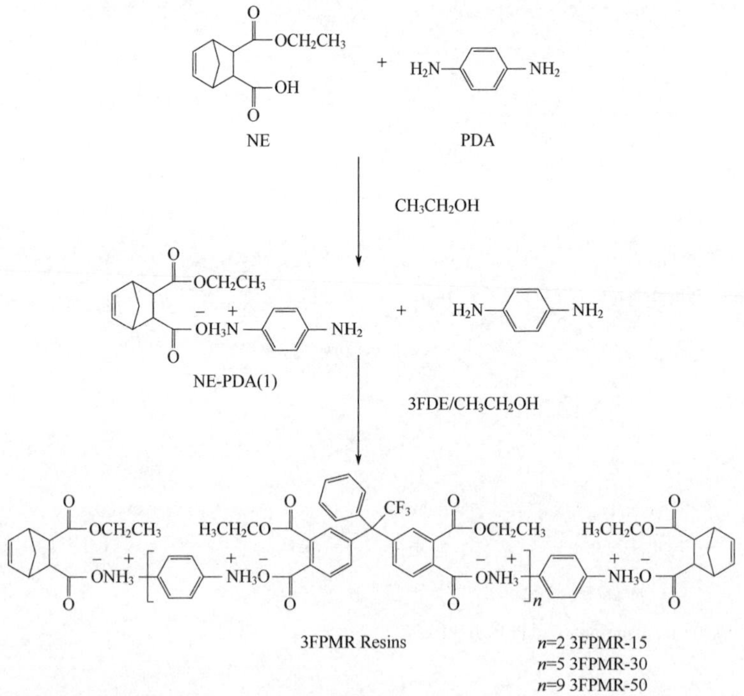

NE + PDA

CH_3CH_2OH

NE-PDA(1) + H_2N—C₆H₄—NH_2

3FDE/CH_3CH_2OH

3FPMR Resins

n=2 3FPMR-15
n=5 3FPMR-30
n=9 3FPMR-50

FIGURE 3.12 Chemistry of the fluorinated 3FPMR polyimide resins.

The homogeneous resin solution with 50%±2% solid contents shows absolute viscosities in the range of 70-85 mPa·s, which is stable for storage of 3-4 weeks at RT and more than 2 months at <4 °C. Neither phase separation nor decrease in viscosity was observed upon storing. The 3FPMR resins exhibit good adhesion to carbon fiber surfaces and make it easy to produce carbon fiber prepregs. The prepregs display good processing charcteristics in the fabrication of composite laminates with high quality.

The thermally cured 3FPMR neat resins at 320 °C followed with postcuring at 371 °C/10 h in nitrogen show T_g values of 375 °C for 3FMR-30, and 370 °C for 3FPMR-50, respectively. About 10 °C of T_g increases was observed if the postcuring time extended from 10 to 20 h. Moreover, 3FPMR cured resins exhibit T_gs of 30-50 °C higher than the corresponding PMR-15 resins with the same Calc'd M_w. It was found that the cured resins with higher Calc'd M_w exhibit lower T_g values. These may be explained by the thermoplasticity of the long linear polymer chains in the higher Calc'd M_w resins. The high T_g values of the 3FPMR resins imply that these materials may be potentially useful for 350 °C long-term applications.

The 3FPMR cured neat resins show excellent thermal stabilities with weight losses of only 4.0%-5.2% of their original weights after isothermal aging at 350 °C for more than 500 h (Fig. 3.13A). The weight loss values increase linearly with the extension of isothermal aging time, averaging about 0.01% of weight loss per hour. It is clear that 3FPMR-50 exhibits better thermal stability than 3FPMR-30. The thermo-oxidative stabilities of the 3FPMR cured neat resins lost 10%-12% of their original weights after isothermal aging at 371 °C for 100 h (Fig. 3.13B), much lower than that of the corresponding PMR-15 resins with the same Calc'd M_w. The 3FPMR-50 always lost less weight than 3FPMR-30 under the same conditions, indicating that the former exhibits better thermo-oxidative stability than the latter. This might be attributed to the lower concentration of the endcaps in 3FPMR-50 due to its high Calc'd M_w.

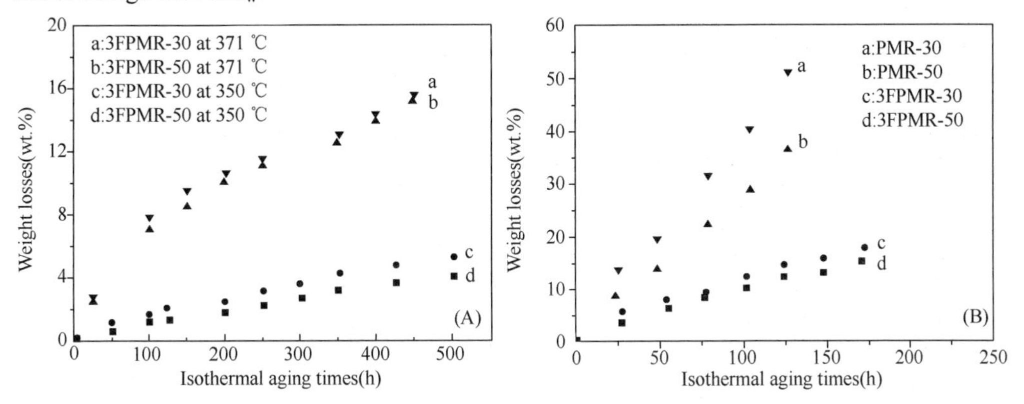

FIGURE 3.13 Thermal and thermos-oxidative stabilities of 3FPMR resins.

In order to improve the processing and mechanical properties of the carbon-fiber-reinforced PMR-II resin composites laminates, Hao et al. [22] prepared a series of mono-endcapped PMR-15-type polyimide resins derived from monoethyl ester of 5-norbornene-2,3-dicarboxylic acid (NE); a mixture of diethyl ester of BTDE and diethyl ester of hexafluoroisopropylidene-bis(phthalic acid) (6FDE), *p*-PDA using ethyl alcohol as solvent. The monomer mixture was stirred at RT for 3-5 h to yield a homogeneous resin solution with solid content of 50 %(w). The mole ratio of NE:(6FDE+BTDE):*p*-PDA in which BTDE/(6FDE+BTDE)=0.50 was 1:n:(n+1), producing a polyimide matrix resin (PMR3-M1) with Calc'd M_w of 5000 (n=10) and BTDE/(6FDE+BTDE) of 0.50. Similarly, PMR1-M1(n=8.0) and PMR2-M1(n=9.0) with Calc'd M_w of 5000 and BTDE/(6FDE+BTDE) of 0.0 and 0.33, respectively, were also prepared.

The matrix resins were low-viscosity and stable homogeneous solutions, possessing good characteristics for carbon fiber prepregs. The thermally cured polyimide materials exhibited glass transition temperature (T_g) of >470 °C and onset temperature of the storage modulus of >440 °C. Continuous carbon fiber composites laminates exhibited high mechanical strength and mechanical property retention at 371 °C, superior to those of commercial materials (CF/PMR-II-50). The thermal and thermo-oxidative stability of the fully cured polyimide resins determined by isothermal aging at 371 °C showed a loss of their original weights of only <3.0 %(w) after isothermal aging for 550 h in nitrogen, and 5.0%-8.0 %(w) after 220 h in air (Fig. 3.14). The weight losses increased almost linearly with increasing isothermal aging time, indicating that the fully cured polyimides possess excellent thermal and thermo-oxidative stability, suitable for 371 °C long-term servicing application.

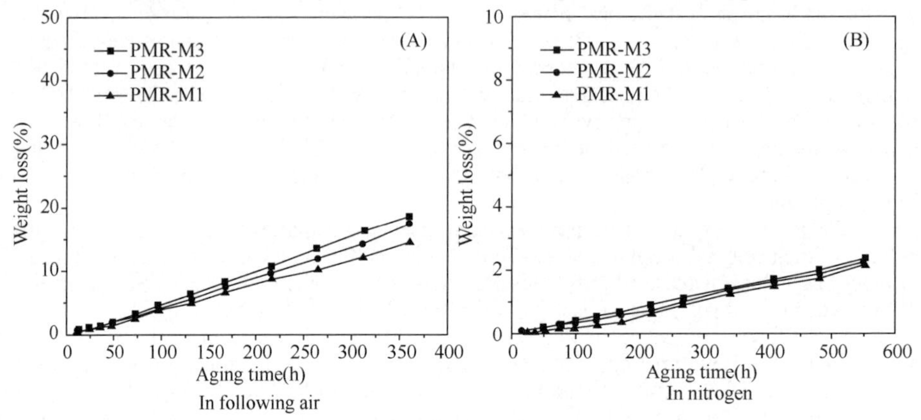

FIGURE 3.14 Thermal and thermos-oxidative stabilities of the thermally cured polyimides.

High-quality unidirectional T300/PMR3-1 composite laminated with void contents of 1%-2% were fabricated by hot press at 350 °C. The nonpostcured laminate showed RT flexural strength of 1057 and 505 MPa at 371 °C, the strength retention was 47.7%; The RT flexural modulus is 106.7 GPa and 99.7 GPa at 371 °C, the modulus retention was 93.4%, superior to that of T300/PMR-II-50 composite (Table 3.1).

Besides polyimides, other heteroaromatic polyimides such as polybenzimidazopyrrole and polypyrrolones have been also considered as promising high-temperature polymer materials because of their excellent thermal and chemical stabilities. However, the poor melt processability caused by the rigidity of the polymer backbones limited their practical applications. The PMR process (in situ PMR) was employed to synthesize endcapped pyrrolone-imide oligomers (PPI) by reaction of 6FDE, NE, and a mixture of *p*-PDA and 3,3-diaminobenzidine (DAB) using ethyl alcohol/*N*-methylpyrrolinone (NMP) mixture as solvent [23]. Thus, PPI-15 with Calc'd M_w of 1500, PPI-20 with Calc'd M_w of 2000, PPI-30 with Calc'd M_w of 3000 were obtained with monomers of 6FDE, NE, a mixture of *p*-PDA and DAB at mole ratio of 2:n:(n+1). The thermally cured neat resins cured at 340 °C/2 h showed excellent thermal stabilities with initial decomposition temperature of >530 °C. After postcuring at 350 °C/10 h in N_2, PPI-20 showed the onset temperatures of the DMA storage modulus curve of >400 °C. It was found that the postcuring caused remarkable increases in modulus retention at elevated temperatures.

Due to the PPI resin's limited solubility in the low-boiling-point solvent, strong bipolar organic solvents such as NMP must be employed as cosolvents. The high-boiling-point NMP was difficult to remove completely in the following procedure, remarkably influencing the quality of the cured polymeric composites as well as their mechanical and thermal performances at high temperatures. Shang et al. prepared a series of novel PMR poly(pyrrolone-imide) resins derived from 6FDE, NE, and a mixture of *p*-PDA and 3,3,4,4-tetraaminodiphenyl ether (TADPE) using low-boiling-point methyl alcohol as solvent [24]. The PPI processability was apparently improved by employing the aromatic tetra-amine monomer, which has a flexible ether bond in the chemical structure. The fully cured PPI at 340 °C/2 h exhibited good thermal stability with T_g of >400 °C and T_d of 525-537 °C, respectively. The onset temperatures of the DMA storage modulus curve for PPI-20 were as high as 395 °C. The short carbon fiber-reinforced PPI composites (C_f/PPI) showed excellent thermal stability and good mechanical properties with T_g of 413-421 °C and the storage modulus retention of 76%-87% at 370 °C.

TABLE3.1 Mechanical Properties of the T300/PMR3-M1 Composites

Mechanical Property	Conditions	T-300/PMR-II-50	T-300/PMR3-M1
Flexural strength (MPa)	RT	1050	1057
	371 °C	398	505
	Retention	37.9%	47.7%
Flexural modulus (GPa)	RT	80.0	106.7
	371 °C	43.0	99.7
	Retention	53.8%	93.4%

The chemical preparation of the thermosetting of NA-endcapped PMR resins has been an area of intensive research activities for many years, but the image of the thermosetting process and thermal degradation mechanism is still not clear due to lack of direct experimental evidence. Jin et al. investigated the thermosetting process and thermal degradation mechanism of the PMR-II resin by high-resolution pyrolysis-gas chromatography-mass spectrometry (HR PyGCMS) in the pyrolysis temperature range of 315-590 °C [25].

PMR polyimide is the reaction product of the three monomers: *p*-PDA, HFDE, and NE. The first step, condensation polymerization, involves heating the monomers at 120-200 °C, which causes imidization to yield a PMR-II prepolymer (A) as shown in Table 3.2. The second step occurs at elevated temperatures (330-390 °C) and allows the norbornene endcaps to cross-link. Hence, the PMR-II samples cured for various temperatures and periods were subjected to pyrolysis in the temperature range of 315-590 °C. The partial TIC pyrograms of samples with different thermosetting stages are obtained by pyrolyzing at temperature of 485 °C for 5 s. The major pyrolyzate of sample A thermally baked at 204 °C is cyclopentadiene, which is the product of a reverse Diels-Alder reaction of the norbornene endcaps of the prepolymer without cross-linking. On the other hand, the pyrolyzates of cyclopentene and small amount of norbornene together with cyclopentadiene were observed in the pyrograms for samples B-G cured at 330-370 °C. These additional pyrolyzates are formed from the thermal scission via carbon-carbon and carbon-carbonyl bond scissions on the polymer ends. But for the samples H, I, J, and K, cured finally at 390 °C, both cyclopentadiene and cyclopentene were not detected in their pyrograms.

The yields of cyclopentene gradually increased to a maximum, then decreased and finally disappeared when the thermosetting temperature and period increased. The pyrolysis of PMR-II prepolymer (A) produced a significant amount of cyclopentadiene, but no cyclopentene was detected, indicating that cyclopentene is not the secondary reaction product of cyclopentadiene. It is confirmed by presence of cyclopentene segments in the cross-linking bridges after their preliminary cure, i.e., cyclopentene is the degradation product of the corresponding cross-linking bridges for the preliminary cross-linked polymer. Therefore, the yield of cyclopentene also reflects the relative cross-linking level on one side.

TABLE3.2 Thermosetting Conditions for PMR II 50 Polyimide Resin

Code	Thermosetting Temperature and Period
A	Prepolymer
B	330 °C/2 h
C	330 °C/2 h + 350 °C/2 h
D	330 °C/2 h + 350 °C/2 h + 370 °C/2 h
E	330 °C/2 h + 350 °C/2 h + 370 °C/2 h
F	330 °C/2 h + 350 °C/2 h + 370 °C/2 h
G	330 °C/2 h + 350 °C/2 h + 370 °C/2 h
H	330 °C C/2 h + 350 °C/2 h + 370 °C/2 h + 390 °C/2 h
I	330 °C /2 h + 350 °C/2 h + 370 °C/2 h + 390 °C/5 h
J	330 °C/2 h + 350 °C /2 h + 370 °C /2 h + 390 °C/10 h
K	330 °C/2 h + 350 °C/2 h + 370 °C/2 h + 390 °C/20 h

The image of the thermosetting process of PMR-II is then presented as shown in Fig. 3.15. In the first thermosetting stage, a reverse Diels-Alder reaction of the norbornene endcaps takes place, resulting in cyclopentadiene and maleimide groups on the end of the macromolecules. The resulting cyclopentadiene was then condensed with the double bonds of maleimide groups or/and unreacted norbornene end groups to produce a cross-linked structure. There are still uncross-linked norbornene end groups in the polymer because cyclopentadiene was identified after the preliminary thermosetting at an elevated temperature. In other words, no cyclopentadiene would be released and the norbornene end groups disappeared for the completely cross-linked polymer.

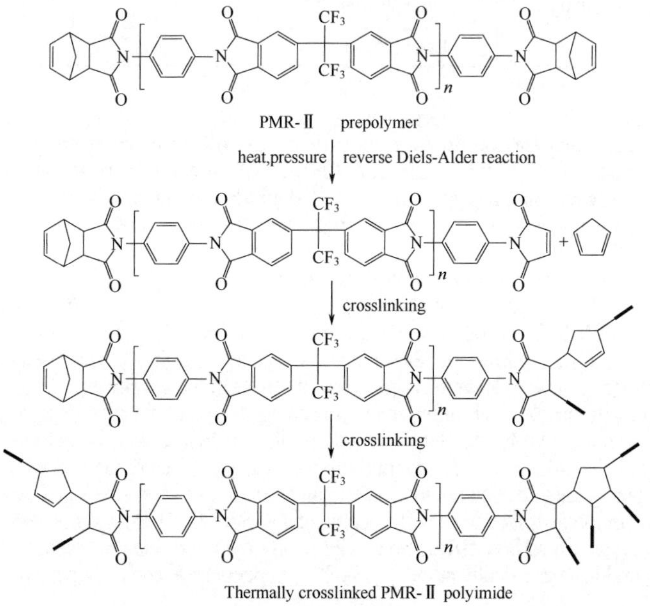

PMR-Ⅱ prepolymer

heat,pressure | reverse Diels-Alder reaction

crosslinking

crosslinking

Thermally crosslinked PMR-Ⅱ polyimide

FIGURE 3.15 Image of the thermosetting process of PMR-II polyimides.

The pyrolysis products of the samples with various curing stages were identified at different pyrolysis temperatures of 315 °C, 386 °C, 485 °C, and 590 °C, respectively. Some characteristic pyrolyzates, including cyclopentadiene and norbornane related to the norbornene end-capped groups were identified in the pyrograms of samples B-G cured at 330-370 °C; however, these pyrolyzetes disappeared in the pyrograms of samples H-K cured at 390 °C, indicating that samples B-G cured at 330-370 °C contained some uncross-linked norbornene segments and those (H-K) cured at 390 °C were cross-linked completely. The cross-linking level from samples B-G cured at 330-370 °C increased gradually, and the pyrolyzates were related to the cross-linking of PMR-II polyimide. It was found that the relative yield of cyclopentene increased with increasing the curing period and reached a maximum in 1.0-2.0 h.

Considering the composition and distribution of the major pyrolyzates of the polymer, it was concluded that all of the uncross-linked and partially cross-linked samples (samples A-G) pyrolyzed at lower temperatures (315 °C and 386 °C) produced some end-group related compounds such as cyclopentadiene and norbornane, and small amounts of the main chain decomposition products, 2,2-diphenylhexafluoropropane, at the pyrolysis temperature of 485 °C. At a higher pyrolysis temperature (590 °C), besides the scission of the norbornene end-capped groups, the scissions of the carbon-nitrogen bonds and carbon-carbon bonds in the imide rings as well as the carbon-carbon bonds along the main chain of the polymer took place. These backbone cleavages gave rise to some characteristic pyrolyzates such as 2,2-diphenyl-hexafluoropropane, 2-phenyl-2-phththalimide-hexafluoropropane, and 2-phenyl- 2-(*N*-phenyl) phththalimide- hexafluoro-propane, reflecting the structural features of the polymer backbone. As a result, the thermal degradation mechanism of PMR-II polyimide is suggested to be as shown in Fig. 3.16.

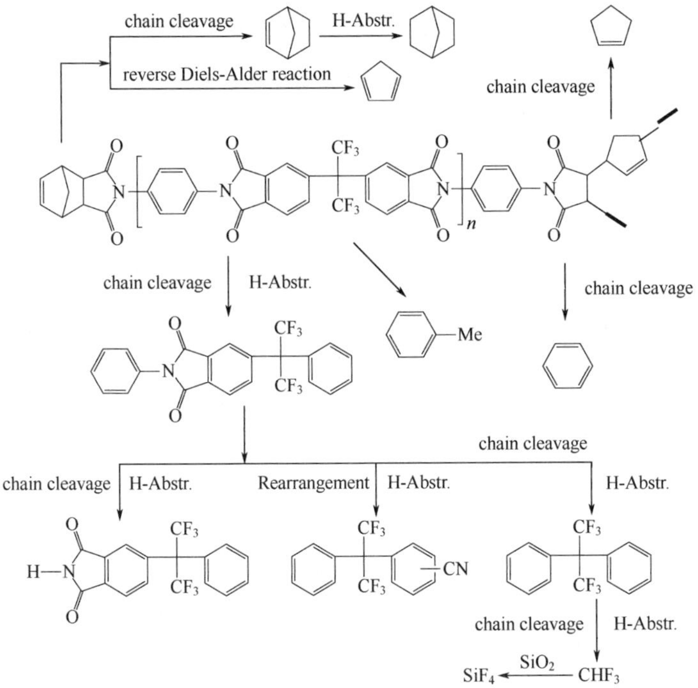

FIGURE 3.16 Thermal degradation mechanism of PMR-II polyimide.

3.3 PE-endcapped Oligoimide Resins

3.3.1 Chemistry

Over the past three decades, ethynyl-endcapped oligomers have been developed and used in various applications. In the 1990s, a large number of research works were directed towards phenylethynyl (PE)-endcapped oligomers, especially PE-endcapped imide oligomers or oligoimides [26,27]. The PE endcap offers distinct advantages over the simple ethynyl endcap. The former is more chemically and thermally stable than the latter, allowing the PE-endcapped imide oligomers to remain unchanged in chemical structures during harsh synthetic conditions, and having great processing windows. The original intensive research was initiated jointly by industry and NASA to develop adhesives and composite matrices for a Mach 2.4 high-speed civil transport (HSCT). These materials must possess a great combination of processing, mechanical, and thermal properties. Good handlability, long shelf life and safety (nontoxicity) are also required. The matrices must exhibit good processability in fabrication of large carbon fiber composite structures, which should be volatile-less or low organic volatiles systems, or the materials must be amenable to acceptable volatile management during the production of large composite parts. Finally, the carbon fiber composite parts must exhibit high mechanical properties over a temperature range of -54 °C to 177 °C and upon exposure to cyclic temperature and stress while being exposed to aircraft fluids and moisture. The long-term servicing life at 177 °C was designed to be 6.0×10^4 h or 6.7 years. No microcracking in the composite parts is allowed because of the long-term servicing at 177 °C. The matrix resins should be amenable to the conventional high temperature autoclave curing and RTM process. At the beginning of the screening material work, all commercially available and some advanced experimental resins were found to exhibit some shortcomings. In experience, thermoplastic polymer resins with moderate to high molecular weights such as linear polyimides were generally difficult to melt-flowing process because of the high melt viscosity. Moreover, linear thermoplastics under stress are also usually sensitive to aircraft fluids. On the other hand, thermosetting materials such as the PMR-15 type polyimides have poor toughness and damage tolerance, and microcracking upon cyclic stress and temperature exposure are always produced. Hence, PE-endcapped oligoimide mers were chosen because of the availability of monomers and the excellent mechanical and thermal properties of the fully cured polyimides.

The PE group was selected because the past work indicated that this group could be placed along the backbone of various polymers and subsequently thermally reacted to provide cross-links [28-31]. PE groups could also be placed on the ends of imide oligomers and subsequently thermally reacted to afford thermoset polyimides with attractive properties [32,33]. PE endcap offers certain distinct advantages over other reactive endcaps. It has adequate chemical and temperature stability to withstand relatively harsh synthetic conditions. PE-endcapped imide oligomers afford excellent shelf life and exhibit broad processing windows without volatile evolution during curing. After thermal-curing, thermoset polyimides with excellent combinations of properties can be obtained.

3.3.2 Structures and Properties

3.3.2.1 PE-endcapped Oligoimides for Autoclave Processing

A variety of imide oligomers and polymers containing phenylethynyl groups have been prepared and evaluated. Hergenrother and Smith have synthesized a series of PE-terminated imide (PETI) oligomers by the reaction of aromatic dianhydride(s) with a stoichiometric excess of aromatic dianmine(s) at Calc'd number average molecular weights (M_n) of 1500-9000 g·mol^{-1} and endcapped with phenylethynylphthalic anhydrides in NMP [34]. Unoriented thin films cured in flowing air to 350 °C exhibited T_g of 251-274 °C, and mechanical properties including tensile strengths of 105.5-139.3 MPa, moduli of 2.8-3.2 GPa, and elongation at breakage of about 40% at 23 °C, and good retention of properties at 177 °C. Stressed film specimens exhibited excellent resistance to a variety of solvents after a 2-week exposure period at ambient temperature.

Hou et al. [35] and Smith et al. [36] have evaluated PETI oligomer as the resin matrix of carbon fiber composite. The PETI-5 oligoimide was prepared by the reaction of 3,3′,4,4′-biphenyltetracarboxylic diahydride (s-BPDA), an aromatic diamine mixture of 3,4′-oxydianiline(3,4′-ODA) and 1,3-bis(3-aminophenoxy)benzene (1,3,4-APB)(85/15 mole ratio), and 4-phenylethynylphthalic anhydride (PEPA) as the endcapper at Calc'd M_n of 5000 g·mol^{-1} (Fig. 3.17). Unidirectional prepreg was made by coating an NMP solution of the amide acid oligomer on unsized carbon fiber (IM7).

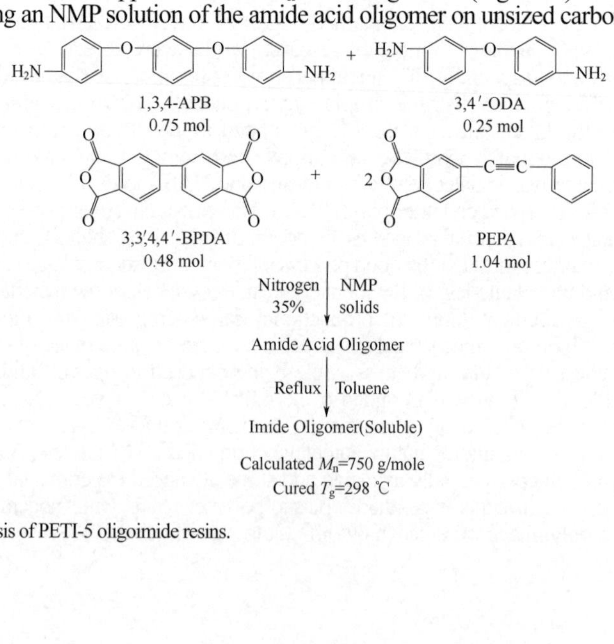

1,3,4-APB 0.75 mol + 3,4′-ODA 0.25 mol

3,3′,4,4′-BPDA 0.48 mol + 2 PEPA 1.04 mol

Nitrogen 35% NMP solids

Amide Acid Oligomer

Reflux Toluene

Imide Oligomer(Soluble)

Calculated M_n=750 g/mole

Cured T_g=298 ℃

FIGURE 3.17 Synthesis of PETI-5 oligoimide resins.

There is a drawback with PETI-5 oligoimide in the fabrication of complex composite structures, primarily skin stringer structures. A resin matrix with improved flow and equivalent or better mechanical properties is desired. In order to improve the melt processability, PETI oligoimides with lower Calc'd M_n of 1250 and 2500 g·mol^{-1} were prepared and their NMP solutions (35 %(w)) were used to prepare unidirectional carbon fiber prepreg. The unoriented polyimide films prepared by casting the NMP solutions of the amic acid oligomers with Calc'd M_n of 2500 (PETI-2.5) and cured at 350 °C/1 h exhibited tensile strength of 151.7 MPa, higher than that of PETI-5 (129.6 MPa). However, the elongation at breakage of PETI-2.5 (14%) was much lower than that of PETI-5 (32%).

The minimum melt viscosity was decreased with lowering of the Calc'd M_n. The minimum melt viscosity of PETI-5 with Calc'd M_n of 5000 was measured at 10,000 Pa·s at 371 °C, compared with PETI-2.50 (900 Pa·s at 335 °C) and PETI-1.25 (50 Pa·s·at 335 °C), respectively. The IM7/PETI composites finally cured at 371 °C/1 h showed great combination of mechanical properties (Table 3.3). The 2500 and 5000 g·mol^{-1} PETI-5 composites exhibited comparable open hole compression (OHC) strengths, compressive strength after impact (CAI) at 6.67 kJ·m^{-1}, and microstrain regardless of the stiffness of the lay-up. Notable better retention of OHC (177 °C dry) properties of the 58/34/8 lay-up were exhibited by the 2500 g·mol^{-1} PETI-5 as compared to the 5000 g·mol^{-1} version. The 1250 g·mol^{-1} PETI-5 exhibited comparable OHC properties to the 2500 and 5000 g·mol^{-1} PETI-5, however, the CAI strength and microstrain were reduced. The 1250 g·mol^{-1} PETI-5 laminates had low resin contents (26%-28%) due to excessive resin flow during fabrication. Crossply laminate (58/34/8) specimens of the 1250 and 2500 g·mol^{-1} PETI-5 composites were thermally cycled from −54 °C to 177 °C at a heating rate of 8.3 °C min^{-1} with a 1 h hold at −54 and 177 °C. The specimens were examined after 100 and 200 cycles under a microscope for microcracks. After 200 cycles, neither sample exhibited any microcracks.

TABLE3.3 Comparison on Mechanical Properties of IM7/PETI-5 Composite Laminates With Different Calc'd M_ws

Property	Lay-up	1250 g·mol^{-1}	2500 g·mol^{-1}	5000 g·mol^{-1a}
OH tension str. (MPa)				
RT (dry)	$(+45,0,-45,90)_{4S}$		444	461
177 °C (wet)	(25/50/25)	—	436	452 (dry)
OH tension str. (MPa)				
RT (dry)	$(+45,-45,90,0,0,+45,-45,0)_S$		578	557
177 °C (wet)	(38/50/12)	—	566	565 (dry)
OHC str. (MPa)				
RT (dry)	$(+45,0,-45,90)_{4S}$		342	335
177 °C (wet)	(25/50/25)	—	219	238
OHC str. (MPa)	$(+45,-45,90,0,0,+45,-45,0)_S$			
RT (dry)	(38/50/12)	—	377	369
177 °C (wet)			263	296
OHC str. (MPa)				
RT (dry)		431.6	458.6	450.3
177 °C (dry)	$(\pm45,90,0,0,\pm45,0)_{2S}$	366.2	395.2	342.7
177 °C (wet)	(58/34/8)	368.5	344.1	344.6
CAI str. (MPa)	$(+45,0,-45,90)_{4S}$ (25/50/25)	244.6	334.5	331
CAI mod (GPa)	$(+45,0,-45,90)_{4S}$ (25/50/25)	55.8	57.9	55.9
Microstrain (×10^{-6})	$(+45,0,-45,90)_{4S}$ (25/50/25)	4377	5908	5986

[a] *Normalized to 62% fiber volume. RT, room temperature.*

Overall, the 1250 g·mol^{-1} PETI-5 exhibited high resin flow which prohibited the fabrication of quality composites. The 2500 g·mol^{-1} version had better processability than the 5000 g·mol^{-1} PETI-5 oligomer while the properties of the cured resins were similar. The similarity in properties is attributed to the supposition that the nature of the cured resins is similar (low cross-link density). Thus, the 2500 g·mol^{-1} PETI should afford improved processability in the fabrication of complex composite structures while exhibiting equivalent properties to that of the 5000 g·mol^{-1} version.

Chen et al. [37] have prepared a series of PE-endcapped fluorinated aromatic oligoimides with Calc'd M_n of 1250-10,000 by thermal polycondensation of an aromatic dianhydride, 3,3′,4,4′-oxydiphthalic anhydride (ODPA), with an aromatic diamine mixture of 1,4-bis(4-amino-2-trifluoromethylphenoxy)-benzene (1,4,4-6FAPB) and 3,4′-oxydianiline (3,4′-ODA) in the presence of 4-phenylethynylphthalic anhydride (PEPA) as reactive endcapping agent (Fig. 3.18). Effects of chemical compositions and molecular weights of the aromatic oligoimides on their meltability as well as the thermal and mechanical properties of the thermal-cured polyimide resins were systematically investigated.

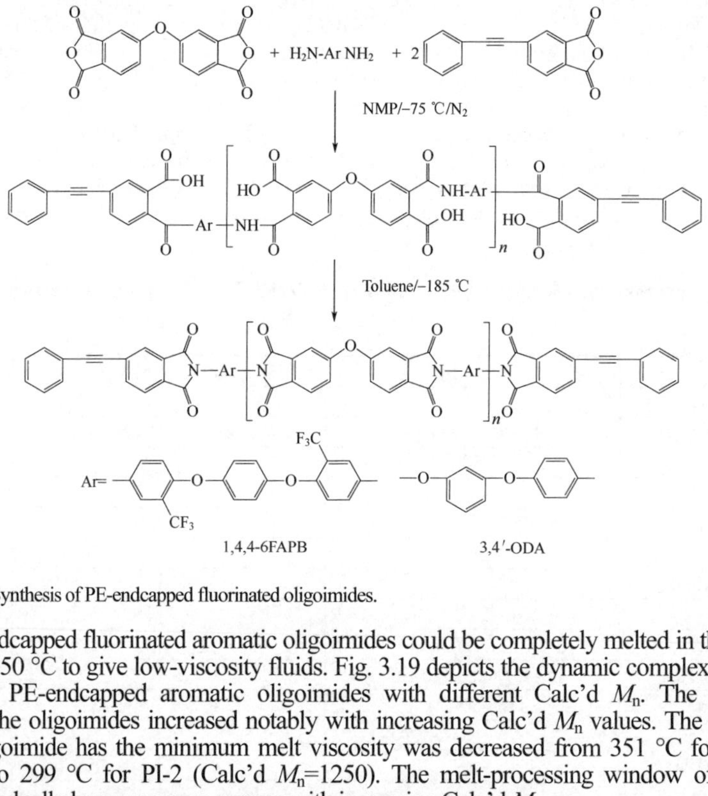

FIGURE 3.18 Synthesis of PE-endcapped fluorinated oligoimides.

The PE-endcapped fluorinated aromatic oligoimides could be completely melted in the temperature range of 250-350 °C to give low-viscosity fluids. Fig. 3.19 depicts the dynamic complex melt viscosity curves of the PE-endcapped aromatic oligoimides with different Calc'd M_n. The complex melt viscosities of the oligoimides increased notably with increasing Calc'd M_n values. The temperature at which the oligoimide has the minimum melt viscosity was decreased from 351 °C for PI-6 (Calc'd M_n=10,000), to 299 °C for PI-2 (Calc'd M_n=1250). The melt-processing window of the aromatic oligoimides gradually becomes more narrow with increasing Calc'd M_n.

Isothermal viscosity measurements at high temperatures of 280 °C or more were also conducted. Fig. 3.20 shows two representative curves of melt viscosity versus isothermal standing time at 280 °C. The aromatic oligoimides with Calc'd M_n=2500 could be completely melted at 250 °C to give molten fluid with melt viscosity of lower than 200 Pa·s, and the melt viscosity decreased gradually with increasing temperature, and then increased at a certain temperature range. PI-2 exhibited a melt viscosity variation at 280 °C from 0.56 to 2.09 Pa·s after isothermal aging for 2 h, compared with PI-4 with Calc'd M_n=2500 (from 25.0 to 178.6 Pa·s), implying that the melt stabilities of the phenylethynyl-endcapped aromatic oligoimides were closely related to their molecular weights and distributions. The oligoimides with the same Calc'd M_n but different chemical structures also show different melt stabilities, of which PI-3 Calc'd M_n=1250 shows the best melt stability (0.35-1.25 Pa·s) at 280 °C/2 h.

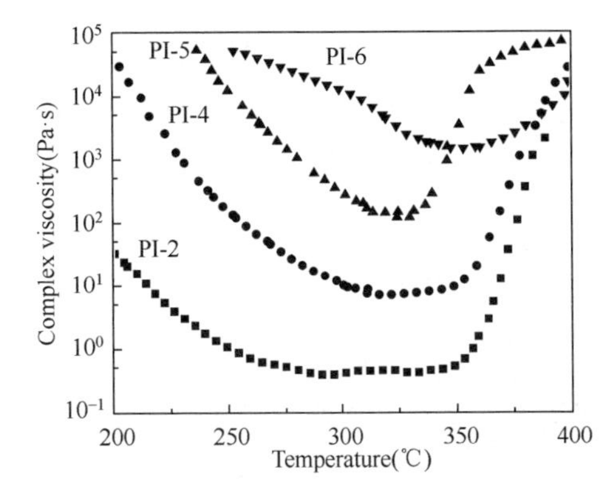

FIGURE 3.19 Dynamic complex melt viscosities vs temperature for PE-endcapped fluorinated oligoimides with different Calc'd M_n.

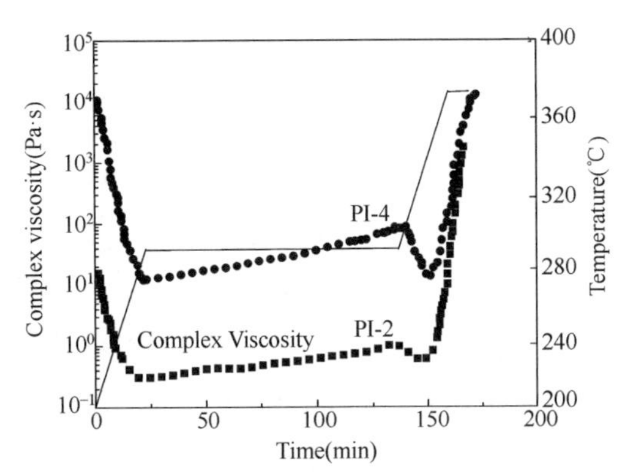

FIGURE 3.20 Dependence of the complex melt viscosities on isothermal aging time For the representative PE-endcapped fluorinated oligoimides.

After thermal curing at 371 °C for 1 h, all of the thermoset polyimide resins exhibited excellent mechanical properties with tensile strengths of 65.7-95.6 MPa, tensile moduli of 1.4-1.9 GPa, elongations at breakage of 3.9%-14.0%, flexural strengths of 135.5-153.8 MPa, and flexural moduli of 3.1-3.5 GPa. It seems that the tensile strengths and elongations at breakage were improved gradually with increasing the Calc'd M_n from 1250 to 5000. PI-5 with Calc'd M_n of 5000 exhibited the best mechanical properties with tensile strength 95.6 MPa, elongation at breakage of 14.0%, flexural strength of 144.7 MPa and flexural moduli of 3.1 GPa. No more benefits in the improvements of the tensile strength and elongation at breakage were observed after the Calc'd M_n of 10,000.

Overall, the PE-endcapped fluorinated oligoimide with Calc'd M_w of 1250 (PI-2) showed good thermal stability with a T_g value of 301 °C (DSC) and 338 °C (DMA), comparable to PETI-298 (T_g=298 °C by DSC) and PETI-330 (T_g=330 °C by DSC), and combined mechanical properties with flexural strength of 135.5 MPa, flexural modulus of 3.2 GPa, tensile strength of 65.7 MPa, and elongation at breakage of 3.9%, probably suitable for RTM application. Additionally, the version with Calc'd M_w of 5000 (PI-5), due to its relatively higher melt viscosity and high mechanical properties, might be more appropriate for infusion or extrusion molding processes.

Cho and Drzal [38] have investigated the imidization and reaction of phenylethynyl end-group of PETI-5 with Calc'd M_n of 2500 g·mol^{-1} by FT-IR spectra. There was an obvious temperature effect on formation of the polyimide structures. Fig. 3.21 depicts FT-IR spectra of the characteristic absorption bands during imidization and thermal curing. In the as-received sample, the absorption band from phenylethynyl-terminated amide acid oligomer in the range of 3000-2800 cm^{-1} is largely screened because of NMP solvent contained in the neat resin. The solvent is detected at a significant level at 100 °C and completely disappeared at 250 °C. It is likely that the resin begins to imidize below 200 °C and completes its imidization reaction at around 250 °C. The two peaks at 3254 and 3194 cm^{-1} in the as-received resin in Fig. 3.21A are attributed to the N—O—H stretching, strongly associated with hydrogen bonding in the amide acid oligomer. Typical C—O—H stretching bands from the solvent are shown between 3100 and 2800 cm^{-1}. The O—O—H stretching of the carboxylic acid group in the oligomer can be also characterized by a broad band between 2650 and 2400 cm^{-1}.

As seen in Fig. 3.23B, the band at 1660 cm^{-1} is attributed to the C═O stretching of amide bonds in the oligomer, which was almost disappeared due to the progressed imidization reaction after heating at 200 °C for 1 h. The absorption band resulting from the C—O—N bending of amide bonds is seen at 1545 cm^{-1} below 200 °C, which has also mostly disappeared at 200 °C and completely at 250 °C, indicating that the phenylethynyl-terminated amide acid oligomer is converted into the corresponding imide polymer through imidization reaction. The reaction is started and activated below 200 °C and then proceeds at a slower rate above 200 °C.

There are several characteristic absorption bands representing the imide polymer in Fig. 3.21B. The most useful bands for identification of the corresponding imide polymer are at 1777 cm^{-1} from C–A–O asymmetrical stretching, at 1374 cm^{-1} from C–O–N stretching, and at 739 cm^{-1} from C=O bending from imide groups. The band at 1777 cm^{-1} is also slightly interfered with by absorption of anhydrides occurring around 1780 cm^{-1}. It can be concluded that the heat treatment of PETI-5 above 200 °C for 1 h is sufficient to virtually develop the structure of imide polymer, especially above 250 °C for complete imidization.

Interestingly, the weak absorption band seen at 2213 cm^{-1} definitely results from stretching of alkyne groups, which are present in the reactive phenylethynyl group at the imide polymer chain ends of PETI-5. This peak is hardly observed in the presence of NMP solvent in the neat resin. The band is largely screened by the presence of a significant amount of solvent below 200 °C but it is apparently shown above the temperature because most of the solvent has been removed. The absorption peak slightly decreases with increasing temperature and it finally disappears completely at 350 °C. This indicates that at 350 °C the alkyne bonds are broken to be converted into C=C bonds that are ready to cross-link three-dimensionally with intermolecular chains in proximity. In fact, it has been found that the terminal phenylethynyl group is definitely responsible for the cure reaction of LaRC PETI-5. Accordingly, it can be concluded that a temperature above 350 °C for 1 h in air is needed to fully cure the PETI-5 resin through complete reaction of its phenylethynyl end groups.

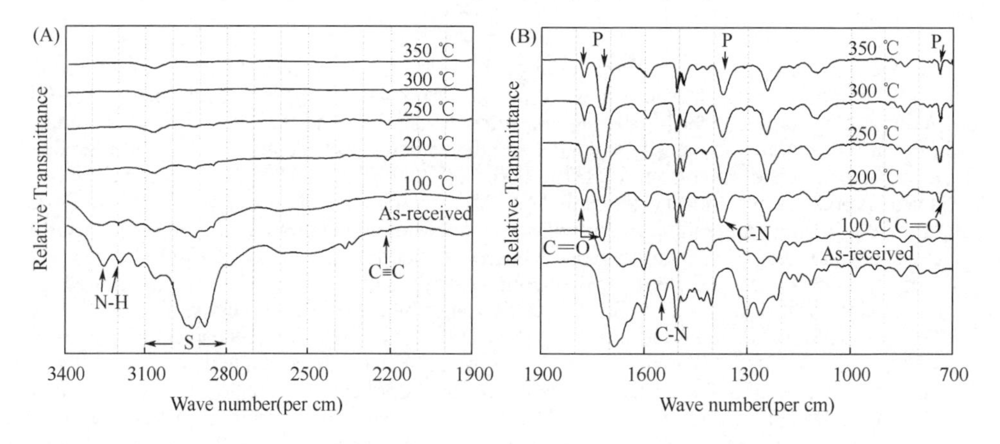

FIGURE 3.21 Variations of FTIR spectra between 3400 and 1900 cm^{-1}(A) and 1900 to 700 cm^{-1} (B) of PETI-5 oligoimide thermally treated at different temperatures for 1 h in air.

Overall, heat treatment of PETI-5 above 200 °C for 1 h is sufficient to develop the structure of the imide polymer, and heat treatment above 250 °C results in complete imidization. The imidization reaction becomes active at around 100 °C, the rate increases linearly up to about 200 °C, and then proceeds more slowly to 250 °C. The activation energy of cure reaction based on the phenylethynyl end group of PETI-5 is about 36.9 kcal·mol^{-1}. The cure reaction involving cross-linking and chain extension by a reaction of phenylethynyl end groups is completed at 350 °C for 60 min or at 370 °C for 10 min in air. The rate of disappearance of the C–C triple bond increases with increasing temperature from 300 °C to 370 °C but does not completely disappear even after curing for 4 h at 300 °C.

The PMR approach has also been used in synthesis of PE-endcapped prepolyimide resins. Originally, the PE-endcapped oligoimides, such as PETI-5 with Calc'd M_w of 5000 g·mol^{-1}, have been used to produce C_f/PETI-5 composites. It was reported that the composite laminate showed very high CAI value (~320 MPa) [17]. However, one problem for this composite was its poor processing properties. High boiling point and strong polar solvents (such as *N*-methyl-2-pyrrolidinone (NMP)) must be employed to prepare the carbon fiber prepreg due to the poor solubility of PETI-5 in common solvents. Another problem is the high melt viscosity of PETI-5 oligoimide. Hence, polyimide matrix resins with both high impact toughness and good processabilities are desired for high-temperature carbon fiber composite fabrication.

Liu et al. [39] developed a modified PMR method to prepare the PE-endcapped thermosetting poly(amic ester) resins using ethyl alcohol as the major solvent, which are the high solid content solutions with low solution viscosity, suitable for impregnating with carbon fiber to produce high-quality prepregs. The PE-endcapped poly(amic ester) resins were prepared by the reaction of the diesters of aromatic dianhydrides (s-BPDA) and/or (α-BPDA), an aromatic diamine mixture of 3,4′-ODA and 1,3,4-APB, and the monoester of 4-phenylethynyl phthalic anhydride (PEPE) as endcapping agent using ethanol as solvent (Fig. 3.22). The PMR matrix resins showed the characteristics of high resin concentration and low viscosity (solid content of 50% and Brookfield viscosity of 40-50 mPa·s at 25 °C), suitable for impregnating carbon fibers (C_f) to give high-quality prepreg.

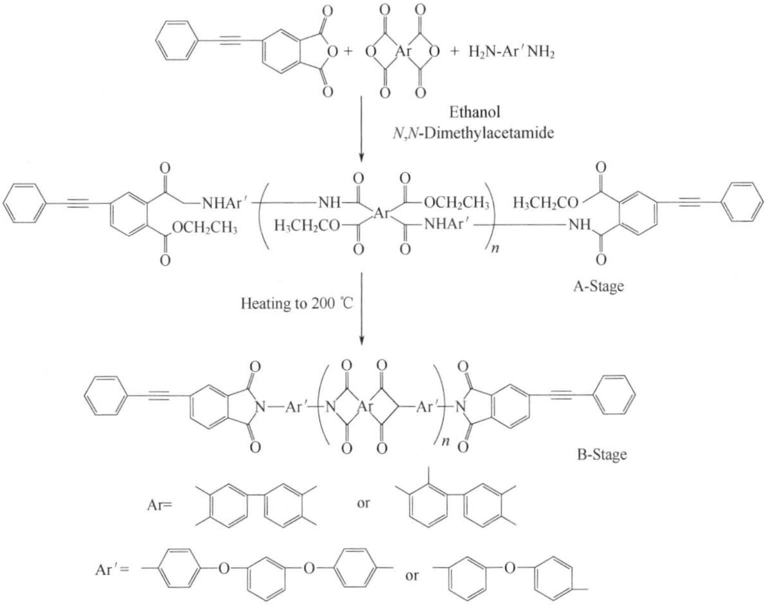

FIGURE 3.22 Preparation of the PE-endcapped PMR poly(amic ester) matrix resins.

The solubility of the diester of aromatic dianhydrides in low-boiling-point alcohol solvent has been considered a key issue for the successful preparation of PMR resins. To prepare a solution of PMR resin, aromatic dianhydrides and an endcapping agent have to be esterified and dissolved in alcohol solvents, and aromatic diamines are then dissolved in this blend by the reaction of acid groups and amino groups. In order to obtain a homogeneous solution of the diester of BPDA (s-BPDA and a-BPDA), 20% of DMAc was blended with ethyl alcohol to form a solvent mixture, which was then employed as the solvent for the esterification of s-BPDA and a-BPDA to provide a homogeneous solution of s-/a-BPDE. The solution of s-/a-BPDE was then mixed with the alcohol solution of PEPE to give a homogenous solution of s-/a-BPDE and PEPE, to which the solid mixture of 1,3,4-APB and 3,4′-ODA was added with stirring for 2 h to give a homogenous A-staged PMR poly(amic ester) resin solutions.

Fig. 3.23 depicts the rheological behaviors of the B-staged oligoimide resins, which were performed by ramping the temperature from 250 °C to 400 °C at a heating rate of 4 °C·min^{-1}. for the powder molded discs. The B-staged oligoimide resin with semicrystallized state (PI-1) exhibited very high melt viscosity at temperatures as high as 330-340 °C. However, the minimum melt viscosities of the B-staged oligoimide resins with amorphous state were significantly reduced by increasing the mole ratios of a-BPDA/(a-BPDA+s-BPDA) in the resin chemical compositions, especially when the mole ratio was over 20%. Meanwhile, the temperatures at the minimum melt viscosity (T_m) decreased with the increase in the a-BPDA loadings, demonstrating that the incorporation of a-BPDA in the B-staged oligoimide resins reduced not only the resin's melt viscosity but also the melting temperatures, thus improving the melt processability of the resins.

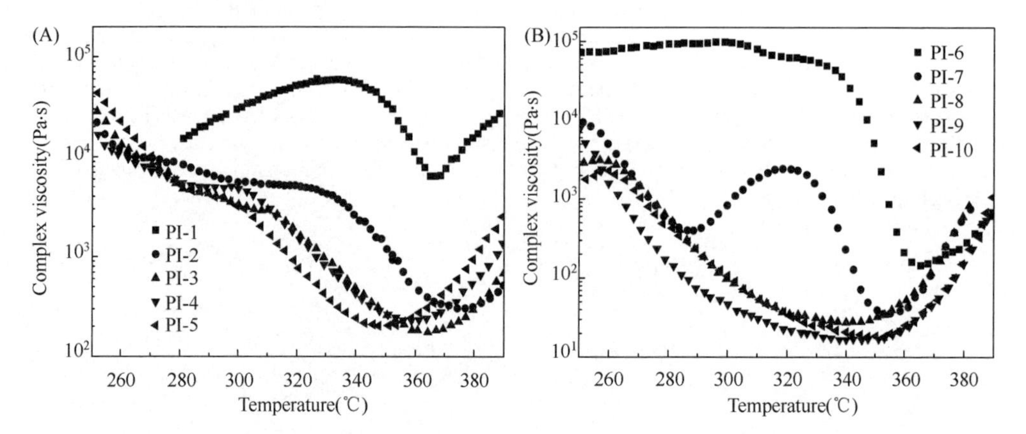

FIGURE 3.23 Melt viscosity changes on temperature of the B-staged oligimides with Calc'd M_w of 5000 (A) and 2500 (B).

Another improvement by incorporating α-BPDA in the polymer backbone of B-staged oligoimide resin was the extended melting processing windows. With the increasing α-BPDA loadings in the resin's backbones, the B-staged oligoimide resins with Calc'd M_w=5000 g·mol^{-1} showed extended temperature scale where the resin's melt viscosity was lower than 500 Pa·s, implying that melt processing widows were improved by increasing the α-BPDA loadings in the B-staged resin's composition.

Additionally, reducing the Calc'd M_w of the oligoimides was also an effective approach to reducing the melt viscosities. The B-staged oligoimide resins with Calc'd M_w of 2500 g·mol^{-1}. showed much lower melt viscosities than those with Calc'd M_w of 5000 g·mol^{-1}. Hence, copolymerization of s-BPDA and a-BPDA with aromatic diamines not only efficiently reduced the melt viscosities of the B-staged oligoimide resins but also widened the melt processing windows. In addition, reducing the Calc'd M_n of the oligoimide was also an effective way to improve the melt processability. However, this is accomplished with a compromise in mechanical properties.

The cured neat resins with Calc'd M_w of 5000 g·mol^{-1} show T_gs of >280 °C determined by DMA, and excellent mechanical properties, including tensile strengths of >124 MPa, elongation at breakage of >16.2%, flexural strength of >154 MPa, and flexural modulus of >3.2 GPa. It was found that lowering the Calc'd M_w could obviously reduce the mechanical strength and toughness. During the thermal curing of the B-staged PE-endcapped oligoimide resins, polyimide chain extension and cross-linking occurred simultaneously at the elevated curing temperatures [40]. In this process, the chain extension yielded the resins with highmolecular-weights, beneficial to improving the strength and toughness while the chain cross-linking led to the resins with high modulus and high T_g.

In comparison with the series of the PI resins with different chemical structures and Calc'd M_w, as discussed above, PI-3 had the best combination of mechanical and thermal properties, along with good melt processability, and was selected as a candidate to prepare the C_f/PI composites.

3.3.2.2 PE-endcapped Oligoimides for RTM Processing

In the late 1990s, seeking thermoset polymer resins suitable for processing of carbon fiber composites by RTM technique while possessing thermos-oxidative stability superior to BMIs began [41,42,27]. The research project was supported by NASA HSCT Program in which the lightweight structural composites were required to have high retention of mechanical properties after 60,000 h at 177 °C. The fabrication of support structures, such as frames, ribs, stringers, etc., in the HSCT program was preferably performed by RTM process due to its low manufacturing costs. However, RTM process required a thermosetting resin with low (preferably < 3Pa·s) and stable (>2 h at the injection or infiltration temperatures) melt viscosity during composite part fabrication. Although PETI-5 exhibited good autoclave processability, the melt viscosity was too high (about 6000 Pa·s at 370 °C) to fabricate composites by RTM or RI processes.

Initial attempts at lowering the melt viscosity of PETI-5 to make it amenable to RTM through molecular weight reduction [43] and the incorporation of a reactive additive [42] met with limited success. In the first approach, melt viscosity of lower-molecular-weight versions of PETI-5 decreased relative to that of the Calc'd M_w=5000 g·mol^{-1} version, but were still too high for RTM. The second approach reduced the melt viscosity of PETI-5 to acceptable levels; unfortunately, >80% of the additive was required to be effective. Additionally, the minimum melt viscosities under both approaches occurred at temperatures where the phenylethynyl group reacted appreciably, leading to an unstable melt. Modification of the oligomer composition with the same monomers of PETI-5 were performed by adjustment to the diamine ratio as well as the molecular weight of the oligomers, affording resins that were processable by RTM [42,27,44]. These volatile-free oligomers exhibited stable, low melt viscosities at temperatures of <290 °C where the phenylethynyl group was slow to react. One composition, designed PETI-RTM, was scaled-up to 30 kg quantities and successfully used in the fabrication of high quality 2.4 m-long curved F-frames by RTM. The cured resin has a T_g of 256 °C, making it an attractive material for the HSCT program.

Research was continued to develop RTM resins which could produce the cured thermoset materials with higher T_gs and better mechanical properties at elevated temperatures for application in advanced aerospace vehicles such as reuasable launch vehicles (RLVs). The operational environment for RLV composite structures differs significantly from that of the HSCT in that higher-temperature performance is needed for short periods of time. Numerous resin compositions were evaluated which looked at other diamines than those used in the PETI-5. In the past two decades, many researchers have carried out extensive works to improve the resin's RTM processability and thermo-oxidative stability. RTM resins with higher T_gs [45] then emerged, including the PETI-298 (T_g of 298 °C), PETI-330 (T_g=330 °C), and PETI-375 (T_g=375 °C). The chemistry, physical, and mechanical properties of the polyimides resins are described herein.

The PETI-298 was prepared by the reaction based on α-BPDA, PEPA, and a diamine mixture of 1,3,4-APB/3,4′-ODA (75/25 mole ratio) with Calc'd M_w of 750 g·mol^{-1} [46,47]. The cured neat resin at 371 °C/1 h exhibited a T_g of 298 °C by DSC. The melt viscosity of PETI-298 was generally in the range of 1-10 poise at 280 °C and did not change appreciably at this temperature, indicating excellent melt stability. This is because the phenylethynyl group reacts very slowly at this temperature. The melt stability at the minimum viscosity was critical for processing of large parts by RTM, VARTM. These PETI materials have an excellent processing window (i.e., temperature difference between where the minimum melt viscosity occurs and onset of cure) and consequently exhibits excellent robustness in processing. The carbon fiber composite laminate processed by RTM showed good initial mechanical properties and excellent retention of RT properties after aging at 288 °C in air for 1000 h [36]. However, the void content in composite laminate was about 4%.

Smith et al. [47] prepared a series of PE-endcapped oligoimides with Calc'd M_w of 750 g·mol^{-1} by the reaction of the appropriate amount of aromatic dianhydride(s) with the appropriate amount of aromatic diamines and endcapped with PEPA to first yield a PE-terminated amide acid, which was then imidized to produce the oligoimides by azeotropic distillation with toluene under a Dean Stark trap (Fig. 3.24). It was found that an oligoimide prepared by the reaction of α-BPDA, a diamine mixture of 1,3,4-APB (50 %(mol)) and m-PDA (50 %(mol)) and PEPA showed excellent melt viscosity stability (0.06-0.09 Pa·s at 280 °C/2 h, and the thermally cured resin at 370 °C/0.5 h exhibited a T_g of 330 °C, which was designed as PETI-330.

The C_f/PETI-330 composite laminates by RTM have ⩽2% porosity with fiber volume of about 57% as determined by acid digestion. The RT OHC strength and short shear (SBS) strength were comparable to those of C_f/PETI-298 composite. The retention of OHC and SBS strengths at 288 °C was significantly better for PETI-330 than for PETI-298 due to its higher T_g. The RT unnotched compressive strength was lower for PETI-330 as compared to PETI-298.

As a part of an ongoing effort to develop materials for RTM of high-performance/high-temperature composites, the PETI-375 was prepared using α-BPDA, an aromatic diamine mixture of 1,3,4-APB and 2,2′-bis(trifluoromethyl) benzidine (TFBZ) using PEPA as endcapper. The PETI-375 oligoimide exhibited a stable complex melt viscosity of 0.1-0.4 Pa·s at 280 °C, suitable for RTM injection processing. High-quality, void-free laminates were fabricated by high-temperature RTM using unsized T-650 carbon fabric. The dried oligomer powder was loaded into the injector and heated to 280 °C under vacuum to degas the resin. The mold was placed in a platen press, heated to about 288 °C, and the resin injected under about 1.34 MPa of hydrostatic pressure. The mold was then heated to 371 °C for 1 h, subsequently cooled to ~100 °C, and the pressure released. Overall, PETI-375 exhibited excellent RTM processability.

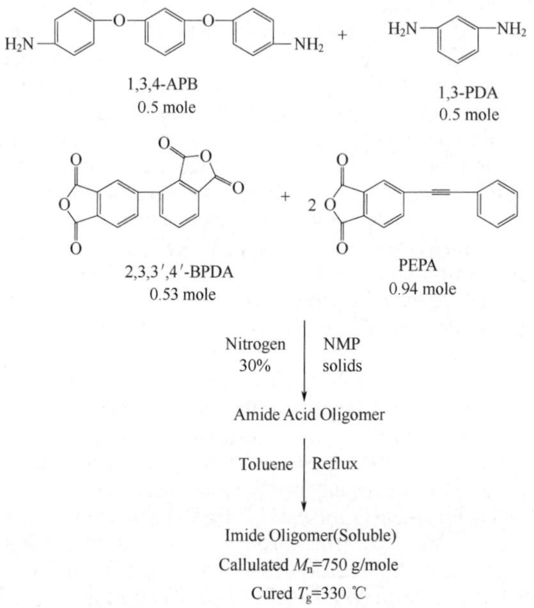

FIGURE 3.24 Synthesis of PE-endcapped oligoimide (PETI-330).

The T650/PETI-375 laminates showed excellent quality, as evidenced by the C-scans and microscopic analysis, had fiber volumes of 57%-59%, and void contents less than 0.75% as determined by acid digestion. T_g was measured at ~375 °C by TMA. The PETI-375 laminates exhibited relatively high RT mechanical properties with greater than 50% retention of RT properties when tested at 316 °C.

Table 3.4 compares the melt viscosities of the RTM oligoimides, including PETI-298, PETI-330, and PETI-375. The complex melt viscosities (η*) initially and after 2 h aging at 280 °C were measured in the range of 0.6-1.4 Pa·s for PETI-298, 0.06-0.9 Pa·s for PETI-330, and 0.1-0.4 Pa.s for PETI-375, successively. All oligoimides were prepared at the same Calc'd M_w of 750 g·mol^{-1}, so the cross-link densities of the thermally cured resins are comparable. It is significantly noted that the melt viscosities of both PETI-330 and PETI-375 are lower than that of PETI-298, but the thermally cured resin shows higher T_g. This might be primarily due to the contribution of α-BPDA since it is known to provide imides with lower melt viscosity and high T_gs as compared to similar materials based on s-BPDA. In addition, because of the unique properties of α-BPDA, more rigid diamine such as 1,3-PDA or TMBZ can be used to increase the overall rigidity of the oligoimide, consequently the thermally cured resins with higher T_gs and lower melt viscosities are successfully achieved.

TABLE3.4 Melt Viscosities of RTM Oligoimides With Different T_gs

Oligomer	Diamine Composition (%)	BPDA	η*@280 °C, Initial (Pa·s)	η*@280 °C, after 2 h (Pa·s)
PETA-298	1,3,4-APB (75), 3,4-ODA (25)	s	0.6	1.4
PETA-330	1,3,4-APB (50), 1,3-PDA (50)	a	0.06	0.9
PETA-375	1,3,4-APB (50), TFMBZ (50)	a	0.1	0.4

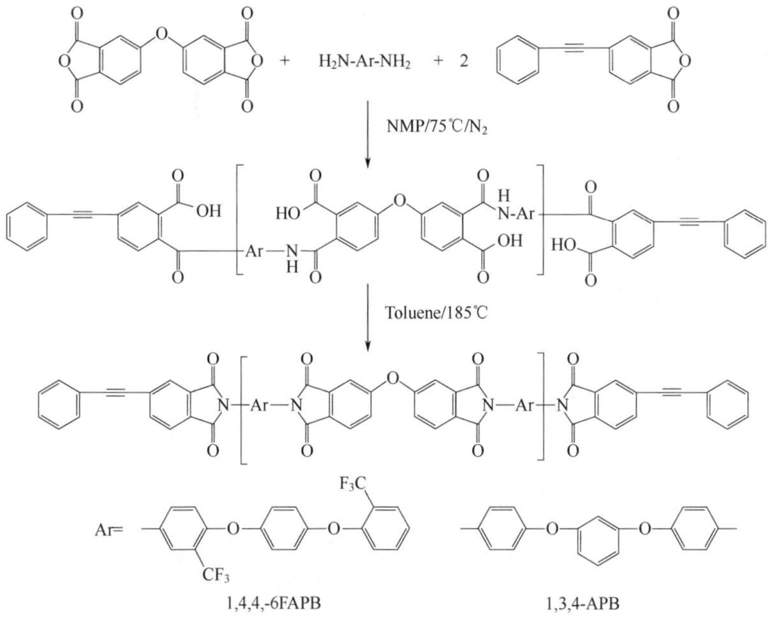

H_2N-Ar-NH_2 + 2

NMP/75℃/N_2

Toluene/185℃

Ar=

1,4,4,-6FAPB 1,3,4-APB

FIGURE 3.25 Synthesis of the fluorinated RTM oligoimide.

Zuo et al. [48] have prepared a series of molecular-weight-controlled PE-endcapped fluorinated oligoimides by the polycondensation of a mixture of 1,4-bis(4-amino-2-trifluoromethylphenoxy) benzene (6FAPB) and 1,3-bis(4-aminophenoxy) benzene (134-APB) with 4,4′-oxydiphthalic anhydride using PEPA as endcapping agent (Fig. 3.25, Table 3.5).

The melt processability of the PE-endcapped oligoimides was affected by the resin's chemical structures and their molecular weights. The minimum melt viscosities and the melt viscosity variations for 2 h at 280 °C are shown in Table 3.6. The resin's minimum melt viscosities also increased gradually with increases in the Calc'd M_n values.

Fig. 3.26 compares the dynamic thermal rheological curves of a series of imide resins with different Calc'd M_w values. In which PI-2 could completely melt at a temperature as low as 250 °C, and its melting temperature width (D) was 100 °C (250-350 °C), compared with those of PI-4 (300-340 °C, Δ=40 °C). The resin meltability and melt processability deteriorated with increases in the Calc'd M_w values.

The isothermal melt stabilities are also shown in Table 3.6, and two representative curves of the melt viscosity against the isothermal standing time at 280 °C are depicted in Fig. 3.27. The melt viscosity variations at 280 °C or 290 °C for 2 h of standing depend on the Calc'd M_w values as well as the polymer chemical structures. PI-2 exhibited a melt viscosity variation at 280 °C of 0.38-0.95 Pa·s, in contrast to PI-4 (11.5-81.79 Pa·s), implying that the melt stabilities of the PE-endcapped imide resins were closely related to their molecular weights and distributions as well as the chemical structures. In addition, the imide resins with the same Calc'd M_w values but different chemical structures (PI-1, PI-2, and PI-3) also showed different melt stabilities. Of which PI-2 showed the lowest minimum viscosity at 288 °C and good melt stability at an elevated temperature.

Fig. 3.27 depicts DSC (A) and DMA (B) curves of the thermally cured polyimides. The T_g values of the thermally cured polyimides were changed with both the polymer chemical backbones and the calculated molecular weights. The T_g value of the thermally cured polyimide increased with decreases in the Calc'd M_w values, and this resulted from the high polymer chain cross-linking density. The thermoset polyimides showed T_g values in the range of 224-341 °C, which decreased with increases in the Calc'd M_w values. The storage modulus did not decrease until the temperature was scanned to 275 °C (Fig. 3.12B), indicating that the thermally cured polyimide had outstanding thermomechanical properties.

TABLE3.5 Stoichiometry of the Fluorinated RTM Oligoimides

Sample	Aromatic Diamine (mmol)	ODPA (mmol)	PEPA (mmol)	Calcuated Molecular Weight
PI-1	1,4,4-6FAPB (13.11)	22.45	60.00	1250
	1,3,4-APB (39.33)			
PI-2	1,4,4-6FAPB (25.21)	20.42	60.00	1250
	1,3,4-APB (25.21)			
PI-3	1,4,4-6FAPB (42.54)	21.73	70.00	1250
	1,3,4-APB (14.18)			
PI-4	1,4,4-6FAPB (45.60)	66.20	50.00	2500
	1,3,4-APB (45.60)			
PI-5	1,4,4-6FAPB (37.88)	65.76	20.00	5000
	1,3,4-APB (37.88)			
PI-6	1,4,4-6FAPB (50.00)	93.50	13.00	10,000
	1,3,4-APB (50.00)			

TABLE3.6 Complex Melt Viscosities of the Fluorinated Oligoimides

Sample	Complex Melt Viscosity (Pa·s)					Minimum Melt Viscosity (Pa·s/ °C)	Melt Viscosity Variation at 280 °C for 2 h (Pa·s)
	250 °C	275 °C	300 °C	325 °C	350 °C		
PI-1	54.1	0.60	0.50	0.57	0.69	0.50 at 300 °C	0.57-4.01
PI-2	0.58	0.32	0.31	0.36	0.44	0.25 at 288 °C	0.38-0.95
PI-3	0.61	0.36	0.30	0.33	1.78	0.28 at 308 °C	0.48-3.30
PI-4	63.31	15.37	7.17	7.18	29.61	6.39 at 313 °C	11.56-81.79
PI-5	3801	491.8	121.3	70.48	1162	69.88 at 322 °C	264.8-1563[a]
PI-6	27,010	8349	1389	392.6	418.8	328.1 at 334 °C	896.4-537 8[b]

[a]*Melt viscosity variation at* 290 °C *for 2* h.
[b]*Melt viscosity variation at* 310 °C *for 2* h.

Table 3.7 compares the mechanical properties of the thermoset polyimides. After thermally curing at 371 °C for 1 h, all the thermoset polyimide sheets exhibited excellent mechanical properties. The tensile strengths and elongations at break apparently improved gradually with increases in the Calc'd M_w from 1250 to 5000. No more benefit for improving the mechanical properties was observed when the Calc'd M_w values were over 5000. The lower tensile strength and elongation at breakage of PI-2 and PI-4 versus those of PI-5 were probably due to the higher concentrations of low-molecular-weight resin species in the imide resins. In addition, X-ray diffraction patterns of the thermoset polyimides demonstrated that the thermoset polyimides did not contain any crystalline-like phase; they were amorphous polymer materials with low water uptakes in the range of 0.68%-0.85%.

Chen et al. [37] developed a PE-endcapped imide-oligomer (PI-2) for RTM processing by the reaction of ODPA and a mixture of 6FAPB and 3,4-ODA (50/50 mole ratio) using PEPA as endcapping agent. It was found that the PE-endcapped imide-oligomers exhibited excellent melt processability, thermal stability, and mechanical properties. Moreover, the thermally cured PI-2 at 371 °C/2 h exhibited a T_g of 301 °C, 11.4 °C higher than that derived from ODPA, 6FAPB/134-APB (50/50 mole ratio), and PEPA at the same Calc'd M_n (1250), implying that diamine 3,4′-ODA was more rigid than 1,3,4-APB, which resulted in the T_g increase.

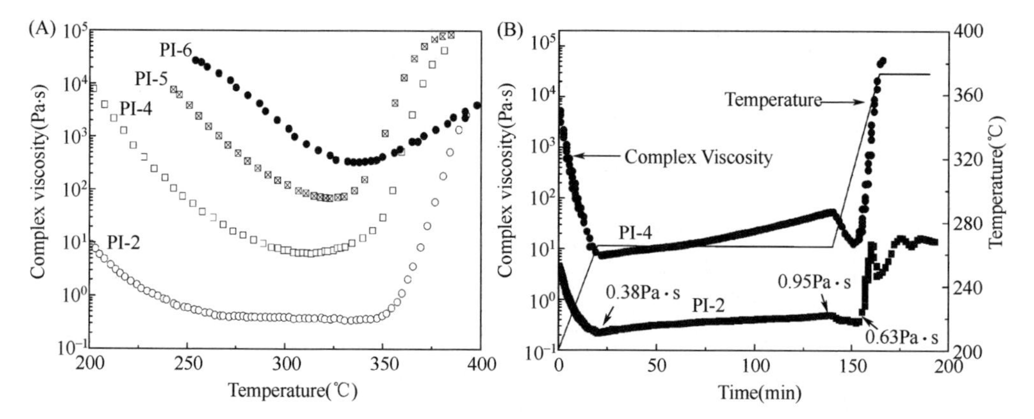

FIGURE 3.26 Complex melt viscosities and melt stabilities of the fluorinated oligoimides.

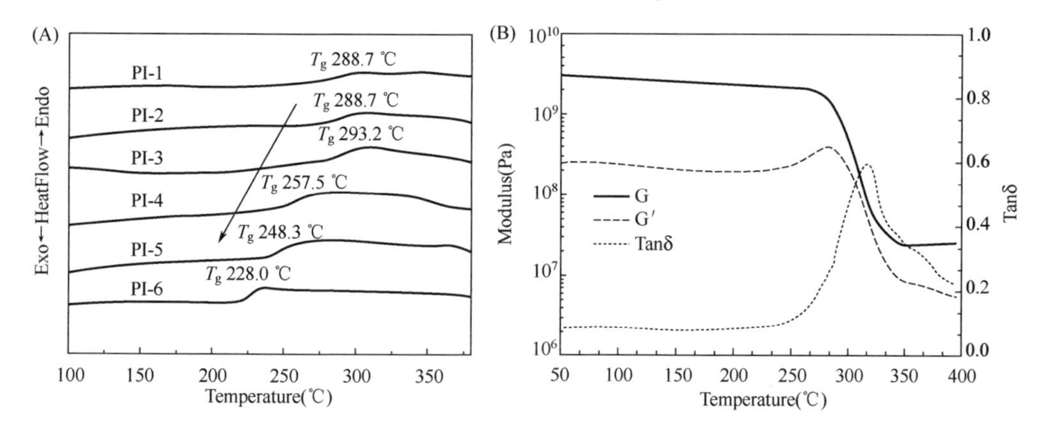

FIGURE 3.27 DSC and DMA curves of the thermoset polyimides.

Su et al. [49] investigated the effects of the molecular weights on the thermal stability of PE-endcapped oligomides with low melt viscosities in order to find a pathway to increase the thermal stability of the RTM oligomide resins. A series of oligomides with Calc'd M_w of 714-1226 g·mol^{-1} were prepared by the high-temperature polycondensation of α-BPDA, 4,4′-methylenebis (2,6-dimethylaniline) (MBDA) using PEPA as endcapping agent (Fig. 3.28). MALDI-TOF results indicated that the oligoimides were mixtures of biphenylethynyl-endcapped imide oligomers with different degrees of polymerization (DP=0, 1, 2, 3, 4), corresponding to the molecular weights (M_w) of 737-2787. The melt viscosities of the oligomides were reduced by lowering the Calc'd M_w, but the viscosity stability was poor. The thermally cured polyimide neat resins showed a T_g of >400 °C and 5% weight loss temperature (T_5) of >499 °C. The mechanical properties of the thermally cured polyimide neat resins were dependent on the Calc'd M_w values and improved gradually with increasing Calc'd M_n, but the tensile strength and elongation at breakage were too low (Table 3.8).

Yang et al. [50] have prepared a series of PENA-endcapped imide oligomers using 4-(2-phenylethynyl)-1,8-naphthalic anhydride (PENA) as endcapping agent (Fig. 3.29). The PENA-endcapped imide oligomers were mixtures of mono- and double-endcapped imide oligomers with polymerization degree (P_n) of 1-5 and M_n of 2515-3851 g·mol^{-1} determined by GPC. They studied the effect of chemical structures on the curing behaviors of two model compounds: PENA-m based on PENA and PEPA-m derived from PEPA revealed that PENA-m showed the cure temperature of 50 °C lower than PEPA-m and the activity energy of thermal curing reaction for PENA-m was also lower than that of PEPA-m. The PENA-endcapped imide oligomers could be melted at temperatures of >250 °C with minimum melt viscosities of 1.2-230 Pa·s at 275-301 °C and the widened melt processing windows, along with 10-40 °C lower cure temperatures than the PEPA-endcapped analog. The PENA-endcapped oligoimides could be thermally cured at 350 °C/1 h to afford the thermoset polyimides with good combined thermal and mechanical properties including T_g of 344-397 °C by DMA, T_d of 443-513 °C, tensile strength of as high as 54.7 MPa, flexural strength of as high as 126.1 MPa and modulus of as high as 2.3 GPa, respectively.

TABLE3.7 Comparison of the Mechanical Properties of the Thermoset Polyimides

Sample	Tensile Strength (MPa)	Tensile Modulus (GPa)	Elongation at Break (%)	Flexural Strength (MPa)	Flexural Modulus (GPa)	Water Uptake (%(*w*))
PI-2	60.7	1.7	4.3	163.6	3.0	0.75
PI-4	68.8	1.1	8.4	147.3	3.3	0.85
PI-5	93.8	1.2	14.7	140.1	3.1	0.76
PI-6	88.0	1.1	12.1	154.1	3.6	0.68

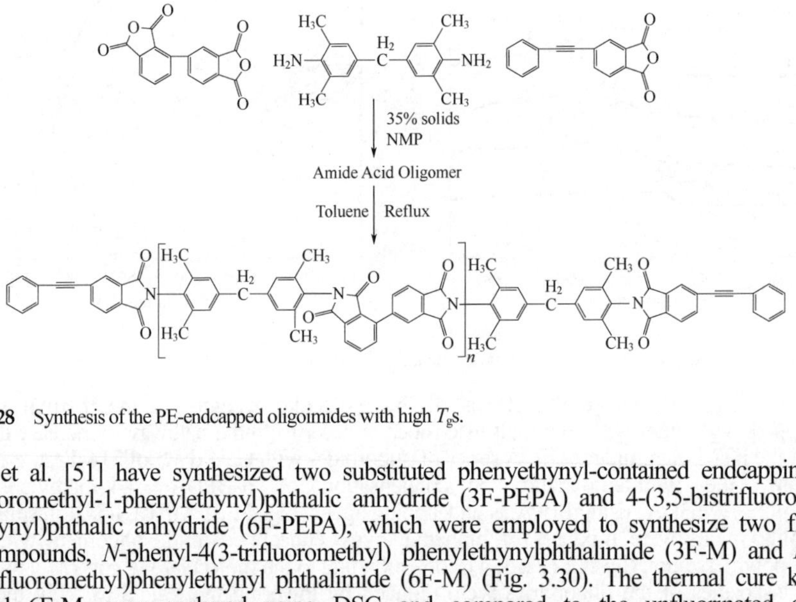

FIGURE 3.28 Synthesis of the PE-endcapped oligoimides with high T_gs.

Yang et al. [51] have synthesized two substituted phenyethynyl-contained endcapping agents, 4-(3-trifluoromethyl-1-phenylethynyl)phthalic anhydride (3F-PEPA) and 4-(3,5-bistrifluoromethyl-1-phenylethynyl)phthalic anhydride (6F-PEPA), which were employed to synthesize two fluorinated model compounds, *N*-phenyl-4(3-trifluoromethyl) phenylethynylphthalimide (3F-M) and *N*-phenyl-4(3,5-bitrifluoromethyl)phenylethynyl phthalimide (6F-M) (Fig. 3.30). The thermal cure kinetics of 3F-M and 6F-M were analyzed using DSC and compared to the unfluorinated derivative, *N*-phenyl-4-phenylethynylphthalimide (PEPA-M). The thermal cure temperatures of 3F-M and 6F-M were 399 and 412 °C, which were 22 and 35 °C higher than that of PEPA-M, respectively. The thermal cure kinetics of 3F-M and 6F-M best fit a first-order rate law, although 3F-M and 6F-M reacted slower than PEPA-M. However, the exothermic enthalpy of 3F-M and 6F-M were only half that of PEPA-M. Based on the model compounds study, a series of fluorinated phenylethynyl-terminated imide oligomers (F-PETIs) with different Calc'd M_n were synthesized by thermal polycondensation of a-BPDA and 3,4′-ODA using 3F-PEPA or 6F-PEPA as the endcapping agent. The substituent effects of the trifluoromethyl (CF_3) groups on the thermal cure behavior and melt processability of F-PETIs were systematically investigated. Experimental results reveal that the melt processability of F-PETI was apparently improved by the reduced resin melt viscosities and the enhanced melt stability due to the incorporation of the CF_3 groups in the imide backbone. All of those F-PETIs exhibit outstanding thermal and mechanical properties.

TABLE 3.8 Mechanical Properties of the Thermally Cured Polyimide Resins

Sample[a]	Calc'd M_n	Tensile Strength (MPa)	Tensile Modulus (GPa)	Elongation at Break (%)
PI-1	1226	35.5±2.7	1.72±0.06	2.98±0.18
PI-2	1103	29.2±4.5	1.72±0.04	2.36±0.36
PI-3	980	26.1±4.6	1.69±0.08	2.32±0.50
PI-4	858	23.4±5.7	1.68±0.06	2.21±0.36

[a]*PI-5 was too brittle to prepare the testing sample.*

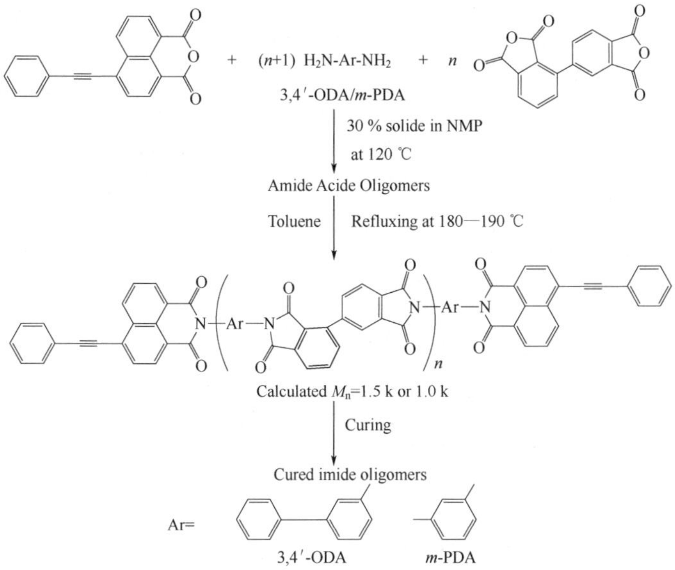

FIGURE 3.29 Synthesis of PENA-endcapped oligoimides.

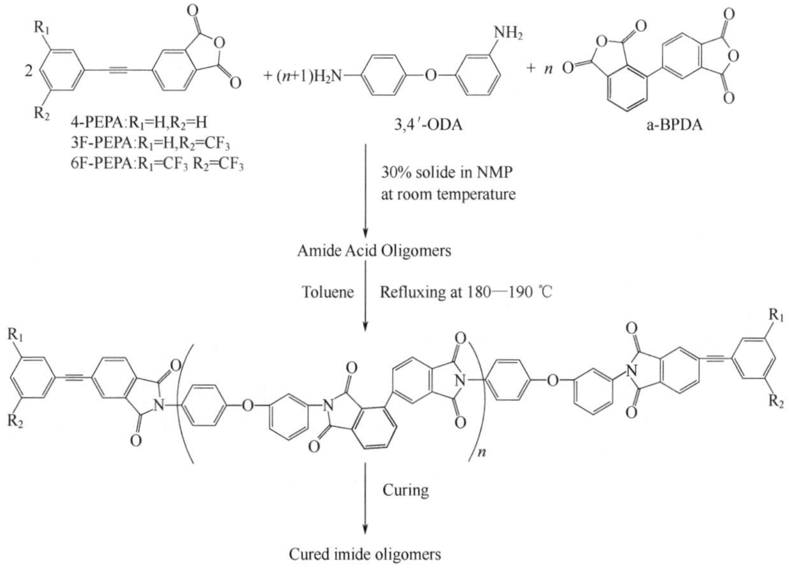

FIGURE 3.30 Synthesis of substituted PE-endcapped oligoimides.

3.4 Properties of Polyimide/Carbon Fiber Composites

3.4.1 Preparation of PMR-type Resin Prepregs

The NA-endcapped PMR poly(amic ester) resin solutions are usually employed as matrix to impregnate carbon fiber, giving fabric prepregs (unidirectional fibers or tapes). The fabric prepreg is prepared by a solution-impregnation route (wet method) using PMR resin solutions with 50 %(*w*) of solid resin content in order to obtain carbon fiber composite components with high enough resin loading. After impregnation, the fabric is then passed through a heated tower or dried in a clean room at RT to reduce the solvent content to about 5%-10 %(*w*) by weight. The residual solvent (methyl alcohol or ethyl alcohol, etc.) can provide the prepreg with appreciable tack, so that individual plies of prepreg can be easily laid up to the desired pattern. Then, the rolls of prepreg are interleaved with polyethylene release film and stored at -18 °C to prevent any reaction of the prepolyimides.

Production of PMR matrix resin prepreg is not easy. Close control must be maintained over impregnation parameters, in which thermal history is critical. In order to reduce the unwanted by-products, the amount of heat exposure to the resin must be minimized. Solid resin content of the prepreg is also important which is usually specified to fine limits by the customer.

The amount of residual solvent is an important parameter. A threshold residual solvent level exists, above which the prepreg is tacky and below which it becomes dry. Hence, the residual solvent levels should be considered in specifying PMR-type prepregs and cure cycles. In addition, the viscous stability of the prepregging solution is also of great importance. It was found that even short storage at RT could affect the chemistry of the PMR-type resins to form unwanted by-products. It was possible to dry the prepregs in a clean room at RT. Additionally, the shelf life (out-time) of the prepreg must be long enough to ensure the lay-up of large complex parts before resin viscosity becomes too high.

3.4.2 Fabrication of Carbon Fiber Composites

Carbon fiber composite are usually made from woven fabrics or parallel arrays of the prepregs by either compression molding (hot press) or autoclave route to produce high-quality composite components. In fabrication of a composite component, the first step is laying up of individual plies to the prescribed orientation. In this process, the physical characteristics of the prepreg, such as tack and drape, etc., are important. The composites are usually fabricated at the maximum cure temperatures of 310-330 °C for 2-3 h, followed by a postcuring at the same temperatures for about 4 h.

3.4.2.1 Compression Molding

Prepreg plies are first cut and assembled; the stack is then placed inside a vacuum oven. The prepreg stack is "staged" in a vacuum at lower temperatures to remove residual solvent or volatiles arising from any chemical reaction prior to placing in the mold. This process converts the PMR-type monomeric resin to the reactive endcapped imide oligomers by heating at various temperatures and times under full vacuum. The aim is to remove all solvent and condensation volatiles arising from the reaction.

The prepreg stack is then transferred to a matched die mold. Matching tool-steel male and female dies are closed to form a cavity of the shape of the component. The prepreg stack is placed in the lower mold section, which is then covered by the upper section. The two halves of the mold are brought together in a press. The prepreg stack is then heated under high pressure (up to 7 MPa) at the required temperature and for the required time to produce the composite laminate or component.

3.4.2.2 Autoclave Molding

The prepreg stack is placed in a flexible bag made from polyimide film. The lay-up is consolidated against an open die or mold, then a vacuum bag is formed over the lay-up and mold surface to allow the simultaneous application of both pressure and vacuum. The vacuum bag is then placed into an autoclave and evacuated. Heat and pressure (usually 1.4-2.0 MPa of nitrogen) are applied at the desired times. The close control of temperature and pressure is imperative, ensuring all of the fibers are completely wetted out by the resin and all of organic volatiles are completely removed before the laminate is consolidated by the cross-linking reaction. Otherwise, highly voided composite laminates will be produced with poor mechanical properties. Usually, thermal cure cycles are highly proprietary and complex.

PMR-15 is relatively easy to process compared with other high-temperature thermosetting resin systems. The major chemical reactions are widely spaced in temperatures. DSC results indicated that the melting of the PMR-type resins was below 100 °C; in situ polycondensation of the monomeric reactants was at about 140 °C; the melting of the NA-endcapped imide oligomers was measured in the range of 176-254 °C (melt flow temperature range); and the cross-linking reaction occurred rapidly at 310 °C. The fabrication of PMR-15/carbon fiber composite components is complicated. The thermal cure processing of PMR-15 laminates can be monitored by dielectric analysis and electromagnetic sensors.

Other high-temperature thermosetting resins cannot be processed into useful articles because the chemical transitions described above are too close in temperature, and the processing window is too narrow to yield high-quality composite laminates or components.

3.4.2.3 Resin Transfer Molding

Prior to laminate fabrication, the fabric sizing was removed by bagging in the fabric in Kapton® film, sealing with a high-temperature sealant, and heating to 400 °C/2 h under vacuum. Laminates of PE-endcapped oligomides were made by RTM using carbon fabric. The resins were injected into the unsized carbon fiber fabric positioned on an Invar tool using a high-temperature injector. The maximum processing parameters of the injector were a temperature of 288 °C, flow rate of 500 $cm^3 \cdot min^{-1}$, and pressure of 2.75 MPa. The tool containing the fabric was loaded into a press, heated to 316 °C/1.5 h prior to resin injection. The PETI resins were degassed in the injector by heating to 288 °C and holding for 1 h prior to injection. The degassing step was required primarily to remove moisture, residual solvent, and air from the resin. The molten resin was used to infiltrate eight-ply stacks of unsized fabric in an Invar tool under 1.34 MPa hydrostatic pressure during the entire process cycle. After the resin was injected at 288 °C at a rate of 200 $cm^3 \cdot min^{-1}$, the part was held under 1.34 MPa of hydrostatic pressure while heating to 371 °C and holding for 1 h. The composite laminates were cooled under pressure.

The laminates were ultrasonically scanned, cut into specimens, and tested for mechanical properties. The laminates were examined for microcracks by a microscope up to 400× magnification. Resin content, fiber volume, and void content were determined by acid digestion using a 1:1 (*w/w*) solution of concentrated sulfuric acid and 30% hydrogen peroxide. OHC properties (Northrup Grumman Test) were determined on specimens of 25.4×3.81 cm with a 0.64 cm hole in the center. Unnotched compression properties (Boeing Test, BSS 7260) were determined on specimens of 8.08×1.27 cm. Short beam shear (SBS) strength (ASTM D2344-84) was determined on specimens of 0.64×1.91 cm. Unstressed specimens were aged at 288 °C in a forced air oven. Five specimens were tested under each condition.

3.4.3 Properties of NA-Endcapped Polyimide Composites by Autoclave

The mechanical properties of carbon fiber (C_f)/PMR-type composites (C_f/PMR) are as good as those of state-of-the-art carbon fiber/epoxy composites (C_f/Ep). However, the thermal stability of C_f/PMR composites is much better than that of C_f/Ep and carbon fiber/bismaleimide (BMI) composites. Many studies have been conducted to determine the effects of various hostile environments on the physical and mechanical properties of PMR-type composites. Excellent retention of interlaminar shear properties of PMR-15/C_f composites was observed on aging at 316 °C (600°F) for 1600 h. Table 3.9 shows the typical mechanical properties for PMR-15/carbon fiber composite laminates [52]. The RT tensile and flexural strengths of T300/KH-304 composite were 1484 and 1619 MPa, which reduced to 1410 and 1128 MPa at 320 °C, respectively, corresponding to 94.9% and 69.7% of strength retentions. The tensile and flexural moduli were not obviously reduced at elevated temperature. The ISSL strength was 108 MPa at RT, reducing to 56 MPa at 320 °C. AS4/PMR-15 shows the comparable mechanical properties.

RP-46 is an improved PMR polyimide resin matrix system developed by NASA Langley Research Center in order to solve performance problems in extreme heat and stress environments without microcracking or blistering. RP-46 is available as liquid for traditional composite prepreg or as tow-prepreg for filament winding. Powder versions are suitable for compression molding process that combine functional fillers.

Table 3.10 shows the mechanical properties of the C_f/RP-46 composites, including the unidirectional IM7/RP-46 composite laminate and woven carbon T650-35/RP-46 laminate. The flexural strength of the unidirectional IM7/RP-46 laminate was 1723.7 MPa at 25 °C, which was decreased to 917.0 MPa at 316 °C (53.2% of property retention) and 793.0 MPa at 371 °C (46.0% of property retention). Meanwhile, the interlaminar shear strength was 131 MPa at 25 °C, reducing to 51.0 MPa at 316 °C (38.9% of property retention), and 32.4 MPa at 371 °C (24.7% of property retention). The tensile strength and compression strength at RT were 2627.0 MPa and 1799.6 MPa, respectively. The composite exhibited excellent thermo-oxidative stabilities. After isothermal aging at 316 °C for 1500 h, the weight loss was 7.6%. Similarly, the weight loss was only 3.7% after isothermal aging at 371 °C for 100 h. The woven carbon T650-35/RP-46 composite also showed outstanding mechanical properties and hygrothermal stabilities.

Pan et al. [53] have carried out research on the processing of C_f/ KH-420 composite for 420 °C application. The KH-420, a PE-endcapped PMR prepolyimide resin, showed a minimum melt viscosity of 242 Pa·s at 305 °C and fine processability. The resin and void contents could be effectively controlled by step by step pressure application. The void content in the carbon fiber composite of <1% could be obtained by applying 0.05 MPa at 90 °C, followed by 0.3 MPa at 120 °C, giving a C_f/KH-420 composite with 65% retention of flexural strength and 89% retention of flexural modulus at 420 °C.

TABLE3.9 Mechanical Properties of T300/KH-304 and AS4/PMR-15 Composites

Laminate	Test Temperature (°C)	0° Tensile Strength (MPa)	0° Tensile Modulus (GPa)	0° Flexural Strength (MPa)	0° Flexural Modulus (MPa)	ISSL (MPa)
T300/KH304	25	1486	139	1619	117	108
	320	1410	138	1128	114	56
AS4/PMR-15	25	—	—	1850	118	98
	320	—	—	930	104	65

TABLE3.10 Mechanical Properties of C_f/RP-46 Composites

Properties	Value for		Specification
	Unidirectional IM7/RP-46	Woven Carbon T650-35/RP-46	
Cure cycle temperature at 1.4 MPa/2 h (°C)	315-325		
Postcure temperature (°C)	349		
T_g (°C)	345		TMA
Flexural strength (°C)			ASTM D-790
25	1723.7		
316	917.0		
371	793.0		
23 wet		1155.7	
288wet		821.0	

TABLE3.10 (Continued)

Properties	Value for		Specification
	Unidirectional IM7/RP-46	Woven Carbon T650-35/RP-46	
288		487.7	
343^{wet}		599.7	
Interlaminar shear Strength (°C)			ASTM D-2344
25	131.0		
316	51.0		
371	32.4		
23 ^{wet}		67.8	
288^{wet}		50.3	
288		49.1	
343^{wet}		31.0	
Tensile strength, (MPa), 25 °C	2627.0	608.2	ASTM D-3039
Compression strength (MPa)			
25 °C	1799.6		ASTM D-3410 ASTM D-695
25 °C^{wet}		589.9	
288 °C^{wet}		364.7	
348 °C^{wet}		304.1	
Thermo-oxidative stability, % weight loss after			
10,000 h/232.2 °C	0.9		
5000 h/288 °C	8.6		
1500 h/316 °C	7.6		
200 h/371 °C	5.6		
100 h/371 °C	3.7		
Moisture absorption, weight gain % after 30 days at 85 °C and 85% relative humidity	0.62	0.51	ASTM D-570

Gao et al. [54] have investigated the high-temperature mechanical properties of carbon fiber (MT300)/PMR polyimides (KH420). The MT300/KH420 composite laminates with 60%±5% fiber volume were fabricated by autoclave at 350 °C and 1.5 MPa followed by postcuring at 420 °C/2 h. The laminates have a T_g value of 443 °C and decomposition temperature at 5% of original weight loss of 590 °C.

The tensile strength of the $[0°]_7$ undirectional laminate was 1056.1 MPa at ambient temperature (Table 3.11), which slightly increased to 1172.0 MPa at 350 °C, then decreased to 931.0 MPa at 420 °C (88.2% of strength retention), and 695.8 MPa at 500 °C (65.9% of strength retention), respectively. The RT tensile modulus was 110.7 GPa, which did not reduce even at 420 °C. The RT interlaminar shear strength (ISSL) was 68.0 MPa, which reduced gradually to 50.4 MPa at 350 °C, 35.9 MPa at 420 °C, and 38.3 MPa at 500 °C, successively, indicating excellent property retentions at high temperatures. The RT flexural strength was 1428 MPa, which reduced to 1060 MPa at 350 °C, 729 MPa at 420 °C, and763 MPa at 500 °C, successively. The flexural modulus did not show an obvious reduction up to 500 °C where the modulus retention was still as high as 92.5%.

The longitudinal tensile strength of the [±45°/0°/90°/+45°/$0°_2$]s multidirectional laminate was 232.6 MPa at ambient temperature (Table 3.12), which slightly increased to 253.5 MPa at 350 °C, then lowered to 221.5 MPa at 420 °C (95.2% of strength retention) and 169.6 MPa at 500 °C (72.9% of strength retention), respectively. The transversal strength was 110.5 MPa at 25 °C, which slightly increased to 117.4 MPa at 350 °C, then lowered 93.2 MPa at 420 °C (84.4% of strength retention) and 63.6 MPa at 500 °C (57.6% of strength retention), respectively.

TABLE3.11 Mechanical Properties of [0 °C] MT300/KH420 Composite Laminates at High Temperatures

Test Temp (°C)	Tensile Strength (MPa) and retention (%)	Tensile Modulus (MPa) and Retention (%)	ISSL (MPa) and Retention (%)	Flexural Strength (MPa) and Retention (%)	Flexural Modulus (MPa) and Retention (%)
25	1056.1	110.7	68.0	1428	118.3
200	1134.1/107.4	108.5/98.0	66.3/97.5	1233/86.4	117.4/99.3
350	1172.0/111.0	115.3/104.2	50.4/74.1	1060/74.2	117.0/98.9
420	931.0/88.2	103.7/93.7	35.9/52.8	729/51.1	105.8/89.5
500	695.8/65.9	92.6/83.7	38.3/56.3	763/53.4	109.4/92.5

TABLE3.12 Mechanical Properties of [±45 °/0 °/90 °/+45 °/$0°_2$]s MT300/KH420 Composite Laminates at High Temperatures

Test temperature (°C)	Longitudinal tensile strength (MPa) and retention (%)	Transversal tensile strength (MPa) and retention (%)	Longitudinal flexural strength (MPa) and retention (%)	Transversal fFlexural strength (MPa) and retention (%)	Longitudinal ISS (MPa) and retention (%)	Transversal ISS (MPa) and retention (%)
5	232.6	110.5	292	209	12.3	8.0
200	235.5/101.2	117.5/106.4	315/107.8	199/95.2	12.2/99.2	9.0/112.5
350	253.5/109.0	117.4/106.3	344/117.7	209/99.6	13.3/108.1	8.5/106.3
420	221.5/95.2	93.2/84.4	302/103.3	167/79.7	12.7/103.3	8.8/110.0
500	169.6/72.9	63.6/57.6	153/52.4	104/49.5	8.5/69.1	4.5/56.3

The longitudinal flexural strength was 292 MPa, which reduced to 344 MPa at 350 °C, 302 MPa at 420 °C, and 153 MPa at 500 °C, successively. The transversal flexural strength was 209 MPa, which reduced to 209 MPa at 350 °C, 167 MPa at 420 °C, and 104 MPa at 500 °C, successively. The longitudinal and transversal flexural strengths did not change up to 420 °C, showing excellent thermal stabilities.

Fig. 3.31 compares the effect of temperature on the retention of mechanical strengths [55]. MT300/KH420 composite laminate showed flexural strength retention of 86.4% at 200 °C, 74.2% at 350 °C, and 53.4% at 500 °C, successively. Compared with the commercially available epoxy composite MT300/603 with flexural strength retention of 20.4% at 220 °C, and BMI composites T300/GW300 and T800/803 with retentions of 67%-73% at 230 °C, demonstrating the PMR polyimide composite MT300/KH420 exhibited excellent high-temperature performance.

PMR-type prepolyimide resins, including prepregs, molding compounds, and adhesives, have been commercially available since about the mid-1970s. A variety of components ranging from small compression-molded bearings to large autoclave-processed cowls and ducts have been produced. The first composite component using PMR-15 matrix resin and carbon fiber by autoclave was the duct for GE's F404 air-engine. The duct, approximately 20,000 mm long and 1000 mm in diameter, was installed on the F404 engine and successfully withstood the accelerated mission testing cycles.

Other components made from C_f/PMR-15 composites include fan blades, inner cowls, swirl frames, nozzle flaps, and the shuttle orbiter aft body flap. Apart from engines and nacelles, other applications are in the missile field such as fins and bodies, and in lightweight ducting for aircraft. PMR prepolyimide resins are the current leading matrix resins for structural composites for high-temperature (300-500 °C) applications. Fig. 3.32 shows the photograph of T300/KH-304 composite component by autoclave, which is 800 mm long and 550 mm in diameter.

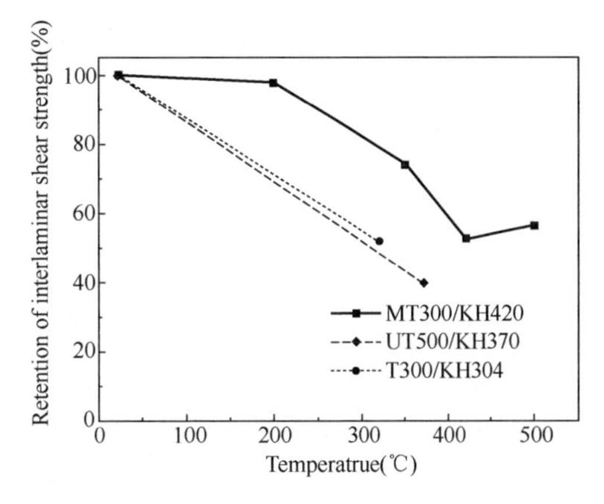

FIGURE 3.31 Effect of temperature on the retention of mechanical strengths.

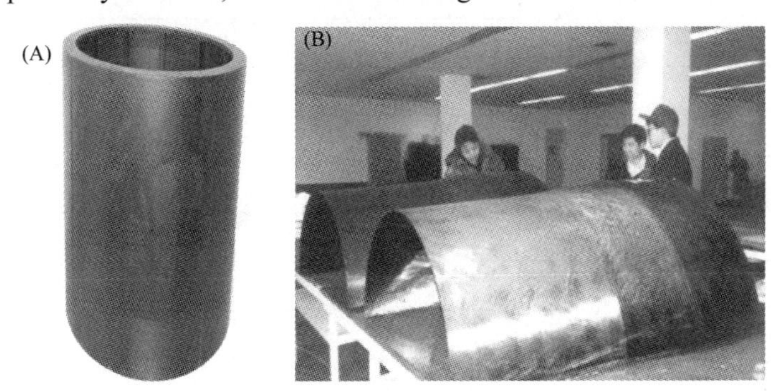

FIGURE 3.32 T300/KH-304 composite components fabricated by autoclave at 310-320 °C/1.5 MPa. (A) 800 mm long and 550 mm in diameter; (B) 1100 mm long and 890 mm in diameter.

3.4.4 Properties of PE-Endcapped Polyimide Composites by Autoclave

Hou et al. [35] have investigated the mechanical properties of carbon fiber (IM7)/PETI-5 composites. Unidirectional prepreg was made by coating an NMP solution of the amide acid oligomer on unsized carbon fiber (IM7), and IM7/PETI-5 composites were molded under 1.38 MPa at 371 °C. Composite mechanical properties are summarized in Table 3.13. The 0° flexural strength of 1787.5 MPa and 0° flexural modulus of 144.7 Gpa were measured, respectively. The 0° tensile strength at RT was measured at 2928.7 MPa, and the 0° tensile modulus was 175 GPa, indicating very good translation of fiber properties. The longitudinal compression strength was measured at 1298.3 MPa, higher than average thermoplastic composites, e.g., 1169 MPa for IM7/LaRC-IA polyimide and 1140 MPa for IM7/PEEK. RT SBS strength was measured at 106.5 MPa, and RT OHC strength was 429.3 MPa, respectively, implying that the PETI composites have good damage tolerance. The 73% retention of the RT OHC strength at 177 °C was excellent.

Liu et al. [39] developed a modified PMRs method to prepare the PE-endcapped thermosetting poly(amic ester) resins using ethyl alcohol as the major solvent. The PE-endcapped MPR matrix resin was used to fabricate continuous T300/PI prepreg. The prepreg was cut and plied up into the desired size and directions, and then cured in a hot press under 1.5 MPa. A high-quality unidirectional and quasi-isotropic composite laminate was obtained, which shows no voids or defects in any layer of the quasi-isotropic laminate, and the C_f volume fraction was about 57%-58%, as calculated from these images.

In order to determine whether the B-staged oligoimide resins were completely cured in the carbon fiber composites, the C_f/PI composite laminates prepared by thermal curing at 370 °C/2 h were further postcured at 380 °C for another 2-4 h. Table 3.14 shows the mechanical properties of the T300/PI-3 unidirectional laminates tested at different elevated temperatures. The flexural strength of the composite was 1536.8±86.7 MPa at RT, and 819.1 ± 36.2 MPa at 250 °C (53.3%), respectively. The flexural modulus was 143.8±4.6 GPa at RT, only 2.6% of loss at 250 °C, exhibiting great modulus retention at elevated temperatures. The interlaminar shear strength was 108.1 ± 2.7 MPa, lowering to 62.8 ± 3.9 MPa at 230 °C (58.1%), and 52.8 ± 2.6 MPa at 250 °C (48.8%), respectively. These results indicated that the T300/PI-3 composite might be long-term served at a temperature as high as 250 °C.

TABLE3.13 Mechanical Properties of IM7/PETI-5 Composite Laminates

Mechanical Properties	Temperature (°C)	IM7/PETI (5000 g·mol^{-1})
SBS strength (MPa)	RT	106.5
	93	97.8
	150	80.8
	177	62.8
0° flexural strength (MPa)	RT	1787.5
	93	1818.5
	150	1547.0
	177	1441.6
0° flexural modulus (GPa)	RT	144.7
	93	151.6
	150	143.3
	177	133.7
0° tensile strength (MPa)	RT	2928.7
0° tensile modulus (GPa)	RT	175.0
0° compress strength (MPa)	RT	1298.3
0° compress modulus (GPa)	RT	138.5
OHC strength (MPa)	RT	429.3
	177(wet)	317.7

TABLE3.14 Mechanical Properties of T300/PI-3 Composite Laminates

	Lay-up	Properties
0° flexural strength (MPa)		
RT		1536.8±86.7
177 °C		1137.9±71.8
230 °C		927.1±45.8
250 °C		819.1±36.2

TABLE3.14 (Continued)

	Lay-up	Properties
0° flexural modulus (GPa)		
RT		143.8±4.6
177 °C		144.4±3.3
230 °C		143.7±1.8
250 °C		140.0±0.9
0° Interlaminar shear (MPa)		
RT		108.1±2.7
177 °C		81.0±4.6
230 °C		62.8±3.9
250 °C		52.8±2.6
CAI strength (MPa)	$(-45°,0°,45°,90°)_{3\%}$	259.8±13.2
OHC strength (MPa)	$(-45°,0°,45°,90°)_{2\%}$	303.2±17.5
OHT strength (MPa)	$(-45°,0°,45°,90°)_{2\%}$	366±49.9

The CAI strength measured on quasi-isotropic laminates after impacted at an energy of 6.7 $kJ\cdot m^{-1}$ was 259.8 ± 13.2 MPa, and the OHC and OHT were 303.2 ± 17.5 MPa and 366.9 ± 49.9 MPa, respectively, indicating that the composite had high impact toughness. When comparing with the study of Smith et al. [36], the OHC and CAI strength showed a little decrease in this work, indicating that incorporating asymmetric structures has certain negative effects on the toughness of polyimide composites. But there are some other factors affecting these results such as the types of carbon fiber layup, the processing condition of PI composites, the differences of testing details between different works, and so on. Therefore, the effect of asymmetric structure on toughness might need a more systematic and comprehensive work that will be part of our following research.

In order to further understand the relationship between the impact energy and the CAI, four energy levels of 2.3, 4.5, 6.7, and 8.9 $kJ\cdot mL^{-1}$ were employed in the impact testing experiments. As expected, higher-impact energies resulted in larger damage areas and lower CAI values. There was a positive correlation between the impact energies and the damage areas, while there was a negative correlation between the impact energies and the CAI values (Fig. 3.33). When the impact energy was increased to 8.9 $kJ\cdot mL^{-1}$, the CAI value of the testing laminate was also as high as 230 MPa, indicating very high impact toughness.

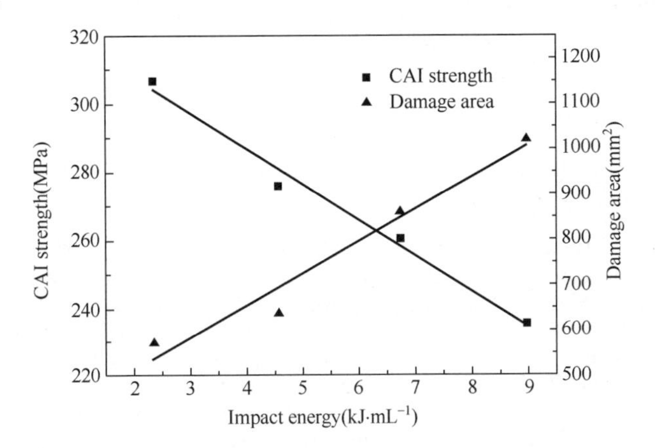

FIGURE 3.33 Relationship between the CAI values and damage areas versus the impact energies.

3.4.5 Properties of PE-Endcapped Polyimide Composites by RTM

Connell et al. [46] investigated the mechanical properties of the carbon fiber composites of PETI-298 and PETI-330 by RTM process. The laminates were made with eight plies of unsized 8HS carbon fabric with a (0/90) lay-up (quasi-isotropic). Fabric sizing was removed by heating the fabric at 400 °C/2 h under vacuum prior to insertion in the tool. The PETI powder was charged to the resin chamber, heated to 280 °C and degassed prior to injection into the tool. Laminate fabrication involved injecting the molten resin at ~280 °C into the preheated tool followed by a cure at 371 °C/1 h under ~ 1.4 MPa hydrostatic pressure.

All of the laminates were of high quality as determined by C-scan and photomicroscopy; the void contents were 0.7%-1.7% for PETI-298 composites and 1.7%-2.5% for PETI-330, respectively. Acid digestion was used to determine the resin content of each panel, the fiber volumes were 57.3%-60.5% for PETI-298 composite and 56.8%-58.1% for PETI-330 composites. The panels were machined into specimens and tested according to ASTM or other well-documented procedures. The T_gs were 275-287 °C for PETI-298 composites and 310-320 °C for PETI-330 composites.

Since the thermo-oxidative stability of AS-4 carbon fibers was poor [25], T650-35, a carbon fabric often used for high-temperature applications, was selected for use in subsequent laminate evaluation. Thus, PETI-298, PETI-330, and BMI-5270/T650-35 laminates were fabricated by RTM. Figs. 3.34 and 3.35 show the OHC properties determined at RT and 288 °C. The OHC strengths of all three resins on T650-35 fabric were comparable at RT. When tested at 288 °C, the PETI-330/T650 laminates exhibited the highest OHC strength, followed by PETI-298 and the BMI-5270. When tested at 288 °C, the PETI-330 exhibited a retention of RT OHC strength of ~74%, followed by PETI-298 (~69%) and then BMI-5270 (~60%). The retention of OHC modulus at 288 °C was higher for PETI-298 (91% of RT modulus) and PETI-330/T650 (92% of RT modulus) laminates as compared to the BMI-5270/T650 (75% of RT modulus) laminates.

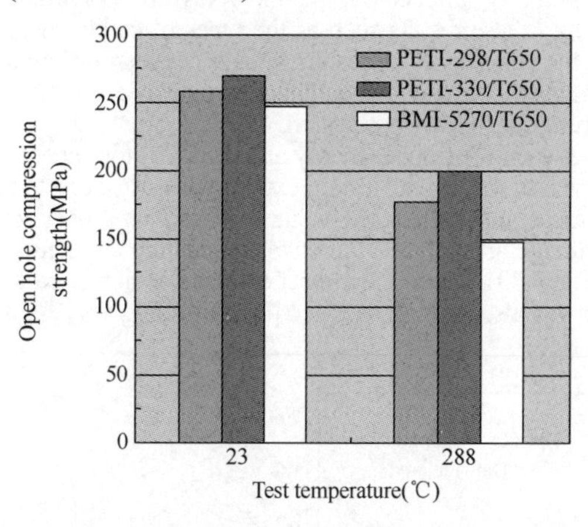

FIGURE 3.34 The OHC strengths of the PETI/ T650 composites at room temperature and at 288 °C.

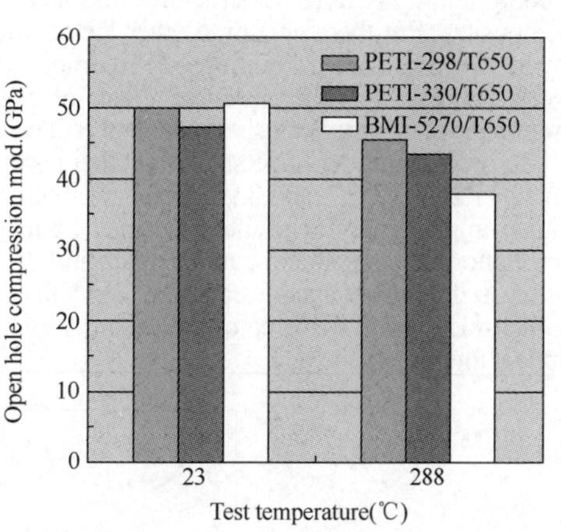

FIGURE 3.35 The OHC moduli of the PETI/ T650 composites at room temperature and 288 °C.

Fig. 3.36 shows the SBS strengths as a function of temperature. The SBS strengths of PETI-330/ T650 and the BMI-5270/T650 are compared. As expected, the PETI-330/T650 exhibited significantly higher SBS strengths at RT and elevated temperature. These specimens retained 62% of their RT SBS strength at 288 °C. This is significantly less than that for the PETI-330/AS-4 SBS specimens, which retained 88% of the RT SBS strength at 288 °C, although the absolute strength values were lower. The RT UNC strengths for PETI-298, PETI-330, and BMI-5270 T650 laminates were 457, 520, and 356 MPa, respectively.

Figs. 3.37 and 3.38 compare the effects on RT OHC strength and modulus of PETI-298/T650 and PETI-330/T650, respectively. For comparison purposes, BMI-5270/AS-4 laminate properties are also included. The effect of aging on OHC strength is comparable for both PETI-298 and PETI-330 T650 laminates with both retaining about ~78% of their unaged OHC strength after 1000 h. The effect of isothermal aging on the OHC modulus (Fig. 3.14) was minimal with PETI-298 and PETI-330/T650 laminates retaining ~85% and ~92%, respectively, of their unaged moduli after 1000 h at 288 °C in air.

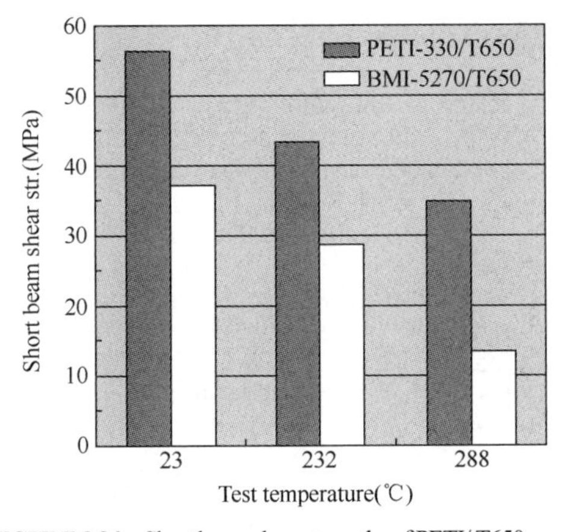

FIGURE 3.36 Short beam shear strengths of PETI/ T650 composites at room temperature, 232, and 288 °C.

FIGURE 3.37 Effect of isothermal aging at 288 °C in air on OHC strength.

Fig. 3.39 depicts the effect of isothermal aging on the SBS strength. In this case, PETI-298/T650 specimens were not available for aging, thus the data for PETI-298/AS-4 SBS specimens were included for comparison. The PETI-330/T650 specimens exhibited a retention of ~93% and ~75% of unaged RT SBS strength after 500 and 1000 h aging at 288 °C in air, respectively. In contrast, the BMI-5270/T650 specimens retained ~73% and 40% of unaged RT SBS strength after 500 and 1000 h aging at 288 °C in air, respectively. On comparing the properties of PETI-298 and PETI-330 on the two different fabrics (AS-4 and T650), the OHC strengths at RT and 288 °C were comparable, however, the absolute values for the PETI-330 were higher, particularly at 288 °C for the AS-4 laminates (87% retention on AS-4 vs 73% on T650). In comparing the OHC moduli, the absolute values were higher on T650 at RT, but when tested at 288 °C, the absolute values were comparable for both the AS-4 and T650 laminates. With regard to the SBS strengths as a function of temperature, only the PETI-330 and BMI-5270 laminates are discussed because no SBS strength data was available on the PETI-298/T650 laminates. The PETI-330/T650 specimens exhibited higher SBS strength at RT and 232 °C; however, when tested at 288 °C, the strengths were comparable for both the AS-4 and T650 PETI-330 specimens (~34 MPa).

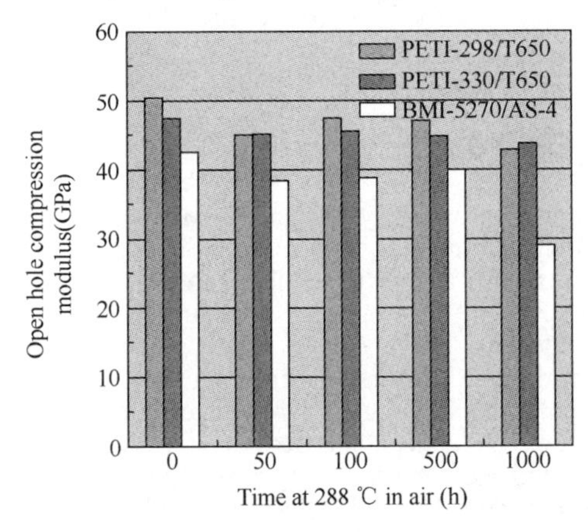

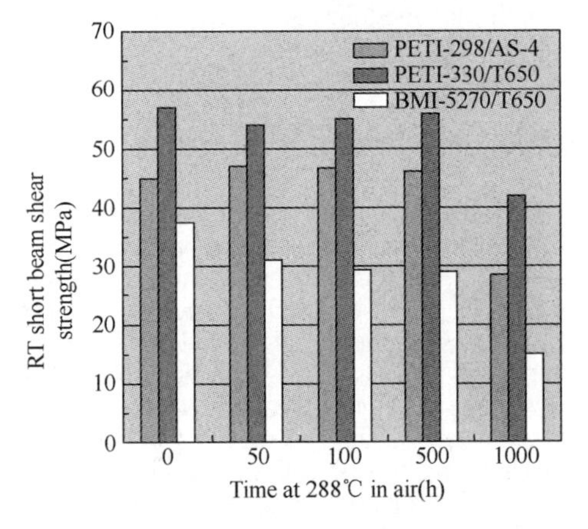

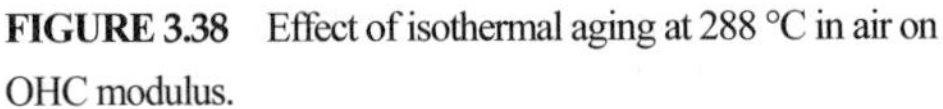

FIGURE 3.38 Effect of isothermal aging at 288 °C in air on OHC modulus.

FIGURE 3.39 Effect of isothermal aging at 288 °C in air on RT short beam shear strength.

The PETI-330/AS-4 and T650 laminates exhibited higher SBS strengths than the comparable BMI-5270 laminates at all of the temperatures tested. Based on this data, no conclusions can be drawn to determine which fabric offers better mechanical properties. It was anticipated that some differences would become apparent upon isothermal aging of the laminates.

The isothermal aging was conducted on unstressed, specimens at 288 °C in a forced air oven. The specimens were removed at various time intervals and subsequently tested at RT for OHC strength and modulus, and SBS strength. In comparing the OHC properties of PETI-298/AS-4 and PETI-298/T650 after aging, the results indicated no significant difference in the OHC strengths. They are very close in absolute strengths and in the retention of strength at every exposure interval. The OHC moduli were comparable except that the moduli of the T650 laminates were slightly higher. However, the retention of unaged modulus after aging for 1000 h was better on the PETI-298/AS-4 laminates (98% vs 85%). No obvious difference was observed in the mechanical properties as a function of aging that was attributable to the type of carbon fabric.

FIGURE 3.40 Photo of T300/KHRTM-350 composite demonstrator part by RTM injected at 260 °C and finally cured at 370 °C/2 h.

Overall, the PETI-330 laminates exhibited higher mechanical properties than the PETI-298 laminates, most notably at elevated temperature. Both PETI-298 and PETI-330 laminates consistently exhibited higher mechanical properties than the BMI-5270 laminates at elevated temperatures. No significant difference in the retention of mechanical properties after aging was apparent for the PETI-298 and PETI-330 laminates, although the PETI-330 laminates typically had higher absolute values. PETI-298 and PETI-330 transfer molding resins offer an unprecedented combination of processability, high-temperature performance, and toughness.

Fig. 3.40 shows a photo of T300/KHRTM-350 composite demonstrator part fabricated by RTM injected at 260 °C and finally cured at 370 °C/2 h. The KHRTM-350 is a PE-endcapped oligoimide resin for RTM processing. The unidirectional laminate showed great RT mechanical properties, including tensile strength of 1500 MPa, tensile modulus of 124 GPa, compression strength of 895 MPa, compression modulus of 124 GPa, flexural strength of 1450 MPa, flexural modulus of 113 GPa, interlaminar shear strength of 86.6 MPa, successively. At 350 °C, flexural strength and modulus are 1050 MPa and 100 GPa, respectively, corresponding to 72.4% and 88.5% of property retention.

3.5 Applications of Polyimide/Carbon Fiber Composites

High-temperature polyimide composites are required in a variety of applications in air engine, advanced aerospace vehicles, and propulsion systems, etc. Aromatic polyimide/carbon fiber composites (C_f/PI) are leading candidates for these applications due to their excellent thermo-oxidative stability and the combination of physical and mechanical properties. The C_f/PI composite components have usually been processed from solvent-laden prepreg by autoclave processing. A variety of composite components such as cowls and ducts have been fabricated. The first C_f/PI composite component, produced using PMR-15/carbon fiber, is the ducts for General Electric's F404 air engine, which is approximately 2000 mm long and 1000 mm in diameter. The fabrication is performed by an autoclave route and involves several stages. The shell is first fabricated and tested by nondestructive testing, which is then adhesive bonded with ply build-ups using PMR-15 as the adhesive. The build-ups are then drilled and the shell is cut into two parts. Split-line stiffeners and titanium end-flanges are then attached. The duct was successfully installed on the F404 air engine. Compared with the titanium version the PMR-15 duct is about 3 kg (15%) lighter with a significant cost reduction. PMR-type polyimide/carbon fiber composites are currently the leading composites for high-temperature applications. Other components fabricated from PMR-15/carbon fiber composites include inner cowls, the core cowl for the CF6 aero-engine, swirl frames, nozzle flaps, shuttle orbiter aft body flaps, and fan blades.

Apart from engines and nacelles, there are many applications in the missile and aircraft field, including the fins and bodies of missiles, the lightweight ducting of aircrafts, and the airframe of the high-speed aircraft. Advanced aerospace vehicles such as high-speed aircraft and RLV need high-temperature carbon fiber composites for structural applications. PMR-15/carbon fiber composites have been used for many years due to their good processability and high quality.

PE-endcapped polyimide/carbon fiber composites have been extensively evaluated in the HSCT program [56,45]. The main operational issues for the HSCT included high temperature, high stress, object impact, thermal cycling and exposure to moisture and aircraft fluids. The composite components were required to retain acceptable material properties at 177 °C for 6×10^4 h of long-term servicing. One PE-endcapped oligimide, designed PETI-5, was selected as the matrix resin of the carbon fiber composite. PETI-5 resin exhibited good autoclave processability during the fabrication of large bonded and composite structures at 350-371 °C for 1 h under 1.41 MPa of pressure. The cured resin with a T_g of 270 °C exhibited excellent combinations of properties including high toughness, high strength, moderate modulus, good moisture, and solvent resistance.

The fabrication of support structures such as frames, ribs, or stringers in the HSCT program was preferably performed by RTM process due to their low manufacturing costs, requiring the resin with low (preferably <3.0 Pa·s) and stable (>2 h at the injecting temperature) melt viscosity during the part fabrication. Although PETI-5 exhibited good autoclave processability, the melt viscosity was too high (about 6×10^4 Pa·s at 370 °C) to produce composite component by RTM. In order to satisfy this requirement, PETI-RTM resins were developed, which were scaled-up to 30 kg quantities and successfully used in the fabrication of high-quality 2400 mm-long curved F-frames by RTM.

After the closing of the HSCT program in 1999, research continued to develop RTM resins with higher cured T_gs and better mechanical properties than PETI-RTM for the use in composite structures for RLVs. The operational environment for RLV composites differs significantly from that of the HSCT in the higher temperature performance is needed for shorter periods of time. Hence, a series RTM resins, including PETI-298 with cured T_g of 298 °C, PETI-330 with cured T_g of 330 °C and PETI-350 with cured T_g of 350 °C were successfully developed by NASA [46]. Good mechanical properties were obtained on the composite laminates and excellent retentions of RT properties were observed after aging at the elevated temperatures in air for as long as 1000 h.

A series of high-temperature polyimide/carbon fiber composites have been developed for aerospace applications at the Aerospace Research Institute of Materials & Processing Technology in Beijing, China, which have servicing temperatures of 320 °C, 370 °C, and 500 °C, respectively. Fig. 3.41 shows photographs of typical structural and functional components of polyimide composites [57]. The PMR-type KH-304/carbon fiber composites have been successfully used for the fabrication of the out cowls for air-engines at the Aviation Research Institute of Material Processing in Beijing, China (Fig. 3.41).

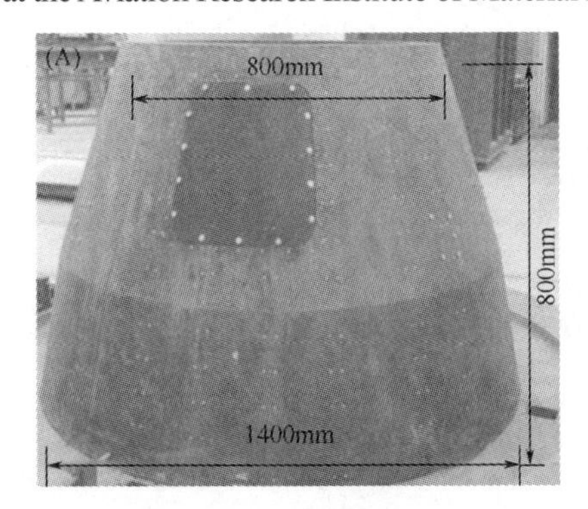

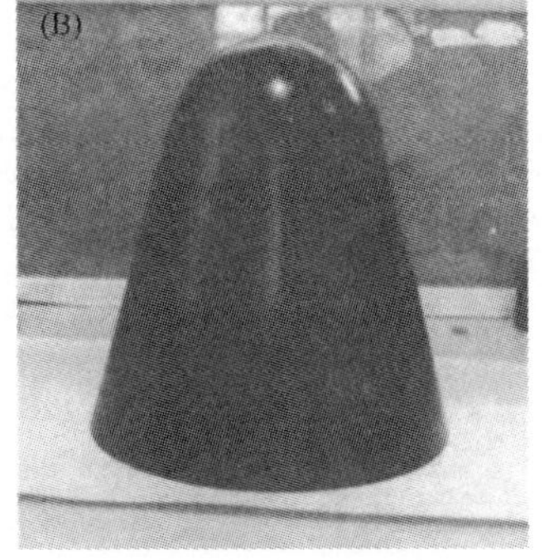

FIGURE 3.41 Typical components of polyimide/carbon fiber composites for high-temperature applications (>300 °C).

3.6 Summary

Polyimide matrix resins play a key role in the fabrication of composite components, not only determining the component's thermal servicing and mechanical properties, but also its processing methods and quality. However, it is very difficult to design and synthesize polyimide matrix resins for high-temperature composite component fabrications, because the matrix resins are required to have not only enough thermal and mechanical properties, but also appropriate melt processability. The NA-endcapped thermoset polyimides have been developed for 316 °C carbon fiber composites applications. The PE-endcapped thermoset polyimides have been demonstrated the suitability for different processing methods such as autoclave and RTM. The carbon fiber composites produced have exhibited excellent combined properties for 370 °C applications.

REFERENCES

[1] G. Lubin, S.J. Dastin, Aerospace applications of composites, *Handbook of composites*, edn, Springer, 1982, pp. 722-743.

[2] D. Wilson, H.D. Stenzenberger, P.M. Hergenrother, Polyimides., Springer, Dordrecht, 1990.

[3] R, L.H. Polyimide polymers. *USA*. US3528950, 1970.

[4] P. Delvigs, T.T. Serafini, G.R. Lightsey. Addition-type polyimides from solutions of monomeric reactants, 1972.

[5] T.T. Serafini, P. Delvigs, G.R. Lightsey, Thermally stable polyimides from solutions of monomeric reactants, J. Appl. Polymer Sci. 16 (4) (1972) 905.

[6] T.T. Serafini, P. Delvigs, W.B. Alston. Pmr polyimides-review and update, 1982.

[7] P.J. Dynes, R.M. Panos, C.L. Hamermesh, Investigation of the crosslinking efficiency of some additional curing polyimides, J. Appl. Polymer Sci. 25 (6) (1980) 1059-1070.

[8] A.C. Wong, W.M. Ritchey, Nuclear magnetic-resonance study of norbornene end-capped polyimides. 1. Polymerization of n-phenylnadimide, Macromolecules 14 (3) (1981) 825-831.

[9] D. Wilson, Pmr-15 processing, properties and problems - a review, Bri. Polymer J. 20 (5) (1988) 405-416.

[10] P. Young. Characterization of initial cure reactions in propargyl and nadic end capped model compounds. NASA Technical Reports Server (NTRS), 1981.

[11] V. Jigajinni, P. Preston, V. Shah, S. Simpson, I. Soutar, N. Stewart, Structure-property relationships in pmr-15-type polyimide, High Perform. Polym 5 (1993) 239-258.

[12] I. Soutar, B. Woodfine, P.N. Preston, V.B. Jigajinni, N.J. Stewart, J.N. Hay, Structure-property relationships in pmr-type polyimide resins. 1. Novel anthraquinone-based pmr resins, Polymer 34 (24) (1993) 5048-5052.

[13] B. Woodfine, I. Soutar, P.N. Preston, V.B. Jigajinni, N.J. Stewart, J.N. Hay, Structure-property relationships in polymerization of monomeric reactant type polyimide resins. 2. New polyimides incorporating alkylenedianilines, Macromolecules 26 (24) (1993) 6330-6334.

[14] R.H. Pater. The 316 °C and 371 °C composite properties of an improved pmr polyimide: LaRC-RP46. The 36th International SAMPE Symposium and Exhibition, San Diego, CA, 1991, pp. 15-18.

[15] T.H. Hou, S.P. Wilkinson, N.J. Johnston, R.H. Pater, T.L. Schneider. Processing and properties of im7/laRC-RP46 polyimide composites. The 39th International SAMPE Symposium and Exhibition, Anaheim, CA, 1994.

[16] B.D. Potter, F.G. Yuan, R.H. Pater, The effect of molecular-weight on transverse microcracking in high-temperature laRC-RP46t polyimide composites, J. Adv. Mater. 25 (1) (1993) 30-34.

[17] T.J. Xiao, S.Q. Gao, A.J. Hu, X.C. Wang, S.Y. Yang, Thermosetting polyimides with improved impact toughness and excellent thermal and thermo-oxidative stability, High Perform. Polymers 13 (4) (2001) 287-299.

[18] R.D. Vannucci, Pmr polyimide compositions for improved performance at 371-degrees-c, Sampe Quarterly-Soc. Adv. Mater. Process Eng. 19 (1) (1987) 31-36.

[19] R.D. Vannucci, D. Cifani. The 700 f properties of autoclave cured pmr-2 composites, 1988.

[20] J.Y. Hao, A.J. Hu, S.Q. Gao, X.C. Wang, S.Y. Yang, Processable polyimides with high glass transition temperature and high storage modulus retention at 400°C, High Performance Polymers 13 (3) (2001) 211-224.

[21] A.J. Hu, J.Y. Hao, T. He, S.Y. Yang, Synthesis and characterization of high-temperature fluorine-containing pmr polyimides, Macromolecules 32 (24) (1999) 8046-8051.

[22] J.Y. Hao, A.J. Hu, S.Y. Yang, Preparation and characterization of mono-end-capped pmr polyimide matrix resins for high-temperature applications, High Performance Polymers 14 (4) (2002) 325-340.

[23] Y.F. Li, A.J. Hu, X.C. Wang, S.Q. Gao, S.Y. Yang, Synthesis and properties of pmr type poly(benzimidazopyrrolone-imide)s, J. Appl. Polymer Sci. 82 (7) (2001) 1600-1608.
[24] Y.M. Shang, L. Fan, S.Q. Gao, A.J. Hu, S.Y. Yang, Processable pmr poly(pyrrolone-imide)s matrix resins and their short carbon fiberreinforced composites, High Performance Polymers 16 (1) (2004) 39-54.
[25] X.G. Jin, L.Y. Huang, S. Yi, S.Y. Yang, A.J. Hu, Thermosetting process and thermal degradation mechanism of high-performance polyimide, J. Analy. Appl. Pyrol. 64 (2) (2002) 395-406.
[26] J.W. Connell, J.G. Smith, P.M. Hergenrother. International SAMPE Technical Conference Series. Vol. 30, p. 545, 1998.
[27] J.M. Criss, C.P. Arendt, J.W. Connell, J.G. Smith, P.M. Hergenrother, Resin transfer molding and resin infusion fabrication of high-temperature composites, Sampe J. 36 (3) (2000) 32-41.
[28] F.L. Hedberg, F.E. Arnold, Phenylethynyl-pendant polyphenylquinoxalines curable by an intramolecular cycloaddition reaction, J. Polymer Sci. Part A: Polymer Chem. 14 (11) (1976) 2607-2619.
[29] A. Banihashemi, C.S. Marvel, Aromatic polyether, aromatic-ketone, aromatic-sulfones as laminating resins. 11. Polymers derived from 2,2′-diiododiphenyl-4,4′-dicarboxylic acid, J. Polymer Sci. Part A: Polymer Chem. 15 (11) (1977) 2653-2665.
[30] F.W. Harris, S.M. Padaki, S. Vavaprath, Polym. Prepr. 21 (1) (1980) 3.
[31] P.M. Hergenrother, Poly(phenylquinoxalines) containing phenylethynyl groups, Macromolecules 14 (4) (1981) 898-904.
[32] F.W. Harris, A. Pamidimukkala, R. Gupta, S. Das, T. Wu, G. Mock, Synthesis and characterization of reactive end-capped polymide oligomers, J. Macromol. Sci. Chem. A21 (8-9) (1984) 1117-1135.
[33] F.W. Harris, K. Sridhav, S. Das, Polyimide oligomers terminated with thermally-polymerizable groups, Polym. Prepr. 25 (1) (1984) 110.
[34] P.M. Hergenrother, J.G. Smith, Chemistry and properties of imide oligomers end-capped with phenylethynylphthalic anhydrides, Polymer 35 (22) (1994) 4857-4864.
[35] T.H. Hou, B.J. Jensen, P.M. Hergenrother, Processing and properties of im7/peti composites, J. Composite Mater. 30 (1) (1996) 109-122.
[36] J.G. Smith, J.W. Connell, P.M. Hergenrother, The effect of phenylethynyl terminated imide oligomer molecular weight on the properties of composites, J. Composite Mater. 34 (7) (2000) 614-628.
[37] J. Chen, H. Zuo, L. Fan, S. Yang, Synthesis and properties of novel meltable fluorinated aromatic oligoimides endcapped with 4-phenylethynylphthalic anhydride, High Performance Polymers 21 (2) (2009) 187-204.
[38] D.W. Cho, L.T. Drzal, Characterization, properties, and processing of LaRC^{TM} peti-5 as a high-temperature sizing material. I. Ftir studies on imidization and phenylethynyl end-group reaction behavior, J. Appl. Polymer Sci. 76 (2) (2000) 190-200.
[39] B. Liu, M. Ji, F. Lin, S. Yang, Phenylethynyl-endcapped polymerizable monomer reactants poly(amic ester) resins for high impact-toughened carbon fiber composites, High Performance Polymers 25 (2) (2013) 225-235.
[40] X.M. Fang, D.F. Rogers, D.A. Scola, M.P. Stevens, A study of the thermal cure of a phenylethynyl-terminated imide model compound and a phenylethynyl-terminated imide oligomer (peti-5), J. Polymer Sci. Part A:Polymer Chem. 36 (3) (1998) 461-470.
[41] J.W. Connell, J.G. Smith, P.M. Hergenrother, M.L. Rommel, Neat resin, adhesive and composite properties of reactive additive peti-5 blends, in: B.A. Wilson, B.J. Hunter, L.L. May Rand Clements (Eds.), Materials - The Star at Center Stage, edn, Vol. 30, 1998, pp. 545-556.
[42] J.M. Criss, J.W. Connell, J.G. Smith. Resin transfer molding of phenylethynyl imides. In: Wilson B.A., Hunter B.J., May Rand Clements L.L. (eds.), Materials-the Star at Center Stage, edn, Vol. 30: 341-350, 1998.
[43] J. Smith Jr, J. Connell, P. Hergenrother, Thermal and mechanical properties of phenylethynyl-containing imide oligomers, Sci. Adv. Mater. Proc. Eng. Ser. 43 (1998) 93-102.
[44] J.W. Connell, J.G. Smith, P.M. Hergenrother. Composition of and method for making high performance resins for infusion and transfer molding processes. USA. US6359107B1, 2002.
[45] J.G. Smith, J.W. Connell, P.M. Hergenrother, J.M. Criss, Resin transfer moldable phenylethynyl containing imide oligomers, J. Composite Materi. 36 (19) (2002) 2255-2265.
[46] J.W. Connell, J.G. Smith, J.M. Criss, High temperature transfer molding resins: Laminate properties of peti-298 and pett-330, High Performance Polymers 15 (4) (2003) 375-394.
[47] J.G. Smith, J.W. Connell, P.M. Hergenrother, L.A. Ford, J.M. Criss, Transfer molding imide resins based on 2,3,3′,4′-biphenyltetracarboxylic dianhydride, Macromol. Symp. 199 (2003) 401-418.
[48] H.J. Zuo, J.S. Chen, H.X. Yang, A.J. Hu, L. Fan, S.Y. Yang, Synthesis and characterization of melt-processable polyimides derived from 1,4-bis(4-amino-2-trifluoromethylphenoxy)benzene, J. Appl. Polymer Sci. 107 (2) (2008) 755-765.
[49] C.-n Su, M. Ji, L. Fan, S.-y Yang, Phenylethynyl-endcapped oligomides with low melt viscosities and high t(g)s: Effects of the molecular weights, High Performance Polymers 23 (5) (2011) 352-361.
[50] Y. Yang, L. Fan, M. Ji, S. Yang, Phenylethynylnaphthalic endcapped imide oligomers with reduced cure temperatures, Europ. Polymer J. 46 (11) (2010) 2145-2155.

[51] Y. Yang, L. Fan, X. Qu, M. Ji, S. Yang, Fluorinated phenylethynyl-terminated imide oligomers with reduced melt viscosity and enhanced melt stability, Polymer 52 (1) (2011) 138-148.

[52] W. Zhao, Pmr-type polyimide matrix composites and their applications, Aeros. Mater. Technol. 39 (4) (2009) 1-5.

[53] L. Pan, W. Zhao, H. Liu, C. Cui, N. Lin, Processing method of polyimide matrix composites for 420°C application, Aeros. Mater. Technol. 46 (4) (2016) 52-55.

[54] Y. Gao, Y. Shi, K. Wang, Y. Yang, L. Long, L. Pan, High-temperature mechanical properties of carbon fiber reinforced polyimide resin matrix composites mt300/kh420(i) - tensile and interlaminar shear properties, Acta Mater. Compos. Sinica 33 (6) (2016) 1206-1213.

[55] Y. Gao, Y. Shi, K. Wang, Y. Yang, L. Long, L. Pan, High-temperature mechanical properties of polyimide resin matrix mt300/kh420 composits (ii)-Flexural properties, Acta Mater. Compos. Sinica 33 (12) (2016) 2699-2705.

[56] P.M. Hergenrother, Development of composites, adhesives and sealants for high-speed commercial airplanes, Sampe J. 36 (1) (2000) 30-41.

[57] W. Zhao, L. Wang, L. Pan, H. Liu, C. Zhao, Recent advances in polyimides matrix structural composites, Aeros. Mater. Technol. 43 (4) (2013) 14-19.

Chapter 4

Super Engineering Plastics and Foams

Shi-Yong Yang, Hai-Xia Yang and Ai-Jun Hu
Institute of Chemistry, Chinese Academy of Sciences, Beijing, China

4.1 Introduction

The market requirement for high-performance polymers (HPPs) is currently experiencing about 9%-10% annual growth. There are many challenging applications in military and aerospace, microelectronics, optoelectric display, healthcare, and down-hole oil and gas industries, etc., which need the HPP resins that combine very high heat and chemical resistance with good melt processability. There are three major categories of HPPs on the market: amorphous thermoplastics, semicrystalline thermoplastics, and nonmelting and imidized materials. The continuous long-term servicing temperature and the chemical resistance are the primary considerations for the selection of HPP materials by customers. The secondary considerations include wear resistance and flammability, etc.

Amorphous thermoplastics such as poly(ether-imide) and polyethersulfone are typically characterized by relatively high glass transition temperature (T_g) values (217-230 °C), which have good retention of strength, stiffness, and dimensional stability at elevated temperatures. However, the practical applications are always restricted by their poor chemical resistance, particularly to highly alkaline materials, and the limited continuous servicing temperature of lower than 200 °C. Semicrystalline polymers such as poly(phenylene-sulfide) (PPS) and poly(ether ether ketone) (PEEK) exhibit high melting temperatures (285-334 °C) as well as good chemical resistance, but they are susceptible to dimensional change and have relatively low T_gs (85 °C for PPS and 150 °C for PEEK, respectively). Currently, imidized polymers such as polyamide-imide (PAI) and aromatic polyimide (PI) offer a good combination of amorphous and semicrystalline materials with excellent thermal, mechanical, and electrical properties, but they fall short in mass productivity and high cost due to the long thermal curing cycles, postmold curing or crystallization, or milling parts from extruded stock shapes. For instance, although PAI can be injection molded as a thermoplastic, it must be postcured for up to 15 days in order to achieve its maximum mechanical properties. In its uncured state, PAI is a brittle, low-strength material. After a long-duration thermal curing cycle, the imidization process is completed, which raises the mechanical strengths, wear, and thermal stabilities.

Aromatic polyimides can be molded with long cure cycles by compression molding. The parts can be fabricated by cutting or milling the thermally cured shapes such as sheets and slab stocks. Hence, the geometry of the final parts is limited. Although there are some other melt-processable thermoplastic polyimides (TPI) on the market, they usually have low T_gs and require postmold crystallization to achieve maximum properties.

In this chapter, thermally stable polyimide materials, especially the super engineering plastics and heat-resistant foams will be discussed in detail, including the chemical structures, molding processability, mechanical properties, and applications.

Advanced Polyimide Materials. DOI: https://doi.org/10.1016/B978-0-12-812640-0.00004-4

4.2 Compression-Molded Polyimide Materials

Aromatic polyimides, due to their rigid/stiff backbone structures, are usually high-melting-temperature materials, which cannot melt, giving an easily flowing molten fluid suitable for injection or infusion like common polymers. For instance, the polyimides derived from PMDA and ODA have T_g of about 385 °C and a theoretical melting point of 592 °C [1]. They do not melt at temperatures up to its theoretical melting point and directly decomposes either anaerobically or in air before the temperature reaches its theoretical melting point. Hence, a molding method for fabrication of inherently intractable aromatic polyimides was developed by Dupont in the 1960s. The shaped polyimide sheets or disks were first prepared by structuring of the intractable polyimide resin powders followed by high-temperature sintering. In this process, aromatic polyimide powder derived from PMDA and ODA is placed in a mold and heated without exertion of pressure to about 300 °C/10 min. Pressure is then applied and maintained at a level of 20 MPa for about 2 min. A shaped article is produced, which is then further heated in a vacuum oven at 450 °C/5 min. These molded polyimide shapes can be ground, cut, and drilled to form an amazing variety of shaped articles, including bushings, roller guides, gears, and bearings, which are useful over a wide temperature range and suitable for continuous servicing at 315 °C. Besides parts prepared from polyimide resin alone, filled resins can be also used to prepare parts; fillers include graphite, molybdenum sulfide, and a mixed filler of graphite and Teflon, etc. This process was used to produce the Vespel family of products, including SP-1, SP-21, SP-22, SP-3, etc. Table 4.1 compares the typical properties of Vespel products.

An improvement in the process for preparation of Vespel products involves the use of a ram extruder for compacting the polyimide resin. Polyimide resin was extruded at temperatures below its T_g into a rod; the rod was then sintered under nitrogen by heating cycles up to a maximum of 400 °C, followed by cooling. Although the basic preparation process of Vespel molded products is successful, it is very complex and not readily adaptable to conventional molding machinery. Hence, a great deal of research was performed on the modification of polyimide backbone structures to improve resin's melt flow ability while concurrently retaining the thermal stability and mechanical properties at elevated temperatures of the molded parts.

The moldable processability of polyimide resins were improved by polycondensation of oxydiphthalic anhydride (ODPA) with different aromatic diamines. ODPA was selected because of its flexible ether-linked chemical structure, being beneficial to decreasing the melting point and T_g of polyimide resins. The polyimide resin derived from ODPA and ODA (PI-ODPA/ODA) was moldable at 370-390 °C, and then crosslinkable by heating in air at 400-450 °C [2]. The possibility of preparing shaped polyimide articles by pressuring below 400 °C and subsequently crosslinking above 400 °C was considered as an interesting potential route to compete with Vespel molded products.

An alternative approach to obtain melt-processable polyimides is the modification of BTDA-based polyimides using sulfone-containing diamine as monomer [3]. The polyimides derived from BTDA and 3,3′-diaminodiphenylsulphone can be melt processed at 250-350 °C. Moldings were prepared by compression molding at 220-280 °C/45 min both for neat resin and for filled resins with 40 %(w) of carbon fibers.

Significant research at Dupont on melt-fusible polyimides was also performed. Codiamine compositions which, when polymerized with PMDA, gave polyimides which could be readily melt-pressed in film at 400 °C [4,5]. The related diamines included 2.2-bis[3,5-dichloro-4-(4-aminophenoxy)phenyl]propane (4Cl-BAPP) with T_g of 314 °C and T_m of 413 °C for the PMDA-based polyimides. In addition, PMDA-based polyimides which were melt-processable included those derived from the 50/50 codiamines of 1,3-bis[3-aminophenoxyl]benzene (1,3,3-APB) and 2,2′-bis[4-(4-aminophenoxyl) phenylpropane] (BAPP) with T_g of 257 °C and T_m of 354 °C. Similarly, polyimide prepared derived from the codiamine of BAPP and 1,3,3-APB (1/3), exhibit T_g of 222 °C and T_m of 308 °C, respectively [5,6].

Modification of the aromatic dianhydride segment of the polyimides was also an effective pathway to improve the melt processability. 2,2-Bis(3,4-diacarboxy-phenyl) hexafluoropropane dianhydride (6FDA) confers melt flowability on polyimides with a variety of diamines [7]. The polyimide derived from 6FDA and ODA prepared via stoichiometric equivalence was not melt-flowable; however, the use of 4 %(mol) excess of diamine resulted in a polyimide which was melt processable at 390-420 °C.

Substantially improved melt flowability was obtained when 6FDA formed a polyimide with 5% excess of 1,3-bis(4-aminophenoxy) benzene (1,3,4-APB); the resulting polyimide has a melt index of 11.3 g·min^{-1} at 390-420 °C and a T_g of 229 °C. The 40/60 coploymer with ODA has a T_g of 295 °C and a melt index of 1.4 g·min^{-1} at 390-420 °C, respectively.

TABLE4.1 Typical Properties of Vespel Products

Property	SP-1	SP-21	SP-22	SP-3
CTE (×10^{-6} °C^{-1})	54	49	38	52
Thermal conductivity (W·m^{-1}·°C^{-1})	0.35	0.87	1.73	0.47
Deflection temperature at 2 MPa (°C)	360	360	—	—
Tensile strength at 23 °C (MPa)	86.2	65.5	51.7	58.5
Elongation at breakage at 23 °C (%)	7.5	4.5	3.0	4.0
Flexural strength at 23 °C (MPa)	110.3	110.3	89.6	75.8
Flexural modulus at 23 °C (GPa)	3.1	3.8	4.8	3.3
Notched Izod impact strength at 23 °C (J·m^{-1})	42.7	42.7	—	21.3
Coefficient of friction PV=0.875 MPa·m·s^{-1}	0.29	0.24	0.30	0.25

TABLE4.2 Typical Properties of Neat Avimid *N*

Property	Values
Coefficient of thermal expansion (×10^{-6} °C^{-1})	56
Density (g·cm^{-3})	1.43-1.45
T_g (°C)	340-370
Tensile strength at 23 °C (MPa)	110
Elongation at breakage at 23 °C (%)	6.0
Flexural strength at 23 °C (MPa)	117
Flexural modulus at 23 °C (GPa)	4.2
Fracture toughness at 23 °C (kJ·m^{-2})	2.45
Notched Izod impact strength at 23 °C (J·m^{-1})	42
Rockwell hardness (E-scale)	70
Weight loss (%)	
After 100 h at 371 °C in N_2	0.3
After 100 h at 371 °C in air	1.5

Avimid N polyimide product was prepared derived from 6FDA and a MPD/PDA mixture, which is melt processable at the temperature of its T_g region (340-370 °C) [8]. The Avimid N can be used neat or in solution in the preparation of composites. Table 4.2 shows the typical properties of Avimid N.

4.3 Injection and Extrusion Processed Polyimide Materials

TPI offer several potential advantages over thermoset materials. First, TPI has an indefinite shelf life, low moisture absorption, excellent thermal stability and chemical resistance, high toughness and damage tolerance, short and simple processing cycles and potential in reductions of manufacturing costs. Second, TPI can be remelted and reprocessed, thus the damaged structures can be repaired by applying heat and pressure. Thirdly, TPI offer advantages with respect to environmental concerns. Usually, TPI has very low toxicity because it is the completely imidized polymer, and does not contain any reactive chemicals. Due to the remelt possibility by heating and redissolvability in solvents, TPI can be recycled or combined with other recycled materials in the market to make new products.

Conventional aromatic polyimides do not have enough melt flow properties for injection or extrusion moldings; only limited fabrication processes such as compression plus sintering methods could be employed. Hence, significant efforts have been devoted to developing melt-processable polyimide resins. Much effort has been focused on exploiting the relationship between chemical structures and combined properties of aromatic polyimides. Based on the experimental results to improve TPI's melt processability, some commercial TPI materials have been developed, such as amorphous LaRC-TPI resin ($T_g \sim 250$ °C), Ultem resin ($T_g \sim 217$ °C), and semicrystalline Aurum resins ($T_g \sim 250$ °C and $T_m \sim 380$ °C).

LaRC-TPI derived from BTDA and 3,3′-diaminobenzophenone was developed by NASA, which has a T_g of 260 °C and could be melt-molded by conventional techniques at temperatures of <350 °C from neat polyimide resin or from filled resins with filler or fiber [9]. The molded resins are normally amorphous and have a transient crystalline form which melts at 272 °C. LaRC-TPI shows excellent thermal stability, no weight loss was detected below 400 °C.

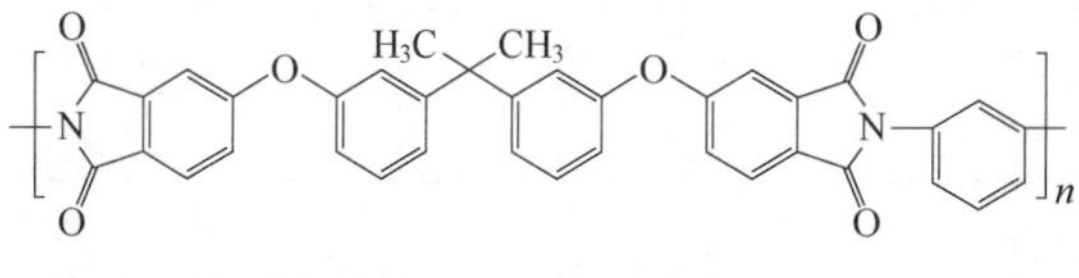

FIGURE 4.1 Chemical structure of Ultem polyimides.

Under the licensing arrangement with NASA, Mitsui-Toatsu has developed a series of products based on LaRC-TPI [10] with much lower melt viscosities (< 20 Pa·s) than the conventionally NASA-prepared version. It was found that the low viscosity is temporary and attributed to the transient form of crystallinity. Upon melting, the polymer was reordered to a higher melting form with increasing the viscosity to a level of 10^4-10^5 Pa·s. Although the melt viscosity is high, it is still melt processable. This melt-processable polyimide resin is in reality a *co*-poly(imide/amic acid) which can convert into polyimide on heating. The transient crystalline form appears to be the melting point of the copolymers. In addition, Rogers has also developed a series of LaRC-TPI-based polyimide materials under the tradename Durimid, including varnish, films, and adhesives. Engineering plastic parts can be produced from the neat polymer or filled resins with graphite (Gr), Teflon, or molybdenum sulfide (MoS_2).

In recent years, a newly developed TPI was commercialized including Ultem and Extem resins by SABIC Innovative Plastics. With its high temperature capability ($T_g \sim 267$ °C and 311 °C) and high melt flow abilities, Extem resin differentiates its position within TPIs as well as other HPPs. Ultem polyimide was amorphous with T_g of 215 °C [11], which has reasonable thermal stability with 1% of original weight loss in air of 446 °C and 5% of original weight loss in air of 528 °C [12]. Ultem polyimide (Fig. 4.1), attributed to the aliphatic and ether-bridge segments in its backbone structure, is soluble in polar and chlorinated solvents, but is unharmed by hydrocarbons such as oil and gasoline. At equilibrium, Ultem polyimide absorbs 1.2% of water. The yield stress of the wet polyimide is reduced by 30% compared with the dry polymer at room temperature [13]. The lack of crystallinity and low T_g resulted in obvious decreases in mechanical strengths on heating; the tensile strength decreased about 50% for dry polymer and about 30% for wet polymer.

Ultem polyimide can be reinforced by a variety of materials such as glass or carbon fibers. The polymer can be injection-molded, compression-molded, and has important applications in automotive, aerospace, and electric industries.

Extem resins have a great combination of melt processability and outstanding long-term servicing performance, which narrows the performance vs. processability gap of the current high-heat thermoplastic and thermoset materials. Extem resin has the highest T_g of any amorphous thermoplastic (311 °C), offers high chemical resistance, high strength and stiffness, robust dimensional stability and high inherent flame resistance. Meanwhile, it can be mass manufactured with high productivity, suitable for melt processing such injection, extrusion, or thermoforming without requiring any postcure or crystallization steps to achieve its maximum performance (Table 4.3).

Extem resin exhibits extreme heat stability and improved chemical resistance to hydrocarbons and chlorinated solvents, and has been extensively used as a component in the down-hole oil and gas industry. Customers in aerospace and military industries also benefit from its high as-molded performance without additional postcuring time or the long cycle times of thermoset polyimide resins. Extem resin's potential lead-free solder capability-along with its balance strength, stiffness and dimensional stability makes it an excellent candidate for electrical, electronic, and semiconductor applications. In the automotive industry, Extem resin could offer opportunities for metal and alloy replacement. And in healthcare, it offers an excellent combination of physical and chemical properties showing significant benefits for hemocompatible membranes.

The Extem family includes two series of products, Extem UH and Extem XH. Extem UH products have the highest heat resistance with T_g up to 311 °C and broad chemical resistance, while Extem XH shows a T_g up to 267 °C along with higher melt flowability. Additionally, Extem resins exhibit good dimensional stability, high strength and stiffness, high creep resistance at elevated temperatures, outstanding flame and smoke, as well as smoke-toxicity performance with limiting oxygen index (LOI) of up to 45%-47%.

TABLE4.3 The Typical Properties of Extem and Ultem Resins

Property	Ultem 1000	Ultem XH6050	Extem XH	Extem UH
Physical				
Density ($g \cdot cm^{-3}$)	1.27	1.31	1.31	1.37
Water Abs. (%) Equilib. (23 °C)	1.25	1.75	2.36	1.37
Moisture Abs. (%) 24 h/50% RH/23 °C	0.25	0.60	0.60	—
MFR ($g \cdot (10\ min)^{-1}$)				
337 °C/6.6 kgf 367 °C/6.6 kgf	8	12.5	7.1	—
Mechanical				
Tensile strength (MPa)	114	95	90	120
Tensile modulus (GPa)	3.6	3.5	3.3	3.8
Elongation (%)				9
Flexural strength (MPa)				175
Flexural modulus (GPa)				3.5
Izod impact (ft-ib/in)				
Notched (23 °C)	1	1.3	0.82	1.5
Unnotched (23 °C)	34	35	37	—
Reverse notched (23 °C)	24	35	27	33
Thermal				
T_g (°C)	217	247	267	280
Vicat softening (B) (°C)	209	242	260	—
HDT (°C)				
0.45 MPa, 3.2 mm	206	232	249	263
1.82 MPa, 3.2 mm	191	216	235	240
CTE ($\times 10^{-6}$ °C^{-1}) (23 °C-150 °C)				
$CTE_{Flow\ Direction}$, 10^{-5} mm·mm^{-1}· °C^{-1}	5.40	5.30	5.00	4.50

Extem UH products have tailorable T_g from 260 °C to 311 °C. The first commercial product can be melt processed on standard injection-molding equipment. This product is a fully amorphous thermoplastic with a T_g of 280 °C, the highest of any amorphous thermoplastic available today. The HDT was measured at 265 °C. The primary drawback compared with other semicrystalline resins is its relatively low T_g, resulting in the losses in stiffness, strength, and creep resistance above that temperature. Extem UH products also offer continuous-servicing temperatures as high as 230 °C. In addition to very high heat resistance, Extem UH also possesses superior chemical resistance. For example, Ultem resin, though is amorphous, has good resistance to chemicals such as organics and acids, in which Ultem CRS5001 resin currently offers the best chemical resistance in that family. Extem UH resins radically outperform Ultem CRS5001 resin in resistance to hydrocarbons such as toluene and aggressive solvents like methyl ethyl ketone and chlorinated solvents such as methylene chloride, demonstrating excellent chemical resistance.

Extem XH products show extreme heat capability balanced with greater processability. While maintaining its high creep resistance and strength at elevated temperatures, Extem XH resin can be injection molded to fill thin-wall, complex parts. Extem XH resin is a high-strength material for short-term heat resistance above 200 °C, providing potential for use in lead-free soldering of electronics. At room temperature, many high-performance amorphous and semicrystalline thermoplastics exhibit very high tensile strength (over 100 MPa). Although Extem XH resin is not the strongest material at ambient temperature, it outperforms many others at temperatures up to 240 °C, especially with higher retention of tensile strength and creep resistance. Extem XH resins, with the best flow properties in the Extem family, exhibit about 20% greater melt flowability than Ultem 1000 resin at process temperatures of 412 °C and 365 °C, respectively. Extem XH resin is suitable for complex, multicavity tools to injection molding parts with 0.010 inch wall thickness at flow lengths of over 0.250 in. Longer flows up to 4 in. for wall sections 0.030 in. thick have been achieved at injection pressures of 32,000 psi. Molding cycle times are comparable to those for Ultem resin. Minimal warpage is observed only at high ejection temperatures.

Extem resin was designed to be completely melt-processable on high-temperature injection molding and extrusion equipment without postcuring steps. The Extem resins must be fully dried to < 0.2% moisture uptake prior to injection molding. This can be accomplished in standard dryers at 150-175 °C with a dew point of –29 to –40 °C for 4-6 h. Injection molding equipment must be operated over the resin melt temperatures of 390-420 °C. Oil or electric-heated molds that are capable of working in the range of 160-200 °C must be handled at the temperatures required to maximize flow length and minimize molded-in stress. Molds must be built by P20 or stronger steel to accommodate necessary thermal loads, and designed to accommodate a stiffer overall flow, incorporating multiple gates and flow runners when challenged with a high flow-length/wall-thickness ratio. Generous gates, such as fan gates, should be employed when possible to minimize internal stresses. Subgates have also been successfully used to mold the Extem resins. Grades containing mold release will be offered to minimize sticking and improve ejection efficiency. Mold designs should incorporate typical amorphous isotropic shrinkage rates of 0.5%-0.7% for unfilled grades. Glass-filled materials will exhibit lower but anisotropic shrinkage. In general, barrels and screws of conventional materials of construction are acceptable, though bimetallic screws are suggested. Screw designs should have 16:1 to 24:1 L/D and low compression ratios of 1.5:1 to 3:1. Compression should be accomplished with a gradual and constant taper without sudden transitions in order to minimize excessive shear and material degradation. Barrels should be sized to accommodate 40%-80% of the shot size to minimize residence time.

Extem resin meets the international standards for recyclability. This is demonstrated by single-pass 100% regrind studies, which indicate properties of over 95% retentions of tensile and elongation. The amount of regrind utilized was governed by performance requirements of the application. Extrusion of stock shapes, profiles, and thin films is also achievable with Extem resin. Melt temperature was from 370 °C to 405 °C.

The scientists at the Institute of Chemistry, Chinese Academy of Sciences, have carried out a systematical research on developing melt-processable TPI. The TPI resins have designed polymer backbones and controlled molecular weights, which were prepared by a one-step thermal polycondensation procedure as shown in Fig. 4.2. The offset of the aromatic dianhydride (ODPA) to the aromatic diamine (6FAPB) and end-capping agent (PA) was used to adjust the polymer molecular weights. The water evolved during the thermal imidization was removed simultaneously from the reaction system by azeotropic distillation. The TPI resins showed excellent melt flow capability, which can be reinforced by chopped fiber or filler to give high-quality TPI composites. The pure TPI resin powder was dried at 205 °C for 6 h in a vacuum dryer to completely remove the moisture in the resin, which was then extruded at elevated temperatures with carbon fiber, glass fiber, MoS_2 or PTFE powders to afford the TPI molding particulates, which were then melt processed at 350-370 °C with molding temperature at 150-160 °C.

Fig. 4.3 depicts the melt viscosities of carbon fiber-filled TPI resins with different loadings at different temperature. The molten viscosities of molding particulates decreased gradually with increasing the temperature scanned at 200-400 °C, primarily attributed to the melting of TPI resin in the molding particulates. The minimum melt viscosities of the molding particulates increased with increasing of the carbon fiber loadings. Meanwhile, the melt viscosity at the processing temperature (360 °C) was increased from 4.7×10^3 Pa·s for CF-TPI-10 to 9.4×10^3 Pa·s for CF-TPI-30, indicating that the addition of carbon fiber in TPI resins increased the melt viscosities of the molding particulate, thus lowering their melt processabilities. The glass fiber-filled TPI molding particulates showed much higher complex melt viscosities than the pure TPI resin.

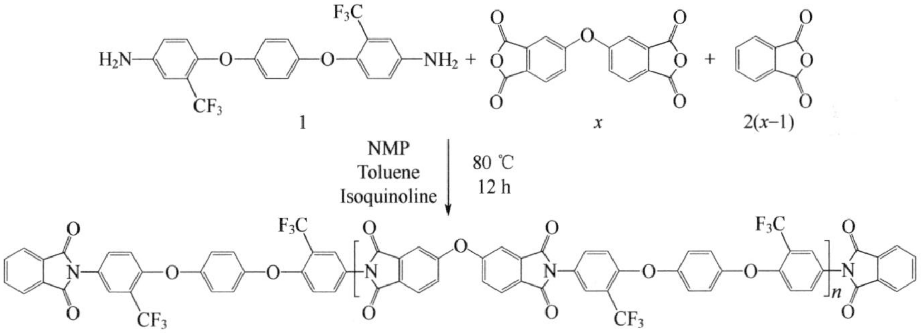

FIGURE 4.2 Preparation of thermoplastic polyimide resins.

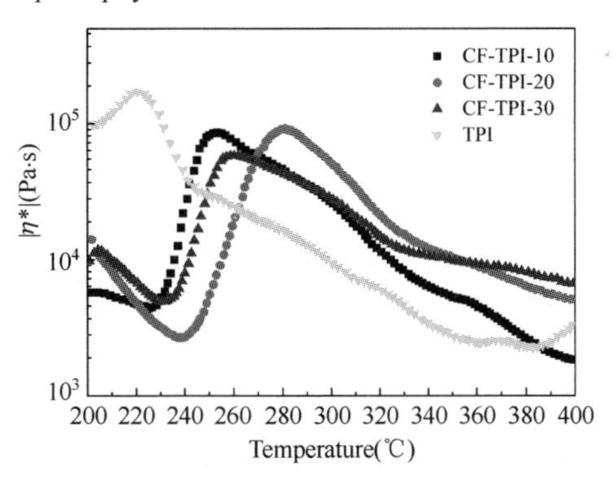

FIGURE 4.3 Dynamic rheological behaviors of the carbon fiber-filled TPI molding particulates.

Table 4.4 shows the melt flow index of the molding particulates. The pure TPI resin showed highest melt flow index of 10.1 g·(10 min)$^{-1}$ at 360 °C under a pressure of 10 kg, which could be employed to inject very thin-walled complex parts due to its excellent melt flow ability. Fig. 4.4 shows a representative thin-walled complex part which has a wall thickness of 0.2 mm. In general, the TPI resins filled with different fillers such as graphite, molybdenum disulfide, and poly(tetrafluoroethylene), etc., all showed good melt flow properties with a melt flow index of > 2.0 g·(10 min)$^{-1}$. In comparison, the carbon fiber-filled TPI molding particulates showed better melt flow properties than the glass fiber-filled ones. For instance, CF-TPI-30 has a melt flow index of 2.4 g·(10 min)$^{-1}$ at 360 °C under 10 kg, compared with GF-TPI-30 (0.5 g·(10 min)$^{-1}$ at 360 °C under 21.6 kg). Meanwhile, the filler loadings also have obvious effect on lowering the melt flow index. For instance, MoS_2-TPI-30 has a melt flow index of 0.8 g·(10 min)$^{-1}$ at 360 °C under 10 kg, much lower than MoS_2-TPI-15(3.2 g·(10 min)$^{-1}$ at 360 °C under 10 kg).

TABLE4.4 Melt Flow Index at 360 °C of the Molding Particulates

	Temperature (°C)	M·kg^{-1}	t·s^{-1}	Melt Flow Index (g·(10 min)$^{-1}$)
TPI	360	10	10	10.1
CF-TPI-10	360	10	30	3.8
CF-TPI-20	360	10	30	2.5
CF-TPI-30	360	10	30	2.4
GF-TPI-15	360	21.6	60	0.9
GF-TPI-30	360	21.6	60	0.5
Gr-TPI-15	360	10	10	8.9
Gr-TPI-40	360	10	30	6.0
MoS_2-TPI-15	360	10	30	3.2
MoS_2-TPI-30	360	10	60	0.8
PTFE-TPI-20	360	10	60	2.4

Fig. 4.5 shows the DMA curves of a representative molded composite (CF-TPI-20), in which the peak temperature in the Tan δ curve was at 211 °C. The storage modulus curve did not turn down until the temperature was scanned up to 201 °C, demonstrating that the carbon fiber-filled TPI molded composites have outstanding thermomechanical properties.

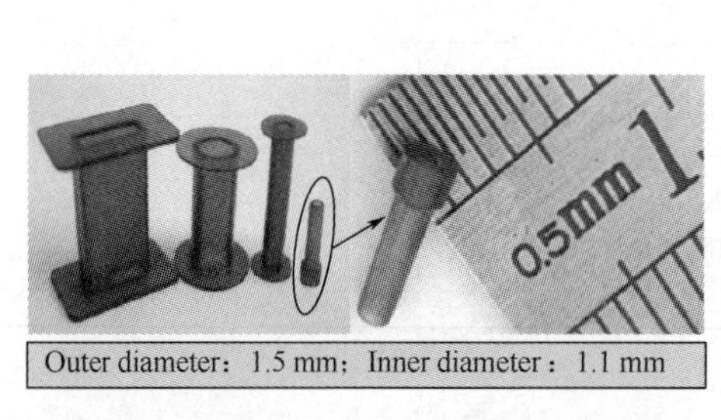

FIGURE 4.4 The injection-molded TPI thin-walled parts.

FIGURE 4.5 DMA curves of CF-TPI-20 molded composite.

Table 4.5 compares the mechanical properties of the injection-molded TPI composites. The pure TPI resin exhibited good combined mechanical properties with tensile strength of 100 MPa, tensile modulus of 5.6 GPa, elongation at break of 57.6%, flexural strength of 154 MPa, flexural modulus of 3.8 GPa, and izod impact (unnotched) of 156 kJ·m^{-2}. The carbon-fiber-filled TPI molded composites possess mechanical properties better than the pure TPI resin with tensile strength of 177-219 MPa, tensile modulus of 7.3-12.4 GPa, flexural strength of 241-327 MPa, and Izod impact of 20.8-24.4 kJ·m^{-2}, demonstrating that carbon fiber has a significant reinforcing effect. The glass fiber-filled TPI molded composites also showed good mechanical strength, but lower modulus than the carbon-fiber-filled ones. The other TPI molded composites filled with graphite, molybdenum disulfide, and poly(tetrafluoroethylene) all showed good combined mechanical properties, demonstrating that the addition of filler did not deteriorate the TPI mechanical properties.

Overall, the thermoplastic polyimide resins with designed polymer backbones and controlled molecular weights can be reinforced with carbon fiber, glass fiber, or modified by adding of solid lubricants such as graphite, poly(tetrafluoroethylene) (PTFE) or molybdenum disulfide (MoS_2) to give

TPI molding particulates. The molding particulates can be injection-molded at elevated temperature to give the TPI composites. Thin-walled molded parts were successfully fabricated. The TPI composites show an excellent combination of thermal and mechanical properties.

4.4 Structures and Melt Processabilities of Aromatic Polyimide Resins

Aromatic polyimides are usually infusible and insoluble due to their long, rigid polymer backbone, and unique chemical structures. It is very difficult to be melt molded by conventional melt-injection or melt-extrusion methods. In recent years, many researchers have made great efforts to improve the melt processability of aromatic polyimides [13-19]. The improvements in melt processability can be to some extent obtained either by introduction of flexible backbones, fluorine-containing groups, *meta*-linkages, bulky pendant groups, or noncoplanar moieties into aromatic polyimide backbones in order to reduce the stiffness of polymer chains and polymer chain-chain interactions, to loosen the polymer chain packing, or to weaken the charge-transfer electronic polarization interactions. However, the improvements in melt processability of aromatic polyimide usually resulted in the sacrificing of mechanical properties at elevated temperatures. For instance, although commercial TPI (Ultem resins) were processed by melt-extrusion or injection methods, their long-term service temperatures were limited to temperatures of lower than 170 °C. Moreover, the thermos-oxidative stability of Ultem polyimides were also relatively poor because of the presence of thermally unstable isopropylidene groups in the polymer backbones [20,21]. Hence, there is still a big concern for developing melt-processable aromatic polyimides with less sacrificing of their long-term service temperature and mechanical properties at elevated temperatures.

TABLE4.5 Mechanical Properties of the TPI Molded Composites

	Tensile Strength (MPa)	Tensile Modulus (GPa)	Elongation at Breakage (%)	Flexural Strength (MPa)	Flexural Modulus (GPa)	Compressive Strength (MPa)	Izod Impact (Unnotched) ($kJ \cdot m^{-2}$)
TPI	100	5.6	57.6	154	3.8	159	156
CF-TPI-10	177	7.3	3.40	241	8.8	168	20.8
CF-TPI-20	177	10.1	2.14	278	13.5	215	20.8
CF-TPI-30	219	12.4	2.34	327	17.8	205	24.4
GF-TPI-15	121	4.8	3.42	191	5.7	145	20.4
GF-TPI-30	137	7.5	2.20	211	9.2	182	23.1
CF-TPI-45	106	8.8	1.62	242	14.6	207	16.6
Gr-TPI-15	94	4.4	4.37	150	5.9	106	15.4
Gr-TPI-40	64	8.2	1.71	114	11.2	86	6.7
MoS_2-TPI-15	102	8.9	13.8	148	3.8	116	23.7
MoS_2-TPI-30	83	4.1	3.36	145	5.6	115	13.1
PTFE-TPI-20	98	3.0	6.64	156	4.3	99	26.9

In this section, the influences of backbone structures, molecular weights, and end-cappings on the melt processabilities of aromatic polyimides will be discussed in detail.

4.4.1 Polyimide Backbone Structures

In order to investigate how the backbone structures affect the melt processability of aromatic polyimide, two aromatic diamines, 4,4′-bis(3-amino-5-trifluoromethyl phenoxy)biphenyl (*m*-6FBAB) and 4,4′-bis(4-amino-5-trifluoromethylphenoxy) biphenyl (*p*-6FBAB) were employed to react with various aromatic dianhydrides including PMDA, BPDA, 6FDA, and ODPA to prepare a series of aromatic polyimides. The polyimides were prepared with a given offset (3.0%) of aromatic dianhydride to the fluorinated diamines through the high-temperature polycondensation procedure (Fig. 4.6). Thus, PI-2a (*m*-6FBAB/PMDA) was prepared derived from m-6FBAB, PMDA and PA as end-capping agent. Similarly, PI-2b (*m*-6FBAB/BPDA), PI-2c (*m*-6FBAB/6FDA), and PI-2d (*m*-6FBAB/ODPA) were prepared from *m*-6FBAB with BPDA, 6FDA, and ODPA. Similarly, PI-4a (*p*-6FBAB/PMDA), PI-4b (*p*-6FBAB/BPDA), PI-4c (*p*-6FBAB/6FDA), and PI-4d (*p*-6FBAB/ODPA) were also prepared from *p*-6FBAB with PMDA, BPDA, 6FDA, and ODPA by the similar procedure.

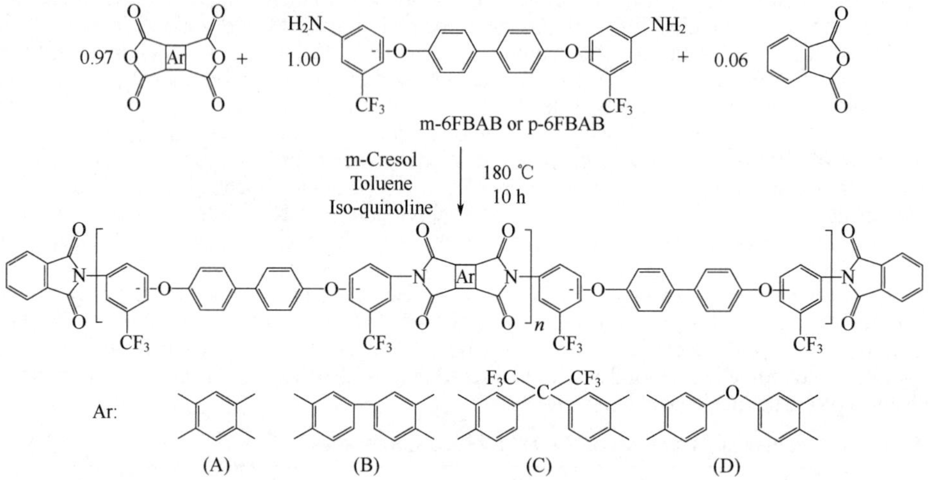

FIGURE 4.6 Synthetic pathways of the fluorinated aromatic polyimides.

Fig. 4.7 shows the melt viscosities of the fluorinated aromatic polyimides as a function of heating temperature. The melt rheological behaviors of the PI-2 series depended greatly on the chemical structures of the polymer backbones. The melt viscosity of PI-2c (*m*-6FBAB/6FDA) decreased abruptly with increasing of temperature greater than 250 °C and reached 1×10^2 Pa·s at 300 °C, which was probably due to the very low molecular weight. PI-2d (*m*-6FBAB-ODPA) exhibited similar melt rheological behavior to PI-2c, with a drop in melt viscosity at 360 °C. PI-2a (*m*-6FBAB/PMDA) and PI-2b (*m*-6FBAB/BPDA) showed the highest melt viscosities of $0.3\text{-}0.8\times10^4$ Pa·s at temperatures above 360 °C. In comparison, PI-4d showed much higher melt viscosity than PI-2d. Meanwhile, PI-4a, PI-4b, and PI-4c were not fusible in the same conditions, probably due to the *para*-substituted aromatic diamine.

The melt stabilities of the fluorinated aromatic polyimides were also investigated by rheology at a suitable shear rate at 330 °C. Fig. 4.8 shows the melt viscosities of PI-2d and PI-4d as a function of time in air and N_2, respectively. It was clear that PI-4d possessed good melt stability in both air and N_2. However, the melt viscosity of PI-2d increased gradually during the measurement.

The fluorinated aromatic polyimides showed excellent thermal properties, with T_gs in the range of 205-250 °C for the PI-2 series and 236-264 °C for the PI-4 series. The T_g was decreased with increasing of polyimide backbone flexibility. Moreover, PI-4 series had higher T_g values than PI-2 series, which is attributed to the linkage style of imide segments in the polymer chains. PI-4 series were *para*-linked imides compared to *meta*-linked ones in PI-2 series. The *para*-linked PI-4 series had more well-packed structures than the *meta*-linked PI-2 series.

Table 4.6 compares the mechanical properties of the solution-cast polyimide-films and melt-molded polyimide sheets. Except for PI-2c (*m*-6FBAB/6FDA), other polyimides could be melt molded at 260-310 °C to give tough and transparent polymer sheets with thicknesses of 1.0-4.0 mm. The tensile strengths were measured in the range of 84.3-102.8 MPa, and the tensile moduli were measured in the range of 1.81-1.98 GPa. The flexural strengths were determined in the range of 126.5-127.6 MPa, and the flexural moduli were determined in the range of 2.60-2.98 GPa. The mechanical properties of the solution-cast polyimide films were also shown in Table 4.6 except PI-4a due to its difficulty being dissolved in common solvents. The tensile strengths and elongations at break for the PI-2 series ranged from 82.3 to 95.5 MPa and 4.8% to 7.0%, respectively, lower than those for the PI-4 series (102.1-104.9 MPa and 7.8%-11.4%).

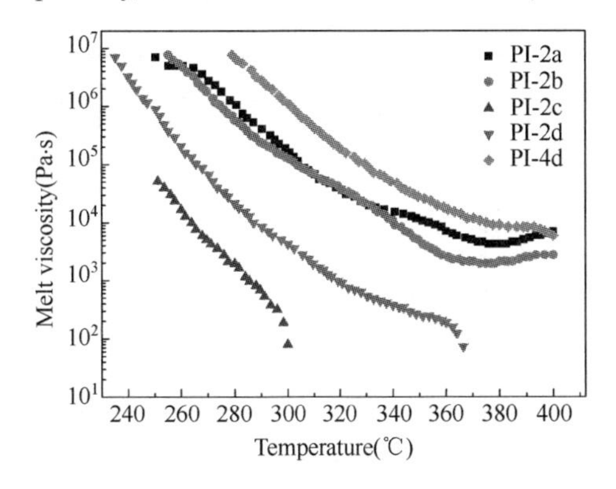

FIGURE 4.7 Melt viscosities of the fluorinated aromatic polyimides as a function of temperature.

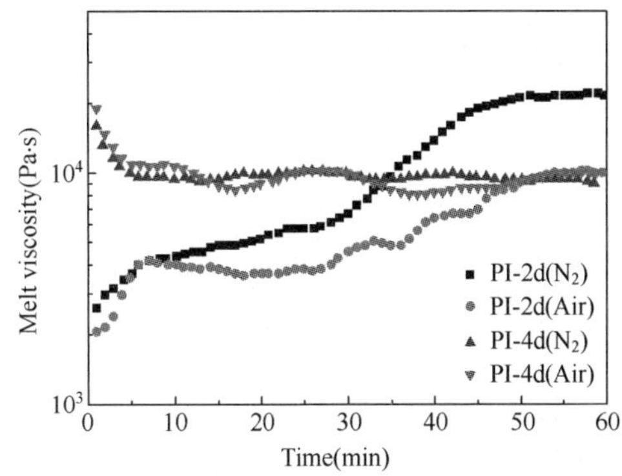

FIGURE 4.8 Melt viscosities of PI-2d and PI-4d in air and N_2 as a function of time at 330 °C

TABLE4.6 Mechanical Properties of the Solution-cast Polyimide Films and Melt-Molded Sheets

PI	Solution-cast Films			Melt-pressed Sheets				
	Tensile Strength (MPa)	Tensile Modulus (GPa)	Elongation (%)	Tensile Strength (MPa)	Tensile Modulus (GPa)	Elongation (%)	Flexural Strength (MPa)	Flexural Modulus (GPa)
2a	82.3	2.79	4.8	85.1	1.81	8.4	127.6	2.75
2b	92.5	2.74	7.0	84.3	1.98	5.1	127.9	2.65
2c	88.9	2.90	4.8	—[a]	—	—	—	—
2d	95.5	2.72	5.1	92.7	1.95	7.5	127.0	2.98
4a	—[a]	—	—	—	—	—	—	—
4b	103.2	2.35	11.4	—	—	—	—	—
4c	102.1	2.42	7.8	—	—	—	—	—
4d	104.9	2.38	10.5	102.8	1.81	19.6	126.5	2.60

[a]*Not detected.*

4.4.2 Controlled Molecular Weights

The molecular weight control was an effective pathway to improve the melt viscosity of polyimides. A series of fluorinated aromatic polyimides with different controlled molecular weights were prepared derived from 1,4-bis(4′-amino-2′-trifluoromethylphenoxy)benzene (6FAPB), ODPA, and PA as the molecular-weight-controlling and end-capping agent (Fig. 4.2). Experimental results indicated that the molecular-weight-controlled aromatic polyimides exhibited a great combination of melt processability, mechanical properties, and thermal stability.

TABLE4.7 The Molecular-weight-controlled Polyimides Prepared With Different Stoichiometric Imbalances

Polyimide	Offset (%)	M_n^T ($g \cdot mol^{-1}$)	M_n ($g \cdot mol^{-1}$)	M_w ($g \cdot mol^{-1}$)	Polydispersity Index (M_w/M_n)	M_n/M_n^T	Inherent Viscosity ($dL \cdot g^{-1}$)
PI-1	4.6	1.5×10^4	3.7×10^4	5.8×10^4	1.57	2.41	0.56
PI-2	2.8	2.5×10^4	5.3×10^4	8.3×10^4	1.57	2.11	0.69
PI-3	2.0	3.5×10^4	6.1×10^4	9.4×10^4	1.53	1.75	0.84
PI-4	1.4	5.0×10^4	7.8×10^4	11.7×10^4	1.49	1.56	1.06

Table 4.7 shows the molecular weights determined by GPC, which were controlled by the adjustment of the offsets of ODPA to 6FAPB from 4.6% to 1.4%. Both M_n and M_w increased with a reduction of the stoichiometric imbalances. For instance, the polyimide (PI-1) prepared at a 4.6% offset had M_n=3.7×10^4 and M_w=5.8×10^4, whereas PI-4, prepared at the lowest offset (1.4%), showed the highest molecular weights (M_n=7.8×10^4 and M_w=11.7×10^4), more than twice those of PI-1. There was no obvious difference in the polydispersity indices in the range of 1.49-1.57. The inherent viscosities of the polyimides increased gradually with increasing molecular weights, which were increased from 0.56 for PI-1 to 1.06 for PI-4.

The melt viscosities of the polyimides were dependent on both the temperature and the molecular weight (Fig. 4.9). The melt viscosities of the polyimides decreased abruptly at temperatures above 250 °C and then stabilized. The onset points in the curves of the melt viscosity versus the heating temperature were 260 °C for PI-1, and 310 °C for PI-4. Clearly, the molecular weights had an obvious effect on the lowest melt viscosity of the polyimides.

Table 4.8 shows the melt viscosities at 360 °C, which increased from 950 (PI-1) to 0.6×10^4 (PI-2), 2.4×10^4 (PI-3), and 9.6×10^4 Pa·s (PI-4). The melt viscosity thought to be suitable for injection molding should be lower than 0.1×10^4 Pa·s [22], hence PI-1 might be processed by melt injection. The temperature at which the melt viscosity was 0.5×10^6 Pa·s ($T_{0.5MPa \cdot s}$) was increased from 267 °C for PI-1 to 320 °C for PI-4.

The melt stability was determined by the melt viscosity ratio (MVR), which is the ratio of the MV30 to MV5 (the melt viscosities measured at 30 min (MV30) and 5 min (MV5), respectively, after the solid sample was completely molten). The polymer could be considered as having good melt stability if its MVR value was located in the range of 1.0-1.5.

Fig. 4.10 shows the melt viscosities of PI-2 and PI-4 measured as a function of time at 330 °C in air and N_2, respectively. In the experiment, the test specimen was loaded when the chamber was preheated, and

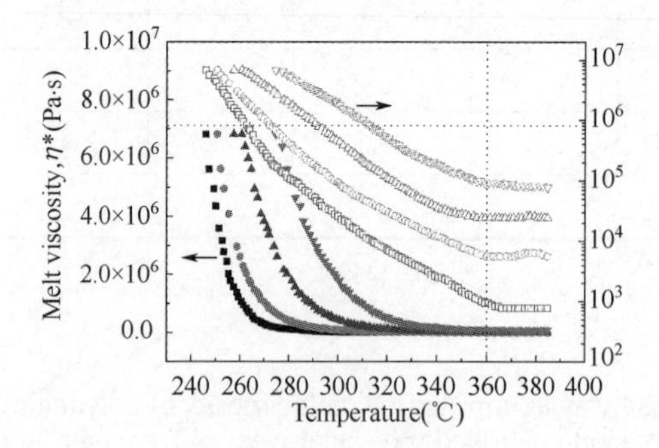

FIGURE 4.9 Melt viscosities of the molecular-weight-controlled polyimides as a function of temperature.

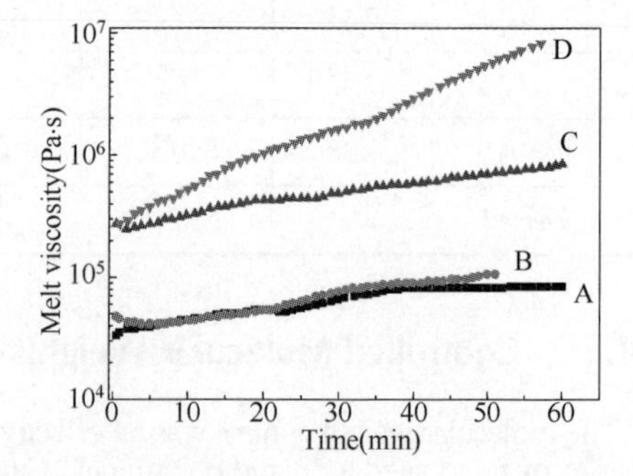

FIGURE 4.10 Dependence of the melt viscosity of the molecular-weight-controlled polyimides on the heating time: (A) I-2 in N_2; (B) PI-2 in air; (C) PI-4 in N_2; and (D) PI-4 in air.

the measurement was started when the temperature of the specimen and chamber was stabilized at 330 °C. The melt viscosity ratio measured in nitrogen (MVR1) and the melt viscosity ratio measured in air (MVR2) are summarized in Table 4.9. PI-I and PI-2 showed good melt stabilities, compared with PI-3 and PI-4, demonstrating that the melt stability decreased with increasing molecular weights. Moreover, the MVR2 data were higher than the MVR1 data, implying that air had a negative effect on the melt stability.

TABLE4.8 Rheological Behavior Data of the Molecular-weight-controlled Polyimides at 360 °C

Polyimide	η_{360}(Pa • s)[a]	$T_{0.5MPa\cdot s}$(°C)[b]	MVR_1	MVR_2
PI-1	950	267	0.94	1.23
PI-2	5800	279	1.59	1.83
PI-3	23890	299	1.66	3.78
PI-4	96210	320	1.74	4.31

[a] *Melt viscosity of the polyimide at 360 °C.*
[b] *Determined with a rheometer.*

TABLE4.9 Mechanical Properties of the Solution-cast Films and the Melt-molded Sheets

	Solution-cast Polyimide Films			Melt-molded Polyimide Sheets				
Polyimide	Tensile Strength (MPa)	Tensile Modulus (GPa)	Elongation (%)	Tensile Strength (MPa)	Tensile Modulus (GPa)	Elongation (%)	Flexural Strength (MPa)	Flexural Modulus (GPa)
PI-1	—[a]	—	—	89.3	1.92	5.9	137.8	3.38
PI-2	98.9	2.56	7.5	105.9	1.82	31.6	127.8	2.71
PI-3	103.5	2.64	7.7	108.5	1.85	18	126.5	2.84
PI-4	104.4	2.67	8.2	113.5	1.98	14.9	115.7	3.85

[a]*The film was too brittle to prepare the tensile test specimen.*

The T_g values were measured in the range of 216.5-226.2 °C, T_5 was in the range of 562.6-586.7 °C, and T_{10} was in the range of 595.4-610.6 °C. The thermal stabilities of the molecular-weight-controlled polyimides were improved slightly with increasing polymer molecular weights. For instance, PI-4, which had the highest molecular weight, exhibited the best thermal stability with T_g=226.2 °C, T_5=586.7 °C, and T_{10}=610.6 °C.

Table 4.9 summarizes the mechanical properties of the solution-cast polyimide films and the melt-molded sheets. The solution-cast polyimide films showed good mechanical properties, including tensile strengths of 98.9-104.4 MPa, tensile moduli of 2.56-2.67 GPa, and elongations at break of 7.5%-8.2%. PI-4 exhibited the highest tensile strength (104.4 MPa) and elongation at break (8.2%), while PI-1 film was too brittle, probably due to its limited molecular weight.

Melt-molded polyimide sheets were prepared by the melt pressing of the polyimide powders in a mold. In a typical experiment, PI-1 powder was charged into a matched mold, which was then placed in a hot press. The mold was heated to 320 °C for 15 min, and then a pressure of 5-6 MPa was applied. After being kept at an elevated temperature for 30 min, the mold was cooled to room temperature. A strong and tough polyimide sheet, light brown in color, was obtained, and it was void- and defect-free, and translucent in appearance. The melt-molded polyimide sheets showed excellent mechanical properties, with tensile strengths of 89.3-113.5 MPa, tensile moduli of 1.82-1.98 GPa, elongations at break of 5.9%-31.6%, flexural strengths of 125.7-137.8 MPa, and flexural moduli of 2.71-3.38 GPa. PI-2 sheets exhibited the best mechanical strengths with a flexural strength of 127.8 MPa, a tensile strength of 105.9 MPa, and an elongation at break of 31.6%.

Fig. 4.11 shows typical stress-strain curves of the molecular-weight-controlled polyimide sheets. PI-2, PI-3, and PI-4 showed distinct yield behaviors and plastic failure while PI-1 showed a brittle rupture, implying that the molecular-weight-controlled polyimide had a critical molecular weight or

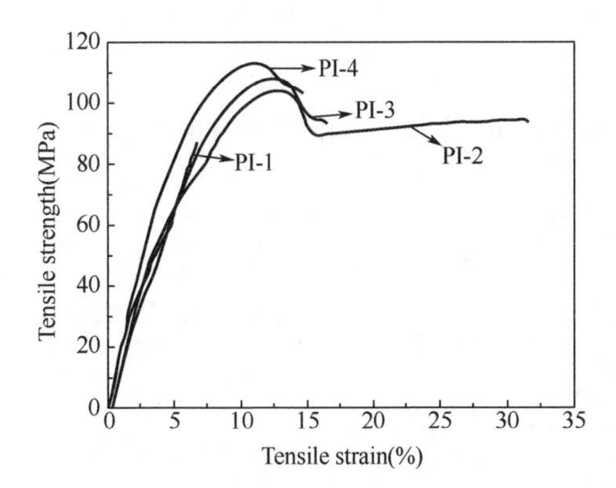

FIGURE 4.11 Strain–stress curves of the melt-molded polyimide sheets at 320 °C in air.

polymer chain length that ensured its good mechanical properties. However, the elongations at break of melt-molded sheets were reduced in the order of PI-2 (31.6%) > PI-3 (18.0%) > PI-4 (14.9%). This might be explained by the deteriorated melt flowing characteristic of the higher-molecular-weight polyimides at the fixed melt-molding temperature (320 °C), which was probably not their optimal melt-molding condition.

Overall, the molecular-weight-controlled aromatic polyimides showed good melt processability for melt pressing. The melt processability depended significantly on the polymer chain length as well as its chemical structure. The melt-processed sheets exhibited a good combination of mechanical and thermal properties, in which the PI-2 derived from ODPA and 6FAPB with 2.8% showed excellent mechanical properties with a flexural strength of 127.8 MPa, a tensile strength of 105.9 MPa, and an elongation at break of 31.6%.

4.5 Meltable Thermoplastic Polyimide Composites

In recent years, the microelectronic packaging technology has been focused on better performance, higher reliability, more miniaturization and lower weight, etc. One of the critical issues in advanced microelectronic packaging is to develop high-density multilayer packaging substrates [23,24]. Currently, the conventional packaging substrates are usually produced based on the E-glass cloth-reinforced polymer matrix composite laminates as the substrate core, in which the E-glass cloths are employed as the reinforcing materials due to their outstanding electrical insulation and dielectric properties, good mechanical strength and modulus, etc., and the polymer resins are used as the composite matrices. Multifunctional epoxy resins (FR-4), cyanate esters or bismaleimide triazine (BT) resins are usually adopted as the composite matrices due to their great balance of mechanical, electrical, processing properties and cost-effectiveness. However, the conventional packaging substrates have always showed limited thermal resistance which could not endure the harsh thermal environment in the electronic packaging production line [25-29]. Hence, advanced packaging substrates with high thermal stability and good processability are required for high-density microelectronics packaging.

Aromatic polyimides, due to their excellent thermal stability, mechanical and electrical properties, have been extensively employed in the microelectronic manufacturing and packaging industry such as interlayer dielectrics in multilayer structures, flexible and rigid substrates in high-density electronic packaging [30-34]. However, the conventional aromatic polyimides are generally insoluble and infusible due to the rigid chemical backbone and the strong interaction of polymer chains, which makes them impossible for use as the matrix resins of the E-glass cloth-reinforced composites. Hence, polyimide precursors, i.e., poly(amic acid or ester), instead of aromatic polyimides themselves are usually employed to impregnate the reinforcing glass-fiber cloth to give the resin/glass prepregs, which are then thermally imidized to yield the fully imidized polymers [15,35-37]. During thermal imidization, the organic volatiles including water and alcohol as the thermal condensation byproducts as well as the solvent residue must be carefully handled to prevent the void or defect formation in the composite [38-40]. The voids and defects in the laminate could deteriorate the mechanical and electrical properties, resulting in the early failure of packaged device. Therefore, the development of highly soluble and meltable aromatic polyimides, which can be used as the matrices of glass cloth-reinforced composite is still a major technological issue.

The meltable thermoplastic polyimides (MTPI) with designed molecular structures and controlled molecular weights derived from 6FAPB or 6FBAB, ODPA and PA were evaluated as the composite matrices to impregnate E-glass cloth (EG) to yield MTPI/EG prepregs. After complete removal of the organic volatiles, the MTPI/EG prepregs were plied up and melt processed at elevated temperature to produce high-quality composite laminates as packaging substrate core. The melt processability as well as mechanical and electrical properties of the composite laminates have been systematically investigated. MTPIa (ODPA/6FAPB/PA, the offset of ODPA to 6FAPB=2.8%) was prepared as shown in Fig. 4.2. MTPIb (ODPA/6FBAB/PA, the offset of ODPA to 6FBAB=2.2%) was also synthesized by a similar procedure.

Table 4.10 compares the thermal and mechanical properties of the neat MTPI resins. The MTPI resins exhibited good thermal stability with T_g of 218 °C for MTPIa and 238 °C for MTPIb, thermal decomposition temperature at 10% weight loss (T_{10}) of 593 °C for MTPIa and 601 °C for MTPIb in nitrogen, and coefficients of thermal expansion (*CTE*) of 48 and 47×10^{-6} °C^{-1}, respectively. The MTPI polyimide films, which were prepared by casting of the MTPI resin solutions on glass plates followed by thermal baking, showed outstanding mechanical properties with tensile strengths of 104-121 MPa, tensile moduli of 1.98-2.53 GPa, and elongations at breakage of 13.5%-34.5%, respectively. Obviously, MTPIb showed better mechanical and thermal properties than MTPIa, probably due to the biphenyl group contribution in the polyimide backbone [41]. In addition, the MTPI polyimide resins exhibited low dielectric constants (3.0) and low moisture absorptions (1%).

TABLE4.10 Thermal and Mechanical Properties of the Neat MTPI Resins

Properties	Methods	MTPIa	MTPIb
T_g (°C)	DSC	218	238
T_{10} (°C)	TGA	593	601
[a]*CTE* ($\times10^{-6}$ °C^{-1})	TMA	48	47
Tensile strength (MPa)	GB/T 1447—2005	121	104
Tensile modulus (GPa)		1.98	2.54
Elongation at breakage (%)		13.5	34.5
Water absorption (%)	25 °C, 24 h	0.97	0.81

[a]*Calculated from TMA curves between 50 °C and 200 °C.*

TABLE4.11 Melt Complex Viscosity of the Neat MTPI Resins

	Melt Complex Viscosity (Pa·s) at		Minimum Complex Viscosity (Pa·s)	Melt Viscosity Variation at 300 °C for 1 h (Pa·s)
	280 °C	300 °C		
MTPIa	2.0×10^4	9.1×10^3	0.48×10^4 at 346 °C	1.2×10^4 to 2.0×10^4
MTPIb	3.0×10^4	2.6×10^4	1.8×10^4 at 348 °C	4.5×10^4 to 5.3×10^4

Fig. 4.12 shows the melt processabilities of the neat TPI. The MTPI resins exhibit low melt viscosities at > 300 °C, in which MTPIa was $(0.9\text{-}2.0)\times10^4$ Pa·s at 280-300 °C, lower than MTPIb ($(2.6\text{-}3.0)\times10^4$ Pa·s) (Table 4.11). The lowest melt viscosity (0.48×10^4 Pa·s) for MTPIa was measured at 346 °C, compared with that of MTPIb (1.8×10^4 Pa·s at 348 °C). The lower melt viscosity of MTPIa could be explained by its more flexible polymer backbone.

Fig. 4.13 compares the melt resin viscosity of MTPI resins as a function of isothermal standing time at 300 °C. The melt viscosity variation of MTPIa was measured in the range of $(1.2\text{-}2.0)\times10^4$ Pa·s, in comparison with $(4.5\text{-}5.2)\times10^4$ Pa·s for MTPIb (Table 4.11), implying that the melt MTPI resins were very stable at 300 °C. Fig. 4.14 shows the melting behaviors of the MTPI/EG prepregs after being thermally baked at 200 °C for 1 h. The melt viscosities of the MTPI/EG cloth prepreg were decreased gradually with temperature increasing. The melt viscosity of MTPIa/EG cloth prepreg was 7.5×10^7 Pa·s at 280 °C and 5.0×10^7 Pa·s at 300 °C, respectively, compared to 1.5×10^8 Pa·s at 280 °C for MTPIb/EG cloth prepreg. The melt viscosity of MTPIb prepreg then remained at the same level until the temperature was scanned up to 350 °C.

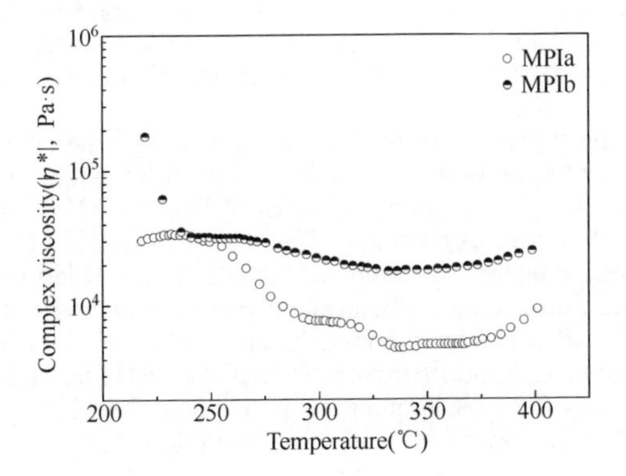

FIGURE 4.12 Melt complex viscosities versus temperature for the neat MTPI resins.

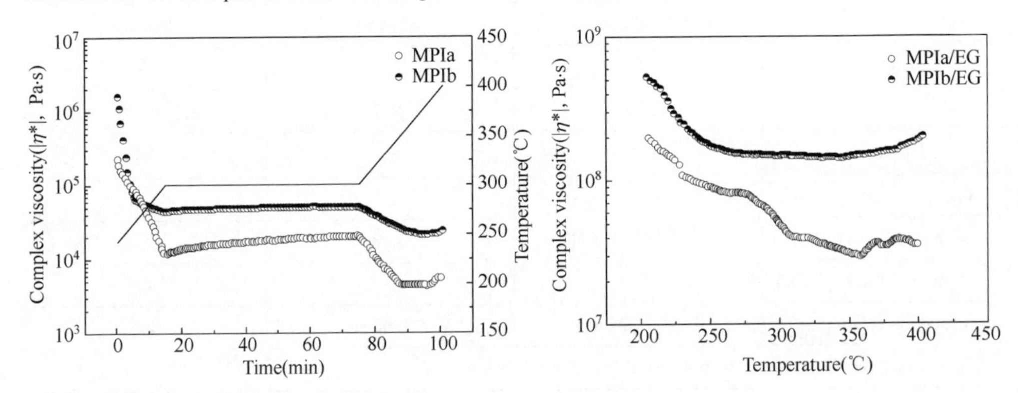

FIGURE 4.13 Melt complex viscosity stability versus isothermal aging time at 300 °C for the neat MTPI resins.

FIGURE 4.14 Melt complex viscosity versus temperature for the MTPI/EG prepregs.

TABLE4.12 Thermal Properties of the MTPI/EG Composite Laminates

	T_g (°C)	T_d (°C)	[a]T_5 (°C)	[b]T_{10} (°C)	CTE ($\times10^{-6}$ °C^{-1})		[e]G' (°C)	[f]G'' (°C)	[g]tan δ (°C)
MTPIa/EG	215	420	489	602	[c]19	[d]61	206	208	214
MTPIa/EG	233	445	521	647	[c]15	[d]64	217	218	224

[a]*Temperature at 5% weight loss.*
[b]*Temperature at 10% weight loss.*
[c]*XY-axis direction.*
[d]*Z-axis direction.*
[e]*The onset temperature in the storage modulus curve.*
[f]*The peak temperature in the loss modulus curve.*
[g]*The peak temperature in the tan δ curve.*

The melt complex viscosities of the MTPI/EG prepregs (Fig. 4.14) were about three orders of magnitude higher than those of neat MTPI resins (Fig. 4.9). This could be interpreted by the modulus contribution to viscosity. Obviously, the prepregs with 40 %(w). of glass-fiber cloths have larger modulus than neat polyimide resins, resulting in higher complex viscosity than neat resins.

Based on the melt rheology of the prepregs, the MTPI/EG composite and copper foil-clad

MTPI/EG laminates were produced by melt processing of the MTPI/EG prepregs or copper foil-clad MTPI/EG prepregs at 280/300 °C by the thermal diagram.

The weight fractions of the MTPI resins in the composite laminates were determined at 57%± 2%(*w*). Table 4.12 compares the thermal properties of the E-glass cloth-reinforced polyimide laminates (MTPI/EG). The T_g values of MTPIa/EG and MTPIb/EG laminates were determined at 215 °C and 233 °C, respectively. The thermal decomposition temperatures of the laminates were measured in the range of 420-445 °C. The dimensional changes of the MTPI/EG laminates in *XY*- and *Z*-axis directions as a function of temperature are also listed in Table 4.12. The in-plane (*XY*-axis) and the *Z*-axis *CTE* of MTPIa/EG laminate were measured at 19 and 61×10^{-6} $°C^{-1}$, respectively, which were comparable to that of MTPIb/EG laminate (15×10^{-6} and 64×10^{-6} $°C^{-1}$, respectively).

The MTPI/EG showed excellent thermal stabilities. No weight loss was observed until the temperature was scanned up to 400 °C by TGA. DMA results indicated that the storage modulus curves did not turn down until the temperature was scanned up to 206 °C, demonstrating that the MTPI/EGs have outstanding thermomechanical properties.

TABLE4.13 Mechanical Properties of the MTPI/EG Composite Laminates

Property	Test Method	MTPIa/EG	MTPIb/EG
Tensile strength (MPa)	GB/T 1447—2005	231	268
Tensile modulus (GPa)		9.5	8.1
Elongation at breakage (%)		3.3	5.6
Flexural strength (MPa)	IPC-TM-650	432	451
Flexural modulus (GPa)		19.5	19.1
Impact strength ($kJ\cdot m^{-2}$)	GB/T 1451—2005	84	103
Peel strength ($N\cdot mm^{-1}$)	IPC-TM-650	1.5-1.8	1.5-1.7

Table 4.13 summarizes the mechanical properties of the MTPI/EG composite laminates. The MTPIa/EG laminates showed tensile strength of 231 MPa, tensile modulus of 9.5 GPa, elongations at breakage of 3.3%, flexural strength of 432 MPa, flexural modulus of 19.5 GPa, and impact strength of 84 $kJ\cdot m^{-2}$, respectively, demonstrating great combined mechanical properties. In comparison, MTPIb/EG laminates showed better mechanical strength than MTPIa, probably due to the more rigid biphenyl-contained polymer backbone. Although the mechanical strength of composites is highly dependent on the reinforced materials (E-glass cloth), the experimental results indicate that the toughness of the matrices also affect the composite toughness and impact strength. Additionally, the peel strengths of the Cu/MTPI/EG laminates were measured at 1.5-1.8 $N\cdot mm^{-1}$, showing good adhesion between copper foil and the MTPI/EG composite.

In summary, meltable thermoplastic polyimide resins (MTPIs) with designed polymer backbones and controlled molecular weights were employed to impregnate EG to yield high-quality MTPI/EG prepregs. After complete removal of the organic volatile, the MTPI/EG prepregs were plied up and melt processed to produce high-quality MTPI/EG laminates. Experimental results indicated that the MTPI/EG composite laminates exhibited a good combination of thermal, mechanical, electrical properties, as well as good melt processability. The copper foil-clad MTPI/EG laminates showed good peel strength between copper foil and the composite laminate surface, demonstrating good potential for packaging substrate core laminate applications.

4.6 Reactive End-Capped Meltable Polyimide Resins

As discussed above, the melt-processable thermoplastic polyimide resins were end-capped by phthalic anhydride (PA), which is an unreactive end-capper. Although the resulting polyimide resins can be melt processed, their melt viscosities were very high and T_gs were limited to < 250 °C, hampering their applications in the high-density microelectronics packaging industry. Hence, NA-and

PEPA-end-capped molecular weight-controlled polyimides have been investigated to satisfy the requirements of the microelectronics manufacturing and packaging industry.

4.6.1 NA-end-capped Meltable Polyimide Resins

An NA-end-capped thermoset polyimide resin was prepared derived from 1,4-bis(4- amino- 2-trifluoro-methylphenoxy)benzene (1,4,4-6FAPB), *p*-phenylenediamine (PDA), diethyl ester of 3,3′,4,4′-benzopheno netetracarboxylic acid (BTDE), and monoethyl ester of *cis*-5- norbornene-endo-2,3-dicarboxylic acid (NE). The thermosetting polyimide solutions with solid content of 40 %(*w*). in anhydrous alcohol were homogeneous, with absolute viscosity of 5-20 mPa·s at 30 °C, and showed color changes from yellow to red-brown, depending on the *p*-PDA's concentration, and they were all stable both in viscosity and in color after storing for more than 3 weeks at room temperature (25 °C), and more than 3 months in a refrigerator (4 °C) (Fig. 4.15).

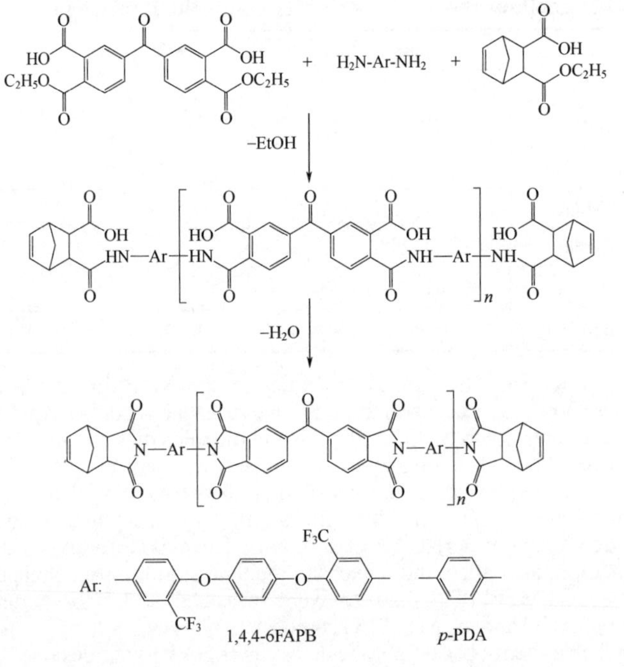

HO, C_2H_5O, OH, OC_2H_5 + H_2N-Ar-NH_2 + OH, OC_2H_5

−EtOH

OH, HN—Ar—HN, HO, OH, NH—Ar—NH, HO, *n*

$-H_2O$

N—Ar—N, N—Ar—N, *n*

Ar: F_3C, O, O, CF_3 1,4,4-6FAPB *p*-PDA

FIGURE 4.15 Synthesis of the thermosetting polyimide resins.

The PMR resin solutions were used to impregnate EG cloth to give high-quality prepregs. The EG/HTPI prepregs were baked up to 240 °C before composite melt processing to remove the organic volatiles as completely as possible. The low volatiles of the B-staged prepregs could reduce the possibility of voids and defects formation in the composite processing. Generally, the PMR polyimide prepregs are treated at temperatures lower than 200 °C to avoid the B-staged resin with too-high melt viscosity. After baking by the same procedure, the B-staged resins had almost the same volatile content. Fig. 4.16 depicts the melt viscosities of typical HTPI-2 resin baked at different temperatures. The results indicated that when the baking temperature was raised from 200 to 240 °C, the minimum resin melt viscosities increased from 121 to 839 Pa·s, but they were still low enough for composite processing. Moreover, the total weight loss in air (50-300 °C) of the B-staged resin treated at 240 °C was only 1.3 %(*w*), compared to 5.5 %(*w*) for the B-staged resin treated at 200 °C. It should be noted that the low volatile outlet was critical for the B-staged prepregs to fabricate void-free copper-clad composite laminates.

TABLE4.14 Mechanical Properties of the EG/HTPI Composite Laminates Thermally Cured at 320 °C/2 h

	Tensile Strength (MPa)	Tensile Modulus (GPa)	Elongation (%)	Flexural Strength (MPa)	Flexural Modulus (GPa)	Impact Strength ($kJ \cdot m^{-2}$)	Peel Strength ($N \cdot mm^{-1}$)
EG/HTPI-0	253	10.9	4.2	724	23.5	47.9	1.24
EG/HTPI-1	270	11.2	4.2	730	24.2	53.5	1.23
EG/HTPI-2	266	11.7	3.3	534	20.0	46.9	1.18
EG/HTPI-5	235	10.4	3.2	419	16.9	—[a]	—[b]
EG/PMR-15	269	10.5	3.3	570	17.9	38.0	NA

[a]*No valid data.*
[b]*No specimens available.*

Fig. 4.17 compares the melt viscosities of the B-staged thermosetting polyimide resins with different *p*-PDA concentrations at different temperatures. The molten resin viscosities decreased with the temperature increase at the beginning stage and then started to increase after a specific temperature. The temperatures, at which the resin had the minimum melt viscosity, increased from 279 °C(91 Pa·s) for HTPI-0 to 284 °C (223 Pa·s) for HTPI-1 and 301 °C (839 Pa·s) for HTPI-2, respectively. In comparison, HTPI-5 had remarkably higher melt viscosity than other resins, showing the minimum melt viscosity of 3728 Pa·s at 267 °C. Therefore, the melt processability of the thermosetting polyimide resins could be controlled by adjusting the molar ratios of the rigid phenyl groups to the flexible ether segments in the resin chemical backbone.

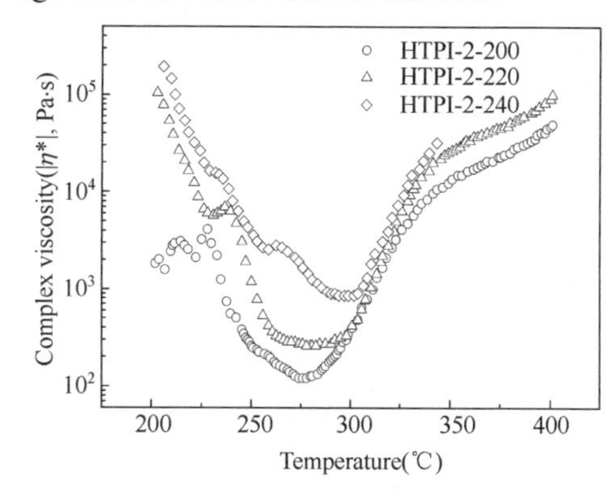

FIGURE 4.16 The melt viscosities versus temperature for HTPI-2 resin baked at different temperature.

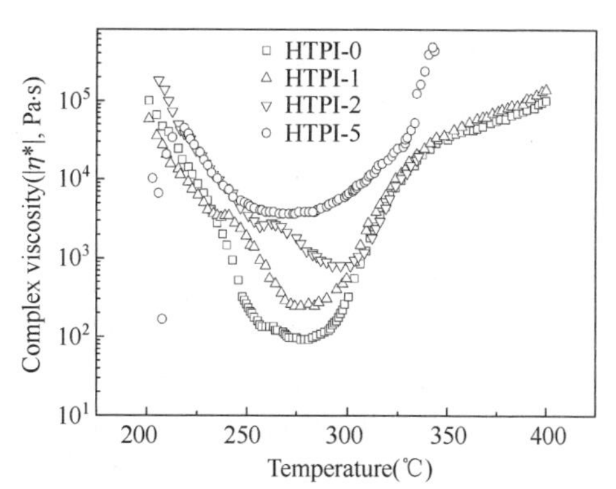

FIGURE 4.17 The melt viscosities versus temperature for the B-staged polyimide resins.

Table 4.14 compares the mechanical properties of the EG/HTPI composites and EG/PMR-15 composite. The EG/HTPI composites possess tensile strength in the range of 253-270 MPa, tensile modulus of 10.9-11.7 GPa, flexural strength of 534-730 MPa, and flexural modulus of 20.0-24.2 GPa. Moreover, the composites showed high impact strength in the range of 46.9-53.5 $kJ \cdot m^{-2}$, the impact toughness of EG/HTPI composites are improved without sacrificing the flexural and tensile strength. The EG/HTPI-5 composite could not give good mechanical properties and valid data in impact strength because of its poor quality derived from the high *p*-PDA loading. It can be concluded that the mole ratio of *p*-PDA in diamines should not exceed 0.2, otherwise the composite will have deteriorated mechanical properties. The peel strength of copper foil from the composite surface in Cu/EG/HTPI laminates was measured in the range of 1.18-1.24 $N \cdot mm^{-1}$, implying that the composite surface has good adhesion to copper foil.

The T_g values increased from 284 °C for EG/HTPI-0 to 322 °C for EG/HTPI-5 and an increase of

38 °C was observed, which was attributed to the increasing of *p*-PDA loadings in the composite. For instance, EG/HTPI-1 composite showed T_g of 288 °C and its storage modulus did not go down until the temperature was scanned to 266 °C, implying that the thermally cured polyimides possess outstanding thermomechanical properties. Fig. 4.18 depicts the dimension changes versus time in a fast temperature ramp and isothermal process at 288 °C for the EG/HTPI composite laminates except for EG/HTPI-5. No delamination in laminates was observed after isothermal aging for 60 min at 288 °C. Then, one of Cu/EG/HTPI laminates (Cu/EG/HTPI-1) was selected to investigate its failure time at different temperatures (Fig. 4.19). The results show that the laminates expanded rapidly as the temperature increased at the heating rate of 40 °C·min^{-1}. The higher the temperature, the larger the expansion was detected. When the Cu/EG/HTPI-1 laminates were isothermally aged at different temperatures, the failure times were drastically changed. When the laminate was isothermally aged at 260 °C, no obvious failure was detected after 60 min aging. However, when the isothermal aging temperature was increased to 288 °C, the "popcorning" phenomenon was observed in 16 min. Moreover, the Cu/EG/HTPI-1 laminate could stand for 4 min at 300 °C, indicating that the composite laminates could withstand the harsh environments in lead-free microelectronics packaging production lines where the temperature could be ramped to 260 °C in a short time. Hence, the EG/HTPI and Cu/EG/HTPI laminates have outstanding thermal reliability, showing good potential for application in high-density packaging substrates.

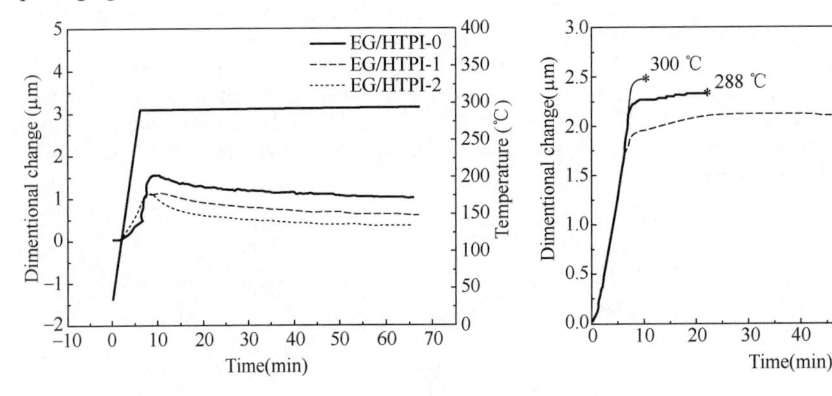

FIGURE 4.18 Dimension changes versus time at 288 °C or the EG/HTPI composite laminates.

FIGURE 4.19 Failure time at different temperature for the Cu/EG/HTPI-1 composite laminates.

Low-dielectric-constant materials play an important role in the development of advanced integrated circuit (IC) manufacturing. A rigid substrate is the main requirement in high-speed signal transmission areas, such as high-end IC packaging and communication equipment in networks, because of its better dimensional stability and impact resistance compared to a flexible substrate. A thinner substrate and higher-frequency signal transmission have been required by the packaging substrate industry in recent years. Therefore, new requirements, such as a low dielectric constant and dielectric loss, good heat resistance, and low warp, have been proposed for package substrate materials. To lower the warping rate of the IC package substrate, the elastic modulus and heat resistance of the substrate material must be enhanced, and the coefficient of thermal expansion (*CTE*) must be reduced.

In the conventional electricity industry, glass-fiber cloth is used as a reinforcement to fabricate substrates [42]. With the rapid development of the dielectric industry, to improve the dielectric properties, various types of glass fibers have been produced to reduce the amount of impurities. E-type glass-fiber cloth was applied, but the composite showed a high dielectric constant of 4.2. Quartz fiber, which has a high content of SiO_2 (99.9%), shows better dielectric properties and thermal properties than glass fiber. Therefore, chopped quartz fiber has been used as a reinforcement in composites. Gao et al. [43] used chopped quartz fiber to reinforce PI, but the *CTE* of PI was (50-52)×10^{-6} °C^{-1}, and the *CTE* of PI/QF was (43-48)×10^{-6} °C^{-1}. It was found that the enhancements of the mechanical properties and thermal properties through the incorporation of chopped fiber were limited. However, there has been little reported on quartz-fiber-cloth implications in the rigid substrate packaging area. In this study, quartz-fiber cloth was applied to lower the dielectric constant and *CTE*.

In order to develop electronic packaging substrates with low dielectric constant, a new polymerization-of-monomer reactant (PMR)-type PI resin with low melt viscosity was prepared as shown in Fig. 4.20, and was used to prepare a series of novel quartz-fiber-cloth-reinforced polyimide substrates with low dielectric constants. The thermosetting PMR solutions have solid content of 37.5 %(w) in anhydrous alcohol with absolute viscosities of 5-15 mPa·s at 25 °C, showing golden colors and stable viscosity after being stored for more than 3 months at room temperature (25 °C).

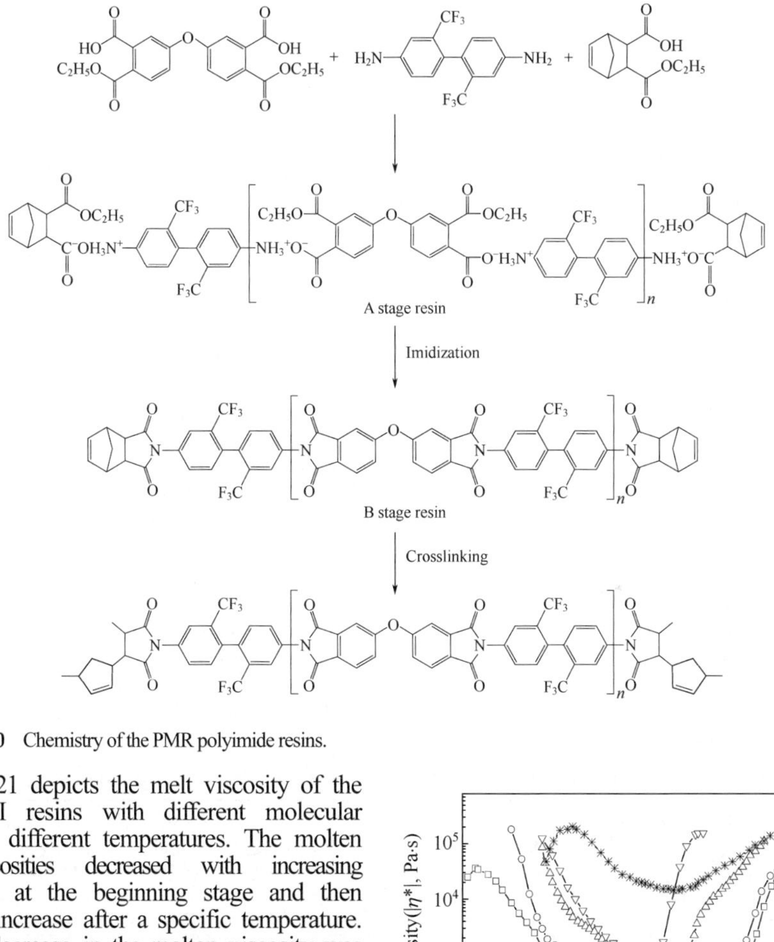

A stage resin

Imidization

B stage resin

Crosslinking

FIGURE 4.20 Chemistry of the PMR polyimide resins.

Fig. 4.21 depicts the melt viscosity of the B-stage PI resins with different molecular weights at different temperatures. The molten resin viscosities decreased with increasing temperature at the beginning stage and then started to increase after a specific temperature. The first decrease in the molten viscosity was primarily attributed to the melting of the B-stage PI resin, and the following increase in molten viscosity was due to the crosslinking reaction of the NA-end-capping groups. The minimum melt viscosities of these five pure resins rose with increasing molecular weight, and the range of temperature at low viscosity diminished. The smaller molecular weight was attributed to the lower melt viscosity. In the preparation of the multilayer composite, the lower melt viscosity of

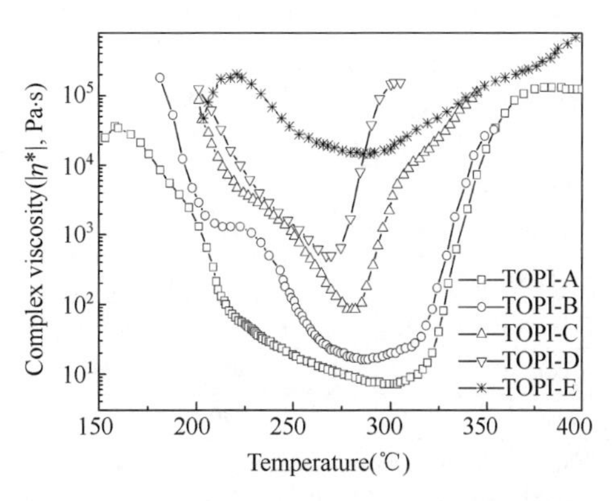

FIGURE 4.21 The melt viscosity of the B-stage PI resins with different molecular weights at different temperatures.

the resin was attributed to a high-quality product with little void.

Table 4.15 shows the mechanical properties of the pure resins cured at 320 °C/2 h. The tensile modulus and tensile strength increased with increasing molecular weight, and the flexural modulus and strength also had the same tendency.

TABLE4.15 Mechanical Properties of the Pristine TOPIs With Different Molecular Weights

Specimen	Tensile Modulus (GPa)	Tensile Strength (MPa)	Elongation (%)	Flexural Modulus (GPa)	Flexural Strength (MPa)
TOPI-A	1.6	59.5	6.1	3.4	142.5
TOPI-B	1.8	75.2	6.8	3.8	147.1
TOPI-C	2.1	91.1	9.9	3.7	157.4
TOPI-D	2.1	95.7	6.0	4.5	184.8

Based on the rheological properties (Fig. 4.21) and mechanical properties (Table 4.15), TOPI-B was chosen as the matrix resin for preparing the laminate. The QF/TOPI prepregs were baked up to 220 °C to remove the organic volatiles as much as possible before melt processing. The reduction of low-boiling-point volatiles of the B-stage prepregs decreased the odds of voids and defect formation in composite processing. Three different curing temperatures were used: after a curing procedure at 320 °C/2 h, QF/TOPI-B-1 was cured at 320 °C/1 h, QF/TOPI-B-2 at 330 °C/1 h, and QF/TOPI-B-3 at 340 °C/1 h.

Table 4.16 listed the mechanical properties of the composite laminates, in which the tensile strength of QF/TOPI-2 was 567 MPa, the flexural strength was 845 MPa, the impact strength was 141 $kJ \cdot m^{-2}$, and the interlaminate strength was 62 MPa. The adhesion properties between PI and the Cu foil were measured at 1.06 $N \cdot mm^{-1}$.

Fig. 4.22 shows the dimensional changes of TOPI-B and QF/TOPI-B in different directions. The *CTE* of QF/TOPI composite laminate in a direction parallel to the interlaminate of the quartz cloth (CTE_{XY}) was 8.4×10^{-6} $°C^{-1}$, and that in the perpendicular direction (CTE_Z) was 43.0×10^{-6} $°C^{-1}$.

Fig. 4.23 shows the dimensional changes vs. time at different temperatures for QF/TOPI-B. Both dimensional changes of QF/TOPI-B at 288 °C were lower than those at 300 °C; this was due to the fact that the chain movement at 300 °C was more active than that at 288 °C, so the interchain distance was larger. No delamination took place in the QF/TOPI composite laminate after 60 min of isothermal conditions at 288 °C and 300 °C, demonstrating that the QF/TOPI-B can withstand the tough circumstances of lead-free electronics processes to 260 °C and has the potential to be used as a high-density packaging substrate.

TABLE4.16 Mechanical Properties of the QF/TOPI-B and EG/HTPI-1 Laminates

Specimen	Resin Content % (*w*)	Resin Content % (vol)[a]	Tensile Strength (MPa)	Tensile Modulus (GPa)	Elongation (%)	Flexural Strength (MPa)	Flexural Modulus (GPa)	Interlaminate[c] Strength (MPa)	Peel Strength ($N \cdot mm^{-1}$)
QF/TOPI-B-1	-c	-c	363.9	7.4	6.2	670.8	14.2	61.6	-c
QF/TOPI-B-1	46	38	567.6	10.1	6.4	845.9	19.5	62.1	1.06
QF/TOPI-B-1	-c	-c	557.1	10.8	6.0	835.0	19.5	58.9	-c
QF/TOPI-B-1	53	45	389.5	9.9	7.2	745.8	17.9	60.7	-c
QF/TOPI-B-1	49	41	547.2	11.4	6.2	834.1	20.0	63.7	-c
EG/HTPI-1[b]	—	—	270	11.2	4.2	730	24.2	—	1.23

[a]*Estimated from scanning electron microscopy photographs.*
[b]*Taken from Xu et al. [44] study.*
[c]*Not tested.*

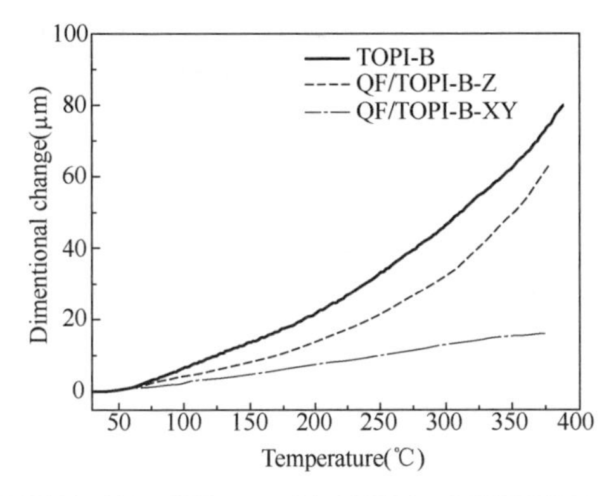

FIGURE 4.22 *CTE* curves of the TOPI-B and QF/TOPI-B in different directions.

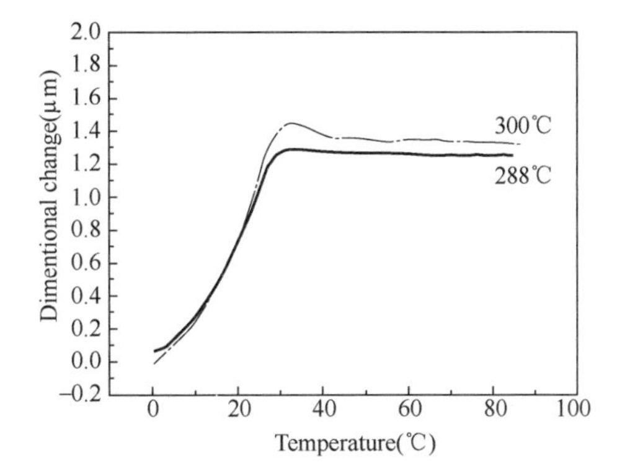

FIGURE 4.23 Dimensional changes verse time for QF/TOPI-B at 288 °C and 300 °C.

Fig. 4.24 shows the dielectric constant and dissipation factor changes of the TOPI-B and QF/TOPI-B composite laminates under different test frequencies. The dielectric constant of QF/TOPI-B was stable in the range of 3.17-3.20, although the frequency was changing, and these data are slightly higher than those of pure TOPI-B. In contrast, the dissipation factor (tan δ) of TOPI-B was higher than that of QF/TOPI-B, which was in the range of $(2.3\text{-}5.5)\times10^{-3}$. The excellent electrical stability indicated that QF/TOPI-B could be applied as a high-frequency, high-speed microelectronic packaging substrate material.

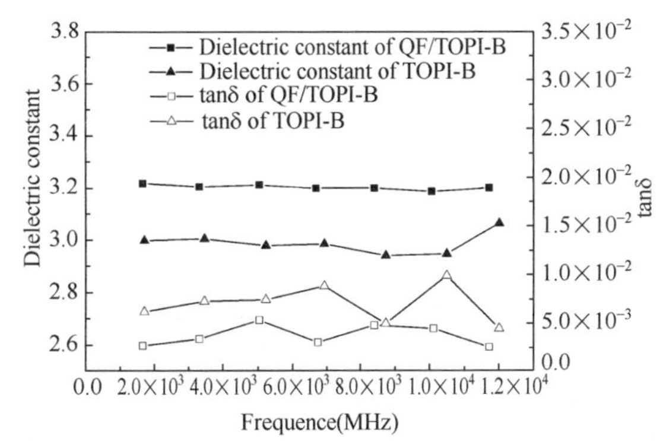

Figure 4.24 Dielectric constant and dissipation factor of TOPI-B and QF/TOPI-B.

Overall, QF/TOPI composite laminates exhibited high glass-transition temperature of 360 °C, low and steady dielectric constant of <3.2 at a test frequency of 1 GHz-12 GHz, and high volume resistance over 1.8×10^{17} Ω·cm. The laminates also showed excellent mechanical properties with flexural and impact strengths in the range of 845 MPa-881 MPa and 141 kJ·m^{-2}-155 kJ·m^{-2}, respectively. The excellent mechanical, thermal properties and good dielectric properties indicated that the laminate composites are good candidates for high-density IC packaging substrates.

4.6.2 PE-end-capped Meltable Polyimide Resins

Phenylethynyl (PE)-end-capped fluorinated aromatic oligoimides with calculated molecular weights (Calc'd M_w) of 1250-10,000 were prepared by thermal polycondensation of an aromatic dianhydride, 4,4′-oxydiphthalic anhydride (ODPA), with aromatic diamine mixtures of 1,4-bis (4-amino-2-trifluoromethylphenoxy) -benzene (1,4,4-6FAPB) and 3,4′-oxydianiline (3,4-ODA) in the presence of 4-phenylethynylphthalic anhydride (PEPA) as reactive end-capping agent (Fig. 4.25). The PE-end-capped fluorinated aromatic oligoimides could be completely melted in the temperature range of 250°C-350°C to give low viscosity fluids. The melt viscosity of the oligoimide increased with increasing molecular weight. After thermally curing at 371 °C, the thermoset polyimide resins produced showed a good combination of thermal and mechanical properties.

PI-1 to PI-3 were prepared by reacting ODPA with the diamine mixtures of 1,4,4-6FAPB and 3,4-ODA at different mole ratios of 25:75, 50:50, and 75:25, respectively, with the same Calc'd M_w of

1250. PI-4, PI-5 and PI-6 were prepared by reacting ODPA with the diamine mixture of 1,4,4-6FAPB and 3,4-ODA at mole ratio of 50:50 with the Calc'd M_w of 2500, 5000, and 10,000, respectively.

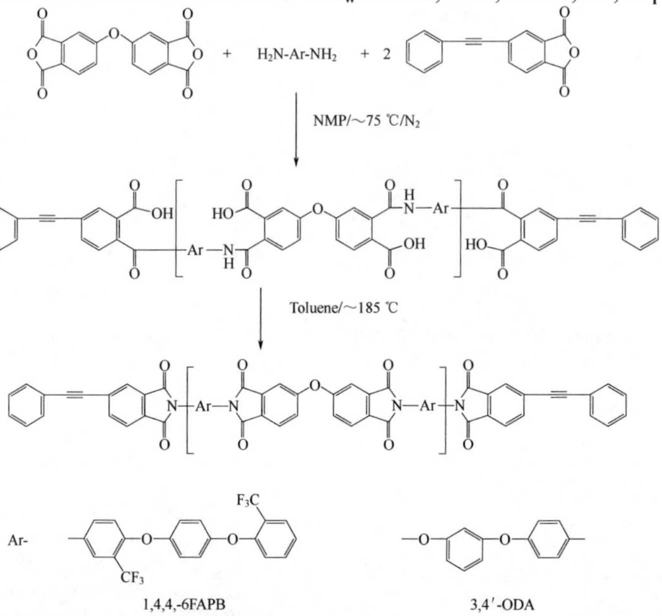

H_2N-Ar-NH_2

NMP/~75 ℃/N_2

Toluene/~185 ℃

Ar-

1,4,4,-6FAPB

3,4′-ODA

FIGURE 4.25 Synthesis of PE-end-capped aromatic oligoimides.

Fig. 4.26 shows the dynamic complex melt viscosity curves of the PE-end-capped aromatic oligoimides with different Calc'd M_w. The complex melt viscosities of the oligoimides increased obviously with increasing Calc'd M_w values. The temperature at which the oligoimide has the minimum melt viscosity was decreased from 351 °C (1451 Pa·s) for PI-6 (Calc'd M_w=10,000), to 327 °C (131 Pa·s) for PI-5 (Calc'd M_w=5000), 325 °C (7.54 Pa·s) for PI-4 (Calc'd M_w=2500), and 294-301 °C (0.27-0.67 Pa·s) for PI-1 to PI-3 (Calc'd M_w=1250), respectively. The melt viscosity of PI-1 at 350 °C is higher than PI-2 and 3. In addition, the melt-processing window of the aromatic oligoimides gradually becomes more narrow with increasing Calc'd M_w.

Isothermal viscosity measurements at 280 °C or more high temperatures were also conducted. Fig. 4.27 shows two representative curves of melt viscosity versus isothermal standing time at 280 °C

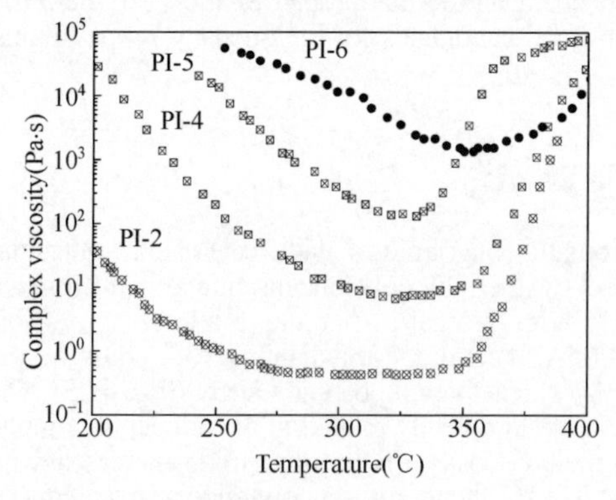

FIGURE 4.26 Plot of complex melt viscosities against temperature for the phenylethynyl-end-capped oligoimides.

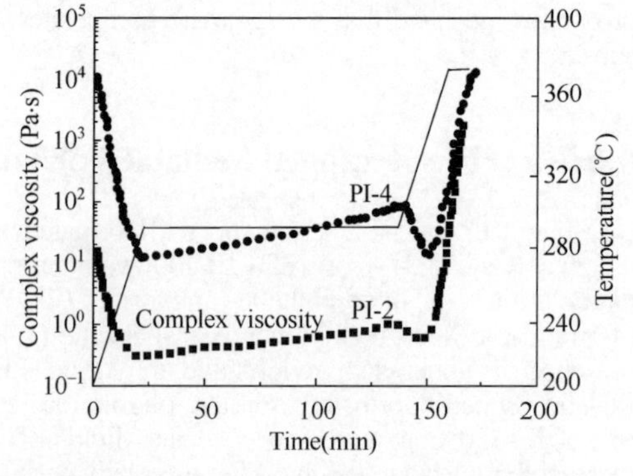

FIGURE 4.27 Dependence of the complex melt viscosity on isothermal aging time for the representative PE-end-capped oligoimides.

and Table 4.17 summarizes the isothermal melt viscosity data of the aromatic oligoimides with different Calc'd M_w values. The aromatic oligoimides with Calc'd M_w of < 2500 could be completely melted at 250 °C to give molten fluid with melt viscosity of lower than 200 Pa·s, and the melt viscosity decreased gradually with increasing temperature and then increased at a certain temperature range. PI-2 exhibited a melt viscosity variation at 280 °C from 0.56 to 2.09 Pa·s after isothermal aging for 2 h, compared with PI-4 (from 25.0 to 178.6 Pa·s), implying that the melt stabilities of the PE-end-capped aromatic oligoimides were closely related to their molecular weights and distributions. The oligoimides with the same Calc'd M_w but different chemical structures (PI-1, PI-2 and PI-3) also show different melt stabilities, of which PI-3 shows the best melt stability at 280 °C (Table 4.17).

TABLE4.17 Complex Melt Viscosity of the PEPA-end-capped Aromatic Oligoimides

Sample	Complex Melt Viscosity (Pa·s) at					Minimum Melt Viscosity (Pa·s)	Melt Viscosity Variation at 280 °C for 2 h (Pa·s)
	250 °C	275 °C	300 °C	325 °C	350 °C		
PI-1	19.8	0.84	0.69	0.82	1173	0.67 at 294 °C	0.65-11.9
PI-2	1.0	0.39	0.29	0.33	0.57	0.27 at 299 °C	0.56-2.09
PI-3	1.0	0.70	0.71	0.64	0.70	0.60 at 301 °C	0.35-1.25
PI-4	156	32.0	10.9	7.54	10.2	7.54 at 325 °C	25.0-178.6
PI-5	12430	1421	323	135	3449	131 at 327 °C	479-3145[a]
PI-6	NA	27590	12475	3030	1472	1451 at 351 °C	8268-59300[b]

[a] *Melt viscosity variation at 290 °C for 2 h.*
[b] *Melt viscosity variation at 310 °C for 2 h.*

Table 4.18 summarizes the thermal data for all of the thermally cured resins (PI-1 to PI-6). The T_g values of the thermally cured resins increased as the Calc'd M_w value decreased. For instance, PI-2 exhibited a T_g value of 301 °C by DSC, 34 °C higher than PI-4 (267 °C) and 41 °C higher than PI-6 (260 °C), respectively. In addition, there was no obvious difference in the T_g values for the thermally cured resins with the same Calc'd M_w value. For instance, PI-2 with the mole ratio of 1,4,4-6FAPB to 3,4-ODA=50:50 had a T_g value of 301 °C, which was only 1 °C higher than PI-1 (300 °C, 1,4,4-6FAPB/3,4-ODA=25:75), about the same as PI-3 (301 °C, 1,4,4-6FAPB/3,4-ODA=75:25), respectively.

TABLE4.18 Thermal Properties of the Thermally Cured Polyimide Resins

Resin	T_g (°C) by DSC	T_g (°C) by DMA		T_5 (°C)[c]	T_{10} (°C)[d]	Char (%) at 700 °C
		E''[a]	tanδ[b]			
PI-1	300	315	343	576	617	72
PI-2	301	294	338	548	595	66
PI-3	301	313	345	542	589	64
PI-4	267	258	278	584	617	70
PI-5	252	241	260	587	617	68
PI-6	260	252	272	583	620	68

[a] *E": The peak temperature in the loss modulus curve.*
[b] *tanδ: The peak temperature in the tanδ curve.*
[c] *Temperature at 5%(w) loss.*
[d] *Temperature at 10%(w) loss.*

The T_g values defined as the tan δ in DMA were measured in the range of 260-345 °C, which decreased as the Calc'd M_w increased. For instance, PI-2 (Calc'd M_w=1250) showed a T_g value of 338 °C, which was 66 °C higher than PI-6 (272 °C, Calc'd M_w=10,000). The thermally cured polyimide resins showed excellent thermal properties with T_5 of 542-587 °C and T_{10} of 589-620 °C, respectively.

Table 4.19 summarizes the mechanical properties of the thermally cured resins with different Calc'd M_w values. After thermal curing at 371 °C for 1 h, all of the thermoset polyimide resins exhibited excellent mechanical properties with tensile strength of 65.7-95.6 MPa, tensile moduli of 1.4-1.9 GPa, elongation at breakage of 3.9%-14.0%, flexural strength of 135.5-153.8 MPa, and flexural moduli of 3.1-3.5 GPa. The tensile strengths and elongations at breakage were improved gradually with the increasing of Calc'd M_w from 1250 to 5000. PI-5 with Calc'd M_w of 5000 exhibited the best mechanical properties with tensile strength 95.6 MPa, elongation at breakage of 14.0%, flexural strength of 144.7 MPa and flexural moduli of 3.1 GPa. No more benefits in improvement of the tensile strength and elongation at breakage were observed after the Calc'd M_w values were > 10,000. The reason that PI-2 and PI-4 have the lower tensile strength and elongation at breakage than PI-5 and PI-6 was probably because their lower Calc'd M_w resulted in limited polymer chain lengths.

TABLE4.19 Mechanical Properties of the Thermally Cured Polyimide Resins

Resin	Tensile Strength (MPa)	Tensile Modulus (GPa)	Elongation Breakage (%)	Flexural Strength (GPa)	Flexural Modulus (GPa)
PI-2	65.7	1.9	3.9	135.5	3.2
PI-4	74.5	1.7	5.6	146.8	3.4
PI-5	95.6	1.4	14.0	144.7	3.1
PI-6	92.8	1.5	13.2	153.8	3.5

Overall, PI-2 showed good thermal stability with a T_g value of 301 °C (DSC) and 338 °C (DMA) and combined mechanical properties with flexural strength of 135.5 MPa, flexural modulus of 3.2 GPa, tensile strength of 65.7 MPa and elongation at breakage of 3.9%. PI-5 and PI-6, due to their relative higher melt viscosity and high mechanical properties, might be appropriate for infusion or extrusion molding processes.

4.7 Heat-Resistant Polyimide Foams

4.7.1 Introduction

Polyimide foams, due to their ultralow weight, excellent thermal insulation and acoustic absorbing properties, high strength-to-weight ratio, and cost-effectiveness, etc., have been extensively used in many high-technology fields such as aerospace and aviation industries. Polyimide foams are usually employed as thermal and acoustic insulation materials as well as structural support materials [45-49]. In the 1960s, Hendrix's patent describes a process to prepare polyimide foams using poly(amic acid) resin solution (the precursor of polyimide) derived from 4,4′-oxydianiline (ODA) and pyromellitic dianhydride (PMDA) in organic solvent [50]. Currently, commercial polyimide foams such as Solimide, Solrex, TEEK, and so on have been developed by different methods [51-55]. Solimide foams manufactured by In-spec Corporation Private Ltd (Singapore) have many outstanding mechanical and thermal properties, but are limited to the opened-cell (i.e., low closed-cell content (C_c)) and low densities (ρ=8 kg·m^{-3}), being unsuitable for use as insulation on future reusable launch vehicles. TEEK foams developed by the NASA (Washington, District of Columbia, USA) have been prepared by the reaction of a unique polyimide precursor residuum to yield foams with densities from 8 to 320 kg m^{-3} and were open-cell structures with C_c of only 32%.

In 2002, Weiser et al. [56] reported a method using intermediate precursor termed "friable balloons" to prepare polyimides with densities from 16 to 128 kg·m^{-3}, and C_c of the foam with a density of 48kg·m^{-3} was as high as approximately 78%. However, the C_c of the foams was obviously reduced with the changes in foam density. For instance, the foams with a density of 96 kg·m^{-3} showed C_c of only approximately 42%. Chu et al. [48,57] first reported a method to prepare the thermally crosslinked polyimide foams using 3,3′,4,4′-benzophenonetetracarboxylic dianhydride (α-BTDA), 4,4′-ODA as monomers, and 2,4,6-triaminopyrimidine as a crosslinking agent via a poly(ester-amine salt) precursor process. The resultant crosslinked polyimide foams are open-cell structures and showed enhanced mechanical strength and modulus.

Recently, the aerospace and aviation industries have growing requirements for closed-cell polyimide foams with high thermal stability and high compression properties for structural support material applications. However, the currently developed open-celled polyimide foams usually had low C_c, and their mechanical properties and thermal stabilities did not satisfy the future needs of aerospace and aviation industries. Dutruch et al. [58] reported a chemical approach to prepare thermostable foams by thermal foaming of the nadimide end-capped oligobenzhydrolimides with a Calc'd M_w of 1000 and 1500 g·mol^{-1}. However, the foams prepared from the neat oligomers were too brittle to prepare the testing samples. The fracture toughness of the crosslinked foams must be improved by blending with thermoplastic aromatic polymers. The foams toughened by thermoplastic polymers with a density of 200 kg·m^{-3} showed compression strength of > 2.0 MPa at room temperature, which was reduced to 1.3 MPa at 300 °C.

Polyimide foams can be classified into opened-cell soft and closed-cell rigid foams according to the cell structures. In this chapter, the synthesis, characterization, and properties of both opened-cell soft and closed-cell rigid polyimide foams will be described.

4.7.2 Opened-Cell Soft Polyimide Foams

Opened-cell soft polyimide foams have the solid phase as the cell edges with void space or gas phase connected through the cell faces. The material is permeable since the cavities are connected. Closed-cell foams have cell faces with isolated cavities filled with trapped gases [59]. Many types of polyimide foams contain a ratio or a percentage of open and closed cells. The polyimide foam properties can be adjusted by controlling of the ratios of opened to closed cell contents.

Researchers at NASA Langley Research Center (LaRC) have developed a method of preparing low-density polyimide foams [45,47]. The salt-like foam precursor is first synthesized by mixing monomer reactants of an aromatic diamine with a foaming agent (tetrahydrofuran) in methanol at room temperature (Fig. 4.28). The aromatic diamine is dispersed in a mixture of tetrahydrofuran (THF) and methanol (MeOH) at room temperature. To the stirring aromatic diamine solution, an aromatic dianhydride is added gradually at 15 °C to yield a homogenous solution. To this solution, an aromatic tetra-acid is added gradually and the mixture is stirred for 24 h at 30 °C to yield a homogenous precursor solution. The solution is then charged into a stainless-steel vat and treated at 70 °C for 14 h in order to evaporate the solvents. The resulting material is cooled and crushed into a fine powder (2-500 μm). The polyimide precursor solid residuum is then treated for an additional amount of time (0-300 min) at 80 °C to further reduce the residual solvents to about 1%-10%(w). depending on the final foam density desired. The polyimide precursor powders are further treated at 100 °C to expand the powders without thermal imidization so that the apparent density of the precursor is decreased without thermal imidization. Thus, a friable balloon was obtained.

The process of foaming the friable balloons into a solid neat piece of foam or a foam-filled honeycomb is accomplished by a closed mold foaming technique (Fig. 4.29). The friable balloons are placed in a mold containing a piece of honeycomb core, which was placed in an oven for curing. In order to obtain a specific density from the friable balloons a simple back calculation is utilized. The desired density is multiplied by the mold volume and a specific weight is obtained. To this weight an additional 20% is added to account for solvent removal and water formation during precursor imidization.

Fig. 4.30 shows the chemical structures of the open-celled polyimide foams, including TEEK-HH (0.082 g·mL^{-1}) and HL (0.032 g·mL^{-1}) derived from ODPA, and 3,4′-ODA, TEEK-LH (0.082 g·mL^{-1}), and LL(0.032 g·mL^{-1}) derived from BTDA and 4,4′-ODA, and TEEK-CL(0.032 g·mL^{-1}) derived from BTDA and 4,4′-DDSO$_2$, respectively.

Fig. 4.31 shows that, as the density is increased from 0.02 to 0.048 g·cm^{-3}, the closed cell content of the foams increases from 32% to a maximum of 78%. The low closed cell content at 0.02 g·cm^{-3} can be explained in part by the lack of sufficient material present to completely fill the interstitial voids and thus a lower closed cell value is obtained. As the density is increased, more material is present and a greater percentage of the interstitials are filled. As the density of the foam is further increased from 0.048 to 0.096 g·cm^{-3}, the closed cell percentage decreases from 78% to 42%. Thus, the greater the foam density, the greater the open cell percentage. For the lowest-density foams, the friable balloon experiences the largest degree of expansion. The closed cell fraction is proportional to the degree of friable balloon inflation and thereby, inversely proportional to foam density.

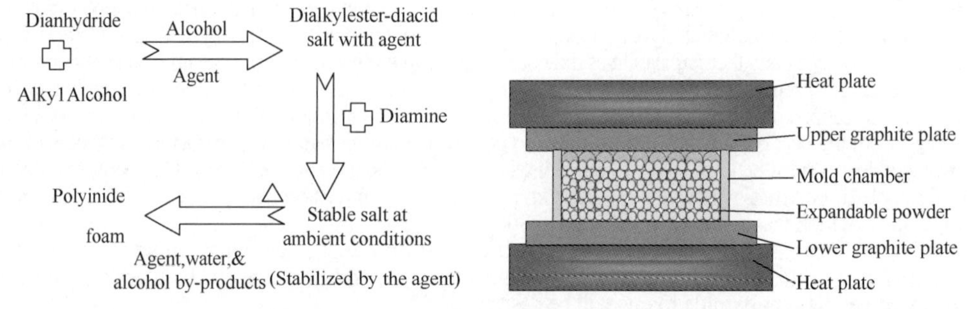

FIGURE 4.28 Chemical process of the open-celled polyimide foams.

FIGURE 4.29 Cartoon of friable balloon processing lay-up.

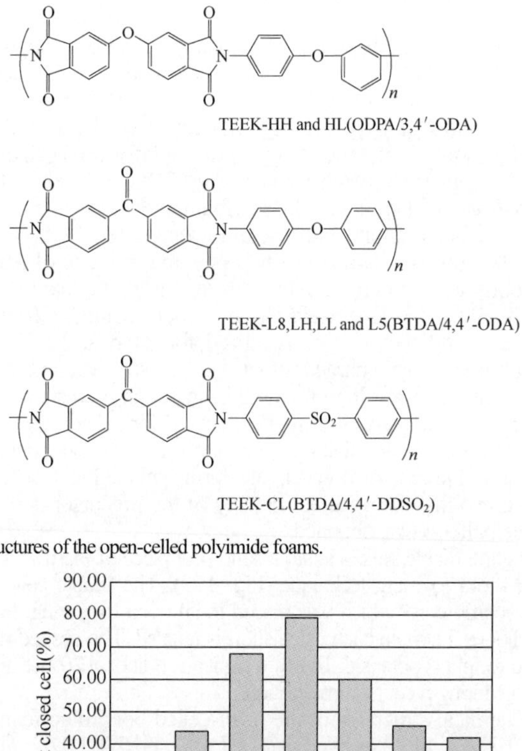

TEEK-HH and HL(ODPA/3,4′-ODA)

TEEK-L8,LH,LL and L5(BTDA/4,4′-ODA)

TEEK-CL(BTDA/4,4′-DDSO$_2$)

FIGURE 4.30 Chemical structures of the open-celled polyimide foams.

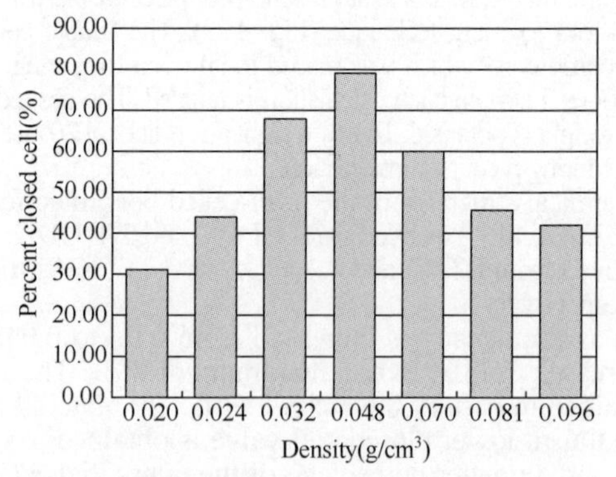

FIGURE 4.31 Closed cell content of friable balloon foams.

Table 4.20 shows the mechanical properties of the polyimide foams. The materials in the TEEK H series have the highest tensile strengths of the three series and TEEK L the highest compressive strength and modulus. The compression specimens performed adequately for all three polyimide foams. The TEEK HH foam had a compression strength of 0.84 MPa at a 10% deflection and a modulus of 6.13 MPa. The TEEK HL, LL, and CL foams had values between 0.098 and 0.30 MPa. Compressive strength was also measured at 177 °C and −253 °C. The compressive strength of the 0.08 g·mL^{-1} foam was 0.31 and 0.72 MPa, respectively. The TEEK HL and LL had values ranging from 0.06 to 0.10 MPa and 0.13 to 0.46 MPa, respectively.

TABLE4.20 Mechanical Properties of the Open-Celled Polyimide Foams

Property	Test Method	TEEK HH	TEEK HL	TEEK LL	TEEK CL
Density	ASTM D-3574(A)	0.08 g·cm^{-3}	0.032 g·cm^{-3}	0.032 g·cm^{-3}	0.032 g·cm^{-3}
Tensile strength	ASTM D-638-97	1.68 MPa	0.28 MPa	0.26 MPa	0.09 MPa
Compressive strength	ASTM D-3574(C)	0.84 MPa at 10% deflection	0.19 MPa at 10% deflection	0.30 MPa at 10% deflection	0.098 MPa at 10% deflection
Compressive modulus	ASTM D-3574(C)	6.13 MPa	3.89 MPa	11.03 MPa	In test
Compressive strength at 177 °C	ASTM D-3574(C)	0.31 MPa	0.06-0.10 MPa at 10% deflection	0.06-0.09 MPa at 10% deflection	In test
Compressive strength at -253 °C	ASTM D-3574(C)	0.72 MPa	0.14-0.46 MPa at 10% deflection	0.13-0.40 MPa at 10% deflection	In test

Fig. 4.32 shows the tensile strengths of the polyimide foams ranging in density from 0.040 to 0.11 g·cm^{-3}. The tensile strength is measured in the range of 0.074 to 0.081 g·cm^{-3} foam density at 0.88 MPa. The foam density varies within the specimen from 0.080 to 0.155 g·cm^{-3}. The failure occurs at the interface between the two density regions. The foaming process can lead to a laminar structure wherein layers of different foam densities are developed near the surfaces.

Fig. 4.33 displays the compressive strengths. The compressive strength properties are measured for samples of 10 different densities, ρ, ranging from 0.039 to 0.097 g·cm^{-3}. These results show that as the density is increased, the compressive strength also increased, showing a linear relationship between foam compressive strength and density as opposed to the results obtained by Gibson et al. [11] where foam strength is shown to be a cubic function of density. Values were obtained on foam specimens with densities ranging from 0.039 to 0.097 g·cm^{-3}. In all cases, the 50% deflection strength increased with increasing density from a lower value of 0.21 MPa to a higher value of 0.74 MPa.

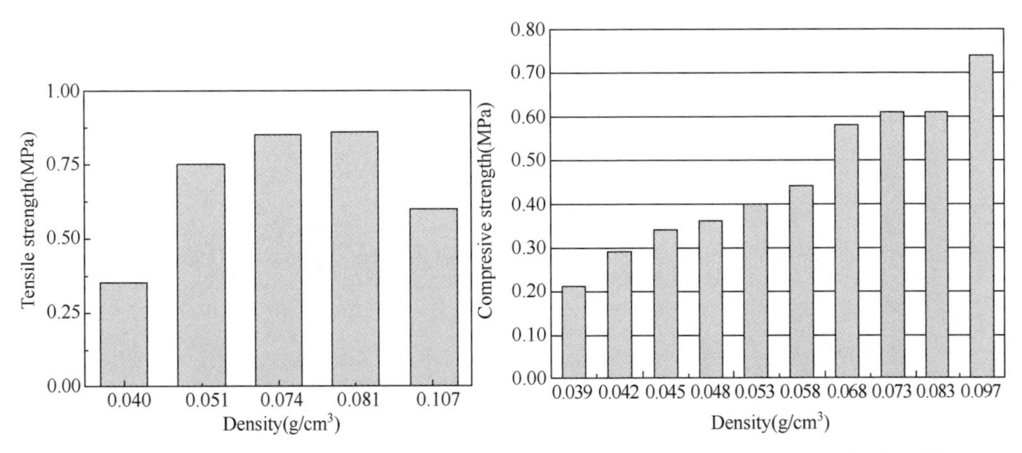

FIGURE 4.32 Tensile strength of the polyimides foams with different density.

FIGURE 4.33 Compressive strengths (50% deflection) of the polyimide foams with different density.

All of the polymers used in this study were aromatic polyimides which exhibit excellent stability up to 320 °C. The specimens' glass transition temperatures ranged from 237 °C to 321 °C. The TEEK-HH and-HL foams exhibited the lowest glass transition temperatures (237 °C), while the TEEK-LL (300 °C) and TEEK-CL (321 °C) had higher T_gs. Both TEEK-HH and -HL showed good thermal stability with weight loss values of 0.6% and 1.07%, respectively. TEEK-LL had the best thermal stability with no appreciable weight loss at 204 °C. TEEK-CL had excellent thermal stability with a weight loss of 0.5%. However, it had a significant amount of moisture absorption (~1.5%), while the TEEK-HH, -HL, and -LL had little water absorption (~0.25%).

The conductivity data for the TEEK-HH, TEEK-HL, and TEEK-CL foams exhibit good values at room temperature and show decreasing values at subambient temperatures, which indicates that these foams would be good cryogenic insulators. The cryogenic temperatures are −253 °C for hydrogen and −193 °C for LO_x. The outer surface of the exterior insulated cryogenic tanks must be at a temperature high enough to prevent air liquefaction, frost build-up or the condensation of moisture. All of the foams have conductivity data which demonstrate that the tank would be sufficiently insulated to prevent frost build-up and air liquefaction with a relatively small thickness (< 0.025 m). The elevated thermal conductivities are not as efficient as the room temperature and subambient values. This indicates that these foams are better suited for cryogenic insulation than elevated-temperature insulation. However, in the case of next-generation launch vehicles, TPS will be the main high-temperature insulator and the foam will provide adequate protection of the cryotank wall.

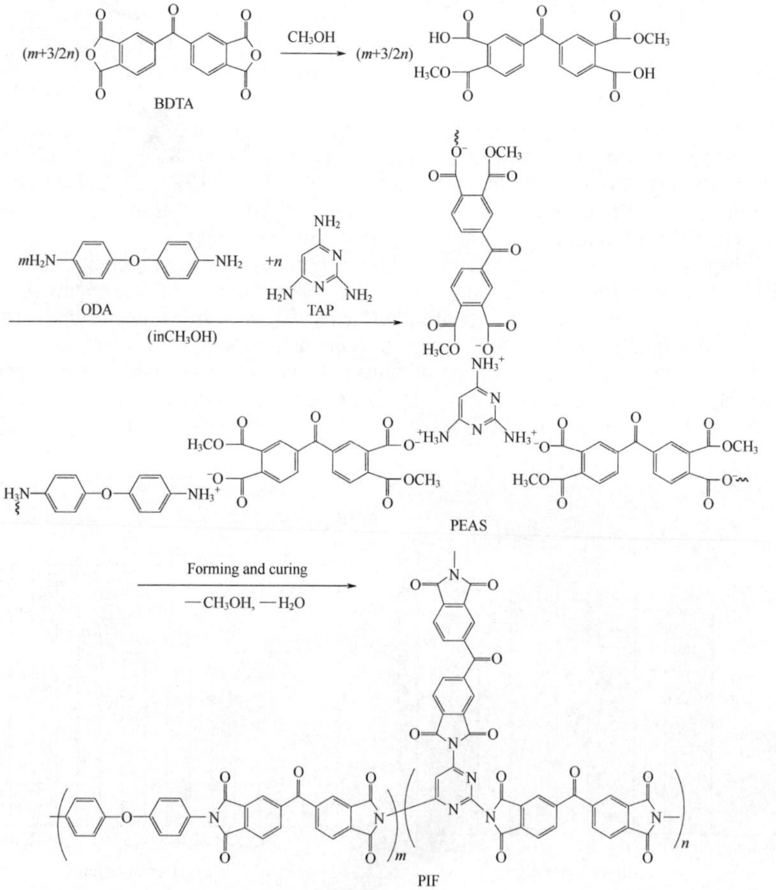

FIGURE 4.34 Preparation of (BTDA/ODA/TAP) polyimide foams.

Resistance to an oxygen atmosphere, smoke generation, and flammability are critical properties that polyimide foams will be required to have if they are to be used on the next-generation RLV. The three polyimide foams tested for the LOI show excellent resistance to combustion in an oxygen-rich environment. The foams of lower density, 0.032 $g \cdot mL^{-1}$, exhibit lower values of about 42%-49% as compared to the polyimide foam at a higher density of 0.08 $g \cdot mL^{-1}$ which had a LOI of 51%.

A series of crosslinked polyimide foams have been prepared by polycondensation of 3,3′,4,4′-benzophenonetetracarboxylic dianhydride (BTDA), 4,4′-oxydianiline (ODA), and 2,4,6-triaminopyrimidine (TAP) via a poly(ester-amine salt) (PEAS) process (Fig. 4.34) [48].

Table 4.21 shows the chemical compositions and properties of the polyimide foams. The crosslinking of TAP could effectively improve the mechanical properties of PIFs. PIF-4 with the largest content of TAP had the tensile and compressive modulus as high as 13.41 and 17.62 MPa, respectively. The void cell size in these polyimide foams ranged from 40 to 200 μm. The average cell size found from mercury porosimetry measurements for these four PIFs was 141, 149, 153, and 162 μm, respectively.

Fig. 4.35 shows the morphologies of polyimide foams with different contents of TAP and comparable porosities. All of the PEASs were converted into polyimide foams with uniform porous structure. The inner of samples showed foam structure mainly in open cells, while the surface of samples was dense.

TABLE4.21 Compositions and Properties of BTDA/ODA/TAP Based Polyimide Foams

Molar Ratio of TAP/(ODA+TAP)	0.0	0.034	0.069	0.105
PEAS number	PEAS1	PEAS2	PEAS3	PEAS4
PIF number	PIF-1	PIF-2	PIF-3	PIF-4
Density ($g \cdot cm^{-3}$)	0.156±0.007	0.146±0.009	0.137±0.006	0.122±0.01
Porosity (%)	87±1	88±1	88±1	90±1
Average cell size (μm)	141±5	149±5	153±5	161±5
Tensile strength (MPa)	1.03±0.06	1.29±0.05	1.36±0.05	1.39±0.07
Tensile modulus (GPa)	5.49±0.45	9.23±0.57	12.12±0.73	13.41±0.68
Compressive strength (MPa) (at 10% deflection)	0.91±0.07	1.12±0.08	1.28±0.06	1.44±0.09
Compressive modulus (MPa)	10.76±0.76	13.24±0.63	15.07±0.65	17.62±0.57

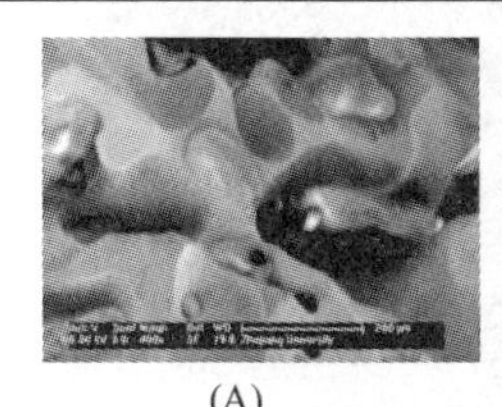
(A)
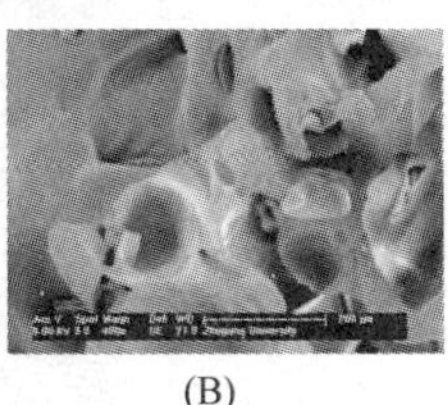
(B)
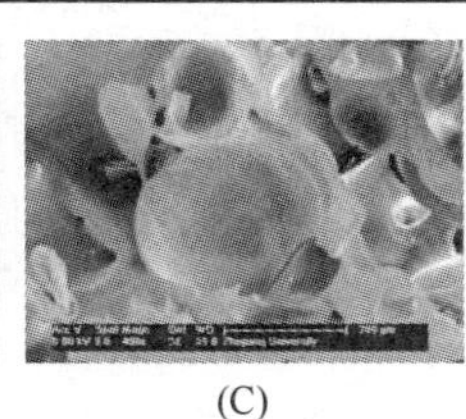
(C)
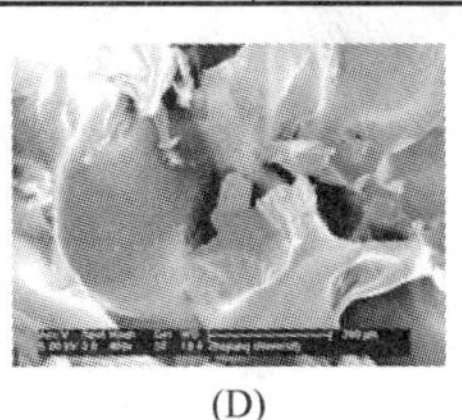
(D)

FIGURE 4.35 SEM images of BTDA/ODA/TAP based polyimide foams: (A) PIF-1; (B) PIF-2; (C) PIF-3; and (D) PIF-4.

Fig. 4.36 shows the dielectric constants and the dielectric loss of PIFs in 100, 1000, and 10,000 Hz fields, respectively. The dielectric constants decreased with the increasing frequency of the applied field. The lower dielectric constants are attributed to their higher porosity. The dielectric constant of PIF-4 containing 10.5 %(mol) TAP was as low as 1.77 at 10,000 Hz. The crosslinking structures and the increasing linkages in polyimide backbones can reduce the chain packing density. The density of the dense polyimide film prepared from poly(amic acid) precursor with monomer composition similar to PIF-1, PIF-2, PIF-3, and PIF-4 was 1.25, 1.21, 1.19, and 1.17 $g \cdot cm^{-3}$, respectively. The dielectric losses of all the polyimide foams were smaller than 3×10^{-2} in the field with frequency higher than 100 Hz.

The dielectric loss increased gradually with the TAP content in the polyimide foams.

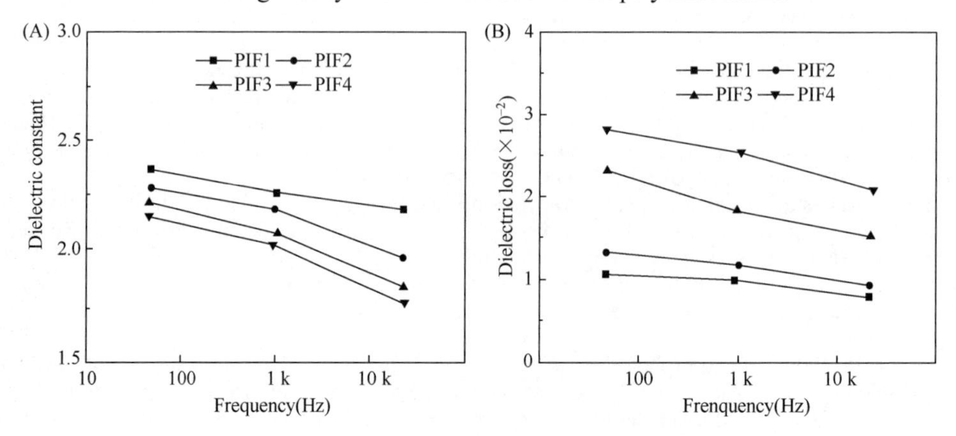

FIGURE 4.36 Dependence of (A) dielectric constant and (B) dielectric loss on frequencies (25 °C).

Fig. 4.37 shows the variation of dielectric constant and dielectric loss of the crosslinked polyimide foams at different temperatures. The increases in both dielectric constant and dielectric loss with increasing temperature were generated from the enforcement of polarization and the intensification of the chaotic thermal oscillations of molecules at higher temperature. When the temperature was elevated from 25 °C to 150 °C, the dielectric constants of all PIFs were still smaller than 2.4, and the increase of absolute dielectric constant was smaller than 0.26 in 10,000 Hz field. To investigate the dependence of the dielectric property on porosity, several PIF-3 samples were prepared with porosity of 68%, 75%, 80%, and 88%, respectively. For PIF-3 with porosity of 68% and 88%, the dielectric constant was 2.30 and 2.04, respectively. When porosity increased by 29%, the reduction in dielectric constant was only 11%.

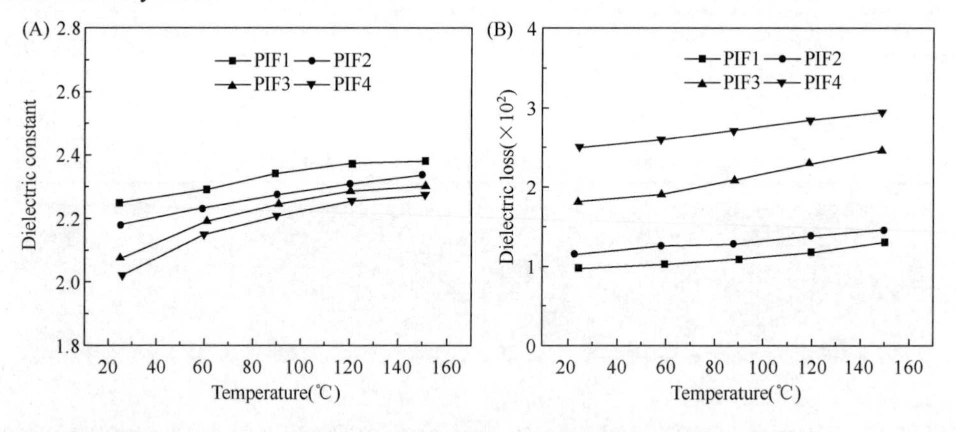

FIGURE 4.37 Influences of temperature on (A) dielectric constant and (B) dielectric loss of PIFs (1000 Hz).

The BTDA/ODA/TAP based PIFs exhibit outstanding thermal stability. The pure BTDA/ODA PIF-1 has a degradation temperatures (T_d) at 5% weight loss of 557 °C, compared with the TAP-crosslinked foams (PIF-2, PIF-3, and PIF-4) of 551 °C, 549 °C, and 527 °C, respectively. In the field with frequency higher than 100 Hz, the dielectric constants of the obtained PIFs ranged from 1.77 to 2.4, and the dielectric losses were smaller than 3×10^{-2} at 25-150 °C. The thermal stability and dielectric property can satisfy the requirements of high-performance polymeric foam candidates for most potential applications.

4.7.3 Closed-Cell Rigid Polyimide Foams

A novel chemical approach to preparing rigid closed-cell polyimide foams (PIFs) was developed through thermal foaming of nadimide end-capped imide oligomers (NAIOs) obtained by thermally treating an in situ polymerization of monomeric reactant (PMR)-type poly(amide-ester) resin solution (Fig. 4.38). Recently, the aerospace and aviation industries have had growing requirements for closed-cell PIFs with high thermal stability and high compression properties for structural support material applications. However, the currently developed PIFs usually had low C_c, and their mechanical properties and thermal stabilities did not satisfy the future needs of aerospace and aviation industries.

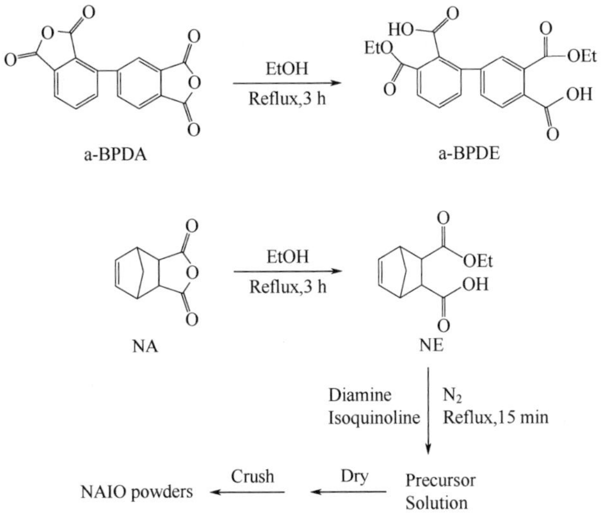

FIGURE 4.38 Synthetic route of nadimide-end-capped imideoligomers.

4.7.3.1 Preparation of NAIO Powders

The PMR poly(amide-ester) was synthesized by the reaction of diethyl ester of 2,3,3′,4′-biphenyltetracarboxylic dianhydride (α-BPDE) and aromatic diamines using monoethyl ester of *cis*-5-norbornene-endo-2,3 -dicarboxylic acid (NE) as reactive end-capping agent in ethyl alcohol. The resulting viscous liquid was dried by baking in a vacuum oven at a temperature of < 200 °C to give a solid, which was then crushed and sieved with a mesh to produce the nadimide end-capped imide oligomer (NAIO) powder (PI (polyimide)-1). Similarly, PI-2, PI-3, PI-4, PI-5 were prepared by the same procedure as PI-1 except that *p*-PDA was replaced by 3,4′-ODA for PI-2, 4,4′-ODA for PI-3, MDA for PI-4, and 4,4′-DDS for PI-5.

4.7.3.2 Preparation of Rigid Closed-cell PIFs

The rigid PIFs were prepared by thermal foaming of the NAIO powders in a closed mold. Thus, the imide oligomer powder was placed at the bottom of the mold at room temperature and then heated in a hot press. After the mold was heated stepwise from 300 °C to 330 °C and held for 3 h, the mold was cooled down to room temperature. The resulting PIF was removed from the mold and cut into the designed dimensions for testing. The amount of the NAIO powders placed in the mold was determined by the designed foam density and the mold volume as well as the calculated amount of organic volatiles produced in the thermal foaming process. Thus, a series of rigid PIFs (PIF-1, PIF-2, PIF-3, PIF-4, and PIF-5) with the same densities ($100\ kg \cdot m^{-3}$) were prepared.

4.7.3.3 Foam Formability of NAIO Powders

In the thermal foaming process (200-330 °C), the NAIO powders first melt to give a molten fluid, then the nadimide end-cap thermally decomposes to evolve a small molecule gas, cyclopentadiene (CPD), which acts as both foaming agent as well as crosslinking agent to give thermoset PIF. Only a small amount of organic volatiles was evolved from powders at a temperature of < 270 °C, which was a mixture of ethanol and water gas produced during the imidization reaction. The evolving ethanol and water gas from the melt fluid could act as the initial point of gas bubble, making the bubble growth much easier when large amounts of CPD evolved in the following thermal foaming stage. When the temperature was increased to over 270 °C, CPD produced by the thermal decomposition of nadimide end-cap in the imide oligomers acted as a foaming agent to yield foams with cellular structures.

Fig. 4.39 depicts the rheological behaviors of the NAIO powders. Except for PI-1, other NAIO powders began to melt after the temperature was increased to over 200 °C. At 235-290 °C, the melt resins showed the minimum melt viscosities (1-10 Pa·s/230-275 °C), which were then increased gradually due to the thermal crosslinking of CPD. The minimum melt viscosities of NAIOs depend on their chemical structures (Table 4.22). PI-4 showed the minimum melt viscosity of 0.6-0.8 Pa·s at 235-270 °C. The NAIO's meltabilities could be improved by adjusting the resin's chemical structures by employing different aromatic diamine monomers in the order of MDA > 4,4′-ODA > 4,4′-DDS > 3,4′- ODA > *p*-PDA. PI-1, due to its high rigid polymer backbone derived from the aromatic diamine monomer (*p*-PDA), showed much higher melting temperature and higher melt viscosity than other NAIO resins. It melted at > 225 °C with three molten stages, in which the first one has a melt viscosity of 1.2×10^3 Pa·s at 242-255 °C, the second has 1.5×10^2 Pa·s at a very sharp temperature scale of 277-282 °C, and the third has 3.6×10^2 Pa·s at a temperature of 320-330 °C.

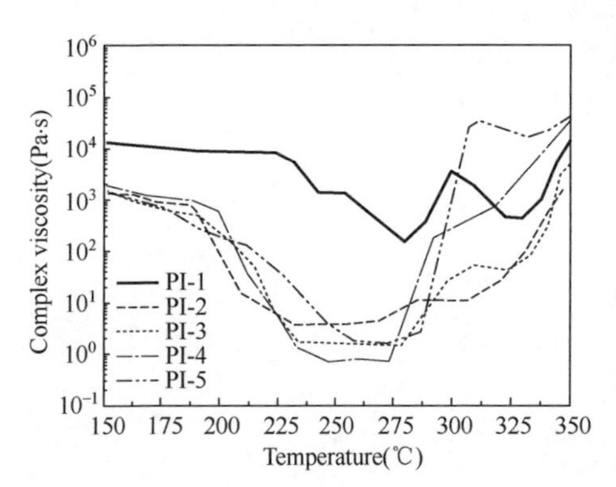

FIGURE 4.39 Rheological behaviors of the nadimide end-capped imide oligomer powders.

TABLE4.22 The Minimum Melt Viscosities of the Nadimide-end-capped Imide Oligomer Powders

	Minimum Melt Viscosity	
	value (Pa·s)	**Temperature (°C)**
PI-1	147-150	277-282
PI-2	5.0-6.0	235-265
PI-3	1.0-2.0	235-290
PI-4	0.6-0.8	235-270
PI-5	1.0-2.0	250-290

In order to observe the foaming process more visually, an improved parallel plate test mode was employed, in which the specimen disk was loaded on the lower plate and a 0.5 mm gap between the specimen disk on the lower plate and the upper parallel plate surface was placed. In this case, when the NAIO powders began their thermal foaming process, the melt fluid expanded to contact the upper plate, thus the melt viscosity could be detected by rheometer. In this way, we can detect at which temperatures the NAIO powders start to foam by testing the melt viscosities. Fig. 4.40 compares the rheological behaviors of the NAIO powders. All of the NAIO powders start to foam quickly at 280-290 °C, of which PI-4 starts to foam at 280 °C, about 10 °C lower than the others (290 °C).

Visual observation tests indicated that the NAIO powder began to melt at 280 °C and foam at 300 °C, respectively. This observation is well in accordance with the rheological testing as shown in Fig. 4.40. By extending the standing time to 300 °C, the foams were expended gradually and kept unbroken even after standing for 30 min at 300 °C. The foam expansion was attributed to the evolved CPD released gradually from the thermal decomposition of the end-cap and acted as the blowing agent. Meanwhile, the following chain extension and crosslinking reaction of CPD with the nadimide end-cap strengthened the bubble wall, preventing it from breaking to give closed-cell structures. Hence, PIFs with high C_c could be obtained.

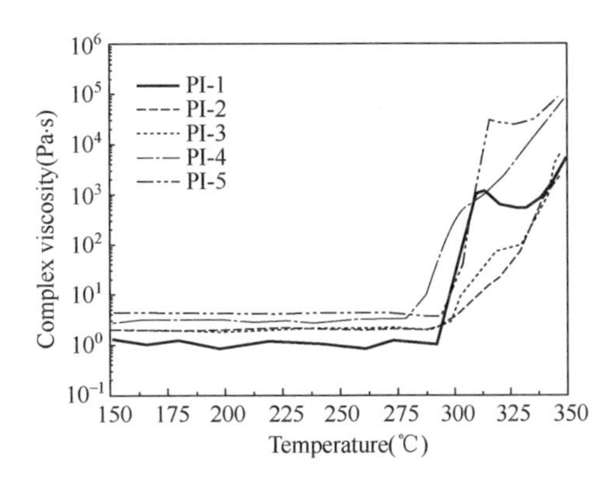

FIGURE 4.40 Rheological behaviors of the nadimide end-capped imide oligomer powders by the gaped testing mode.

4.7.3.4 Combined Properties of the Rigid PIFs

The NAIO powders could be converted to rigid PIFs with uniform cellular structures, as shown in Fig. 4.41. The PIFs showed homogeneous cell structures with an average diameter of about 400 mm. Fig. 4.42 compares the C_c of the PIFs with the same calculated M_n (1500 $g{\cdot}mol^{-1}$) and same foam density (100 $kg{\cdot}m^{-3}$). The PIFs showed a C_c of 70%-89%. The C_c of the PIF (4,4′-oxydiphthalic anhydride/3,4′-ODA) with a density of 96 $kg{\cdot}m^{-3}$ was approximately 42%, much lower than that of PIF-1 derived from α-BPDE/p-PDA/NE (C_c=89%). The C_c of the PIFs was controlled by several factors such as melt viscosity, melt fluid strength, concentration of blowing agent, and so on. In the process of thermal foaming, the concentration of blowing agent CPD in the melt resin produced by the thermal decomposition of nadimide end-cap in the imide oligomers continued to increase, leading to bubble generation, and then growing in size. As resin melt flowed, the bubble wall became thinner and thinner with the bubble size growing, resulting in the cell walls destabilizing and being more likely to rupture, giving an open-cell structure. On the other hand, the melt viscosity could make melt resin more difficult to flow, preventing bubbles from being ruptured, yielding a closed-cell structure. Because NAIOs have the same calculated M_n and density, the concentration of blowing agent in the different NAIO powders should be constant. Hence, the melt resin viscosity has a determinative effect on the resin's thermal-foaming ability as well as the morphology of the cellular structures. Due to the thermal crosslinking of melt resins at elevated temperatures (>280 °C), the melt viscosity increased gradually to produce polymer gel while the bubble was foamed and growing, which could aid the bubbles' stability and lead to high Cc (>70%). Among the five NAIOs prepared, PI-1 has the highest melt viscosity at the thermal foaming temperatures (approximately 270-300 °C), giving PIF with higher $C_{c(89\%)}$ than others.

FIGURE 4.41 Appearance of the rigid polyimide foams.

Table 4.23 summarizes tensile properties of the rigid PIFs with the same density (100 kg m^{-3}) and the same calculated M_n (1500 $g{\cdot}mol^{-1}$). PIF-2 has the best tensile properties (δ_m: 29.4 MPa, δ_b:

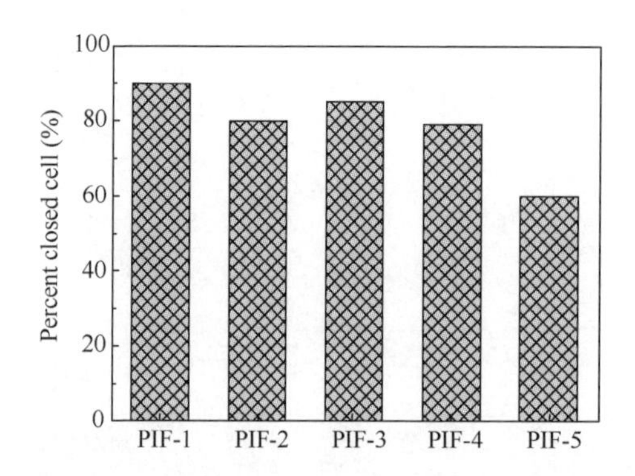

FIGURE 4.42 Closed-cell content of the rigid polyimide foams.

1.34 MPa), comparable to PIF-1 (δ_m: 27.4 MPa, δ_b: 0.96 MPa) and much higher than PIF-3 (δ_m: 19.3 MPa, δ_b: 1.13 MPa) and PIF-4 (δ_m:12.0 MPa, δ_b: 0.66 MPa). This might be interpreted by the NAIO's asymmetrical chemical structure of the PI-2 polymer backbone derived from the asymmetrical diamine monomer (3,4′-ODA), which not only ensured the NAIO resin with lower melt viscosity beneficial for thermal foaming, but also gave the PIF with T_g of 289 °C, which was lower than the final thermal foaming temperature (300-330 °C), resulting in the more spherical cellular structures, which could strengthen the foam produced. In addition, the elongation at breakage (ε) increased from 4.6% (PIF-1) to 6.2% (PIF-2) and 8.2%-8.7% (PIF-3 and PIF-4) with the changes in molecular structures of aromatic diamines from rigidity to flexibility and nonlinear to linear types. The flexural properties of the rigid PIFs were also shown in Table 4.22. PIF-2 showed better flexural properties than others. However, it was found that PIF-5 was too brittle to prepare the test samples for flexural property testing.

TABLE4.23 Mechanical Properties of the Rigid PIFs

	Tensile Properties			Flexural Properties	
	σ_m (MPa)	σ_b (MPa)	ε (%)	f_m (MPa)	f_s (MPa)
PIF-1	27.4 ± 2.6	0.96 ± 0.08	4.6 ± 0.5	20.1 ± 4.2	1.19 ± 0.20
PIF-2	29.4 ± 1.9	1.34 ± 0.11	6.2 ± 0.6	24.6 ± 0.9	1.82 ± 0.09
PIF-3	19.3 ± 1.7	1.13 ± 0.20	8.7 ± 1.2	14.7 ± 3.1	1.29 ± 0.22
PIF-4	12.0 ± 2.0	0.66 ± 0.06	8.2 ± 1.2	13.1 ± 2.9	1.09 ± 0.27
PIF-5	12.6 ± 2.2	0.15 ± 0.04	1.4 ± 0.2	—[a]	—

σ_m, tensile modulus; σ_b, tensile strength; f_m, flexura modulus; f_s, flexural strength; PIF, polyimide foam.
[a]*PIF-5 was too brittle to prepare the test sample.*

Table 4.24 shows the compression properties of the rigid PIFs with the same density (100 kg·m^{-3}) and the same calculated M_n (1500 g·mol^{-1}). The c_m and c_s at 10% compressive deformation reduced in the order of PIF-1 > PIF-2 > PIF-3 > PIF-4 > PIF-5. PIF-1 exhibited the compression strength of 1.34 MPa and modulus of 37.1 MPa, about 1.9- to 2.3-fold increase in PIF-3 and 3.7- to 3.9-fold increase in PIF-5, respectively. Importantly, the foams showed good compression performance at 300 °C. PIF-1 exhibited compression modulus of 22.5 MPa and strength of 0.68 MPa at 300 °C, 60.6% and 50.7% retention of the properties at room temperature. However, PIF-2 showed 32.2% retention in modulus and 26.0% of retention in strength at 300 °C, probably due to the flexibility of the polymer backbone derived from 3,4′-ODA.

TABLE4.24 Compression Properties of the Rigid PIFs

	c_m (MPa)		c_s (MPa) at 10% Deformation		Compression Creep[a] (%)
	RT	300 °C	RT	300 °C	RT
PIF-1	37.1 ± 7.1	22.5 ± 3.3	1.34 ± 0.23	0.68 ± 0.06	0.18 ± 0.02
PIF-2	21.1 ± 5.8	6.8 ± 0.8	0.96 ± 0.17	0.25 ± 0.03	0.44 ± 0.13
PIF-3	16.3 ± 0.8	10.3 ± 0.4	0.70 ± 0.05	0.28 ± 0.02	0.64 ± 0.15
PIF-4	14.5 ± 0.8	10.4 ± 0.2	0.69 ± 0.06	0.28 ± 0.01	5.83 ± 2.39
PIF-5	10.1 ± 2.1	6.9 ± 0.2	0.34 ± 0.04	0.24 ± 0.01	—[b]

PIF, polyimide foam; RT, room temperature; c_m, compression modulus; c_s, compression strength.
[a]*Creep test was conducted under 0.4 MPa at room temperature for 2 h.*
[b]*The compression strength of PIF-5 was lower than 0.4 MPa, so PIF-5 cannot endure the creep condition.*

The compression creep resistance of the rigid PIFs was measured at 0.4 MPa for 2 h at room temperature (Table 4.24). PIF-1 showed the best creep resistance with creep strain of 0.18%, much better than PIF-2(0.44%), PIF-3 (0.64%), and PIF-4 (5.83%), respectively. Furthermore, the compression creep resistance of PIF-1 was measured at elevated temperatures under the pressure of 0.4 MPa and 2 h standing.

Fig. 4.43 shows the plot of creep strain versus temperature. It can be seen that the creep strain was kept at very low level (<2.0%) and did not change obviously at temperatures below 250 °C, then increased abruptly with the testing temperature increase. Fig. 4.44 shows the dependence of the creep strain on the applied pressures. The compression creep strain of PIF-1 was measured at a low level (<2.0%) under the applied pressures of <0.4 MPa, indicating that the rigid PIFs with density of 100 kg·m^{-3} could be serviced at 250 °C without obvious dimensional change, suitable as structural foams for high-temperature applications. Resistance to oxygen atmosphere, smoke generation, and flammability properties were also measured. It was found that the PIFs we prepared showed high LOI in the range of 33%-41%, indicating excellent resistance to combustion in an oxygen-rich environment. The vertical burn test results also indicated that this PIF has no after-flame, no dripping, and low smoke.

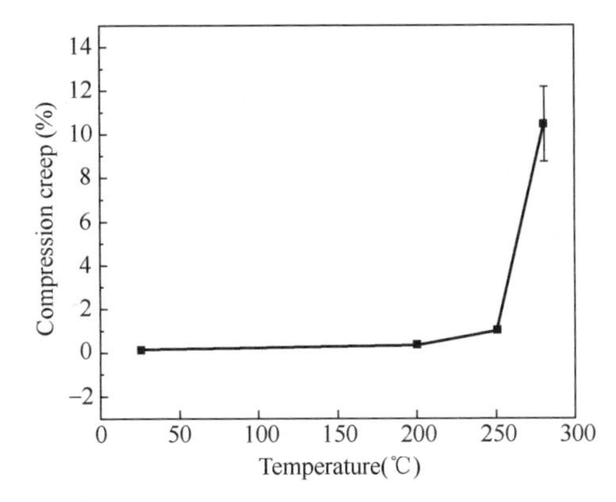

FIGURE 4.43 Compression creep strain of PIF-1 versus temperature.

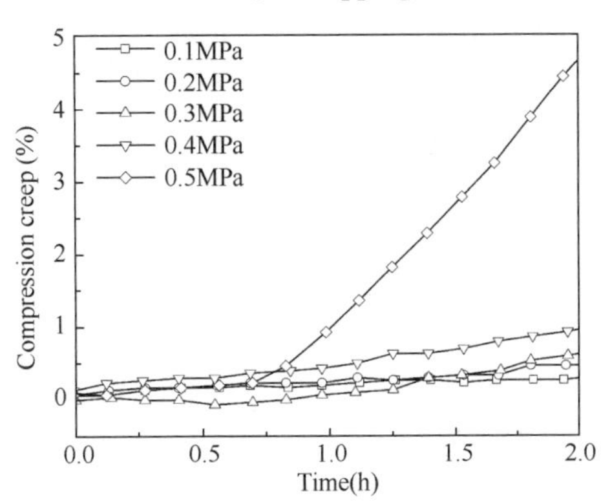

FIGURE 4.44 Compression creep strain of PIF-1 versus applied pressures.

Overall, the thermal foaming of the NAIOs was attributed to CPD evolved from the thermal decomposition of the nadimide end-caps by the reverse Diels-Alder reaction during the crosslinking of the bisnadimide oligomers, which acted as blowing agents. The resulting rigid polyimide foams showed uniform cellular structure with high C_c as well as high thermal stability and mechanical properties. The polyimide foams with C_c of 89% exhibited high compression strength at 10% compressive deformation of 1.34 MPa, good creep resistance, and outstanding thermal stabilities at temperatures as high as 250 °C, making it a desirable candidate for high-temperature structural foam materials for aviation and spacecraft applications.

Furthermore, the chemical structures and properties of the closed-cell polyimide foams were systematically investigated.

4.7.3.5 Thermal Foaming Properties of NAIOs With Different Backbone Structures

If α-BPDA in PI-2 (α-BPDA/3,4′-ODA/NA) was replaced by an other dianhydride (4,4′-oxydiphthalic anhydride (ODPA), for example) to give a new NAIO (ODPA/3,4′-ODA/NA) or NA replaced by other end-caps such as 3-ethynylaniline (3-APA) or 4-phenylethynylphthalic anhydride (PEPA) to give NAIO (α-BPDA/3,4′-ODA/3-APA, or α-BPDA/3,4′-ODA/PEPA) the NAIOs could lead to collapsed or uniformed cellular structures (Table 4.25), implying that the rigid and asymmetrical molecule structure of a-BPDA could not only improve foaming capability as the asymmetrical structure increase the “free volume” among molecule chains which could act as the initial foaming point, but also rise the T_g of polymer matrix to endure the high curing temperature of NA which decomposed CPD as blowing agent that the other two crosslinking agent cannot generate.

TABLE4.25 Thermal Foaming Results of 3,4′-ODA-based NAIOs With Different Dianhydrides and End-capping Agents

Dianhydride	NA	3-APA	PEPA
s-BPDA	Uniform, closed-cell	Collapsed	Collapsed seriously
ODPA	Collapsed	Collapsed seriously	Collapsed completely
BTDA	Nonuniform	Collapsed seriously	Collapsed completely
PMDA	Nonuniform	Collapsed seriously	Collapsed completely

4.7.3.6 Effect of Calc'd M_w on NAIO's Thermal Foaming Properties

On the basis of the observations described above, NAIO (α-BPDA/p-PDA/NA) was selected to investigate the effect of Calc'd M_w on the mechanical and thermal properties of the polyimide rigid foams. Thus, a series of NAIOs powders with different Calc'd M_w were prepared. The M_w of the oligomers determined by GPC were in accordance with the trend of the designed molecular weight but a little greater than the calculated ones with polydispersities of 1.48-2.16. The thermal foaming properties of NAIO powders were changed by either increasing or decreasing of Calc'd M_w. It can be seen that NAIO powders with Calc'd M_w=1000-2500 g·mol^{-1} (PI-1000, PI-1500, PI-2000, and PI-2500) could be easily thermally foamed to give homogeneous polyimide foams (Fig. 4.45). However, PI-3000 could be only partly foamed. It also can be seen from Fig. 4.46 that the pore of foams became smaller and less uniform, due to decreasing amounts of blowing agent CPD with Calc'd M_w increasing. Fig. 4.47 compares the rheological behaviors of NAIO powders with different Calc'd M_w. The NAIO's minimum melt viscosities were increased gradually with increasing Calc'd M_w, i.e., 78 Pa·s at 254 °C for PI-1000, 144 Pa·s at 280 °C for PI-1500, and 1280 Pa s at 279 °C for PI-2500, respectively. PI-3000 showed melt viscosity of as high as 2374 Pa·s at 306 °C, much higher than others, resulting in poor thermal foaming performance, probably due to the insufficient amount of blowing agent (CPD), and low melt flowability.

FIGURE 4.45 Thermal foaming results of NAIOs with different Calc'd M_w.

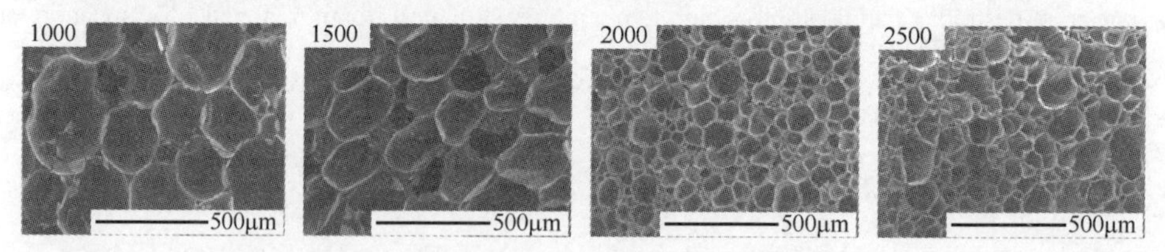

FIGURE 4.46 SEM photographs of the polyimide rigid foams derived from different Calc'd M_n.

4.7.3.7 Closed Cell Percent and Thermal Properties of Polyimide Rigid Foams

Fig. 4.48 compares the C_c of the polyimide rigid foams with different Calc'd M_w. PIF-2000 showed the highest C_c of 92%, compared with 85% for PIF-1000, 88% for PIF-1500, and 86% for PIF-2500, respectively. The C_c of polyimide foams was controlled by factors such as melt viscosity, melt strength, and amount of blowing agent, etc. The melt viscosity determined the thermal-foaming ability of the melt resins and the melt strength and toughness were critical factors in whether opened- or closed-cell structures could be produced. The concentration of blowing agent was also a key factor for the production of a closed-cell structure. PIF-1000 has the lowest melt viscosity due to its lowest Calc'd M_n, meanwhile, its melt strength and toughness are relatively poor compared with PIF-1500 and PIF-2000 due to the short polyimide backbone chains, resulting in polyimide foams with opened-cell structures.

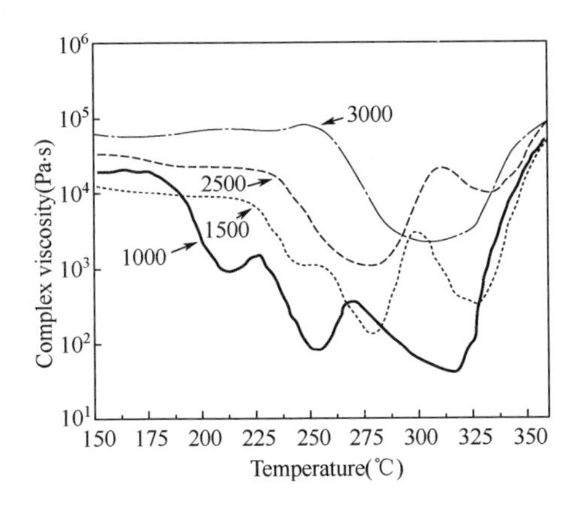

FIGURE 4.47 Rheological behaviors of NAIO powders with different Calc'd M_n.

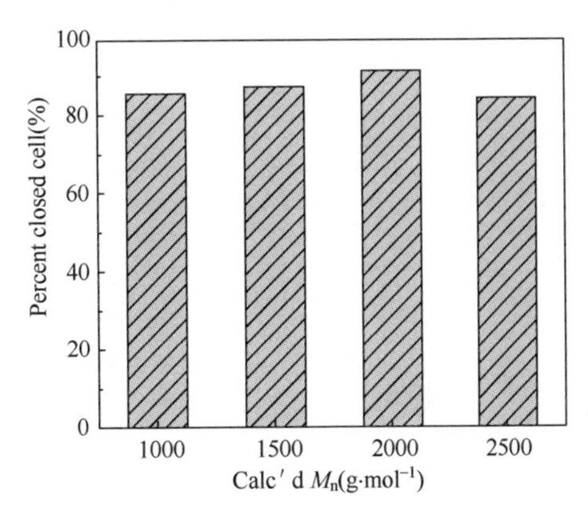

FIGURE 4.48 Closed-cell percent (C_c) of the polyimide rigid foams with different Calc'd M_n.

4.7.3.8 Mechanical Properties of the Polyimide Rigid Foams

Table 4.26 summarizes the mechanical properties of the polyimide rigid foams. The flexural modulus (f_m) decreased from 28.2 MPa for PIF-1000 to 20.1 MPa for PIF-1500, 18.7 MPa for PIF-2000, and 6.4 MPa for PIF-2500. The flexural strength (f_s) was also reduced with increasing Calc'd M_w. The tensile strength (δ_b) was measured in the range of 0.44-0.96 MPa and tensile modulus (δ_m) of 7.23-27.4 MPa. The elongations at breakage (ε) ranged from 4.3% to 4.9% for the polyimide foams with Calc'd M_w=1000-2000 g·mol^{-1}, and reached 9.0% for PIF-2500, indicating that the polyimide rigid foams have good toughness.

TABLE4.26 Mechanical Properties of the Polyimide Rigid Foams With Different Calc'd M_w

Calc'd M_n (g·mol^{-1})	Tensile Properties			Flexural Properties	
	σ_m (MPa)	σ_b (MPa)	ε (%)	f_m (MPa)	f_s (MPa)
1000	18.6±2.2	0.62±0.09	4.3±0.7	28.2±4.8	1.36±0.19
1500	27.4±2.6	0.96±0.08	4.6±0.5	20.1±4.2	1.19±0.20
2000	18.5±1.4	0.75±0.04	4.9±0.6	18.7±3.18	1.17±0.19
2000	7.23±0.9	0.44±0.06	9.0±1.4	6.4±1.10	0.59±0.07

Table 4.27 shows the compression properties of the polyimide rigid foams both at room temperature and elevated temperatures. In general, polyimide foams with low molecular weight, and high crosslinking-point densities had better compression properties because the crosslinking structures confined the movability of the molecule chain and made the foams much stiffer (PIF-1500 > PIF-2000 >

PIF-2500). However, increasing the crosslinking points could not only increase the stiffness of foams, but also increased the fragility, which could strongly affect the properties. As a result, PIF-1000 showed inferior compression properties compared with PIF-1500. PIF-1500 which reached balance between the stiffness and fragility showed the best compression properties at room temperature (c_m=37.1 MPa and c_s=1.34 MPa), compared with PIF-1000 (c_m=29.0 MPa and c_s=1.24 MPa), and PIF-2500 (c_m= 22.7 MPa and c_s=0.89 MPa), respectively.

TABLE4.27 Compression Properties of the Polyimide Rigid Foams With Different Calc'd M_w

Calc'd M_n (g·mol^{-1})	c_m (MPa)		c_s (MPa) at 10% Deformation		Compression Creep[a] (%)	
	RT	300	RT	300	RT	250
1000	29.0 ± 2.1	17.5 ± 1.1	1.24 ± 0.06	0.70 ± 0.02	0.18	1.28
1500	37.1 ± 7.1	22.5 ± 1.3	1.34 ± 0.23	0.71 ± 0.02	0.18	1.03
2000	32.0 ± 9.7	20.6 ± 2.3	1.27 ± 0.08	0.61 ± 0.02	0.23	3.87
2500	22.7 ± 1.9	11.9 ± 4.6	0.89 ± 0.05	0.50 ± 0.04	0.22	10.31

[a]*Creep test was conducted under 0.4 MPa for 2 h.*

At 300 °C, the compression properties were reduced by 48% for modulus and 52% for strength. For instance, PIF-1500 had c_m=37.1 MPa and f_s=1.34 MPa at room temperature, which was reduced to 22.5 MPa (61%) and 0.71 MPa (53%) at 300 °C (higher than the others). The polyimide rigid foams also showed excellent compression creep properties (0.18%-0.23%) under 0.4 MPa for 2 h at room temperature. But a quite different phenomenon was shown at 250 °C, PIF-1500, and PIF-1000 still exhibited very low compression creeps (1.03% and 1.28%, respectively) as the crosslinking networks could stabilize the molecule chains to ensure the structural integrity of foams at high temperatures. However, the polyimide foams with higher Calc'd M_w (PIF-2000 and PIF-2500) which lacked enough crosslinking points, showed increased compression creep (3.87% for PIF-2000 and 10.3% for PIF-2500, respectively).

4.7.3.9 Mechanical Properties of Polyimide Foams With Different Densities

Foam density can be controlled in the range of 50-400 kg·m^{-3}. Making subtle changes in chemistry, density, and opened or closed cell content, the physical properties of foams can be tuned to a specific application. It has been shown experimentally and theoretically that density is the most important parameter that affects physical properties. PIF-1500 showed the most excellent properties (relatively high T_g, tensile and flexural property, the highest class of compression properties) compared with the other polyimide foams with different Calc'd M_w and the same density of 100 kg·m^{-3}, NAIO powder with Calc'd M_w of 1500 g·mol^{-1} was selected to prepare foams with different densities and to illustrate how density affects the mechanical properties and closed-cell content of the rigid polyimide foams.

Fig. 4.49 depicts the dependence of the c_m of the polyimide foams on their densities (ρ), both at room temperature and at 300 °C. It can be seen that c_m of the polyimide rigid foams at room temperature increased gradually with ρ increasing. At ρ < 100 kg·m^{-3}, c_m increased from 3.3 MPa at ρ=50 kg·m^{-3} to 37.1 MPa at ρ=100 kg·m^{-3}, which was then increased obviously from 37.1 MPa at ρ=100 kg·m^{-3} to 72.3 MPa at ρ=150 kg·m^{-3} and 108.7 MPa at ρ=200 kg·m^{-3}, respectively. The polyimide foams with ρ=300 kg·m^{-3} and 400 kg·m^{-3} showed c_m of as high as 144.5 and 177.8 MPa, respectively. It was noteworthy that the polyimide foams showed very high modulus retention at 300 °C. For instance, the polyimide foams with ρ=400 kg·m^{-3} and 100 kg·m^{-3} showed modulus retention of 60%-70%.

The c_s of polyimide foams also showed obvious dependence on its ρ, both at ambient temperature and at 300 °C (Fig. 4.50). The c_s at room temperature for the polyimide foam with ρ=100 kg·m^{-3} was measured at 1.3 MPa, increased to 3.1 MPa at ρ=150 kg·m^{-3} and 8.0 MPa at ρ=300 kg·m^{-3}, respectively, indicating that polyimide foams could be strengthened by increasing their density.

At 300 °C, the polyimide foams also showed reasonable compression strength. For instance, the polyimide foams with ρ=100 and 400 kg·m^{-3} exhibited strength retention of > 50%, demonstrating that the polyimide foams have outstanding thermal stabilities.

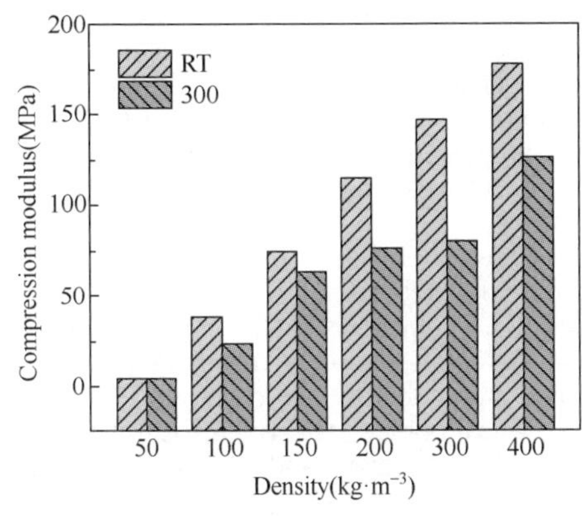

FIGURE 4.49 Compression modulus of the polyimide rigid foams with different densities.

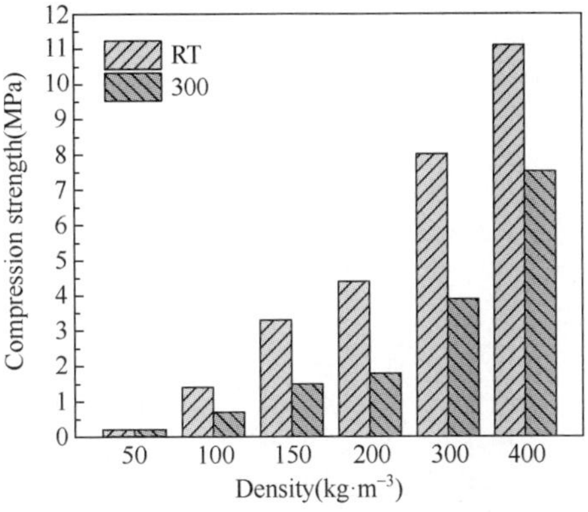

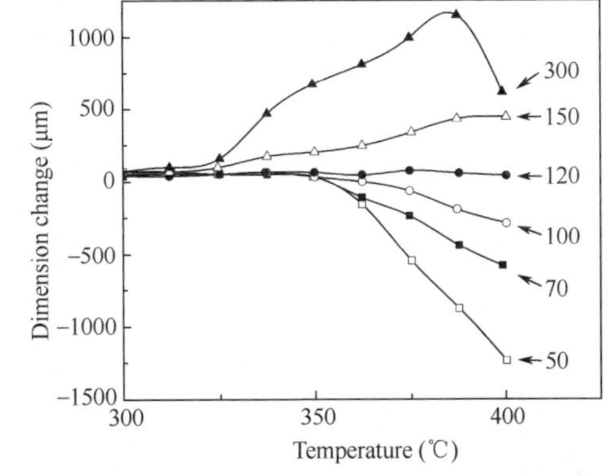

FIGURE 4.50 Compression strength of the polyimide rigid foams with different densities.

4.7.3.10 Closed-cell Percent and Thermal Properties of Polyimide Foams With Different Densities

Fig. 4.51 reveals the dependence of the C_c on the foam densities at two levels of testing pressures. At a lower testing pressure (2.9 psi), C_c was measured at 72% for polyimide foam with ρ=550 kg·m^{-3}, then increased gradually to 85% at ρ=570 kg·m^{-3}, and finally reached 89% at ρ=100 kg·m^{-3}. However, at a higher testing pressure of 17.0 psi, much lower C_c was observed for polyimide foams with lower densities (ρ < 100 kg·m^{-3}). For instance, polyimide foam with ρ=570 kg·m^{-3} showed C_c of 85% at 2.9 psi and 40% at 17.0 psi, respectively. However, C_c values of 90% have been reached for polyimide foams with higher densities (ρ > 100 kg·m^{-3}) both at 2.9 and 17.0 psi. This indicated that the low-density polyimide foams have weaker cells in the cellular foam structures, which could be damaged under high testing pressures.

Fig. 4.52 shows TMA curves of the polyimide foams with different densities. At temperature of < 320 °C, the polyimide foams showed linear thermal dimension expansion with increasing of temperatures. The *CTE* was measured at about (38-53) × 10^{-6} °C^{-1} (Fig. 4.53), and high-density polyimide foams had relatively high *CTE*. Abrupt changes in thermal expansions have been observed at > 320 °C (Fig. 4.45B). The polyimide foams with ρ < 120 kg·m^{-3} showed dimension shrinkages and that with ρ > 120 kg·m^{-3} expanded in dimension with increasing temperatures.

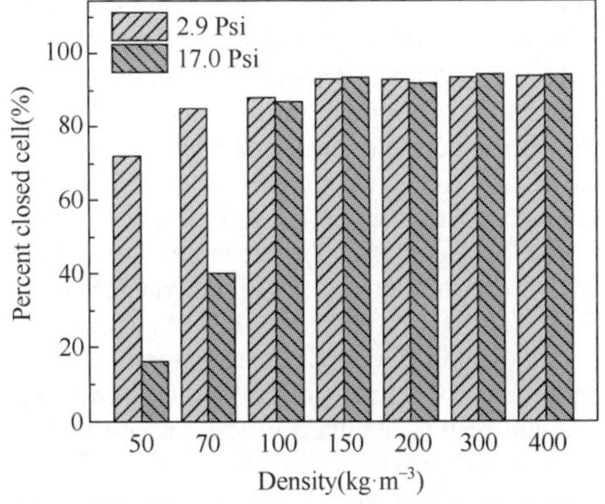

FIGURE 4.51 Closed-cell percent of the polyimide rigid foams with different densities.

FIGURE 4.52 TMA curves of the polyimide rigid foams with different densities.

This could be interpreted by the loading force (20 mN) on the testing samples, resulting in the foam cell shrinkage at temperatures near T_g for the low-density foams due to their large size and weak cells. The high-density foams with $\rho > 120$ kg·m^{-3} were strong enough to endure the preloading force on the testing samples and exhibited the same thermal expansion behaviors as solid resin without cellular structures.

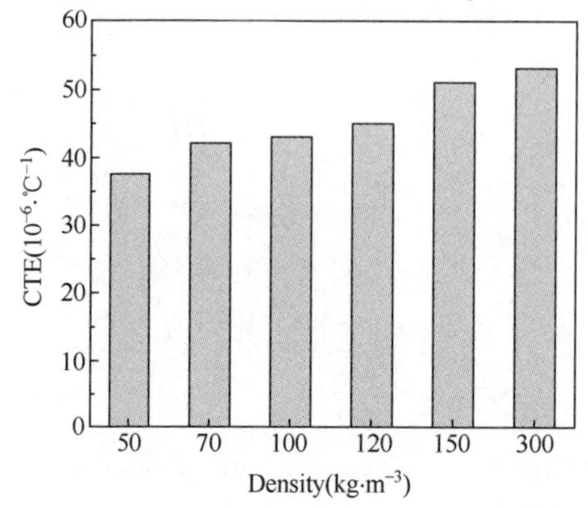

FIGURE 4.53 *CTE* of the polyimide rigid foams with different densities.

This observation was well in accordance with the conclusion derived from the compression property and closed-cell content testing that high-density foams have strong small size cells, resulting in thermally stable and stiffer foams. The density of 120 kg·m^{-3} was the critical point above which foam would have high closed-cell content and excellent mechanical properties.

In summary, there are obvious effects of Calc'd M_w and foam densities on mechanical and thermal properties of the polyimide rigid foams. Generally, low Calc'd M_w could yield foams with improved T_g, flexural, and compression properties, but excessively low Calc'd M_w could also make foams fragile and opened-cell. Thus, only the appropriate Calc'd M_w could yield polyimide foams with both high closed-cell percent and outstanding mechanical properties. Moreover, the mechanical properties could be improved gradually by increasing the foam densities which decreased the cell size and meanwhile increased the thickness of the cell vertices. The density of 120 kg·m^{-3} was the critical point above which foam would have high closed-cell content and excellent mechanical properties.

4.8 Thermally Stable Flexible Polyimide Aerogels

Organic aerogels have attracted increasing attention in recent years due to their intrinsic low density, low thermal conductivity, high thermal insulation, and low dielectric constants compared with their inorganic counterparts [60]. Various organic aerogels such as polyurethane, polyurea, polystyrene, aramide, and polydicyclopentadiene have been widely investigated in literature and have found applications in high-tech fields [61-66]. However, common organic aerogels usually suffer from low thermal and dimensional stability at elevated temperatures; therefore, they cannot meet the severe demands of high-temperature applications such as thermal protective coatings or interlayer thermal insulation systems for microelectronic device fabrication [67-69]. In order to prevent thermal deformation of the aerogels in the above applications, high-temperature-resistant organic aerogels are highly desired.

Aromatic polyimides (PIs) were considered as the most important high-temperature-resistant polymers and have been widely used in the electronic, microelectronic, and optoelectronic industries owing to their excellent combined thermal, mechanical, and dielectric properties. Thus, it can be anticipated that PI aerogels might exhibit good comprehensive properties. Some pioneering work on PI aerogels has been reported very recently [70-75]. Several PI aerogels based on aromatic dianhydrides, 3,3′,4,4′-biphenyltetracarboxylic dianhydride (BPDA), 3,3′,4,4′-benzophenone tetracarboxylic dianhydride (BTDA), aromatic diamines, paraphenylenediamine (PPD), 4,4′-oxydianline (ODA), bisanilinep-xylidene (BAX), 2,2′-dimethylbenzidine (DMBZ), and multifunctional end-cappers, 1,3,5-tris (4-aminophenyl)benzene (TAPB), 1,3,5-tris(4-aminophenoxy)benzene (TAB), octa(aminophenyl) silsesquioxane (OAPS), have been developed. Experimental results obtained in the above-mentioned reports suggest that PI aerogels derived from flexible dianhydride BTDA, flexible diamine ODA, and TAB crosslinker usually exhibited T_g below 280 °C, while their volume shrinkages were as low as 20%. Conversely, those derived from rigid dianhydride BPDA, rigid diamine PPD or DMBZ, and TAB showed high T_g values up to 346 °C; however, their volume shrinkages were as high as 48%. Thus, it is a challenge to achieve a good balance between the thermal and dimensional stability of the PI aerogels at elevated temperatures.

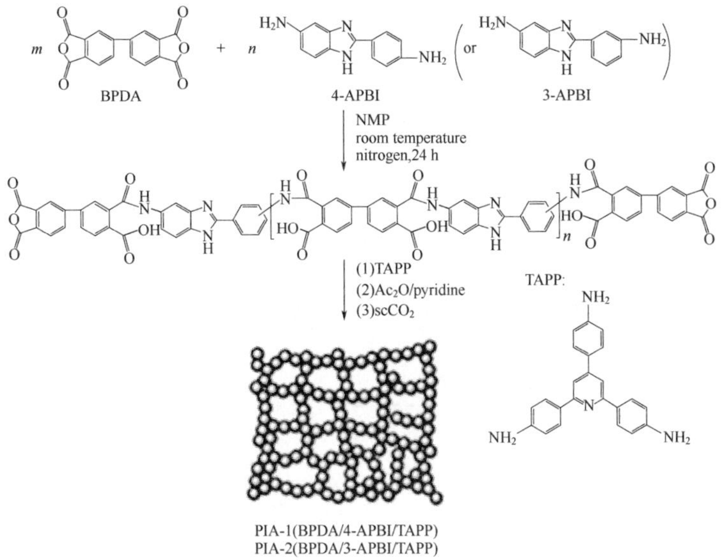

m BPDA + n 4-APBI (or 3-APBI)

NMP
room temperature
nitrogen,24 h

(1)TAPP
(2)Ac_2O/pyridine
(3)$scCO_2$

TAPP:

PIA-1(BPDA/4-APBI/TAPP)
PIA-2(BPDA/3-APBI/TAPP)

FIGURE 4.54 Synthesis of PIA aerogels.

A series of novel flexible PI aerogels with excellent thermal stability at elevated temperatures. The aerogels were synthesized based on a well-established route for polyimide aerogel preparation, as shown in Fig. 4.54. First, a triphenylpyridine-containing triamine compound, 2,4,6-tris(4-aminophenyl)-pyridine (TAPP) was synthesized. Then, two polyimide aerogels, PIA-1 (BPDA/4- APBI/TAPP) and PIA-2 (BPDA/3-APBI/TAPP) were prepared by the polycondensation of BPDA dianhydride, benzimidazole-containing diamines (4-APBI) and TAPP, respectively, followed by drying in supercritical carbon dioxide ($scCO_2$).

Both 4-APBI and TAPP monomers contain rigid components (benzimidazole or triphenylpyridine segments), which are very beneficial for improving the thermal resistance of the derived polyimide aerogels.

There is obvious influence of the rigid molecular structures of the polyimide aerogels on their properties, including density, porosity, surface area, volume shrinkage, and thermal and mechanical properties. Fig. 4.55 shows the morphology of the polyimide aerogels by field emission scanning electron microscopy (FESEM). Both polymers produced open-pore structures consisting of three-dimensional networks of tangled nanofibers, with nanoscale diameters. This three-dimensional networked microstructure results from the highly crosslinked molecular chains in the aerogels. The solvents present in the networks were continuously extracted by the solvent exchange process, followed by supercritical drying procedures, leaving nanopores in the final aerogels.

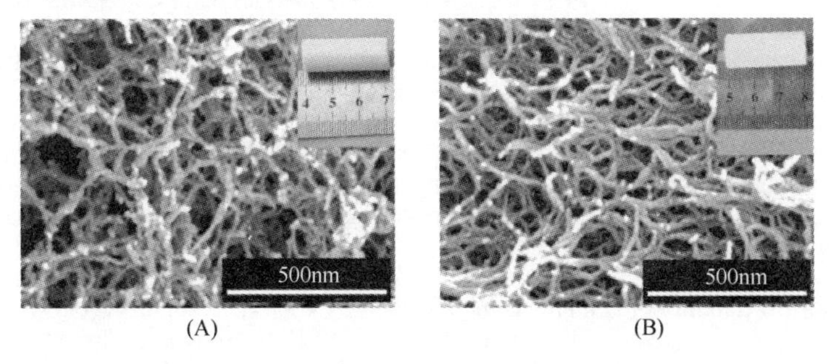

FIGURE 4.55 SEM images of PI aerogels. (A) PIA-1 and (B) PIA-2.

The porous structures of the current PI aerogels can be further probed by N_2 adsorption and desorption isotherm plots, measured by the Brunauer Emmett Teller (BET) method. Both isotherm plots showed rapid increase in adsorbed volumes at relative pressures above 0.9, which in combination with a narrow desorption loop indicates the existence of both meso- and macro-porous structures in the PIAs. Meanwhile, both isotherms showed an early rise in the adsorbed volumes ($P/P_0 < 0.1$), indicating the microporous structure of the aerogels. This porous structure of varying size is further confirmed by the pore size distribution measurements. Both aerogels had nanopores with sizes in the range of several to 100 nm. The average pore diameters of the aerogels are 18.2 nm for PIA-1 and 40.5 nm for PIA-2 (Table 4.28). From the BET measurements, the surface areas (δ) of PIA are determined to be 204 $m^2 \cdot g^{-1}$ for PIA-1 and 251 $m^2 \cdot g^{-1}$ for PIA-2. The bulk densities (ρ_b) of the PI aerogels, determined by mercury intrusion porosimetry, were 0.21 and 0.11 $g \cdot cm^{-3}$, respectively. The porosities of the aerogels calculated from the density values were 85.9% for PIA-1 and 92.3% for PIA-2. Owing to the bulky molecular packing in PIA-2, caused by the meta-substituted molecular structures, PIA-2 exhibited lower bulky density and higher porosity values.

TABLE 4.28 Properties of the PIA Aerogels

PI	Porosity[a] (%void)	σ^b ($m^2 \cdot g^{-1}$)	$\rho_b{}^c$ ($g \cdot cm^{-3}$)	$\rho_s{}^d$ ($g \cdot cm^{-3}$)	d^e (nm)	$T_{5\%}{}^f$ (°C)	Weight Loss After Aging for 24 h (%)			
							300 °C	400 °C	450 °C	500 °C
PIA-1	85.9	204	0.21	1.51	18.2	542	2.7	2.9	8.5	21.3
PIA-2	92.3	251	0.11	1.47	40.5	541	3.1	5.2	9.2	29.9

[a] *Calculated by* $1-(\rho_b/\rho_s)$.
[b] *Brunauer-Emmett-Teller surface area.*
[c] *Bulk density.*
[d] *Skeletal density.*
[e] *Average pore diameter from the BJH desorption plot.*
[f] *Temperature at 5% weight loss.*

The stress-strain curves obtained from the compression tests for the PI aerogels are shown in Fig. 4.56. The appearance of a thin sheet of PIA-1 is shown in the inset. The curves exhibited linear elastic regions below 10% strain and yielded in a relatively low slope. Young's modulus of PI aerogels is 9.32 MPa for PIA-1 and 2.38 MPa for PIA-2, indicating good flexibility and the tough nature of the polymers. The inset picture shows that PIA-1 aerogel can be folded without visible cracks. The good flexibility of the current PI aerogels is very beneficial for their utility in high-tech applications.

Thermal properties of the PI aerogels were evaluated by thermogravimetric analyses (TGA), thermomechanical analyses (TMA), and isothermal aging measurements. TGA measurements of the PI aerogels were performed in nitrogen from 50 °C to 750 °C. Both aerogels showed good thermal stability up to 400 °C, and they began decomposing at temperatures greater than 450 °C. The smooth plots around 300 °C indicated the complete removal of residual NMP solvent by the supercritical drying procedure and the complete imidization in the systems. The 5% weight loss temperatures (T_5%) are 542 °C for PIA-1 and 541 °C for PIA-2. The char yields of both aerogels exceed 70% in nitrogen, which indicated the good thermal stability of PIAs.

Changes in aerogel dimensions were investigated by TMA measurements, as depicted in Fig. 4.57. The aerogels showed different dimension change behaviors at elevated temperatures. PIA-2 shrank sharply when the temperature exceeded its T_g; however, PIA-1 showed expansion behavior beyond the T_g. The plots revealed the T_g values to be 360 °C for PIA-1 and 356 °C for PIA-2. Owing to the more symmetric molecular structure of PIA-1, it exhibited a slightly higher T_g value, compared with PIA-2. Nonetheless, both PI aerogels showed good dimensional and thermal stability. This is mainly due to the presence of rigid-rod substituents in the aerogels.

The thermal stability of the PI aerogels was further determined by isothermal aging measurements carried out at four different temperatures (300 °C, 400 °C, 450 °C, and 500 °C) in nitrogen for 24 h. Graphs of weight retention versus aging time of both aerogels are shown in Fig. 4.58. PIA-1 showed weight loss values of 2.7% at 300 °C, 2.9% at 400 °C, 8.5% at 450 °C, and 21.3% at 500 °C, indicating good thermal stability of the polymer below 400 °C. However, the corresponding thermal loss data of

PIA-2 at those specific aging temperatures are all greater than those observed for PIA-1. This is mainly due to the more rigid molecular structure of PIA-1. The SEM images of PIA-1 and PIA-2 after heating at 300 °C for 24 h showed little change. However, after aging for 24 h at 450 °C, the mesoporous structures of the PIA-1 aerogel partly collapsed; in contrast, the porous structure of PIA-2 was completely destroyed. This also reflects the higher thermal stability of PIA-1 compared to its analog PIA-2.

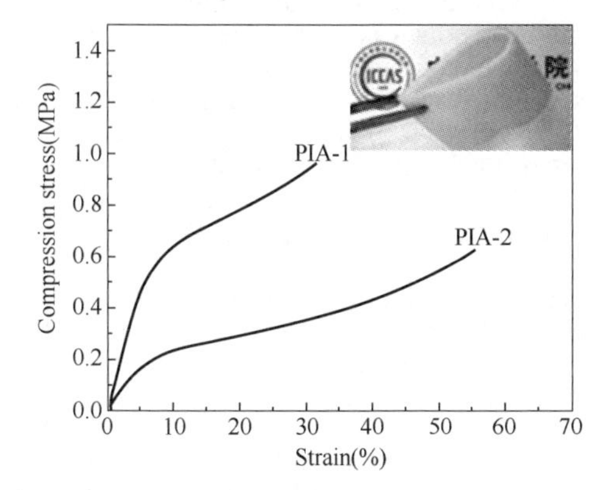

FIGURE 4.56 Stress-strain curves of PIA aerogels (inset: appearance of PIA-1).

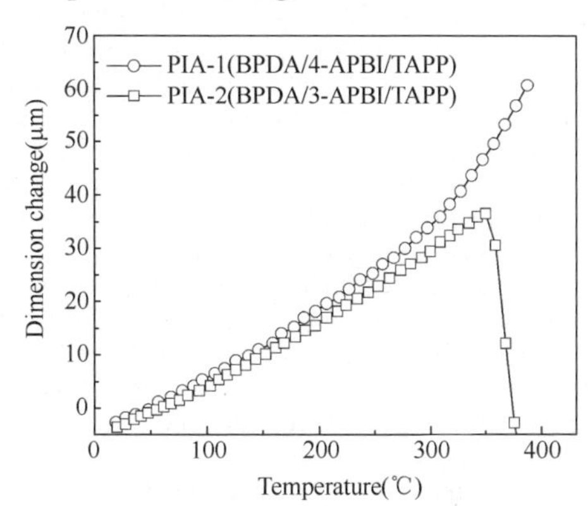

FIGURE 4.57 TMA curves of PIA aerogels.

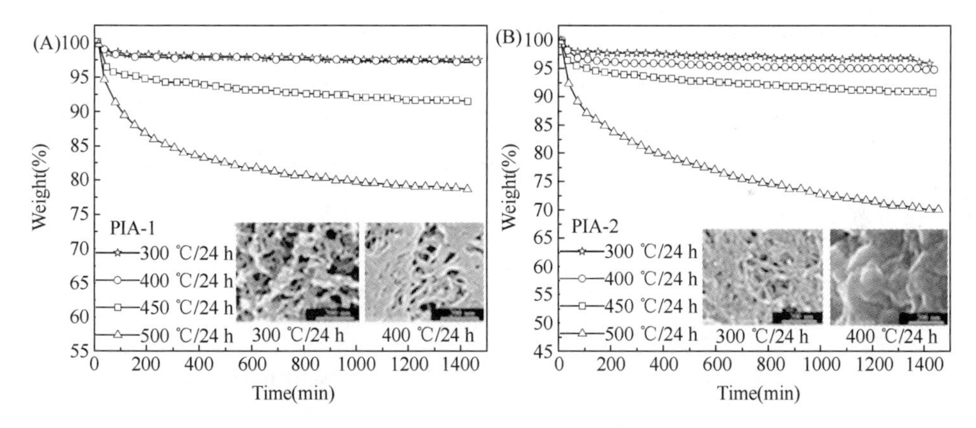

FIGURE 4.58 Isothermal thermal aging plots of PIA aerogels. (A) PIA-1; (B) PIA-2 (inset: SEM images of PIA aerogels after aging at 300 °C and 400 °C for 24 h. The scale number is 200 nm).

Overall, rigid-rod biphenyl, benzimidazole, and triphenylpyridine moieties endow polyimide aerogels with excellent thermal stability ($T_g > 350$ °C). The mesoporous structures could retain their initial morphology even after heating at 300 °C for 24 h, and the weight losses for both aerogels after aging for 24 h at 450 °C were lower than 10%(*w*).

Furthermore, intrinsically highly hydrophobic semialicyclic fluorinated polyimide aerogels with ultralow dielectric constants have been developed at the Institute of Chemistry, Chinese Academy of Sciences (ICCAS).

In recent years, polymer materials with low dielectric constants (usually abbreviated as low-*k* polymers, where *k* refers to the dielectric constant) and dielectric dissipation factors have been playing an increasingly important role in the interconnect fabrication of ultra large-scale integrated circuits (ULSIs) because of their ability to achieve a more rapid signal transport speed and a lower signal crosstalk. Various low-*k* polymers have been extensively investigated in the past decades. Among the low-*k* polymers, polyimides (PIs) have received great attention because of their intrinsic excellent combined properties such as high thermal and chemical resistance, high mechanical and dielectric strength, and more importantly, their relatively lower dielectric constant (k=3.0) compared to the common inorganic dielectrics. Thus, several wholly aromatic PIs have been commercialized and widely used as dielectrics for semiconductor chips. However, with the rapid development of ULSIs, dielectrics with lower k-values are highly desired. According to the International Technology Roadmap for Semiconductors (ITRS), by 2016, a k-value below 2.0 at 1 GHz and a dielectric loss below 0.003 must be met for the dielectrics [76]. It has been proven that the common methodologies for reducing the dielectric constants of conventional PIs, including introduction of groups with low molar polarizability (fluorinated groups, etc.) or with large molar volumes, could only achieve a lowest k-value of around 2.5. In recent years, it was found that incorporation of air voids (k=1.0) into PIs might be a promising method for achieving a k-value lower than 2.5. By this methodology, various ultralow-k PIs such as nanoporous PIPMMA [poly(methyl methacrylate)] films (k=2.1) [77] porous PI nanoparticles (k=1.9) [78] and PI-POSS (polyhedral oligomeric silsesquioxane) nanocomposites (k=2.4) [79] have been prepared and reported. However, attempts to reduce the k-values of polyimides to below 2.0 seem to be more challenging because the conventional methods could only achieve very limited air loading.

In the continuous efforts toward developing ultralow-k dielectrics for ULSI fabrication [80-82], attention was turned to aerogels because of their highly porous nature. The porosity of an aerogel can usually reach over 80% (volume ratio). The trapped air efficiently decreases the k-values of the aerogels to an extremely low level. Hence, aerogels based on PIs might be an efficient solution to achieving k-values below 2.0. In order to develop PI aerogels with k-values below 2.0, several limitations for conventional PI aerogels have to be overcome. First, wholly aromatic PIs usually possess polar molecular structures which can absorb humidity, thus increasing the dielectric constants of the polymers. Second, the nanoporous structure of the aerogels usually leads to condensation of water in the pores, resulting in the collapse of the pore structure by capillary forces. In order to overcome the above-mentioned limitations, a less polar molecular structure should be contained for the PI matrix and a hydrophobic surface is preferred for the PIs.

An intrinsically highly hydrophobic polyimide aerogel with ultralow dielectric constants has been synthesized from an alicyclic dianhydride, 1,2,3,4-cyclobutane tetracarboxylic dianhydride (CBDA), an aromatic fluoro-containing diamine, 2,2′-bis(trifluoromethyl)-4,4′-diaminobiphenyl (TFDB), and an *octa*-(aminophenyl) silsesquioxane (OAPS) end-capper (Fig. 4.59). As a POSS compound having eight amino

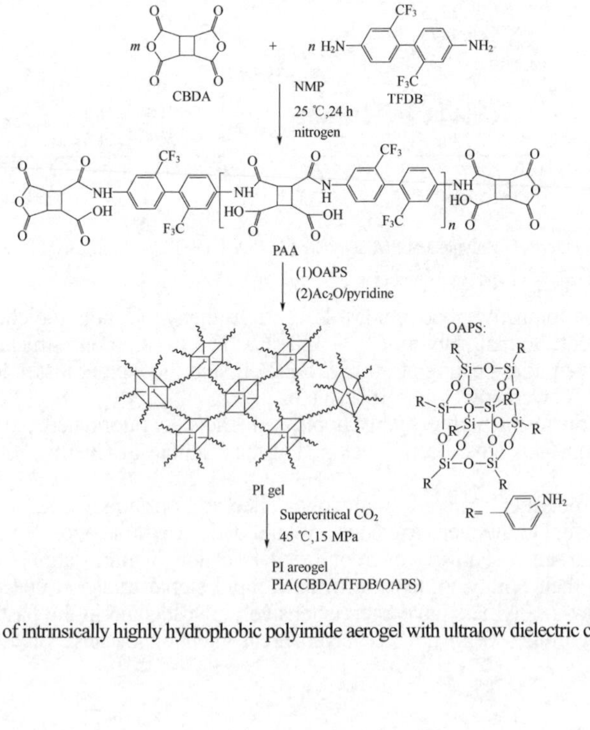

FIGURE 4.59 Synthesis of intrinsically highly hydrophobic polyimide aerogel with ultralow dielectric constants.

groups in its structure, OAPS works as not only a crosslinker but also as a component increasing the hydrophobicity and reducing the *k*-value for the PI aerogel. Excessive CBDA (the molar ratio of CBDA to TFDB equals to 1.03:1 and the number of the repeating units n is 30) was first reacted with TFDB to give an anhydride-capped PI precursor, poly(amic acid) (PAA) solution. Then, OAPS (the molar ratio of total amino to anhydride=1:1) was added to the PAA solution, inducing a crosslinking reaction of PAA.

The dehydrating agent (acetic anhydride/pyridine system) was then added to the solution, affording a PI wet gel. Then, the wet gel was immersed in ethanol to extract off the NMP solvent. Finally, the residual NMP and ethanol trapped in the gel were dried with supercritical carbon dioxide ($scCO_2$) to give the final PI aerogel (PIA). Young's modulus of the PIA according to the stress-strain curve was 10.4 MPa, reflecting the flexible and tough nature of the aerogel. The prepared PIA has several structural features. First, the PI matrix has relatively low polarity because of the introduction of the low-molar-polarizability trifluoromethyl and cyclobutane units. Second, the PI has a highly crosslinked three-dimensional network structure caused by the multifunctional POSS end-capper. Third, the supercritical drying process results in a high air content in the PI. The synergistic effects of these factors might provide the polymer with good hydrophobicity and a low dielectric constant.

Fig. 4.60 shows the morphology of the polyimide aerogel by field emission scanning electron microscopy (FESEM) measurements. The PIA consisted of tangled nanofibers with diameters in the range of dozens of nanometers. This three-dimensional network might be due to the highly crosslinked structure of the PI gel. Air occupies the free empty space in the PIA with a porosity of up to 85.8% (volume ratio), as calculated from the bulky density (ρ_b) and skeletal density (ρ_s) of the PIA. The ρ_b and ρ_s values of the PIA measured by the mercury intrusion porosimetry method are 0.22 and 1.55 $g\cdot cm^{-3}$, respectively. The surface area (δ) and pore size distribution of the PIA were measured by N_2 adsorption and desorption isotherms, using the Brunauer-Emmett-Teller (BET) method. As deduced from the N_2 adsorption desorption isotherm, the PIA should have a BET surface area of 407 $m^2\cdot g^{-1}$. In addition, the PIA shows a rapid increase in adsorbed volume at relative pressures above 0.9 and that the desorption plot exhibits a narrow loop. All this information indicates that both meso- and macroporosity exist in the PIA.

Fig. 4.61 shows the highly hydrophobic behavior of PIA using water contact angle measurements. The surface of the PIA is highly hydrophobic with a contact angle value of 135 degrees. This value is 62 degrees higher than that of the analogous PI aerogel derived from an aromatic dianhydride, 3,3′,4,4′-biphenyltetracarboxylic dianhydride (BPDA), unfluorinated aromatic diamine, 2,2′-dimethyl-4,4′-diaminobiphenyl (DMBZ), and OAPS. Furthermore, the PIA surface shows no sign of water seepage even after 24 h. This intrinsically hydrophobic behavior for the PIA is very beneficial for its application as ILDs in ULSIs because the *k*-value of the polymer would not be affected by atmospheric moisture.

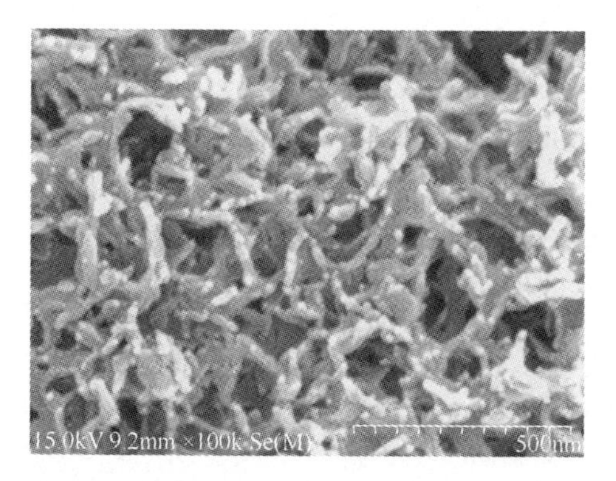

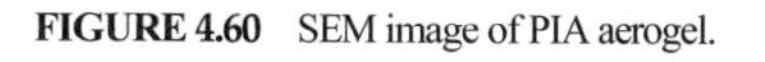

FIGURE 4.60 SEM image of PIA aerogel.

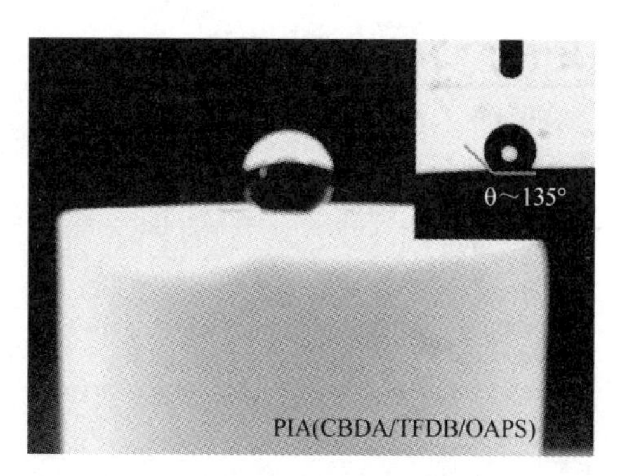

FIGURE 4.61 Status of water droplet on PIA aerogel. Inset: contact angle (θ) of water droplet.

Fig. 4.62 depicts the broadband dielectric constant (*k*) and dissipation factor (δ) of the PIA in the frequency range of 2-12 GHz measured at room temperature. The monolithic PIA sample (50×30×2mm^3) exhibited an almost constant *k*-value (1.17-1.19) over the entire frequency range. It has been reported that the pristine PI film (CBDA/TFDB) has a *k*-value of 2.5 [78]. Thus, introduction of air into the PI reduced the *k*-value significantly.

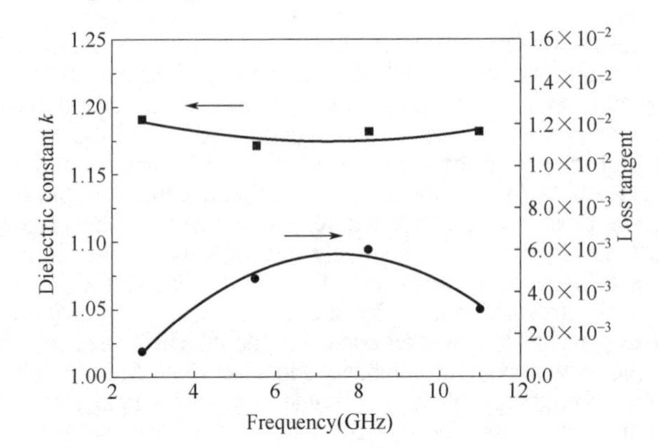

FIGURE 4.62 Dielectric constant (*k*) and dissipation factor (loss tangent) for PIA as a function of frequency.

The loss tangent (δ) value of the PIA also changed in a narrow range of 1.1×10^{-3} to 6.1×10^{-3}. The frequency-independent feature of the *k* and values for the PIA, on the one hand, is attributed to the high air loading (porosity: 85.8%) in its structure and, on the other hand, to the relatively low polarity of the fluoro-containing alicyclic PI matrix. Correspondingly, the ultralow-*k*-value of PIA is also due to the high volume ratio of the trapped air in PIA and the low ratio of the molar polarizability over molar volume by the trifluoromethyl groups and alicyclic units in the PIA. To the best of our knowledge, the *k*-value of around 1.19 at a frequency of 2.75 GHz is one of the lowest values for PI materials. Thus, the polymer might be a good candidate as an interlayer dielectric for advanced semiconductor chip interconnections.

Table 4.29 summarizes the typical properties of the PIA. The PIA was highly porous (porosity: 85.8%, volume ratio) and hydrophobic (water contact angle: 135 degrees) with a large surface area (407 $m^2\cdot g^{-1}$) and an ultralow dielectric constant (*k*=1.19 at 2.75 GHz). It is worth noting that the water contact angle of the PIA did not reach a super hydrophobic level (CA > 150 degrees). The experimental results indicated that a super hydrophobic surface might be achieved by adjusting the fluorine content of the PI matrix.

TABLE4.29 Properties of the PIA Aerogel

Porosity[a] (% void)	σ^b ($m^2\cdot g^{-1}$)	ρ_b^c ($g\cdot cm^{-3}$)	ρ_s^d ($g\cdot cm^{-3}$)	d^e (nm)	$T_{5\%}^f$ (°C)	k^g	δ^h
85.8	407	0.22	1.55	32.3	442	1.19	0.0011

[a]*Calculated by 1-(ρ_b/ρ_s).*
[b]*Brunauer-Emmett-Teller surface area.*
[c]*Bulk density.*
[d]*Skeletal density.*
[e]*Average pore diameter from the BJH desorption plot.*
[f]*Temperature at 5% weight loss.*
[g]*Value at 2.75 GHz.*
[h]*Loss tangent at 2.75 GHz.*

Overall, flexible polyimide aerogels with an intrinsically hydrophobic nature and a dielectric constant as low as 1.19 at a frequency of 2.75 GHz was synthesized. The low-*k* polyimide aerogels exhibit excellent combined properties such as high thermal and chemical resistance, high mechanical and dielectric strength, and more importantly, have potential for applications in microelectronics.

4.9 Summary

Conventional aromatic polyimides, due to their rigid/stiff backbone structures, are usually high-melting polymer materials, which cannot melt giving an easily flowing molten fluid suitable for injection or infusion, such as common polymers. Hence, a great deal of research has been performed on the modification of polyimide backbone structures to improve polyimide resin's melt flow ability while concurrently retaining the thermal stability and mechanical properties at elevated temperatures. Compression-molded polyimide engineering plastics and injection/extrusion processed engineering plastics have been developed. There are obvious effects of chemical structures of the aromatic polyimide resins on their combined properties including the melt processability, mechanical and thermal properties.

In addition, polyimide foams have been extensively used in many high-technology fields such as aerospace and aviation industries as the thermal and acoustic insulation materials as well as structural support materials due to their ultralow weight, excellent thermal insulating and acoustic absorbing properties, high strength-to-weight ratio and cost-effectiveness, etc. Two types of polyimide foams have been successfully developed, including the opened-cell soft polyimide foams and the closed-cell rigid polyimide foams according to the cell structures. The closed-cell polyimide rigid foams exhibited high thermal stability and high compression properties for structural support material applications in aerospace, while the opened-cell polyimide soft foams showed low thermal stability and low mechanical modulus.

REFERENCES

[1] M. Bessonov, Polyimides-thermally stable polymers, Plenum Publishing Corp, New York, 1987.

[2] W. Volksen, Recent advances in polyimides and other high performance polymers. Workshop sponsored by American Chemical Society, Polymer Division, 1990.

[3] T.L.S. Clair, D.A. Yamaki, Solvent resistant, thermoplastic aromatic poly(imidesulfone) and process for preparing same. US4398021, 1983.

[4] T.P. Gannett, H.H. Gibbs, R.J. Kassal, Melt-fusible polyimides. US4485140, 1984.

[5] T.P. Gannett, R.J. Kassal, R.S. Ro, J. Uradnisheck, Melt-fusible co-polyimide from diamine mixture. US4725642, 1988.

[6] T.P. Gannett, H.H. Gibbs, Melt-fusible polyimides. US4576857, 1986.

[7] F.E. Rogers, Melt-fusible linear polyimide of 2,2-bis(3,4-dicarboxyphenyl)-hexa fluoropropane dianhydride. US3959350, 1976.

[8] D. Wilson, H.D. Stenzenberger, P.M. Hergenrother, Polyimides., Springer, 1990.

[9] V.L. Bell, B.L. Stump, H. Gager, Polyimide structure-property relationships. 2. Polymers from isomeric diamines, J. Polym. Sci. Part A:Polym. Chem. 14 (9) (1976) 2275-2291.

[10] H.D. Burks, T.L. Saint Clair, T.-H. Hou, Characterization of crystalline LaRC-TPI powder, 1986.

[11] I.W. Serfaty, Polyetherimide: a versatile, processable thermoplastic, In: K.L. Mittal (Ed.), Polyimides: Synthesis, Characterization, and Applications Volume 1, MA: Springer US, Boston, 1984, pp. 149-161.

[12] T. Takekoshi, J.E. Kochanowski, J.S. Manello, M.J. Webber, Polyetherimides. 2. High-temperature solution polymerization, J. Polym. Sci.- Polym. Symposia 74 (1986) 93-108.

[13] S. Tamai, A. Yamaguchi, M. Ohta, Melt processible polyimides and their chemical structures, Polymer 37 (16) (1996) 3683-3692.

[14] R.G. Bryant, LaRC™-Si: A soluble aromatic polyimide, High Perform. Polym. 8 (4) (1996) 607-615.

[15] T.H. Hou, R.G. Bryant, Processing and properties of IM7/LaRC™-Si polyimide composites, High Perform. Polym. 9 (4) (1997) 437-448.

[16] T.H. Hou, T.L. St Clair, IM7/LaRC™-IAX-3 polyimide composites, High Perform. Polym. 10 (2) (1998) 193-206.

[17] V. Ratta, E.J. Stancik, A. Ayambem, H. Pavatareddy, J.E. McGrath, G.L. Wilkes, A melt-processable semicrystalline polyimide structural adhesive based on 1,3-bis(4-aminophenoxy)benzene and 3,3′,4,4′-biphenyltetracarboxylic dianhydride, Polymer 40 (7) (1999) 1889-1902.

[18] V. Ratta, A. Ayambem, J.E. McGrath, G.L. Wilkes, Crystallization and multiple melting behavior of a new semicrystalline polyimide based on 1,3-bis(4-aminophenoxy)benzene (TPER) and 3,3′,4,4′-biphenonetetracarboxylic dianhydride (BTDA),

Polymer 42 (14) (2001) 6173-6186.

[19] S. Tamai, T. Kuroki, A. Shibuya, A. Yamaguchi, Synthesis and characterization of thermally stable semicrystalline polyimide based on 3,4′-oxydianiline and 3,3′,4,4′-biphenyltetracarboxylic dianhydride, Polymer 42 (6) (2001) 2373-2378.

[20] Z.M. Shi, M. Hasegawa, Y. Shindo, R. Yokata, F.F. He, H. Yamaguchi, et al., Thermo-processable polyimides with high thermo-oxidative stability as derived from oxydiphthalic anhydride and bisphenol a type dianhydride, High Perform. Polym. 12 (3) (2000) 377-393.

[21] M. Hasegawa, Z. Shi, R. Yokata, F.F. He, H. Ozawa, Thermo-processable polyimides with high t-g and high thermo-oxidative stability as derived from 2,3,3′,4′-biphenyltetracarboxylic dianhydride, High Perform. Polym. 13 (4) (2001) 355-364.

[22] T. Kuroki, A. Shibuya, M. Toriida, S. Tamai, Melt-processable thermosetting polyimide: Synthesis, characterization, fusibility, and property, J. Polym. Sci. Part A:Polym. Chem. 42 (10) (2004) 2395-2404.

[23] Y.H. Song, S.G. Kim, S.B. Lee, K.J. Rhee, T.S. Kim, A study of considering the reliability issues on asic/memory integration by sip (systemin-package) technology, Microelectron. Reliab. 43 (9-11) (2003) 1405-1410.

[24] T. Shimoto, K. Baba, K. Matsui, J. Tsukano, T. Maeda, K. Yachi, Ultra-thin high-density lsi packaging substrate for advanced csps and sips, Microelectron. Reliab. 45 (3-4) (2005) 567-574.

[25] F.P. McCluskey, L. Condra, T. Torri, J. Fink, Packaging reliability for high temperature electronics: A materials focus, Microelectron. Int. 13 (3) (1996) 23-26.

[26] D.Y. Chong, B. Lim, K.J. Rebibis, S. Pan, S. Krishnamoorthi, R. Kapoor, et al., Development of a new improved high performance flip chip bga package. Electronic Components and Technology Conference, 2004. Proceedings. 54th. IEEE, 2004, 1174-1180.

[27] P.S. Ho, G. Wang, M. Ding, J.-H. Zhao, X. Dai, Reliability issues for flip-chip packages, Microelectron. Reliab. 44 (5) (2004) 719-737.

[28] D.R. McGregor, G.S. Cox, T.D. Lantzer, Recent developments in polyimide-based planar capacitor laminates. IPC/FED Conference on Embedded Passives, 2004.

[29] R. Tummala, P.M. Raj, V. Sundaram, Next-generation packaging materials, Adv. Packaging 13 (6) (2004).

[30] E.D. Feit, C.W. Wilkins Jr, Polymer Materials for Electronic Applications, ACS Publications, 1982.

[31] P. Buchwalter, A. Baise, K. Mittal, Polyimides: Synthesis, Characterization and Applications, *Plenum, New York*, 1984, p. 537.

[32] V. Smirnova, M. Bessonov, C. Feger, M. Khojasteh, J. McGrath, Polyimides: Materials, Chemistry and Characterization, Elsevier Science Publishers BV, Amsterdam, 1989.

[33] J. Coburn, M. Pottiger, in: M.K. Ghosh, K.L. Mittal (Eds.), Polyimides: Fundamentals and Applications, Marcel Dekker, New York, 1996.

[34] Y. Watanabe, Y. Shibasaki, S. Ando, M. Ueda, Synthesis and characterization of polyimides with low dielectric constants from aromatic dianhydrides and aromatic diamine containing phenylene ether unit, Polymer 46 (16) (2005) 5903-5908.

[35] T.H. Hou, R.G. Bryant, Processing and properties of IM7/LaRC^{TM}-Sai composite, High Perform. Polym. 8 (2) (1996) 169-184.

[36] R.J. Cano, T.H. Hou, E.S. Weiser, T.L. St Clair, Polyimide composites from 'salt-like' solution precursors, High Perform. Polym. 13 (4) (2001) 235-250.

[37] J.S. Chen, S.Y. Yang, Z.Q. Tao, A.J. Hu, L. Fan, Processing and properties of carbon fiber-reinforced pmr type polyimide composites, High Perform. Polym. 18 (3) (2006) 377-396.

[38] J.B. Schutz, Hybrid process for resin transfer molding of polyimide matrix composites, in: P.J. Adams, S.A. Elsworth, T.C. Petkauskos and M J. Walton (Eds.), Revolutionary Materials: Technology and Economics, Vol. 32, 2000, pp. 319-328.

[39] J.E. Lincoln, R.J. Morgan, E.E. Shin, Fundamental investigation of cure-induced microcracking in carbon fiber/bismaleimide cross-ply laminates, Polym. Compos. 22 (3) (2001) 397-419.

[40] T.H. Hou, B.J. Jensen, Double-vacuum-bag technology for volatile management in composite fabrication, Polym. Compos. 29 (8) (2008) 906-914.

[41] J. De Abajo, J. De la Campa, Processable aromatic polyimides, Prog. Polyimide Chem. I (1999) 23-59.

[42] S. Bhuvana, M.S. Devi, Bisphenol containing novel polyimides/glass fiber composites, Polym. Compos. 28 (3) (2007) 372-380.

[43] S.Q. Gao, X.C. Wang, A.J. Hu, Y.L. Zhang, S.Y. Yang, Preparation and properties of pmr-ii polyimide/chopped quartz fibre composites, High Perform. Polym. 12 (3) (2000) 405-417.

[44] H.Y. Xu, H.X. Yang, L.M. Tao, J.G. Liu, L. Fan, S.Y. Yang, Preparation and properties of glass cloth-reinforced meltable thermoplastic polyimide composite for microelectronic packaging substrates, High Perform. Polym. 22 (2010) 581-597.

[45] E.S. Weiser, T.F. Johnson, T.L. St Clair, Y. Echigo, H. Kaneshiro, et al., Polyimide foams for aerospace vehicles, High Perform. Polym. 12 (1) (2000) 1-12.

[46] C. Resewski, W. Buchgraber, Properties of new polyimide foams and polyimide foam filled honeycomb composites, Materialwissenschaft Und Werkstofftechnik 34 (4) (2003) 365-369.
[47] M.K. Williams, D.B. Holland, O. Melendez, E.S. Weiser, J.R. Brenner, G.L. Nelson, Aromatic polyimide foams: Factors that lead to high fire performance, Polym. Degrad. Stab. 88 (1) (2005) 20-27.
[48] H.-J. Chu, B.-K. Zhu, Y.-Y. Xu, Polyimide foams with ultralow dielectric constants, J. Appl. Polym. Sci. 102 (2) (2006) 1734-1740.
[49] C.-J. Chang, M.-H. Tsai, G.-S. Chen, M.-S. Wu, T.-W. Hung, Preparation and properties of porous polyimide films with tio2/polymer double shell hollow spheres, Thin Solid Films 517 (17) (2009) 4966-4969.
[50] W.R. Hendrix, A process for the production of foamed polyimide articles comprises adding an anhydride of a lower fatty acid (as defined) or an anhydride of an aromatic monoba. US3249561-A, 1966.
[51] E. Lavin, I. Serlin, Preparing poly-imide foams from polycarboxylic acids or. US3483144-A, 1968.
[52] E. Lavin, I. Serlin, Polyimide foam prepn. US3554939-A, 1971.
[53] J.R. Barringer, H.E. Broemmelsi, C.W. Lanier, R. Lee, H.E. Broemmelsiek, Polyimide foam provided in a range of densities|prepd. By incorporating a polar protic foam-enhancing material as an additive to a foamable polyimide precursor compsn. US5096932-A, 1991.
[54] R. Lee, Polyimide foam precursor, liq. Within specified temp.-range|comprising tetra: Carboxylic acid alkyl:Di-half-ester, di:Amine-cpd. And volatile, polar, protic foam-enhancer, useful for mg. Fire-resistant insulating foam. US5122546-A, 1992.
[55] E. Weiser, T. St. Clair, Y. Echigo, H. Kaneshiro, T. St. Clair, E.S. Weiser, et al., 2000. Solid polyimide precursor, used to make e.G. Foams for aerospace applications. US6180746-B1.
[56] E.S. Weiser, Synthesis and characterization of polyimide residuum, friable balloons, microspheres and foams, 2004
[57] H.-J. Chu, B.-K. Zhu, Y.-Y. Xu, Preparation and dielectric properties of polyimide foams containing crosslinked structures, Polym. Adv. Technol. 17 (5) (2006) 366-371.
[58] L. Dutruch, M. Senneron, M. Bartholin, P. Mison, B. Sillion, Preparation of thermostable rigid foams by control of the reverse diels-alder reaction during the cross-linking of bis-nadimide oligomers, in: K.C. Khemani (Ed.), Polymeric Foams: Science and Technology, Vol. 669, 1997, pp. 37-53.
[59] D.J. Green, R. Colombo, Cellular ceramics: Intriguing structures, novel properties, and innovative applications, Mrs Bull. 28 (4) (2003) 296-300.
[60] L. Ratke, Monoliths and fibrous cellulose aerogels, Aerogels Handbook, Springer, 2011, pp. 173-190.
[61] R. Petricevic, M. Glora, J. Fricke, Planar fibre reinforced carbon aerogels for application in pem fuel cells, Carbon 39 (6) (2001) 857-867.
[62] A. Rigacci, J.C. Marechal, M. Repoux, M. Moreno, P. Achard, Preparation of polyurethane-based aerogels and xerogels for thermal superinsulation, J. Non-Crystal. Solids 350 (2004) 372-378.
[63] J.K. Lee, G.L. Gould, Polydicyclopentadiene based aerogel: a new insulation material, J. Sol-Gel Sci. Technol. 44 (1) (2007) 29-40.
[64] C. Daniel, D. Sannino, G. Guerra, Syndiotactic polystyrene aerogels: adsorption in amorphous pores and absorption in crystalline nanocavities, Chem. Mater. 20 (2) (2008) 577-582.
[65] N. Leventis, C. Sotiriou-Leventis, N. Chandrasekaran, S. Mulik, Z.J. Larimore, H. Lu, et al., "Multifunctional polyurea aerogels from isocyanates and water. A structure-property case study, Chem. Mater. 22 (24) (2010) 6692-6710.
[66] X. Wang, S.C. Jana, Syndiotactic polystyrene aerogels containing multi-walled carbon nanotubes, Polymer 54 (2) (2013) 750-759.
[67] J.O. Simpson, A.K. St Clair, Fundamental insight on developing low dielectric constant polyimides, Thin Solid Films 308 (1997) 480-485.
[68] T. Homma, Low dielectric constant materials and methods for interlayer dielectric films in ultralarge-scale integrated circuit multilevel interconnections, Mater. Sci. Eng. R-Rep. 23 (6) (1998) 243-285.
[69] A. Kuntman, H. Kuntman, A study on dielectric properties of a new polyimide film suitable for interlayer dielectric material in microelectronics applications, Microelectron. J. 31 (8) (2000) 629-634.
[70] C. Chidambareswarapattar, Z. Larimore, C. Sotiriou-Leventis, J.T. Mang, N. Leventis, One-step room-temperature synthesis of fibrous polyimide aerogels from anhydrides and isocyanates and conversion to isomorphic carbons, J. Mater. Chem. 20 (43) (2010) 9666-9678.
[71] H. Guo, M.A.B. Meador, L. McCorkle, D.J. Quade, J. Guo, B. Hamilton, et al., Polyimide aerogels cross-linked through amine functionalized polyoligomeric silsesquioxane, ACS Appl. Mater. Interfaces 3 (2) (2011) 546-552.
[72] N. Leventis, C. Sotiriou-Leventis, D.P. Mohite, Z.J. Larimore, J.T. Mang, G. Churu, et al., Polyimide aerogels by ring-opening metathesis polymerization (romp), Chem. Mater. 23 (8) (2011) 2250-2261.
[73] H. Guo, M.A.B. Meador, L. McCorkle, D.J. Quade, J. Guo, B. Hamilton, et al., Tailoring properties of cross-linked polyimide aerogels for better moisture resistance, flexibility, and strength, ACS Appl. Mater. Interfaces 4 (10) (2012) 5422-5429.

[74] M.A.B. Meador, E.J. Malow, R. Silva, S. Wright, D. Quade, S.L. Vivod, et al., Mechanically strong, flexible polyimide aerogels cross-linked with aromatic triamine, ACS Appl. Mater. Interfaces 4 (2) (2012) 536-544.

[75] M.A.B. Meador, S. Wright, A. Sandberg, B.N. Nguyen, F.W. Van Keuls, C.H. Mueller, et al., Low dielectric polyimide aerogels as substrates for lightweight patch antennas, ACS Appl. Mater. Interfaces 4 (11) (2012) 6346-6353.

[76] P.A. Kohl, Low-dielectric constant insulators for future integrated circuits and packages. In: Prausnitz JM (ed). Annual review of chemical and biomolecular engineering, vol. 2, pp. 379-401, 2011.

[77] G.D. Fu, B.Y. Zong, E.T. Kang, K.G. Neoh, Nanoporous low-dielectric constant polyimide films via poly(amic acid)s with raft-graft copolymerized methyl methacrylate side chains, Ind. Eng. Chem. Res. 43 (21) (2004) 6723-6730.

[78] G. Zhao, T. Ishizaka, H. Kasai, M. Hasegawa, T. Furukawa, H. Nakanishi, et al., Ultralow-dielectric-constant films prepared from hollow polyimide nanoparticles possessing controllable core sizes, Chem. Mater. 21 (2) (2009) 419-424.

[79] C.M. Leu, Y.T. Chang, K.H. Wei, Polyimide-side-chain tethered polyhedral oligomeric silsesquioxane nanocomposites for low-dielectric film applications, Chem. Mater. 15 (19) (2003) 3721-3727.

[80] H.S. Li, J.G. Liu, K. Wang, L. Fan, S.Y. Yang, Synthesis and characterization of novel fluorinated polyimides derived from 4,4′- 2,2,2-trifluoro-1-(3,5-ditrifluoromethylphenyl) ethylidene diphthalic anhydride and aromatic diamines, Polymer 47 (4) (2006) 1443-1450.

[81] L. Tao, H. Yang, J. Liu, L. Fan, S. Yang, Synthesis and characterization of highly optical transparent and low dielectric constant fluorinated polyimides, Polymer 50 (25) (2009) 6009-6018.

[82] X.-J. Zhao, J.-G. Liu, H.-S. li, L. Fan, S.-Y. Yang, Synthesis and properties of fluorinated polyimides from 1,1-bis(4-amino-3,5,-dimethylphenyl)-1-(3,4,5-trifluorophenyl)-2,2,2 -trifluoroethane and various aromatic dianhydrides, J. Appl. Polym. Sci. 111 (5) (2009) 2210-2219.

Chapter 5

Polyimides for Electronic Applications

Qing-Hua Lu and Feng Zheng
Shanghai Jiao Tong University, Shanghai, China

5.1 Introduction

Polyimides (PIs) can be directly formed from the solution of a dianhydride and a diamine in a phenolic solvent with high boiling point [1,2]. Alternatively, with the more common two-step approach as shown in Fig. 5.1, a precursor poly(amic acid) (PAA) is firstly synthesized from the reaction of a dianhydride and a diamine in polar organic solvents (e.g., DMF, DMAC, and NMP), and then converted into PI product via thermal/chemical imidization of PAA. They are characterized by the repeating imide structure as a linear or heterocyclic unit in the main chain of the polymer backbone, and classified into aliphatic, semiaromatic, and aromatic groups.

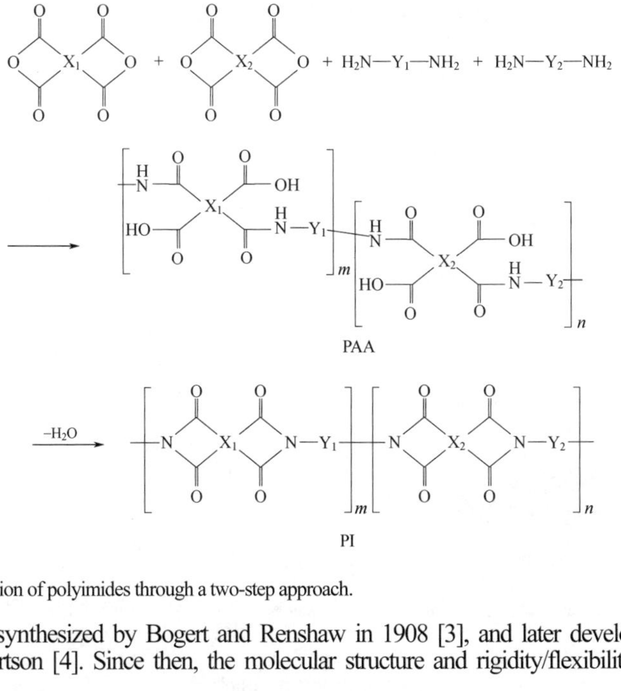

FIGURE 5.1 Preparation of polyimides through a two-step approach.

PIs were first synthesized by Bogert and Renshaw in 1908 [3], and later developed as material by Edwards and Robertson [4]. Since then, the molecular structure and rigidity/flexibility of PIs have been

Advanced Polyimide Materials. DOI: https://doi.org/10.1016/B978-0-12-812640-0.00005-6

judiciously designed due to the diversity of both dianhydrides and diamines; they become a class of high-performance polymers with unique chemical, physical, mechanical and electrical properties [5]. These properties include (1) temperature stability from -190 °C to 540 °C, (2) outstanding mechanical properties with tensile and compressive strengths over 100 MPa, and tensile elongation greater than 10%, (3) great radiation resistance, (4) chemical stability with high resistance to solvents and moisturize, (5) good electrical insulation with volume resistivity at $10^{14}\,\Omega$-$10^{15}\,\Omega$ and surface resistivity at $10^{15}\,\Omega$-$10^{16}\,\Omega$, (6) excellent dielectric properties, (7) high purity, (8) good adhesive properties for normal inorganic, metallic, and dielectric materials, (9) easy casting of thin or thick films, and (10) good processibility. As a consequence, PIs are used in a variety of applications including aerospace, automotive, electrical, tribological and photonics applications [6]. The suitability of PIs for such applications is attributed to their solubilities in the PAA precursor form and excellent combination of their thermal and mechanical properties after cure.

The electrical properties of PIs are very stable over a wide range of temperature [5], and are ideally suited for electrical and electronics applications. In the 1960s, PIs were widely used in electrical industries as wire insulation materials, high heat connectors, switches, housings, and controls. However, the history of PIs for electronics dates back to the early 1970s, when highly purified and high-heat-resistant PI was first developed for microelectronics as the interlayer dielectric of dual metal semiconductor devices [7]. After this point, the application of PI to microelectronics rapidly expanded. In the late 1970s, with the development of semiconductor integration, PI has been transferred from electrical insulating materials to electronic materials. In the last few decades, PIs for microelectronics have been reviewed from different applications [8-12]. In the following sections, a brief description of such applications and the recent progress will be discussed.

5.2 Polyimide Materials for Microelectronics

5.2.1 Combined Property Requirements

A package is a structure that interconnects the semiconductor chips to each other and to the peripherals such as input and output devices [13]. Basically, packaging can support the chip for subsequent processing, handling and performance, can protect the chip from moisture, dust, and gases. Packaging can also dissipate heat produced by the chip, and maintain the integrity of the electrical signals traveling to and from the chips. With the development of integrated circuits, the higher levels of integration emerge as the times require, such as very-large-scale integration (VLSI), ultralarge-scale-integration (ULSI), very-high-speed integrated circuits (VHSIC), etc. This requires more and more input and output counts and improved thermal and electrical performance, and consequently places more demands on the circuit packaging. Hence, packaging needs to be cost-effective, and to achieve the desired function with minimal adverse effects on the performance, efficiency, and life of packaged electronic circuits.

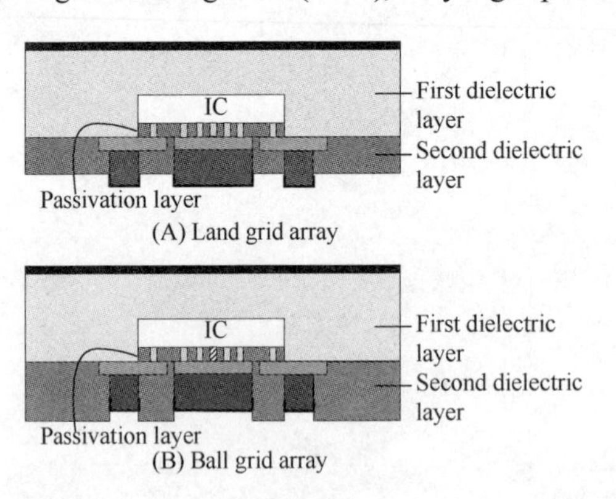

FIGURE 5.2 Schematic representations of microelectronic packages.

In advanced microelectronic packaging, one of the critical issues is to develop polymer substrates with high thermal, chemical, and mechanical stresses associated with microelectronics fabrication and good processability [8,9]. Fig. 5.2 shows the simplified schematic representations of an electronic package with one embodiment that enables coupling a packaged chip to a substrate by either a land grid array (LGA) or a ball grid array (BGA) [14]. It can be seen that in the electronic chip packages, a die or chip with aluminum pillars in a passivation layer is encapsulated by laminating with the first dielectric layer. The pads and pillars may be encapsulated in a second dielectric layer. A thin layer of a dark dielectric (typically black) may be laminated over the top of the die package. Herein, the passivation layer encapsulates material and the dielectric layer may be provided as a polymer film such as PI.

TABLE 5.1 Commercially Available Polyimides for Microelectronic Packaging

PMDA-ODA		BPDA-PDA		BTDA-ODA		Fluorinated PIs	
Trade name	Manufacturer	Trade name	Manufacturer	Trade name	Manufacturer	Trade name	Manufacturer
PI-2545 (Kapton)	DuPont	UpilexS	Hitachi	Pyralin 2525	DuPont	Ultradel 5000	Amoco
PI-2540	DuPont	PIQL100	Hitachi	Pyralin 2555	DuPont	Ultradel 9000	Amoco
		PIXL100	Hitachi	Pyralin 2722	DuPont	SIXEF	Hoechst Celanese
		Pyralin 2610	DuPont	Thermid IIP (6001, 6010, 1015, 6030)	National Starch		
		Pyralin 2611	DuPont				
		PI 2730	DuPont				
		PI 2135	DuPont				
		PIQ 13	Hitachi				

Compared to other organic substrates, polymers used in microelectronic packaging have several advantages, including their ability to form thicker layers (with much lower stress), low cost and high speed of deposition, and better planarization [5]. Among the polymers explored for the potential as electronics packaging materials, PIs, epoxies, poly(*p*-xylylenes), polyquinolines, benzocyclobutenes (BCBs), and cynate ester formulations (CEs) have been commercially available [15]. Semirigid aromatic PIs such as PMDA-ODA, PMDA-PDA, and BPDA-PDA (Table 5.1) have emerged as a favored class of materials for such applications [5]. The suitability of PIs as packaging materials in microelectronics depends on three major factors, including (1) dielectric properties, (2) thermal properties including curing behavior (imidization) and thermal expansions, and (3) mechanical properties including thin-film deposition from precursor solution and adhesion characteristics.

5.2.1.1 Dielectric Properties

Dielectric properties include dielectric constant and dissipation factor, in which the dielectric constant is usually used to describe the dielectric characterization of a material. The lower the dielectric constant (k) and dissipation factor, the less energy is absorbed from an electric field. In microelectronic packaging, a low-k dielectric material is required, and the low dielectric constant (ε=2.0 to 3.5) should be almost constant over a wide frequency range from dc to GHz and low dissipation factor (tanδ) that should be frequency-independent. PIs fulfilled such dielectric requirements and were initially introduced in the early 1980s in electronics packaging to counteract the package delays that were caused by the interaction of the signal with the dielectric medium. The dielectric constant of PI films has been generally measured by the parallel-plate capacitance method that covers a frequency range of 0.1 kHz to 1 MHz and allows variation of temperature over a wide range [16,17]. In most PIs, the dielectric constant does not vary significantly with frequency, except for some fluorine-containing PIs, which exhibit significant (of the order of 0.3-0.5) variations.

However, several sources of error which can cause changes in the values of dielectric constant have been reported. (1) As dielectric constants of thin PI films depend critically on the thickness and uniformity of the film, the measured values in literature are normally not same. (2) The insufficient contact between sample and electrodes can lead to discrepancy in the values of dielectric constant. (3) Water adsorption has significant effects on dielectric properties of PIs, the values of dielectric constant can be varied by the order of 0.5 depending on the overall water uptake of PI [18,19]. (4) Contact resistance, edge effects, and pinholes can also lead to variations (up to 0.7) in dielectric constant for identical PIs [5]. For PIs with anisotropic properties such as BPDA-PDA, the dielectric constant in the plane is significantly higher than that perpendicular to the plane [20-22]. Therefore, efforts are needed to control or reduce these errors, such as surface modification or molecular design to improve the values of dielectric constant.

In addition, dielectric strength of a material is also used as a critical factor in chip applications and high-power electrical applications. Low-dielectric-strength materials (implicitly with low dissipation factors) are of high interest for electronic packaging.

5.2.1.2 Thermal Properties

Besides the dielectric constant, the second most important requirement for materials used in microelectronics is the thermal stability. It has been suggested that PIs should be cured at or above the highest subsequent process temperature to minimize outgassing and subsequent degradation in packaging. The curing sequence of PIs comprises firstly evaporation of solvent, the imidization of PAA precursor, and the subsequent cooling of the film. The rate of curing and the temperature affects the final film properties. If PI is cured partially or subjected to fast curing, the presence of certain amounts of unevaporated solvent results in higher values of thermal expansion coefficient (*CTE*) of the film. The difference in *CTE* of PI, metal(s) and substrate(s) causes thermal mismatch and thermal stresses in the cured PI films, and these stresses become more pronounced depending on the cure cycle. Therefore, thermal expansion behavior of PIs plays an important role in the package structure consisting of PI-substrate/metal combinations. Thermal stress can be controlled at low levels using PIs with similar *CTE* as those of the substrates or metals.

5.2.1.3 Mechanical Properties

In general, the PIs used in electronic applications should be sufficiently strong, have elongation of at least 10%, with tensile strength greater than 103 MPa to survive stresses induced during processing and/or thermal cycling. In packaging applications, PIs should be able to form pinhole-free, uniform thin films with good planarization. The thin films of PIs can be deposited on a variety of substrates by spin coating, spray coating, and screen printing of the PAA precursor solutions. The desired level of uniformity and film thickness can be achieved by adjusting the formulations of PAA solutions (e.g., with different viscosities). In addition, adhesion characteristics decide the suitability of a polymer for packaging applications; the thin-film dielectric is required to have very good adhesion with the substrate and the metal layers deposited thereon. The adhesion strength of PIs with different substrates such as alumina, silicon and copper can be improved with the use of adhesion promoters.

5.2.2 Typical Applications

5.2.2.1 Encapsulates and Coatings

The purpose of encapsulation is to protect the electric IC devices from moisture, mobile ion contaminants, ultraviolet-visible light, and hostile environmental conditions [23]. Among organic polymeric materials, PIs have been accepted as encapsulates because they can withstand IC processing temperatures (up to 500 °C) without significant degradation. PIs could be easily either spun-on or flow-coated and imaged by conventional photolithography and etch processes. They have been widely used as coatings with enamels and flexible circuits, or as protective coatings on semiconductor devices, and have been used as a VLSI encapsulate with great success [24].

The mechanical properties of PIs should be sufficiently excellent if the coatings are to survive annealing, wire bonding, encapsulation, thermal cycling, and so forth. During the soldering process of a surface mount of semiconductor package, the difference in thermal expansion of mold resin and semiconductor chip causes mold stress, which results in fatal damage to the package such as passivation crack, aluminum electrode slide, and package crack. Therefore, coating on the chip surface with PIs is very important to reduce the mold stress and to prevent damage to the package during soldering or heat cycle. The PIs for chips are required to have residual stress less than 20 MPa, moisture absorption less than 0.5%, T_g greater than 260 °C, and so on [25].

The reliability of the chip coated device is strongly affected by the adhesion of mold resin and PI film. It has been found that PI having dispersion temperature lower than curing temperature shows good adhesion to the mold resin. In addition, low thermal expansion PIs are reported to be the most promising material for the chip coat application, and poor adhesion of these PIs can be improved by the plasma treatment [25]. In addition, advances in PI synthesis have reduced the material moisture absorption and improved the adhesion of the materials. This new type of PI will have significant implications in device of packaging.

With the emergence of the flexible electrics, such as soft, skin-like electric devices, the encapsulation materials are required to be capable of bending, stretching, and be very thin [26]. Thin-film encapsulation (TFE) technology using PI thin film barriers has been developed along with the development of flexible organic light emitting diode (OLED). The advantages of the TFE method include (1) a thinner/lighter form factor and (2) a higher flexibility of device form during in-flex use of the display. Because a polymer substrate does not offer the same barrier performance as a glass one, the TFE should be developed on both the bottom and top sides of the device layers, as shown in Fig. 5.3, to provide the highest degree of protection from moisture and oxygen penetration in electronic devices [27].

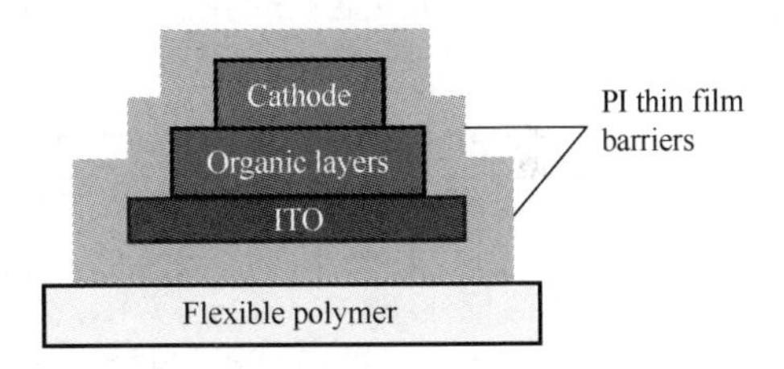

FIGURE 5.3 Schematic diagram for thin-film encapsulation for OLED.

5.2.2.2 Passivation Layers

Passivation involves the creation of an outer layer of shield material that is applied as a microcoating in semiconductor devices of metallization films to reduce the environmental effects [28]. Passivation layers can be organic or inorganic in nature. Among organic passivation layers, PI materials have been widely used in electronic devices due to their good high-temperature resistance, excellent mechanical and electrical properties, and high chemical stability. In order for PI to be useful for passivation, additional property requirements must be satisfied. The material must be an excellent electrical insulator, must adhere well to the substrate, and must provide a barrier for transport of chemical species that could attack the underlying device. The use of PI passivation layers can reduce the damages to electronic devices from various natural environments and operating environments and thus improve the reliability and stability of the resultant devices.

In the electronics industry, PI films are used as high-temperature insulation materials and passivation layers in the fabrication of integrated circuits and flexible circuitry. For example, the PI passivation layer has been found to be effective for protecting the organic semiconductor of organic thin-film transistors (OTFTs) from ambient moisture and to increase device lifetime [29]. Recently, inorganic/organic dual layers as a passivation layer for high field effect transistors (HFETs) was reported [30]. The photosensitive PI (PSPI) as the organic layer in the dual passivation layers is a well-known organic material in the semiconductor devices process. It was found that the SiN_x/PSPI dual layers are promising for the suppression of output power drop of HFETs.

5.2.2.3 Interlayer Dielectrics

Insulating materials are known to be attacked and degraded by energy resulting from voltage stresses. This energy may be produced by UV radiation or kinetic energy from moving ions, electrons, and molecular species. PIs are good insulators and are used in a variety of electrical and electronic applications. They show very low electrical leakage in surface or bulk and form excellent interlayer dielectric insulators and excellent stop coverage in multilayer IC structures. When used as interlayer dielectrics, even greater demands are placed on PIs. The interlayer dielectric film must withstand the temperature of metal sinter at 400 °C, which is the final step of integrated circuit processing, without degradation of electrical, chemical, or mechanical properties. In addition, the deposition, cure, and etch process must provide for reliable interconnection between the metal layers above and below the film [10,31,32]. From the early 1970s to the late 1980s, PIs were used as interlayer dielectrics [33]. Some early progress on PIs as dielectric materials in microelectronics has been reviewed [10,32].

The use of low dielectric constant (*k*) interlayers can greatly reduce the resistance-capacitance (RC) time delays, cross-talks, and power dissipation in the new generation of high-density and high-speed integrated circuits [34]. PIs are known to reveal relatively low-*k*, which ranges 2.6-3.0, depending on the chemical compositions [10]. However, with *k* values of about 3.1-3.5, the conventional PIs are insufficient in meeting the requirement of $k < 2.5$ for the dielectrics of the near future, and the ultralow-*k* of less than 2.2 for the technology nodes below 130 nm. Therefore, there has been considerable interest in the synthesis of ultralow-*k* materials based on PIs in recent years [32]. Multilevel interconnection technology hasbeen essential to realizing high-density and high-performance ULSIs. The interlayer dielectric film technologies are one of the most important keys for the multilevel interconnection fabrication. PI films are one of the most attractive interlayer dielectric films, because of their good surface planarization characteristics, low film stress and lower relative dielectric constant than 3.5 [35].

Recently, through-silicon-vias (TSVs) are regarded as one of the key techniques for three- dimensional (3D) integration technology. 3D integration has been recognized as a promising technological paradigm to overcome further miniaturization obstacles faced by conventional 2D ICs [36]. PI dielectric materials have been reported as TSV liners due to their excellent inherent properties. MIS trench capacitors with PI dielectric liners as shown in Fig. 5.4 was formed by vacuum-assisted spin coating technique on the silicon substrate. Such TSVs could be used in future "via-last/backside via" 3D integration applications.

5.2.2.4 Alpha-particle Barrier

The substrate material and/or the interconnect solder used in integrated circuit chips normally contain impurity elements such as polonium, thorium, and uranium. The radioactive decay of these impurities can emit alpha particles, which cause changes in both logic and memory functions of the device in close proximity to the alpha particle source. The overall disruption of the normal operation of the semiconductor devices is described as the soft error rate (SER). As integrated circuit devices continue to be scaled, capacitor cell sizes and operating voltages continue to decrease while circuit density increases. This causes an increase in the probability of integrated circuits (ICs), especially large multi-chip modules (MCM), experiencing soft errors. Therefore, the peripheral solder connections and conductive pads are encapsulated by PI alpha particle barrier layer as shown in Fig. 5.5 [37].

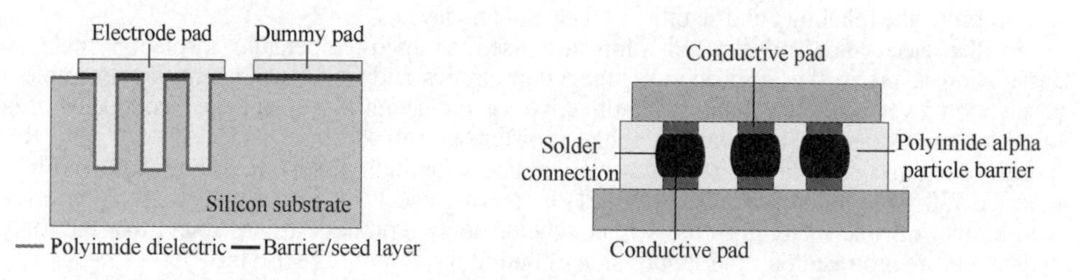

FIGURE 5.4 Schematic diagram for MIS trench capacitors.

FIGURE 5.5 Schematic diagram for PI based alpha-particle barrier in ICs.

PI provided the unintended benefit of absorbing the alpha radiation from the substrate in earlier MCM packages. In recent years, this application for PIs has attracted more and more interest. There are two approaches that can be used to shield alpha particle by PIs: (1) by thin film technology, i.e., to adhere a PI thin film to the active surface of integrated circuit before packaging; and (2) by using PAA solutions, i.e., filling the electronic component within a container with a PAA solution and then curing PAA as an integral part of the component before packaging. The second approach has become a common technique used in IC industries. For example, Motorola used the precursor solution of PI from Dupont, Pyralin PI-2562, in the 64-k memory devices as alpha particle barrier layers. It was found that when the film thickness was around 25 μm, there was no more than one soft error rate occurring in the 10000h test.

5.2.3 Structures and Properties of Polyimide Materials for Microelectronics

5.2.3.1 Conventional Polyimides

Unlike the photosensitive PIs, conventional PIs do not have photosensitive moieties in their polymer backbones. According to the dianhydride and diamine monomers from which PIs are derived, conventional PIs can be broadly classified as fully aromatic PIs, semiaromatic PIs, and fully aliphatic PIs [38], of which aromatic PIs have been extensively used in microelectronics applications. Aromatic PIs are typical of most commercial PIs, such as Ulterm from G.E. and Kapton from DuPont. The most utilized PIs in microelectronic applications are based on aromatic PIs due to their incredible mechanical and thermal properties. However, the charge transfer (CT) interactions between the five-membered ring of the imide group and the aromatic ring in aromatic PIs are so strong that these PIs are often insoluble in their fully imidized form and have low processibility. In addition, these interactions make aromatic PIs able to absorb visible light intensely and have high dielectric constant. Despite all its excellent properties, low processibility, high dielectric constant, and yellow coloration hinders its successful application in opto-electric materials and high-speed multilayer printed wiring boards [38]. Therefore, monomers containing siloxane [39-44], fluorinated [45-48], and aliphatic [38] moieties, as well as side groups [49], have been used in PI structures to solve these problems. For example, PI-siloxanes (PISiO), the segmented copolymers containing siloxane segments along the intractable backbone chain of aromatic PIs, maintain some of the excellent properties of PIs, e.g., high thermal stability and mechanical strength, and some of the desirable properties of siloxanes, such as ductility and adhesion as well as low moisture permeability. As fully imidized PISiOs have superior solubility, even in organic solvents with low boiling point, their processibility can be meliorated by replacing a prolonged polymer-curing step at elevated temperature with a baking procedure at a lower temperature.

However, for microelectronic processing such as solder masks, protective coatings, and intermetal dielectrics (IMD), photoimageable PIs are required to produce a pattern on a substrate. In such process, conventional PI can be used as a dielectric, and an additional photoresist is required as the overcoat. The photolithographic patterns of the photoresist are then transferred to the PI layer. Therefore, pattern formation of the photoresist and the following etching procedure are required (Fig. 5.6); in order to reduce the patterning steps and complexity, increase accuracy while maintaining good dielectric, mechanical, and thermal properties, it is considered to be a promising way to incorporate photosensitivity into PIs (or PI precursors). When a PI precursor is used for pattern formation, it is possible to be convert it into its corresponding PI film by thermal imidization.

5.2.3.2 Photosensitive Polyimides (PSPI)

The history of PSPI research dates back to 1971, when Kerwin and Goldrick [51] from Bell's Lab developed first PSPI by adding a small amount of potassium dichromate to a PAA solution. However, this PSPI was too unstable to have any application values. It was not until 1976 that the first negative-working PSPI based on poly(amic ester) was produced, which laid the foundation for the development of PSPIs [1]. In 1978, Loprest et al. [52] reported the first positive-working PSPI which composed of PAA and a diazonaphthoquinone photoactive compound (DNQ PAC). As the PSPIs are able to produce pattern formation, and are finally mounted on the electronic devices as PI films, the design of PSPIs demands two factors: (1) photolithographic properties including sensitivity, resolution, development system, imidization, etc., and (2) PI film properties such as mechanical strength, electrical characteristics, dimensional stability, adhesive properties, etc. In comparison with conventional photoresists, PSPIs remain in the final manufactured products whereas the photoresists would be removed after pattern transfer to underlaid materials, as shown in Fig. 5.6. Therefore, the advantages of PSPIs are not only demonstrated by the outstanding properties of PIs, but also by the significant simplification of the pattern formation process.

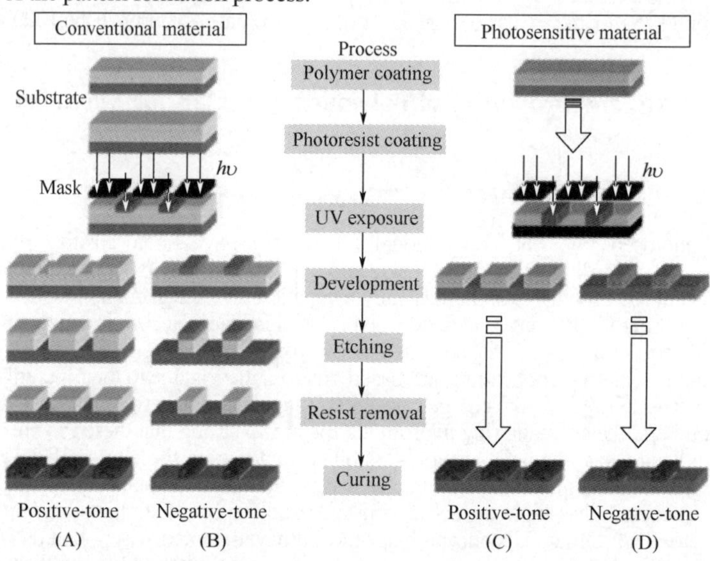

FIGURE 5.6 Comparison on photolithographic patterning process of conventional polyimides and photosensitive polyimides. *Reproduced from K.-I.Fukukawa, M.Ueda, Recent progress of photosensitive polyimides. Polym.J.40(2008)281—296(Ref.[50]).*

The general patterning process of PSPI involves four steps: (1) spin-coating a PSPI onto a substrate of interest (e.g., a silicon wafer); (2) exposing this coated film to a UV light through a mask to transfer the pattern information to the film; (3) developing, either the exposed or unexposed area is selectively removed to form a pattern; and (4) curing, to convert the precursor pattern into the PI one. It should be noted that wavelength at 436 nm (g-line) or 365 nm (i-line) is usually chosen as the source of UV light for the lithography with PSPIs. The photoirradiation process allows the PSPI to undergo a chemical change such as deprotection, polarity change, chain-scission, cross-linking, etc. PSPIs can be categorized into positive- and negative-working types, depending on the washed-out area. The exposed area of positive-tone PSPIs (*p*-PSPIs) is washed out, while that of negative-tone PSPIs (*n*-PSPIs) is insoluble in the developer (as seen in Fig. 5.6).

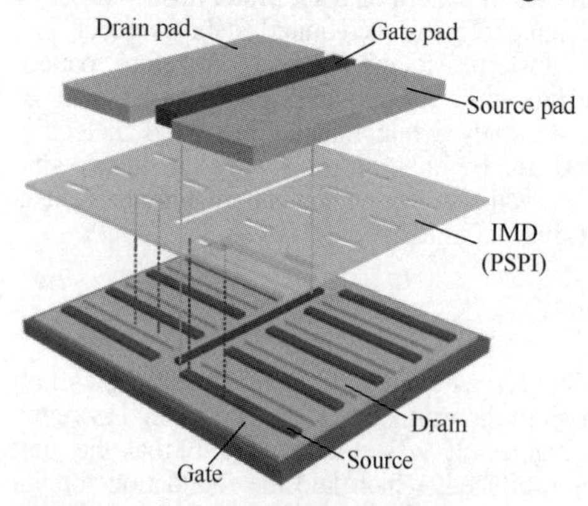

FIGURE 5.7 Schematic diagram for high-power hetero-structure field-effect transistors.

Many electronics packages are built with PSPIs, and PSPIs have significantly enhanced the development of microelectronic devices because they eliminate the need for a photoresist. In recent years PSPIs have greatly contributed to the progress of microelectronics. They are widely used in interconnects, multichip modules, protection layers, insulating films, optical interconnects, and resists because of their excellent thermal and chemical stabilities, low dissipation factors, and reasonably low dielectric constants. Quite recently, PSPIs have been employed as the intermetal dielectric (IMD) layers in high-power hetero-structure field-effect transistors (Fig. 5.7) [53,54], and the applications in microelectronics still continue to expand.

After a brief introduction to positive- and negative-tone PSPIs, this section highlights the recent progress of *p*-PSPIs composed of PAA or PI and a DNQ PAC, which are regarded as the most popular PSPIs. The previously developed PSPIs have been reviewed in the literature [6,50,55-57].

5.2.3.2.1 Negative PSPIs

When casting an *n*-PSPI film to specific areas of microelectronic devices, the exposed areas become insoluble through a crosslinking mechanism, and the masked portions (unexposed areas) are washed away with solvent (developer) to form a negative image. The first commercial PSPI was a negative-tone type, in which a photoreactive methacryl group was introduced to the side chain of a PI precursor [58,59]. According to the composition of photoresists, *n*-PSPIs can be generally classified into two catalogs, i.e., with and without photoinitiator, as shown in Fig. 5.8.

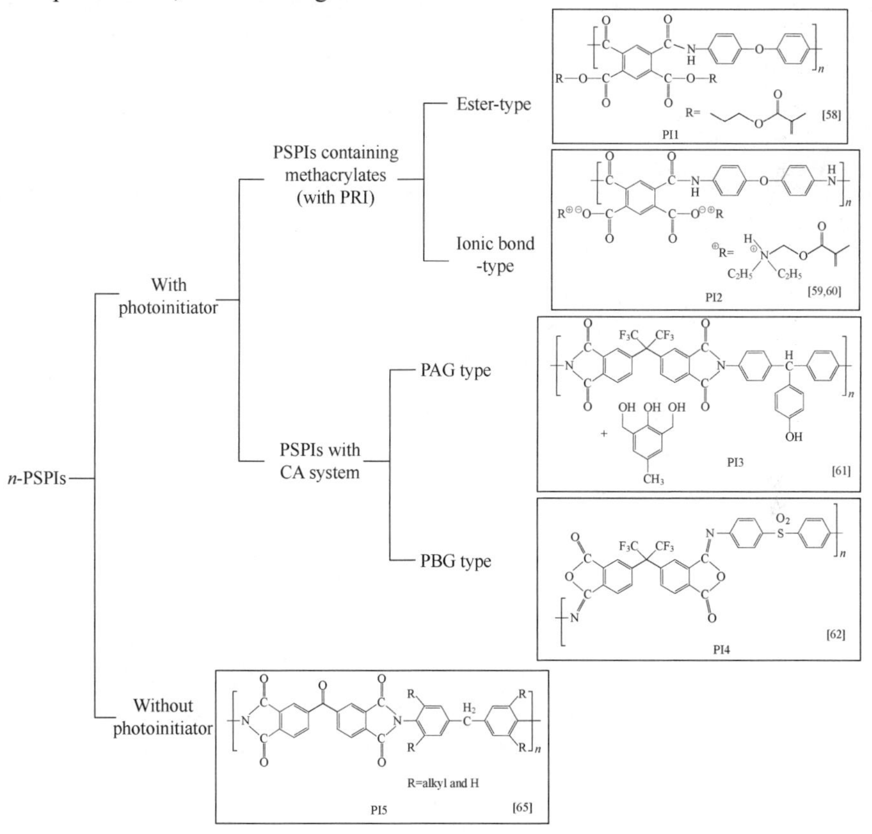

FIGURE 5.8 Classifications of *n*-PSPIs.

Ester/or Ionic-type *n*-PSPIs: Most commercial *n*-PSPIs are based on a PAA in which photoreactive methacrylate groups are linked to carboxylic acids through ester linkage via the covalent bond (ester-type *n*-PSPI, **PI1** in Fig. 5.8) [58], or acid-amine ion linkages via the inonic bond (ionic-type *n*-PSPI, **PI2** in Fig. 5.8) [59,60]. The photochemistry mechanisms of pattern formation are different. In the case of ester-type PSPIs, methacrylate groups can react in a radical polymerization fashion by UV irradiation. Conversely, a "charge separation" mechanism is regarded to pattern the ionic-bond type PSPIs, in which charge-transfer complexes are formed between PAA and a photoradical initiator (PRI) in the exposed area upon UV exposure.

The typical photoinitiators for *n*-PSPIs are usually PRI, photoacid generator (PAG), and photobase generator (PBG). PRI is known as photosensitizer for PSPIs containing a photoreactive acryloyl group, such as *N*-phenyl-diethanol amine [60]. PAG and PBG can photochemically generate a strong acid or a base upon UV irradiation, and promote thermal conversion from precursor polymers such as PAA to robust structures, PI, upon thermal treatment at a relatively lower temperature [57]. For example, the PAG [(5-propylsulfonyloxyimino-5*H*-thiphen-2-ylidene)-(2-methylphenyl)acetonitrile, PTMA] and the PBG of N-[(4,6-dimethoxy-2-nitrobenzyl)oxy]carbonyl-2,6-dimethylpipridin (DNCDP) as depicted in Fig. 5.9 are mainly used in literature. Therefore, both PAG and PBG can be employed in a chemical amplification (CA) fashion, and have been applied in the preparation of PSPIs for both negative- and positive-working systems.

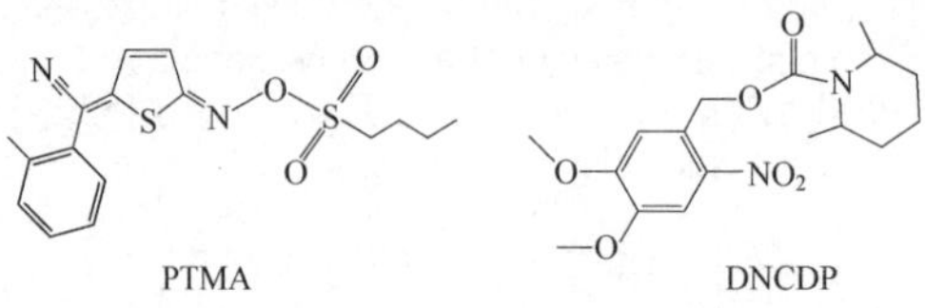

PTMA DNCDP

FIGURE 5.9 Chemical structures of PTMA and DNCDP.

Chemically Amplified (CA) *n*-PSPIs: CA *n*-PSPIs are known to show very high sensitivity as compared to nonchemically amplified ones, and will continue to be imaging materials for microlithography in the future. Form this point of view, the concept of CA has been successfully introduced to the design of PSPIs for improving the photosensitivity. CA can lead to a change in the structure and physical properties of PSPI material with a minimum amount of UV exposure energy.

The CA-PSPIs are mainly cataloged in two types according to the additional photoinitiators: PAG type and PBG type. The PAG type *n*-PSPIs are composed of a PAA or PI, a benzylic alcohol-type cross-linker and a PAG. The benzylic alcohol-type cross-linker, such as 2,6-bis(hydroxymethyl)-4-methylphenol was effective in making a negative pattern with an alkaline developer. An acid-catalyzed cross-linking mechanism was proposed for this kind of *n*-PSPI to obtain negative patterns. For example, for the poly(hydroxyl imide) CA system (**PI3**) [61] as shown in Fig. 5.8, a photogenerated acid catalyzes the protonation of a benzylic alcohol group to form a benzylic carbocation species, which then undergoes electrophilic substitution (alkylation reaction) on the aromatic ring and the hydroxy group to produce *C*- and *O*-alkylated polymers (Scheme 5.1). The formation of network polymers decreases the solubility of the exposed area and enables the formation of the negative image pattern.

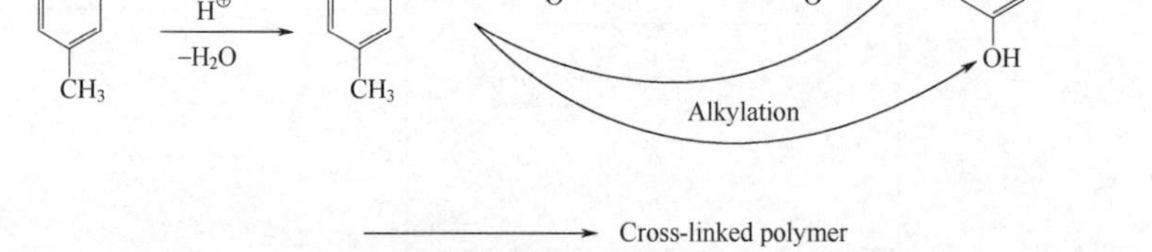

Photoacid generator (PAG) $\xrightarrow{h\upsilon}$ $H^{\oplus}R^{\ominus}$

Alkylation

$\longrightarrow$ Cross-linked polymer

Scheme 5.1 Acid-catalyzed cross-linking mechanism for the formation of network polymers for PAG type *n*-PSPI.

PI precursors such as poly(isoimide) [62], poly(amic alkylester) [63], and PAA [64] may be utilized to prepare PBG type *n*-PSPIs. For example, in the poly(isoimide)/PBG system (**PI4** in Fig. 5.8), PBG photochemically generates an imine compound as shown in Scheme 5.2, and promotes thermal imidization of poly(isoimide) into the corresponding PI. This results in a different solubility between the exposed and unexposed regions in this resist and the formation of a negative pattern [62].

Intrinsic *n*-PSPIs: The addition of large amount of low-molecular weight additives such as PRI, PAG, and PBG, etc., into a PI system is found to have the loss of thermal- and/or mechanical properties of the PI. This problem can be avoided by using pure PI contents without the additional of photoinitiators. It is well-known that benzophenone is a photoreactive group, and can cross-link without photoinitiator [65]. Therefore, photoinitiator-free *n*-PSPI reported so far are all benzophenone-containing PSPIs. An example of this type PSPI (**PI5**) is depicted in Fig. 5.8, which was prepared from *ortho*-multialkyl-substituted aromatic diamines and 3,3′4,4′-benzophenone tetracarboxylic dianhydride (BTDA) [66]. In the UV-exposed area, a radical cross-linking reaction occurred between alkyl substituents in diamines and carbonyl groups in BTDA, and made the polymer become insoluble in an organic developer. As shown in Scheme 5.3, this radical cross-linking contributed to hydrogen abstraction from the *o*-alkyl group and triplet state of benzophenone moiety excited by UV exposure.

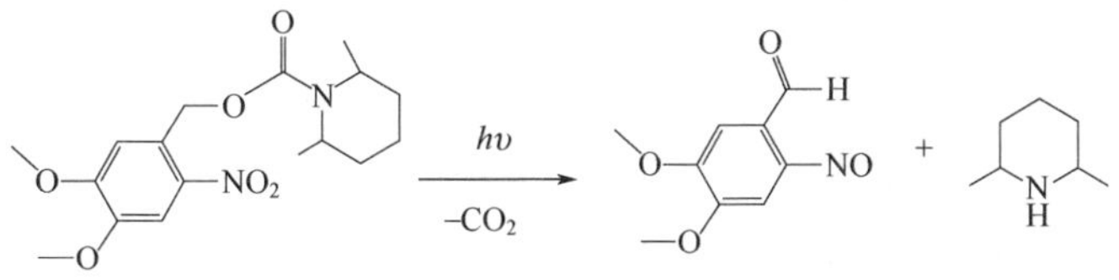

Scheme 5.2 Photochemical generation of amine from PBG (DNCDP).

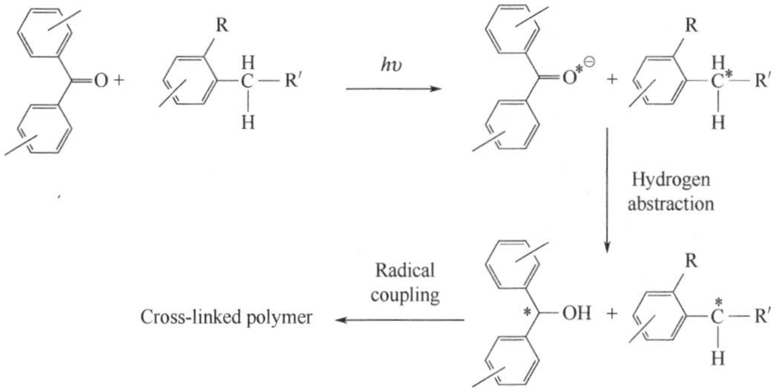

Scheme 5.3 Mechanism of crosslinking via photoradical generation.

5.2.3.2.2 Positive PSPIs

The major disadvantage of *n*-PSPIs is the inevitable use of organic solvents during pattern development, which can cause a swelling problem of the resist as well as environmental and safety issues at the industrial level. To avoid these problems, positive-tone PSPIs (*p*-PSPIs), which use aqueous alkaline developers, become more desirable. Similar to *n*-PSPIs, *p*-PSPIs can be divided into two groups, with-and without photoinitiators, as shown in Fig. 5.10. Herein, we give a brief introduction to the photoinitiator-free *p*-PSPIs and then a detailed description on the systems with photoinitiators.

***Instrinsic p*-PSPIs:** From the viewpoint of molecular design, photosensitive moieties which can provide acid groups via photochemical reactions, or moieties containing hydroxy groups are chemically bonded to PIs. Thus, this type of *p*-PSPIs can develop with an aqueous alkaline solution and form a positive pattern without the addition of photoinitiators.

Upon UV irradiation the photosensitive compound, *o*-nitrobenylester, can be converted to *o*-nitrosobenzaldehyde and a carboxylic acid, as shown in Scheme 5.4. Based on this photochemical reaction, *p*-PSPIs containing *o*-nitrobenylester groups as side substituents were developed. An example is shown in Fig. 5.10 based on poly(amide *o*-nitrobenzylester) (the o-Nitrobenylester type *p*-PSPI, **PI7**), where the ester bond was cleaved into PAA containing a carboxylic acid group and *o*-nitrosobenzaldehyde upon UV exposure [67]. However, this type of *p*-PSPI had synthetic difficulty and low photosensitivity.

Diazonaphthoquinone (DNQ) moieties can be directly introduced into PAAs or PIs as side-chain substituents to prepare *p*-PSPIs. It was reported that the high solubility of PAA in an aqueous alkaline solution can be avoided by an ester to a carboxylic acid group. Thus a DNQ-contained *p*-PSPI (**PI8**) was developed based on a poly(amic ester) bearing a phenolic hydroxyl group, and DNQ moiety was attached partially to the phenol group as shown in Fig. 5.10 [68]. This resist system became soluble after UV exposure through the photolysis of DNQ.

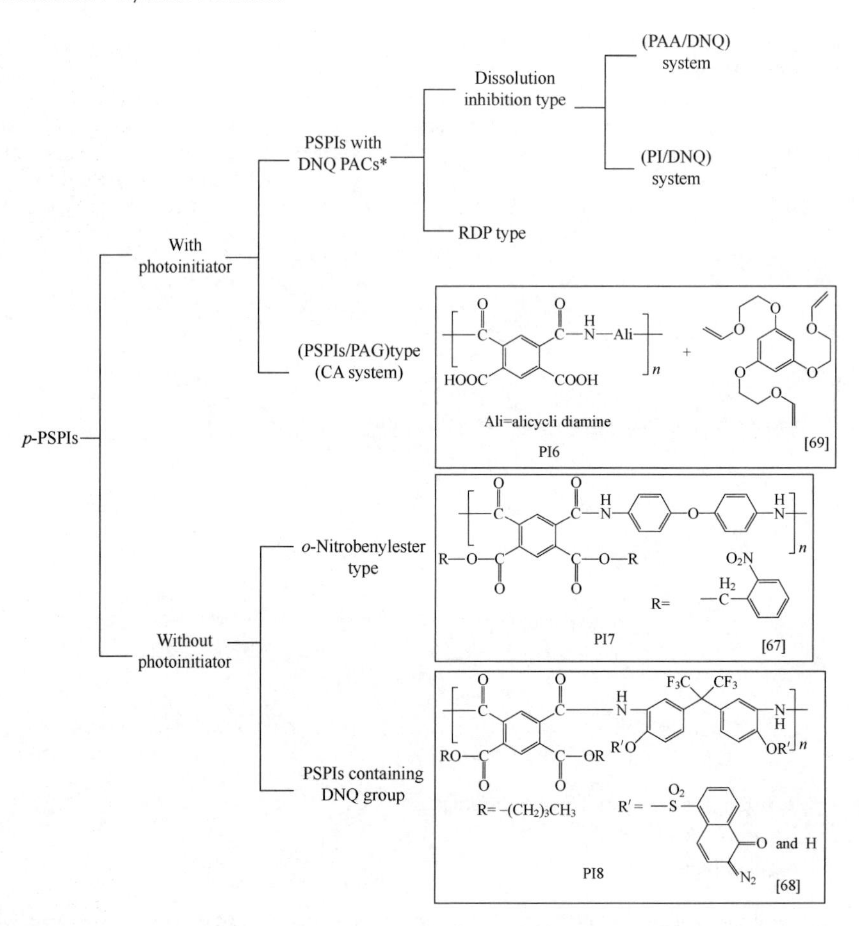

FIGURE 5.10 Classifications of *p*-PSPIs.

Scheme 5.4 Photochemical decomposition of *o*-nitrobenylester.

***p*-PSPIs with photoinitiators:** Photoinitiators for *p*-PSPIs are mainly DNQ photoactive compounds (DNQ PACs), PAG, and PBG. DNQ PACs are known as dissolution inhibitor additives, which are typical and widely used for Novolac-type classical positive-working photoresists for microlithography [67]. According to the different photoinitiators employed, there are mainly two categories for this type of *p*-PSPIs: (1) the CA systems, i.e., PSPI in combination with PAG or PBG, and (2) PSPI in combination with DNQ PACs, where the latter category can be divided into two groups based on different pattern formation mechanisms, dissolution inhibition type, and reaction development patterning (RDP) type.

Chemically Amplified (CA) p-PSPIs: It is known that PAG and PBG can be applied in the preparation of PSPIs for both negative- and positive-working systems in CA fashion, which exhibits a large change in solubility by polar modification occurring upon irradiation and in the following postexposure baking step. The most reported structures for *p*-PSPIs with CA system are those containing an acetal linkage, which have been formulated with vinyl ether derivatives as thermally crosslinking agents to form acetal linkages. As shown in Scheme 5.5, vinyl ester reacts with both phenolic hydroxyl and carboxylic groups upon thermal treatment, and gives alkaline aqueous soluble polymers via acidolysis with photogenerated acid.

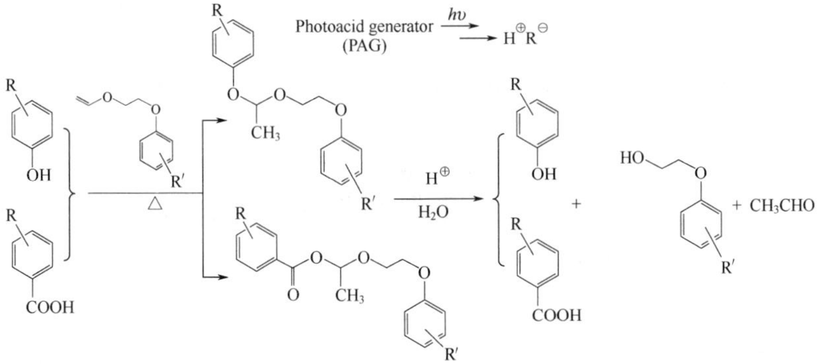

Photoacid generator (PAG) $\xrightarrow{h\upsilon}$ $H^{\oplus}R^{\ominus}$

Scheme 5.5 Radical generation from benzophenone moiety upon UV irradiation.

Fig. 5.10 illustrates an example of chemically amplified positive-working resists based on this chemistry **(PI6)**, which contains a three-component formulation including a semialicyclic PAA, a multifunctional vinyl ether as a crosslinker, and a PAG [69].

p-PSPIs with DNQ PACs (dissolution inhibition type): PSPIs based on alkaline-soluble polymeric matrices and certain dissolution controllers such as DNQ PACs have been widely studied and reported. For the dissolution inhibition type PSPIs, DNQ works as dissolution inhibitor, and photoresists bearing acidic functional groups such as phenolic hydroxyl group, carboxyl group, etc., are absolutely required to make patterns. As shown in Scheme 5.6. DNQ PACs photochemically transfer into indenecarboxylic acid derivatives upon UV exposure. Therefore, DNQs play two important roles in the pattern formation process for *p*-PSPI systems; i.e., DNQs themselves are hydrophobic to depress the dissolution of PSPIs, while indenecarboxylic acid derivatives accelerate the solubility of PSPIs in aqueous alkaline solutions.

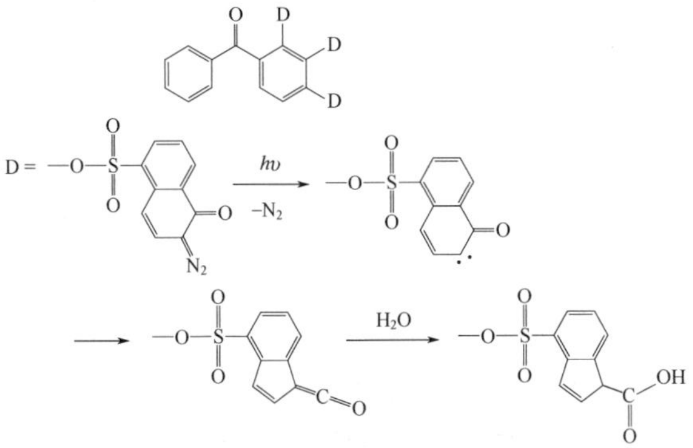

Scheme 5.6 Photorearrangement of DNQ compound upon UV irradiation.

From the literature, either the precursors of PI including poly(amic acid) (PAA) [70], poly(amic ester) (PAE) [71] and poly(isoimide) (PII) [72], or alkaline soluble PIs with acidic functional groups in side chains [73], can combine with DNQ compounds to achieve positive-image patterns. These types of PSPIs have many advantages over the other types, because they have a relatively simple synthetic route and controllable relative molecular weight of the polymers, and they can be directly utilized by adjusting the ratio of polymer to DNQ. Most commercialized *p*-PSPIs are based on this type.

It is well known that PAAs are too soluble in an aqueous alkaline solution to obtain a large dissolution contrast between the exposed and unexposed areas. The solubility of PAA is able to be regulated by the ratio of the hydrophilic groups such as carboxylic groups, and the hydrophobic groups such as aromatic groups and fluorine groups. A variety of *p*-PSPIs composed of PAA and a DNQ compound are summarized in Table 5.2 and described in this section.

The introduction of more hydrophobic structures into PAA by increasing the aromatic and fluorine contents is an effective way to depress the solubility of PAA in alkaline aqueous solution. Haba et al. [74] developed a *p*-PSPI (**PI9**) based on a simple PAA by controlling the hydrophobicity of the main chain. PAAs were prepared from several diamines with various hydrophobicity and 4,4′-(hexafluoroisopropylidene)diphthalic anhydride (6FDA). It was found that, the resists composed of PAA prepared form the more hydrophobic diamine, 1,3-bis(4-aminophenoxy)benzene (TPE-R) and DNQ compounds gave the high sensitivity and contrast of 100 mJ·cm^{-2} and 7.1, respectively.

Based on this work, Sakayori et al. [75] developed an alkaline-developable PSPI (**PI10**) consisting of PAA derived from TPE-R and 2,2′,6,6′-biphenyltetracarboxylic dianhydride (2,2′,6,6′-BPDA), with a DNQ compound as a dissolution inhibitor. The twisted conformation of 2,2′,6,6′-BPDA gave the PAAs good solubility and excellent transparency at 365 nm.

TABLE5.2 (PAA+DNQ) Type *p*-PSPIs

	X_1	X_2	Y_1	Y_2	*m*:*n* (molar ratio)	References
PI9, PI10	PI9: 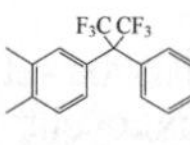 F_3C CF_3	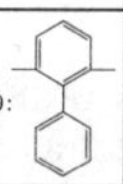 PI10:	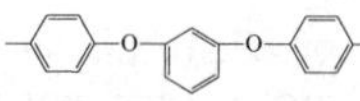 O O			[74,75]
PI11	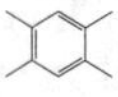	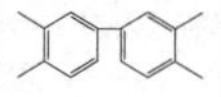	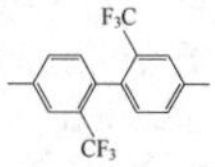 F_3C CF_3		1:1	[70]
PI12	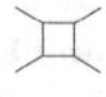		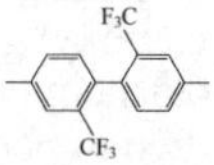 F_3C CF_3	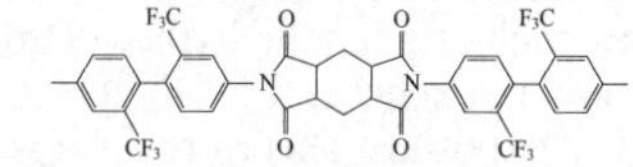 F_3C O O F_3C N N O O CF_3 CF_3	9:1	[76]
PI13	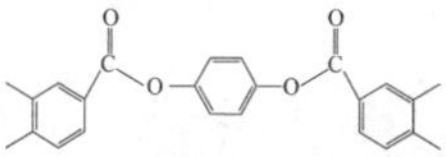 O O C O O C		 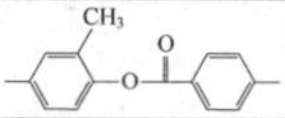 a: ; b: CH_3 O O	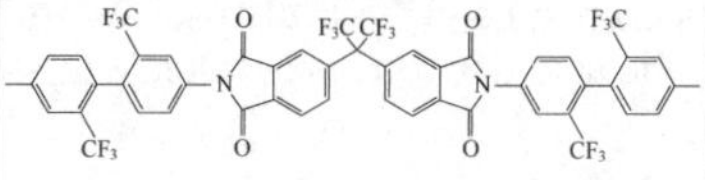 F_3C O F_3C CF_3 O F_3C N N O O CF_3 CF_3; c: F_3C CF_3	a: 80: 20 (*w*) b: 85: 15(*w*)	[77]
PI14			O or O or F_3C CF_3			[78]
PI15	O		O			[79]
PI16	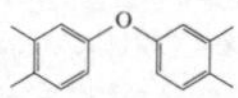 O		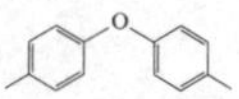 O	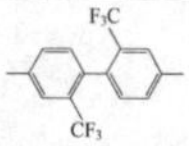 F_3C CF_3	1:1	[80,81]
PI17	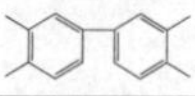 ; F_3C CF_3		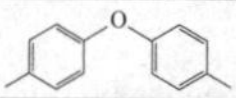 O; F_3C CF_3 O O		PAA: FPAA=85:15 (*w*)	[82]
PI18	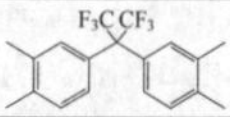 F_3C CF_3		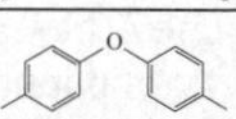 O			[83]
PI19	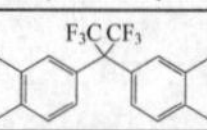 F_3C CF_3		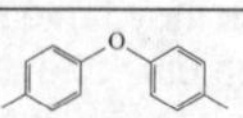 O	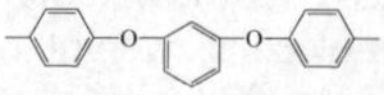 O O	4:1	[84]

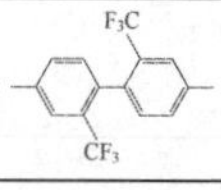

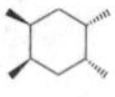

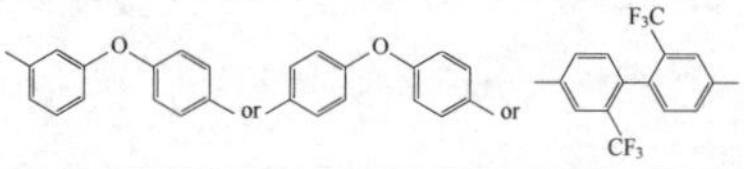

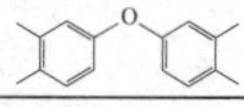

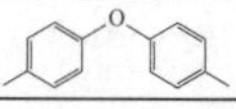

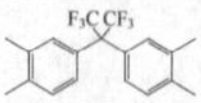

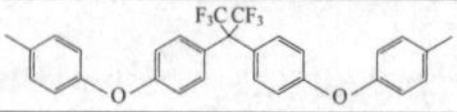

In addition, the fluorine contents increase not only the transparency of PAAs, but also the hydrophobicity to decrease their solubility. Therefore, the fluorinated dianhydrides such as 6FDA [74,82-84], and fluorinated diamines such as 2,2′-bis(trifluoromethyl)benzidine (BTFB) [70,76-78] are promising candidates for the preparation of the desired PAAs. Seino et al. [70] prepared a fluorinated poly(aminc acid) (FPAA) (**PI11**) by ring-opening polyaddition of dianhydrides, pyromellitic dianhydride (PMDA), and biphenyltetracarboxylic dianhydride (BPDA), with diamine BTFB in methanol. FPAA containing 30 %(*w*) DNQ compound was found to be a *p*-PSPI precursor and showed a sensitivity of 80 $mJ \cdot cm^{-2}$ and a contrast of 7.8 with 365 nm. The corresponding FPI film had a low coefficient of thermal expansion of 10.3×10^{-6} $°C^{-1}$ and a low dielectric constant of 3.04. On the other hand, Inoue et al. [82-84] developed fluorinated DNQ (FDNQ) compound by the reaction of 1,2-naphthoquinone-2-diazido-5-sulfonyl chloride (215-DNQ) with 4,4′-(hexafluoroisopropylidene) diphenol (6F-BPA). FDNQ was reported to act as a dissolution inhibitor and successfully develop *p*-PSPIs based on FPAA and PAA (**PI17-PI19**). The PSPI exhibited an excellent sensitivity of 60 $mJ \cdot cm^{-2}$ and a good contrast of 3.3 with i-line exposure.

The solubility of PAA in alkaline aqueous developer can be resolved in another way, i.e., by reducing the content of carboxylic acid in the repeating unit of PAAs to decrease the hydrophilicity. This can be achieved by partial esterification (imidization) of PAAs.

Although partial imidization can be obtained by simple heating of PAA solutions, the extent of imidization is difficult to control. In addition, the decomposition of DNQ occurs above 105 °C. Therefore, Hasegawa et al. [76,77] introduced imide-containing diamines into the (CBDA/TFMB) PAA backbones (CBDA=1,2,3,4-cyclobutanetetracarboxylic dianhydride, TFMB=2,2′-bis(trifluoromethyl) benzidine), to precisely control the extent of imidization and control the PAA solubility. It was found that the partially imidized PAA (**PI12**) (10%-15%) in combination with DNQ enabled the formation of fine positive patterns. After being developed in 0.1% tetramethylammonium hydroxide (TMAH) aqueous solution, this *p*-PSPI film showed a sensitivity of 250 $mJ \cdot cm^{-2}$ and a contrast of 0.65 at 365 nm [76]. Based on this work, Hasegawa et al. [77] incorporated aromatic ester linkage into PAA and developed ester-containing PIs (PEsIs) (**PI13**) with both low *CTE* and high modulus. PEsIs with an imide content of 15%-20% processed higher hydrophobicity and formed fine positive patterns using DNQ as photoactive compound [77].

Later on, Hasegawa and Nakano [78] employed a cyclodehydration reagent (acetic anhydride/ pyridine) into PAA solutions to control the extents of imidization. It was reported that novel colorless PIs (**PI14**) derived from 1R,2S,4S,5R-cyclohexanetetracarboxylic dianhydride (a new hydrogenated pyromellitic dianhydride (PMDA) with a controlled steric strucutre, H”-PMDA) and ODA provided good positive-tone patterns with the use of DNQ when the imidization extent reached 80%. Tomikawa et al. from Toray Industries [79-81] reported a novel quantitative esterification in PAAs (**PI15** and **PI16**) by reacting with *N,N*-dimethylformamide diethyl acetal (DMFDEA), as shown in Scheme 5.7. The partial esterification of PAA was controlled by the amount of DMFDEA, and allowed the strict control of the PAA solubility; by using it, *p*-PSPI could be obtained by mixing with DNQ compound.

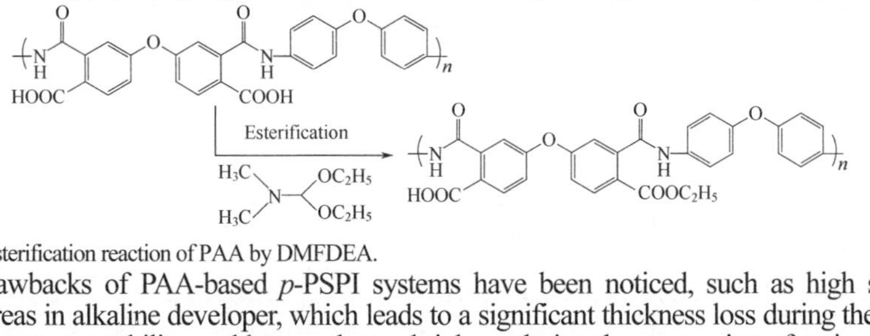

Scheme 5.7 Esterification reaction of PAA by DMFDEA.

Some drawbacks of PAA-based *p*-PSPI systems have been noticed, such as high solubility of unexposed areas in alkaline developer, which leads to a significant thickness loss during the developing process, poor storage stability, and large volume shrinkage during the conversion of amic acid to imide unit [85]. To overcome these problems, PSPIs of fully imidized type have also been investigated and summarized in Table 5.3. After the imidization process, water molecules eliminated from PAA chain to form PI, and the solubility of PI dramatically decreased in alkaline aqueous solutions. Therefore, the molecular design of PI structures focuses on the promotion of the PI solubility by introduction of phenolic hydroxyl group and carboxylic group (–COOH) into the PI backbones.

In the early 1990s, Abe et al. [86] developed *p*-PSPI (**PI20**) composed of solvent-soluble PIs having weak acidic groups and DNQ compound. These soluble PIs were synthesized by copolymerization of

3,5-diaminobenzoic acid (DBA), bis[4-(3-aminophenoxy)phenyl]sulfone (BAPS), and cyclobutane tetracarboxylic dianhydride (CBDA). They found the photosensitivity properties of PIs were verified by the fraction of —COOH groups. When the mole-fraction of DBA was 30%, the *p*-PSPI showed a high sensitivity of 80 mJ·cm^{-2} and the contrast of 4.96. Fukushima et al. [87] replaced CBDA with 6FDA to prepare soluble PI (**PI21**) by a direct one-pot polycondesation of 6FDA with DBA and BAPS. The solubility of the PIs was controlled by the diamine composition, and the corresponding *p*-PSPI systems with the addition of DNQ gave the positive pattern with 10 μm line/space resolution with about 15 μm of film thickness. A recent US patent (US8722758) [88] also reported a water-soluble *p*-PSPI material (**PI22**) having pendant carboxyl groups. These PIs were prepared from the copolymerization of DBA or 3,5-diamino-4-hydroxybenzoic acid (DAPHBA), and diamine having a siloxane group (PLSX) with ODA and BTDA. The corresponding PSPIs contained 12%(*w*) of DNQ compound and formed positive patterns with 3 μm line/space resolution after developing in 1%(*w*) Na_2CO_3 solution.

TABLE5.3 (PI+DNQ) type *p*-PSPIs

	X_1	X_2	Y_1	Y_2	*m*:*n* (molarratio)	References
PI20, PI21	PI20:	PI21: F₃C CF₃	COOH	O O S O O	PI20, 30:70 (*w*); PI21, 1:1	[86,87]
PI22	a: O		Si—O(—Si—O)—Si y n x ; OH COOH	O	3:4:2.8	[88]
	b: O		Si—O(—Si—O)—Si y n x ; OH COOH	COOH ; O	3:1:2:3.8	
PI23	F_3C CF_3		SO_3H	O	9:1	[89]
PI24	F_3C CF_3		SO_3H S HO_3S	O	1:1	[90]
PI25			F_3C CF_3 HO OH	O O S O O	1:1	[91]
PI26	a:		F_3C CF_3	F_3C CF_3 HO OH	4:1	[92]
	b: O O S		O O S HO OH	R Si—O(—Si—O)$_n$—Si 3 R 3	1:1.44	
PI27	F_3C CF_3		Si—O—Si 3 3	O O S HO OH	26:6	[93]
PI28	H N O O S H N O HO OH O		O	F_3C CF_3 O O	25:75(% (*w*))	[94]
PI29	O		CH_3	H_3C CH_3 O O	1:1	[73]
PI30		F_3C CF_3	F_3C CF_3	F_3C CF_3 O O	1:1	[95]
PI31	O O R O O ; R=$(CH_2)_4$– ; O S O ; O C			O O S O O		[96]

PIs bearing sulfonic acids (PIS) are found to be good candidates for alkaline developable PSPI precursors as the cleavage of sulfo-groups from aromatic rings easily occurs by heating. Morita et al. [89] developed a *p*-PSPI (**PI23**) based on PIS and a sulfo-containing DNQ compound (S-DNQ). In this system, PIS was prepared by ring-opening polyaddition of 1,3-phenylenediamine-4-sulfonic acid (PDAS), ODA and 6FDA, and followed by thermal cyclization in *m*-cresol. With combination of 30 %(*w*) S-DNQ, this *p*-PSPI showed a sensitivity of 100 $mJ\cdot cm^{-2}$ and a contrast of 1.7. Afterwards, the same authors prepared another PIS (**PI24**) using 2,2-oxybis(5-aminobenzensulfonic acid) (OBAS) or 2,2′-thiobis(5-aminobenzenesulfonic acid) (TBAS). The TBAS system containing 30%(*w*) S-DNQ showed a sensitivity of 100 $mJ\cdot cm^{-2}$ and a contrast of 1.7 [90].

p-PSPIs with hydroxyl groups (–OH) are the most reported soluble PIs, and the –OH group can be introduced into PI backbones via the molecular design of both diamines and dianhydrides. Jin et al. [91] synthesized a soluble block copolyimide (Bco-PI) (**PI25**) with a hydroxyl group by polycondensation of cyclohexanetetracarboxylic dianhydride (H-PMDA), hydroxyl group containing diamines 2,2-bis(3-amino-4-hydroxyphenylhexafluoropropane) (Bis-AP-AF) and bis(4-(3-aminophenxoy)phenyl) sulfone (*m*-BAPS) in the presence of catalyst. The resulting PI film was colorless and transparent, showed a sensitivity of 250 $mJ\cdot cm^{-2}$ and a contrast of 2.56 using DNQ as photoactive compound. Ishii et al. [92] developed two types of soluble PIs (**PI26**) from hydroxyl group containing diamines, the (PMDA/BPDA+BisApAf/TFMB) system and the siloxane-containing (DSDA+BSDA/PLSX) (BSDA=bis(3-amino-4-hydroxyphenyl)sulfone) system. Both *p*-PSPIs containing 30%(*w*) DNQ showed fine positive patterns and were applied as cover layer materials for flexible printed circuit boards. A (6FDA+BSDA/PLSX) based soluble PI (**PI27**) was also reported in a patent (WO2009/110764) [93]. The relative *p*-PSPI materials based on the mixture of soluble PI (60%) and a conventional PAA (40%), and 30%(*w*) DNQ, showed a sensitivity of 60 $mJ\cdot cm^{-2}$ at 365 nm.

The polymerization reactions of hydroxyamide-containing dianhydrides and diamines to prepare soluble PIs are commonly reported. Hasegawa et al. [94] synthesized hydroxyamide-containing tetracarboxylic dianhydride from trimellitic anhydride chloride (TA) and BAPS as shown in Scheme 5.8. This dianhydride was then reacted with 4,4′-ODA and 2,2′-bis[4-(4-aminophenoxy) phenyl]-hexafluoropropanane (HFBAPP) to form the desired soluble OH-pendant PI (**PI28**). The alkaline-solubility of the PI was adjusted by the ratio of 4,4′-ODA to HFBAPP. When the ratio was 4:1, photoirradiation of the PI/DNQ film at g-line allowed the formation of a fine positive-type pattern by development using 2.38%(*w*) TMAH containing 2-propanol.

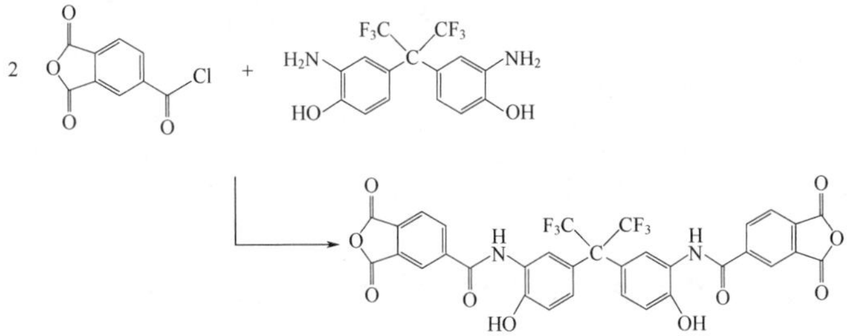

Scheme 5.8 Synthetic route of hydroxyamide-containing tetracarboxylic dianhydride.

p-PSPIs with DNQ PACs (RDP type): The other type of *p*-PSIPs with DNQ based on a new positive-imaging technique, reaction development patterning (RDP). The most obvious characteristic of the RDP principle is the unnecessary alkaline-reactive and photosensitive groups (such as phenolic hydroxyl group, carboxyl group) in the polymer structure [73,95]. Both the (BPDA/DAT+BPDA/BAPP) PI system (**PI29**) [73] (DTA=2,4-diaminotoluene, BAPP=2,2-bis-[4-(4-aminophenoxy) phenyl]propane), and the (6FDA/HFBAPP+BPDA/2-DMB) PI system (**PI30**) [95] (2-DMB=2,2′-dimethylbenzidine) developed by Tomoi et al. are classified as RDP type. These PIs with flexible contents in the backbone were found to be easily soluble in ethanolamine-containing solvents such as NMP. By mixing with DNQ compounds, the corresponding PI films gave positive-tone behavior by UV irradiation, followed by development in a mixture of ethanolamine/NMP/H_2O (1/1/1 by weight).

This process differs from the conventional dissolution inhibition system as well as the chemically amplified system. In the RDP technique, the amine-containing developer causes chemical chain-scissions of PIs in the UV-exposed area during the development step and generates positive patterns. The pattern formation mechanism for RDP is illustrated in Scheme 5.9 and can be described in three steps: [73] (1) in the UV-exposed area, an acid-base reaction occurs between indencarboxylic acid, which is generated from DNQ compound through photorearrangement, and ethanolamine; (2) the ammonium salt generated from the acid-base reaction promotes the permeation of the developer into the exposed film; and (3) the nucleophilic attack of the permeated amine induces a ring-opening of the imide linkage, resulting in the depolymerization of the PI and promotes the PI dissolution; consequently, the dissolution difference between the exposed and unexposed area of the film causes the formation of a positive-working pattern.

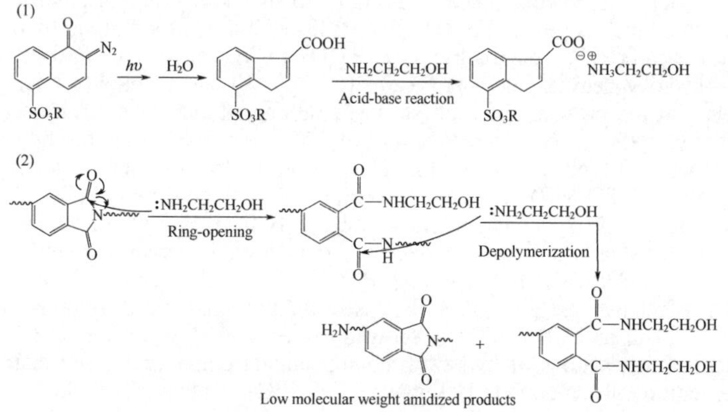

Scheme 5.9 RDP mechanism of positive PSPI system with DNQ treated with ethanolamine-containing developer.

As there are no structural restrictions on polymers, RDP can apply to polymers without specific functional groups for dissolving these polymers in alkaline solution, such as **PI31** [96] in Table 5.3 as well as commercially available engineering plastics like polycarbonates [97] and polyarylates [98]. In microelectronic packaging such as dicing, bonding and encapsulations, PSPIs are being used as interlevel dielectrics and high-density interconnects. They can provide a straightforward patterning of the bond pad windows and the scribing or dicing paths, and also provide protection overcoats for electronic devices during packaging or during the life of the device. With the development of microelectronic devices, properties and functionalities of PSPIs still need to be improved. New PSPIs with high sensitivity, high transparency, low dielectric constant, low *CTE*, low thermal imidization temperature and environmentally friendly production, and so on are greatly demanded.

5.3 Polyimide Materials for Optoelectric Planar Displays

5.3.1 Liquid Crystal Alignment Layers

Liquid crystal displays (LCD) have established a firm foothold on the market as flat displays for computers, transportation, communication, instrumentation, and television. LCDs are composed of a LC, a color filter, and a LC alignment layer. In LCD picture image is obtained through the change of the transmission of light by orienting the LC molecule under the electrical field, whereas the LC alignment layers orient the LC molecules uniformly to give a desired optical effect. Therefore, surface-induced alignment of LC molecules is a key technology for fabricating high-performance LCDs. The LC alignment on the cell wall surface is macroscopically characterized by the surface anchoring energy coefficient and the pretilt angle of LC molecules [99]. The surface anchoring energy is the surface excess free energy when the LC director on the cell wall surface deviates from the "easy direction." The pretilt angle is defined by the tilt angle of the easy direction measure from the cell wall surface plane as shown in Fig. 5.11 [25]. In order to obtain uniform LC alignment, the selective of LC alignment films is quite important as they have significant effects on the performance of LC devices by their characteristic parameters, including pretilt angle, alignment stability, transmittance and birefringence.

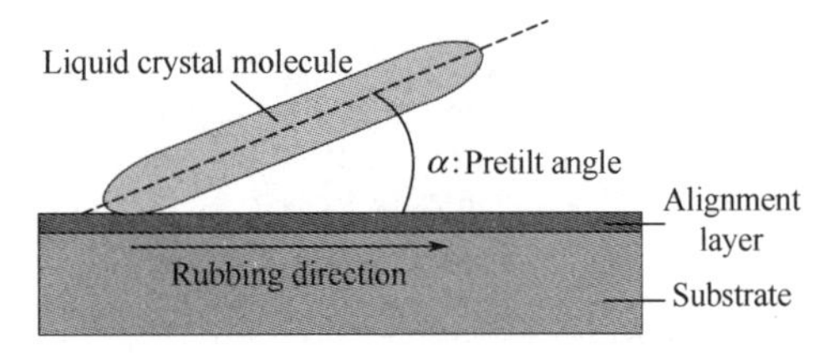

FIGURE 5.11 Generation of pretilt angle by rubbing the surface of alignment layer.

PIs are good candidates for LC alignment layers because of their advantageous properties, including excellent optical transparency, adhesion, heat resistance, dimensional stability, and insulation. In addition, the suitability of polymers as LC alignment materials also depends on the following factors: (1) uniformly align the LC molecules, (2) ease of processing (to form thin films on the substrate surface by several conventional methods such as spin coat, rolling coat, dipping coat, and so on; (3) not reacting with the LC molecules; (4) strong adhesion with the substrate; and (5) to be hydrophobic. Therefore, one of the most recently developed and important electronics applications of PIs is their use in liquid crystal (LC) alignment layers for LC flat-panel display devices [100,101].

One of the key technologies in LCDs is the surface alignment of LCs. Such PI film surfaces need to be treated if they are to produce a uniform alignment of LC molecules with a defined range of pretilt angle values. The pretilt angle of LC molecules is a sector angle between the LC molecule and substrate that influences the electro-optic characteristics of LCDs, such as bias voltage, threshold voltage, and response time [102]. At present, there are two methods have been developed to treat PI film surfaces for the control of LC alignment, one is a rubbing process using velvet fabrics, and the other is a rubbing-free process based on irradiation of the PI film surfaces with linearly polarized ultraviolet light (LPUVL) [10].

5.3.2 Mechanical Rubbing Alignment Polyimides

The rubbing process rubs an organic-polymer-coated substrate, such as PI films, with a rotating drum that is covered with a rayon velvet fabric cloth. Unidirectional LC alignment with a controlled pretilt angle is typically produced using the rubbing techniques. This is the only technique adopted in the LC industry in the mass-production of flat-panel LC display devices because of its simplicity and controllability of both the LC anchoring energy and pretilt angle [10].

An understanding of the LC alignment mechanism is not only interesting from a scientific point of view, but is also of great importance from the industrial point of view. This is because rubbed PI films are widely used as alignment layers for LC molecules in present LC displays, and also because determination of the alignment mechanism is very helpful in developing an alignment method that does not have the drawbacks associated with mechanical rubbing. Different mechanisms have been proposed for the alignment of LC molecules on the rubbed polymer surface. Although the exact mechanism of LC alignment is still not clear, two models of the alignment process have received the most attention. They are known as the "microgroove effect" and the "molecule-molecule interaction" [103]. The former suggests that the mechanical rubbing creates microgrooves or scratches on the polymer surfaces, and the LC aligns along the grooves to minimize the energy of elastic distortion [104]. Alternatively, the latter mechanism suggests that the rubbing process aligns surface polymer chains, which in turn, align the LCs through the anisotropic van der Waals interaction between the LC molecules and the oriented surface polymer molecules or segments [105].

As a result of extensive efforts to understand the alignment mechanism, it is understood in the following two steps: (1) molecular alignment of the first LC monolayer is induced through a short-range interaction between the LC molecules and the underlying PI film. The LC molecules in the first monolayer are strongly anchored to the PI film surface; (2) the molecular alignment of the first LC monolayer propagates into the bulk via a long-range (elastic) interaction among LC molecules [106].

Pretilt angle is an important parameter in the design of LC devices. A suitable pretilt angle depends on the driving mode of LCDs: e.g., several degrees for twisted nematic (TN)-mode LCDs, 5-10 degrees for optically compensated bend (OCB)-mode LCDs, ~50 degrees for bistable ben-splay LCDs, and ~88 degrees for the multidomain vertical alignment (MVA)-mode LCDs [107]. Therefore, the LC alignment techniques that can continuously control the pretilt angle from 0 to 90 degrees are desired.

One of the advantages of PI is to be able to control the pretilt angle. In rubbed PI layers, the pretilt angle is thought to result from the inclination of polymer backbone by unidirectional rubbing [108]. The LC pretilt angle depends on both the polymer inclination angle and the nature of the surface-LC interaction. In addition to the uniform, unidirectional alignment of LC molecules, the pretilt angle of alignment film plays the most important role in determining the optical and electrical performance of LCD devices. If the pretilt angle is not constant throughout the panel, the orientation of LC is dispersed, resulting in an ununiformed image.

Much effort has also been made to achieve the desired LC pretilt angle and to understand the mechanism behind LC pretilt phenomena. Based on reported investigations, some mechanisms have been proposed [10], including (1) LC pretilt is governed by van der Waals interactions between the LC molecules and the alkyl side chains of the PI, and (2) LC pretilt angle is determined by the polymer backbone structure but is independent of the length of the side chain.

The control of pretilt angle of LC molecules has been intensively investigated [109]. These studies implied that pretilt angle is strongly related to the surface polarity. The pretilt angle can be increased not only along with the rubbing strength of the rubbing process, but also by simply introducing long, linear alkyl side-chains or other nonpolar groups to the PI alignment layers, which is due to the decrease in the surface polarity. In contrast, the pretilt angle can be decreased by applying the surface treatment of UV exposure or O_2 plasma to the PI alignment layers, which would increase the surface polarity.

5.3.3 Photoinduced Alignment PIs

The rubbing process with velvet fabrics is simple, convenient, inexpensive, and widely used in the LC industry. However, the shortcomings of this process have also been disclosed. This process generates dust, has electrostatic problems, and it is difficult to control the rubbing strength and uniformity in the production of TFT-LCDs [110]. Moreover, this process has a serious limitation in the fabrication of very large areas of PI films which are required in the mass production of large LC televisions. Therefore, alternative methods which do not suffer from the shortcomings of the rubbing process were developed [12], including photoalignment, and the use of microgrooved surfaces and Langmuir-Blodgett membranes.

The photoalignment technique is the treatment of the alignment film surfaces by irradiation of the polymer with LPUVL to generate an anisotropic distribution of alignment material molecules [110]. This is a noncontact alignment technology with potential capability for two-dimensional alignment patterning. Such patterning capability is useful for fabricating LCDs with a large viewing angle, for realizing multistable in-plan LC alignment over a large area. Because there is no need for a vacuum environment, photoalignment is suitable for mass production. In 2010, Sharp Corp. first adopted a photoalignment method in the commercial production line of multidomain vertical alignment mode LCDs [111].

The photoalignment can be induced by three types of photochemical reactions: (1) reversible *cis-trans* isomerization of azo-containing polymers [112], (2) photodimerization of crosslinkable prepolymers [113], and (3) photodegradation of polymers [114]. Among the polymer materials being used, photoinduced alignment PIs are of great interest because of their superior properties such as unique photo- and thermal stability.

5.3.3.1 Photodegradation of PIs

An advantage of the alignment methods based on photoinduced decomposition reactions is that there is no requirements on polyimide structures. Therefore, this method is applicable to a variety of polyimides. Different polyimides [108, 114-118] as shown in Table 5.4 have been studied and shown an enormous range of properties. The first photodegradable PI alignment material was reported by Hasegawa and Taira [114]. They exposed positive-type PI (**PI32**) to LPUVL at 257 nm and induced homogeneous alignment of nematic LCs. They believed that the anisotropic photodegradation of the PI main chains causes the LC alignment, i.e., the PI chains parallel to the exposed UV polarization direction are selectively decomposed by the UV light, the LPUVL-exposed film has an anisotropy orientation of the remaining PI molecules [119-121]. The anisotropic orientation of PI molecules induces uniform alignment of LC molecules. It was pointed out by West et al. [116] that those nondecomposed PI chains, which are perpendicular to the exposed UV polarization, cause the anisotropic van der Waal forces to align the LC along its optical axis. The average orientation direction of LC molecules is parallel to the film surface and perpendicular to the polarization direction of LPUVL.

A PI with cyclobutane structure, **PI33**, was reported to require a fluence of 720 $mJ \cdot cm^{-2}$ to reach saturated alignment. When the cyclobutane ring replaced by a benzene ring to give **PI34**, the required fluence increased by a factor of 10 [115]. The LC alignment direction of **PI35** was found to depend on the excitation wavelength [108]. When **PI35** was excited at 254 nm, the parallel direction was obtained, while at 313 nm it was the perpendicular direction. The PI containing a fluorene moiety, **PI36**, showed parallel alignment to the UV polarization [116]. **PI37** is a polyamic acid-azo copolymer with a better temperature stability at low UV exposure [117], in which the alignment that arose from the *trans/cis* transition of the azo segment was locked in by simultaneous imidization of the PAA. **PI38** is a tailored PI structure to minimize photodegradation, which gave a strong anchoring energy with minimal change of spectroscopic properties upon UV irradiation [118].

TABLE5.4 Polyimides Showing Photoalignment Because of Anisotropic Photodegradation

	Molecular Structure	References
PI32	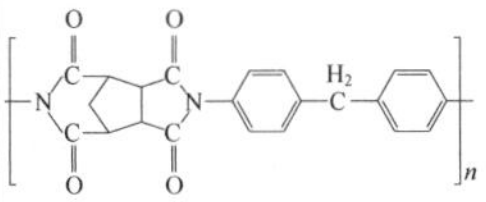	[114]
PI33	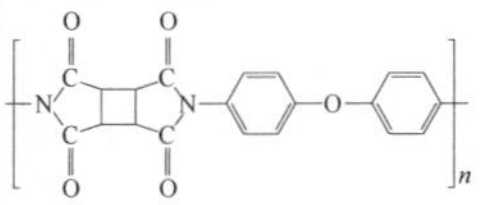	[115]
PI34	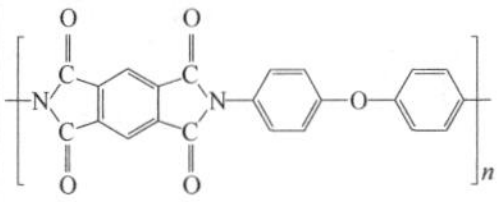	[115]
PI35	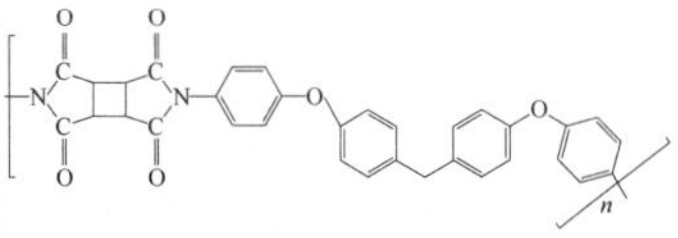	[108]
PI36	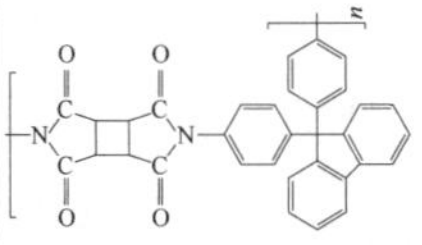	[116]
PI37	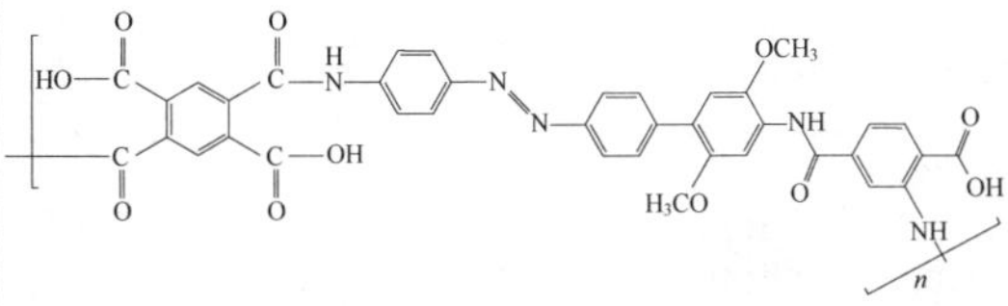	[117]
PI38	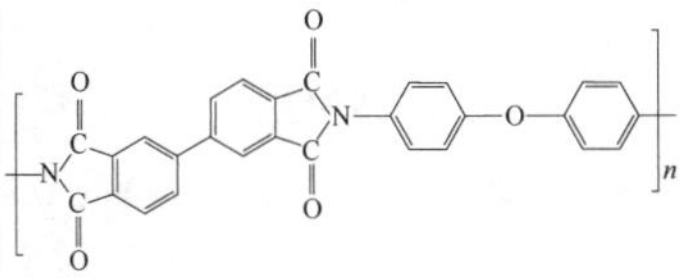	[118]

However, one may be concerned about the lowering of the thermal, chemical, and mechanical stability of the PI film due to the photoinduced decomposition. Some other disadvantages have also been revealed for such photodecomposable PI materials. (1) The photosensitivity of PIs are low, so inconveniently long exposure times are usually required to achieve photoalignment. (2) Photodegradation products are thought to negatively affect LCD performance. (3) The polar anchoring energy is quite low, possibly because of the lowering of the surface energy by decomposition [122]. (4) The surface anisotropy and dichroic ratio of the PI films induced by the LPUV light irradiation are quite small [117,119]. (5) The anisotropic orientation of the alignment units can be damaged during the thermal process in manufacturing LC devices and lead to a deterioration of the LC aligning properties of the alignment layers [123]. (6) The microscopic structure of a polymer surface is sensitive to heating, the thermal treatment of the LC-filled cells could modify the fine morphology of the PI surface [123]. (7) Photochemical reactions in PIs would produce charges that could induce noticeable image sticking and flicker, which can be a severe defect when trying to achieve high reliability of LCD panels [124]. (8) The photopolymer film is also optically unstable. Therefore, the design of the chemical structure of PI with a large a/b ratio (a and b stand for two decomposition rates of PI chains oriented parallel and perpendicular to the polarization direction of the LUPV light) is important to obtaining a large surface anisotropy by LPUV light irradiation. Moreover, maintenance of the alignment is more important for applications compared to the induction of the alignment. Therefore, it is very necessary to develop PI films which can provide excellent and stable LC alignment.

5.3.3.2 Photopolymerization of PIs

Polymers containing photodimerizable chromophores, such as cinnamate, coumarin, and chalcone units, at the side groups or in the polymer main chain, have been investigated more extensively due to their advantageous properties compared with other photochemical systems. They are photochemically irreversible, and chemically more stable after UV irradiation than azobenzene-type *cis-trans* isomerizable polymers. They are also more sensitive to UV light, requiring a much lower exposure dose to achieve a saturated alignment than the photodegradation-type polymers. Several polyvinyl derivatives containing these photoreactive groups have thus been reported [2,6,125,126]. However, these polymers have poor thermal stability due to their high chain flexibility, and this hinders their commercial application. PI is the most attractive material for use as alignment films owning to its verified properties. Therefore, many attempts have been made to improve the thermal stability using PI backbones with crosslinkable chromophores in the structure.

5.3.3.2.1 Cinnamate-contained Polyimides

In the cinnamate-containing polymer film, the cinnamate side-chain groups undergo angular- selective photodimerization of □CH═CH□ double bonds, as shown in Scheme 5.10, and general surface anisotropy when the film is irradiated by LPUV light [127]. Thus, the polymer film homogeneously aligns the LCs. Table 5.5 summarizes the cinnamate-containing PIs for photoalignment layers.

UV

Head to tail cycloaddition

UV

Head to head cycloaddition

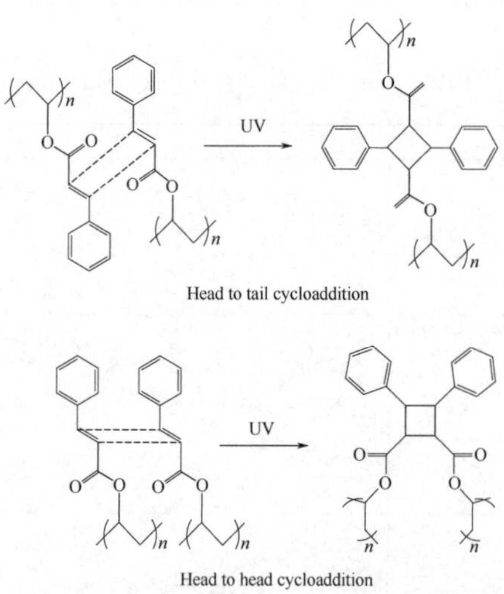

Scheme 5.10 (2+2) Cycloaddition for a cinnamate side-chain polymer.

The homogeneous, uniaxial LC-alignment ability of 6F-HAB-CI (CI=cinnamate) PSPI (**PI39**) was revealed by Lee et al. [128]. This photosensitive PI with cinnamate side groups was synthesized to have a high T_g and was thus stable both thermally and dimensionally. Its LC alignment test suggested that the LC alignment process was principally governed in the irradiated PSPI films by the orientation of the polymer main chains and of the unreacted CI side groups, whose directionally anisotropic interactions contributed to the alignment of the LC molecules. Therefore, this PSPI is a promising candidate material for use as an LC alignment layer in advanced LCD devices.

TABLE5.5 Cinnamate-Containing Polyimides Showing Photoalignment Because of Photopolymerization

	Molecular Structure	References
PI39	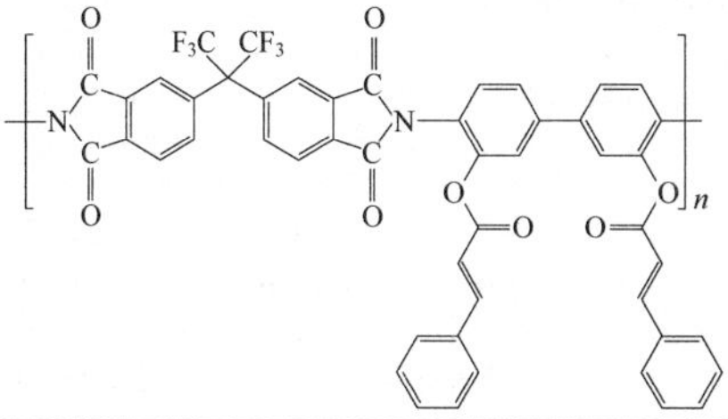	[128]
PI40	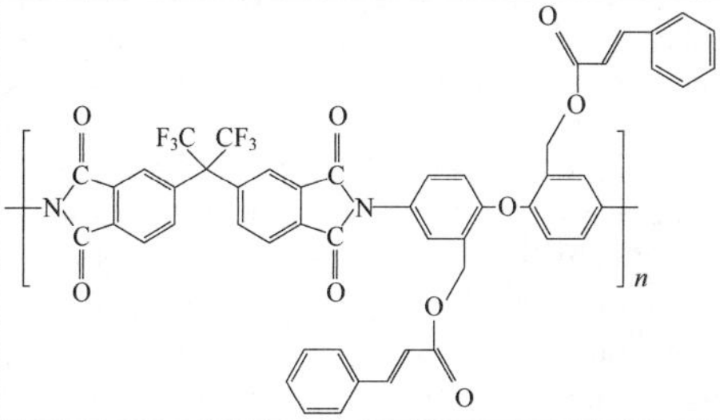	[129]
PI41	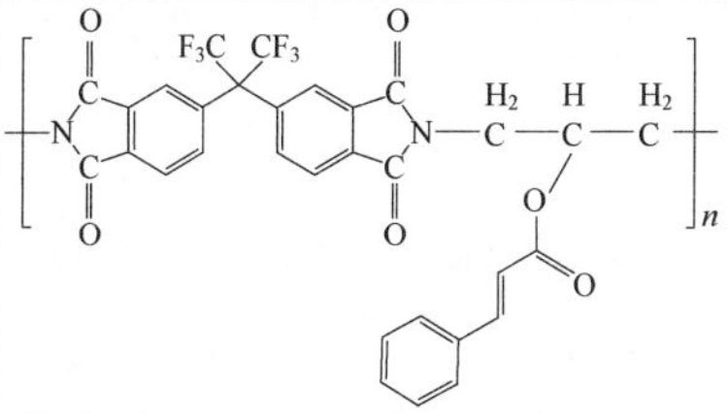	[130]
PI42	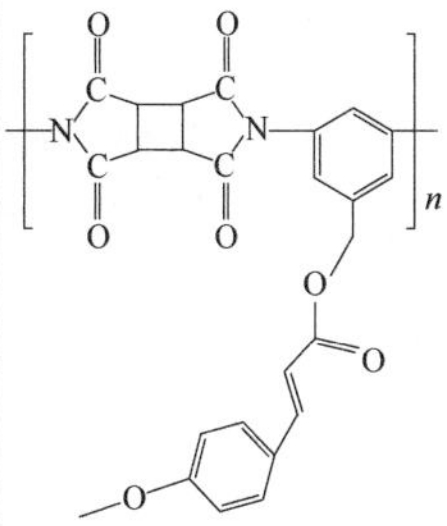	[131]
PI43	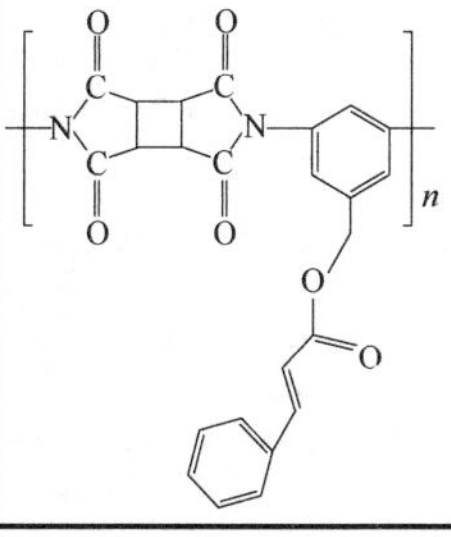	[127]

However, a high-exposure dose of 0.5 $J \cdot cm^{-2}$ was required for the saturated alignment of LCs in this system. To improve photosensitivity and thermal stability, a soluble fluorinated PI bearing methylene cinnamate side groups, PIMC (**PI40**), was developed as a photoalignment material [129]. **PI40** was firstly developed as a negatively working photosensitive PI for making a PI pattern by a lithographic process. Its alignment behavior showed homogenous LC alignment was successfully induced on the PIMC film by LPUV light irradiation with small exposure energy of 0.045 $J{\cdot}cm^{-2}$. Moreover, the LC cells retained their alignment even after severe thermal treatments.

For 6F-DAHP-CI (**PI41**) [2+2] photodimerization was found to occur mainly in the solid thin-film state [130]. The directionally selective photoreaction of the cinnamate chromophores in the PSPI films due to LPUVL exposure, leaves cinnamate chromophores unreacted along the direction perpendicular to the polarization of the LPUVL. Kim et al. compared photoinduced alignment behavior of PSPIs containing 2-, 4- and 2,5-methoxy cinnamate chromophores. They found that 2-methoy cinnamate acid was the most photosensitive and UV irradiation of CBDA-2MCI (**PI42**) film proceeded predominantly by the photoisomerization [131]. They also investigated the effect of polarized UV light on rubbed CBDA-2MCI PSPI film. It was reported that irradiation of depolarized UV light suppressed anchoring energy, while irradiation of linearly polarized UV light controlled anchoring energy effectively [132].

However, photoalignment methods have an image sticking problem with unreacted photosensitive functional groups, which can be a severe defect in the attempts to achieve high reliability of LCD panels. To overcome this problem, Hwang et al. [127] provided a method to react bromine and ethanethiol with CDBA-CI PSPI (**PI43**) film after irradiation of LPUV light at 350 nm - 360 nm. The unreacted photoreactive functional groups were effectively deactivated by forming carbon-bromine or carbon-sulfur bonds from carbon-carbon double bonds.

It has been noticed that several photoreaction pathways (i.e., *trans-cis* photoisomerization, photopolymerization, and photocyclization) are available for cinnamate chromophores, making it difficult to predict the effects of LPUV light on cinnamate-containing polymers [133]. Therefore, it is worth exploring the photoinduced LC alignment properties of polymers with other chromophores.

5.3.3.2.2 Coumarin-contained Polyimides

One promising alternative to the cinnamate chromophore is the coumarin chromophore, which undergoes a simple [2+2] photodimerization because of its fused-ring structure (Scheme 5.11) [133]. Taking this property of the coumarin chromophore into account, and considering the advantageous properties of PIs, photoreactive PI containing the coumarin moieties have been developed and are listed in Table 5.6.

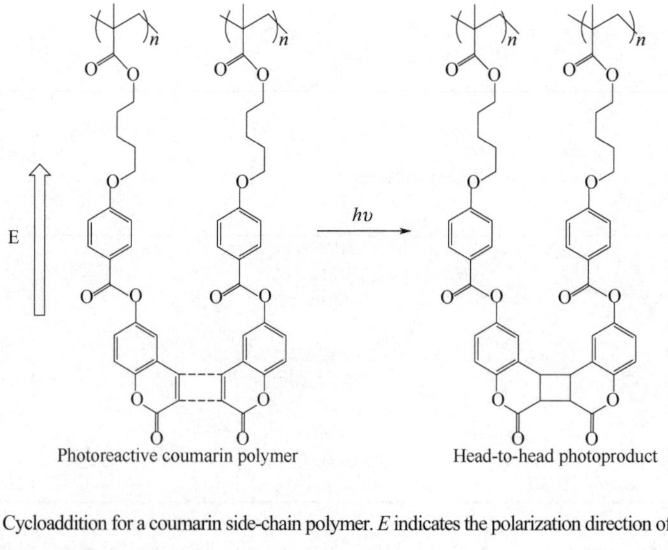

Photoreactive coumarin polymer   Head-to-head photoproduct

Scheme 5.11 (2+2) Cycloaddition for a coumarin side-chain polymer. *E* indicates the polarization direction of the incident beam.

TABLE5.6 Coumarin-contained Polyimides Showing Photoalignment Because of Photopolymerization

	Molecular Structure	References
PI44	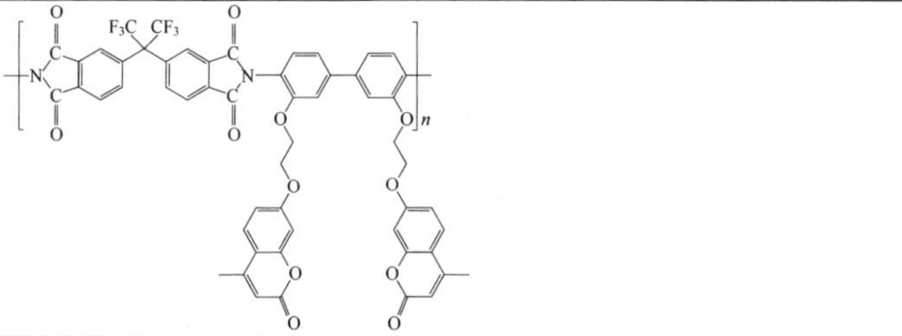	[133,134]
PI45	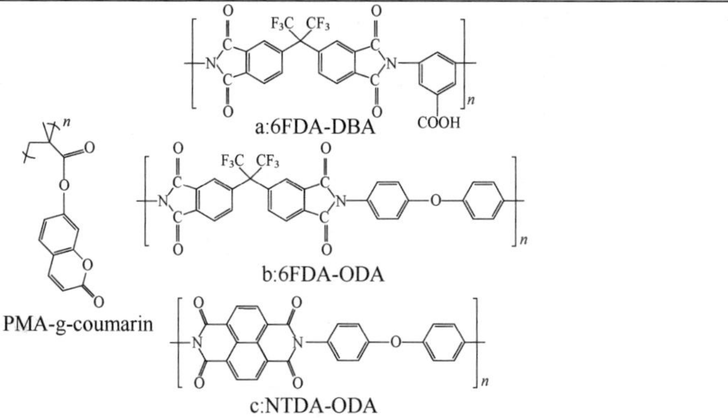 a:6FDA-DBA b:6FDA-ODA PMA-g-coumarin c:NTDA-ODA	[135,136]

A new photoreactive PI containing the coumarin moiety in a side group, 6F-HAB-COU (COU=coumarin) PSPI (**PI44**), was developed [133,134]. This PSPI was stable up to 300 °C. Irradiation of the PSPI films with LPUV light induced anisotropic reorientations of the polymer main chains and the unreacted COU side groups, whose directionally anisotropic interactions with the LC molecules contribute to the homogeneous LC alignment at an angle of 100 degrees.

The alternative approach to improving the thermal stability of the LC alignment is to introduce PI into a photopolymer based on coumarin (**PI45**) [135,136]. Park et al. prepared blend alignment layers from polymethacrylate with coumarin side chains (PMA-*g*-coumarin) and PIs for the orientation of LCs using LPUV irradiation. Four different PIs, 6FDA-DBA (**PI45a**), PMDA-ODA (**PI34**) [135] and 6FDA-ODA (**PI45b**), NTDA-ODA (**PI45c**) [136] were used. It was found that the thermal stability of the LC alignment layers was enhanced regardless of the type of PI while the direction of LC orientation was dependent on the type of PI. When using 6FDA-ODA with high photoreactivity, the direction of LC orientation was changed from parallel to perpendicular to the polariztion direction of LPUV light. On the contrary, no directions changes were observed when using PMDA-ODA and NTDA-ODA, which have low photoreactivity. Therefore, it is clear that photoreactivity is an important factor related to the LC orientation of a photoinduced alignment layer based on blend systems.

5.3.3.3 Photoisomerization of Polyimides

In the photoisomerizable polymer films with azobenzenes, the azobenzenes oriented parallel to the polarization direction of the exposed UV light are selectively isomerized from the *trans* form to the *cis* from (Scheme 5.12) due to the selective absorption of the linearly polarized light.

Therefore, the LC molecules unidirectionally align perpendicular to the polarization direction of the exposed UV after a prolonged irradiation with an LPUV light [117]. But these kinds of polymers are rebuilt on the molecular level while the induced direction remains stable. Photoisomerizable PIs are summarized in Table 5.7.

TABLE 5.7 Polyimides Showing Photoalignment Because of Photoisomerization

	Molecular structure	References
PI46	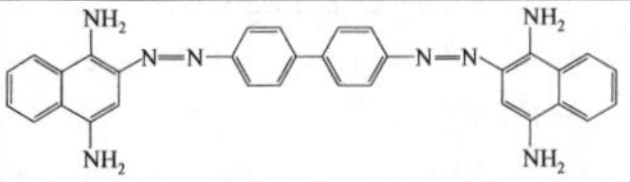	[117]
PI47	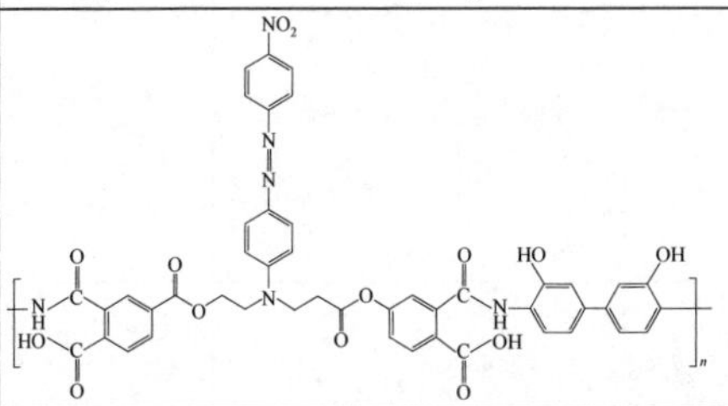	[117]
PI48	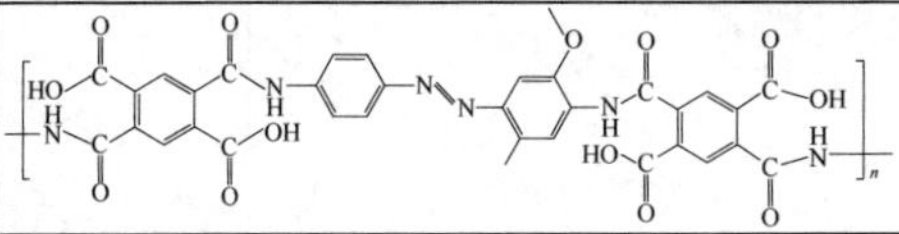	[117]
PI49	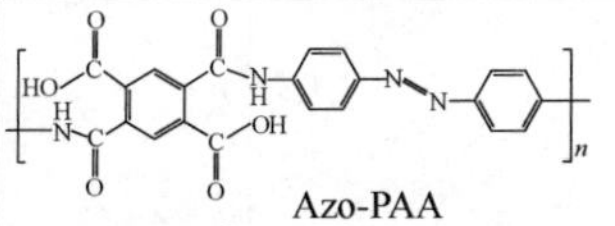 Azo-PAA 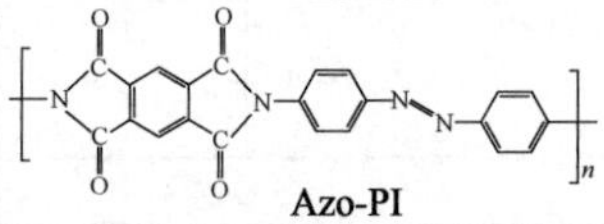 Azo-PI	[137-140]
PI50	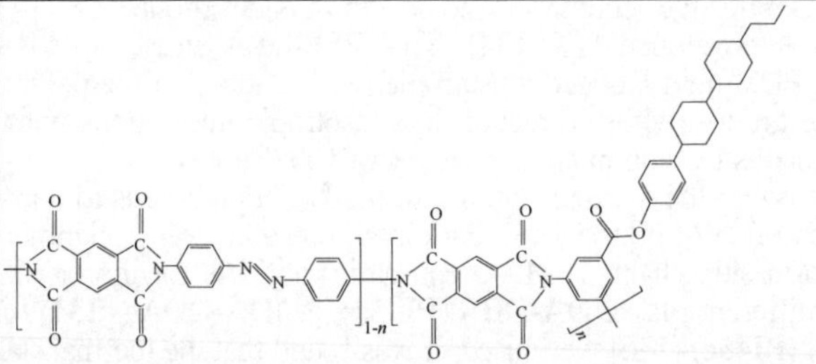	[141, 142]
PI51	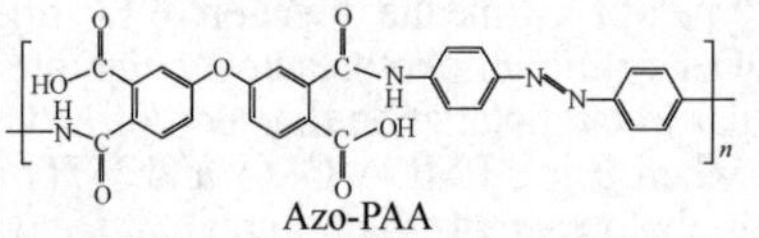 Azo-PAA 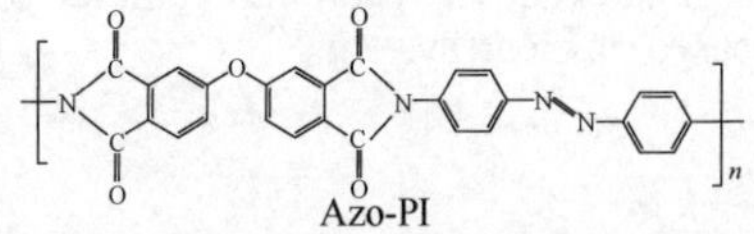 Azo-PI	[143]

In order to obtain photoisomerizable PI films used as LC alignment layers, Park et al. [117] first introduced the photosensitive function, azobenzene, into the polyamic acid backbone (Azo-PAA) and prepared three polyaimic acid systems modified with azobenzene molecules: a polyamic acid doped with azobenzene molecules (**PI46**), a side-chain-substituted polyaimic acid with azobenzene units (**PI47**), and a main-chain-substituted polyamic acid with azobenzene units (**PI48**). It was found that the main-chain PI system exhibited good unidirectional LC alignment and excellent thermal (300 °C for 1 h) and optical stability, while the doped system and the side-chain system were thermally unstable for use as an LC alignment layer. In this method, photoalignment is carried out on the film of polyaimic acid containing azobenzene in the backbone structure (Azo-PAA). The alignment of the Azo-PAA backbone structure is caused by the photoinduced *trans-cis-trans* isomerization of the azobenzene molecule. After the photoalignment treatment, the Azo-PAA is converted into the corresponding PI (Azo-PI) film by thermal imidization. Since the photoresponsivity disappears after conversion, the resulting alignment layer shows high optical stability along with the inherent high thermal and chemical stability of PI [137,144].

Scheme 5.12 The *cis* (*Z*) and *trans* (*E*) isomeric forms of azobenzene.

The photoinduced alignment performance of Azo-PI films, synthesized from ODPA and 4-[4′-propylbi(cyclohexane)-4-yl]phenyl 3,5-diamino-benzoate (DABA) (ODPA-DABA, **PI49**), were investigated by Sakamoto et al. [137-140]. It was found that the alignment of Azo-PI backbone structures can be induced by either a double-exposure method of UV/visible light [139] or an oblique angle irradiation of unpolarized light (UP-L) on the Azo-PAA films [137]. The pretilt angle was confirmed to be generated by single oblique angle irradiation with UP-L at an incidence angle of 45 degrees [140]. However, as long as the main-chain-type Azo-PAA was used, it was hard to generate a pretilt angle greater than ~3 degrees from the cell wall plane [140]. The controllable range of the pretilt angle by oblique angle irradiation is between 2 and 3 degrees.

In order to introduce a side-chain structure into Azo-PI, random copolymers (**PI50**) were synthesized from pyromellitic dianhydride and a mixture of 4,4′-diaminoazobenzene and DABA (PBCP-DABA). When used as alignment layers, PBCP-DABA can induce any desired pretilt angle between 0 and 90 degrees by varying the molar fraction (n) from 0 to 0.5 [141,142], and the pretilt angle was found to increase with the increasing side-chain content. Defect-free uniform LC alignment was obtained for the pretilt angle range less than 5 degrees and greater than 85 degrees, which is sufficient for practical applications of the TN-mode and MVA-mode LCDs. Furthermore, the defect-free pretilt angle range can be slightly extended using the blend films containing main-chain-type Azo-PI and PBCP-DABA instead of the pure PBCP-DABA films. The azimuthal anchoring strength of the blend film is stronger than that of the pure films, and the pretilt angle for the blend systems was less than 11 degrees and greater than 78 degrees [107]. The controllable pretilt angle range of main-chain-type Azo-PIs without the appearance of threadlike disclination loops can be further extended at least up to 38 degrees by a simple method, which is an alkyl-amine (octadecylamine (ODA)) vapor treatment and requires no chemical synthesis of specially designed PIs. This is attributed to the fact that the photoalignment efficiency of Azo-PIs is enhanced and results in the extended pretilt angle.

For the conventional Azo-PI, a relatively large light exposure ($\geqslant 0.5\ J \cdot cm^{-2}$) is required to induce uniform LC alignment [106,137,145,146]. To obtain the in-plane anisotropy comparable to that induced by rubbing, an exposure of linearly polarized light more than ~10 $J \cdot m^{-2}$ is needed. The backbone structure of Azo-PI films plays an important role in the photoalignment efficiency. Azo-PIs with PMDA-DABA backbone structures, **PI51**, were synthesized and their photoalignment efficiencies were compared with ODPA-DABA Azo-PI films, **PI49** [143]. It was found that the photoalignment efficiency can be significantly improved using ODPA-DABA (**PI49**). The in-plane anisotropy of the Azo-PI film is induced by the pure photoalignment effect of Azo-PAA and the self-alignment effect during thermal imidization. For the ODPA-DAAB film, both effects were enhanced. The photoalignment efficiency can be enhanced by treatment with ODA vapor. The 18 nm-thick Azo-PAA film swelled by ~300% after the ODA vapor treatment at 80 °C for 1 h. The photoinduced in-plane anisotropy was found to be correlated very closely with the swelling ratio. For enhancing the photoalignment efficiency, this alkyl-amine vapor treatment must be performed prior to photoalignment.

Photoaligned films of **PI51** are reported to be excellent alignment layers for light-emitting polymer, poly (9,9-dioctylfluorenyl-2,7-diyl) (PFO), which is in liquid crystalline phase at above 165 °C. A 30 nm-thick PFO layer was spin coated on an Azo-PI film, and thermally treated to its liquid crystalline phase. After quenching it to room temperature, highly oriented PFO layers that showed a polarization ratio of approximately 30 in photoluminescence were formed. This excellent alignment ability is attributed to the uniformity and the sufficiently large in-plan anisotropy of the photoaligned film surface [147]. In addition, two-dimensional alignment patterns of PFO can be formed on the photoaligned

Azo-PI film by performing a double exposure of polarized light through a photomask [148].

In summary, among a variety of photoalignment methods, the photoisomerization method based on PI containing azobenzene in the backbone structure (Azo-PIs) is the most attractive. Azo-PIs have pronounced advantages including: (1) In addition to inherent high thermal and chemical stability, Azo-PI films are optically stable, as their photoresponsivity disappears after conversion from their corresponding Azo-PAA precursor [137,144]. (2) The in-plane molecular orientation of the photoaligned Azo-PI film can be controlled over a wide range, and the in-plan anisotropy is much larger than that induced by rubbing. (3) The azimuthal anchoring energy coefficient of main-chain-type Azo-PI films for 4-*n*-pentyl-4′-cyanobiphenyl (5CB) are ~1.5×10^{-3} J·m^{-2} [145], and is comparable to or larger than that of conventional rubbed PI alignment layers (> 10^{-4} J·m^{-2}) [149]. Although the low photoalignment efficiency of Azo-PIs is the major drawback, it can be overcome by exposing the Azo-PAA film to alkylamine vapor prior to photoalignment. Therefore, the azobenzene-containing PIs are promising alternative materials for conventional rubbed PI alignment layers.

5.3.4 The Microgroove Polyimide Surfaces

It was reported in 1973 by Berremance that LC molecules were able to align along microgrooves on the surface of alignment films [150]. Some researchers have attempted to replace the PI alignment layers in the TN-LC cell with microgroove grating structures [151]. Several processes to produce microgrooved structures have been proposed [12]. They are (1) reactive ion-etching on glass surfaces with chromium masks, (2) pattern formation by laser-induced periodic structures on a polymer surface (LIPSS), (3) holographic light exposure or exposure of photocurable polymer films to UV through masks with a grating pattern, and (4) the soft embossing method.

Among these methods, the LIPSS has been widely investigated. Lu et al. reported the studies on molecular orientation in LIPSS formed on a PI surface [152], and the alignment behavior of LC molecules on such laser-induced microgroove surfaces [103]. The PI used was prepared from the solution polycondensation between 4,4′-diamino-3,3′-dimethyldiphenylmethane and benzophenone-3,3′,4,4′-tetracarboxylic dianhydride (**PI52**). They found that the PI chains tended to orient in a direction perpendicular to the surface microgrooves [152], while the alignment direction of LC molecules was dependent on the depth of the microgrooves [103].

PI52

Chiou et al. proposed a nonrubbing soft-embossing method (Fig. 5.12A) to fabricate reliable periodical microgrooves on the PI surfaces as shown in Fig. 5.12B to align the LC molecules [109]. It

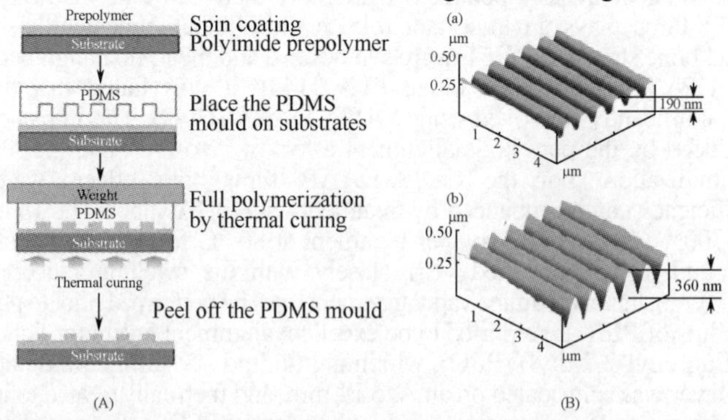

FIGURE 5.12 (A) The schematic illustration of the soft embossing method. (B) AFM images of the microgrooved polyimide surfaces obtained by the soft embossing method. *Reproduced from D.-R. Chiou, L.-J. Chen, C.-D. Lee. Pretilt angle of liquid crystals and liquid-crystal alignment on microgrooved polyimide surfaces fabricated by soft embossing method. Langmuir 22 (2006) (22) 9403-9408.*

was found that the director of LC molecules uniformly align along the groove direction even when the groove width is as high as 3 μm. The anchoring energy of these microgrooved PI surfaces was found to be higher than that of the typical rubbed surfaces.

5.3.5 Langmuir-Blodgett Polyimide Films

In 1999, Koo et al. [153] fabricated polyimide alignment layers by combining the rubbing and Langmuir-Blodgett (LB) methods to investigate the relationship between the pretilt angle of LC molecules and the surface structure of the alignment layer. In a rubbed PI film with two LB layers the contact angle was shown to be the same as the LB film, while the pretilt angle of LB layer on rubbed film was reduced. At the same time, a kind of silicon-contained PI LB film was fabricated to align ferroelectric liquid crystal (FLC) [154]. The aligning effect on the FLC molecules was found to be affected by different imidization temperatures, i.e., high-temperature-imidized LB films can provide the energy barrier for the realization of excellent bistability, while the ultrathinness of LB films is helpful for the quick response of the FLC cell.

The LC alignment capability in 4-*n*-pentyl-4′-cyanobiphenyl (5CB) on PI LB film with alkyl chain lengths was investigated by measuring the pretilt angle, the induced optical retardation, and the extrapolation length of 5CB [155]. The characteristics of the PI, such as chain numbers, were found to have strong influences on the LC aligning capability in NLC. The orientational ordering process of 5CB [156] was also studied during evaporation onto N-layer PI LB films (N=0, 1, 3, 5, 11) using polar and nematic order parameters S1 and S2. It was shown that S1 decreases but S2 is constant as the PILB layer increases [156].

5.4 Polyimide Materials for Optoelectronic Flexible Displays

5.4.1 Combined Property Requirements

Fabrication of electronic devices on flexible substrates is attracting increasing interest for many applications, such as displays, sensors for biomedical, robotics and environmental control applications. As the structural support and optical signal transmission pathway and medium, flexible substrates are playing ever-increasingly important roles in advanced optoelectronic display devices. Currently, there are mainly three types of candidates for flexible substrates: ultrathin glass, metal foil, and plastic (polymer) films. As summarized in Table 5.8 [157], plastic substrates, especially transparent plastic substrates, possess good optical transmittance similar to that of thin glass, in addition, the good flexibility and toughness comparable to those of metal foils. Thus they are ideal for flexible optoelectronic devices, including flexible displays (TFT-LCDs or AMOLEDs, etc.), FPCBs, touch panels, electronic paper, and thin photovoltaic cells [158,159].

TABLE5.8 Comparison of Plastic, Thin Glass, and Metal-Foil Substrate

	Advantages	Disadvantages
Plastic substrate	Rugged	Poor dimension stability
		Low process temperature
	Bendable, rollable	High H_2O and O_2 permeation
	Transparent	Low chemical resistance
Thin glass	Conformable	Low mechanical stability
	Transparent	
	Low H_2O and O_2 permeation	Low process temperature

TABLE5.8 (Continued)

	Advantages	Disadvantages
Metal foil	Rugged conformable	Opaque
	Low H_2O and O_2 permeation	Rough surface
	High process temperature	Capacitive effect
	Good dimension stability	

The plastic film candidates for flexible display devices include polyethylene terephthalate (PET), polyethylene naphthalate (PEN), and PI, among which PIs have been considered one of the promising plastic substrates for flexible optoelectronic devices due to their good thermal stability, excellent mechanical and dielectric properties. Except flexible displays such as AMOLEDs, PI substrates have also been applied to fabricate microelectrodes [160], solar cells [161], and sensors [162], etc. In this section, we focus on PI substrates for fabricating flexible displays.

Of the critical components in personal mobile devices, the display is perhaps the most important because it facilitates interactions between human and machine. It is believed that the next generation of displays will be "flexible," because the use of flexible substrates, such as PIs, instead of conventional glass substrates can significantly reduce the thickness and weight of displays.

Generally speaking, PIs for fabricating flexible displays can be classified as coating-type substrate and film-type substrate according to different fabrication methods of flexible displays. For example, active-matrix OLED (AMOLED) displays on a flexible PI film can be fabricated by either a typical coating/debonding method, or a transfer method as shown in Figs. 5.13 and 5.14 [163]. The essential process steps used in the coating/debonding method are schematically illustrated in Fig. 5.13: (1) coating a PI solution directly onto the glass substrate, (2) fabricating a thin-film transistor (TFT) device on the PI substrate, and (3) the final debonding of the PI substrate from the glass carrier by either mechanical delamination [164] or laser delamination [165,166]. The coating-type PI substrate is basically a PAA solution. As the first step a temporary glass carrier substrate is spin-coated with PAA solution, which turns to a PI film after thermal imidization.

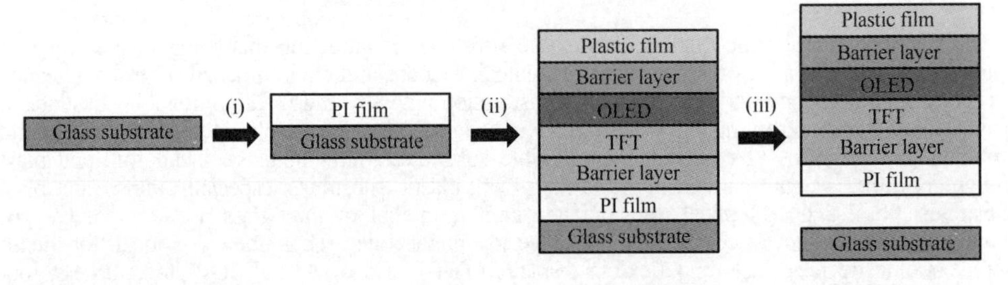

FIGURE 5.13 Schematic illustration of the coating/debonding method: (i) coating PI film, (ii) fabricating TFT device, (iii) debonding of glass substrate.

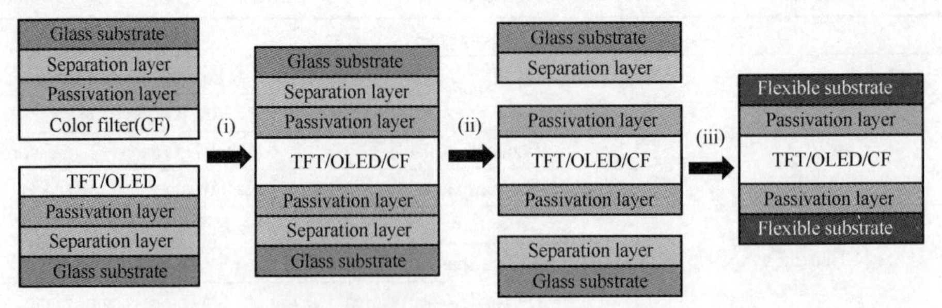

FIGURE 5.14 Schematic illustration of the transfer method.

In the transfer method, which favors the top emission OLED with a color filter [167-169], film-type PI substrates, i.e., PI films are used as flexible substrates. The process of transfer method is schematically illustrated in Fig. 5.14, in which the TFT/OLED substrate and the color filter substrate are prepared separately and then bonded together. After separating the glass substrates from the passivation layers by physical force, PI films are finally bonded to the AMOLED device.

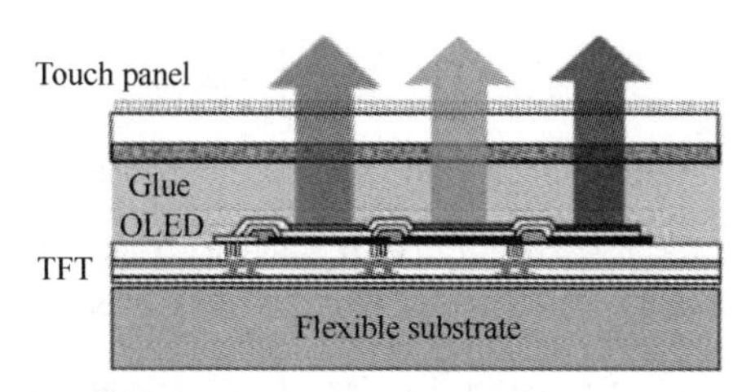

FIGURE 5.15 Schematic diagram for the structure of top-emission AMOLDS.

According to the structure of devices, AMOLEDs can be divided into two groups: top-emission AMOLEDs and bottom-emission AMOLEDs. Fig. 5.15 schematically illustrated the structure of the top-emission AMOLDS. From this viewpoint, all PI substrates can be applied for top-emission AMOLEDs, while only colorless PIs can be used for fabricating bottom-emission AMOLEDs. The colorless PI substrates for flexible displays have been recently reviewed [159].

The structure support, the flexible substrate, is one of the challenging materials that need to be focused on [170]. The characteristics and functionalities of flexible substrates have become the important factors that affect the quality of flexible devices. In order to achieve a practical application for PI film substrates in flexible displays, several issues have to be addressed.

5.4.1.1 High Thermal Stability

The thermal stability of the flexible substrate should have tolerance to the thin-film transistor (TFT) process, because fabrication of TFTs on flexible substrates is one of the most important procedures, for instance, for flexible display devices such as AMOLED processing. As shown in Table 5.9, the most popularly used producing technologies for TFT fabrications in AMOLE include: (1) amorphous silicon (a-Si) TFTs [171], (2) low-temperature polycrystalline silicon (LTPS) TFTs [157], (3) oxide TFTs [172], and (4) organic TFTs [173].

Among these TFTs, LTPS TFTs technique exhibits the highest field-effect mobility (>50 $cm^2 \cdot V \cdot s^{-1}$) and stable electrical performance [174]. In this procedure, an a-Si film is initially deposited on the substrate as an active layer, and then a high-temperature annealing step is performed to crystallize the a-Si to be polycrystalline silicon (Poly-Si). The temperature associated with solid-state crystallization of a-Si is about 300-500 °C [174]. Alternatively, the excimer laser crystallization process can also convert the a-Si precursor into poly-Si with a thermal budget compatible with plastic substrates such as PI films [175-177]. In contrast, a-Si TFTs process has been widely used to produce AMOLED devices, which shows uniform electrical characteristics over large areas, reasonable field-effect mobility, low-temperature process (<300 °C), and low cost compared to the other techniques. Hence, colorless and transparent polymer substrates with good thermal resistance above 300 °C are highly desired in advanced flexible display engineering.

PI films have tolerance to TFT process because their glass transition temperatures (T_g) can be higher than 300 °C. To achieve a good performance of the flexible displays, PIs with high T_g (>450 °C) for flexible substrates have been a research direction.

TABLE5.9 Key Features for a-Si TFTs, LTPS TFTs, OTFTs, and IGZO TFTs

	a-Si TFTs	LTPS TFTs	OTFTs	IC20 TFTs
Field-effect mobility ($cm^2 \cdot V \cdot s^{-1}$)	<1	50-100	0.1-1	10-30
Process temperature (°C)	<300	300-500	<300	<300
Device stability	Challenging	Good	Challenging	OK
Uniformity	Good	Challenging	OK	OK
Manufacturability	Excellent	Maturing	Developing	Developing
Cost	Low	Medium	To be determined	

5.4.1.2 High Thermal Dimensional Stability

An excellent thermal dimensional stability is also required for flexible substrate in order to apply it to the LTPS process [157]. The quality and reliability of the flexible displays largely depend on the dimensional stability of flexible substrates. The *CTE* is the key parameter dominating the dimensional stability. Conventional polymer substrates usually have a *CTE* value higher than 30×10^{-6} °C^{-1}; however, the inorganic components, such as SiN_x gas barrier layer have *CTE* value below 20×10^{-6} °C^{-1} [159]. The difference in the *CTE* values of the plastic substrate and Si film can cause thermal stress upon cooling from the deposition temperature to room temperature. This stress relaxation can result in convex bending of the substrate, and consequently lead to delamination, cracking, and other failures in the devices [157]. Therefore, comparable *CTE* value to the inorganic or metal components in display devices is required when developing PI substrates. For lowering the *CTE* values, PI composites containing inorganic nanoparticles, such as silica, titania, via sol-gel route have been proven to be an effective approach [178,179].

5.4.1.3 High Flexibility

As shown in Fig. 5.16, the flexible devices have progressed through the "unbreakable, curved, wearable, and foldable" development stages. Consequently, the development of flexible substrates is experiencing a roadmap from plane (2013) to bended (2015) then to rollable (2018) finally to foldable (2020) in the coming years [170]. The radius of curvature of the highly flexible substrates might reach below 3 mm in the year of 2020. Transparent PIs with high flexibility will be undoubtedly demanded in the future to fabricate such foldable display devices.

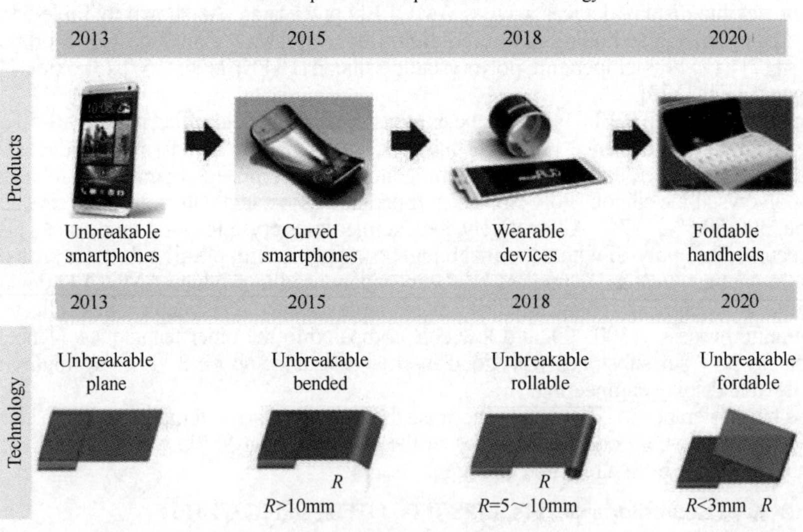

FIGURE 5.16 Roadmap of flexible AMOLED technology and products [170].

5.4.1.4 Low Water-vapor Transmission Rate and Oxygen Transmission Rate

When exposed to environmental moisture, most high-performance semiconductor organic compounds built on plastic substrate show degraded performance [11,174]. This requires the plastic substrate to have low water vapor transmission rate (WVTR) and oxygen transmission rate (OTR) features. For OLED, WVTR, and OTR of the substrates are limited to be below 10^{-6} g·m^{-2} per day and 10^{-4} cm^3·m^{-2} per day, respectively [27]. However, conventional polymer film substrate have WVTR and OTR values of around 10^{-1} to 40 g·m^{-2} per day and 10^{-2} to 10^{2} g·m^{-2} per day, respectively, dependent on the aggregation structures of their molecular chains (Table 5.10) [180]. It is therefore preferred to overcome the insufficient protection to the water and oxygen permeants when developing new polymer substrates. For PI substrates, their moisture barrier properties might be improved to some extent by the incorporation of specific additives, such as grapheme [181,182] and organoclay [158,183,184].

5.4.1.5 Surface Smoothness

Surface qualities such as roughness and cleanliness are also important issues for flexible substrates. Unlike glass and metal foil, polymer substrates cannot be smoothed by mechanical surface polishing. The roughness of the polymer surface can cause deterioration of the OLED performance, such as a short-circuited pixel or the formation of dark spots and device degradation [185,186]. Therefore, a smooth diffusion barrier film is required on the surface of flexible polymer substrates to provide a defect-free surface.

TABLE5.10 WVTR and OTR for Various Polymers and Coating

Polymer	WVTR[a] (g·m^{-2} per day) (@37.8 - 40 °C)	OTR[c] (cm^3(STP)·m^{-2} per day) (@20 - 23 °C)
Polyethylene (PE)	1.2-5.9	70-550
Polypropylene (PP)	1.5-5.9	93-300
Polystyrene (PS)	7.9-40	200-540
Poly(ethylene terephthalate) (PET)	3.9-17	1.8-7.7
Poly(ethersulfone) (PES)	14[b]	0.04[b]
Poly(ethylene naphthalate) (PEN)	7.3[b]	3.0[b]
Polyimide (PI)	0.4-21	0.04-17
OLED requirement (estimated)	1×10^{-6}	1×10^{-5} to 1×10^{-6}

[a] *Calculated assuming 100 μm polymer film.*
[b] *Temperature not given.*
[c] *Calculated assuming a 100 μm polymer film and 0.2 atm O_2 pressure gradient.*

5.4.2 Polyimides for Flexible Electronic Substrates

Excellent comprehensive properties of PIs make them good candidates for advanced optoelectronic devices. With the increasing demands of optoelectronic fabrication, PIs, especially colorless PIs have attracted more and more attention from both academia and industry. In the past decades, PI substrates have been developed along with the development of flexible displays, such as OLED. As can be seen from Table 5.11, the development of commercialized PI substrates is trending towards higher T_g and lower *CTE* to tolerate the TFT fabrication process. For example, ITRI (Industrial Technology Research Institute, Taiwan) have recently developed a unique flexible-universal-plane (FlexUP) solution for flexible display applications [195]. This new technique relies on two key innovations: flexible substrate and a debonding layer (DBL). As for the flexible substrate, ITRI developed a CTPI substrate, which contains a high content (>60%(w)) of inorganic silica particles in the PI matrix. The CTPI substrate exhibits good optical transmittance (90%), high T_g (>300 °C), low *CTE* (28×10^{-6} °C^{-1}), and good chemical resistance. In addition, the CTPI substrate with additional barrier treatment shows a WVTR value of less than 4×10^{-5} g·m^{-2} per day. Moreover, this barrier property suffered only to a minor drop, to 8×10^{-5} g·m^{-2} per day, after the flexible panel had been bent 1000 times at a radius of 5 cm. A 6 in. flexible color AMOLED display device was successfully fabricated using this substrate. By using this CTPI substrate, a flexible touch panel was successfully prepared.

TABLE 5.11 Polyimide Flexible Substrates in Industries

	Affiliation	Materials	T_g (°C)	*CTE* ($\times 10^{-6}$ °C^{-1})	Transmittance (%)	References
2005	General Electric	Colorless PI	240		90	[187]
2008	ITRI	Colorless PI	230	60	90	[170]
2009	ITRI	Colorless PI	350	40	90	[188]
2009	Samsung	Colorless PI	360	3.4		[172]
2010	ITRI	Colorless PI (MCL-PIH2)	>400	20		[178]
2011	Ube	Colored PI (U-varnish-S)	>400	12		[189]
2012	ITRI	Yellowish PI	450	7		[170]

Some recent examples of commercialized flexible AMOLED displays with a PI film are summarized in Table 5.12. In 2007, a 7 in. flexible VGA transmissive-type active matrix TFT-LCD display with a-Si TFT was successfully fabricated on CTPI film substrate by ITRI [195]. The CTPI film substrate has the features of high T_g (>350 °C) and high light transmittance (>90%), which ensure the successful fabrication of 200 °C a-Si:H TFT in the flexible device, as shown in Fig. 5.17. The flexible panel showed resolution of 640×RGB×480, pixel pitch of 75×225 mm, and brightness of 100 nit. This technique is fully a-Si TFT backplane compatible, making it very attractive for applications in high-performance flexible displays.

TABLE 5.12 Examples of Flexible AMOLED Displays With a Polyimide Film

Affiliation	Size (Diagonal)	Format	Resolution	TFT	RGB	Emitting Direction	Process	References
ITRI	7″	640×480		a-TFT	RGB	Top		[195]
Toshiba	11.7″	qHD	94 ppi	a-IGZO	White/CF	Bottom	Coating/debonding	[190]
Toshiba	10.2″	WUXGA (1920×1200)		a-IGZO	White/CF	Bottom	Coating/debonding	[191]
SEL	5.9″	720×1280	249 ppi	CAAC-IGZO	White/CF	Top	Transfer	[167]
LG	5.98″	720×1280	245 ppi	ELA-TFT	RGB	Top	Coating/debonding (by laser)	[166]
LG	18″	WXGA		IGZO	RGB	Top	Coating/debonding (by laser)	[192]
AUO	4.3″	qHD	257	ULTPS		Top	Debonding (mechanical)	[164]
Holst center	6 cm	QQVGA	85 dpi	IGZO solution	Monochrome	Top	Debonding	[193]
BOE	9.55″	640×432	81 ppi	a-IGZO	FMM	Top	Debonding	[194]

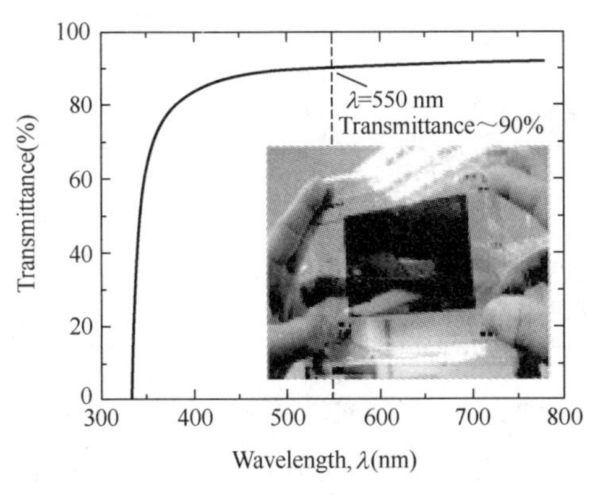

FIGURE 5.17 Colorless PI substrate and the color VGA flexible TFT-LCD.

Similarly, a-Si TFTs deposited on clear plastic substrates (from DuPont) at 250 °C-280 °C was reported [196]. The free-standing clear plastic substrate has a T_g value of higher than 315 °C and a *CTE* value of lower than 10×10^{-6} $°C^{-1}$. The maximum process temperature of 280 °C is close to the temperature used in industrial a-Si TFT production on glass substrates (300 °C-350 °C).

Toshiba Corp. Japan used the coat/debond method to successfully fabricate an 11.7 in. flexible AM-OLED having qHD format (960×540 dots) with 94 ppi on a PI film of 100 mm thickness [190]. The display has α-IGZO TFTs with two transistors and one capacitor (2Tr+1 °C), white emission with color filters (WOLED+CF), and bottom emitting device structure. Afterwards, Toshiba fabricated a flexible 10.2 in. WUXGA (1920×1200) bottom-emission AMOLED display device driven by amorphous indium gallium zinc oxide (IGZO) TFTs on a CTPI film substrate [191]. The threshold voltage shifts of these amorphous IGZO TFTs on the PI substrates under bias-temperature stress have been successfully reduced to less than 0.03 V, which is equivalent to those on glass substrates.

Semiconductor Energy Laboratory (SEL) developed a folding screen type OLED display driven by CAAC (crystalline oxide semiconductor)-IGZO TFT, using the transfer method [167]. The device structure of the OLED is top emission combined with a color filter substrate having RGBW (R, red; G, green; B, blue; W, white) pixels. The flexible OLED display has a pixel number of 720×1280 and a resolution of 249 ppi.

In 2013, LG Display commercialized flexible AM-OLED displays with PI substrates [166]. These curve-shaped displays are 5.98 in. diagonal, pixel number 720×1280, resolution 245 ppi, ELA-TFT on a PI backplane, radius of curvature 700 mm. Later on, LG Display developed the world's first large 18 in. flexible OLED display [192]. The OLED display has resolution 810×RGB×1200 (WXGA) of top emission type, an a-IGZO TFT on a yellowish PI substrate, and can be bendable up to bending radius of 30 mm. Pictures of the developed flexible 18 in. OLED displays are shown in Fig. 5.18.

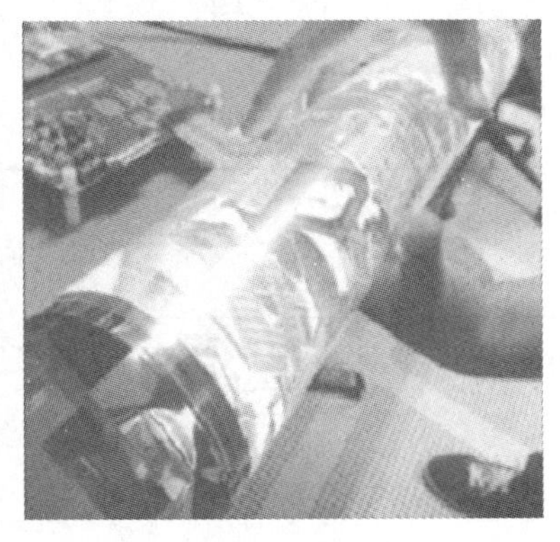

FIGURE 5.18 Pictures of the developed flexible 18-in. OLED display.

A 4.3 in. AMOLED panel was fabricated by AU Optronics Corporation (AUO) on the ultralow-temperature poly-crystalline silicon TFT (ULTPS TFT) backplane on PI substrate with a reliable ultra-high gas barrier layer [164]. This is a top-emission type OLED with a true resolution of 257 ppi. The total thickness of the display is merely 0.2 mm. In the same year, Holst Centre developed a top-emitting OLED display which was fabricated on a PI film deposited on a glass wafer by spin-coating [193]. The display has QQVGA format and a resolution of 85 dpi. BOE Technology from China developed a full-color flexible AMOLED display using a-IGZO TFT backplane fabricated on PI film, and top-emission OLED structure [194]. The specifications of the display are 9.55 in. diagonal, pixel number 640×432, resolution 81 ppi and color gamut greater than 100%.

Furthermore, low-temperature TFT technologies have also been developed. For example, IGZO TFTs have been reported to prove at the highest temperature of 160 °C by Fruehauf et al. of the University of Stuttgart (Germany). Such TFT technologies can be applied to various polymer films including PI films [197].

The structure design of PI substrates is more towards the requirement of optical transparent, high T_g and low *CTE*. However, as the trade secrets, the molecular structures of such PI materials have not been reported much in the literature. There are only a few examples of molecular design of PI flexible structures, as shown in Table 5.13.

TABLE5.13 Polyimides for Flexible Displays

	Molecular Structure	References
PI53	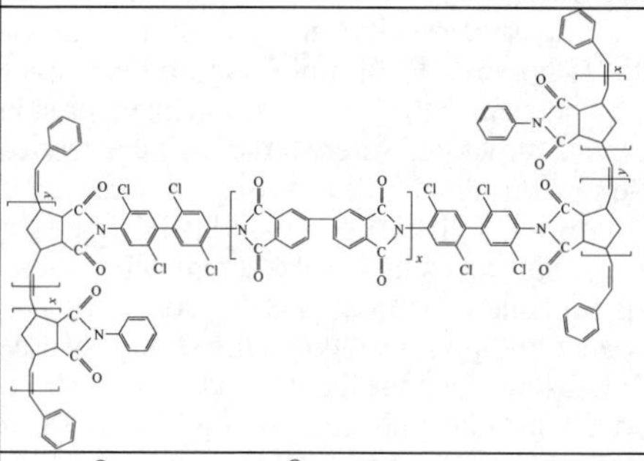	[198]
PI54	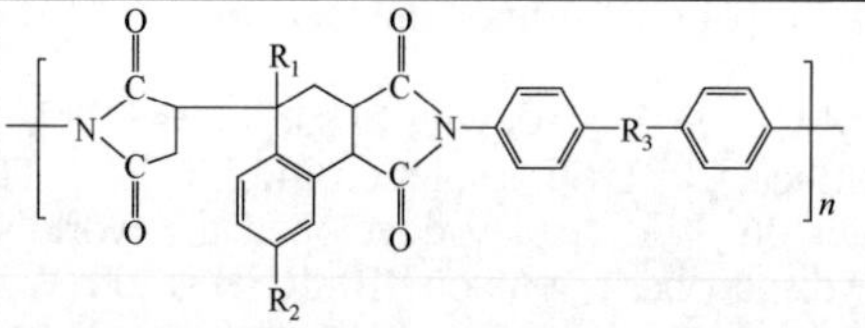 R_1 = H, Me; R_2 = H, C_mH_{2m+1}(m=1-16) R_3 = O, O-C_6H_4-O, O-C_6H_4-C$(CH_3)_2$-C_6H_4-O, O-C_6H_4-C$(CF_3)_2$-C_6H_4-O	[199]
PI55	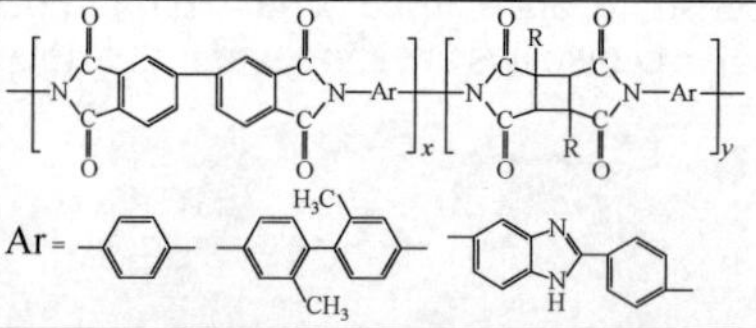 Ar =	[200]
PI56	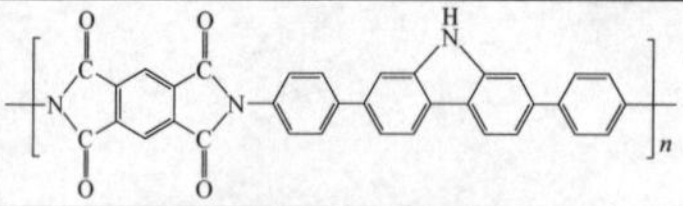	[201]
PI57	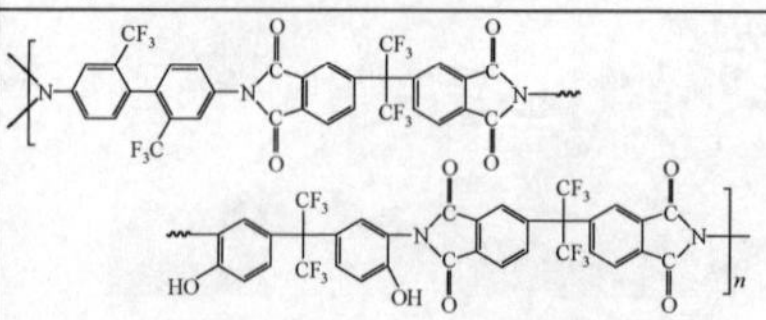 +graphene	[202]

TABLE5.13 (Continued)

	Molecular Structure	References
PI58	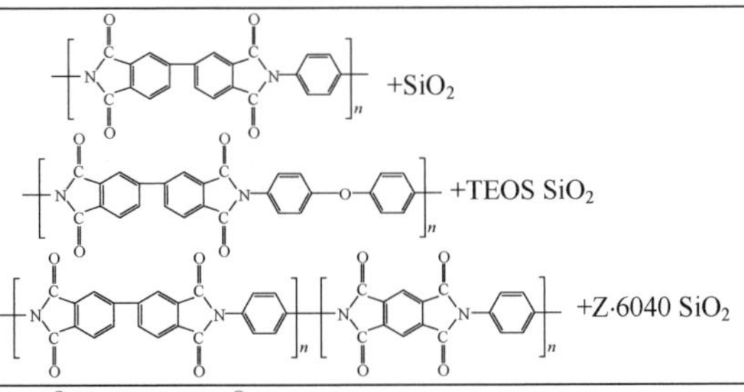 $+SiO_2$ +TEOS SiO_2 +Z-6040 SiO_2	[203]
PI59	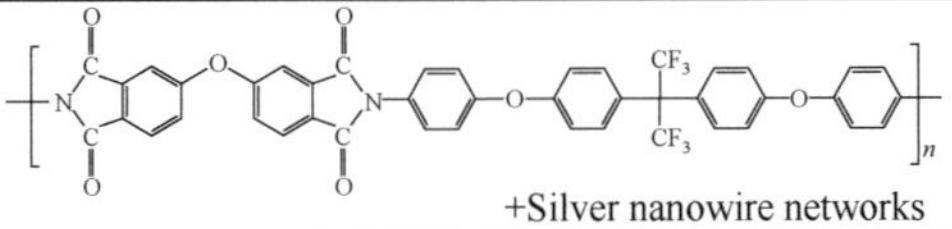 +Silver nanowire networks	[204]

Choi et al. [198] reported novel polynorbornene (PNB)-PI copolymers (**PI53**) based on poly(*N*-phenyl-*exo*-norbornene-5,6-dicarboximide) (PPhNI) and chlorinated PI (BPDA/TCDB). These copolymer films exhibited good optical transparency with a transmittance of around 70% at 400 nm and a good thermal stability with a T_g at 276 °C-300 °C, as well as good chemical resistance. Based on such flexible films, an OLED was fabricated with a structure of PNIC08/ITO (anode)/hole-transporting layer (HTL)/emitting and electron-transporting layer (EM&EFL)/aluminum (cathode), and exhibited a performance that was comparable to corresponding ITO-grown substrates.

PIs (**PI54**) containing an alicyclic component, an aromatic component, and a flexible ether chain fraction, were patented by Yang et al. [199]. These PIs were reported to have enhanced thermostability, optical transparence and organic solvent dissolvability, and can be applied to fabricate TFT-LCD photoelectronic devices or as coating of fiber communication and solar cells. This was followed by another patent from Yang et al. [200]. PIs (**PI55**) were prepared using aromatic diamines and a combination of aromatic dianhydride (sBPDA) and alicyclic dianhydride. The thermal stability and optic transparency of **PI52** can be adjusted by adjusting the ratio of sBPDA and alicyclic dianhydride, which are reported as promising candidates for flexible substrates to fabricate flexible display devices such as flexible OLED and flexible TFT-LCD.

Liu et al. [201] developed an intrinsic high-barrier and thermally stable PI (**PI56**) by polymerization of PMDA and a novel diamine (2,7-CPDA), which consists of rigid planar carbazole moiety. **PI56** materials were found to exhibit excellent barrier properties with low oxygen transmission rates and water vapor transmission rates at 0.2 $cm^3 \cdot m^{-2}$ per day and 0.1 $g \cdot m^{-2}$ per day, respectively. **PI56** also presents outstanding thermal stability and mechanical properties with a T_g of 437 °C, 5%(*w*) loss temperature of 556 °C under N2, *CTE* of 2.89×10^{-6} K^{-1} and tensile strength of 143.8 MPa. Owing to such excellent properties, **PI56** shows attractive potential applications in flexible electronics and high-grade packaging areas, especially for using as flexible substrates for top-emitting OLED displays.

On the other hand, nanocomposites have been introduced to PI films to add benefits of ease of fabrication from solution, chemical stability, and so on. Such PI composite films are also reported to be flexible substrates for the fabrication of flexible OLEDs. **PI57** were synthesized by Kim et al. [202] using selected monomers containing multiple CF_3 groups. The corresponding PI-graphene composite films (PGCFs) were prepared by film coating of mixture of polyamic acid solution with reduced grapheme oxide. The mechanical properties of the GPI films were found to increase up to 0.7%(*w*) of graphene. The flexible OLEDs were fabricated using these PGCFs on which the ITO/Ag/ITO thin films were deposited, and exhibited better performance than the one based on commercial PI films.

PI-silica composite films (PSCFs, **PI58**) were patented by Liu et al. [203] as flexible substrates for manufacturing flexible electronic devices. These PSCFs were fabricated by mixing polyamic acid solution with silica sol gel, tetraethoxy silane (TEOS) and *r*-glycidoxy propyl trimethoxy silane (Z-6040), respectively.

Spechler et al. [204] developed a transparent, smooth, thermally robust, and conductive PI (**PI59**) by combing a colorless PI with the embedded silver nanowire (AgNW) networks. It was found that the addition of a titania coating on the nanowires increases their thermal stability and allows for **PI59** to be thermally imidized at 360 °C. This PI material was used as a substrate for a thermally deposited, flexible, organic light-emitting diode, which showed improved device performance compared to a control device made on ITO-coated glass.

Due to the attractive features of flexible OLEDs and the current technological progress induced by the efforts of scientists and engineers, we believe that the next generation of displays is going to be "flexible." It is worth noting that PI material is beginning to be applied in flexible fields due to its good mechanical strength, higher T_g, electrochemical stability and flexibility in polymer materials. As a promising candidate to be flexible substrates, PI will be more and more feasible to apply for fabricating OLED displays in future.

5.5 Polyimide Materials for Electronic Memories

5.5.1 Introduction

The electrical switching phenomena in polymers and the corresponding polymer electronic memories is an emerging area in organic electronics over the past two decades. PI-based materials have received much attention due to their simple preparation method, chemical inertness, mechanical and thermal stability, and high biocompatibility. As shown in Fig. 5.19, polymer-based memory devices can be classified into three types. PI materials have been mainly applied in resistor-type [12,205-207] and transistor-type memories [208]. As detailed reviews have been published from different aspects of PI-based memory devices, especially for the resistor-type memories, herein we give a brief introduction on this topic.

5.5.1.1 Resistor-type Memories

The Resistor-type memory devices are those incorporating switchable resistive materials, which need two distinct (bistable) electronic states, i.e., "0" and "1," or "OFF" and "ON," in terms of storing data. Resistive-type memory devices based on PI materials store data using the conductivity response of the active layer to the applied voltage, in which the low- and high-conductivity states are assigned to "0" and "1," respectively. This is completely different to current memory devices using electric circuits, which stores data by means of discharge (0) and charge (1). From this point of view, polymeric film can be a memory element, which acts for the combination of a minute and a complicated electric circuit in current memory device. The schematic structure of polymer-based resistive memory device is shown in Fig. 5.20 [205], including bottom electrode such as ITO on substrate, PI active layer, and top electrode such as AL.

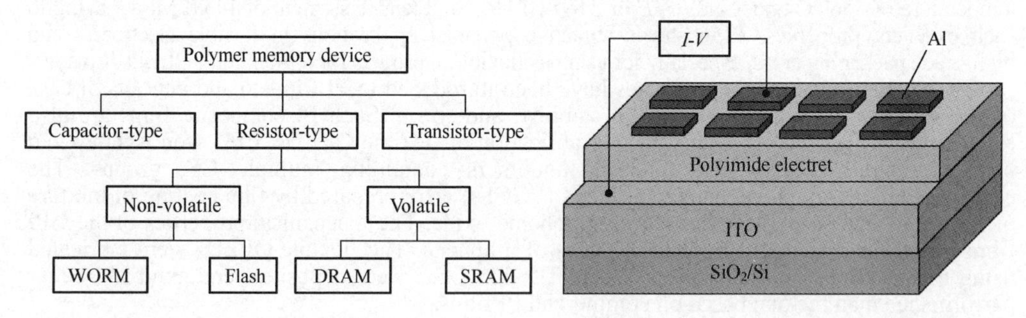

FIGURE 5.19 Different types of polymer memory device.

FIGURE 5.20 Polyimide-based resistive memory devices with the configuration of ITO/PI/Al.

Resistive-type memories can be divided into two primary classifications according to the retention time after removing power: volatile memories (dynamic random access memory (DRAM), static random access memory (SRAM)) and nonvolatile memories (rewritable type (flash), write once read many (WORM) type). A volatile memory could erase the stored data as soon as the system is powered off, which cannot sustain the two distinct electronic states without an external electronic power supply. In contrast, the nonvolatile memory can retain the stored data even after removing the electrical power supply, which can sustain the two distinct states without the power supply. In these memory devices, PIs act as switchable resistive materials, which is known as active layer in the memory.

5.5.1.1.1 Device Performance

Device behavior for each type of memory is different, as shown in Fig. 5.21. DRAM and SRAM differ in the retention time of the ON state after removal of the applied external power, while flash (rewritable) memory and WORM differ in terms of whether a suitable voltage can switch the ON state to the OFF state.

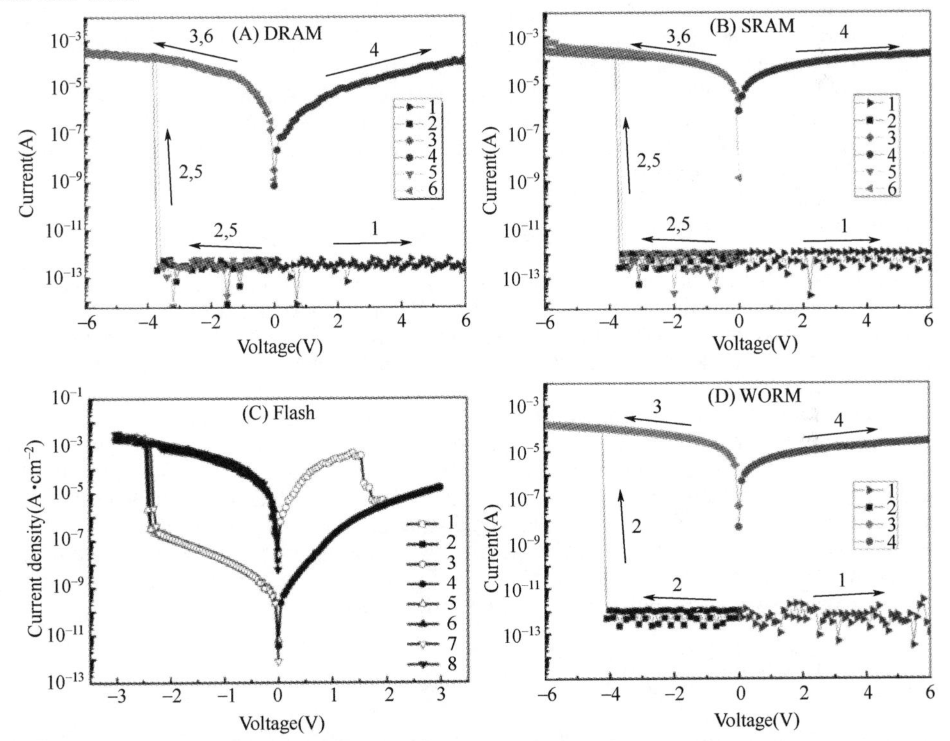

FIGURE 5.21 Current-voltage (*I*-V) characteristics of the ITO/p、olymer/Al memory device demonstrating (A) DRAM behavior (the fifth sweep was conducted approximately 15 s after turning off the power), (B) SRAM behavior (the fifth sweep was conducted approximately 3 min after turning off the power), (C) flash behavior, and (D) WORM behavior.*Reproduced from Refs, [209,210] with permission from the Royal Society of Chemistry.*

As shown in Fig. 5.21A for DRAM behavior, only a short time (less than 1 min) of the ON state can be retained after removing the applied voltage [209]. On the contrary, the device can stay in the ON state for a longer time after turning off the power for SRAM behavior (Fig. 5.21B) [209]. For nonvolatile memory devices, the ON state of flash behavior can be switched back to the OFF state by applying a suitable voltage, which is known as an erasing process (Fig. 5.21C) [210]. It is contrast to WORM behavior as shown in Fig. 5.21D, which is capable of maintaining the ON state (holding data) permanently even after the application of a reverse voltage, and the data can be read repeatedly [209].

5.5.1.1.2 Mechanism

Mechanisms underlying the switching phenomena of memory devices have been conducted by many studies. The most widely proposed ones in PI-based resistive memory devices are charge transfer, space charge trapping, and filamentary conduction [206,207].

Charge Transfer Electronic charge can be induced to partially transfer from the donor (D) to the acceptor (A) moiety in an electron D-A system by applying a suitable voltage. This phenomenon is called charge transfer (CT), which can result in a sharp increase in conductivity [211]. CT is the primary mechanism responsible for the memory characteristics. In aromatic PIs, charge transfer occurs from the electron-donating diamine segments to the electron-accepting dianhydride segments in the polymer chains. By tuning the electron-donating or electron-accepting capabilities of PIs, memory behaviors can be adequately adjusted. There several factors that have been pointed out which are beneficial to maintaining the conductive CT state:

1. Conformational change [212]. Ueda et al. discovered that the difference in linkage conformation plays an important role in the DRAM and SRAM properties of PI-based memories, in which PI with a more twisted conformation would prefer SRAM characteristics.

2. LUMO [213] and HOMO [209] energy levels. A lower LUMO energy level or a higher HOMO energy level can provide a more stable CT state, and consequently result in an increase in retention times.

3. Dipole moment [209]. A higher dipole moment in the PI structure can stabilize the CT state and lead to nonvolatile memory properties.

4. High conjugation [214]. PIs with highly conjugated or high-electron-affinity units can give nonvolatile characteristics to the corresponding memory devices.

Space Charge Traps The electrical switching behaviors of some polymer materials have been reported to be associated with space charges and traps. When the interface between the electrode and polymer is ohmic and the polymer is trap-free, a space charge can be built up by the accumulation of the carries near the electrode. Traps may be present in the bulk material or at interfaces in which they will reduce carrier mobility.

Filament Conduction It is termed as "filamentary" conduction for the phenomenon when the ON state current is highly localized to a small area of the memory device. In polymer-resistive memory devices, two types of filament conduction have been reported: (1) the carbon-rich filaments, which are formed by the local degradation of polymer films; and (2) the metallic filaments, which result from the migration of electrodes through polymer films. The filamentary conduction mechanism has often been proposed in many polymer memory devices; it is, however, suggested to have arisen from device physical damage in resistive random access memories. To overcome the filament effect of resistor-type memories, transistor-type memories were subsequently developed.

5.5.1.2 Transistor-type Memories

Transistor-type memory devices based on PI materials consist of conventional transistors, and an additional PI dielectric layer (referred to as PI electret) which is placed between a semiconductor layer and a gate contact. Generally, these devices are fabricated with the configuration of n+Si/SiO_2/PI/semiconductor/Au, as shown in Fig. 5.22.

Electret is known as a dielectric material exhibiting a quasi-permanent electric field which is either caused by trapping of electrostatic changes (homocharge electret) or by macroscopically oriented dipoles (heterocharge electrets) (Fig. 5.23).

The charge storage capability of a polymer electret material is to a large extent determined by the chemical structure of the polymer. In addition, parameters such as end groups, catalyst residues, concentration, the type of processing aids and additives, processing conditions, and thermal history also have a great impact on the electret behavior of the polymer materials. Some early works on PI and thermoplastic polyetherimide (PEI) have been reviewed by Erhard et al. [215].

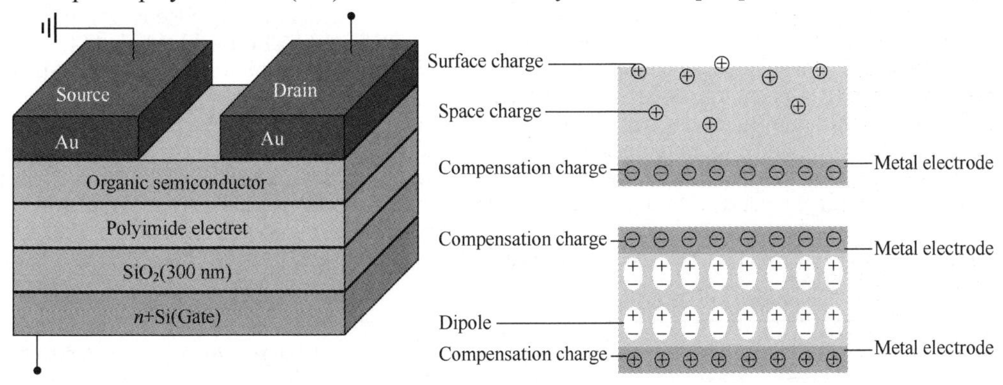

FIGURE 5.22 Polyimide memory devices with the configuration of n+Si/SiO_2/PI/Semiconductor/Au.

FIGURE 5.23 Schematic representation of a cross-section of a real-charge electret film with one electrode (top) and a dipolar electret film with two electrodes (bottom).

5.5.2 Polyimides for Resistive-type Memory Devices

Functional PIs containing both electron donor and acceptor moieties within a repeating unit contribute to electronic transition between the ground and excited states, which could be manipulated by the induced CT from donor (D) to acceptor (A) under applied electric fields. The memory properties can be tuned through the molecular design of the PI materials. As shown in Fig. 5.24, the electron donor or acceptor for functional donor-acceptor type PIs, the use of linkage groups in the polymer chain, and conformation changes are intrinsic factors to influence the memory properties. In addition, the change of film thickness of the active PI layer within the devices is the external factor.

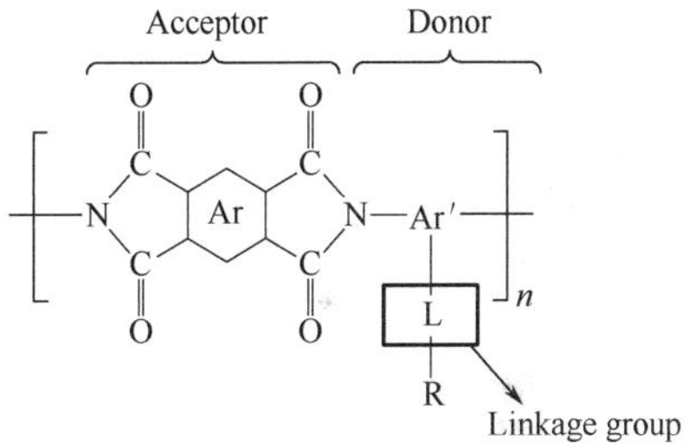

FIGURE 5.24 The electron donor and acceptor for functional donor–acceptor type polyimides.

5.5.2.1 Volatile Memory Devices

The volatile memory devices reveal DRAM and SRAM behavior depending on the retention time of the ON state after removal of the external voltage. For the DRAM effect, the retention time is quite short, normally less than 1 min, while for the SRAM effect, a longer retention time is observed. The general molecular structure of PIs for volatile memory devices is shown in Fig. 5.25, in which the electron acceptor segment is based on 6FDA. By varying the electron-donating diamine segments, the DRAM properties of memory devices are varied. PI materials for fabricating DRAM-type devices are summarized in Table 5.14.

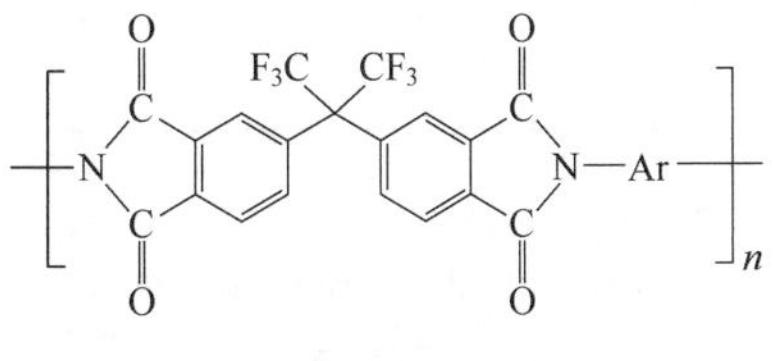

FIGURE 5.25 The molecular structure of polyimides for volatile memory devices.

The majority PIs for DRAM memory are aromatic PIs containing a triphenylamine (TPA) or TPA-substituted diamine moiety. PIs containing a TPA moiety that is directly bonded to (**PI60** [209]) or incorporated via a mono-mediated phenoxy linkage into the backbone (**PI61**

[212]) exhibit DRAM properties. PIs containing TPA-substituted diphenylpyridine moieties (**PI62** [216]) and PIs of TPA derivatives bearing an electron-accepting bis(trifluoromethyl)phenyl group (**PI63** [219]), a pendant anthraquinone attached to backbone (**PI64** [218]) were also found to exhibit DRAM characteristics. **PI65** was synthesized from 6FDA and *N,N'*-bis(4-aminophenyl)-*N,N'*-di(4-methoxy-phenyl)1,4-phenylenedi- amine ($(OMe)_2$TPPA), and revealed DRAM behavior [213]. For polymer **PI66**, the film thickness was reported to influence the memory behavior. It reveals DRAM memory properties under a film thickness of 100 nm, because only local filaments, which are responsible for the DRAM memory performance, were formed in such thick films [219]. On the other hand, various kinds of functional PIs containing the electron-donating structures apart from the TPA derivatives were also found to exhibit DRAM characteristics. These donating structures include carbazole (**PI67**) [220], 2,8- and 3,7-phenylenesulfanyl-substituted dibenzothiophenes (**PI68** and **PI69**) [221].

TABLE 5.14 Polyimides for DRAM

	Structure of Ar	References		Structure of Ar	References
PI60	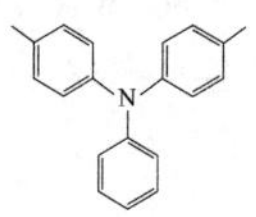	[209]	**PI65**	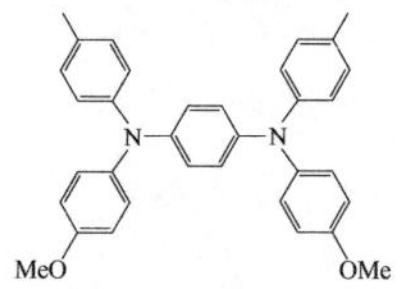 MeO OMe	[213]
PI61	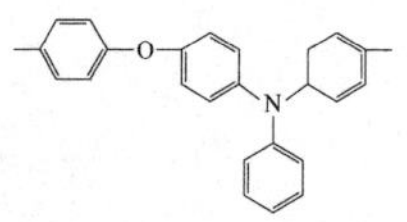	[212]	**PI66**	Ar: 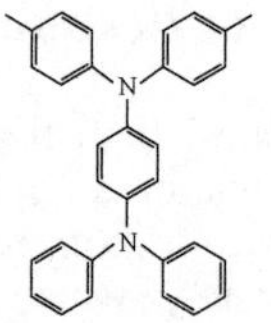	[219]
PI62	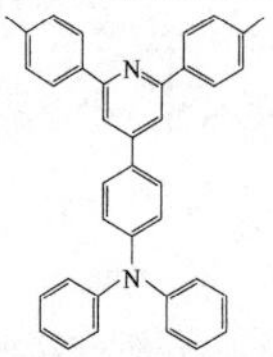	[216]	**PI67**	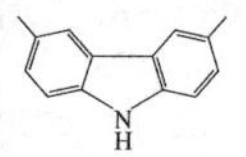	[220]
PI63	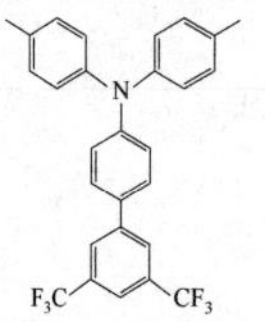 F_3C CF_3	[217]	**PI68**	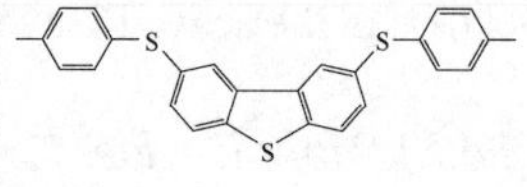	[221]
PI64	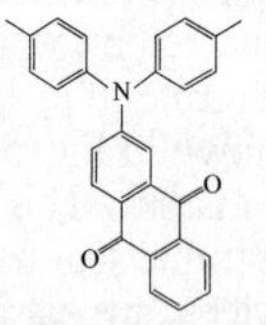	[218]	**PI69**	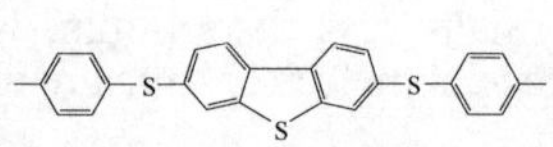	[221]

Similar to those for DRAM memory, the basic structures of PIs for SRAM type memory consist of 6 FDA and different diamines (as seen in Fig. 5.25), in which the TPA derivative are predominant (see Table 5.15). The TPA-based aromatic PI, **PI70**, has a similar structure to **PI61** but has dual-mediated phenoxy linkages in the repeating unit and shows SRAM properties [212]. TPA-based PI, **PI71**, containing anthraquinone which was incorporated via ether linkages into polymer backbone (analog to **PI64**) was also found to exhibit SRAM characteristics [218]. In polymer **PI70** and **PI71**, the linkage effect was demonstrated. It was suggested by Chen et al. [209] that the retention time of the memory device shows a systematic increase with the intensity of electron donation increasing. Therefore, the functional PI **PI72** with starburst triarylamine units was found to show SRAM properties [209], while its corresponding 3Ph-PI (**PI60**) and 5Ph-PI (**PI65**) revealed DRAM behavior. Volatile SRAM memory properties were also observed in the PI system containing a carbazole-tethered TPA unit, **PI73**, in which carbazole-tethered TPA has the strong electron-donating ability. In addition, the non-TPA-type diamines, such as triphenylethylene (**PI74** [223,224]) and oxadiazole (**PI75** [225]) exhibited SRAM properties as well.

5.5.2.2 Nonvolatile Memory Devices

The nonvolatile memory devices can be divided into two classes depending on whether a suitable voltage can switch the ON state to the OFF state or not. They are flash-type memory and WORM memory; the latter is capable of maintaining the ON state permanently. PIs showing nonvolatile flash or WORM properties have mostly been developed from the viewpoint of field-induced CT effects. The structures of such PIs show more diversities compared to those for volatile memory, i.e., the electron-accepting moiety can be different dianhydrides other than 6FDA, especially in the case of PIs for WORM memory.

TABLE5.15 Polyimides for SRAM

	Structure of Ar	References
PI70	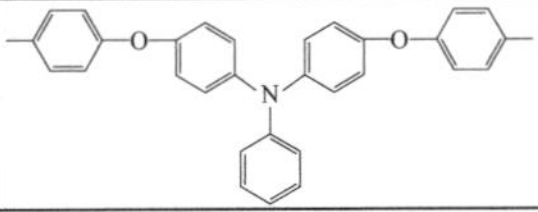	[212]
PI71	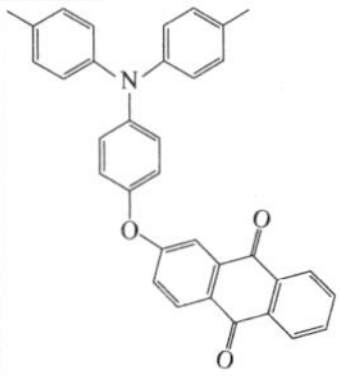	[218]
PI72	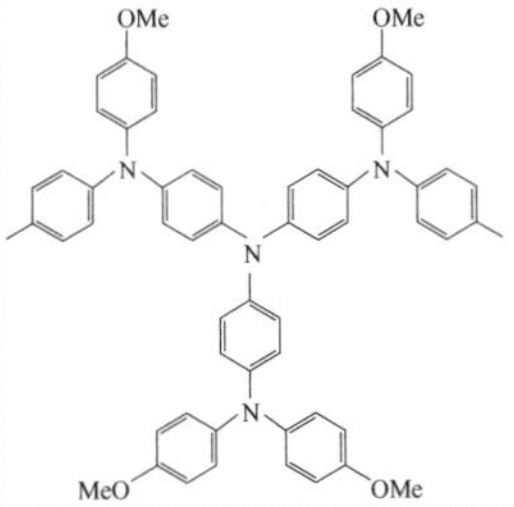	[209]
PI73	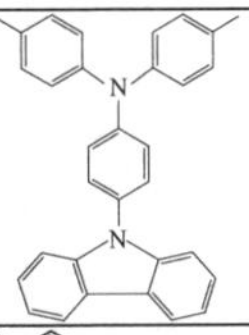	[222]
PI74	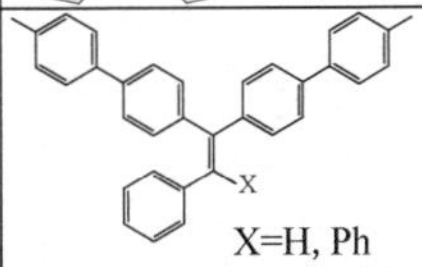 X=H, Ph	[223,224]
PI75	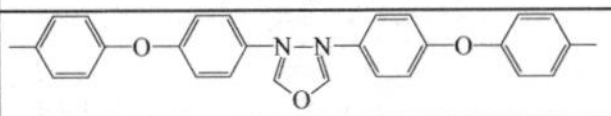	[225]

There are generally three types of structures for PIs exhibiting flash properties: (I) 6FDA-based aromatic PIs with different electron-donating diamines, mainly TPA-based diamines (Table 5.16); (II) PIs containing the 2,2′-position aryl-substituted tetracarboxylic dianhydride and diamine (Table 5.17), and (III) other PIs apart type I and II.

TABLE5.16 Polyimides for Flash Memory (Type I)

	Molecular Structure	References
	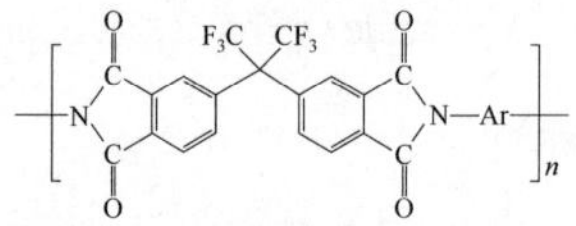 (general structure: 6FDA-based polyimide, –N–Ar–, n)	
PI76	Ar: 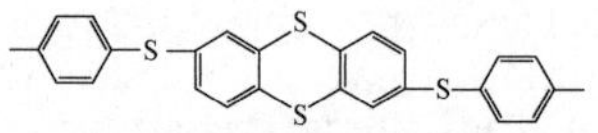	[226]
PI77	Ar: 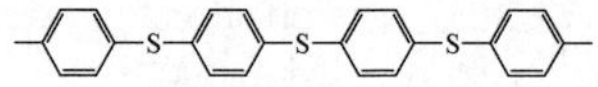	[226]
PI78	Ar: 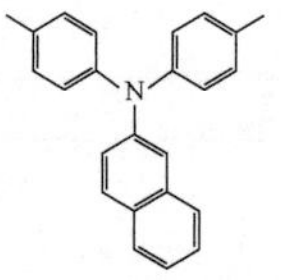	[227]
PI79	Ar: 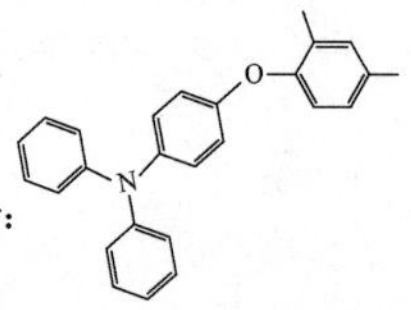	[228]
PI80	Ar: 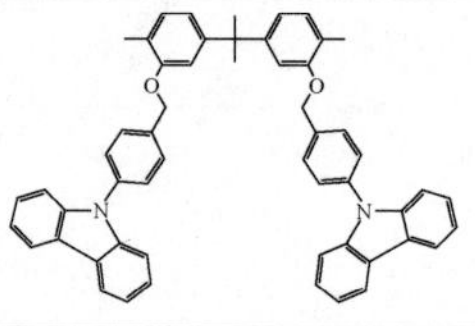	[229]
	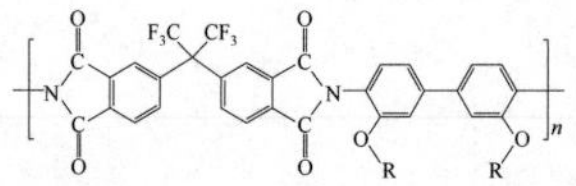 (general structure: 6FDA-based polyimide with O–R substituted biphenyl, n)	
PI81	R: 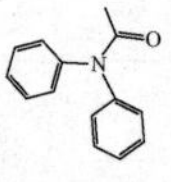	[230]
PI82	R: 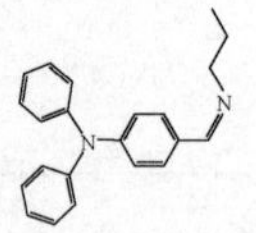	[231]
PI83	R: 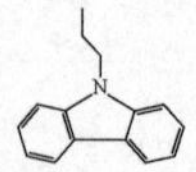	[232]
PI84	R: 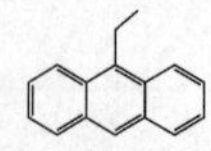	[233]

For type I 6FDA-based PIs, the hole-transporting and electron-donating moieties, such as sulfur-containing, TPA, and phenyl groups, could be introduced into the main chain or the side chain of PI derivatives. Electron-rich sulfur-containing moieties were incorporated into the PI main chain by Ueda and Chen [226]. These two PIs **PI76** and **PI77**, with 2,7-bis(phenylenesulfanyl)thianthrene and 4,4′-thiobis(*p*-phenylenesulfanyl) moieties, respectively, showed flash properties. The diphenylnaphthylamine, in which the naphthyl group is β-position substituted, was introduced to the main chain of **PI78** [227], and its corresponding memory devices revealed programmable flash-type performance.

By changing the hole-transporting and electron-donating moieties on the side chains of PIs have also been reported for application in excellent rewritable memory devices. PI derivatives contain TPA (**PI79** [228]), carbazole (**P80** [229] and **PI83** [232]), diphenycarbamyloxyl (**P81** [230]), diphenylamino- benzylidenylimine (**P82** [231]), and anthracene (**PI84** [233]) on the side chains, respectively. Their corresponding memory devices exhibit flash characteristics, in which a filamentary mechanism was demonstrated for polymers **PI80** and **PI82**.

The type II structure of PIs shown in Table 5.17 was designed by Shen and coworkers to induce a high ring torsion in the PI backbone using the 2,2′-position aryl-substituted tetracarboxylic dianhydride and diaime [234-238]. The resulting PIs have the same main chain but different groups on the 2,2′-position of the biphenyl moieties in electron-donating dianhydride segments and/or electron-accepting diamine segments. These PIs have comparable solubility to that of the PI using 6FDA unit and the variety of design had been expanded. The fabricated memory devices of Al/PI/ITO using these PIs were determined to present flash-type memory behavior.

Apart from type I and II PIs, **PI91** and **PI92** (Fig. 5.26) with different chain structures were also reported to exhibit nonvolatile flash characteristics. Ferrocene moiety was grafted in the repeat units of **PI91**, and phthalimide acted as an electron acceptor [239]. **PI92** was recently reported by Tsai et al. It was found that the switching properties of **PI92** are related to the chain lengths of **PI92** and the thickness of the PI layer [240].

TABLE5.17 Polyimides for Flash Memory (Type II)

	Molecular Structure	References
	(main chain structure with R_1, R_1, R_2, R_2 substituents)	
PI85	R_1: phenyl R_2: –C$_6$H$_4$–N(Ph)$_2$	[234]
PI86	R_1: phenyl R_2: –C$_6$H$_4$–N (carbazolyl)	[234]
PI87	R_1: –C$_6$H$_4$–C$_6$H$_2$F$_3$ R_2: –C$_6$H$_4$–N (carbazolyl)	[235]
PI88	R_1: phenyl R_2: –C$_6$H$_4$–C$_6$H$_2$F$_3$	[236]
PI89	R_1=H R_2: –C$_6$H$_4$–C$_6$H$_2$F$_3$	[237]
PI90	R_1: –C$_6$H$_4$–C$_6$H$_2$F$_3$ R_2: phenyl; –C$_6$H$_4$–naphthyl; –C$_6$H$_4$–C$_6$H$_5$	[238]

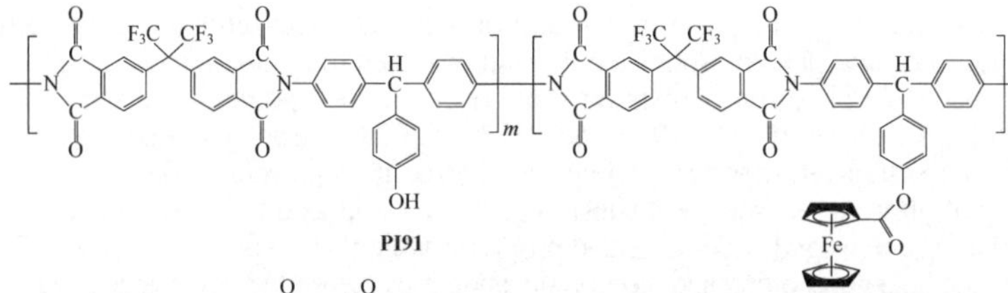

PI91

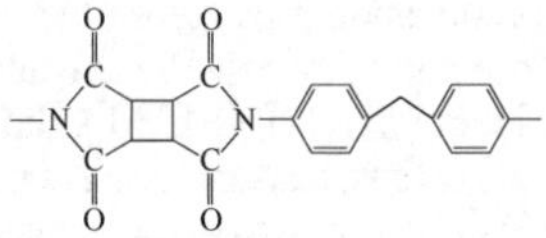

PI92

FIGURE 5.26 The molecular structures of polyimides **PI91** and **PI92.**

PIs for WORM-type devices are summarized in Table 5.18, and can be divided into four types according to their different structures: (I) 6FDA-based aromatic PIs with different electron-donating TPA-based diamines (Table 5.18); (II) DSDA-based (DSDA=3,3′,4,4′-diphenylsulfonyltetracarboxyimide) aromatic PIs with different electron-donating TPA-based diamines (Table 5.19); (III) PIs with same electron-donating structure but different electron-accepting groups (Table 5.20); (IV) other PIs apart from type I, II, and III (Table 5.21).

In the chemical structures of type I PIs (Table 5.18), 6FDA serves as the electron-accepting species and TPA moieties function as electron-donating groups. The only difference presented in these PIs is the different substituents on TPA moiety. **PI93** had a pendant hydroxyl group on the TPA [241], **PI94** contained an electron-donating bithiophene group [217] and **PI95** based on an α-substituted diphenylnaphthylamine [227]. All these PIs were reported to reveal WORM memory characteristics. TPA donor in **PI96** was incorporated into the polymer main chain through the oxadizazole linkage [242]. Memory device based on **PI96** exhibited excellent WORM memory behavior due to the linkage effect. **PI97** containing TPA-substituted triazole moieties [243] and the hyperbranched PI **PI98** containing 6FDA and TPA moiety [244] were also reported to exhibit the WORM memory properties. In addition, the film thickness has an effect on the memory behavior for **PI66** as mentioned above. It exhibits excellent WORM memory properties for films with thicknesses in the range of 34 nm-74 nm (Table 5.18), while showing DRAM behavior with thicknesses below 100 nm (Table 5.14) [219].

TABLE5.18 Polyimides for WORM (Type I)

	Molecular Structure	References
	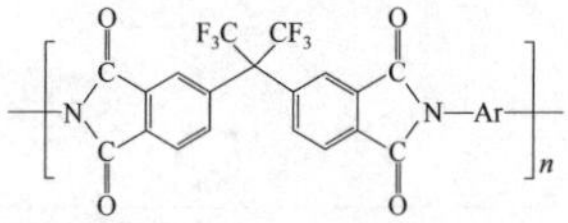	
PI93	Ar: 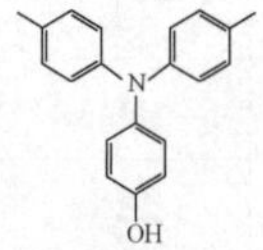	[241]
PI94	Ar: 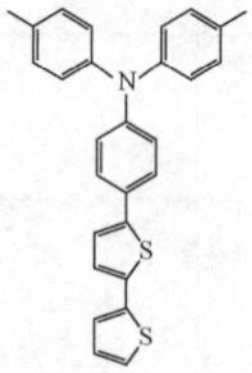	[217]

TABLE5.18 (Continued)

	Molecular Structure	References
PI95	Ar: 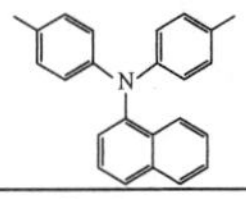 N	[227]
PI96	Ar: 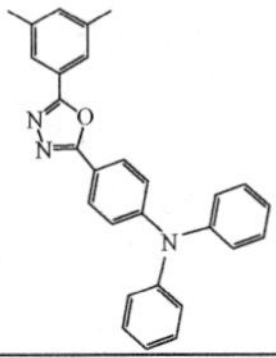 N N O N	[242]
PI97	Ar: 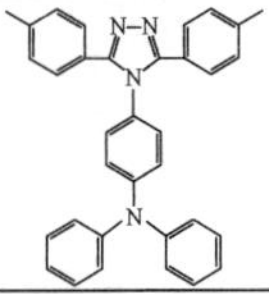 N–N N N	[243]
PI66	Ar: 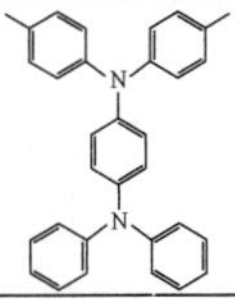 N N Ar: With film thickness of 34-74 nm	[219]
PI98	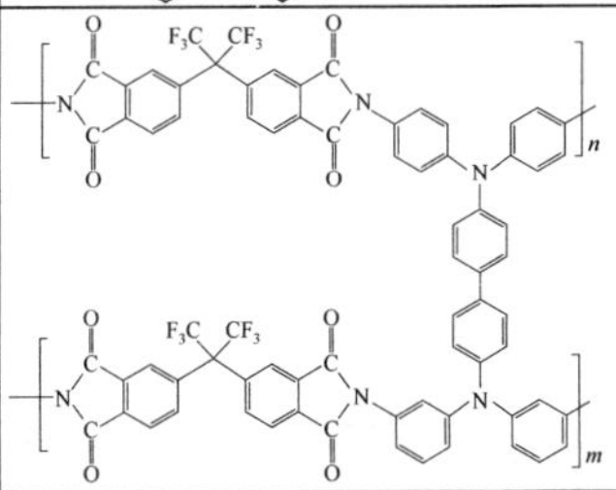 O F_3C CF_3 O C C N C C O O N n O F_3C CF_3 O C C N C C N N O O m	[244]

TABLE5.19 Polyimides for WORM (Type II)

	Molecular Structure	References
	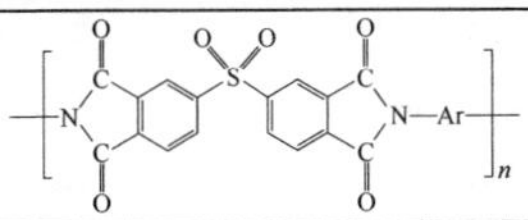 O O O O C S C N C C N–Ar O O n	
PI99	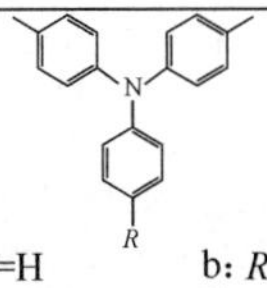 N R Ar: a: R=H b: R=CN c: R=OMe d: R=N(Me)$_2$	[245]
PI100	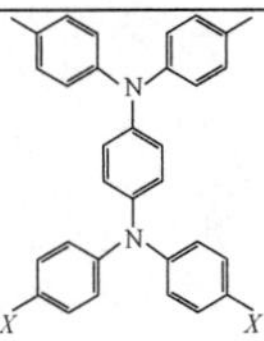 N N X X Ar: X=H, OMe	[246]

Type II PIs have similar main chain structures to Type I PIs but with DSDA as the electron-accepting groups. DSDA-based PIs **PI99** bearing TPA derivatives [245] and **PI100** bearing conjugated bis(triphenylaime) (2TPA) derivations [246] demonstrated unipolar WROM memory behavior. Aromatic PIs, **PI101** were synthesized from tetracarboxylic dianhydride monomers and 2,2-bis[4′-(3″,4″,5″-trifluorophenyl)phenyl]-4,4′-biphenyl diamine (type III structure). Each repeat unit of **PI101** contained different aryl pendants on two imide rings. The memory devices of Al/**PI101**/ITO exhibited WORM memory capability with different threshold voltages. On the other hand, PIs prepared from 2,2-bis[4′-(3″,4″,5″-trifluorophenyl)phenyl]-4,4′-biphenyl tetracarboxylic dianhydride with different electron-donating groups such as **PI102-PI104** were also found to exhibit excellent WORM memory performance. These electron-donating groups include TPA moieties (**PI102**[235]), 3′,′5′-trifluorobiphenyl and phenyl moieties (**PI103**[236]), and dimethoxyphenyl groups (**PI104** [238]).

TABLE5.20 Polyimides for WORM (Type III)

	Molecular Structure	References
	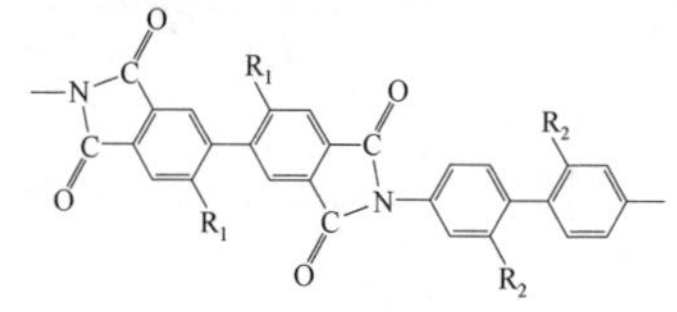 (general structure with R_1, R_2 substituents)	
PI101	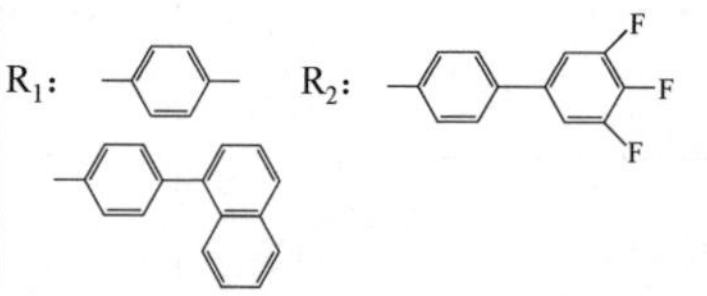 R_1: R_2: F, F, F	[237]
PI102	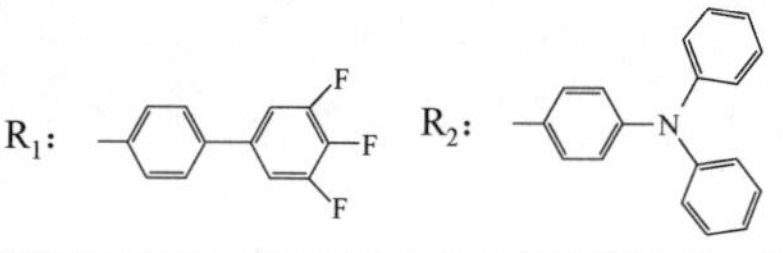 R_1: F, F, F R_2: N	[235]
PI103	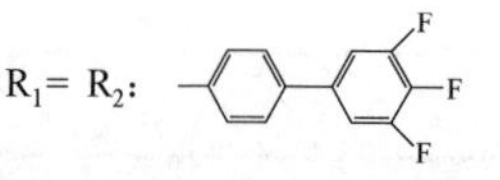 R_1= R_2: F, F, F	[236]
PI104	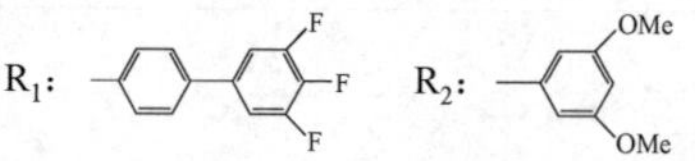 R_1: F, F, F R_2: OMe, OMe	[238]

Other types of PI structures are either PIs contain carbazole as electron donor and different phthalimides as electron acceptor as shown in Table 5.21 [247], or co-PIs with dual or multi A-D systems [214,248,210,249]. The memory devices fabricated by PMDA- and BPDA-based PIs **PI105** revealed excellent unipolar WORM memory behavior [247]. Random coPIs, **PI106**, were prepared by varying the feeding ratio of acceptor units, perylenetetracarboxydiimide (PTI), naphthalenetetracar- boxydiimide (NTI), and benzenetetracarboxydiimide (BTI), respectively [214,248]. It was found that the introduction of a small amount of these high conjugation moieties dramatically changed the memory properties from volatile DRAM to nonvolatile WORM type, and the conjugation length of the acceptor moiety affected the memory characteristics. Memory devices based on coPIs **PI107** with main chain TPA/pyrene groups randomly copolymerized with 6FDA acceptors were also found to have the similar phenomena, i.e., the memory behaviors varied with the APAP molar content to nonvolatile memory [210]. The study on optoelectrical dual-mode memory based on **PI108** with the pendant spiropyran moiety was also reported by Song et al. [249] these properties open up the possibility of potential application of the multimode data storage memory devices with PIs as the functioning medium.

TABLE5.21 Polyimides for WORM (Type IV)

	Molecular Structure	References
PI105		[247]
PI106		[214,248]
PI107		[210]
PI108		[249]

5.5.2.3 PI Hybride Materials for Resistive Memories

Inorganic nanoparticles or fullerene derivatives can be used as supplementary compounds to prepare nanoparticle-embedded PI thin films to introduce the physical electronic transitions with the PI matrix. Such PI hybrid materials applied to fabricate memory devices are summarized in Table 5.22.

A silver-nanoparticle-embedded PI thin film was prepared by incorporating silver(I) complex (AgTFA) in the **PI109** matrix [227]. The corresponding memory device showed nonvolatile memory behavior. Similarly, PI-TiO_2 hybrids were synthesized by introducing titanium butoxide into sulfur- or flurorine-containing poly(*o*-hydroxy-imide), **PI110** [250] and **PI111** [251], with pendant hydroxyl groups via sol-gel reaction. The resulting films with different TiO_2 concentrations from 0 to 50%(*w*). were found to exhibit memory properties from DRAM to SRAM to WORM. Other PI blend films were prepared from different compositions of **PI112** and polycyclic aromatic compounds (cornene or *N*,*N*-bis[4-(2-octyldodecyloxy)-phenyl]-3,4,9,10-perylenetetracarboxylic diimide (PDI-DO)) [252]. As the additive content increases in both blend systems, the memory device characteristics changed from volatile to nonvolatile behavior.

TABLE5.22 Polyimide Hybrid Materials

Molecular Structure	References
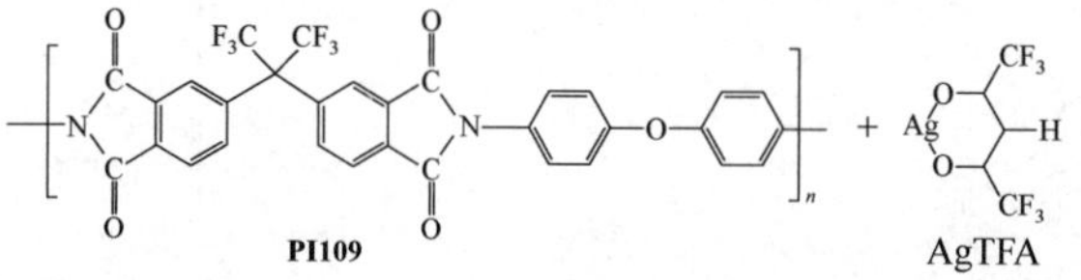 PI109 + AgTFA	[227]
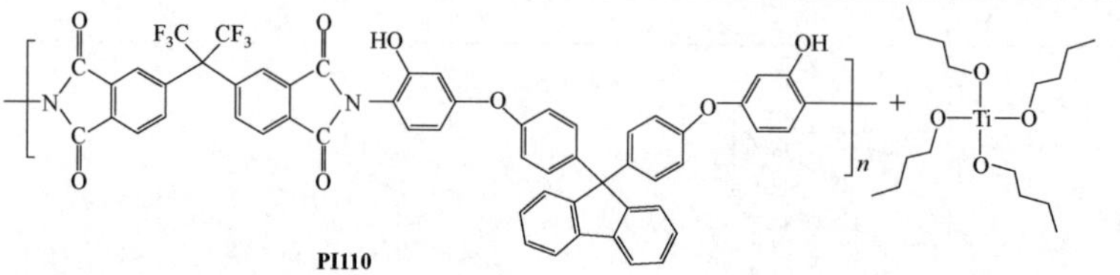 PI110 + Ti(O-butyl)$_4$	[250]
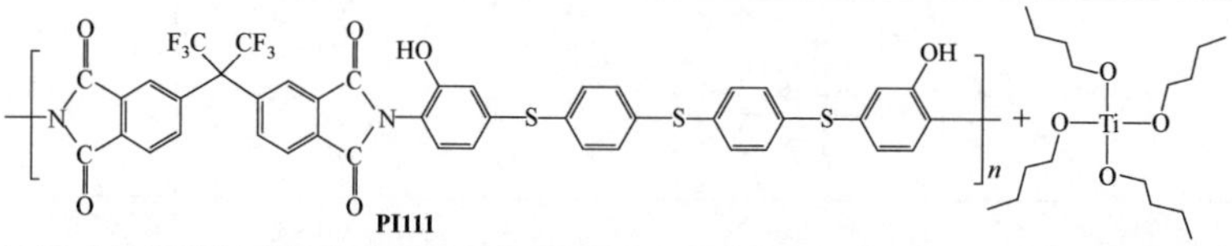 PI111 + Ti(O-butyl)$_4$	[251]
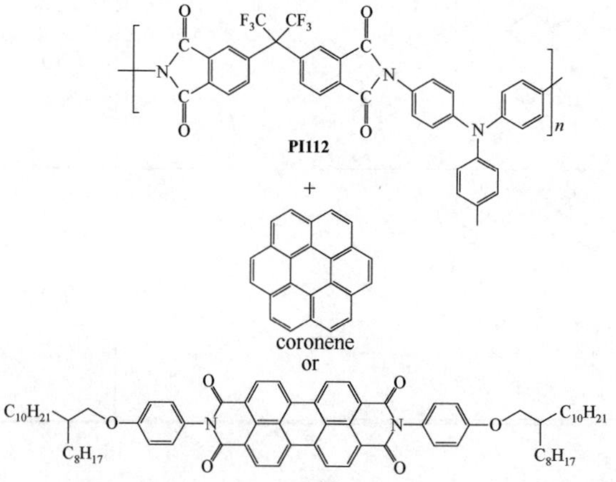 PI112 + coronene or PDI-DO	[252]

5.5.3 Polyimides for Transistor-type Memory Devices

The most important role of PIs in transistor-type memory applications is as electret materials. Some early works on PI and thermoplastic PEI as electrets have been reviewed by Erhard et al. [215]. Organic field-effect transistor (OFET)-based memory devices have been extensively investigated recently. It has been reported that PI derivatives containing electron-donor and electron-acceptor moieties have been successfully used in OFET-based devices [208]. These PI-based OFET memory devices were mainly developed by Chen's group from National Taiwan University, and are summarized in Table 5.23.

It is well known that the electrical properties of OFETs are determined by the interface properties between semiconductors and dielectric layers. Therefore, PI-based polymer dielectrics with quasipermanent charges and immobile electrical dipoles can be used in *n*-type OFETs to achieve highly stable OFETs. The first study on *n*-type *N,N'*-bis(2-phenylethyl)-perylene-3,4,9,10-tetracarboxylic diimide (BPE-PTCDI) OFET memory devices using donor-acceptor (D-A) PI electrets, **PI113** was reported in 2012 [253]. The

PI electrets were consisted of electron-accepting 6FDA and electron-donating 2,5-bis(4-aminophenylenesulfanyl)selenophene (APSP) (**PI113a**) or 2,5-bis(4-aminophenylenesulfanyl)thiophene (APST) (**PI113b**). The order of the band gaps resulted in the of intramolecular charge transfer in PIs, APSP > APST > ODA. The OFET memory device based on **PI113a** electret exhibited the highest field-effect mobility and the largest memory window [253].

The OFET memory devices derived from TPA-based PI electrets with higher dielectric constant (k) demonstrated an enhanced capability for storing the charges [254]. The PI electrets, **PI114** and **PI99b**, consisted of electron-donating TPA-CN and electron-accepting moieties, including BTDA, 6FDA and DSDA. Among the devices, the one using 6FDA-based PI electret, **PI114c**, was found to have the highest field-effect mobility memory window due to its largest dipole moment and torsion angle, which formed a stable CT complex and trapped the charges deeply. The memory devices exhibited a high ON/OFF ratio and the ON or OFF state could be retained over 10^4 s. The write-read-erase-read (WRER) cycles could be operated over 100 cycles.

Two organo-soluble D-A PIs with pendant hydroxyl groups, (poly-[9,9′-bis(4-(4-amino-3-hydroxyphenoxy) phenyl)fluorine-oxydiphthalimide]) **PI110** and (poly[4,4′-bis(4-amino-3-hydroxyph- enylthio) diphenyl sulfide-oxydiphthalimide]) **PI111**, and their PI/TiO_2 hybrids were synthesized as electrets [260]. The BPE-PTCDI OFET memory device based on the **PI110** electret with a conjugated fluorene moiety exhibited the larger memory window compared to **PI111**. The devices using PI/TiO_2 hybrids (20%(w) TiO_2 content) with a higher dielectric constant as electrets revealed the reducing operating voltage, indicating that the charge transfer capability of TiO_2 plays an important role in transferring and storing the charge for OFET memory devices.

At the same time, Chen's group reported five new PIs as electrets for OFET memory devices that have different acceptor conjugation extents, i.e., PI[1,3-diaminopropane (DAP)-BPDA] (**PI115**), PI[DAP-ODPA] (**PI116**), PI[DAP-PMDA] (**PI117a**). PIs with different spacer chain lengths were also synthesized to compare with **PI117a**, i.e., PI[1,6-diaminohexane (DAH)-PMDA) (**PI117b**) and PI[1,12-diaminododecane (DAD)-PMDA] (**PI117c**) [255]. It was found that the acceptor structure and spacer chain length of semiconjugated PI electret could significantly alter the characteristics of OFET memory devices. For the devices based on **PI117a-c**, the memory characteristics changed from flash to WORW behavior with the increase of the aliphatic spacer length. On the other hand, the memory window was found to be reduced as the aliphatic spacer length increased.

TABLE5.23 Polyimide Electrets for OFET Memory Devices

	Molecular Structure	References
PI113	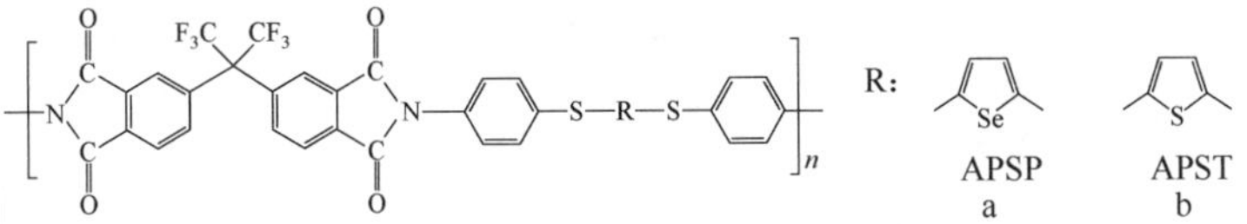 O F₃C CF₃ O C C N N—S—R—S— n O O; R: Se (APSP, a); S (APST, b)	[253]
PI114	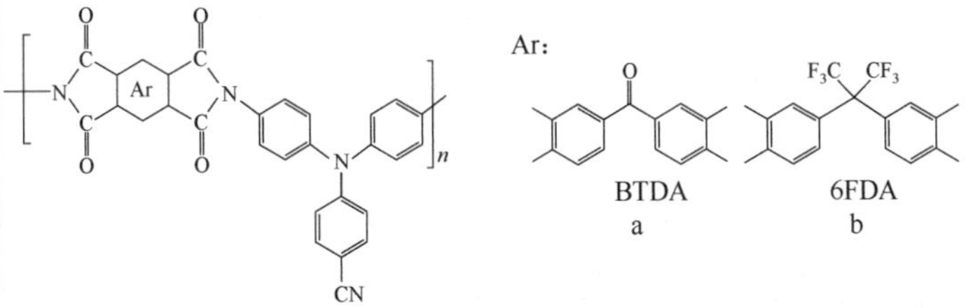 O O C C N Ar N C C O O N CN n; Ar: O (BTDA, a); F₃C CF₃ (6FDA, b)	[254]
PI115	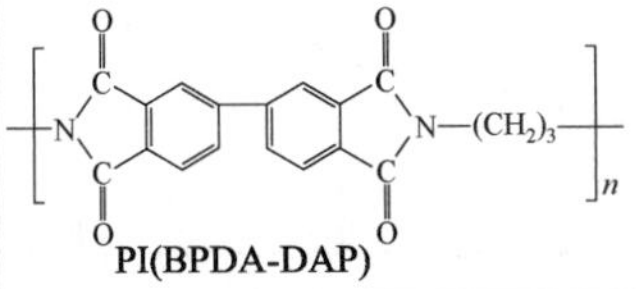 O O C C N N—$(CH_2)_3$— C C O O n; PI(BPDA-DAP)	[255]

TABLE5.23 (Continued)

	Molecular Structure	References
PI116	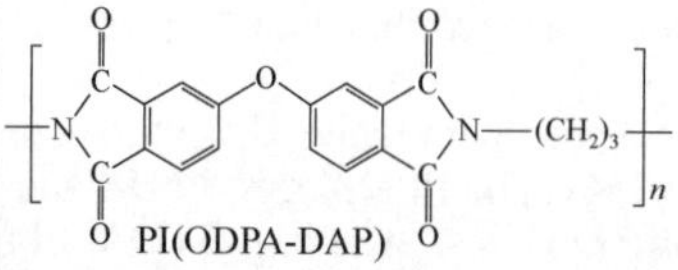 O, C, N, O, C, N, $(CH_2)_3$, n; PI(ODPA-DAP)	[255]
PI117	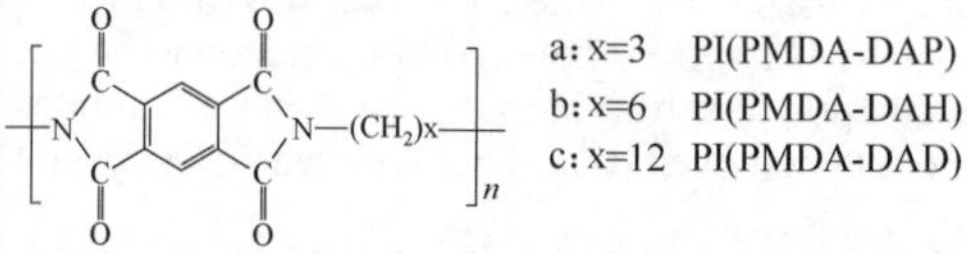 O, C, N, $(CH_2)x$, n; a: x=3 PI(PMDA-DAP); b: x=6 PI(PMDA-DAH); c: x=12 PI(PMDA-DAD)	[255]
PI118	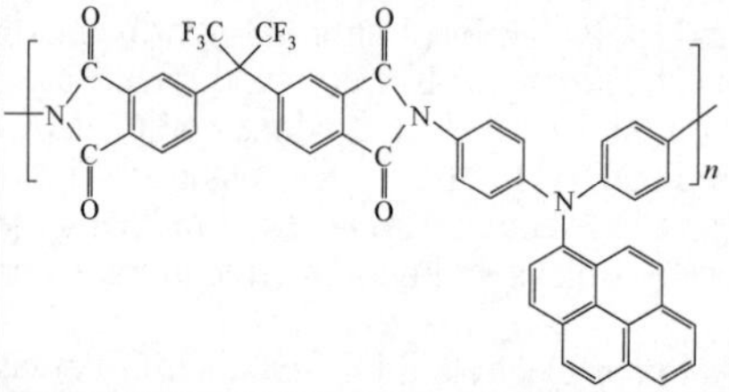 O, C, N, F_3C CF_3, N, n	[256]
PI119	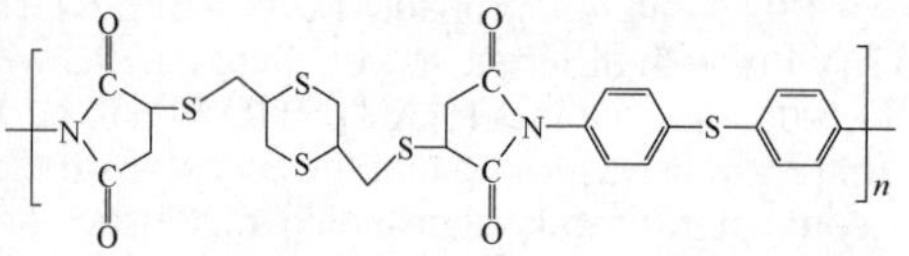 O, C, N, S, S, S, S, N, S, n	[257]
PI120	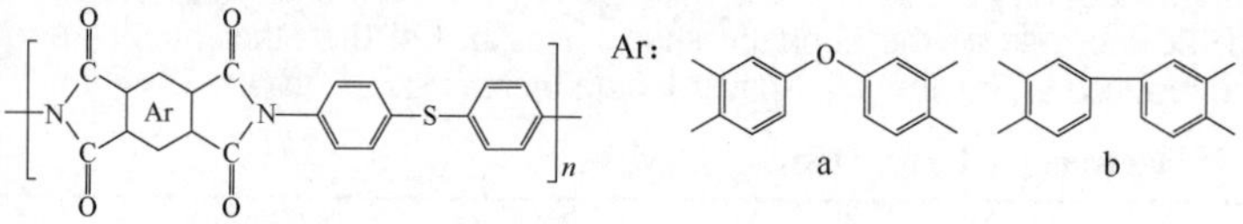 O, C, N, Ar, N, S, n; Ar: a, b	[257]
PI121	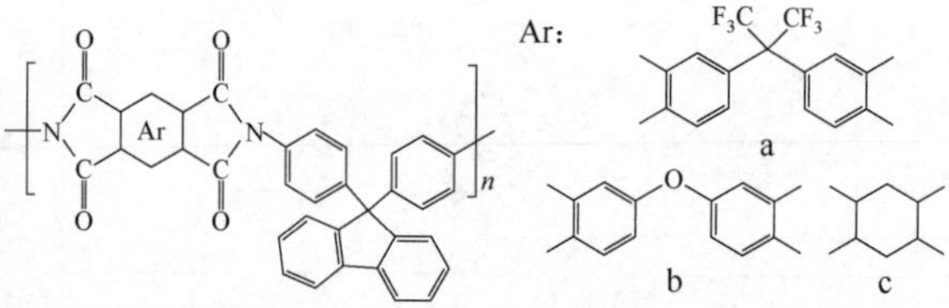 O, C, N, Ar, N, n; Ar: F_3C CF_3 a; O b; c	[258]
PI122	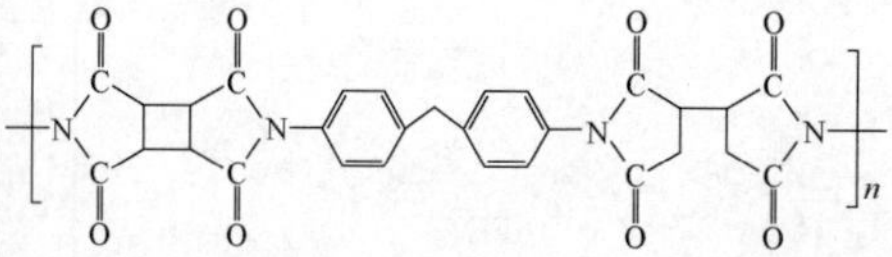 O, C, N, N, N, N, n	
PI123	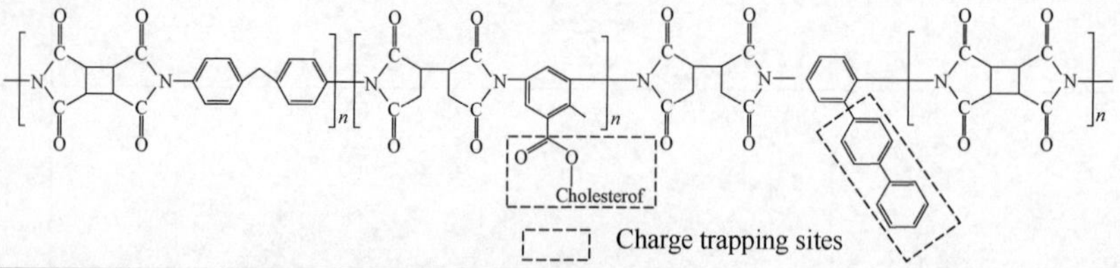 O, C, N, n, N, n, N, n; Cholesterof; Charge trapping sites	[259]

Polycyclic arene-based D-A PI electrets were prepared from different electron donating moieties, 4,4′-diamino-4″-methyltriphenylamine (AMTPA, **PI99a**), *N,N*-bis(4-aminophenyl)aminonaphthalene (APAN, **PI95**), *N,N*-bis(4-aminophenyl)aminopyrene (APAP, **PI118**) and the same electron accepting moiety 6FDA [256]. The OFET memory device based on **PI118** exhibited the largest memory window of 40.63 V and the best charge retention ability. In addition, **PI118**-based OFET memory also performed well in WRER tests for over 100 cycles.

With the aim to study the effect of CT capability of PI electrets on BPE-PTCDI-type OFET memory characteristics, new PIs, polyimidothioether[4,4′-(diaminodiphenylsulfide)bismaleimide- 2,5-bis (mercaptomethyl)-1,4-dithiane] (**PI119**), poly[bis-(4-aminophenyl)-sulfide-oxydiphthalimide] (**PI120a**), and poly[bis-(4-aminophenyl)-sulfile-biphthalimide] (**PI120b**), were prepared [257]. It was found that the device with strong electron-donating **PI119** electret behaved as the inerasable WORW-type memory. For D-A type **PI120a** and **PI120b** electrets, the derived memory devices exhibited programmable flash-type characteristics. It indicated that the performance of transistor memory could be modulated by controlling the CT features of the electrets.

A similar study was carried out by Dong et al. [258] to investigate the effects of acceptor structure in D-A PI electret on pentacene-based OFET memory performance. The electron-donating moiety 9,9-bis(4-aminophenyl)fluorene (BAPF) was used to react with the electron-accepting 6FDA and ODPA, and the neutral 1,2,4,5-cyclohexanetetracarboxylic diimide (HPMDA), respectively, giving **PI121a**, **PI121b** and **PI121c** as electrets. The 6FDA-based memory device exhibited the largest memory window due to its enhanced charge separation effect, which facilitated a stable CT complex and trapped the charges deeply.

Recently, Chou's group synthesized **PI122** from the polymerization of CBDA and mthylenediphenyl diamine (MDA) and **PI123** containing different weight ratios of polar piperazinyl and cholesterol side chains, which have dual-charge (electron and hole) withdrawing ability [259]. Various ratios of **PI123** were further mixed with **PI122** to form composite electrets that were used in OFETs and photo memory devices. It was found that the increased weight ratio of the **PI123** molecules enhanced not only the thermal properties of PI electrets but also the electrical characteristics of the OFET memories such as the field-effect mobility and on/off current ratio. Therefore, **PI123** with polar side chains improves the stability of OFETs and enlarges the organic memory.

It can be concluded from these works that the CT capability of PI electrets, which could be modulated by both the electron accepting moiety [255,258] and the electron donating moiety [256,257], has a predominate effect on the memory performance. In addition, PI electrets with higher dielectric constant (*k*), which could be achieved by introducing a triphenylamine (TPA) moiety to the polymer chain [254] or by doping TiO_2 content to PI matrix [260], can enhance the CT capability and improve the memory performance.

5.6 Summary

The high-performance PIs exhibit excellent thermal stability and mechanical properties in a broad temperature range and have exceptionally high radiation resistance and superior semiconductor properties. These advantages allow PIs to dominate the applications in many high-tech fields. In this chapter, we have briefly described and discussed various functional PIs for microelectronics and optoelectrics applications, including the photosensitive PIs as passivation layers and multilayer dielectrics in microelectronic manufacturing and packaging, LC aligning PI layers in TFT-LCD, the PI-based flexible substrates for OLED, and the PI electret layers for organic memory devices.

Microelectronics is the conventional application for PIs. Improving in the versatility and robustness as well as resolution of the patterns of PSPIs is still crucial for expanding the utility of PSPIs in microelectronics. Although PIs have been widely commercialized in microelectronics, a new developing direction of PIs will be how to reduce dielectric constant and dissipation for high-frequency communication applications. The excellent combined properties make PIs good candidates for optoelectronic planar display devices. Although the mechanical rubbing alignment PIs have successfully used in TFT-LCD and made a great success in commercialization, the mechanical rubbing process still has some shortcomings such as dust generation and electrostatic problems, etc. Thus, photoinduced alignment PIs have been developed recently.

Furthermore, planar displays will be continuously improved in terms of visibility, flexibility, durability, and light weight. PI films can be used as substrates for the development of advanced flexible display devices such as OLED. In this case, low-cost transparent PIs are highly desired, and their combined properties are still further improved for wide applications. Similar to other organic electronics, the development of polymer memory devices is towards low processing cost, flexible, and high-density data storage capacity. Therefore, the development of PI materials for memory devices is another recent emerging area. The relationship between the polymer structure and the resulting memory characteristics still remains unclear. There is still a high demand for the systematical investigation of the switch phenomenon and the mechanism behind it. There is still significant potential to develop the PI-based memory device materials.

REFERENCES

[1] C.E. Sroog, Polyimides, J. Polym. Sci., Macromol. Rev 11 (1976) 161-208.
[2] C.E. Sroog, Polyimides, Prog. Polym. Sci. 16 (4) (1991) 561-694.
[3] M.T. Bogert, R.R. Renshaw, 4-Amino-o-phthalic acid and some of its derivatives, J. Am. Chem. Soc. 30 (1908) 1135-1144.
[4] W.M. Edwards and I.M. Robinson (1955). Polyimides of pyromellitic acid. Copyright r 2017 American Chemical Society (ACS). All Rights Reserved.
[5] M. Fahim, J. Bijwe, H.S. Nalwa, Supramolecular Photosensitive and Electroactive Materials, Academic Press, (2001) 643-726.
[6] M. Ghosh, History of the invention and development of the polyimides, in: M. Ghosh (Ed.), Polyimides: Fundamental and applications, edn., CRC Press, New York, 1996.
[7] K. Sato, S. Harada, A. Saiki, T. Kimura, T. Okubo, K. Mukai, A novel planar multilevel interconnection technology utilizing polyimide, IEEE Transactions on Parts, Hybrids, and Packaging 9 (3) (1973) 176-180.
[8] Y.-H. Song, S.G. Kim, S.B. Lee, K.J. Rhee, T.S. Kim, A study of considering the reliability issues on asic/memory integration by sip (systemin-package) technology, Microelectron. Reliab. 43 (9-11) (2003) 1405-1410.
[9] T. Shimoto, K. Baba, K. Matsui, J. Tsukano, T. Maeda, K. Oyachi, Ultra-thin high-density lsi packaging substrate for advanced csps and sips, Microelectron. Reliab. 45 (3-4) (2005) 567-574.
[10] M. Ree, High performance polyimides for applications in microelectronics and flat panel displays, Macromol. Res. 14 (1) (2006) 1-33.
[11] S. Logothetidis, Flexible organic electronic devices: Materials, process and applications, Mater. Sci. Eng., B 152 (1-3) (2008) 96-104.
[12] D.-J. Liaw, K.-L. Wang, Y.-C. Huang, K.-R. Lee, J.-Y. Lai, C.-S. Ha, Advanced polyimide materials: Syntheses, physical properties and applications, Progress in Polymer Science 37 (7) (2012) 907-974.
[13] M.L. Minges, Electronic materials handbook. edn, vol. 1, Packaging. ASM International, Materials Park, OH, 1989.
[14] D. Hurwitz and A. Huang (2017). Chip package. Copyright r 2017 American Chemical Society (ACS). All Rights Reserved.
[15] G Maier, Polymers for microelectronics, Mater. Today (Oxford, U. K.) 4 (5) (2001) 22-33.
[16] R. Bartnikas, E.J. McMahon, Engineering dielectrics.edn, vol. 2B, ASTM, Philadelphia, PA, USA, 1987.
[17] G Hougham, G Tesoro, A. Viehbeck, J.D. Chapple-Sokol, Polarization effects of fluorine on the relative permittivity in polyimides, Macromolecules 27 (21) (1994) 5964-5971.
[18] J. Melcher, D. Yang, G Arlt, Dielectric effects of moisture in polyimide, IEEE Trans. Electr. Insul. 24 (1) (1989) 31-38.
[19] R. Hirte, Polyimides: Materials, chemistry and characterization. Procedings of the third international conference on polyimides, Ellenville, New York, November 2-4, 1988. Hg. Von C. Feger, M. M. Khojasteh and J. E. McGrath Amsterdam/New York/Oxford/Tokyo: Elsevier 1989. 788 s., geb., dfl. 495.00, Acta Polymerica 42 (2-3) (1991). 135-135.
[20] Boese, D., Herminghaus, S., Yoon, D.Y., Swalen, J.D. and Rabolt, J.F. (1991), "Stiff polyimides: Chain orientation and anisotropy of the optical and dielectric properties of thin films," *Mater. Res. Soc. Symp. Proc*. 227 (Mater. Sci. High Temp. Polym. Microelectron.): 379-386.
[21] Yoon, D.Y., Parrish, W., Depero, I.E. and Ree, M. (1991), "Chain conformations of aromatic polyimides and their ordering in thin films," *Mater. Res. Soc. Symp. Proc*. 227 (Mater. Sci. High Temp. Polym. Microelectron.): 387-393.
[22] S.A. Bidstrup, T.C. Hodge, L. Lin, P.A. Kohl, J.B. Lee, M.G Allen, Anisotropy in thermal, electrical and mechanical properties of spin-coated polymer dielectrics, Mater. Res. Soc. Symp. Proc 338 (1994) 577-587.
[23] C.P. Wong, Application of polymer in encapsulation of electronic parts, *Electronic applications*, edn. Berlin, Heidelberg, Springer, Berlin Heidelberg, 1988, pp. 63-83.

[24] H.F. Mark, N.M. Bikales, C.G. Overberger, G. Menges, J.I. *Kroschwitz, Encyclopedia of polymer science and engineering.* edn, vol. 12, Wiley, New York, 1985.
[25] D. Makino, Recent progress of the application of polyimides to microelectronics polymers for microelectronics, ACS symposium series 537 (Polymer for Microelectronics) (1993) 380-402.
[26] R. Xu, J.W. Lee, T. Pan, S. Ma, J. Wang, J.H. Han, et al., Designing thin, ultrastretchable electronics with stacked circuits and elastomeric encapsulation materials, Advanced Functional Materials 27 (2017) 4.
[27] P. Jin-Seong, C. Heeyeop, C. Ho Kyoon, L. Sang In, Thin film encapsulation for flexible am-oled: A review, Semiconductor Science and Technology 26 (3) (2011) 034001.
[28] T. Walter, M. Lederer, G. Khatibi, Delamination of polyimide/cu films under mixed mode loading, Microelectron. Reliab. 64 (2016) 281-286.
[29] G.W. Hyung, J. Park, J.H. Kim, J.R. Koo, Y.K. Kim, Storage stability improvement of pentacene thin-film transistors using polyimide passivation layer fabricated by vapor deposition polymerization, Solid-State Electron. 54 (4) (2010) 439-442.
[30] S.K. Oh, T. Jang, Y.J. Jo, H.-Y. Ko, J.S. Kwak, Improved package reliability of AlGaN/GaN HFETs on 150 mm si substrates by SiN*x*/polyimide dual passivation layers, Surface and Coatings Technology 307, Part B (2016) 1124-1128.
[31] Senturia, S. (1987), "Polyimides in microelectronics polymers for high technology," ACS *symposium series* 346 (Polym. High Technol.: Electron. Photonics): 428-436.
[32] G. Maier, Low dielectric constant polymers for microelectronics, Progress in Polymer Science (Oxford) 26 (1) (2001) 3-65.
[33] Jensen, R.J. (1987), "Polyimides as interlayer dielectrics for high-performance interconnections of integrated circuits," ACS *Symp. Ser.* 346 (Polym. High Technol.: Electron. Photonics): 466-483.
[34] W.C. Wang, R.H. Vora, E.T. Kang, K.G. Neoh, C.K. Ong, L.F. Chen, Nanoporous ultralow-*k* films prepared from fluorinated polyimide with grafted poly(acrylic acid) side chains, Advanced Materials 16 (1) (2004) 54-57.
[35] T. Homma, M. Yamaguchi, Y. Kutsuzawa, N. Otsuka, Electrical stability of polyimide siloxane films for interlayer dielectrics in multilevel interconnections, Thin Solid Films 340 (1,2) (1999) 237-241.
[36] Y. Yan, M. Xiong, B. Liu, Y. Ding, Z. Chen, Low capacitance and highly reliable blind through-silicon-vias (tsvs) with vacuum-assisted spin coating of polyimide dielectric liners, Science China Technological Sciences 59 (10) (2016) 1581-1590.
[37] Choudhary, R., Fasano, B.V., Iruvanti, S., Reinhardt, D.D. and Sylvester, D.A. (2007). Device including reworkable alpha particle barrier and corrosion barrier to reduce soft error rate. Copyright©2017 American Chemical Society (ACS). All Rights Reserved.
[38] A. Mathews, Synthesis, characterization, and properties of fully aliphatic polyimides and their derivatives for microelectronics and optoelectronics applications, Macromolecular Research 15 (2) (2007) 114-128.
[39] Lee, C.J. (1988). Soluble polyimide-siloxanes and their manufacture and use. Copyright©2017 American Chemical Society (ACS). All Rights Reserved.
[40] Rojstaczer, S.R., Tyrell, J.A. and Tang, D.Y. (1991), "Recent advances in polyimidesiloxanes for microelectronics applications," *Int. SAMPE Electron. Conf.* 5 (Electron. Mater.: Technol. Here Now): 310-314.
[41] N.A. Johnen, H.K. Kim, C.K. Ober, Development of new composite materials based on siloxane-containing polyimides, Polym. Prepr. (Am. Chem. Soc., Div. Polym. Chem.) 34 (2) (1993) 392-393.
[42] Johnen, N.A., Beecroft, L.L. and Ober, C.K. (1996), "Formation of transparent silica-polymer hybrids based on siloxane-containing polyimides," *ACS Symp. Ser.* 624 (Step-Growth Polymers for High-Performance Materials): 392-402.
[43] Y. Yamada, Siloxane modified polyimides for microelectronics coating applications, High Perform. Polym. 10 (1) (1998) 69-80.
[44] J. Li, J. Wu, S. Fan, S. Huang, Progress of synthesis & application of polyimide containing siloxane, Youjigui Cailiao 25 (2) (2011) 121-125.
[45] W. Dong, Y. Guan, D. Shang, Novel soluble polyimides containing pyridine and fluorinated units: Preparation, characterization, and optical and dielectric properties, RSC Adv. 6 (26) (2016) 21662-21671.
[46] P. Zhang, J. Zhao, K. Zhang, R. Bai, Y. Wang, C. Hua, et al., Fluorographene/polyimide composite films: Mechanical, electrical, hydrophobic, thermal and low dielectric properties, Composites, Part A:84 (2016) 428-434.
[47] Q. Li, Y. Wang, S. Zhang, L. Pang, H. Tong, J. Li, et al., Novel fluorinated random co-polyimide/amine-functionalized zeolite MEL50 hybrid films with enhanced thermal and low dielectric properties, J. Mater. Sci. 52 (9) (2017) 5283-5296.
[48] C. Wang, T. Wang, Q. Wang, Controllable porous fluorinated polyimide thin films for ultralow dielectric constant interlayer dielectric applications, J. Macromol. Sci., Part A: Pure Appl. Chem. 54 (5) (2017) 311-315.
[49] L. Yi, W. Huang, D. Yan, Polyimides with side groups: Synthesis and effects of side groups on their properties, J. Polym. Sci., Part A: Polym. Chem. 55 (4) (2017) 533-559.
[50] K.-i Fukukawa, M. Ueda, Recent progress of photosensitive polyimides, Polymer Journal 40 (4) (2008) 281-296.
[51] R.E. Kerwin, M.R. Goldrick, Thermally stable photoresist polymer, Polymer Engineering & Science 11 (5) (1971) 426-430.
[52] Loprest, F.J. and McInerney, E.F. (1978). Positive working thermally stable photoresist composition, article and method of

using. *United States*. 597226.

[53] S.K. Oh, T. Jang, Y.J. Jo, H.-Y. Ko, J.S. Kwak, Bonding pad over active structure for chip shrinkage of high-power algan/gan hfets, IEEE Trans. Electron Devices 63 (2) (2016) 620-624.

[54] S.K. Oh, T. Jang, S. Pouladi, Y.J. Jo, H.-Y. Ko, J.-H. Ryou, et al., Output power enhancement in algan/gan heterostructure field-effect transistors with multilevel metallization, Appl. Phys. Express 10 (1) (2017). 016502/016501-016502/016503.

[55] Horie, K. (1995). *Photosensitive polyimides: Fundamentals and applications*. edn.

[56] A. Mochizuki, M. Ueda, Recent development in photosensitive polyimide (PSPI), Journal of Photopolymer Science and Technology 14 (5) (2001) 677-688.

[57] K.-i Fukukawa, M. Ueda, Chemically amplified photosensitive polyimides and polybenzoxazoles, J. Photopolym. Sci. Technol. 22 (6) (2009) 761-771.

[58] R. Rubner, H. Ahne, E. Kuehn, G Kolodziej, A photopolymer - the direct way to polyimide patterns, Photogr. Sci. Eng. 23 (5) (1979) 303-309.

[59] N. Yoda, H. Hiramoto, New photosensitive high temperature polymers for electronic applications, J. Macromol. Sci., Chem. A21 (13-14) (1984) 1641-1663.

[60] N. Yoda, Recent developments in advanced functional polymers for semiconductor encapsulants of integrated circuit chips and hightemperature photoresist for electronic applications, Polym. Adv. Technol. 8 (4) (1997) 215-226.

[61] M. Ueda, T. Nakayama, A new negative-type photosensitive polyimide based on poly(hydroxyimide), a cross-linker, and a photoacid generator, Macromolecules 29 (20) (1996) 6427-6431.

[62] A. Mochizuki, T. Teranishi, M. Ueda, Novel photosensitive polyimide precursor based on polyisoimide using an amine photogenerator, Macromolecules 28 (1) (1995) 365-369.

[63] J.M.J. Fréchet, J.F. Cameron, C.M. Chung, S.A. Haque, C.G Willson, Photogenerated base as catalyst for imidization reactions, Polymer Bulletin 30 (4) (1993) 369-375.

[64] K.-i Fukukawa, Y. Shibasaki, M. Ueda, Direct patterning of poly(amic acid) and low-temperature imidization using a photo-base generator, Polymers for Advanced Technologies 17 (2) (2006) 131-136.

[65] J.C. Dubois, J.M. Bureau, *Polyimides and other high-temperature polymers*. edn., Elsevier, Amsterdam, 1991.

[66] O. Rohde, P. Smolka, P.A. Falcigno, J. Pfeifer, Novel auto-photosensitive polyimides with tailored properties, Polymer Engineering & Science 32 (21) (1992) 1623-1629.

[67] S. Kubota, T. Moriwaki, T. Ando, A. Fukami, Preparation of positive photoreactive polyimides and their characterization, Journal of Applied Polymer Science 33 (5) (1987) 1763-1775.

[68] S.L.-C. Hsu, P.-I. Lee, J.-S. King, J.-L. Jeng, Novel positive-working aqueous-base developable photosensitive polyimide precursors based on diazonaphthoquinone-capped polyamic esters, Journal of Applied Polymer Science 90 (8) (2003) 2293-2300.

[69] M. Okazaki, H. Onishi, W. Yamashita, S. Tamai, Positive-working photosensitive alkaline-developable polyimide precursor based on semialicyclic poly(amide acid), vinyl ether crosslinker, and a photoacid generator, Journal of Photopolymer Science and Technology 19 (2) (2006) 277-280.

[70] H. Seino, A. Mochizuki, O. Haba, M. Ueda, A positive-working photosensitive alkaline-developable polyimide with a highly dimensional stability and low dielectric constant based on poly(amic acid) as a polyimide precursor and diazonaphthoquinone as a photosensitive compound, Journal of polymer science. Part A: Polymer chemistry 36 (13) (1998) 2261-2267.

[71] S. Hsu, Synthesis and characterization of a positive-working, aqueous-base-developable photosensitive polyimide precursor, Journal of Applied Polymer Science 86 (2) (2002) 352-358.

[72] A. Mochizuki, Positive-working alkaline-developable photosensitive polyimide precursor based on polyisoimide using diazonaphthoquinone as a dissolution inhibitor, Polymer 36 (11) (1995) 2153-2158.

[73] T. Fukushima, T. Oyama, T. Iijima, M. Tomoi, H. Itatani, New concept of positive photosensitive polyimide: Reaction development patterning (rdp), Journal of polymer science. Part A: Polymer chemistry 39 (19) (2001) 3451-3463.

[74] O. Haba, M. Okazaki, T. Nakayama, M. Ueda, Positive-working alkaline-developable photosensitive polyimide precursor based on poly(amic acid) and dissolution inhibitor, Journal of Photopolymer Science and Technology 10 (1) (1997) 55-60.

[75] K. Sakayori, Y. Shibasaki, M. Ueda, A positive-type alkaline-developable photosensitive polyimide based on the polyamic acid from 2,20,6,60-biphenyltetracarboxylic dianhydride and 1,3-bis(4-aminophenoxy)benzene, and a diazonaphthoquinone, Polymer Journal 38 (11) (2006) 1189-1193.

[76] M. Hasegawa, A. Tominaga, Environmentally friendly positive- and negative-tone photo-patterning systems of low-*k* and low-CTE polyimides, Journal of Photopolymer Science and Technology 18 (2) (2005) 307-312.

[77] M. Hasegawa, Y. Tanaka, K. Koseki, A. Tominaga, Positive-type photo-patterning of low-CTE, high-modulus transparent polyimide systems, Journal of Photopolymer Science and Technology 19 (2) (2006) 285-290.

[78] M. Hasegawa, J. Nakano, Colorless polyimides derived from cycloaliphatic tetracarboxylic dianhydrides with controlled steric structures (4). Applications to positive-type photosensitive polyimide systems with controlled extents of imidization,

Journal of Photopolymer Science and Technology 22 (3) (2009) 411-415.
[79] M. Tomikawa, S. Yoshida, N. Okamoto, Novel partial esterification reaction in poly(amic acid) and its application for positive-tone photosensitive polyimide precursor, Polymer Journal 41 (8) (2009) 604-608.
[80] T. Yuba, R. Okuda, M. Tomikawa, J.H. Kim, Soft baking effect on lithographic performance by positive tone photosensitive polyimide, Journal of Photopolymer Science and Technology 23 (6) (2010) 775-779.
[81] M. Tomikawa, M. Suwa, H. Niwa, K. Minamihashi, Novel high refractive index positive-tone photosensitive polyimide for microlens of image sensors, High Performance Polymers 23 (1) (2011) 66-73.
[82] Y. Inoue, Y. Saito, T. Higashihara, M. Ueda, Facile formulation of alkaline-developable positive-type photosensitive polyimide based on fluorinated poly(amic acid), poly(amic acid), and fluorinated diazonaphthoquinone, Journal of Materials Chemistry C: Materials for optical and electronic devices 1 (14) (2013) 2553-2560.
[83] Y. Inoue, T. Higashihara, M. Ueda, Alkaline-developable positive-type photosensitive polyimide based on fluorinated poly(amic acid) and fluorinated diazonaphthoquinone, Journal of Photopolymer Science and Technology 26 (3) (2013) 351-356.
[84] Y. Inoue, Y. Ishida, T. Higashihara, A. Kameyama, S. Ando, M. Ueda, Alkaline-developable and positive-type photosensitive polyimide based on fluorinated poly(amic acid) from diamine with high hydrophobicity and fluorinated diazonaphtoquinone, Journal of Photopolymer Science and Technology 27 (2) (2014) 211-217.
[85] S. Ryu, Synthesis and characterizations of positive-working photosensitive polyimides having 4,5-dimethoxy-o-nitrobenzyl side group, Bulletin of the Korean Chemical Society 29 (9) (2008) 1689-1694.
[86] T. Abe, M. Mishina, N. Kohtoh, Positive photosensitive polyimides with cyclobutane structure, Polymers for Advanced Technologies 4 (4) (1993) 288-293.
[87] T. Fukushima, K. Hosokawa, T. Oyama, T. Iijima, M. Tomoi, H. Itatani, Synthesis and positive-imaging photosensitivity of soluble polyimides having pendant carboxyl groups, Journal of polymer science. Part A: Polymer chemistry 39 (6) (2001) 934-946.
[88] Hwang, K.Y., Tu, A.P., Wu, S.Y., Chen, G.C., Huang, C.J. and Wang, J.F. Novel water soluble polyimide resin, its preparation and use. U.S. Pat. Appl. Publ. (2011), US 20110172324 A1 20110714.
[89] K. Morita, K. Ebara, Y. Shibasaki, M. Ueda, K. Goto, S. Tamai, New positive-type photosensitive polyimide having sulfo groups, Polymer 44 (20) (2003) 6235-6239.
[90] K. Morita, Y. Shibasaki, M. Ueda, New positive-type photosensitive polyimide having sulfo groups 2. Polyimides from 2,20-oxy(or thio)bis(5-aminobenzenesulfonic acid), 4,40-oxydianiline, and 4,4′-hexafluoropropylidene-bis(phthalic anhydride), Journal of Photopolymer Science and Technology 17 (2) (2004) 263-268.
[91] X. Jin, H. Ishii, A novel positive-type photosensitive polyimide having excellent transparency based on soluble block copolyimide with hydroxyl group and diazonaphthoquinone, Journal of Applied Polymer Science 96 (5) (2005) 1619-1624.
[92] J. Ishii, Organo-soluble polyimides and their applications to photosensitive cover layer materials in flexible printed circuit boards, Journal of Photopolymer Science and Technology 21 (1) (2008) 107-112.
[93] Seong, H.-R., Park, C.-H., Oh, D.-H., Shin, H.-I., Kim, K.-J. and Shin, S.-J. (PCT Int. Appl. (2009), WO 2009110764 A2 20090911). Positive photosensitive polyimide composition photoresist for organic led fabrication. WO2009KR01121 20090306.
[94] M. Hasegawa, Hydroxyamide-containing positive-type photosensitive polyimides, Journal of Photopolymer Science and Technology 20 (2) (2007) 175-180.
[95] T. Miyagawa, T. Fukushima, T. Oyama, T. Iijima, M. Tomoi, Photosensitive fluorinated polyimides with a low dielectric constant based on reaction development patterning, Journal of polymer science. Part A: Polymer chemistry 41 (6) (2003) 861-871.
[96] S. Sugawara, M. Tomoi, T. Oyama, Photosensitive polyesterimides based on reaction development patterning, Polym. J. (Tokyo, Jpn.) 39 (2) (2007) 129-137.
[97] T. Oyama, Y. Kawakami, T. Fukushima, T. Iijima, M. Tomoi, Photosensitive polycarbonates based on reaction development patterning (rdp), Polym. Bull. (Berlin, Ger.) 47 (2) (2001) 175-181.
[98] T. Oyama, A. Kitamura, T. Fukushima, T. Iijima, M. Tomoi, Photosensitive polyarylates based on reaction development patterning, Macromolecular Rapid Communications 23 (2) (2002) 104-108.
[99] K. Sakamoto, K. Usami, K. Miki, Photoalignment efficiency enhancement of polyimide alignment layers by alkyl-amine vapor treatment, Appl. Phys. Express 7 (8) (2014). 081701/081701-081701/081704, 081704 pp.
[100] Cognard, J. (1982). Alignment of liquid crystals and their mixtures. edn. Gorden & Breach: London.
[101] P.J. Collings, J.S. Patel, *Handbook of liquid crystal research*. edn, Oxford University Press Inc, New York, 1997.
[102] B. Chae, S.B. Kim, S.W. Lee, S.I. Kim, W. Choi, B. Lee, et al., Surface morphology, molecular reorientation, and liquid crystal alignment properties of rubbed nanofilms of a well-defined brush polyimide with a fully rodlike backbone, Macromolecules 35 (27) (2002) 10119-10130.

[103] L. Qing-Hua, L. Xue-Min, Y. Jie, Z. Zi-Kang, W. Zong-Guang, H. Hiroyaki, Liquid crystal alignment on polyimide surface with laser-induced periodic surface microgroove, Japanese Journal of Applied Physics 41 (7R) (2002) 4635.

[104] D.W. Berreman, Solid surface shape and the alignment of and adjacent nematic liquid crystal, Phys. Rev. Lett. 28 (26) (1972) 1683-1686.

[105] J.M. Geary, J.W. Goodby, A.R. Kmetz, J.S. Patel, The mechanism of polymer alignment of liquid-crystal materials, J. Appl. Phys. 62 (10) (1987) 4100-4108.

[106] K. Usami, K. Sakamoto, Y. Uehara, S. Ushioda, Transfer of the in-plane molecular orientation of polyimide film surface to liquid crystal monolayer, Appl. Phys. Lett. 86 (21) (2005). 211906/211901-211906/211903.

[107] K. Usami, K. Sakamoto, Photo-aligned blend films of azobenzene-containing polyimides with and without side-chains for inducing inclined alignment of liquid crystal molecules, J. Appl. Phys. 110 (4) (2011). 043522/043521-043522/043526.

[108] M. O'Neill, S.M. Kelly, Photoinduced surface alignment for liquid crystal displays, J. Phys. D: Appl. Phys. 33 (10) (2000) R67-R84.

[109] D.-R. Chiou, L.-J. Chen, C.-D. Lee, Pretilt angle of liquid crystals and liquid-crystal alignment on microgrooved polyimide surfaces fabricated by soft embossing method, Langmuir 22 (22) (2006) 9403-9408.

[110] K. Takatoh, M. Hasegawa, M. Koden, N. Itoh, R. Hasegawa, S. *Sakamoto*, *Alignment technologies and applications of liquid crystal devices*. edn, Taylor & Francis, London, 2005.

[111] K. Sakamoto, K. Usami, K. Miki, Photo-alignment property of azobenzene-containing polyimide films swollen by alkyl-amine, Mol. Cryst. Liq. Cryst. 611 (1) (2015) 153-159.

[112] W.M. Gibbons, P.J. Shannon, S.T. Sun, B.J. Swetlin, Surface-mediated alignment of nematic liquid crystals with polarized laser light, Nature (London) 351 (6321) (1991) 49-50.

[113] M. Schadt, H. Seiberle, A. Schuster, Optical patterning of multi-domain liquid-crystal displays with wide viewing angles, Nature (London) 381 (6579) (1996) 212-215.

[114] M. Hasegawa, Y. Taira, Nematic homogeneous photoalignment by polyimide exposure to linearly polarized uv, J. Photopolym. Sci. Technol. 8 (2) (1995) 241-248.

[115] M. Nishikawa, T. Kosa, J.L. West, Effect of chemical structures of polyimides on unidirectional liquid crystal alignment produced by a polarized ultraviolet-light exposure, Jpn. J. Appl. Phys., Part 2 38 (3B) (1999). L334-L337.

[116] M. Nishikawa, B. Taheri, J.L. West, Mechanism of unidirectional liquid-crystal alignment on polyimides with linearly polarized ultraviolet light exposure, Appl. Phys. Lett. 72 (19) (1998) 2403-2405.

[117] B. Park, Y. Jung, H.-H. Choi, H.-K. Hwang, Y. Kim, S. Lee, et al., Thermal and optical stabilities of photoisomerizable polyimide layers for nematic liquid crystal alignments, Jpn. J. Appl. Phys., Part 1 37 (10) (1998) 5663-5668.

[118] Y. Wang, C. Xu, A. Kanazawa, T. Shiono, T. Ikeda, Y. Matsuki, et al., Generation of nematic liquid crystal alignment with polyimides exposed to linearly polarized light of long wavelength, J. Appl. Phys. 84 (1) (1998) 181-188.

[119] K. Sakamoto, K. Usami, M. Watanabe, R. Arafune, S. Ushioda, Surface anisotropy of polyimide film irradiated with linearly polarized ultraviolet light, Appl. Phys. Lett. 72 (15) (1998) 1832-1834.

[120] K. Sakamoto, K. Usami, T. Araya, S. Ushioda, Anisotropic molecular orientation of poly[4,4′-oxydiphenylene-1,2,3,4-cyclobutanetetracarboximide] films irradiated by linearly polarized uv light, Jpn. J. Appl. Phys., Part 2 38 (12A) (1999). L1435-L1438.

[121] K. Usami, K. Sakamoto, S. Ushioda, Influence of molecular structure on anisotropic photoinduced decomposition of polyimide molecules, J. Appl. Phys. 89 (10) (2001) 5339-5342.

[122] O. Yaroshchuk, Y. Reznikov, Photoalignment of liquid crystals: Basics and current trends, J. Mater. Chem. 22 (2) (2012) 286-300.

[123] Y. Wang, C. Xu, A. Kanazawa, T. Shiono, T. Ikeda, Y. Matsuki, et al., Thermal stability of alignment of a nematic liquid crystal induced by polyimides exposed to linearly polarized light, Liq. Cryst. 28 (3) (2001) 473-475.

[124] K.H. Yang, K. Tajima, A. Takenaka, H. Takano, Charge trapping properties of uv-exposed polyimide films for the alignment of liquid crystals, Jpn. J. Appl. Phys., Part 2 35 (5A) (1996). L561-L563.

[125] M.I. Bessonov, V.A. Zubkov, *Polyamic acids and polyimides: Synthesis, transformations, and structure*. edn, CRC Press, Boca Raton, LA, 1993.

[126] M. Ree, K. Kim, S.H. Woo, H. Chang, Structure, chain orientation, and properties in thin films of aromatic polyimides with various chain rigidities, J. Appl. Phys. 81 (2) (1997) 698-708.

[127] Y.J. Hwang, S.H. Hong, S.G Lee, M.H. Cho, D.M. Shin, Surface characteristics of photoaligned polyimide film interfacial reacted with bromine or ethanethiol, Ultramicroscopy 108 (10) (2008) 1266-1272.

[128] S.W. Lee, S.I. Kim, B. Lee, W. Choi, B. Chae, S.B. Kim, et al., Photoreactions and photoinduced molecular orientations of films of a photoreactive polyimide and their alignment of liquid crystals, Macromolecules 36 (17) (2003) 6527-6536.

[129] X.-D. Li, Z.-X. Zhong, G Jin, S.H. Lee, M.-H. Lee, Liquid crystal photoalignment by soluble photosensitive polyimide with methylene cinnamate side units, Macromol. Res. 14 (3) (2006) 257-260.

[130] Y.H. Kim, Y.H. Min, S.W. Lee, Synthesis, characterization, and liquid crystal alignment properties of photosensitive polyimide, Mol. Cryst. Liq. Cryst. 513 (2009) 89-97.
[131] S.Y. Kim, S.E. Shin, D.M. Shin, Photo-sensitive polyimide containing methoxy cinnamate derivatives on photo-alignment of liquid crystal, Mol. Cryst. Liq. Cryst. 520 (2010) 392-397.
[132] Kim, S., Shin, S.E., Shin, D., Nelson, R.L., Kajzar, F. and Kaino, T. (2010), "Anchoring energy of photo-sensitive polyimide alignment film containing methoxy cinnamate," *Proc. SPIE* 7599 (Organic Photonic Materials and Devices XII): 75991L/75991-75991L/75998.
[133] S.W. Lee, S.I. Kim, B. Lee, H.C. Kim, T. Chang, M. Ree, A soluble photoreactive polyimide bearing the coumarin chromophore in the side group: Photoreaction, photoinduced molecular reorientation, and liquid-crystal alignability in thin films, Langmuir 19 (24) (2003) 10381-10389.
[134] M. Ree, S.I. Kim, S.W. Lee, Alignment behavior of liquid-crystals on thin films of photosensitive polymers effects of photoreactive group and uv-exposure, Synth. Met. 117 (1-3) (2001) 273-275.
[135] S.-J. Sung, J. Lee, K.-Y. Cho, W.S. Kim, H. Hah, H.-K. Shim, et al., Effect of photoreactivity of polyimide on the molecular orientation of liquid crystals on photoreactive polymer/polyimide blends, Liq. Cryst. 31 (12) (2004) 1601-1611.
[136] H. Hah, S.-J. Sung, K.Y. Cho, J.-K. Park, Molecular orientation of liquid crystal on polymer blends of coumarin and naphthalenic polyimide, Polymer Bulletin 61 (3) (2008) 383-390.
[137] K. Sakamoto, K. Usami, T. Sasaki, Y. Uehara, S. Ushioda, Inclined alignment of polyimide backbone structures induced by single exposure of unpolarized light, Mol. Cryst. Liq. Cryst. 438 (2005) 1779-1790.
[138] K. Sakamoto, K. Usami, M. Kikegawa, S. Ushioda, Alignment of polyamic acid molecules containing azobenzene in the backbone structure: Effects of polarized ultraviolet light irradiation and subsequent thermal imidization, J. Appl. Phys. 93 (2) (2003) 1039-1043.
[139] K. Sakamoto, K. Usami, T. Sasaki, T. Kanayama, S. Ushioda, Optical alignment control of polyimide molecules containing azobenzene in the backbone structure, Thin Solid Films 464-465 (2004) 416-419.
[140] K. Sakamoto, K. Usami, T. Sasaki, Y. Uehara, S. Ushioda, Pretilt angle of liquid crystals on polyimide films photoaligned by single oblique angle irradiation with unpolarized light, Jpn. J. Appl. Phys., Part 1 45 (4A) (2006) 2705-2707.
[141] K. Usami, K. Sakamoto, J. Yokota, Y. Uehara, S. Ushioda, Polyimide photo-alignment layers for inclined homeotropic alignment of liquid crystal molecules, Thin Solid Films 516 (9) (2008) 2652-2655.
[142] K. Usami, K. Sakamoto, J. Yokota, Y. Uehara, S. Ushioda, Pretilt angle control of liquid crystal molecules by photoaligned films of azobenzene-containing polyimide with a different content of side-chain, J. Appl. Phys. 104 (11) (2008). 113528/113521-113528/113525.
[143] K. Usami, K. Sakamoto, N. Tamura, A. Sugimura, Improvement in photo-alignment efficiency of azobenzene-containing polyimide films, Thin Solid Films 518 (2) (2009) 729-734.
[144] K. Usami, K. Sakamoto, Y. Uehara, S. Ushioda, Stability of azobenzene- containing polyimide film to uv light, Jpn. J. Appl. Phys., Part 1 44 (9A) (2005) 6703-6705.
[145] S. Faetti, K. Sakamoto, K. Usami, Very strong azimuthal anchoring of nematic liquid crystals on uv-aligned polyimide layers, Phys. Rev. E: Stat., Nonlinear, Soft Matter Phys. 75 (5-1) (2007). 051704/051701-051704/051712.
[146] K. Usami, K. Sakamoto, Y. Uehara, S. Ushioda, In-plane order of liquid crystal molecules adsorbed on photoaligned polyimide films: Coverage dependence in submonolayer range, J. Appl. Phys. 101 (1) (2007). 013512/013511-013512/013515.
[147] K. Sakamoto, K. Usami, Y. Uehara, S. Ushioda, Excellent uniaxial alignment of poly(9,9-dioctylfluorenyl-2,7-diyl) induced by photoaligned polyimide films, Appl. Phys. Lett. 87 (21) (2005). 211910/211911-211910/211913.
[148] K. Sakamoto, K. Usami, K. Miki, Light exposure dependence of molecular orientation of glassy polyfluorene layers formed on photoaligned polyimide films, Colloids Surf., B 56 (1-2) (2007) 260-264.
[149] S. Oka, T. Mitsumoto, M. Kimura, T. Akahane, Relationship between surface order and surface azimuthal anchoring strength of nematic liquid crystals, Phys. Rev. E: Stat., Nonlinear, Soft Matter Phys. 69 (6-1) (2004). 061711/061711-061711/061717.
[150] D.W. Berreman, Alignment of liquid crystals by grooved surfaces, Mol. Cryst. Liquid Cryst. 23 (3/4) (1973) 215-231.
[151] K. Tokuoka, H. Yoshida, Y. Miyake, C.H. Lee, Y. Miura, S. Suzuki, et al., Planar alignment of columnar liquid crystals in microgroove structures, Molecular Crystals and Liquid Crystals 510 (1) (2009). 126/[1260]-1133/[1267].
[152] Q.-h Lu, Z.-g Wang, J. Yin, Z.-k Zhu, H. Hiraoka, Molecular orientation in laser-induced periodic microstructure on polyimide surface, Appl. Phys. Lett. 76 (10) (2000) 1237-1239.
[153] Y.-M. Koo, M.-H. Kim, H.-S. Lee, J.-D. Kim, Relationship between pretilt angle of nematic liquid crystal and surface structure of alignment layer, Molecular Crystals and Liquid Crystals Science and Technology. Section A. Molecular Crystals and Liquid Crystals 337 (1) (1999) 515-518.
[154] R. Lu, K. Xu, J. Gu, Z. Lu, Afm investigation of polymer lb films on the alignment of ferroelectric liquid crystal, Physics

Letters A 260 (5) (1999) 417-423.
[155] D.S. Seo, S. Kobayashi, Liquid crystal alignment capability on polyimide langmuir-blodgett surfaces with alkyl chain lengths, Mol. Cryst. Liq. Cryst. Sci. Technol., Sect. A 339 (2000) 1-10.
[156] D. Taguchi, T. Kawate, R. Miyazawa, M. Weis, T. Manaka, M. Iwamoto, Orientational ordering of 4-pentyl-4'-cyanobiphenyl molecules evaporated on multi-layered polyimide film, Thin Solid Films 517 (4) (2008) 1407-1410.
[157] D.-U. Jin, J.-K. Jeong, T.-W. Kim, J.-S. Lee, T.-K. Ahn, Y.-K. Mo, et al., Flexible amoled displays on stainless-steel foil, Journal of the Society for Information Display 14 (12) (2006) 1083-1090.
[158] M.-C. Choi, Y. Kim, C.-S. Ha, Polymers for flexible displays: From material selection to device applications, Prog. Polym. Sci. 33 (6) (2008) 581-630.
[159] H.-j Ni, J.-g Liu, Z.-h Wang, S.-y Yang, A review on colorless and optically transparent polyimide films: Chemistry, process and engineering applications, J. Ind. Eng. Chem. (Amsterdam, Netherlands) 28 (2015) 16-27.
[160] M. Vomero, P. van Niekerk, V. Nguyen, N. Gong, M. Hirabayashi, A. Cinopri, et al., A novel pattern transfer technique for mounting glassy carbon microelectrodes on polymeric flexible substrates, J. Micromech. Microeng 26 (2) (2016). 025018/025011-025018/025010.
[161] X. Mathew, J.P. Enriquez, A. Romeo, A.N. Tiwari, Cdte/cds solar cells on flexible substrates, Sol. Energy 77 (6) (2004) 831-838.
[162] S. Wang, R. Li, C. Zheng, C. Cheng, Y. Yu, S. Bai, et al., Laser direct writing and photonic manufacturing of microball lens for wide-angle imaging, energy devices, and sensors on flexible substrates, J. Laser Micro/Nanoeng. 11 (3) (2016) 285-289.
[163] M. Koden, Flexible oleds, *Oled displays and lighting*, edn, John Wiley & Sons, Ltd, 2016, pp. 166-188.
[164] Y.-L. Lin, T.-Y. Ke, C.-J. Liu, C.-S. Huang, P.-Y. Lin, C.-H. Tsai, et al., 10.4: A delamination method and ultra-high-reliable gas barrier film for flexible oled displays, SID Symposium Digest of Technical Papers 45 (1) (2014) 114-117.
[165] R. Delmdahl, R. Paetzel, J. Brune, Large-area laser-lift-off processing in microelectronics, Phys. Procedia 41 (2013) 241-248.
[166] S. Hong, C. Jeon, S. Song, J. Kim, J. Lee, D. Kim, et al., 25.4: Invited paper: Development of commercial flexible amoleds, SID Symposium Digest of Technical Papers 45 (1) (2014) 334-337.
[167] Y. Jimbo, T. Aoyama, N. Ohno, S. Eguchi, S. Kawashima, H. Ikeda, et al., 25.1: Tri-fold flexible amoled with high barrier passivation layers, SID Symposium Digest of Technical Papers 45 (1) (2014) 322-325.
[168] R. Kataish, T. Sasaki, K. Toyotaka, H. Miyake, Y. Yanagisawa, H. Ikeda, et al., 15.3: Development of side-roll and top-roll panels for an rgbw high-resolution flexible display using a white oled with microcavity structure, SID Symposium Digest of Technical Papers 45 (1) (2014) 187-190.
[169] R. Komatsu, R. Nakazato, T. Sasaki, A. Suzuki, N. Senda, T. Kawata, et al., 25.2: Repeatedly foldable book-type amoled display, SID Symposium Digest of Technical Papers 45 (1) (2014) 326-329.
[170] J. Chen, C.T. Liu, Technology advances in flexible displays and substrates, IEEE Access 1 (2013) 150-158.
[171] R. Ma, K. Rajan, J. Silvernail, K. Urbanik, J. Paynter, P. Mandlik, et al., Wearable 4-in. QVGA full-color-video flexible amoleds for rugged applications, J. Soc. Inf. Disp. 18 (1) (2010) 50-56.
[172] J.-S. Park, T.-W. Kim, D. Stryakhilev, J.-S. Lee, S.-G. An, Y.-S. Pyo, et al., Flexible full color organic light-emitting diode display on polyimide plastic substrate driven by amorphous indium gallium zinc oxide thin-film transistors, Appl. Phys. Lett. 95 (1) (2009). 013503/013501-013503/013503.
[173] Adamovich, V.I., Weaver, M.S. and Brown, J.J. (2008). Improving the performance of pholeds by using dual doping: 70510D-70510D-70518.
[174] Pang, H., Rajan, K., Silvernail, J., Mandlik, P., Ma, R., Hack, M., et al. (2011). Recent progress of flexible amoled displays: 79560J-79560J-79510.
[175] A. Pecora, L. Maiolo, M. Cuscuna, D. Simeone, A. Minotti, L. Mariucci, et al., Low-temperature polysilicon thin film transistors on polyimide substrates for electronics on plastic, Solid-State Electron. 52 (3) (2008) 348-352.
[176] S. An, J. Lee, Y. Kim, T. Kim, D. Jin, H. Min, et al., 47.2: 2.8-inch WQVGA flexible amoled using high performance low temperature polysilicon TFT on plastic substrates, SID Symposium Digest of Technical Papers 41 (1) (2010) 706-709.
[177] M. Kim, J. Cheon, J. Lee, Y. Park, S. An, T. Kim, et al., 16.2: World-best performance LTPs TFTs with robust bending properties on amoled displays, SID Symposium Digest of Technical Papers 42 (1) (2011) 194-197.
[178] J.-M. Liu, T.M. Lee, C.-H. Wen, C.-M. Leu, 61.2: Invited paper: High performance organic-inorganic hybrid plastic substrate for flexible display and electronics, SID Symposium Digest of Technical Papers 41 (1) (2010) 913-916.
[179] C.H. Choi, B.H. Sohn, J.-H. Chang, Colorless and transparent polyimide nanocomposites: Comparison of the properties of homo- and copolymers, Journal of Industrial and Engineering Chemistry 19 (5) (2013) 1593-1599.
[180] J.S. Lewis, M.S. Weaver, Thin-film permeation-barrier technology for flexible organic light-emitting devices, IEEE Journal of Selected Topics in Quantum Electronics 10 (1) (2004) 45-57.
[181] J.-S. Park, J.-H. Chang, Colorless polyimide nanocomposite films with pristine clay: Thermal behavior, mechanical property,

morphology, and optical transparency, Polymer Engineering & Science 49 (7) (2009) 1357-1365.
[182] I.H. Tseng, Y.-F. Liao, J.-C. Chiang, M.-H. Tsai, Transparent polyimide/graphene oxide nanocomposite with improved moisture barrier property, Materials Chemistry and Physics 136 (1) (2012) 247-253.
[183] I.H. Choi, J.-H. Chang, Colorless polyimide nanocomposite films containing hexafluoroisopropylidene group, Polymers for Advanced Technologies 22 (5) (2011) 682-689.
[184] U. Min, J.-C. Kim, J.-H. Chang, Transparent polyimide nanocomposite films: Thermo-optical properties, morphology, and gas permeability, Polymer Engineering & Science 51 (11) (2011) 2143-2150.
[185] H. Chatham, Oxygen diffusion barrier properties of transparent oxide coatings on polymeric substrates, Surface and Coatings Technology 78 (1) (1996) 1-9.
[186] A. Yoshida, A. Sugimoto, T. Miyadera, S. Miyaguchi, Organic light emitting devices on polymer substrates, Journal of Photopolymer Science and Technology 14 (2) (2001) 327-332.
[187] M. Yan, T.W. Kim, A.G. Erlat, M. Pellow, D.F. Foust, J. Liu, et al., A transparent, high barrier, and high heat substrate for organic electronics, Proceedings of the IEEE 93 (8) (2005) 1468-1477.
[188] J.-J. Huang, Y.-P. Chen, Y.-S. Huang, G.-R. Hu, C.-W. Lin, Y.-J. Chen, et al., 58.1: A 4.1-inch flexible QVGA amoled using a microcrystallinesi: H TFT on a polyimide substrate, SID Symposium Digest of Technical Papers 40 (1) (2009) 866-869.
[189] C. Hassler, T. Boretius, T. Stieglitz, Polymers for neural implants, J. Polym. Sci., Part B: Polym. Phys. 49 (1) (2011) 18-33.
[190] H. Yamaguchi, T. Ueda, K. Miura, N. Saito, S. Nakano, T. Sakano, et al., 74.2l: Late-news paper: 11.7-inch flexible amoled display driven by a-IGZO TFTs on plastic substrate, SID Symposium Digest of Technical Papers 43 (1) (2012) 1002-1005.
[191] Miura, K., Ueda, T., Saito, N., Nakano, S., Sakano, T., Sugi, K., et al. (2013). TFT technologies and FPD materials. *Proceedings of 2013 Twentieth International Workshop on Active-Matrix Flatpanel Displays and Devices (AM-FPD 13)*, Kyoto, Japan. Institute of Electrical and Electronics Engineers (IEEE).
[192] J. Yoon, H. Kwon, M. Lee, Y.-y Yu, N. Cheong, S. Min, et al., 65.1: Invited paper: World 1st large size 18-inch flexible oled display and the key technologies, SID Symposium Digest of Technical Papers 46 (1) (2015) 962-965.
[193] B. Cobb, F.G. Rodriguez, J. Maas, T. Ellis, J.-L. van der Steen, K. Myny, et al., 13.4: Flexible low temperature solution processed oxide semiconductor TFT backplanes for use in amoled displays, SID Symposium Digest of Technical Papers 45 (1) (2014) 161-163.
[194] S. Shi, D. Wang, J. Yang, W. Zhou, Y. Li, T. Sun, et al., 25.3: A 9.55-inch flexible top-emission amoled with a-IGZO TFTs, SID Symposium Digest of Technical Papers 45 (1) (2014) 330-333.
[195] Y.-H. Yeh, C.-C. Cheng, K.-Y. Ho, P.-C. Chen, M.H. Lee, J.-J. Huang, et al., 7-inch color VGA flexible TFT LCD on colorless polyimide substrate with 200°C a-Si:H TFTs, Dig. Tech. Pap. - Soc. Inf. Disp. Int. Symp 38 (Bk. 2) (2007) 1677-1679.
[196] K. Long, A.Z. Kattamis, I.C. Cheng, H. Gleskova, S. Wagner, J.C. Sturm, Stability of amorphous-silicon TFTs deposited on clear plastic substrates at 250°c to 280° c, IEEE Electron Device Letters 27 (2) (2006) 111-113.
[197] N. Fruehauf, M. Herrmann, H. Baur, M. Aman, Low temperature processes for metal-oxide thin film transistors, International Workshop on Active-Matrix Flatpanel Displays and Devices (2015) 39-42.
[198] M.-C. Choi, J.-C. Hwang, C. Kim, S. Ando, C.-S. Ha, New colorless substrates based on polynorbornene-chlorinated polyimide copolymers and their application for flexible displays, J. Polym. Sci., Part A: Polym. Chem. 48 (8) (2010) 1806-1814.
[199] Yang, S., Guo, Y., Song, H. and Liu, J. (2012). Alkyl-substituted alicyclic dianhydride compound and polyimide prepared with the same. Copyright r 2017 American Chemical Society (ACS). All Rights Reserved.
[200] Yang, S., Ni, H., Liu, J. and Yang, H. (2015). Polyimide, manufacture method, polyimide film with good thermal stability and optical transparency as flexible substrate for flexible display device. Copyright r 2017 American Chemical Society (ACS). All Rights Reserved.
[201] Y. Liu, J. Huang, J.H. Tan, Y. Zeng, J.J. Liu, H. Zhang, et al., Intrinsic high-barrier polyimide with low free volume derived from a novel diamine monomer containing rigid planar moiety, Polymer 114 (2017) 289-297.
[202] H.H. Kim, H.J. Kim, B.J. Choi, Y.S. Lee, S.Y. Park, L.S. Park, Fabrication and properties of flexible oleds on polyimide-graphene composite film substrate, Mol. Cryst. Liq. Cryst. 584 (1) (2013) 153-160.
[203] Lin, C.-C., Leu, C.-M. and Kuo, Y.-J. (2015). Aromatic polyimide release layer, substrate structure, and method for manufacturing flexible electronic device. Copyright © 2017 American Chemical Society (ACS). All Rights Reserved.
[204] J.A. Spechler, T.-W. Koh, J.T. Herb, B.P. Rand, C.B. Arnold, A transparent, smooth, thermally robust, conductive polyimide for flexible electronics, Adv. Funct. Mater. 25 (48) (2015) 7428-7434.
[205] T. Kurosawa, T. Higashihara, M. Ueda, Polyimide memory: A pithy guideline for future applications, Polymer Chemistry 4 (1) (2013) 16-30.
[206] H.J. Yen, G.S. Liou, Solution-processable triarylamine-based high-performance polymers for resistive switching memory devices, Polymer Journal 48 (2) (2016) 117-138.

[207] Yen, H.J., Wu, J.H. and Liou, G.S. (2016). Chapter 4: High performance polyimides for resistive switching memory devices. In: *RSC Polymer Chemistry Series* Vol. 2016-January, pp 136-166.
[208] C.-L. Tsai, H.-J. Yen, G.-S. Liou, Highly transparent polyimide hybrids for optoelectronic applications, React. Funct. Polym. 108 (2016) 2-30.
[209] C.-J. Chen, H.-J. Yen, Y.-C. Hu, G.-S. Liou, Novel programmable functional polyimides: Preparation, mechanism of ct induced memory, and ambipolar electrochromic behavior, J. Mater. Chem. C 1 (45) (2013) 7623-7634.
[210] A.-D. Yu, T. Kurosawa, Y.-C. Lai, T. Higashihara, M. Ueda, C.-L. Liu, et al., Flexible polymer memory devices derived from triphenylaminepyrene containing donor-acceptor polyimides, J. Mater. Chem. 22 (38) (2012) 20754-20763.
[211] C.W. Chu, J. Ouyang, J.-H. Tseng, Y. Yang, Organic donor-acceptor system exhibiting electrical bistability for use in memory devices, Adv. Mater. (Weinheim, Ger.) 17 (11) (2005) 1440-1443.
[212] T. Kuorosawa, C.-C. Chueh, C.-L. Liu, T. Higashihara, M. Ueda, W.-C. Chen, High performance volatile polymeric memory devices based on novel triphenylamine-based polyimides containing mono- or dual-mediated phenoxy linkages, Macromolecules (Washington, DC, U. S.) 43 (3) (2010) 1236-1244.
[213] C.-J. Chen, H.-J. Yen, W.-C. Chen, G.-S. Liou, Resistive switching non-volatile and volatile memory behavior of aromatic polyimides with various electron-withdrawing moieties, J. Mater. Chem. 22 (28) (2012) 14085-14093.
[214] T. Kurosawa, Y.-C. Lai, T. Higashihara, M. Ueda, C.-L. Liu, W.-C. Chen, Tuning the electrical memory characteristics from volatile to nonvolatile by perylene imide composition in random copolyimides, Macromolecules (Washington, DC, U. S.) 45 (11) (2012) 4556-4563.
[215] D.P. Erhard, D. Lovera, C. von Salis-Soglio, R. Giesa, V. Altstädt, H.-W. Schmidt, Recent advances in the improvement of polymer electret films, in: A.H.E. Mü ller, H.-W. Schmidt (Eds.), *Complex macromolecular systems ii*, edn, Springer Berlin Heidelberg, Berlin, Heidelberg, 2010, pp. 155-207.
[216] Y.-L. Liu, Q.-D. Ling, E.-T. Kang, K.-G. Neoh, D.-J. Liaw, K.-L. Wang, et al., Volatile electrical switching in a functional polyimide containing electron-donor and -acceptor moieties, J. Appl. Phys. 105 (4) (2009). 044501/044501-044501/044509.
[217] D.M. Kim, Y.-G. Ko, J.K. Choi, K. Kim, W. Kwon, J. Jung, et al., Digital memory behaviors of aromatic polyimides bearing bis(trifluoromethyl)-and bithiophenyl-triphenylamine units, Polymer 53 (8) (2012) 1703-1710.
[218] Y.-C. Hu, C.-J. Chen, H.-J. Yen, K.-Y. Lin, J.-M. Yeh, W.-C. Chen, et al., Novel triphenylamine-containing ambipolar polyimides with pendant anthraquinone moiety for polymeric memory device, electrochromic and gas separation applications, J. Mater. Chem. 22 (38) (2012) 20394-20402.
[219] T.J. Lee, C.-W. Chang, S.G. Hahm, K. Kim, S. Park, D.M. Kim, et al., Programmable digital memory devices based on nanoscale thin films of a thermally dimensionally stable polyimide, Nanotechnology 20 (13) (2009). 135204/135201-135204/135207.
[220] G. Tian, D. Wu, S. Qi, Z. Wu, X. Wang, Dynamic random access memory effect and memory device derived from a functional polyimide containing electron donor-acceptor pairs in the main chain, Macromol. Rapid Commun. 32 (4) (2011) 384-389.
[221] C.-L. Liu, T. Kurosawa, A.-D. Yu, T. Higashihara, M. Ueda, W.-C. Chen, New dibenzothiophene-containing donor-acceptor polyimides for high-performance memory device applications, J. Phys. Chem. C 115 (13) (2011) 5930-5939.
[222] G.-S. Liou, S.-H. Hsiao, H.-W. Chen, Novel high-tg poly(amine-imide)s bearing pendant n-phenylcarbazole units: Synthesis and photophysical, electrochemical and electrochromic properties, J. Mater. Chem. 16 (19) (2006) 1831-1842.
[223] Y. Liu, Y. Zhang, Q. Lan, S. Liu, Z. Qin, L. Chen, et al., High-performance functional polyimides containing rigid nonplanar conjugated triphenylethylene moieties, Chem. Mater. 24 (6) (2012) 1212-1222.
[224] Y. Liu, Y. Zhang, Q. Lan, Z. Qin, S. Liu, C. Zhao, et al., Synthesis and properties of high-performance functional polyimides containing rigid nonplanar conjugated tetraphenylethylene moieties, J. Polym. Sci., Part A: Polym. Chem. 51 (6) (2013) 1302-1314.
[225] Y.-L. Liu, K.-L. Wang, G.-S. Huang, C.-X. Zhu, E.-S. Tok, K.-G. Neoh, et al., Volatile electrical switching and static random access memory effect in a functional polyimide containing oxadiazole moieties, Chem. Mater. 21 (14) (2009) 3391-3399.
[226] N.-H. You, C.-C. Chueh, C.-L. Liu, M. Ueda, W.-C. Chen, Synthesis and memory device characteristics of new sulfur donor containing polyimides, Macromolecules (Washington, DC, U. S.) 42 (13) (2009) 4456-4463.
[227] L. Shi, H. Ye, W. Liu, G. Tian, S. Qi, D. Wu, Tuning the electrical memory characteristics from worm to flash by α- and β-substitution of the electron-donating naphthylamine moieties in functional polyimides, J. Mater. Chem. C 1 (44) (2013) 7387-7399.
[228] T. Kurosawa, A.-D. Yu, T. Higashihara, W.-C. Chen, M. Ueda, Inducing a high twisted conformation in the polyimide structure by bulky donor moieties for the development of non-volatile memory, Eur. Polym. J. 49 (10) (2013) 3377-3386.
[229] B. Hu, F. Zhuge, X. Zhu, S. Peng, X. Chen, L. Pan, et al., Nonvolatile bistable resistive switching in a new polyimide bearing 9-phenyl-9hcarbazole pendant, J. Mater. Chem. 22 (2) (2012) 520-526.
[230] S.G. Hahm, S. Choi, S.-H. Hong, T.J. Lee, S. Park, D.M. Kim, et al., Electrically bistable nonvolatile switching devices

fabricated with a high performance polyimide bearing diphenylcarbamyl moieties, J. Mater. Chem. 19 (15) (2009) 2207-2214.

[231] K. Kim, S. Park, S.G. Hahm, T.J. Lee, D.M. Kim, J.C. Kim, et al., Nonvolatile unipolar and bipolar bistable memory characteristics of a high temperature polyimide bearing diphenylaminobenzylidenylimine moieties, J. Phys. Chem. B 113 (27) (2009) 9143-9150.

[232] S.G. Hahm, S. Choi, S.-H. Hong, T.J. Lee, S. Park, D.M. Kim, et al., Novel rewritable, non-volatile memory devices based on thermally and dimensionally stable polyimide thin films, Adv. Funct. Mater. 18 (20) (2008) 3276-3282.

[233] S. Park, K. Kim, D.M. Kim, W. Kwon, J. Choi, M. Ree, High temperature polyimide containing anthracene moiety and its structure, interface, and nonvolatile memory behavior, ACS Appl. Mater. Interfaces 3 (3) (2011) 765-773.

[234] Y. Li, R. Fang, S. Ding, Y. Shen, Rewritable and non-volatile memory effects based on polyimides containing pendant carbazole and triphenylamine groups, Macromol. Chem. Phys. 212 (21) (2011) 2360-2370.

[235] Y.-Q. Li, R.-C. Fang, A.-M. Zheng, Y.-Y. Chu, X. Tao, H.-H. Xu, et al., Nonvolatile memory devices based on polyimides bearing noncoplanar twisted biphenyl units containing carbazole and triphenylamine side-chain groups, J. Mater. Chem. 21 (39) (2011) 15643-15654.

[236] Y. Li, H. Xu, X. Tao, K. Qian, S. Fu, Y. Shen, et al., Synthesis and memory characteristics of highly organo-soluble polyimides bearing a noncoplanar twisted biphenyl unit containing aromatic side-chain groups, J. Mater. Chem. 21 (6) (2011) 1810-1821.

[237] Y. Li, H. Xu, X. Tao, K. Qian, S. Fu, S. Ding, et al., Resistive switching characteristics of polyimides derived from 2,2′-aryl substituents tetracarboxylic dianhydrides, Polym. Int. 60 (12) (2011) 1679-1687.

[238] Y. Li, Y. Chu, R. Fang, S. Ding, Y. Wang, Y. Shen, et al., Synthesis and memory characteristics of polyimides containing noncoplanar aryl pendant groups, Polymer 53 (1) (2012) 229-240.

[239] G. Tian, S. Qi, F. Chen, L. Shi, W. Hu, D. Wu, Nonvolatile memory effect of a functional polyimide containing ferrocene as the electroactive moiety, Appl. Phys. Lett. 98 (20) (2011). 203302/203301-203302/203303.

[240] S.-H. Liu, W.-L. Yang, C.-C. Wu, T.-S. Chao, M.-R. Ye, Y.-Y. Su, et al., High-performance polyimide-based reram for nonvolatile memory application, IEEE Electron Device Lett. 34 (1) (2013) 123-125.

[241] D.M. Kim, S. Park, T.J. Lee, S.G. Hahm, K. Kim, J.C. Kim, et al., Programmable permanent data storage characteristics of nanoscale thin films of a thermally stable aromatic polyimide, Langmuir 25 (19) (2009) 11713-11719.

[242] K.-L. Wang, Y.-L. Liu, J.-W. Lee, K.-G. Neoh, E.-T. Kang, Nonvolatile electrical switching and write-once read-many-times memory effects in functional polyimides containing triphenylamine and 1,3,4-oxadiazole moieties, Macromolecules (Washington, DC, U. S.) 43 (17) (2010) 7159-7164.

[243] K.-L. Wang, Y.-L. Liu, I.H. Shih, K.-G. Neoh, E.-T. Kang, Synthesis of polyimides containing triphenylamine-substituted triazole moieties for polymer memory applications, J. Polym. Sci., Part A: Polym. Chem. 48 (24) (2010) 5790-5800.

[244] F. Chen, G. Tian, L. Shi, S. Qi, D. Wu, Nonvolatile write-once read-many-times memory device based on an aromatic hyperbranched polyimide bearing triphenylamine moieties, RSC Adv 2 (33) (2012) 12879-12885.

[245] Y.-G. Ko, W. Kwon, H.-J. Yen, C.-W. Chang, D.M. Kim, K. Kim, et al., Various digital memory behaviors of functional aromatic polyimides based on electron donor and acceptor substituted triphenylamines, Macromolecules (Washington, DC, U. S.) 45 (9) (2012) 3749-3758.

[246] K. Kim, H.-J. Yen, Y.-G. Ko, C.-W. Chang, W. Kwon, G.-S. Liou, et al., Electrically bistable digital memory behaviors of thin films of polyimides based on conjugated bis(triphenylamine) derivatives, Polymer 53 (19) (2012) 4135-4144.

[247] S. Park, K. Kim, J.C. Kim, W. Kwon, D.M. Kim, M. Ree, Synthesis and nonvolatile memory characteristics of thermally, dimensionally and chemically stable polyimides, Polymer 52 (10) (2011) 2170-2179.

[248] T. Kurosawa, Y.-C. Lai, A.-D. Yu, H.-C. Wu, T. Higashihara, M. Ueda, et al., Effects of the acceptor conjugation length and composition on the electrical memory characteristics of random copolyimides, J. Polym. Sci., Part A: Polym. Chem. 51 (6) (2013) 1348-1358.

[249] Q. Liu, K. Jiang, Y. Wen, J. Wang, J. Luo, Y. Song, High-performance optoelectrical dual-mode memory based on spiropyran-containing polyimide, Appl. Phys. Lett 97 (25) (2010). 253304/253301-253304/253303.

[250] C.-L. Tsai, C.-J. Chen, P.-H. Wang, J.-J. Lin, G.-S. Liou, Novel solution-processable fluorene-based polyimide/TiO_2 hybrids with tunable memory properties, Polym. Chem. 4 (17) (2013) 4570-4573.

[251] C.-J. Chen, C.-L. Tsai, G.-S. Liou, Electrically programmable digital memory behaviors based on novel functional aromatic polyimide/TiO_2 hybrids with a high on/off ratio, J. Mater. Chem. C 2 (16) (2014) 2842-2850.

[252] A.-D. Yu, T. Kurosawa, Y.-H. Chou, K. Aoyagi, Y. Shoji, T. Higashihara, et al., Tunable electrical memory characteristics using polyimide: Polycyclic aromatic compound blends on flexible substrates, ACS Appl. Mater. Interfaces 5 (11) (2013) 4921-4929.

[253] Y.-H. Chou, N.-H. You, T. Kurosawa, W.-Y. Lee, T. Higashihara, M. Ueda, et al., Thiophene and selenophene donor-acceptor polyimides as polymer electrets for nonvolatile transistor memory devices, Macromolecules 45 (17) (2012)

6946-6956.

[254] Y.-H. Chou, H.-J. Yen, C.-L. Tsai, W.-Y. Lee, G.-S. Liou, W.-C. Chen, Nonvolatile transistor memory devices using high dielectric constant polyimide electrets, J. Mater. Chem. C 1 (19) (2013) 3235-3243.

[255] L. Dong, Y.-C. Chiu, C.-C. Chueh, A.-D. Yu, W.-C. Chen, Semi-conjugated acceptor-based polyimides as electrets for nonvolatile transistor memory devices, Polym. Chem. 5 (23) (2014) 6834-6846.

[256] A.-D. Yu, T. Kurosawa, M. Ueda, W.-C. Chen, Polycyclic arene-based d-a polyimide electrets for high-performance n-type organic field effect transistor memory devices, J. Polym. Sci., Part A: Polym. Chem. 52 (1) (2014) 139-147.

[257] A.-D. Yu, W.-Y. Tung, Y.-C. Chiu, C.-C. Chueh, G.-S. Liou, W.-C. Chen, Multilevel nonvolatile flexible organic field-effect transistor memories employing polyimide electrets with different charge-transfer effects, Macromol. Rapid Commun. 35 (11) (2014) 1039-1045.

[258] L. Dong, H.-S. Sun, J.-T. Wang, W.-Y. Lee, W.-C. Chen, Fluorene based donor-acceptor polymer electrets for nonvolatile organic transistor memory device applications, J. Polym. Sci., Part A: Polym. Chem. 53 (4) (2015) 602-614.

[259] Y.-F. Wang, M.-R. Tsai, P.-Y. Wang, C.-Y. Lin, H.-L. Cheng, F.-C. Tang, et al., Controlling carrier trapping and relaxation with a dipole field in an organic field-effect device, RSC Adv. 6 (81) (2016) 77735-77744.

[260] Y.-H. Chou, C.-L. Tsai, W.-C. Chen, G.-S. Liou, Nonvolatile transistor memory devices based on high-k electrets of polyimide/TiO_2 hybrids, Polym. Chem. 5 (23) (2014) 6718-6727.

Chapter 6

Polyimide Gas Separation Membranes

Xiao-Hua Ma and Shi-Yong Yang
Institute of Chemistry, Chinese Academy of Sciences, Beijing, China

6.1 Introduction

A membrane is considered to be a selective barrier between two phases, which can be solid film or liquid of certain thicknesses [1]. The first membrane process was discovered by Nollet [2], who observed that a pig's bladder could preferentially transfer ethanol when used as a barrier between water-ethanol mixtures. The pressure induced by those two chemical potentials is called osmotic pressure. The first large-scale industrial application of membrane-based gas separation was achieved by Permea (Monsanto) for hydrogen recovery from the purge gas stream of ammonia plants by Prism membranes in 1977 [3]. Since then, membrane technology has attracted intensive attention in both industrial and academic areas because of its versatile advantages over other technologies, including: (1) there are no phase changes or chemical additives in the separation process; (2) it is simple in concept and operations resulting in low maintenance costs; (3) it is able to be made into modules and it is easy to scale up; and (4) it has a small footprint, etc. [4]. Baldus and Tillmman [5] described some general rules for the membrane-based gas separation technique which are extremely preferred in situations such as: (1) bulk separation where only moderate purity recovery is required; (2) the feeding gas is under high pressure and the retentive side is the product required. Currently, there are many industrial applications that membranes can be used in, including:

- air separation, including both on-site nitrogen generation and oxygen fluent gas production;
- recovery of hydrogen from product streams of ammonia plants;
- separation of hydrogen from gases such as nitrogen and methane;
- biogas and landfill gas upgrading;
- natural gas sweetening, including removal of CO_2 and H_2S;
- removal of water vapor from natural gas and other gases;
- on-site N_2 generation for long-term shipment and the food industry;
- removal of volatile organic liquids from the air of exhaust streams;
- olefin/paraffin separation;
- isomer separation such as xylene isomers, benzene and its derivatives;
- flue gas treatment including removal of CO_2 and carbon capture.

In the United States, the total energy consumed in industries takes up 32% of the total energy consumption of the whole country (Fig. 6.1). Separation processes take up almost half of the industrial energy consumption. If membrane technology can replace other separation techniques, ~90% of the energy in the separation process will be saved as compared with the conventional distillation process [6].

Driven by so many advantages of the membrane technique, many plants have been installed in the gas separation field in recent years. Fig. 6.2 shows the road map of industrial membrane gas separation that has occurred in the last 30 years. The first commercial industrial gas separation polyimide was

Advanced Polyimide Materials. DOI: https://doi.org/10.1016/B978-0-12-812640-0.00006-8

established in 1989 for H_2 separation by Ube. Later, Medal Company used polyimide membranes for CO_2/CH_4 separation in 1994. However, the practical largest natural gas sweetening membrane was still cellulose acetate at UOP in 2008.

Polymer-based gas separation membrane can be classified to different categories, including dense membranes, anisotropic membranes, Loeb-Sourirajan membranes, composite membranes, and facilitate membranes, etc. Some membrane types are illustrated in Fig. 6.3.

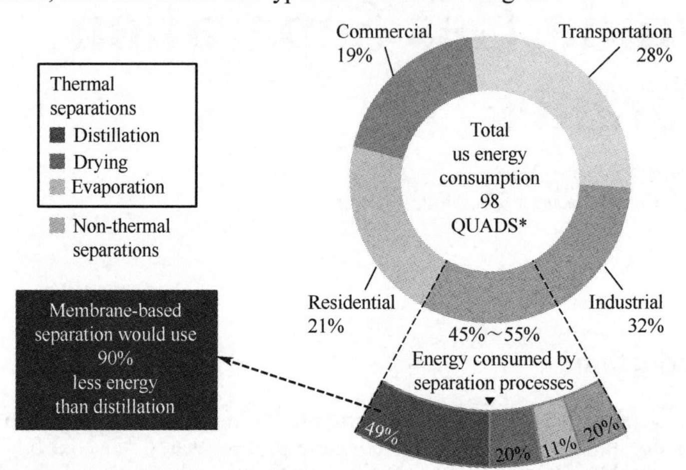

FIGURE 6.1 Energy consumption in the United States and the proposed energy saving by the membrane technique.

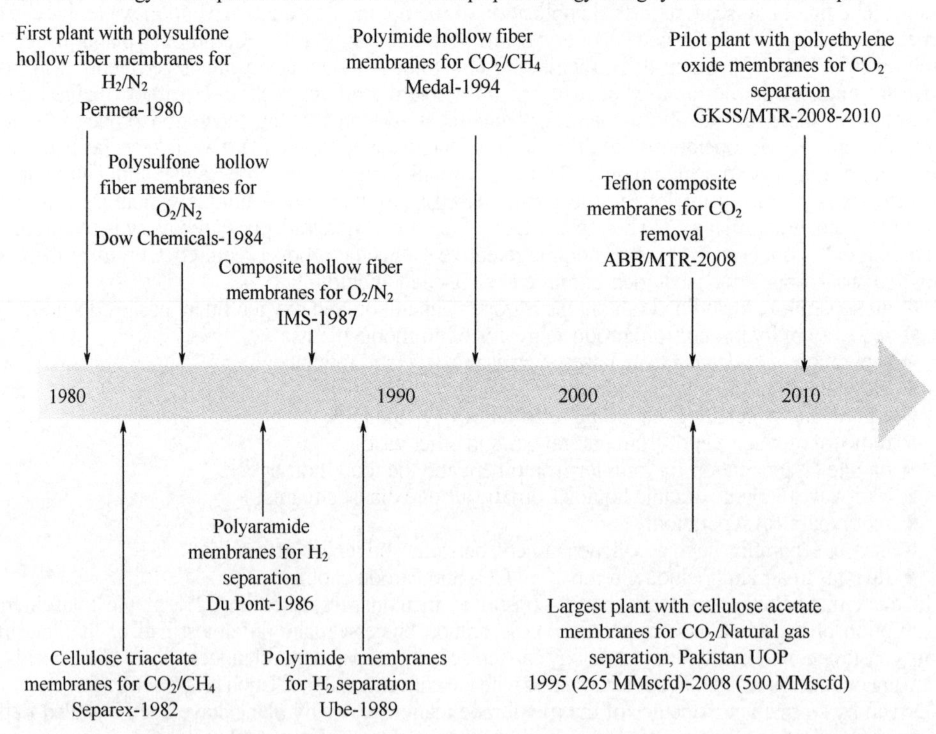

FIGURE 6.2 Road map for industrial applications of membrane-based gas separation.

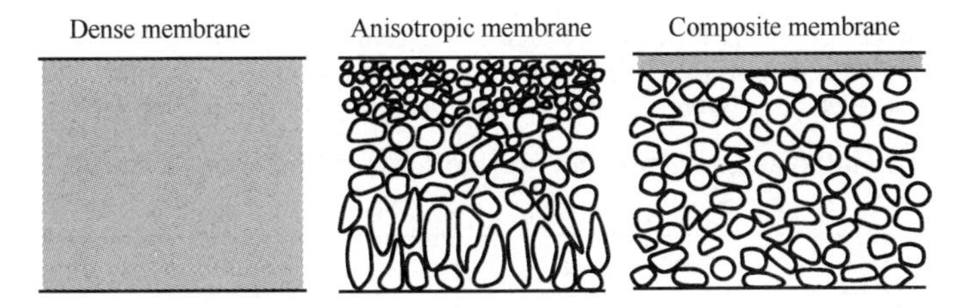

FIGURE 6.3 Schematic structures of polymer-based membranes.

The dense membrane is characterized by no pores between the top layer and the bottom layer, which can be obtained by slow evaporation of the polymeric solution on a special substrate. For better membrane quality, precise control of the casting temperature, solvent polarizability, evaporation of the solvent and moisture in the environment, etc., are required. The dense membranes can also be obtained by the melting-stretching method. The melted polymer resin was extruded and stretched. After cooling down, a thin membrane was obtained.

The anisotropic membrane can be obtained by a phase inversion method, the first anisotropic cellulose acetate membrane was prepared via phase inversion by Loeb and Sourirajan in the 1960s [7]. The membrane was composed of a very thin-skin dense layer with thickness ~ 0.2 μm and a macro-porous support layer with thickness of more than several hundreds of micrometers. This method is especially attractive in industries for large-scale membrane production. Until now, phase inversion is still the most frequently method used for making commercially available polymer membranes.

The composite membrane is a little different from the anisotropic membrane, in which the skin layer and the supported layer are prepared with different polymer materials. In 1963, Riley discovered a technique which could prepare the skin layer and the support layer; the thickness of the skin layer could be controlled down to 30 nm-50 nm [8]. There are several methods used to prepare the active skin layer, including (1) casting the polymer solution on top of the supporting layer; (2) in situ interfacial polymerization on the skin layer; and (3) in situ free radical, UV or thermal polymerization on the top surface of the supporting substrate.

6.2 Mechanisms of Gas Separation and Testing Methods

6.2.1 Gas Transport Mechanism

In principle, membranes separate different gases by the transport speed of gas molecules across the separating membrane. The gas molecule pass-through of the membranes is significantly affected by two factors. One is the dynamic diameter of the gas molecule, another is the porosity and rigidity of the membrane itself. The dynamic diameters of different gas molecules are summarized in Table 6.1.

TABLE 6.1 Gas and Vapor Properties for Sorption and Transport Properties

Gas	He	H_2	N_2	O_2	CO_2	CH_4	C_2H_4	C_2H_6	C_3H_6	C_3H_8	H_2S
T_c^a	5.2	33.2	126.2	154.6	304	191	282.5	305.3	365.2	369.9	373
V_c^b	57.5	64.9	89.3	73.5	91.9	98.6	131.1	147	184.6	200	87.7
d_{L-J}^c	2.55	2.83	3.80	3.47	3.94	3.76	4.16	4.44	4.68	5.11	3.62
d_K^d	2.6	2.89	3.64	3.46	3.3	3.8	3.9	–	4.5	4.3	3.6

[a]*Critical temperature (T_c, K) of the gases [9].*
[b]*Critical volume (V_c, $cm^3 \cdot mol^{-1}$) of the gases [9].*
[c]*Lennard-Jones diameter (d_{LJ}, Å) [10].*
[d]*Kinetic diameter (d_K, Å) of the gas molecules [11].*

The dynamic diameter (d_{L-F}) of the gas molecules range from 2.6Å to 4.5Å, which is in the range of thermal-fluctuation-induced polymer chain rearrangement. In order to get a high-performance membrane, the pore size of the polymers should be in a proper range of ultramicropore (< 7Å, Table 6.2).

TABLE6.2 IUPAC (International Union of Pure and Applied Chemistry) Classification of Pores as a Function of Size

	Micropores		Mesopores	Macropores
	Ultramicropore	Super-micropore		
Range (nm)	<0.7	0.7-2	2-50	>50

Taking the gas molecules and the pore size of the membrane into account, if the pores (r) of the membrane are very large and comparable to the mean free path (λ) of gas molecules (Fig. 6.4), the gas transport though the membrane will follow Knudsen diffusion [12]. In this case, the speed of gas molecules transport though the membrane is determined by the gas molecules frequently colliding with the walls of the pores. The resulting separation selectivity of this mechanism is proportional to the ratio of the inverse square root of molecular weights. In this case, the lighter gas molecule such as H_2 transport much faster than the heavier gas molecules such as CO_2, the separation factor is $(44/2)^{1/2}$ -4.7 (Table 6.3).

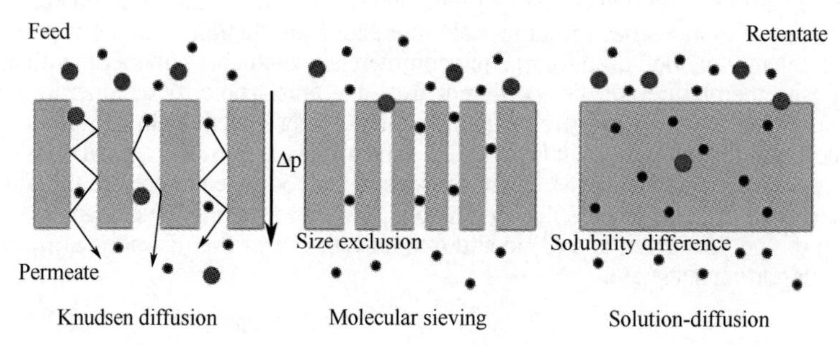

FIGURE 6.4 Schematic transport mechanisms of gas molecules through the membranes: Knudsen diffusion, molecular sieving, and solution-diffusion models.

TABLE6.3 Mean Free Path of the Gases (25 °C, 1 Bar) and Gas Separation Factor Based on Knudsen Flow for Selected Binary Gas Mixtures

Gas	λ (Å)	Gas Pair	Separation Factor
Argon	1017	H_2/N_2	3.73
Hydrogen	1775	H_2/CO_2	3.73
Helium	2809	H_2/H_2S	4.11
Nitrogen	947	H_2/CO_2	4.67
Neon	2005	O_2/N_2	1.07
Oxygen	1039	O_2/CO_2	1.17

If the pore size of the membrane is very small, in the range of ultramicropore region (Fig. 6.4 and Table 6.2), and distributed precisely and uniformly, the membrane can be considered a molecular sieve membrane. The accessibility of the pores can only occur in those specialized gas molecules with smaller dynamic diameters, if the gas molecule is bigger than the pore size, the window size of the membrane will reject such a molecule. This kind of molecular sieve membrane is mostly distributed in inorganic and carbon membranes, and usually shows excellent gas permeability and selectivity, but is limited to special application areas due to its brittle characteristics and high cost.

As for the dense polymeric membranes, the gas molecule transport through the membranes follows the solution-diffusion model (Fig. 6.4), which was first proposed by Graham [13]. The separation process can be divided into several steps: (1) the gas molecules are adsorbed in the upstream of the membrane; (2) the gas molecules diffuse inside the membrane to the downstream; and (3) the gas molecules undergo desorption in the downstream [14]. The permeability of gas molecule pass-through the membrane is determined by the two parameters, diffusivity (D) and solubility (S), as expressed in the below Eq. (6.1):

$$P = D \times S \tag{6.1}$$

The rate control step is the diffusion step which can be expressed in the following Fig. 6.5. This process can be expressed as follows: the gas molecules inside the polymer bounded to part of the polymer chains, under thermal fluctuation or outside pressure driven by the upstream pressure, the gas molecules jumped to the activated state by moving forward for a certain distance, and thereafter, the gas molecules returned to their normal states and were again covered by or attached to the polymer main chain.

The diffusion coefficient was related to the fractional free volume (FFV) [15] of the polymers and can be expressed in the following Eq. (6.2):

$$D = A_0 \exp\left(-\frac{\gamma V^*}{V_f}\right) \tag{6.2}$$

In which, A_0 is a preexponential factor, V_f is the average free volume in the system, γ is a parameter of order unity, and V^* characterizes the size of diffusing molecules. Normally, γV^* is related to the kinetic diameter of the gas molecules. V_f can be calculated using Eq. (6.3) below:

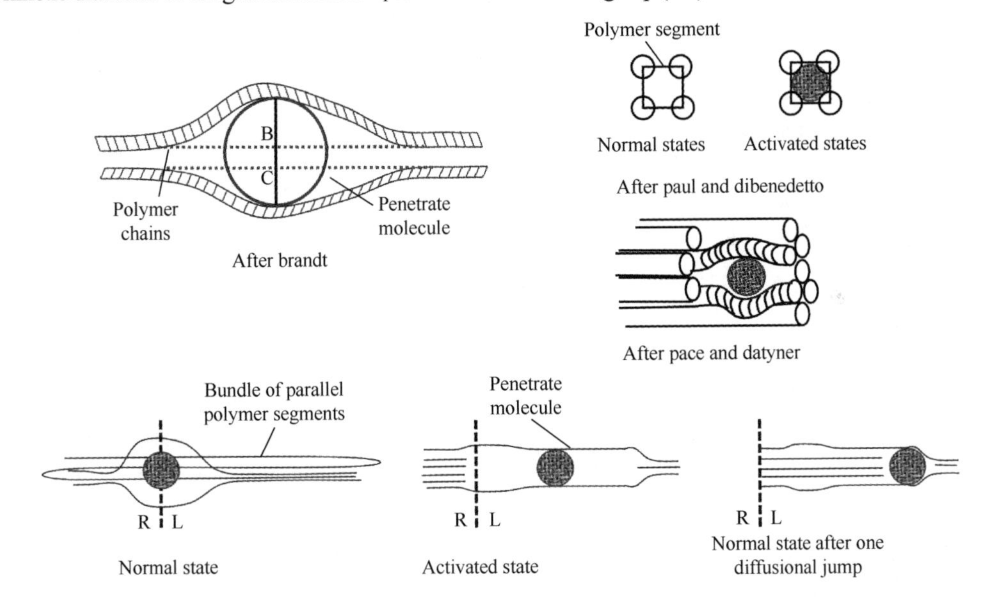

FIGURE 6.5 Mechanism of gas molecule diffusion through polymer membranes.

$$V_f = V_{sp} - 1.3V_w \tag{6.3}$$

$V_{sp}=1/\rho$, where ρ (g·cm^{-3}) is the density of the polymer membrane. V_{oc} is the occupied volume, Bondi proposed calculating V_{oc} as $1.3V_w$ [16] where V_w is the sum of the increments of van der Waals volumes. The V_f/V_{sp} is considered as its FFV of the polymer. Taking this into the Eq. (6.4):

$$D = Ae^{-B/FFV} \tag{6.4}$$

where A and B are constant parameters, the result indicated that the diffusion coefficient is only related to the FFV of the polymers. The higher the fraction free volume, the higher the diffusivity coefficient of the polymer membranes.

The solubility of the gas in the membranes can be expressed by the dual model sorption, which can be expressed in Eq. (6.5):

$$S = k_{\mathrm{D}} p + C'_{\mathrm{H}} \frac{bp}{1+bp} \tag{6.5}$$

The first part ($k_{\mathrm{D}}p$) is called Henry sorption, the second part $C'_{\mathrm{H}}\ bp/(1+bp)$ is the Langmuir sorption. Where k_{D} is the Henry's law parameter characterizing sorption into the densified equilibrium matrix of the glassy polymer, C'_{H} is the Langmuir sorption capacity, which characterizes sorption into the nonequilibrium excess volume associated with the glassy state, and b is the Langmuir affinity parameter. The Langmuir pores inside the polymer main chain both affect the diffusion coefficient (D) and also solubility coefficient (S).

The performance of the polymer membranes for gas separation is often evaluated using Robeson trade-off curves. After analyzing the permeability of thousands of polymers and the selectivity for different gas pairs, he proposed the state-of-the-art Robeson upper bound [17,18]. Some trade-off curves for certain gas pairs are shown in Fig. 6.6 below. There is a correlation between permeability and selectivity of the polymer membranes. The higher the permeability, the lower the selectivity and vice versa.

The permeability and selectivity can be expressed in Eq. (6.6):

$$P_i = K\alpha_{ij}^{n} \tag{6.6}$$

where P_i is the permeability for gas i, K is the preexponential factor, α_{ij} is the selectivity of gas i to gas j. Eq. (6.6) can be switched to Eq. (6.7):

$$\ln\alpha_{ij} = \frac{1}{n} \times \ln P_i - \frac{1}{n} \times \ln K \tag{6.7}$$

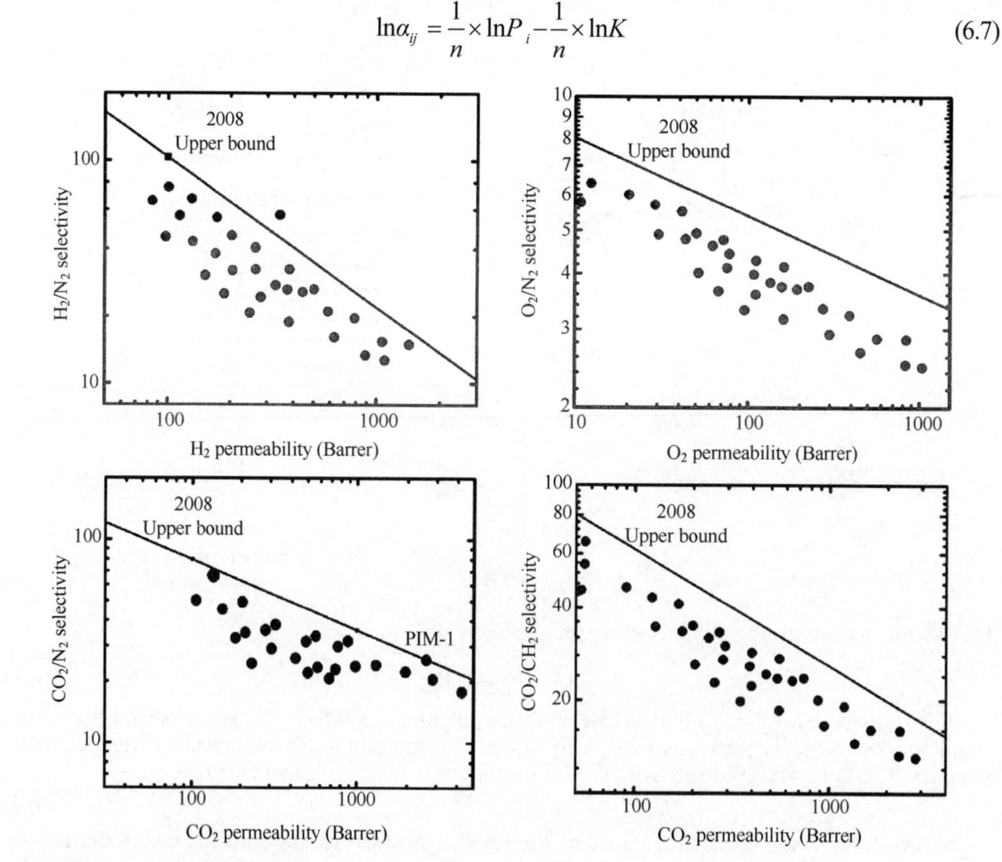

FIGURE 6.6 Robeson upper bounds correlation for O_2/N_2, H_2/N_2, CO_2/N_2, and CO_2/CH_4.

Eq. (6.6) was also theoretically investigated by Freeman [19], and Eq. (6.6) can be expressed in Eq. (6.8):

$$\ln\alpha_{ij} = -\left[\left(\frac{d_j}{d_i}\right)^2 - 1\right]\times\ln P_A + \left\{\ln\left(\frac{S_i}{S_j}\right) - \left[\left(\frac{d_j}{d_i}\right)^2 - 1\right]\times\left(b - f\left(\frac{1-a}{RT}\right) - \ln S_i\right)\right\} \quad (6.8)$$

where d_i and d_j are kinetic diameters of the gas molecules, S_i and S_j are solubility of the gas in polymer membranes; a and b are constant values for different polymers, which are also independent from the gas types; f is a value related to the polymer itself; for different polymers, the f value can range from 0 to 14,000 $kcal\cdot mol^{-1}$ [20].

Compared with the above two Eqs. (6.2) and (6.3), the slopes of the trade-off curves are related to the kinetic diameter of the gas molecules ($1/n=1-(d_i/d_j)^2$). The interceptions change based on the advent of the different polymer itself. The higher the rigidity, the higher the interception of the Robeson upper bond.

6.2.2 Apparatus for Testing Gas Transport Properties

There are generally two methods to measure the gas transport properties for dense membranes. One is constant volume/variable pressure and the other is constant pressure/variable volume method. In the first case, the set-up can be schematically shown as in Fig. 6.7 [21].

The polymer membranes were put in the permeation cell. The speed of the gas transport through the membranes was recorded using a pressure transducer, and the permeability (P) can be obtained by calculation of the permeation rate at the steady state. There is a delay in the pressure versus time because the gases are diffusing in the membrane at the beginning from the upstream to downstream, and this is called "time-lag." This fixed volume/variable pressure method is also called the time-lag method [22]. The schematic time-lag data are shown in Fig. 6.8. At the very beginning of the permeation test, there is a certain time (t) that a very small pressure increase rate is observed as compared to the steady state permeate rate, this delayed time is closely related to the diffusion coefficient of the certain gas to the membrane, which can be expressed as in Eq. (6.9):

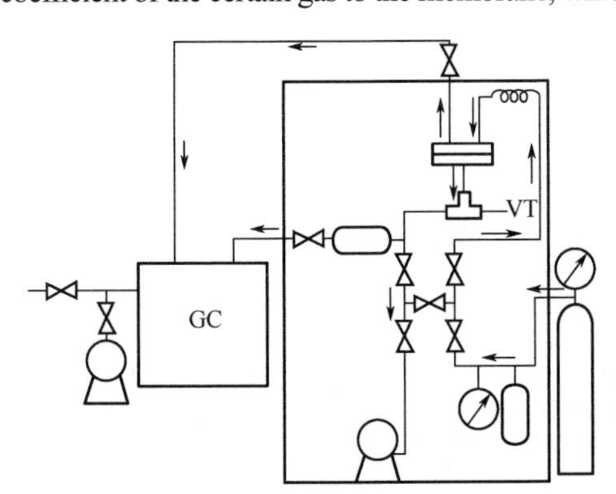

FIGURE 6.7 Schematic set-up for testing the permeability of dense membranes.

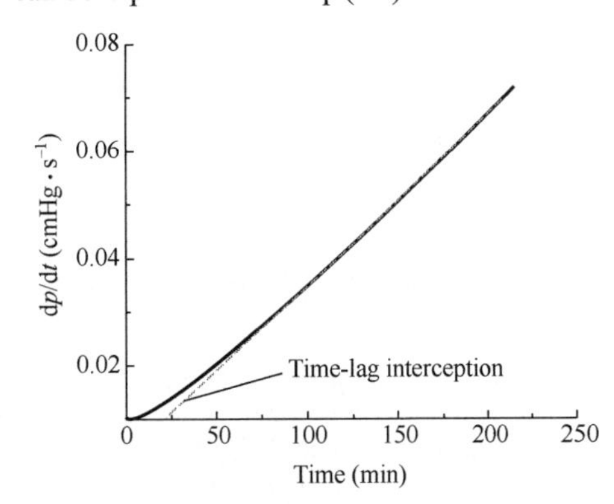

FIGURE 6.8 Pressure response of the downstream on increasing permeation time and the illustration of time-lag interception.

$$D = L^2 / 6\theta \quad (6.9)$$

where L is the thickness of the membrane, θ is the angle between the steady state time-permeation derivative line and the x-axis (Fig. 6.8). The S values can also be obtained and calculated using $S=P/D$.

The schematic diagram of the second method for measuring the membrane permeation properties is shown in Fig. 6.9. The gas transport through the membrane at a fixed upstream pressure and the permeated gases are collected at the downstream. This method is quite good for those membranes with high permeability or flux.

6.3 Structures and Properties of Polyimide Membranes

A good membrane candidate for gas separation should have the following properties: (1) medium to high permeability/flux; (2) high selectivity for the desired gas pairs; (3) environment stability and can tolerate some impurities in the feed gas; (4) mechanically robust and can be easily scaled up for mass production; (5) good thermal stability; (6) producing reproducibility, properties stable from batch to batch; (7) easy packing and can be easily embedded to module; (8) acceptable costs. The current commercially available gas separation membrane materials in industry and those still being intensively investigated in the laboratory are cellulose triacetate, polysulfone, polyamide, polyimide, polycarbonate, and perfluorinated polymers, etc. Among the above-mentioned polymeric materials, aromatic polyimide membranes are one of the desired polymer materials for gas separation due to their excellent thermal, chemical, and mechanical properties, as well as extraordinary film-formation abilities.

Aromatic polyimide was synthesized by condensation of dianhydrides and diamines under certain conditions. Thus, its gas transport performance can be adjusted by tailoring the molecular structures of diamine and/or dianhydride monomers. Pioneering research on the structure/gas transport properties of polyimides were carried out by Hoehn in 1972 [23]. His criteria for polyimides with the chemical structures are shown in Fig. 6.10. Take 6FDA-*m*-PD as an example: (1) the polyimide was classified into four parts separated by carbon-carbon or carbon-nitrogen single bonds; although connected by a single bond, they are not in the same plane as the aromatic structures directly connected to them; (2) these bonds can rotate, however, only with a restricted rotation angle; (3) the predominant building blocks in these polyimides are aromatic structures.

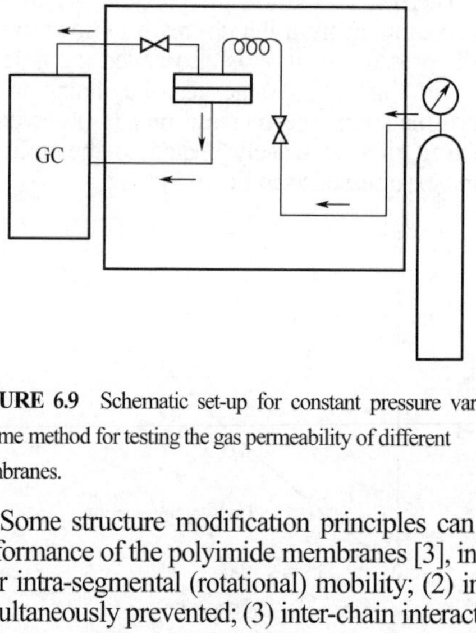

GC

FIGURE 6.9 Schematic set-up for constant pressure various volume method for testing the gas permeability of different membranes.

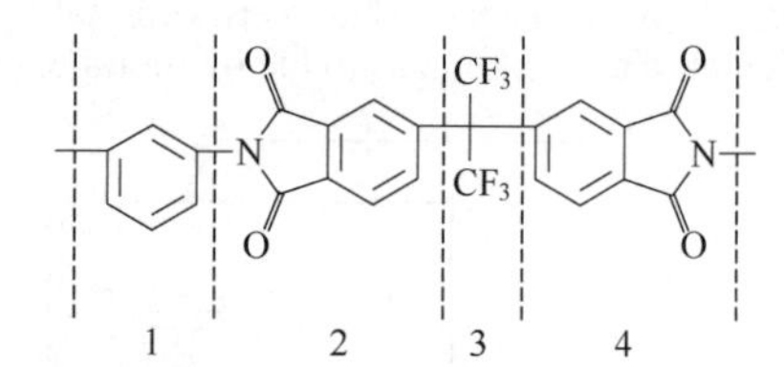

FIGURE 6.10 Structure analysis of 6FDA-*m*PD based on Hoehn's criteria for 6FDA-*m*PD [23].

Some structure modification principles can be applied to achieve higher permeability/selectivity performance of the polyimide membranes [3], including: (1) the main chains must be rigid by inhibiting their intra-segmental (rotational) mobility; (2) inter-segmental packing of the polymer chains must be simultaneously prevented; (3) inter-chain interactions must be weakened.

6.3.1 Isomer Structure Effects From Diamines

Isomers of diamine monomer had significant effects on its macromolecular folding, so as to affect polymer chain packing and fraction free volume. It resulted in different gas transport properties. Pioneering work was carried out by Koros using 6FDA as dianhydride monomer and 6FmDA (3,3′position) and 6FpDA (4,4′ position) as diamine monomers [24] to form PI-(6FDA/6FpDA) and PI-(6FDA/6FmDA) (Fig. 6.11).

The substitution at the *para*-position (6FDA/6FpDA) demonstrated a significantly larger *d*-spacing (5.9Å vs 5.7Å), higher glass transition temperature (T_g) value (320 °C vs 254 °C), *FFV* (0.1897 vs

0.1748) and permeability than the meta-substituted 6FDA/6FmDA (63.9 and 5.1 Barrer for CO_2, respectively), with much lower CO_2/CH_4 selectivity of 39.9 and 63.8, respectively (Table 6.4). It is hypothesized that differences in *sub*-T_g relaxations are responsible for the dramatically higher diffusivities and permeabilities in the *para*-connected materials relative to the *meta*-connected ones. The higher permeability of the polyimides derived from *para*-diamine was also observed in PI-(PMDA/4,4′- ODA) polymer compared with PI-(PMDA/3,3′-ODA) (P_{CO_2} of 1.14 Barrer vs 0.5 Barrer). Stern et al. think the *para*-isomers have very likely a broader free-volume distribution than the *meta*-isomers (Fig. 6.11) [14]. This could be determined by means of molecular dynamics simulations.

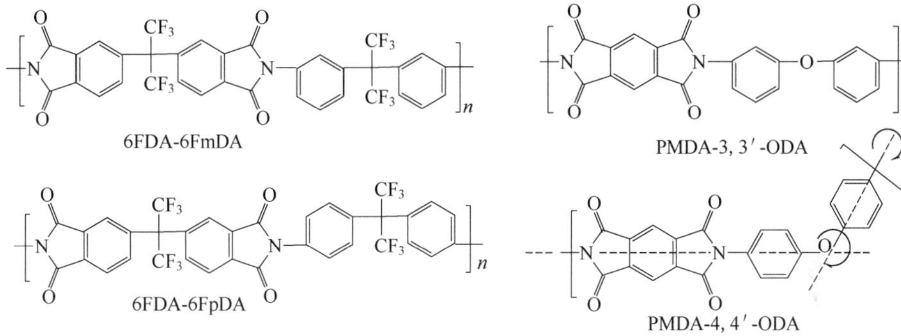

FIGURE 6.11 Structures of the 6FDA-based and PMDA-based polyimide isomers.

TABLE 6.4 Permeability and Selectivity of the Polyimides Derived From Different Diamine Isomers

Polyimides	Permeability[a]					Selectivity ($\alpha_{X/Y}$)			
	H_2	N_2	O_2	CH_4	CO_2	P_{H_2}/P_{N_2}	P_{H_2}/P_{CH_4}	P_{O_2}/P_{N_2}	P_{CO_2}/P_{CH_4}
6FDA/6FpDA[b]	—	3.47	16.3	1.60	63.9	—	—	4.7	39.9
6FDA/6FmDA[b]	—	0.26	1.8	0.08	5.1	—	—	6.9	63.8
PMDA/3,3′-ODA[c]	3.6	0.018	0.13	0.008	0.5	199	450	7.2	62
PMDA/4,4′-ODA[c]	3.0	0.049	0.22	0.026	1.14	61.3	115	4.5	43

[a] *The permeability of the polyimides, scale: Barrer, 1 Barrer=10^{-10} cm^3 (STP)·cm^{-1}·s^{-1}·$cmHg^{-1}$.*
[b] *Tested at the upstream pressure of 10 atm.*
[c] *Tested at the upstream pressure of 4.67 atm.*

6.3.2 Substitution and Geometric Effects From Diamines

The gas transport through membrane is very sensitive to the polyimide chain packing, which is determined by the backbone structure of the polyimide itself. Any modification to the polyimide either from the diamine and/or dianhydride monomers will have a significant effect on the gas transport properties. The effects of methyl substitution on gas transport properties were first investigated in polycarbonate-based membranes [25]. The tetramethyl-substituted polycarbonate demonstrated much higher free volume than its pristine polycarbonate, resulting in almost 1.5 times sorption capacity for CO_2 and twofold diffusion coefficient compared to the conventional PC at 35 °C and 20 atm. The methyl substitution in the diamines for polyimides also resulted in an improved *FFV* by impeding its rotation of the imide bond. Tanaka et al. prepared a series of polyimides by polycondensation of 6FDA with different methyl-substituted diamines, as shown in Fig. 6.12 [26]. The methyl-substituted phenylenediamines included unsubstituted (*m*-PD and *p*-PD), to mono- or bimethyl-substituted (*m*-MPD and *p*-DiMPD) to trimethyl- or tetramethyl-substituted polyimides (*m*-TrMPD and *p*-TeMPD).

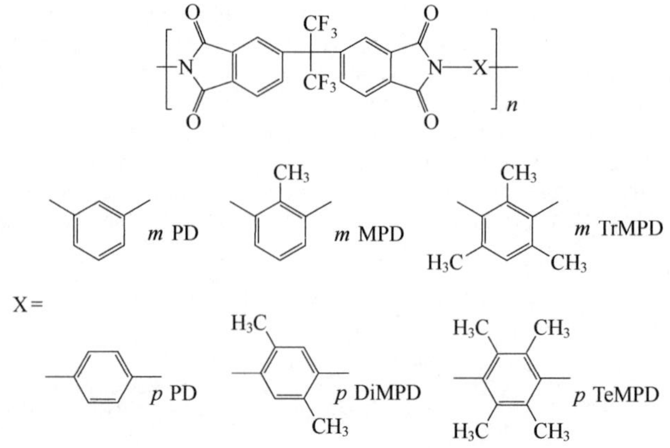

FIGURE 6.12 Polyimides derived from 6FDA and different methyl-substituted diamines.

Similar to the isomer effect, the *meta*-substituted phenylenediamines exhibited a lower T_g, density, and *FFV* (Table 6.5). This is attributed to the greater rotation freedom of the *meta*-substituted isomer in the polymer main chain. The methyl substitution at the *ortho*-position of the diamines also resulted in an increased *FFV*, higher T_g, lower density and solubility parameters. The incorporation of a mono-methyl group resulted in a slightly higher T_g (5 °C-10 °C), while that of multimethyl groups in the diamine resulted in a huge jump in T_g of ~ 80 °C.

The gas permeability and selectivity of the polyimide membranes derived from 6FDA and various methyl-substituted phenylenediamines are summarized in Table 6.6. If there is only one methyl substitution in the *ortho*-position of the polyimides, the permeability for different gases increases by about two to five times, whereas the second methyl substitution leads to another 5-10 times increment of permeability. All of the methyl substitutions result in a decrease in selectivity.

TABLE6.5 Basic Properties of the Polyimide Films

Polyimides	T_g (°C)	Density ($g \cdot cm^{-3}$)	V_F	δ
6FDA/*m*-PD	298	1.474	0.160	27.7
6FDA/*m*-MPD	335	1.416	0.176	27.3
6FDA/*m*-TrPD	377	1.352	0.182	26.4
6FDA/*p*-PD	351	1.473	0.161	27.7
6FDA/*p*-DiMPD	355	1.390	0.175	26.8
6FDA/*p*-TeMPD	420	1.333	0.182	26.1

V_F is the fractional free volume obtained using Bondi's method [16]; δ is the solubility parameter, which is calculated by group contribution method of van Krevelen [20].

TABLE6.6 Permeability and Selectivity of the 6FDA-based Polyimides With Different Methyl Substitutions

Polyimides	Permeability[a]					Selectivity ($\alpha_{X/Y}$)			
	H_2	N_2	O_2	CH_4	CO_2	P_{H_2}/P_{N_2}	P_{H_2}/P_{CH_4}	P_{O_2}/P_{N_2}	P_{CO_2}/P_{CH_4}
6FDA/*m*-PD	40.2	0.447	3.01	0.160	9.20	89.9	251	6.73	57.5
6FDA/*m*-MPD	106	2.24	11.3	0.877	40.1	47.3	121	5.04	45.7

TABLE 6.6 (Continued)

Polyimides	Permeability[a]					Selectivity ($\alpha_{X/Y}$)			
	H_2	N_2	O_2	CH_4	CO_2	P_{H_2}/P_{N_2}	P_{H_2}/P_{CH_4}	P_{O_2}/P_{N_2}	P_{CO_2}/P_{CH_4}
6FDA/*m*-TrPD	516	31.6	109	26	431	16.3	19.8	3.45	16.5
6FDA/*p*-PD	45.5	0.799	4.22	0.286	15.3	56.9	159	5.28	53.5
6FDA/*p*-DiMPD	119	2.67	13.4	1.07	42.7	44.5	111	5.02	39.9
6FDA/*p*-TeMPD	549	35.6	122	28.2	440	15.4	19.4	3.43	15.6

[a]*The permeability of the polyimides, scale: Barrer, 1 Barrer=*10^{-10} cm^3 *(STP)* $\cdot cm^{-1} \cdot s^{-1} \cdot cmHg^{-1}$.

Although there is a huge increase in the fraction free volume, there is a slight change in different gas solubility and gas pair selectivity (Table 6.7), either with mono- or bimethyl-substitution in the *ortho*-position of the diamines.

On the contrary, a huge increase in N_2, O_2, CH_4, and CO_2 diffusion coefficient as well as D_{O_2}/D_{N_2} and D_{CO_2}/D_{CH_4} diffusion selectivity is observed. The results are shown in Table 6.8. The introduction of a methyl group in the *ortho*-position of diamine resulted in an enhancement of diffusion coefficient because of the higher *FFV* and higher rigidity of the polymer main chain. However, the "big" gas molecules such as CH_4 demonstrated a much larger diffusion coefficient increment than the "small" gas molecules, such as H_2, resulting in a sharp decrease in diffusivity selectivity.

The permeability/selectivity of the polyimide membranes are not only affected by the position of methyl substitution, but also by the geometric of the diamines. Fritsch et al. [27] prepared a series of polyimides derived from 6FDA and substituted terphenylene-based diamines (Fig. 6.13). The diamine in PI-1 has four methyl groups substituted in the *ortho*-position, the diamines in PI-2, PI-4 have two methyl groups. The difference between these two diamines is the angle (θ) between the diamine and the terphenylene highlighted in Fig. 6.13, which resulted in the more asymmetric structures in the polyimides. The PI-3 and PI-5 had no methyl group in their *ortho*-position of the diamine segments.

The gas permeability and selectivity of the 6FDA-based polyimide membranes for some gas pairs are listed in Table 6.9. It is clear that the PI-1 with methyl group substituted on both sides exhibited the highest permeability. The permeability of the polyimides derived from dimethyl-substituted diamines is 5 to 10 times larger than that derived from the mono-methyl substituted diamines. On the other hand, the PI-2 with the diamine out of the terphenylene plane showed approximately fivefold higher permeability than the linear PI-4. The linear PI-5, with no methyl substitution in the *ortho*-position of the diamine, exhibited the lowest permeability for most of the gases while its selectivity for different gas pairs such as H_2/N_2, H_2/CH_4, O_2/N_2, and CO_2/CH_4 are the highest among the methyl-substituted terphenylene-based polyimides.

TABLE 6.7 Solubility Coefficient of the Methyl-Substituted Polyimides and Solubility Selectivity for O_2/N_2 and CO_2/CH_4

	S_{N_2}	S_{O_2}	S_{CH_4}	S_{CO_2}	S_{O_2}/S_{N_2}	S_{CO_2}/S_{CH_4}
6FDA/*m*-PD	8.6	12	16	55	1.3	3.4
6FDA/*m*-MPD	12	15	21	65	1.3	3.1
6FDA/*m*-TrPD	13	16	26	80	1.3	3.0
6FDA/*p*-PD	9.9	12	18	59	1.2	3.3
6FDA/*p*-DiMPD	13	16	23	66	1.2	2.9
6FDA/*p*-TeMPD	15	19	29	92	1.2	3.1

The solubility was recorded and the scale is 10^{-3} cm^3 (STP) $\cdot$ cm^{-3} $\cdot cmHg^{-1}$.

TABLE 6.8 Diffusion Coefficient of the Methyl-Substituted Polyimides and Diffusivity Selectivity for O_2/N_2 and CO_2/CH_4

	D_{N_2}	D_{O_2}	D_{CH_4}	D_{CO_2}	D_{O_2}/D_{N_2}	D_{CO_2}/D_{CH_4}
6FDA/*m*-PD	0.52	2.6	0.10	1.7	5.0	17
6FDA/*m*-MPD	1.9	7.5	0.42	6.2	3.9	15
6FDA/*m*-TrPD	25	67	9.9	54	2.7	5.5
6FDA/*p*-PD	0.81	3.5	0.16	2.6	4.3	16
6FDA/*p*-DiMPD	2.1	8.6	0.48	6.5	4.1	14
6FDA/*p*-TeMPD	23	66	9.6	48	2.8	5.0

The diffusion coefficient was recorded with time-lag method and the scale is 10^{-8} $cm^2 \cdot s^{-1}$.

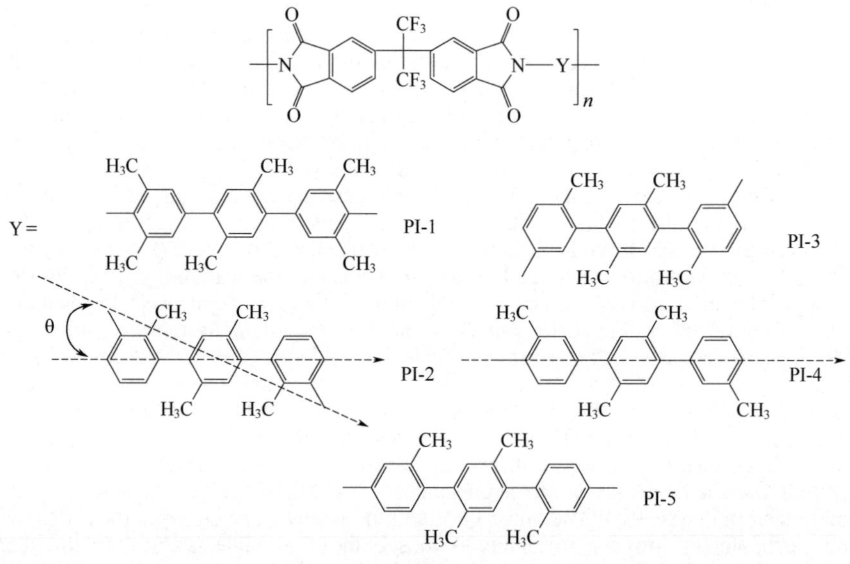

FIGURE 6.13 The structures of the 6FDA-based polyimides derived from the methyl substituted terphenylene-based diamines.

TABLE 6.9 Gas Permeability and Ideal Selectivity of the Polyimide Membranes

Polymer	Permeability (Barrer)						Ideal selectivity ($\alpha_{X/Y}$)			
	He	H_2	N_2	O_2	CH_4	CO_2	H_2/N_2	H_2/CH_4	O_2/N_2	CO_2/CH_4
PI-1	230	350	16.5	67.0	15.0	360	21.2	23.3	4.06	24.0
PI-2	160	210	7.3	32	5.60	190	28.8	37.5	4.38	33.9
PI-3	84	100	2.4	12	1.90	62	41.7	52.6	5.00	32.6
PI-4	47	51	1.6	8.6	1.10	–	31.9	46.4	5.38	–
PI-5	71	72	1.4	7.2	0.88	32	51.4	81.8	5.14	36.4

The permeability of the polyimides was recorded in Barrer, and the scale is 10^{-10} $cm^3 \cdot cm^{-2} \cdot s^{-1} \cdot cmHg^{-1}$.

In 1993, Langsam and Burgoyne [28] investigated the permeability for O_2 and N_2 of a series of 6FDA-based polyimides derived from different bridged diamines, as shown in Fig. 6.14. In which, R_1 and R_2 can be H-, methyl-, or trifluoromethyl-, ethyl-, isopropyl-substituents, etc. These substitutions can be symmetric or asymmetric. The bridging segment can be derived from benzidine, bis(4-aminoaryl)methane, bis(4-aminoaryl)-1-phenyl-2,2,2-trifluoro-ethanes or 9,9-bis(4-aminoaryl) fluorenes, etc.

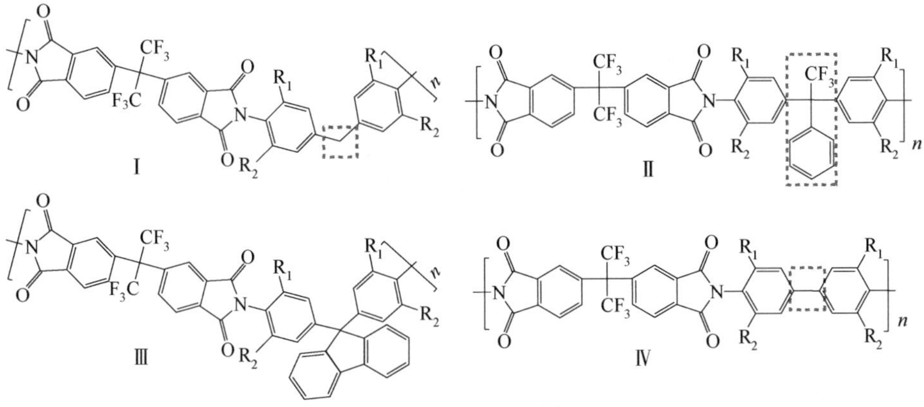

R=H, F, CH_3, C_2H_5, i-C_3H_7, etc.

FIGURE 6.14 Structures of the 6FDA-based polyimides with different bridging structures in the diamine segments.

The O_2 permeability and O_2/N_2 selectivity of the 6FDA-based polyimides is shown in Table 6.10. Based on the comparison of the *ortho*-substitutions and bridging structures, it was found that the larger the substitution at the *ortho*-position of the imide, the higher the O_2 permeability, which follows a logarithmic relationship (Fig. 6.15). If the substitutions are asymmetric, the O_2 permeability will be a little lower than expected, this might be attributed to the interaction of the dipole effect. The effect of substitution on O_2/N_2 selectivity is more complex, it seems that the asymmetric substitution could enhance the O_2/N_2 selectivity coupled with a decrease in O_2 permeability.

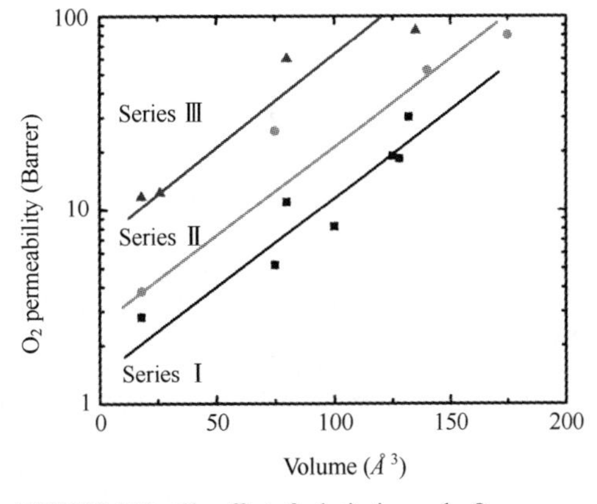

FIGURE 6.15 The effect of substitution on the O_2 permeability of the polyimides.

Bridge segments can affect the gas transport properties of the polyimides, and this was investigated by Coleman and Koros [24]. Polyimides were obtained by reaction of 6FDA and different diamines from MDA, IPDA to 6F*p*DA with similar structures but different bridge heads. The T_g, *FFV* and *d*-spacing increment of the polyimides increase with an increase in the volume of the bridge structure (Table 6.11) (Fig. 6.16).

TABLE 6.10 Permeability of the Polyimides and the O_2/N_2 Selectivity

Polymer	R_1	R_2	P_{O_2} (Barrer)	α_{O_2/N_2}	T_g (°C)	Density ($g \cdot cm^{-3}$)	*d*-spacing (Å)	FFV
Series I	H	H	2.8	5.65	289	1.55	5.51	0.07
	H	*i*-Pr	8.2	5.90	—	—	6.17	—

TABLE 6.10 (Continued)

Polymer	R_1	R_2	P_{O_2} (Barrer)	α_{O_2}/N_2	T_g (°C)	Density (g·cm^{-3})	*d*-spacing (Å)	FFV
Series I	H	*t*-Bu	19.0	4.70	274	1.21	6.20	0.203
	H	CF_3	5.2	5.71	—	1.37	5.91	0.238
	CH_3	CH_3	11.0	4.17	299	1.40	5.93	0.108
	C_2H_5	C_2H_5	18.4	4.20	—	1.29	6.35	0.147
	CH_3	*i*-Pr	30.1	3.82	276	1.27	6.35	0.16
	i-Pr	*i*-Pr	47.1	3.76	275	1.20	6.58	0.181
Series II	H	H	3.80	4.21	—	—	5.70	—
	CH_3	CH_3	25.5	3.00	—	—	5.87	—
	CH_3	*i*-Pr	52.3	3.50	289	1.20	6.10	0.23
	i-Pr	*i*-Pr	80	3.22	274	1.18	—	0.221
Series III	H	H	11.7	3.90	355	1.31	5.80	0.19
	H	F	12.3	5.45	352	1.29	5.82	0.225
	H	*i*-Pr	16.8	5.10	299	1.20	5.98	0.215
	CH_3	CH_3	60.1	3.60	361	1.24	5.87	0.197
	CH_3	*i*-Pr	84.4	4.21	335	1.18	5.93	0.214
Series IV	H	CH_3	5.4	4.61	352	—	—	—
	CH_3	CH_3	53.8	3.71	>375	—	—	—
	CH_3	*i*-Pr	84.4	3.30	—	—	—	—

The permeability of the polyimides increases in the sequence of 6FDA-ODA$<$6FDA-MDA$<$ 6FDA-IPDA $<$ 6FDA-6FpDA (Table 6.12), which is the same trend as the FFV because of the looser packing of the polyimide main chain derived from higher free volume bridge head. However, the selectivities for O_2/N_2 and CO_2/CH_4 are almost constant. These results provided opportunities to design and synthesize high-permeability and -selectivity membranes. The increment of the gas permeability by increase of the size of the bridge group can be originated from the higher diffusion coefficient (Table 6.13). However, the diffusion selectivities for O_2/N_2 and CO_2/CH_4 are almost the same.

TABLE 6.11 Physical Properties of the 6FDA-based Polyimide Films With Different Bridging Segments

Polyimides	T_g (°C)	Density (g·cm^{-3})	V_F	*d*-spacing
6FDA-ODA	304	1.432	0.1635	5.6
6FDA-MDA	304	1.400	0.1600	5.6
6FDA-IPDA	310	1.352	0.1680	5.7
6FDA-6FpDA	320	1.466	0.1897	5.9

V_F, fraction free volume. *d*-spacing is obtained by wide angle X-ray scattering using the Bragg's law.

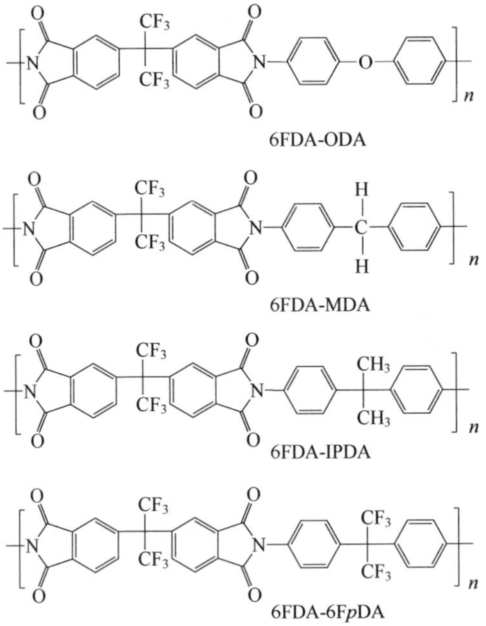

6FDA-ODA

6FDA-MDA

6FDA-IPDA

6FDA-6F*p*DA

FIGURE 6.16 6FDA-based polyimides with different bridges in the diamine part.

TABLE 6.12 Gas Permeability and Selectivity of the Polyimide Membranes

Polymer	Permeability (Barrer)				Ideal Selectivity ($\alpha_{X/Y}$)	
	N_2	O_2	CH_4	CO_2	O_2/N_2	CO_2/CH_4
6FDA-ODA	0.83	4.34	0.38	23	5.2	60.5
6FDA-MDA	0.81	4.6	0.43	19.3	5.7	44.9
6FDA-IPDA	1.34	7.53	0.70	30	5.6	42.9
6FDA-6FpDA	3.47	16.3	1.60	63.9	4.7	39.9

The permeability of the polyimides was measured in Barrer, and the scale is 10^{-10} $cm^3 \cdot cm^{-2} \cdot s^{-1} \cdot cmHg^{-1}$.

The solubility coefficient of the polyimides for different gases are almost the same (Table 6.14) irrespective of the bridge substitution. The introduction of the 6F*p*DA as bridge in the diamines showed approximately twofold higher permeability than the IPDA-based diamine, and almost the same selectivity. Koros et al. attributed this higher permeability of the 6*p*FDA-based polyimides to the presence of the two bulky $-CF_3$ groups. This bulky group can increase the chain stiffness owing to its inhibition of the rotation freedom of the intra-segment mobility. Moreover, it also affects the packing of the polyimide main chain, resulting in much looser packing and reducing the inter-chain interactions. Clair et al. also pointed out [29] that the introduction of $-CF_3$ group into linear polyimide backbone reduce the charge transfer complex (CTC) of the polyimides to a large extent.

Plaza-Lozano et al. prepared a 6FDA-based polyimide [30] by polycondensation of 6FDA with an adamantane-contained diamine (Spiro-(adamantane-2,9′(2′,7′-diamino)-fluorene) (SADAF)). The resulting PI-(6FDA/SABAF) exhibited a relative small permeability of CO_2 of 31 Barrer, and O_2/N_2 and CO_2/CH_4 selectivities of 6.2 and 51, respectively (Table 6.15). This is a much better combination than the 6FDA-ODA, 6FDA-MDA, and 6FDA-IPDA. Their D and S studies clearly indicated that the diffusion selectivity is the major contribution to the overall CO_2/CH_4 and O_2/N_2 selectivity (Fig. 6.17).

TABLE 6.13 Diffusion Coefficient and Diffusivity Selectivity of the Polyimides

Polyimides	D_{N_2}	D_{O_2}	D_{CH_4}	D_{CO_2}	D_{O_2}/D_{N_2}	D_{CO_2}/D_{CH_4}
6FDA-ODA	1.16	3.20	0.22	3.58	2.75	16.3
6FDA-MDA	1.34	4.26	0.28	3.70	3.19	13.4
6FDA-IPDA	2.03	6.36	0.44	5.38	3.14	12.1
6FDA-6FpDA	3.98	12.5	0.84	10.3	3.14	12.2

The diffusion coefficient was recorded with time-lag method and the scale is 10^{-8} $cm^2 \cdot s^{-1}$.

TABLE 6.14 Solubility Coefficient and Diffusivity Selectivity of the Polyimides

Polyimides	S_{N_2}	S_{O_2}	S_{CH_4}	S_{CO_2}	S_{O_2}/S_{N_2}	S_{CO_2}/S_{CH_4}
6FDA-ODA	0.54	1.03	1.32	4.89	1.90	3.70
6FDA-MDA	0.46	0.82	1.18	3.96	1.78	3.36
6FDA-IPDA	0.50	0.90	1.20	4.24	1.81	3.53
6FDA-6FpDA	0.67	0.99	1.44	4.72	1.48	3.28

The diffusion coefficient was recorded with the time-lag method and the scale is 10^{-8} $cm^2 \cdot s^{-1}$.

TABLE 6.15 Gas Permeability and Selectivity of the Polyimide Membranes for Different Gases

	Permeability (Barrer)[a]				Ideal Selectivity ($\alpha_{X/Y}$)	
	N_2	O_2	CH_4	CO_2	O_2/N_2	CO_2/CH_4
6FDA-SADAF	1.21	7.3	0.61	31	6.2	51
D[b]	D_{N_2}	D_{O_2}	D_{CH_4}	D_{CO_2}	D_{O_2}/D_{N_2}	D_{CO_2}/D_{CH_4}
	0.75	3.3	0.12	1.31	4.4	10.9
S[c]	S_{N_2}	S_{O_2}	S_{CH_4}	S_{CO_2}	S_{O_2}/S_{N_2}	S_{CO_2}/S_{CH_4}
	1.6	2.2	5.1	23.0	1.25	4.51

[a] *The permeability of 6FDA-SADAF was recorded in Barrer, and the scale is* 10^{-10} $cm^3 \cdot cm^{-2} \cdot s^{-1} \cdot cmHg^{-1}$.
[b] *Diffusion coefficient was recorded with time-lag method and the scale is* 10^{-8} $cm^2 \cdot s^{-1}$.
[c] *The solubility coefficient was obtained by S=P/D and the scale is* 10^{-2} $cm^3 \cdot cm^{-3} \cdot cmHg^{-1}$.

Two PMDA-based polyimides were obtained by condensation of PMDA with trimethyl (TMPD) and a more bulky cyclic diamine (TMID) substituted meta-phenylenediamine. (Fig. 6.18, [31]). The PMDA-based polyimides showed significant enhancement in permeability and better gas separation efficiency than the 6FDA-based ones for CO_2/N_2 permeability/selectivity (Table 6.16 and Fig. 6.18).

A series of the 6FDA-based polyimides with phenyl-substituted backbones have been prepared by polycondensation of 6FDA with various phenyl-substituted diamines, as shown in Fig. 6.19 [32], and their gas transport properties are shown in Table 6.17. The introduction of the biphenyl groups in the polyimide backbone resulted in a slight increase in the permeability of ~10%. While the more rigid and bulky terphenyl groups were inserted in the backbone, a significant enhancement in the permeability of ~50%-80% was observed. However, the selectivity for most of the interesting gas pairs was decreased. The enhancement in permeability of the terphenyl-substituted polyimides was attributed to the increase in its diffusion coefficient (Table 6.18). Because the biphenyl-substituted diamine is a noncoplanar building block, the D and S of the resulting polyimides are nearly unchanged compared with the 6FDA-phenyl-substituted ones for most of the gas pairs.

A novel phenyl-substituted pyrene dianhydride (DPPTD) was synthesized [33] by Santiago-García et al., and reacted with several substituted diamines to produce a series of DPPTD-based polyimides (Fig. 6.20). The density, FFV, and gas permeabilityies/selectivities are listed in Table 6.19. Polyimide derived from DPPTD and TMPD exhibited very high permeability, and demonstrated a CO_2 permeability of as high as 1600 Barrer and a CO_2/N_2 selectivity of 21.9.

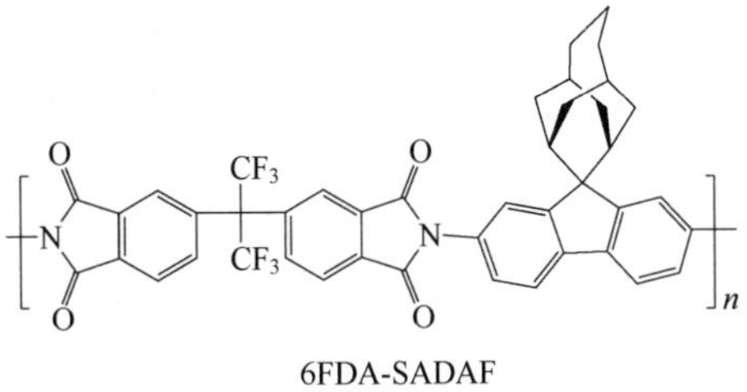

6FDA-SADAF

FIGURE 6.17 Chemical structure of PI-(6FDA/SADAF).

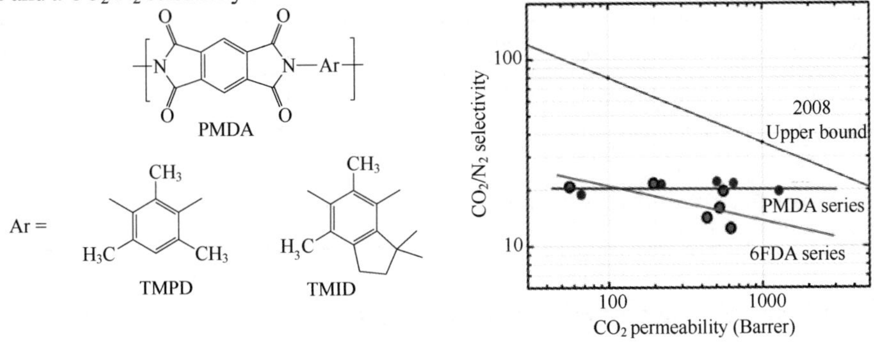

PMDA

Ar =

TMPD

TMID

FIGURE 6.18 The permeability and gas separation efficiency of the PMDA-based polyimides with multimethyl substituted backbones.

6FDA-phenyl

6FDA-biphenyl

6FDA-terphenyl

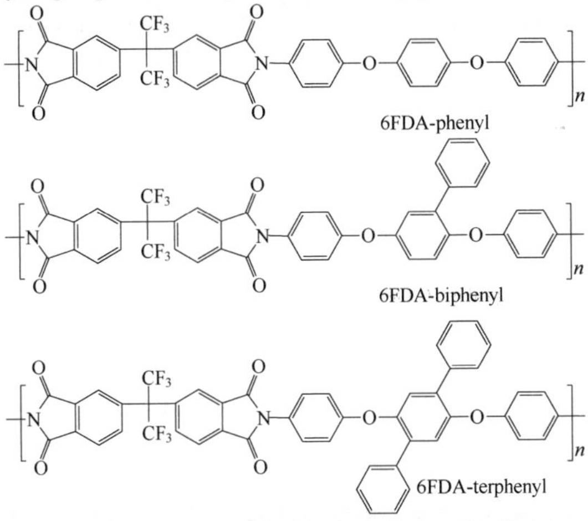

FIGURE 6.19 The structure of the 6FDA-based polyimides with phenyl-substituted backbones.

TABLE 6.16 Gas Permeability and Ideal Selectivity of the Polyimides With Multimethyl Substituted Backbones

Polymer	Permeability (Barrer)[a]			Selectivity ($\alpha_{X/Y}$)		
	N_2	O_2	CH_4	CO_2	O_2/N_2	CO_2/CH_4
PMDA-TMPD	21.9	89	26.6	526	4.0	19.4
PMDA-TMID	58	232	76	1190	4.0	15.7

[a] *The permeability of the polyimides were recorded in Barrer, and the scale is* $10^{-10}\ cm^3 \cdot cm^{-2} \cdot s^{-1} \cdot cmHg^{-1}$.

TABLE6.17 Gas Permeability and Selectivity of the Polyimide Membranes

Polyimides	Permeability (Barrer)					Ideal Selectivity ($\alpha_{X/Y}$)			
	H_2	N_2	O_2	CH_4	CO_2	H_2/N_2	H_2/CH_4	O_2/N_2	CO_2/CH_4
6FDA-Phenyl	33.5	0.56	3.23	0.35	11.9	59.5	94.9	5.74	33.7
6FDA-Biphenyl	34.3	0.62	3.46	0.36	13.0	55.6	95.8	5.62	36.2
6FDA-Terphenyl	45.6	1.02	5.26	0.75	21.5	44.7	61.1	5.16	28.8

The permeability of the polyimides was recorded in Barrer, and the scale is 10^{-10} $cm^3 \cdot cm^{-2} \cdot s^{-1} \cdot cmHg^{-1}$.

TABLE6.18 The Diffusion Coefficient and Solubility Coefficient of the noncoplanar Polyimides for Some Interesting Gases

Polyimides	D_{N_2}	D_{O_2}	D_{CH_4}	D_{CO_2}	S_{N_2}	S_{O_2}	S_{CH_4}	S_{CO_2}
6FDA-Phenyl	1.05	4.4	0.192	1.97	5.35	7.34	18.4	60.3
6FDA-Biphenyl	1.00	4.23	0.192	2.01	6.15	8.19	18.7	64.6
6FDA-Terphenyl	1.66	6.16	0.372	3.37	6.11	8.54	20.0	63.7

The diffusion coefficient was recorded with time-lag method and the scale is 10^{-8} $cm^2 \cdot s^{-1}$. Solubility of the gases were and the scale is 10^{-3} $cm^3 \cdot cm^{-3} \cdot cmHg^{-1}$.

6.3.3 Chemical Structure Effects of Dianhydrides

The effects of dianhydride chemical structures on the gas transport properties have been investigated by Okamoto et al. [34]. A series of polyimides derived from ODA as diamine and different dianhydrides, including PMDA, BPDA, BTDA, and 6FDA. The structures of the polyimides are shown in Fig. 6.21.

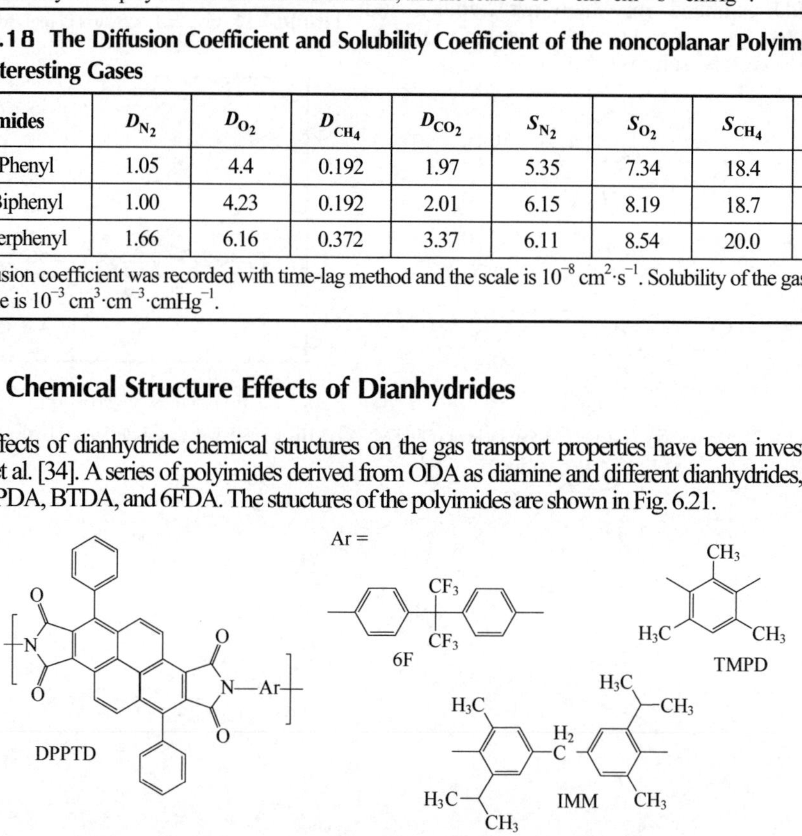

FIGURE 6.20 The chemical structures of the DPPTD-based polyimides.

TABLE6.19 Physical Properties, Gas Permeability and Ideal Selectivity of the DPPTD-based Polyimides

Polymer	Properties		Permeability (Barrer)[a]				Selectivity ($\alpha_{X/Y}$)	
	d[b]	FFV[c]	N_2	O_2	CH_4	CO_2	O_2/N_2	CO_2/N_2
DPPTD-6F	1.312	0.187	11	44	13	261	4	23.7
DPPTD -IMM	1.118	0.190	19	75	24	392	3.9	23
DPPTD-TMPD	1.150	0.205	73	280	108	1600	3.8	21.9

[a] *The permeability of the polyimides was recorded in Barrer, and the scale is 10^{-10} $cm^3 \cdot cm^{-2} \cdot s^{-1} \cdot cmHg^{-1}$.*
[b] *Density of the polyimides ($g \cdot cm^{-3}$).*
[c] *Fractional free volume of the polyimides.*

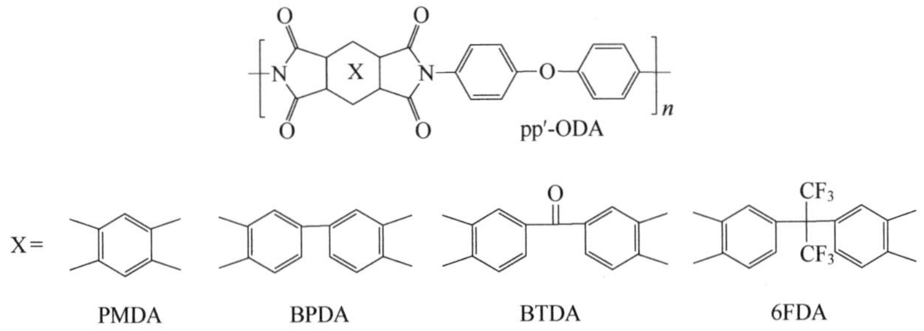

FIGURE 6.21 The structures of ODA-based polyimides.

Introduction of flexible carbon-carbon bonds (BPDA) and carbonyl bonds (BTDA) resulted in sharp decreases in the T_gs, FFV and solubility parameters of the polyimides (Table 6.20). The 6FDA/ODA membrane showed obvious high FFV and lower solubility coefficient, due to the presence of bulky CF_3 groups which restricted the rotation freedom of the polyimide backbone.

TABLE 6.20 Physical properties of the ODA-based polyimides

Polyimides	T_g (°C)	Density ($cm^3 \cdot g^{-1}$)	FFV	δ
PMDA/ODA	420	1.395	0.129	34.6
BPDA/ODA	270	1.366	0.121	32.3
BTDA/ODA	266	1.374	0.124	32.7
6FDA/ODA	299	1.432	0.165	27.2

FFV is the fractional free volume obtained using Bondi's method, δ is the solubility parameter, which is calculated by the group contribution method of van Krevelen.

The gas permeability and some gas pair ideal selectivity of the ODA-based polyimides are shown in Table 6.21. The PMDA/ODA membrane exhibited as much as 2 to 10 times higher permeability for different gases than BPDA/ODA and BTDA/ODA. This is attributed to the relatively higher FFV generated by the short PMDA building unit. The 6FDA/ODA exhibited the highest permeability, which is about three to five times that of the PMDA/ODA.

TABLE 6.21 Gas Permeability and Selectivity of the Polymer Membranes

Polyimides	Permeability (Barrer)					Ideal Selectivity ($\alpha_{X/Y}$)			
	H_2	N_2	O_2	CH_4	CO_2	H_2/N_2	H_2/CH_4	O_2/N_2	CO_2/CH_4
PMDA/ODA	10.6	0.145	0.820	0.094	3.55	73.1	112.8	5.7	38
BPDA/ODA	3.68	—	—	0.010	0.642	—	368.0	—	65
BTDA/ODA	4.79	0.024	0.191	0.011	0.625	199.6	435.5	8.1	57
6FDA/ODA	40.7	0.730	3.88	0.340	16.7	55.8	119.7	5.3	49

The permeability of the polyimides were recorded in Barrer, and the scale is 10^{-10} $cm^3 \cdot cm^{-2} \cdot s^{-1} \cdot cmHg^{-1}$.

The diffusion and solubility coefficients of the polyimides are listed in Table 6.22. The higher permeability of PMDA/ODA than BTDA/ODA and BPDA/ODA originated from its higher solubility and diffusivity coefficients. However, the diffusion coefficient increment (three to four times) is much more significant than solubility enhancement (1.45-1.8 times). Similarly, the higher permeability of 6FDA/ODA was also mostly derived from its much higher diffusion coefficient.

TABLE6.22 Diffusion and Solubility Coefficients of the ODA-based Polyimide

Polyimides	D_{N_2}	D_{O_2}	D_{CH_4}	D_{CO_2}	S_{N_2}	S_{O_2}	S_{CH_4}	S_{CO_2}
PMDA/ODA	0.32	1	0.079	0.8	4.5	8.1	12	45
BPDA/ODA	—	—	0.011	0.18	—	—	9.5	36
BTDA/ODA	0.077	0.42	0.013	0.19	3.1	4.5	8.5	32
6FDA/ODA	1.1	3.4	0.22	3.1	6.6	12	16	54

The diffusion coefficient was recorded with time-lag method and the scale is 10^{-8} $cm^2 \cdot s^{-1}$. Solubility of the gases were in the scale of 10^{-3} $cm^3 \cdot cm^{-3} \cdot cmHg^{-1}$.

Similar research has been performed on a bulky DAI diamine to replace the low free volume ODA (Fig. 6.22) [35], by reacting with different dianhydrides including BTDA, OPDA, BPDA, and 6FDA, four polyimides denoted as BTDA/DAI, OPDA/DAI, BPDA/DAI, and 6FDA/DAI were obtained.

The physical properties and their gas transport performances are shown in Table 6.23. From BTDA, OPDA to BPDA, there is a continuous decrease in packing efficiency of the polymer chains as indicated by the gradual decreases in density, and increases in FFV and *d*-spacing. The changes in gas transport properties show the same trend as their FFV values. 6FDA/DAI membrane with 6FDA as dianhydride showed the highest permeability of all of the tested gases while retaining good selectivity for O_2/N_2 and CO_2/CH_4.

By changing the diamine to a conjugated diamine (DDBT), three different polyimides of 6FDA/DDBT, BPDA/DDBT, and DSDA/DDBT were obtained by reacting with 6FDA, BPDA, and DSDA (Fig. 6.23) [36] dianhydrides. DDBT is a conjugated diamine composed of an isomer of 3,3′-dimethyl or 3,4′-dimethyl-substituted 3,7-diaminodibenzo[b,d]thiophene 5,5-dioxide. The DDBT-based polyimides have high rigidities and there is no T_g detected before decomposition.

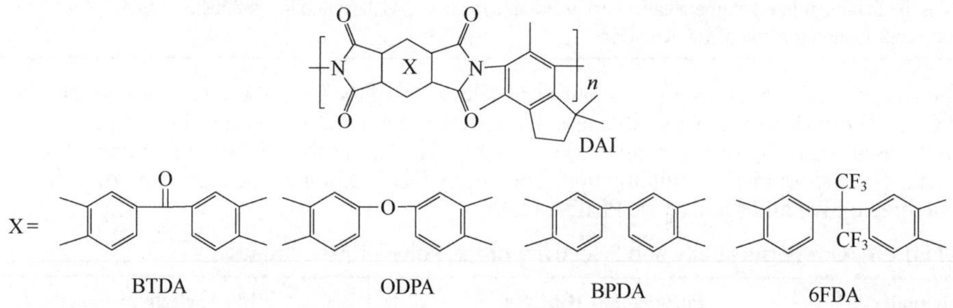

FIGURE 6.22 The structures of the DAI-based polyimides.

TABLE6.23 Physical Properties, Permeability and Ideal Selectivity of the DAI-based Polyimides

Polyimides	Properties			Permeability (Barrer)[a]				Selectivity ($\alpha_{X/Y}$)	
	Density[b]	FFV[c]	d[d]	N_2	O_2	CH_4	CO_2	O_2/N_2	CO_2/CH_4
BTDA/DAI	1.225	0.154	5.74	8.52	37.2	8.12	140.4	4.34	17.28
ODPA/DAI	1.218	0.164	5.89	10.3	43.9	10.3	161.1	4.26	15.6
BPDA/DAI	1.190	0.167	6.00	22.0	83.4	23.9	328.8	3.79	13.73
6FDA/DAI	1.326	0.170	7.02	58.6	198.2	50.5	692.3	3.38	13.72

[a] *The permeability of the polyimides were recorded in Barrer, and the scale is* 10^{-10} $cm^3 \cdot cm^{-2} \cdot s^{-1} \cdot cmHg^{-1}$.
[b] *Density of the polyimides* ($g \cdot cm^{-3}$).
[c] *Fractional free volume of the polyimides.*
[d] *d-spacing of the polymer chains obtained by X-ray scattering results.*

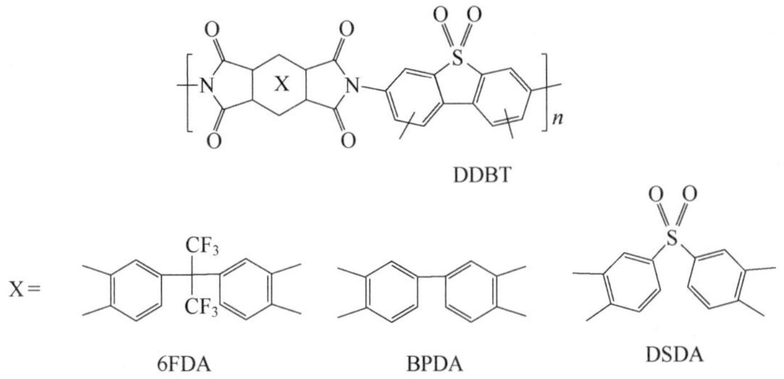

DDBT

X = 6FDA, BPDA, DSDA

FIGURE 6.23 Structures of the DDBT-based polyimides.

The gas permeability and gas pair selectivity of the polyimides are shown in Table 6.24. The FFV and gas permeability are decreased in the order of 6FDA/DDBT > DSDA/DDBT > BPDA/DDBT. The selectivity for CO_2/CH_4 was decreased in the order of DSDA/DDBT > BPDA/DDBT > 6FDA/DDBT. The DSDA/DDBT not only exhibited higher CO_2 permeability than BPDA/ODA, but also higher CO_2/CH_4 selectivity.

Some substituted PMDA dianhydrides were also synthesized, and reacted with different diamines to produce a series of modified PMDA-based polyimides (Fig. 6.24, [37]). The R_1 in modified PMDA can be *t*-trimethylphenyl-or trimethylsilylphenyl-groups. Table 6.25 compares the gas transport performances of the polyimides.

The *t*-trimethylphenyl-substituted polyimides exhibited higher T_g values. The permeability for O_2 was measured in the range of 14-56 Barrer and the O_2/N_2 selectivity in the range of 3.7-5.5, which is slightly higher than the trimethylsilylphenyl-substituted polyimides.

TABLE 6.24 Gas Permeability and Ideal Selectivity of the DDBT-based Polyimides

Polyimides	Properties			Permeability (Barrer)[a]			Selectivity ($\alpha_{X/Y}$)		
	d[b]	FFV[c]	T_g[d]	N_2	O_2	CH_4	CO_2	O_2/N_2	CO_2/CH_4
6FDA/DDBT	1.421	0.169	> 763	5.14	23.5	2.51	91.0	4.6	36.3
BPDA/DDBT	1.372	0.125	> 763	—	—	0.149	7.26	—	48.7
DSDA/DDBT	1.401	0.141	> 763	3.29	0.499	0.258	13.7	6.6	53.1

[a] *The permeability of the polyimides were recorded in Barrer, and the scale is* $10^{-10}\ cm^3 \cdot cm^{-2} \cdot s^{-1} \cdot cmHg^{-1}$.
[b] *Density of the polyimides based on DDBT as diamine* ($g \cdot cm^{-3}$).
[c] *Fractional free volume of the polyimides based on DDBT as diamine.*
[d] *Glassy transition temperature of the polyimides (temperature K).*

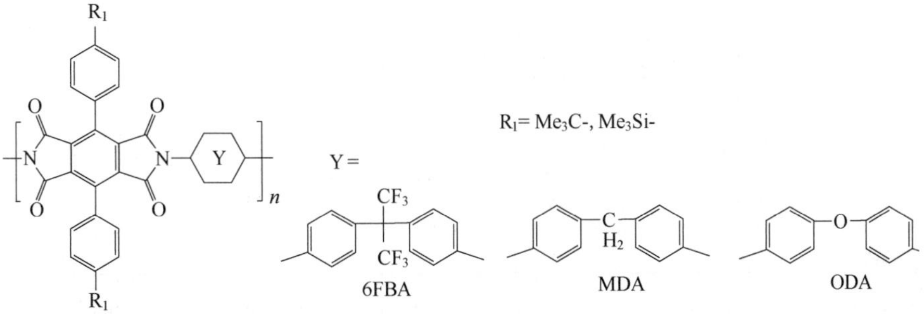

R_1 = Me_3C-, Me_3Si-

Y = 6FBA, MDA, ODA

FIGURE 6.24 Structures of the modified PMDA-based polyimides.

TABLE6.25 Gas Transport Performance of the Modified PMDA-based Polyimides

R_1 Group	Diamine	T_g	P_{O_2}	P_{N_2}	α_{O_2/N_2}
Me_3Si-	6FBA	245.6	56	15	3.7
Me_3Si-	MDA	227.2	14	3.3	4.2
Me_3Si-	ODA	279.6	32	8	4
Me_3C-	6FBA	259.6	50	10.6	4.7
Me_3C-	MDA	291.7	13	2.4	5.5
Me_3C-	ODA	292.3	20	4	5

The diffusion coefficient was recorded with time-lag method and the scale is 10^{-8} $cm^2 \cdot s^{-1}$. Solubility of the gases were in the scale of 10^{-3} $cm^3 \cdot cm^{-3} \cdot cmHg^{-1}$.

Ayala et al. also investigated the substitution effect of the dianhydrides and their gas transport properties [38]. A series of DCDA-based polyimides were prepared by condensation of different substituted DCDA with fluorinated diamines, including HDCDA (*R*=H), PDCDA (*R*=phenyl-), BDCDA (*R*=*t*-butyl-) (Fig. 6.25).

Table 6.26 summarized the physical properties and gas separation performance of the substituted DCDA-based polyimides. The density of the polyimides decreases in the order of H, Phenyl and *t*-Butyl group substitution in the dianhydrides; this is inverse to their *FFV* and T_g orders. Their CO_2 permeabilities are in the range of 4.1 -15.6 Barrer, while the CO_2/CH_4 selectivity is ~ 40. The *t*-butyl group as a substitution showed the best CO_2/CH_4 permeability/selectivity trade-off with 15.6 Barrer of CO_2 permeability and CO_2/CH_4 selectivity of 39.

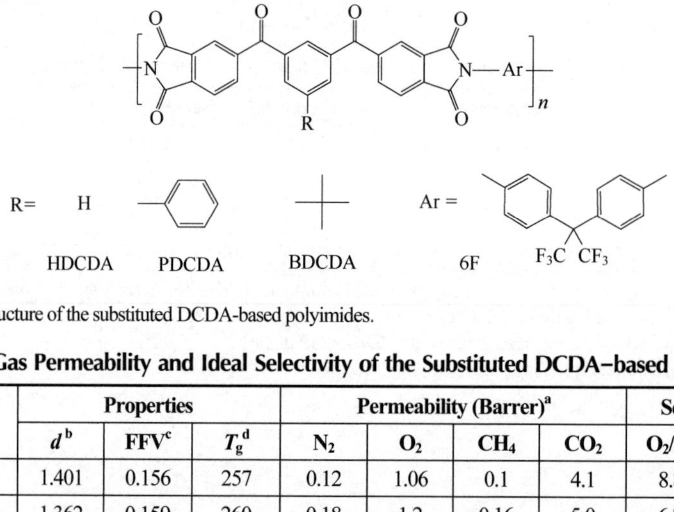

R= H HDCDA; PDCDA; BDCDA; Ar = 6F (F_3C CF_3)

FIGURE 6.25 Structure of the substituted DCDA-based polyimides.

TABLE6.26 Gas Permeability and Ideal Selectivity of the Substituted DCDA-based Polyimides

Polyimides	Properties			Permeability (Barrer)[a]				Selectivity ($\alpha_{X/Y}$)	
	d[b]	FFV[c]	T_g[d]	N_2	O_2	CH_4	CO_2	O_2/N_2	CO_2/CH_4
HDCDA/6 F	1.401	0.156	257	0.12	1.06	0.1	4.1	8.8	41
PDCDA/6 F	1.362	0.159	260	0.18	1.2	0.16	5.0	6.7	31
BDCDA/6 F	1.336	0.164	265	0.64	3.8	0.4	15.6	5.9	39

[a]*The permeability of the polyimides was recorded in Barrer, and the scale is* 10^{-10} $cm^3 \cdot cm^{-2} \cdot s^{-1} \cdot cmHg^{-1}$.
[b]*Density of the polyimides based on 6 F as diamine* ($g \cdot cm^{-3}$).
[c]*Fractional free volume of the polyimides, calculated by Hyperchem.*
[d]*Glass transition temperature of the polyimides (°C).*

A novel series of polyimides were prepared by condensation reaction of the substituted BPDA (BBBPAn and BTSBPAn) with various diamines, including ODA, MDA, 6FDA, and DADBSBF (Fig. 6.26) [39,40]. The BBBPAn and BTSBPAn are aromatic dianhydrides with noncoplanar twisted biphenyl structure with aromatic substitution at the para-position of the dianhydrides in the BPDA structure. The resulting polyimides showed high permeability, with O_2 permeability of 100-120 Barrer and a very low O_2/N_2 selectivity of 2.2-2.84 (Table 6.27).

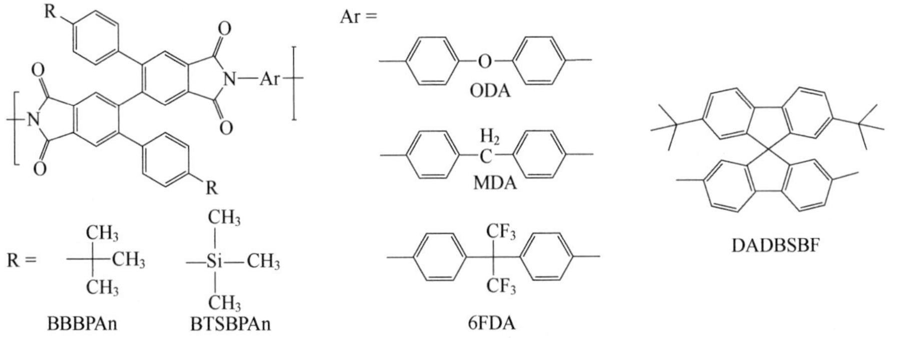

FIGURE 6.26 Structures of the phenyl-substituted BPDA-based polyimides.

TABLE 6.27 Gas Permeability and O_2/N_2 Ideal Selectivity of the Substituted BPDA-based Polyimides

Polyimides	Permeability (Barrer)[a]			Selectivity ($\alpha_{X/Y}$)
	T_g[b]	N_2	O_2	O_2/N_2
BBBPAn/ODA	370	12	43	3.58
BBBPAn/MDA	360	8	31	3.9
BBBPAn/6FDA	381	35	110	3.14
BTSBPAn/ODA	348	18	61	3.4
BTSBPAn/MDA	348	12	52	4.3
BTSBPAn/6FDA	365	37	105	2.84
BBBPAn/DADBSBF	>400	12	52.2	4.4
BTSBPAn/DADBSBF	398	54	121	2.2

[a] *The permeability was recorded in Barrer, and the scale is* 10^{-10} $cm^3 \cdot cm^{-2} \cdot s^{-1} \cdot cmHg^{-1}$.
[b] *Glass transition temperature (°C).*

Another polyimides were also prepared derived from 2,2′-disubstituted-4,4′,5,5′-biphenyl-tetracarboxylic dianhydrides and MDA or ODA (Fig. 6.27) [41]. The polyimide membrane showed the CO_2 permeability in the range of 29-85 Barrer and CO_2/CH_4 selectivity of ~30 (Table 6.28).

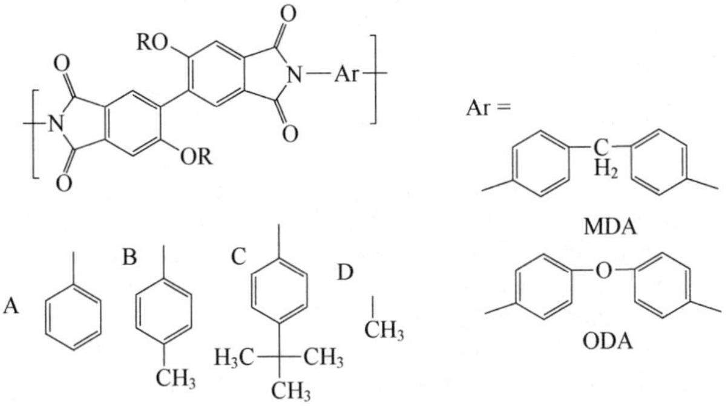

FIGURE 6.27 Structure of the phenyoxy-substituted BPDA-based polyimides.

TABLE6.28 Gas Permeability and Ideal Selectivity of the Substituted BPDA-based Polyimides

Polyimides	Permeability (Barrer)[a]				Selectivity ($\alpha_{X/Y}$)	
	N_2	O_2	CH_4	CO_2	O_2/N_2	CO_2/CH_4
A/ODA	0.83	3.59	0.85	30.1	4.3	35.4
A/MDA	0.73	3.36	0.85	29.0	4.6	34
B/ODA	1.25	4.99	1.45	44.0	4.0	30.3
C/ODA	2.95	11.3	3.16	85.6	3.8	27.1

[a] *The permeability was recorded in Barrer, and the scale is* 10^{-10} $cm^3 \cdot cm^{-2} \cdot s^{-1} \cdot cmHg^{-1}$.

Overall, substitutions on the polyimide backbones either from diamines or dianhydrides showed significant influence in tuning the gas separation properties of the polymer membranes. The *ortho*-methyl substitution provides a more open space to improve the gas permeability and selectivity. Introduction of bulky groups, twisted biphenyl groups, and asymmetric segments also have obvious effects. CF_3 substitution can enhance gas permeability by introducing more volume elements and reduce the interaction between the polymer chains.

6.4 Intrinsically Microporous Polyimide Membranes

The ideal of intrinsically microporous polyimides (PIM-PI) originated from the polymer of intrinsic microporosity (PIM), which is obtained in a ladder-type polymer by introducing the contortion center into the polymer structures. Owning to the inefficient packing of such "kink" structure, the resulting polymer demonstrated remarkable microporosity with a large surface area as tested by N_2 adsorption isotherms. This ideal was successfully extended to polyimides and form intrinsically microporous polyimides, i.e., PIM-PIs. The PIM-PI-based membrane exhibited two orders of magnitude of permeability than conventional polyimide such as Matrimid and modest selectivity for CO_2/CH_4, O_2/N_2, CO_2/N_2, and H_2/N_2. It is considered that PIM-PI membrane is a very strong potential candidate for practical application in gas separation field.

PIM-PI was first confirmed in 2007, which was prepared by reaction of spirobifluorene-based diamine with PMDA [42]. This polyimide is easily soluble in conventional solvent and also maintains a modest surface area of 550 $m^2 \cdot g^{-1}$. The first PIM-PIs reported for gas separation are prepared by adopting the spirobisindane site of contortion as a building block [43,44]. The dianhydride can be obtained in the procedure shown in Fig. 6.28. The site of contortion containing biscatecol was reacted with 4,5-dichlorophthalonitrile. The resulting tetracyano-intermediate hydrolyzed with KOH and further reacted with acetic anhydride to form the spirobisindane-based dianhydrides. A series of PIM-PIs were obtained by reaction of the dianhydride with different diamines, giving PIM-PI-1, PIM-PI-2, PIM-PI-3, PIM-PI-4, PIM-PI-7, and PIM-PI-8, respectively. These polyimides exhibited good solubility in most of the conventional solvents such as $CHCl_3$, THF, DMF, and NMP, etc.

The physical properties of the polyimides are summarized in Table 6.29. The polyimides showed densities of 1.14 $g \cdot cm^{-3}$-1.26 $g \cdot cm^{-3}$, a much lower value than the conventional polyimides (1.3 $g \cdot cm^{-3}$-1.5 $g \cdot cm^{-3}$). Their surface areas were measured in the range of 470 $m^2 \cdot g^{-1}$- 685 $m^2 \cdot g^{-1}$ due to their special site of contortion that make these rigid polyimides unable to fold efficiently. Owning to the high microporous free volume generated in the polymer system, the FFVs are measured in the range of 0.22- 0.23, higher than conventional PIM-PIs (0.11- 0.18). The experimental results indicated that the PIM-PIs have a larger amount of microporosity than conventional polyimides. The gas permeability and selectivity properties are listed in Table 6.30.

The above PIM-PI membranes showed much higher gas permeability than the conventional polyimides. The CO_2 permeability can reach as much as 3700 Barrer, which is about 770 times that of conventional Matrimid. The O_2 permeability can reach as much as 545 Barrer and O_2/N_2 selectivity can reach 3.4. This figure-of-merit is also much larger than conventional polyimides in terms of permeability. The trade-off curves for H_2/N_2 and O_2/N_2 are depicted in Fig. 6.29. Some of the gas pairs are already outperformed the 1991 trade-off curves.

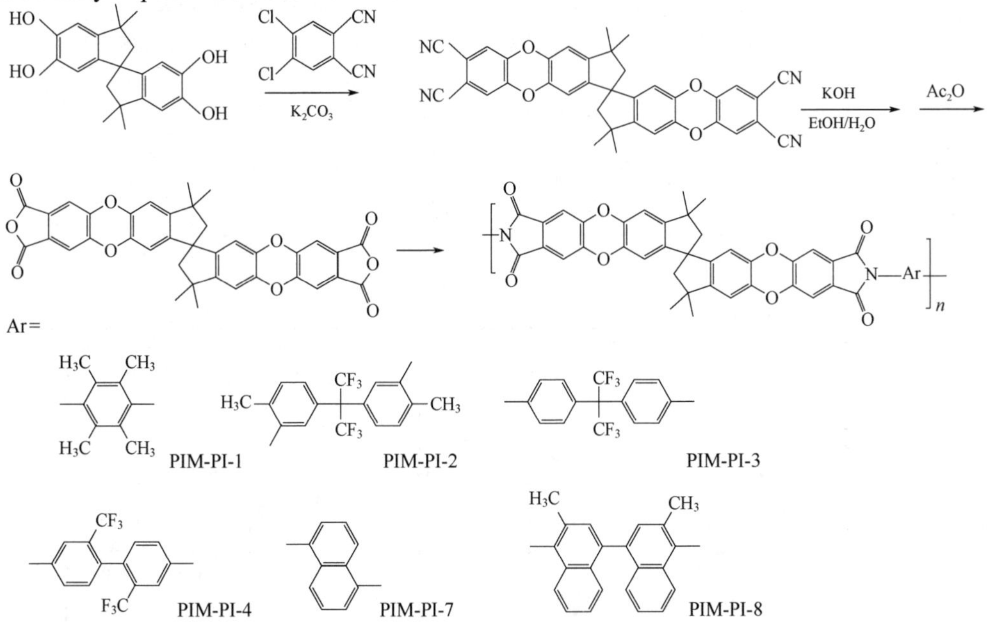

FIGURE 6.28 Synthesis procedure and structure of spirobisindane-based PIM-PIs [44].

TABLE 6.29 Physical Properties of the Intrinsic Microporous Polyimides

Polyimides	M_n ($\times 10^3$ mol·g^{-1})[a]	Density (g·cm^{-3})	S_{BET} (m^2·g^{-1})[b]	FFV[c]
PIM-PI-1	17	1.15	680	0.232
PIM-PI-2	30	—	500	—
PIM-PI-3	22	1.26	471	0.226
PIM-PI-4	11	1.26	486	0.228
PIM-PI-7	11	1.19	485	0.223
PIM-PI-8	54	1.14	683	0.231

[a] *Number-average molar mass of the polyimides, M_n.*
[b] *BET surface area (S_{BET}), determined from N_2 adsorption at 77 K.*
[c] *Fractional free volume (FFV): determined from film density and van der Waals volume [16].*

Other novel PIM-PIs were also synthesized based on different diamines or dianhydrides containing different sites of contortions. Ma et al. reported a spirobifluorene- (SBF) and dibromospirobifluorene- (BSBF) based diamines (Fig. 6.30) [45], which were used to prepare a series of novel PIM-PIs. Significant surface area enhancement was obtained by introducing the bromo-atom next to the imide bond. Moreover, molecular simulations of PMDA-SBF and PMDA-BSBF repeat units indicate that the twist angle between the PMDA and fluorene plane changes from 0° in PMDA-SBF to 77.8° in PMDA-BSBF. The bromine-substituted polyimides showed significantly increased gas permeability but slightly lower selectivity than the SBF-based polyimides (Table 6.31). For example, the CO_2 permeability for PMDA-BSBF (693 Barrer) was 3.5-fold higher than PMDA-SBF (197 Barrer), combined with CO_2/CH_4 selectivity of 19 and 22, respectively.

TABLE6.30 Gas Transport Properties of the Intrinsic Microporous Polyimides

Polyimides	Permeability (Barrer)[a]					Ideal Selectivity ($\alpha_{X/Y}$)[b]				
	P_{H_2}	P_{N_2}	P_{O_2}	P_{CH_4}	P_{CO_2}	H_2/N_2	H_2/CH_4	O_2/N_2	CO_2/N_2	CO_2/CH_4
PIM-PI-1	530	47	150	77	1100	11.3	6.90	3.2	23.4	14.3
PIM-PI-2	220	9	39	9	210	24.4	24.4	4.3	23.3	23.3
PIM-PI-3	360	23	85	27	520	15.7	13.3	3.7	22.6	19.3
PIM-PI-4	300	16	64	20	420	18.8	15.0	4.0	26.3	21
PIM-PI-7	350	19	77	27	510	18.4	13.0	4.1	26.8	18.9
PIM-PI-8	1600	160	545	260	3700	10.0	6.20	3.4	23.1	14.2

[a] *The permeability was recorded in Barrer, and the scale is* $10^{-10}\ cm^3 \cdot cm^{-2} \cdot s^{-1} \cdot cmHg^{-1}$.
[b] *Ideal selectivity of the gas pairs.*

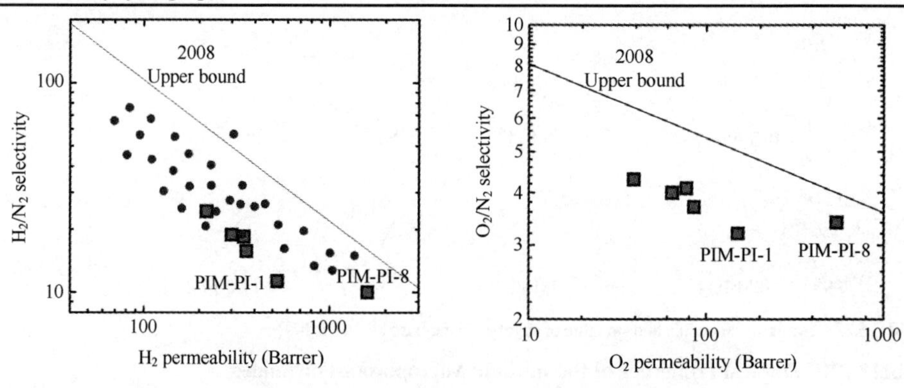

FIGURE 6.29 Trade-off curves for H_2/N_2 and O_2/N_2 gas pairs for the intrinsic microporous polyimides.

FIGURE 6.30 Synthesis of SBF- and BSBF-based PIM-PIs.

TABLE6.31 Permeability and Selectivity of SBF-and BSBF-based PIM-PI Membranes

Polyimides	Permeability (Barrer)[a]					Ideal Selectivity ($\alpha_{X/Y}$)			
	H_2	N_2	O_2	CH_4	CO_2	H_2/N_2	O_2/N_2	CO_2/CH_4	CO_2/N_2
6FDA/SBF	234	7.8	35.1	6.4	182	30.0	4.5	27.3	23.3
PMDA/SBF	230	8.5	35.5	9.1	197	27.1	4.2	21.6	23.2
SPDA/SBF	501	28.6	111	41.1	614	17.5	3.9	14.9	21.5
6FDA/BSBF	531	27.0	107	24.9	580	19.7	4.0	23.3	21.5
PMDA/BSBF	560	28.8	116	36.5	693	19.4	4.0	19.0	24.1
SPDA/BSBF	919	69.0	243	102	1340	13.3	3.5	13.1	19.4

[a] *1 Barrer=*$10^{-10}\ cm^3\ (STP)\ cm\ cm^{-2}\ s^{-1}\ cm\ Hg^{-1}$ *or* $7.5 \times 10^{-18}\ m^3\ (STP) \cdot m\ m^{-2} \cdot s^{-1} \cdot Pa^{-1}$.

Due to the intensive special internal free volume of the triptycene structure, a series of triptycene-contained diamines have been prepared; the triptycene groups acted as contortion centers (Fig. 6.31) [46-50].

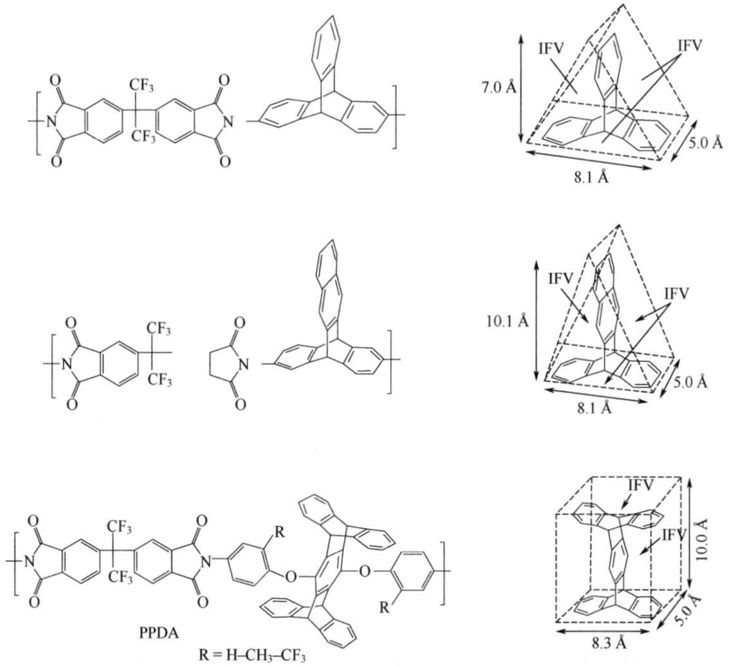

FIGURE 6.31 Structures of the 6FDA-based polyimides derived from different substituted triptycene-contaning diamines.

2,6-Diaminotriptycene was used to react with 6FDA to form the PI-(6FDA/DATRi) (Fig. 6.31). Owing to its special internal free volume between the bridge head and the bottom plane of triptycene, the cavity resulted in a high-speed gas transport while maintaining high selectivity of the gas pairs by the size sieving effect. The gas separation properties of the resulting PIM-PIs are listed in Table 6.32. The permeability of the polyimides decreased in the order of $P_{H_2} > P_{CO_2} > P_{O_2} > P_{N_2} > P_{CH_4}$, which is the same as those low free volume polyimides. The introduction of the benzotriptycene building block in the polyimides resulted in a 40% enhancement in surface area from 320 $m^2 \cdot g^{-1}$ to 450 $m^2 \cdot g^{-1}$, which produced a ~10% higher permeability for all the gases, and the selectivity for different gas pairs such as H_2/N_2, H_2/CH_4, O_2/N_2, and CO_2/CH_4 was almost the same. The pentiptycene-containing polyimide showed a much lower permeability coupled with slightly higher selectivity than the pristine triptycene or benzotriptycene because two more flexible ether groups are introduced.

TABLE 6.32 Gas Permeability and Selectivity of the Triptycene-contained PIM-PIs

Polyimides	Permeability (Barrer)				CO_2	Ideal Selectivity ($\alpha_{X/Y}$)			
	H_2	N_2	O_2	CH_4		H_2/N_2	H_2/CH_4	O_2/N_2	CO_2/CH_4
6FDA/DATRi	257	8.1	39	6.2	189	31	42	4.8	30.5
6FDA/DAT2	281	9	43.3	7.1	210	31.2	39.5	4.8	29.6
6FDA/PPDA(H)	131	3.2	17	2.5	73	41	52.4	5.3	29
6FDA/PPDA(CH_3)	100	2.4	12	2.0	55	42	50.0	5.0	28
6FDA/PPDA(CF_3)	188	7.0	30	5.5	132	27	34.2	4.3	19

The data of 6FDA/DATri from [46]. The data of 6FDA-DAT2 from [49]. The data of 6FDA-PPDA series are from [50]. The permeability of the gases was recorded in Barrer, 1 Barrer=10^{-10} cm^3 (STP) $\cdot cm \cdot cm^{-2} \cdot s^{-1} \cdot cm\ Hg^{-1}$ or 7.5× 10^{-18} m^3 (STP)$\cdot m \cdot m^{-2} \cdot s^{-1} \cdot Pa^{-1}$.

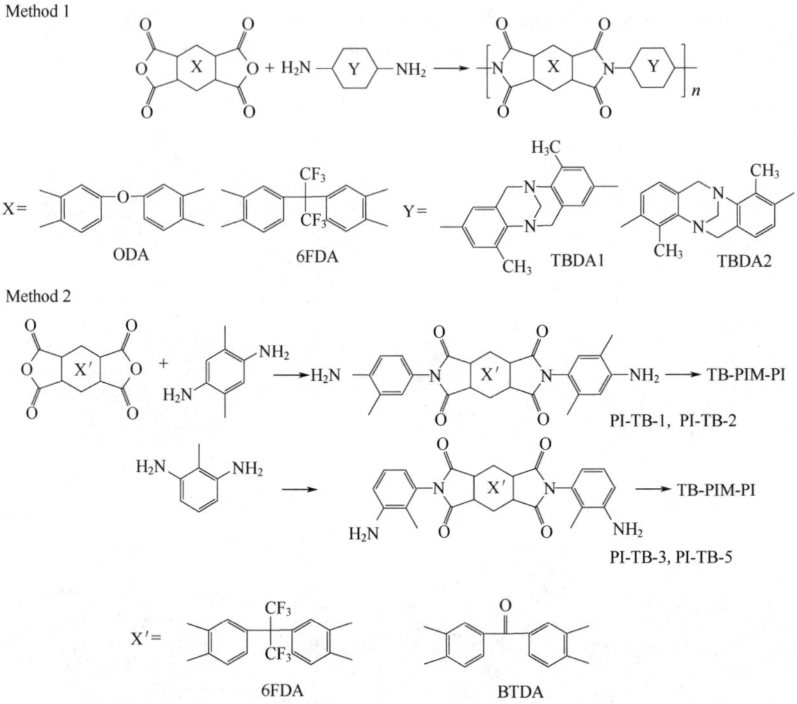

Method 1

X= ODA 6FDA Y= TBDA1 TBDA2

Method 2

PI-TB-1, PI-TB-2

PI-TB-3, PI-TB-5

X′= 6FDA BTDA

FIGURE 6.32 Synthesis of Tröger's base-contained intrinsically microporous polyimides (TB-PIM-PI).

In 2013, an extremely rigid Tröger's base was reported as a building block for ladder-type polymers. The PIM membranes showed extremely high permeability and reasonable selectivity. Their overall performance are located significantly higher than the 2008 trade-off curves for most gas pairs. Thereafter, a series of research results were developed focusing on introducing this building block to polyimide membranes [51-56]. Two different methods have been developed to synthesize intrinsically microporous Tröger's-base-containing polyimides (Fig. 6.32). One method of synthesizing Tröger's-base-based intrinsically microporous polyimides (TB-PIM-PI) is condensation of dianhydrides with Tröger's-base-containing diamines. The second method is synthesis of TB-PIM-PIs using imide-containing diamine as monomer.

The physical properties of the TB-PIM-PIs are listed in Table 6.33 and the gas transport properties of the TB-PIM-PIs are summarized in Table 6.34. High-molecular-weight TB-PIM-PIs have been prepared by the above-mentioned methods. The 6FDA-based polyimides showed higher surface area than the ODPA- and BTDA-based polyimides. Moreover, the *ortho*-methyl group in the polymer backbone significantly enhanced the surface area of the resulting TB-PIM-PI. The TB-PIM-PIs derived from *para*-substituted diamine exhibited higher surface area and FFV than that derived from *meta*-substituted diamine. The PIM-TB-1 showed the H_2 permeability of up to 607 Barrer with the lowest H_2/N_2 selectivity of 19. All the TB-PIM-PIs exhibited the N_2 separation performance over CH_4, regardless of the BET surface area even it is over 500 $m^2 \cdot g^{-1}$. Overall, the gas separation performances of TB-PIM-PIs are quite good, in which the CO_2/CH_4 separation performance was located at a very close position to the 2008 trade-off curves (Fig. 6.33).

The introduction of methyl or halide groups to form totally blocked Tröger's-base-based diamines are also reported, which was used to react with 6FDA to produce a series of blocked Tröger's-base-containing polyimides (Fig. 6.34) [55,57]. The physical properties of the resulting polyimides are listed in Table 6.35. The 6FDA-based PIM-PI-TB membranes have surface areas of 440 $m^2 \cdot g^{-1}$-580 $m^2 \cdot g^{-1}$, which is much larger than the TBDA1-6FDA-PI, TBDA2-6FDA-PI as well as PT-TB. The introduction of bromo-contained groups in PIM-PI-TB-1 resulted in smaller surface areas than PIM-PI-TB-2. This is attributed to the heavy atom of the bromine itself. The surface areas of 4MTBDA-based PIM-PIs are measured in the range of 580 $m^2 \cdot g^{-1}$-740 $m^2 \cdot g^{-1}$, which is in the range of the highest porous PIM-PIs.

TABLE6.33 Physical Properties of the TB-PIM-PIs

Polyimides	M_n (10^4 mol·g^{-1})	PDI	Density (g·cm^{-3})	S_{BET} (m^2·g^{-1})	FFV
TBDA1/6FDA-PI [a]	3	2.6	1.31	89	0.21
TBDA1/ODPA-PI [b]	2.6	2.7	1.27	24	0.18
TBDA2/6FDA-PI [c]	2.3	2.3	1.28	349	0.23
TBDA2/ODPA-PI [d]	2.7	2.7	1.24	38	0.20
PI-TB-1 [e]	5.3	2.9	1.26	544	0.23
PI-TB-2 [f]	4.1	3.0	1.19	270	0.22
PI-TB-3 [g]	5.4	3.41	–	250	0.20
PI-TB-5 [h]	5.3	3.07	–	65	0.17

[a] *TBDA1-6FDA-PI was synthesis by TBDA1 and 6FDA.*
[b] *TBDA1-ODPA-PI was synthesis by ODPA1 and 6FDA.*
[c] *TBDA2-6FDA-PI was synthesis by TBDA2 and 6FDA.*
[d] *TBDA2-ODPA-PI was synthesis by ODPA2 and 6FDA.*
[e] *PI-TB-1 was synthesized when X' is 6FDA.*
[f] *PI-TB-2 was synthesized when X' is BTDA.*
[g] *PI-TB-3 was obtained when X' is 6FDA.*
[h] *PI-TB-2 was synthesized when X' is BTDA.*

TABLE6.34 Gas Permeability and Selectivity of the TB-PIM-PIs

Polyimides	Permeability (Barrer) [a]					Ideal selectivity ($\alpha_{X/Y}$)			
	H_2	N_2	O_2	CH_4	CO_2	H_2/N_2	H_2/CH_4	O_2/N_2	CO_2/CH_4
TBDA1/6FDA-PI [b]	253	6.5	28	3.3	155	35	76.7	3.9	47
TBDA1/ODPA-PI [c]	36.4	0.5	2.5	0.37	13.4	71	107	5.0	36
TBDA2/6FDA-PI [d]	390	12	47	8	285	33	49	4.0	36
TBDA2/ODPA-PI [e]	159	3.8	16.2	2.2	106	42	72	4.3	48
PI-TB-1 [f]	607	31	119	27	457	19	22	3.8	17
PI-TB-2 [g]	134	2.5	14	2.1	55	53	64	5.5	26
PI-TB-3 [h]	299	9.5	42	6.7	218	31	45	4.5	33
PI-TB-5 [i]	54	1.9	4.9	1.7	19.6	29	32	2.6	11.5

[a] *The permeability of the gases are recorded in Barrer, 1 Barrer=10^{-10} cm^3 (STP)·cm·cm^{-2}·s^{-1}·cm Hg^{-1}.*
[b] *TBDA1-6FDA-PI was synthesis by TBDA1 and 6FDA.*
[c] *TBDA1-ODPA-PI was synthesis by ODPA1 and 6FDA.*
[d] *TBDA2-6FDA-PI was synthesis by TBDA2 and 6FDA.*
[e] *TBDA2-ODPA-PI was synthesis by ODPA2 and 6FDA.*
[f] *PI-TB-1 was synthesized when X' is 6FDA.*
[g] *PI-TB-2 was synthesized when X' is BTDA.*
[h] *PI-TB-3 was obtained when X' is 6FDA.*
[i] *PI-TB-2 was synthesized when X' is BTDA.*

As a result, the PIM-PIs exhibit very permeable membrane properties (Table 6.36). The O_2 permeability of 4MTBDA-PMDA membrane with a thickness of 102 μm reaches as high as 1080 Barrer with O_2/N_2 selectivity of 3.72. After aging for 333 days, the O_2 permeability was still measured at 394 Barrer. The overall performance for O_2/N_2, CO_2/N_2 separation against trade-off curves are shown in Fig. 6.35.

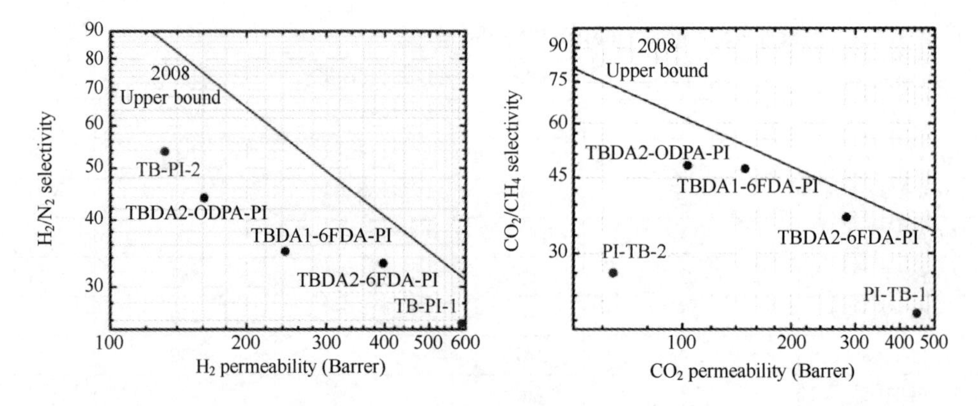

FIGURE 6.33 H_2/N_2 and CO_2/CH_4 trade-off curves for the TB-PIM-PIs.

The introduction of different sites of contortions into the very short dianhydride as part of polyimide backbones had more pronounced effects on improving the gas permeabilities and selectivities. There are generally two methods of introducing these contortion centers as a building block for dianhydrides which are shown in Fig. 6.36.

FIGURE 6.34 Structures of blocked Tröger's base-contained PIM-PIs.

TABLE6.35 Physical Properties of the TB-PIM-PIs

Polyimides	M_n (10^4 mol·g^{-1})	PDI	S_{BET} (m^2·g^{-1})[a]	Pore Volume (cm^3·g^{-1})[b]
PIM-PI-TB-1[c]	3.55	1.57	440	0.350
PIM-PI-TB-2[d]	6.59	1.59	580	0.544
4MTBDA/6FDA[e]	5.9	2.25	584	0.72
4MTBDA/PMDA[f]	4.3	1.79	650	0.57
4MTBDA/SBIDA[g]	7.1	2.25	733	0.71
4MTBDA/SBFDA[h]	5.8	2.23	739	0.54

[a]*BET surface area was obtained by N_2 adsorption isotherms at -196 °C at the relative pressure (p/p_0) from 0.1 to 0.3.*
[b]*Pore volume was measured by N_2 adsorption iostherms method.*
[c] *PIM-PI-TB-1 are obtained by reaction of 6FDA and TB-DA 1.*
[d]*PIM-PI-TB-2 is obtained by reaction of 6FDA and TB-DA 2.*
[e]*The polyimide was obtained by reaction of 6FDA and 4MTBDA.*
[f]*The polyimide was obtained by reaction of PMDA and 4MTBDA.*
[g] *The polyimide was obtained by reaction of SBIDA and 4MTBDA.*
[h]*The polyimide was obtained by reaction of SBFDA and 4MTBDA.*

TABLE6.36 Gas Permeability and Selectivity of the TB–PIM–PIs

Polymer	Permeability (Barrer)[a]				CO_2	Ideal Selectivity ($\alpha_{X/Y}$)			
	H_2	N_2	O_2	CH_4		H_2/N_2	H_2/CH_4	O_2/N_2	CO_2/CH_4
PIM-PI-TB-1[b]	380	19	70	16	361	20.0	23.8	3.68	22.6
PIM-PI-TB-2[c]	582	34	123	31	595	17.1	18.8	3.62	19.2
4MTBDA/6FDA[d]	1446	133	408	116	1672	10.9	12.5	3.07	14.4
4MTBDA/PMDA[e]	3300	290	1080	390	4460	11.4	8.5	3.72	11.4
4MTBDA/PMDA[f]	1531	99	394	114	1689	15.5	13.4	3.98	14.8
4MTBDA/SBIDA[g]	3200	373	1132	591	5140	8.6	5.4	3.03	8.7
4MTBDA/SBFDA[h]	2901	264	941	371	4476	11.0	7.8	3.56	12.1

[a]*The permeability of the gases are recorded in Barrer, 1 Barrer =* 10^{-10} cm^3 *(STP)·cm·*cm^{-2}·s^{-1}·*cm* Hg^{-1} *or* 7.5×10^{-18} m^3 *(STP)·m·*m^{-2}·s^{-1}·Pa^{-1}.
[b]*PIM-PI-TB-1 was obtained by reaction of 6FDA and TB-DA 1.*
[c]*PIM-PI-TB-2 was obtained by reaction of 6FDA and TB-DA 2.*
[d]*The polyimide was obtained by reaction of 6FDA and 4MTBDA.*
[e]*The polyimide was obtained by reaction of PMDA and 4MTBDA with membrane thickness of 102 μm.*
[f]*The membrane e was aged for 333 days.*
[g]*The polyimide was obtained by reaction of SBIDA and 4MTBDA.*
[h]*The polyimide was obtained by reaction of SBFDA and 4MTBDA.*

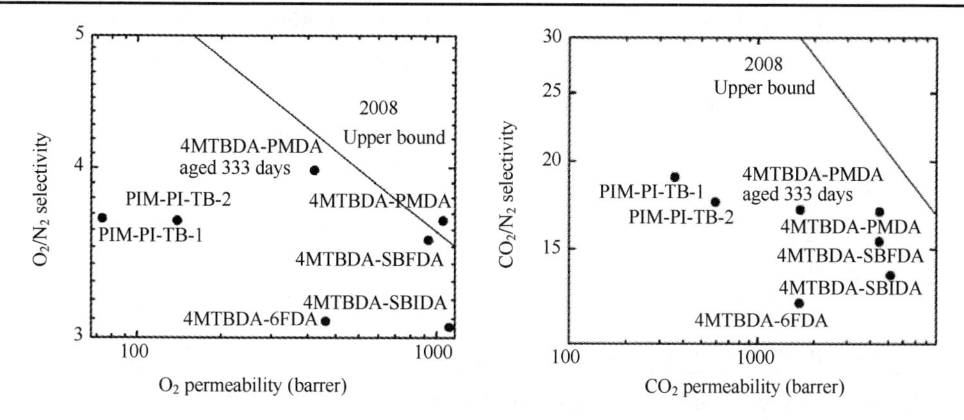

FIGURE 6.35 O_2/N_2 and CO_2/N_2 trade-off curves for the TB-contained PIM-PIs.

A novel dianhydride with spirobisindane and 9,10-dimethylanthrance as site of contortions were synthesized [58,59] according to method 1, by oxidation of their tetramethyl-substituted intermediates to get the tetra-carboxylic acid, and thereafter refluxed with acetic anhydride to get the dianhydrides. A spirobifluorene-based dianhydride was synthesized using method 2, by aromatic substitution to form the tetracyano-intermediate, after hydrolysis and cyclic reaction using acetic anhydride to get the corresponding dianhydrides [60].

These dianhydrides reaction with 3,3-dimethylnaphthadine (DMN) and 2,3,5,6-tetramethyl-p-phenylenediamine (TMPD), four intrinsically microporous polyimides were obtained denoted as SBI-DMN, SBI-TMPD, SBF-DMN and PIM-PI-EA. Their physical properties and gas transport properties are summarized in Tables 6.37 and 6.38. The PIM-PIs showed very high surface areas in the range of 616 $m^2·g^{-1}$-699 $m^2·g^{-1}$ and high pore volume (Table 6.37).

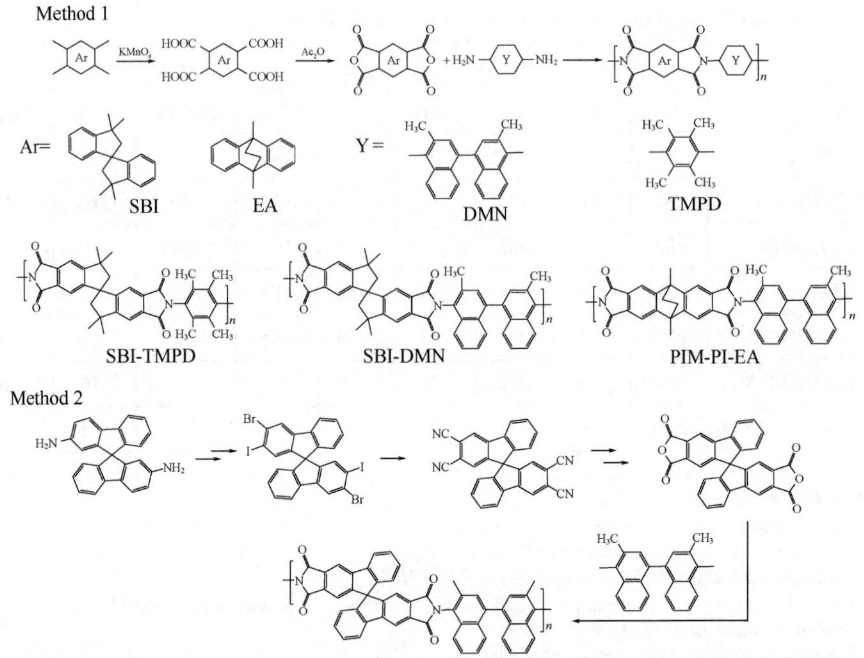

FIGURE 6.36 Synthesis of spirobisindane-, spirobifluorene- and ethanoanthracene-contained PIM-PI membranes [60].

TABLE6.37 Physical Properties of the PIM–PIs Using Site of Contortion as Dianhydrides

Polymer	M_n (10^4 mol·g^{-1})[a]	PDI[b]	S_{BET} (m^2·g^{-1})[c]	Pore Volume (cm^3·g^{-1})[d]
SBI-DMN	22.7	1.8	699	0.67
SBI-TMPD	12.9	1.7	690	0.47
PIM-EA-TB	11.0	3.1	616	0.35
SBFDA-DMN	6.5	1.92	686	—

[a] *Average molecular weight of the polyimide, measured using chloroform as eluent and narrow polydispersity polystyrene as external standard.*
[b] *Polydispersity of the PIM-PIs.*
[c] *BET surface area of the PIM-PIs, obtained by analysis the N_2 adsorption data at -196 °C, the relative pressure (p/p_0) ranged from 0.1 to 0.3.*
[d] *Pore volume of the PIM-PIs, obtained from N_2 adsorption isotherm at -196 °C and the p/p_0=0.98.*

TABLE6.38 Gas Permeability and Selectivity of the PIM–PIs

Polymer	Permeability (Barrer)[a]					Ideal Selectivity ($\alpha_{X/Y}$)			
	H_2	N_2	O_2	CH_4	CO_2	H_2/N_2	H_2/CH_4	O_2/N_2	CO_2/CH_4
SBI-DMN	840	94	295	170	2180	4.3	4.9	3.1	12.7
SBI-TMPD	670	84	270	168	2154	3.5	4.0	3.2	12.8
PIM-EA-TB[b]	4230	369	1380	457	7340	11.5	9.2	3.7	19.0
After aged 273 d	2860	131	659	156	3230	21.8	18.3	5.0	24.6
SBFDA-DMN[a]	2966	226	850	326	4700	13.1	9.1	3.8	14.4
After aged 200 days	878	33	161	40	703	26.6	22	4.9	17.6

[a] *The permeability of the gases are recorded in Barrer, 1 Barrer=10^{-10} cm^3 (STP)·cm·cm^{-2}·s^{-1}·cm Hg^{-1} or 7.5×10^{-18} m^3 (STP)·m·m^{-2}·s^{-1}·Pa^{-1}.*
[b] *The thickness of the membrane is 72 μm.*

The gas transport performance of the above PIM-PIs is shown in Table 6.38. The obtained PIM-PIs show very high permeability. The SBI-based PIM-PIs exhibit high O_2 permeability around 280 Barrer with O_2/N_2 selectivity of 3.1-3.2. The PIM-PI with ethanoanthracene as contortion center have fivefold higher O_2 permeability (~1380 Barrer) than SBI-DMN or SBI-TMPD, with O_2/N_2 selectivity of 3.7. After aging for 273 days, its O_2 permeability still measured at a high level (659 Barrer) with an increase in O_2/N_2 selectivity to 5.0. The O_2 permeability of fresh-made SBFDA-DMN was measured at 850 $m^2 \cdot g^{-1}$, it reduced to 161 Barrer after aging for 200 days.

Triptycene as a building block in the dianhydride part was also synthesized and is shown in Fig. 6.37. By reaction with different diamines, a series of triptycene-containing PIM-PIs were produced, ranging from KAUST-PI-1 to KAUST-PI-7 (Fig. 6.41) [61-63]. The nonextended triptycene-based dianhydrides (TDA1, TDA*i*3) as site of contortions were prepared by substituting at the 9,10 position with a methyl group or isopropyl group, which were reacted with DMN to yield two polyimides, including TDA-DMN and TDA*i*3-DMN, respectively. The physical properties of the triptycene-based PIM-PIs are listed in Table 6.39. All of the polyimides have remarkably high surface areas, and the KAUST-PI-7 exhibits the highest surface area of 840 $m^2 \cdot g^{-1}$. The pore volumes were measured in the range of 0.45 $cm^3 \cdot g^{-1}$-0.61 $cm^3 \cdot g^{-1}$, except KAUST-PI-4 (0.28 $cm^3 \cdot g^{-1}$). The lower pore volume of KAUST-PI-4 was attributed to the introduction of extra-flexible polymer chain segments.

The triptycene-based polyimides demonstrated excellent gas-transport properties (Table 6.40). Very high permeability and high selectivity were simultaneously observed. The KAUST-PI-1 exhibits O_2 permeability up to 627 Barrer with O_2/N_2 selectivity of 5.9. Meanwhile, KAUST-PI-1 shows H_2 permeability of as high as 3983 Barrer with H_2/N_2 and H_2/CH_4 selectivities of 37-38. These excellent results are attributed to the strong size sieve effect of the triptycene structures (Fig. 6.38).

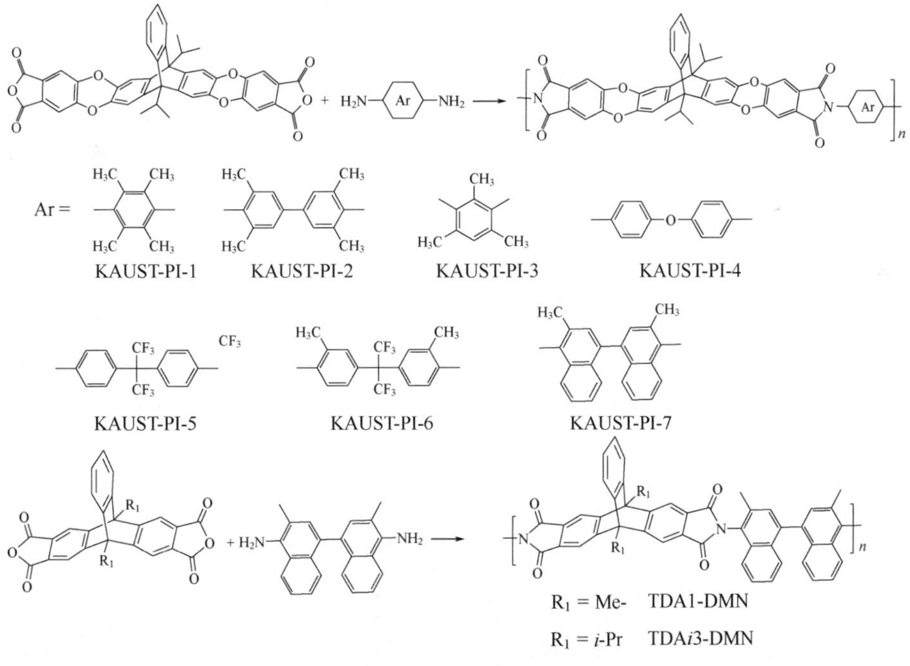

FIGURE 6.37 Synthesis of the triptycene-contained PIM-PIs

Fig. 6.39 depicts the gas separation performance of the triptycene-containing PIM-PIs. Obviously, PIM-PIs derived from triptycene as part of dianhydride demonstrated remarkable gas separation properties. KAUST-PI-1 showed the best gas separation properties, including high permeability and selectivity for separation of H_2/N_2, O_2/N_2, H_2/CH_4, and CO_2/CH_4. Moreover, the chemical structures of the contortion centers in the PIM-PIs also show obvious influence on the gas separation performance.

The triptycene-containing PIM-PIs demonstrated significantly higher selectivity compared with corresponding spirobisindane-, spirobifluorene-, and ethanoanthracene- based PIM-PIs (Fig. 6.43). This is probably due to the high rigidity as well as the special internal free volume generated by the Y-shaped structures of triptycene building blocks (Fig. 6.34).

TABLE6.39 Physical Properties of the Triptycene-contained PIM-PIs

Polymer	M_n (10^4 mol·g^{-1})[a]	PDI[b]	S_{BET} (m^2·g^{-1})[c]	Pore Volume (cm^3·g^{-1})[d]
KAUST-PI-1	158	2	750	0.531
KAUST-PI-2	120	1.8	740	0.49
KAUST-PI-3	82	1.9	760	0.57
KAUST-PI-4	73	2.1	420	0.282
KAUST-PI-5	83	2	650	0.53
KAUST-PI-6	135	1.6	500	0.61
KAUST-PI-7	140	1.7	840	0.59
TDA1-DMN	73	1.6	760	0.53
TDA*i*3-DMN	131	1.50	680	0.48

[a]*Average molecular weight of the polyimide, measured using chloroform as eluent and narrow polydispersity polystyrene as external standard.*
[b]*Polydispersity of the PIM-PIs.*
[c]*BET surface area of the PIM-PIs, obtained by analysis the N_2 adsorption data at −196 °C, the relative pressure (p/p_0) ranged from 0.1 to 0.3.*
[d]*Pore volume of the PIM-PIs, obtained from N_2 adsorption isotherm at −196 °C and the p/p_0=0.98.*

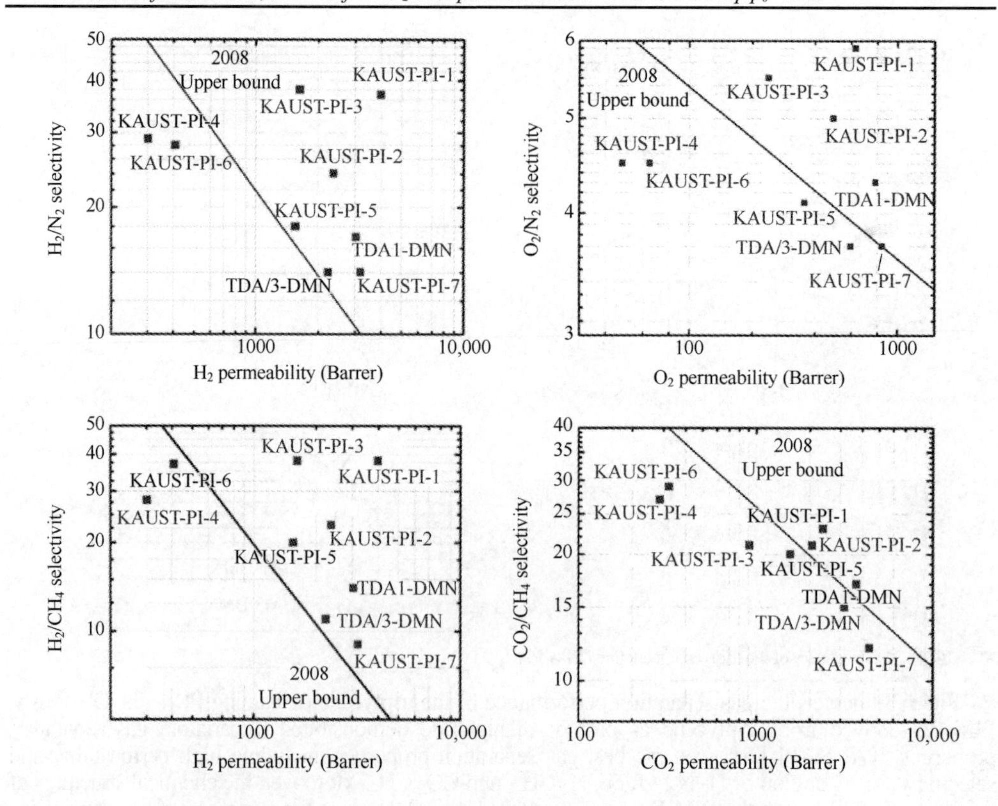

FIGURE 6.38 The trade-off curves on gas separation performance of the triptycene-contained PIM-PIs.

TABLE6.40 Gas Permeability and Selectivity of the Triptycene-contained PIM-PIs

Polymer	Permeability (Barrer)[a]				CO_2	Ideal Selectivity ($\alpha_{X/Y}$)			
	H_2	N_2	O_2	CH_4		H_2/N_2	H_2/CH_4	O_2/N_2	CO_2/CH_4
KAUST-PI-1	3983	107	627	105	2389	37	38	5.9	23
KAUST-PI-2	2368	98	490	101	2071	24	23	5.0	21
KAUST-PI-3	1625	43	238	43	916	38	38	5.5	21
KAUST-PI-4	302	10.6	48	10.7	286	29	28	4.5	27
KAUST-PI-5	1558	87	356	77	1552	18	20	4.1	20
KAUST-PI-6	409	14.4	64.8	11	322	28	37	4.5	29
KAUST-PI-7	3198	225	842	354	4391	14	9.0	3.7	12
TDA1-DMN	3050	182	783	216	3700	17	14	4.3	17
TDAi3-DMN	2233	160	594	211	3154	14	11	3.7	15

[a] *The permeability is recorded in Barrer, 1 Barrer=10^{-10} cm^3 (STP)·cm·cm^{-2}·s^{-1}·cm Hg^{-1} or 7.5×10^{-18} m^3 (STP)·m·m^{-2}·s^{-1}·Pa^{-1}.*

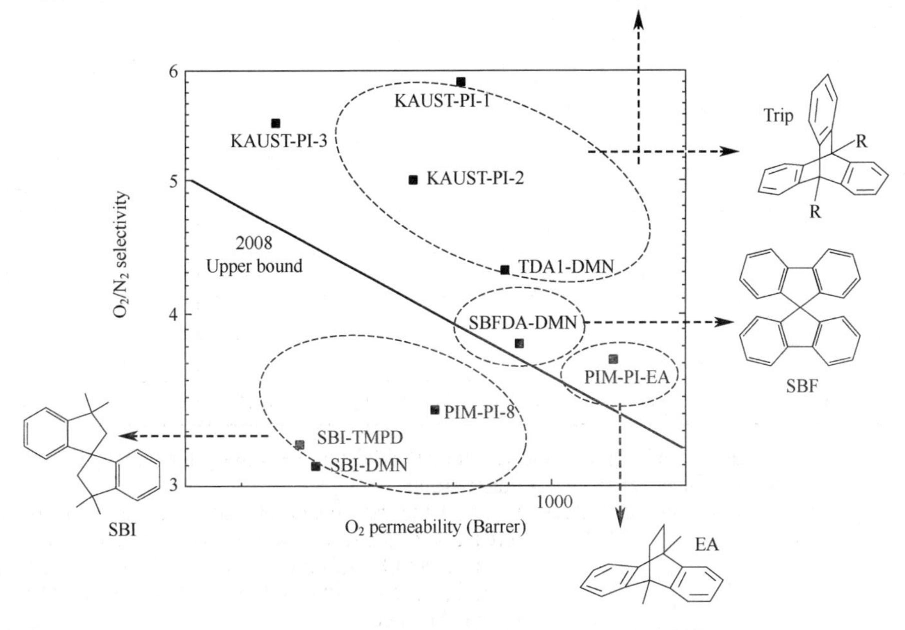

FIGURE 6.39 The trade-off curves on O_2/N_2 separation of the PIM-PIs with different sites of contortions.

6.5 Hydroxyl-functionalized Polyimide and Its Derived Polybenzoxazole Membranes

6.5.1 Hydroxyl-functionalized Polyimide Membranes

Hydroxyl-functionalized polyimide membranes for gas separation applications can be dated back to 1997 [64]. A series of hydroxyl-containing polyimides were prepared derived from 6FDA and different hydroxyl-containing aromatic diamines (Fig. 6.44). Table 6.41 summarizes the gas permeability and some gas pair selectivity of some hydroxyl-functionalized polyimides (Fig. 6.40).

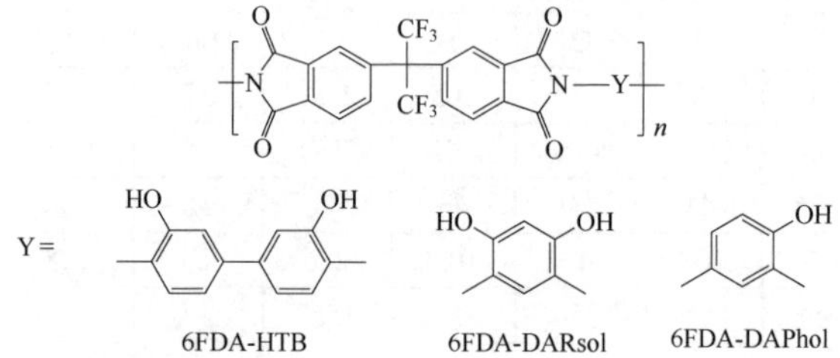

Y =

6FDA-HTB   6FDA-DARsol   6FDA-DAPhol

FIGURE 6.40 Structures of the hydroxyl-contained polyimides.

TABLE6.41 Gas Permeability and Selectivity of the Hydroxyl-contained Polyimides

Polyimides	P_{H_2}	Permeability (Barrer)			P_{CH_4}	Selectivity ($\alpha_{X/Y}$)		
		P_{O_2}	P_{N_2}	P_{CO_2}		P_{O_2/N_2}	P_{CO_2/CH_4}	Ref.
6FDA/*m*-PD	46	3.6	0.62	14	0.2	5.9	70	[65]
	—	5.4	0.87	20.3	0.35	6.2	58	[66]
	—	2.1	0.29	7.3	0.1	7.3	76	[66]
6FDA/DAP	38	2.5	0.37	11	0.12	6.8	92	[65]
	40	2.4	0.34	8.5	0.09	7.0	94	[64]
	—	2.9	0.43	11	0.11	6.8	100	[67]
6FDA/DAR	34	1.9	0.28	8	0.085	6.8	94	[65]
	—	2.2	0.30	8	0.11	7.4	73	[64]
	—	2.1	0.29	8.2	0.08	7.3	102	[66]
Cellulose acetate	—	1.46	0.23	6.56	0.20	6.35	33	[68]

The hydroxyl-containing polyimides showed remarkably high selectivity for CO_2/CH_4 due to the strong polar OH group in the polymer as a pendant group.

Alaslai et al. made a systematic investigation of the gas separation properties of some hydroxyl functionalized polyimides (Fig. 6.41) [65]. By reaction of the 6FDA with meta-phenylenediamine with different OH substitution from 0, 1 to 2 in each repeat unit, 6FDA/*m*-PD, 6FDA/DAP and 6FDA/DAR can be obtained.

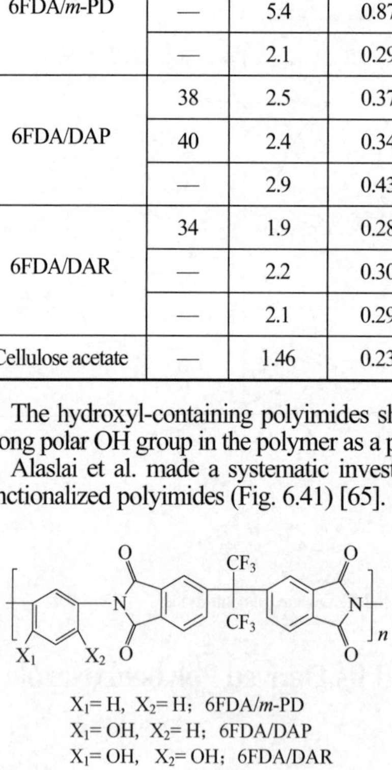

X_1= H, X_2= H; 6FDA/*m*-PD
X_1= OH, X_2= H; 6FDA/DAP
X_1= OH, X_2= OH; 6FDA/DAR

FIGURE 6.41 Structures of 6FDA/*m*-PD, 6FDA/DAP, and 6FDA/DAR.

The gas separation performance of the above OH functionalized polyimides are listed in Table 6.41. The 6FDA/DAP membrane exhibits CO_2 permeability of 8.5-11 Barrer, which is about 1.4 times higher than the cellulose acetate materials. The introduction of one OH group in the polyimides enhanced the CO_2/CH_4 selectivity from ~70 to ~90, and the second OH resulted in a very small change either in the permeability or selectivity. The higher CO_2/CH_4 selectivity was achieved by the higher diffusivity selectivity (Table 6.42). The solubility coefficients and solubility selectivity of these polyimides are almost the same, irrespective of the substitution and functionalization.

The introduction of hydroxyl groups in the polyimides resulted in a decrease with increasing amounts of OH groups, whereas their CO_2 adsorption amount increased with increasing OH group concentration in the repeat unit of the polyimides. Moreover, all polyimides exhibited a very high-pressure resistance to CO_2 pressure. A continuous decrease in permeability was obtained when increase the CO_2 pressure. However, there is a big drop in CO_2/CH_4 selectivity as the pressure of CO_2 increases. The selectivity of CO_2/CH_4, for 6FDA/*m*-PD was 70-76 at the CO_2 partial pressure of 7.3-14 bar, which reduced to 58 when the CO_2 partial pressure increased to 20 bar, indicating the OH functionalized polyimides may be a very promising polyimide in practical natural gas sweetening.

A series of hydroxyl-functionalized polyimides was synthesized by reaction of different dianhydrides with bisAPAF (Fig. 6.42, [69]), including 6FDA, BTDA, ODA, BPDA, and PMDA. The polyimides obtained were denoted as 6F-HPI, P-HPI, B-HPI, Bt-HPI, and O-HPI.

TABLE 6.42 Diffusion Coefficient, Solubility Coefficient, Diffusivity Selectivity and Solubility Selectivity of the Polyimides For CO_2 and CH_4

Polymer	Diffusion Coefficient (10^{-8} $cm^2{\cdot}s^{-1}$)[a]		Solubility Coefficient (10^{-2}) [b]		Diffusivity Selectivity	solubility Selectivity
	CH_4	CO_2	CH_4	CO_2	α_D (CO_2/CH_4)	α_S (CO_2/CH_4)
6FDA/*m*-PDA	0.06	0.78	3.8	18.2	13.0	4.8
6FDA/DAP	0.02	0.42	5.8	27.1	21.0	4.8
6FDA/DAR	0.014	0.29	6.0	29.3	21.0	4.9

[a]*Diffusion coefficient of the gases transport through the membranes, scale: 10^{-8} $cm^2{\cdot}s^{-1}$.*
[b]*solubility coefficient of the membranes, scale: $10^{-2}{\cdot}cm^3 cm^{-3}{\cdot}cmHg^{-1}$.*

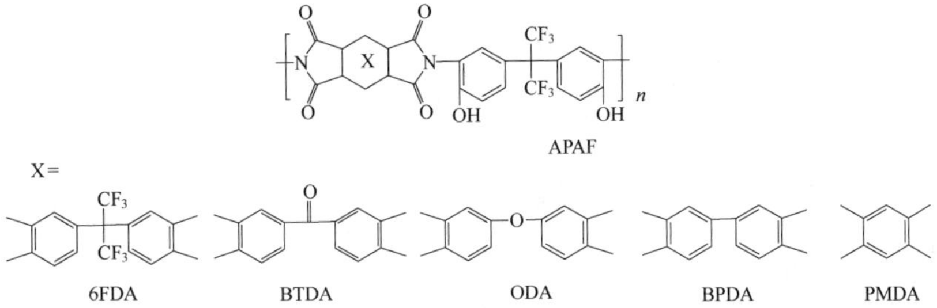

FIGURE 6.42 Structure of the hydroxyl-functionalized polyimides.

Table 6.43 shows the gas separation performances of the hydroxyl-functionalized polyimides. The polyimides exhibited a CO_2 permeability in the range of 1.43—17 Barrer with the CO_2/CH_4 selectivity in the range of 71-121. A rapid decrease in gas permeability was observed from 6F-HPI to Bt-HPI, where the 6F-HPI showed a P_{CO_2} of 17 Barrer and a selectivity of 80.

TABLE 6.43 Gas Separation Properties of the Hydroxyl-functionalized Polyimides

Polyimides	Permeability (Barrer)[a]					Ideal Selectivity ($\alpha_{X/Y}$) [b]				
	P_{H_2}	P_{N_2}	P_{O_2}	P_{CH_4}	P_{CO_2}	H_2/N_2	H_2/CH_4	O_2/N_2	CO_2/N_2	CO_2/CH_4
6F-HPI	42.75	0.55	3.73	0.21	16.95	78	200.1	6.8	31	79.5
P-HPI	35.21	0.36	2.58	0.08	9.87	97.4	433	7.1	27.3	121.4
B-HPI	14.26	0.09	0.72	0.03	2.74	153.5	470.9	7.7	29.5	90.6
Bt-HPI	11.13	0.08	0.6	0.02	1.43	139.1	556.5	6.6	17.9	71.5
O-HPI	13.36	0.06	0.54	0.02	1.81	231.8	792.3	9.3	31.5	107.6

[a]*The permeability was recorded in Barrer, and the scale is 10^{-10} $cm^3{\cdot}cm^{-2}{\cdot}s^{-1}{\cdot}cmHg^{-1}$.*
[b]*The ideal selectivity of the gas pairs.*

Some other hydroxyl functionalized polyimides were also investigated by Pinnau's group. Raja et al. [70] investigated the effect of OH loadings and thermal annealing on the gas transport properties of 6FDA-APAF, TPDA-APAF, and TPDA-ATAF, which were prepared by reacting 6FDA and TPDA with APAF and ATAF (Fig. 6.43A). By changing the dianhydride from 6FDA to TPDA, a much more intensive fluorescence emission was observed, indicating a higher charge transfer complex in TPDA-APAF. The gas transport properties are shown in the Table 6.44. The membranes were annealed at 120 °C and 250 °C, respectively. The higher-temperature-annealed membrane demonstrated a lower permeability for CO_2 and CH_4 coupled with a higher CO_2/CH_4 selectivity. By carefully analyzing the diffusion coefficient and solubility coefficient, the lower gas permeability was thought to originate from the lower diffusion coefficient of the gases (D) while the selectivity increase is from the enhanced diffusivity selectivity (α_D). For instance, the 6FDA-APAF polyimide annealed at 120 °C demonstrated a permeability of 9.6 Barrer for CO_2 and the selectivity for CO_2/CH_4 was 69; whereas the 250 °C annealed membrane only exhibited a CO_2 permeability of 6.8 Barrer and the CO_2/CH_4 selectivity increased to 96. The solubility and the solubility selectivity for CO_2 and CH_4 are almost the same for the polyimides annealed at different temperatures. After annealing at elevated temperatures, the only difference is the diffusion coefficient (D) and the diffusivity selectivity (α_D).

(A)

6FDA-APAF

TPDA-APAF

TPDA-ATAF

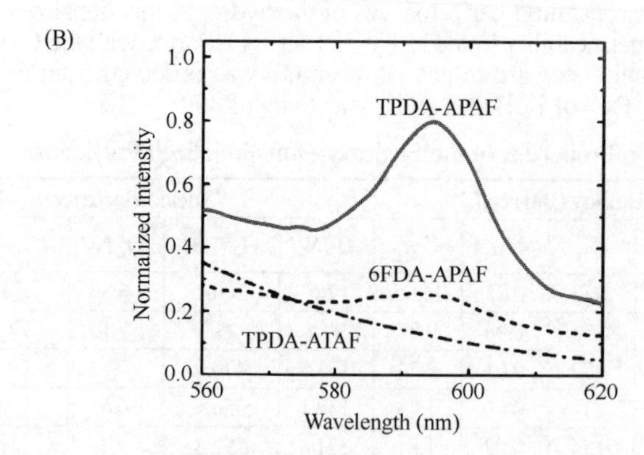

FIGURE 6.43 Polyimides derived from 6FDA and TPDA with different diamines and their fluorescence with the excitation is 395 nm [70].

This is caused by the more dense packing of the polymer chains annealed at higher temperatures, resulting in a decrease in diffusion coefficient and higher diffusivity selectivity. By changing the dianhydride from 6FDA to TPDA, ~10 fold enhanced permeability was observed by introducing this highly contorted triptycene as a kink. The CO_2/CH_4 selectivity decreased from 69 to 38. Compared with TPDA-ATAF, the introduction of the methyl group instead of the OH group resulted in further increase in CO_2 permeability from 99 Barrer to 325 Barrer at the small decrease in selectivity of CO_2/CH_4 from 38 to 30.

Moreover, the hydroxyl-containing polyimides showed significant stability against plasticization resistance (Fig. 6.44). Up to 25 bar of CO_2 pressure, the CO_2 permeability continues to decrease with pressure. The pure gas demonstrated slightly higher permeability as compared to the mixed gas. As for the CO_2/CH_4 selectivity, the TPDA-APAF showed an increased selectivity in mixed-gas separation than the pure-gas selectivity because its ultramicroporous structure is appropriate for rare "blocking" of CH_4 transport by co-permeation of CO_2.

The introduction of hydroxyl functional groups into the intrinsically microporous polyimides was also investigated (Fig. 6.45). The polyimides contain both polar OH group and spiro center, and shows modest surface areas with S_{BET} ~200 $m^2 \cdot g^{-1}$ (Table 6.45). After soaking in a mixture solvent (DCM/Hexane=20/80) and drying in a 120 °C in vacuum oven for 24 h, the PIM-6FDA-OH polyimide exhibited CO_2 permeability of 225 Barrer and CO_2/CH_4 selectivity of 29 (Table 6.46). Compared with PIM-1 and PIM-PI-3, the CO_2 permeability decreased ~10 to ~2 fold and the CO_2/CH_4 selectivity increased from ~18 to 29. Meanwhile, the gas transport properties for CO_2/CH_4 in the high-pressure mixed-gas condition also demonstrated high selectivity and CO_2 pressure resistance. Under 20 bar CO_2 partial pressure, the membrane still exhibited a CO_2/CH_4 selectivity of over 20.

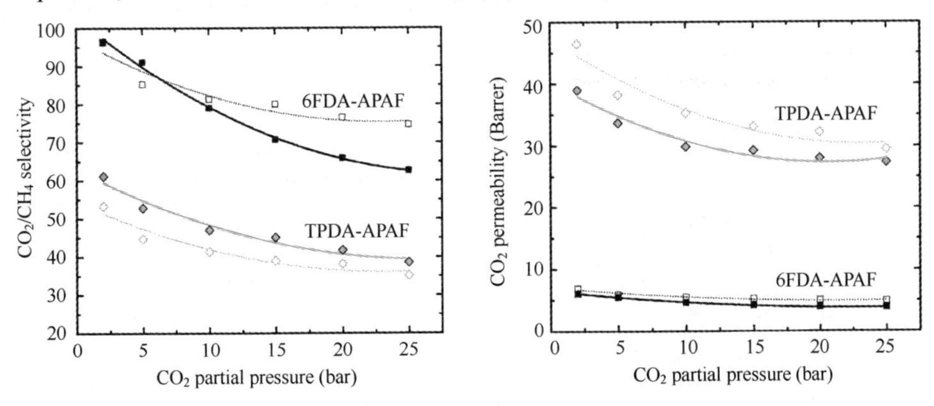

FIGURE 6.44 CO_2 permeability (left) and CO_2/CH_4 mixed-gas selectivity (right) under different CO_2 partial pressure [70].

PIM-6FDA-OH　　PIM-PMDA-OH

FIGURE 6.45 The structure of PIM-6FDA-OH and PIM-PMDA-OH.

TABLE 6.44 Gas Transport Properties of the Hydroxyl Functionalized Polyimides

Polyimides	T (°C) [a]	P		D		S		CO_2/CH_4		
		CH_4	CO_2	D_{CH_4}	D_{CO_2}	S_{CH_4}	S_{CO_2}	α_P	α_D	α_S
6FDA/APAF	250	0.071	6.8	0.026	0.47	2.7	14.7	96	17.7	5.5
	120	0.13	9.0	0.052	0.66	2.5	13.6	69	12.7	5.4

TABLE6.44 (Continued)

Polyimides	T (°C) [a]	P		D		S		CO_2/CH_4		
		CH_4	CO_2	D_{CH_4}	D_{CO_2}	S_{CH_4}	S_{CO_2}	α_P	α_D	α_S
TPDA/APAF	250	0.87	46	0.13	1.9	6.6	24.3	53	14.5	3.7
	120	2.6	99	0.39	3.7	6.6	26.8	38	9.3	4.1
TPDA/ATAF	250	3.8	125	0.59	5.0	6.4	25.0	33	8.4	3.9
	120	11.0	325	1.4	10.9	8.0	29.8	30	7.9	3.7
Matrimid	—	0.28	4.8	—	—	—	—	36	—	—

[a] *Isotherm temperature of the membranes and the time are 24 hrs.*

TABLE6.45 Properties of Functionalized 6FDA and PMDA OH-Based PIM Polymers and Reference PIM Polymers

Polymer	M_n (×10^4) [a]	M_w (×10^4) [a]	PDI [a]	S_B ($m^2 \cdot g^{-1}$) [b]	T_d (°C) [c]	ρ [d]	FFV [e]
PIM-6FDA-OH	8.54	16.5	1.94	225	380	1.22	0.23
PIM-PMDA-OH	13	35.6	2.74	190	380	1.18	0.20
PIM-1	10.1	25.7	2.54	830	350	1.07	0.26
PIM-PI-3	2.2	4.5	2.05	471	420	1.26	0.23

[a] *Molecular weights and polydispersity index (PDI), measured by GPC using THF as solvent, and polystyrene as external standard.*
[b] *BET surface area determined by Ar sorption at 87K.*
[c] *Onset decomposition temperature determined by T_g analysis.*
[d] *Polymer density.*
[e] *Fractional free volume of the polymer membrane calculated based on Bondi's theory [16].*

TABLE6.46 Permeability and Selectivity of Various PIM Membranes for Different Gases at 35 °C

Polymers	Permeability (Barrer) [a]					Ideal Selectivity (α)				
	H_2	N_2	O_2	CH_4	CO_2 [b]	H_2/N_2	O_2/N_2	CH_4/N_2	CO_2/N_2	CO_2/CH_4
PIM-6FDA-OH	259	10.8	45.2	9.1	263	24	4.2	0.83	24	29
PIM-PMDA-OH	190	6.9	30.5	7.7	198	28	4.5	1.1	29	26
PIM-PI-3 [c]	360	23	85	27	520	16	3.7	1.2	23	19

[a] *1 Barrer=10^{-10} cm^3 (STP)·cm·cm^{-2}·s^{-1}·cm Hg^{-1} or 7.5×10^{-18} m^3 (STP)·m·m^{-2}·$s^{-1}$$Pa^{-1}$.*
[b] *Permeability of CO_2 was obtained at the upstream pressure of 1 bar while those of other gases were measured at 2 bar.*
[c] *Data from [44].*

TABLE6.47 Permeability, CO_2/CH_4 and H_2S/CH_4 Selectivity of PIM-6FDA-OH and Other Polymers in Ternary Mixed-Gas Feeds at 35 °C

Polymers	Pressure	Composition	Permeability		Selectivity		References
			CO_2	H_2S	CO_2/CH_4	H_2S/CH_4	
PIM-6FDA-OH	34.5	15/15/70	54.7	36.0	27.8	8.3	[71]
PIM-6FDA-OH	48.3	15/15/70	52.6	63.0	25.0	30.0	[71]
Crosslinked TEGMC	48.3	20/20/60	46.2	33.5	31.2	22.5	[72]
Crosslinked TEGMC	48.3	20/20/60	54.6	38.2	28.4	19.3	[72]
Cellulose acetate	34.5	20/20/60	8.66	8.71	29.5	29.7	[73]
Cellulose acetate	48.3	20/20/60	27.5	39.7	19.1	27.4	[73]
Cellulose acetate	10.1	6/29/65	2.43	2.13	22	19	[74]
6FDA-DAM:DABA (3:2) annealed at 180 °C	48.3	10/20/70	55.6	25.4	32.1	14.7	[72]

TABLE 6.47 (Continued)

Polymers	Pressure	Composition	Permeability		Selectivity		References
			CO_2	H_2S	CO_2/CH_4	H_2S/CH_4	
6FDA-DAM:DABA (3:2) annealed at 230 °C	48.3	10/20/70	50.8	23.6	31.1	14.4	[72]
Pebax 1074	10.1	12.5/18.1/69.4	155	695	11	50	[74]
PU1	10.1	12.5/18.1/69.4	55.8	183	6.9	23	[74]
PU2	10.1	12.5/18.1/69.4	195	618	5.6	18	[74]
PU3	10.1	12.5/18.1/69.4	62.2	280	12	55	[74]
PU4	10.1	12.5/18.1/69.4	50.8	223	15	66	[74]
6F-PAF-1	63.3	10/20/70	8.1	4.2	32	11	[75]

When the PIM-6FDA-OH polyimide was annealed at 250 °C, a decrease in permeability and an increase in CO_2/CH_4 selectivity were observed, a more intensive charge transfer complex was formed and proofed by the fluorescence. Yi et al., tested this annealed membrane for its natural gas separation properties under the condition of high feeding pressure and in the presence of CO_2 and H_2S at the same time (Table 6.47) [71]. Due to the high microporosity and the polar OH effect, both CO_2 and H_2S can be simultaneously separated by this membrane. The permeability of CO_2 and H_2S are 52.6 and 63, coupled with CO_2/CH_4 and H_2S/CH_4 selectivity of 25 and 30 at the feeding pressure of 48.3 bar, respectively. These results are both higher in permeability and selectivity of cellulose acetate (P_{CO_2} ~6 Barrer and CO_2/CH_4 selectivity of 33).

Meanwhile, Ma et al., reported another hydroxyl-functionalized PIM-PI based on spirobifluorene as diamine monomer (Fig. 6.46). When reacted with 6FDA and SPDA, two intrinsically microporous

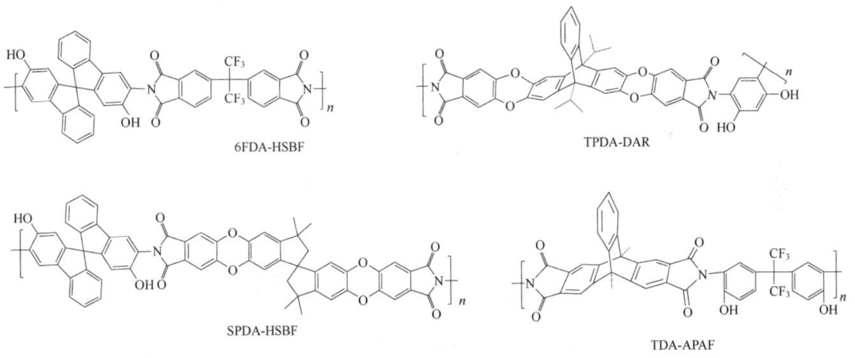

FIGURE 6.46 Structure of the hydroxyl-contained intrinsically microporous polyimides.

polyimides, 6FDA-HSBF and SPDA-HSBF, were obtained [76]. Alaslai and Alghunaimi et al. adopted some triptycene-based dianhydrides and reacted with DAR and APAF, respectively [57,77]. Two intrinsically microporous OH functionalized PIM-PIs are also obtained, denoted as TPDA-DAR and TDA-APAF. Their gas transport properties are shown in Table 6.48.

The basic properties of the hydroxyl functionalized intrinsically microporous polyimides are listed in Table 6.48. The hydroxyl-containing intrinsically microporous polyimides contain sites of contortion either in the dianhydride part or the diamine part. The BET surface areas increased in the order of 6FDA-HSBF (70 $m^2 \cdot g^{-1}$) < TDA-APAF (260 $m^2 \cdot g^{-1}$) < TPDA-DAR (308 $m^2 \cdot g^{-1}$) < SPDA-HSBF (464 $m^2 \cdot g^{-1}$), where the double spiro-containing PIM-PIs exhibited the highest BET surface area of 464 $m^2 \cdot g^{-1}$ (Table 6.48). Compared with spirobisindane as site of contortion, the introduction of spirobifluorene (6FDA-HSBF) showed significant enhanced selectivity compared with spirobisindane-based PIM (6FDA-OH) (42 vs 35) with a slight decrease in selectivity (100 vs 119). The increase in the surface areas resulted in an increase in gas permeability. The H_2 permeability increased in the order of 6FDA-HSBF (162 Barrer) < TPDA-DAR (431 Barrer) < SPDA-HSBF (519 Barrer), respectively.

However, the selectivity for CO_2/CH_4 and H_2/N_2 were continuously decreasing. The trade-off curves of the gas separation performance for CO_2/CH_4 and H_2/N_2 are depicted in Fig. 6.47. The triptycene-containing TDA-APAF showed very good performance. After aging for 250 days, only a slight decrease in permeability and increase in selectivity was observed.

TABLE6.48 Gas Permeability and Selectivity of the Hydroxyl Functionalized Intrinsically Microporous Polyimides at 35 °C at 2 bar Upstream Pressure

Polymers	S_{BET}	T_d	H_2	*P* (Barrer)			CO_2	Ideal Selectivity ($\alpha_{X/Y}$)		
				N_2	O_2	CH_4		H_2/CH_4	O_2/N_2	CO_2/CH_4
6FDA-HSBF[a]	464	380	162	3.8	19.3	2.4	100	67.5	5.1	41.7
SPDA-HSBF	260	400	519	24	98	29	568	17.9	4.1	19.6
TPDA-DAR	308	480	431	13	65	11	349	39.2	5.0	32
TDA-APAF	464	450	94	1.5	8.5	0.73	40	128.8	5.7	55
Aged 250 days	—	—	73	0.90	6.4	0.40	30	182.5	7.1	75

[a]*The membrane was annealed at 250 °C for 24 h. 1 Barrer=10^{-10} cm^3 (STP)· cm·cm^{-2}·s^{-1}·cm Hg^{-1} or 7.5×10^{-18} m^3 (STP)·m·m^{-2}·s^{-1}·Pa^{-1}.*

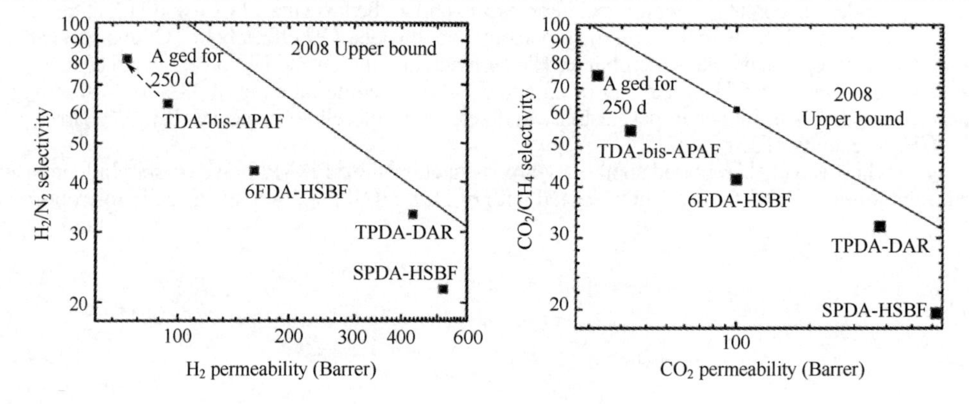

FIGURE 6.47 Performance of the OH functionalized PIM-PIs for H_2/N_2 and CO_2/CH_4.

6.5.2 Thermally Rearranged Polybenzoxazole (TR–PBO) Membranes

In 1999, Tullos et al. found that the *ortho*-hydroxyl-functionalized aromatic polyimide can undergo cyclic reaction to form polybenzoxazole (TR-PBO) under high temperatures of 350 °C-450 °C [78]. The mechanism of this thermal induced arrangement is illustrated in Fig. 6.54. When the *ortho*-hydroxyl-functionalized polyimide was heated to about 350 °C, the hydroxyl group reacted with the imide group to form an intermediate with an evolving CO_2 molecule to form a PBO membrane. This thermal conversion can be investigated in situ by TGA with online mass spectrometry. A typical TGA-MS of the *ortho*-hydroxyl-functionalized polyimide upon heating is highlighted in Figs. 6.48 and 6.49.

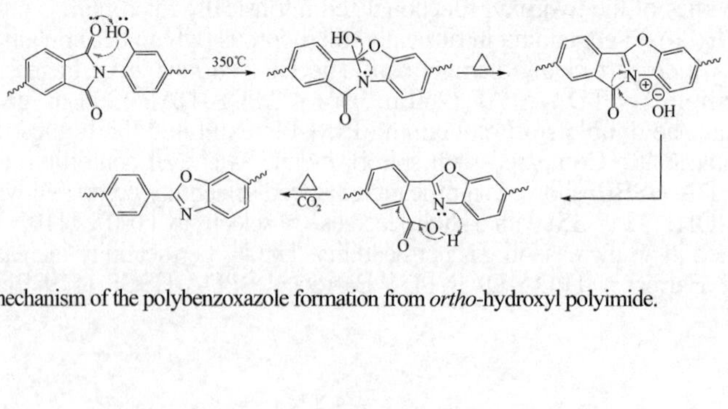

FIGURE 6.48 The mechanism of the polybenzoxazole formation from *ortho*-hydroxyl polyimide.

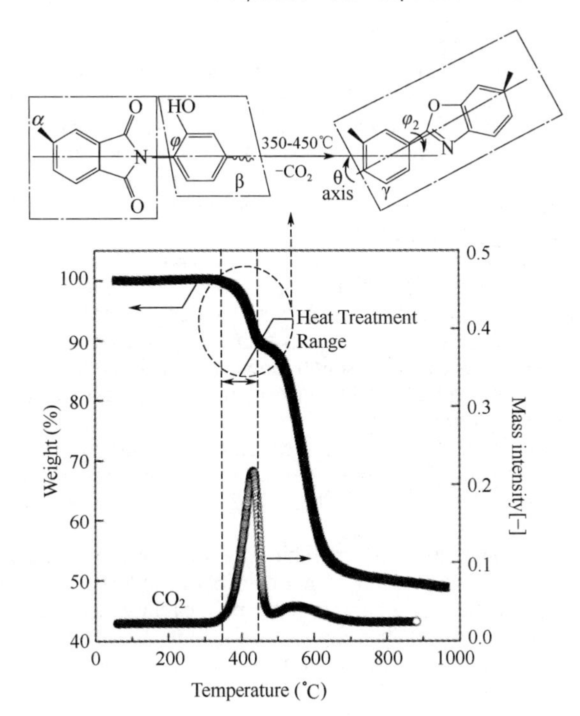

FIGURE 6.49 Weight loss of the *ortho*-hydroxyl-functionalized polyimide under N_2 at elevated temperatures with the simultaneously evolving of CO_2 molecule.

A novel kind of thermally rearranged (TR) PBO membrane derived from *ortho*-hydroxyl or *ortho*-thiol functionalized polyimides as a precursor was obtained [79]. During the TR process, there are some significant changes in the resulting membranes. First, the FFV element of the polymer was enhanced due to the chain conformation rearrangement (~10 to 20 fold). Second, the pore size distribution of the polymers became more uniform after the TR process. Third, the TR-PBO membranes showed excellent gas transport properties with both high permeability and selectivity (Fig. 6.50).

Owing to the outstanding performance of the TR-PBO membranes, a large amount of research effort has been devoted to this application and processing field. A series of *ortho*-hydroxy-functionalized polyimides and their corresponding polybenzoxazoles have been prepared for gas transport studies. Table 6.49 shows the gas permeability and selectivity of the *ortho*-hydroxyl-functionalized poly-imides and the corresponding TR-PBO membranes at 35 °C/2 bar upstream pressure.

Han et al. investigated the effects of the chemical structures of the TR-PBO membranes on the gas separation performance (Fig. 6.51) [80]. The polyimide precursor 6FDA-bisAPAF was prepared by different methods: (1) *tHPI* was formed by direct thermal treatment of the hydroxyl-containing poly-(amic acid) (HPAA) by heating to 60 °C, 100 °C, 150 °C, 200 °C, and 250 °C for 1 h at each isothermal step. (2) *aHPI* was formed using the azeotropic agent o-xylene (30 mL) at 180 °C for 6 h. (3) cAcPI was formed by chemical imidization of the HPAA using acetic anhydride and pyridine as a catalyst at room

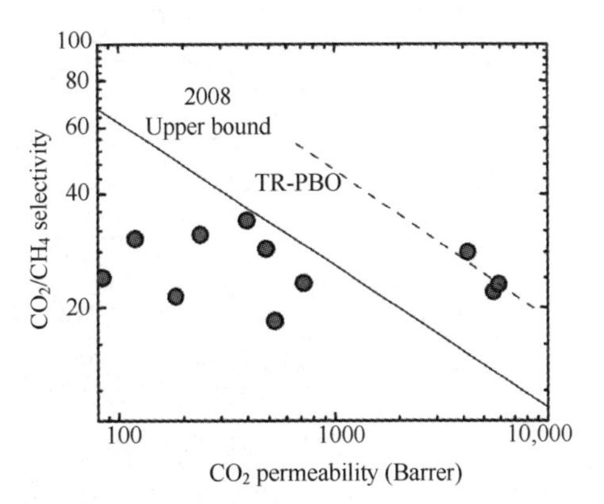

FIGURE 6.50 The trade-off curves of the CO_2/CH_4 performance for the TR-PBO membranes.

temperature for 12 h. (4) sAcPI formation is the same as synthesis of cAcPI except using large amounts of trimethylsilyl chloride to activate the diamine in the formation of the HPAA step.

The polyimide precursor was cast to isotropic films, which was thereafter heated to 300 °C and held for 1 h to eliminate the residual solvent, and further heated to 450 °C and maintained for 1 hr in a high-purity argon atmosphere to obtain the resulting PBO membranes. Table 6.49 summarizes the gas permeation results. The *t*PBO, AcPI, and sAcPI demonstrated good transport performance characterized by very high permeability and reasonable selectivity, whereas the azeotropic method formed aPBO that exhibited the lowest permeability.

Two *ortho*-hydroxyl-functionalized polyimides with biphenyl structures (*m*HAB-6FDA and *p*HAB-6FDA) have been prepared derived from 3,3′-diamino-4,4′-dihydroxybiphenyl (*m*HAB) or 4,4′-diamino-benzendine (*p*HAB) and 6FDA, which were thermally treated at 450 °C for 1 h to give the corresponding PBO membranes (Fig. 6.52). The gas transport properties are also listed in Table 6.49. The *para*-substituted *p*HTB-6FDA exhibited slightly lower permeability than the *m*HTB-6FDA. When they are converted to PBO, a much higher CO_2 permeability (720 Barrer vs 240 Barrer) was observed for the *meta*-substituted diamine *m*HTB [83].

The enhancement in the flexibility of the precursor and the resulting gas transport properties of the PBO membranes were investigated by introducing ether and forming poly(ether-benzoxazole) (PEBO) [81]. After thermally treated at 400 °C and 450 °C for 1, 2, 3 h, respectively, six PBO membranes with different conversions were obtained (Fig. 6.53).

The gas transport performances are listed in Table 6.49. Compared with the precursor HPEI, all of the TR polymer membranes resulted in an increase in permeability and a decrease in selectivity. However, when introducing the flexible ether group, the permeability increment for all of the gases are not as prominent as 6FDA-bisAPAF. When the treatment time is extended at 450 °C, the resulting TR-PBO derived from HPEI demonstrated comparable results to those of the aPBO from 6FDA-bisAPAF.

TABLE6.49 Gas Permeability and Selectivity of the *Ortho*-Hydroxyl-Functionalized Polyimides and the Corresponding TR-PBO Membranes at 35 °C/2 bar Upstream Pressure

Polymers	Permeability (Barrer)					Ideal Selectivity ($\alpha_{X/Y}$)			
	H_2	N_2	O_2	CH_4	CO_2	H_2/N_2	H_2/CH_4	O_2/N_2	CO_2/CH_4
6FDA-bisAPAF[a]									
tPBO	4194	284	1092	151	4201	14.8	28	3.8	28
aPBO	408	19	81	12	398	21.5	35	4.3	34
cPBO	3612	431	1306	252	5568	8.4	14	3	22
sPBO	3585	350	1354	260	5903	10.2	14	3.9	23
*p*HTB-6FDA[b]	35	0.35	2.3	0.16	10	100	219	6.6	62.5
*p*TR450	260	10	45	7.7	240	26.0	33.8	4.5	31.2
*m*HTB-6FDA	46	0.41	2.8	0.18	12	112	255	6.8	66.7
*m*TR450	570	34	130	31	720	16.8	18.4	3.8	23.2
HPEI[c]	29.1	0.46	1.9	0.13	8.6	63.3	224	4.1	66.2
400-1	59.5	1.02	6.2	0.62	23.3	58.3	96	6.1	37.6
400-2	64.5	1.98	10.7	1.35	27	32.6	48	5.4	20
400-3	51.2	1.52	5.6	0.8	22.5	33.7	64	3.7	28.1
450-1	95.3	1.89	10	1.45	41	50.4	66	5.3	28.6
450-2	158	5.5	23.1	3.9	119	28.7	41	4.2	30.4
450-3	439	20	88.5	17	486	22.0	26	4.4	28.6

TABLE 6.49 (Continued)

Polymers	Permeability (Barrer)					Ideal Selectivity ($\alpha_{X/Y}$)			
	H_2	N_2	O_2	CH_4	CO_2	H_2/N_2	H_2/CH_4	O_2/N_2	CO_2/CH_4
PHA[d]	22.1	0.24	1.54	0.13	5.76	92	170	6.5	43
PBO300	94.6	2.28	11.1	1.56	44.7	41	61	4.9	28.6
PBO350	149	4.42	19.9	3.48	84.2	34	43	4.5	24
PBO425	223	10.1	35.2	8.53	183	22	26	3.5	21.4
PBO450	526	30.3	105	28.9	532	17	18	3.5	18.4

[a] *Data from Ref. [80]*
[b] *Data from Refs. [1,83]. 1 Barrer=10^{-10} cm^3 (STP)·cm·cm^{-2}·s^{-1}·cm Hg^{-1}.*
[c] *Data from Ref. [81]*
[d] *Data from Ref. [82]*

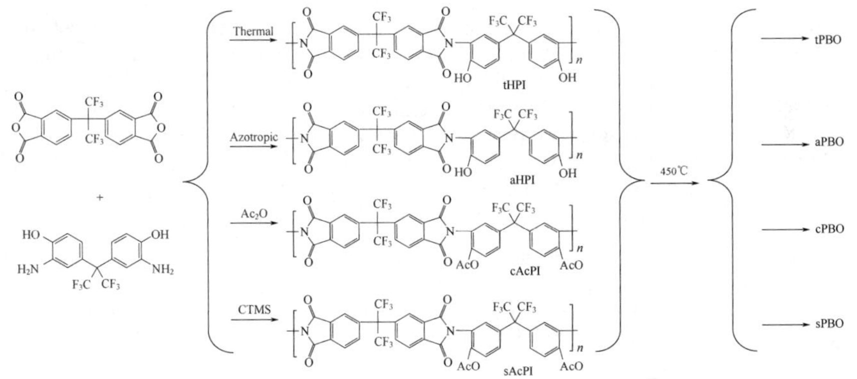

O CF3 O O CF3 CF3 O O
+
HO OH H2N NH2 F3C CF3
Thermal tHPI HO OH → tPBO
Azotropic aHPI HO OH → aPBO
Ac2O cAcPI AcO AcO → cPBO
CTMS sAcPI AcO AcO → sPBO
450℃
n

FIGURE 6.51 Synthesis of the TR-PBO membranes using different HPI precursors and different synthetic conditions.

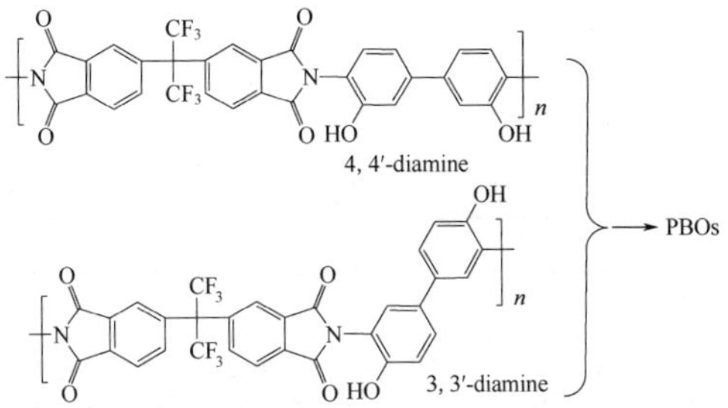

CF3 CF3 N O HO OH n
4, 4′-diamine
3, 3′-diamine
→ PBOs

FIGURE 6.52 Structures of *m*HAB-6FDA and *p*HAB-6FDA and the corresponding TR-PBO membranes.

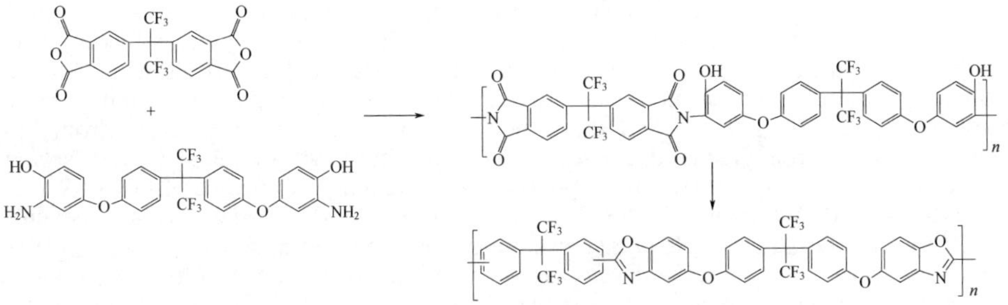

CF3 CF3 O +
HO OH H2N O O NH2
N O OH OH n
CF3 O N O N O n

FIGURE 6.53 Synthesis of HPEI and its corresponding PBOs at different temperatures [81].

Guo et al. investigated two series of *ortho*-hydroxyl-functionalized polyimides, which were then converted into PBOs [84]. The structures of HTB-6FDA and 6FAP-BisADA are shown in Fig. 6.54.

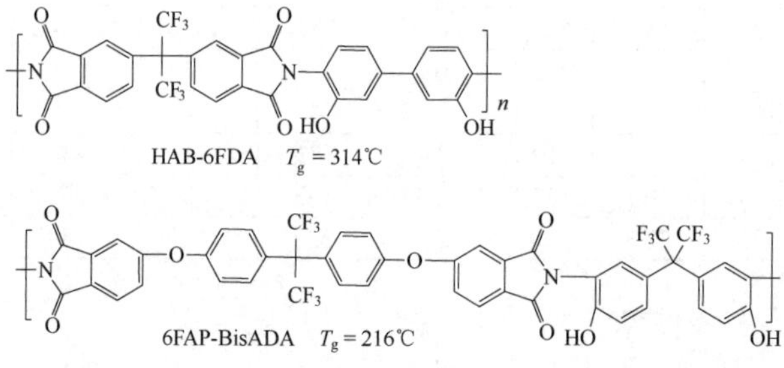

HAB-6FDA T_g = 314℃

6FAP-BisADA T_g = 216℃

FIGURE 6.54 Structures of the hydroxyl functionalized polyimides precursor with different T_g for TR-PBOs.

T_gs of the polyimide precursors have apparent effects on the resulting properties of the PBO membrane. The T_g difference between HAB-6FDA and 6FDP-BisADA is about 100 °C. However, the polyimide precursors exhibited almost the same permeability for CO_2 and H_2 (Table 6.50). The HAB-6FDA-TR350 obtained by thermally treating HAB-6FDA at 350 °C exhibited a slightly higher permeability and selectivity than 6FAP-BisADA-TR350. If higher temperature (400 °C) was chosen for TR temperature, the resulting HAB-6 FDA-TR350 and HAB-6FDA-TR450 exhibited about fourfold higher CO_2 permeability than 6FAP-BisADA and were a little higher in CO_2/CH_4 selectivity. The CO_2/CH_4 performance of TR–PBO derived from HAB-6FDA was much higher than that derived from 6FAP-BisADA.

TABLE 6.50 Gas Permeability and Selectivity of the Hydroxyl Functionalized Intrinsically Microporous Polyimides and the Corresponding PBOs at 35°C and 2 bar Upstream Pressure

Polymers	Permeability (Barrer)					Ideal Selectivity ($\alpha_{X/Y}$)			
	H_2	N_2	O_2	CH_4	CO_2	H_2/N_2	H_2/CH_4	O_2/N_2	CO_2/CH_4
HAB-6FDA	11	0.1	0.64	0.06	2.9	110	183	6.4	48.3
HAB-6FDA-TR350	55	0.62	3.9	0.26	17	88.7	211	6.3	65.4
HAB-6FDA-TR400	115	2.7	13	1.5	58	42.6	76.7	4.8	38.7
HAB-6FDA-TR450	155	4.5	21	3.03	95	34.4	51.2	4.7	31.4
6FAP-BisADA	13	0.11	0.76	0.09	2.9	118	144	6.9	32.2
6FAP-BisADA-TR350	30	0.43	2.6	0.31	11	69.8	96.8	6.0	35.5
6FAP-BisADA-TR400	34	0.54	3.1	0.39	13	63.0	87.2	5.7	33.3

1 Barrer=10^{-10} cm^3 (STP)·cm·cm^{-2}·s^{-1}·cm Hg^{-1}.

A series of thermal crosslinked PBO membranes have been prepared by thermal treatment of the hydroxyl functionalized polyimides with crosslinkable groups at elevated temperatures (Fig. 6.55). The polyimide precursor was first heated at 250 °C to give the crosslinked polyimides, which was then further heated at 400 °C·h^{-1} to produce a thermally rearranged PBO [85].

Table 6.51 compares the gas separation properties of the TR-PBOs. The crosslinking densities have obvious influence of the gas permeabilities. The crosslinked TR-PBOs with 5%-10% of crosslinking contents showed more than two to threefold higher gas permeability than the noncrosslinked TR-PBO (TR-PBO). Further increasing of the crosslinking density from 10% to 15% and 20% resulted in gradual decreases in gas permeability, whereas the CO_2/CH_4 selectivity did not change obviously. The overall CO_2/CH_4 gas separation performance of the XTR-PBOI-5 and XTR-PBOI-10 outperformed the 2008 trade-off curve (Fig. 6.56). This excellent gas permeability may be attributed to the crosslinking that happened at > 250 °C. During the TR process, there is extra free volume developed by the decomposition of the crosslinked polymers. However, when the crosslinking content is over 10%, an apparent decrease in permeability and slight increase in selectivity have been observed.

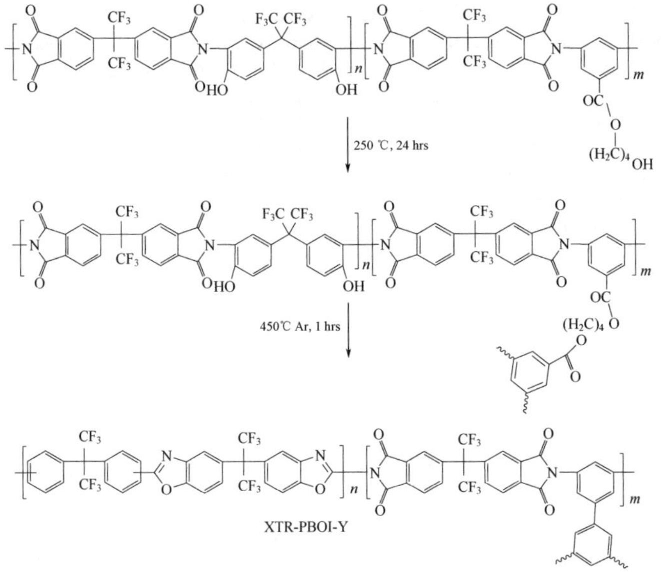

FIGURE 6.55 Preparation of the crosslinked TR-PBO membranes.

TABLE 6.51 Gas Permeability and Selectivity of the Crosslinked PBOs at 35 °C at 2 bar Upstream Pressure

Polymers	Permeability (Barrer)					Ideal Selectivity ($\alpha_{X/Y}$)			
	H_2	N_2	O_2	CH_4	CO_2	H_2/N_2	H_2/CH_4	O_2/N_2	CO_2/CH_4
TR-PBO	294	12.6	52.5	7.5	261	23.3	39.2	4.2	34.8
XTR-PBOI-5	603	29.6	133	19.9	746	20.4	30.3	4.5	37.5
XTR-PBOI-10	763	50.9	193	33	980	15.0	23.1	3.8	29.7
XTR-PBOI-15	515	29.8	119	19.4	668	17.3	26.5	4	34.4
XTR-PBOI-20	421	19.7	81.9	12.4	440	21.4	34	4.2	35.5
aPBO	408	19	81	12	398	21.5	35	4.3	34

Another crosslinked TR-PBO was also prepared [86]. A bismaleimide-based hydroxyl-containing monomer was used as a precursor, which was then heated to produce an oligomer and further heated to the temperature of thermal rearrangement to get the crosslinked PBO (Fig. 6.57).

The resulting crosslinked PBO (TR-BMI- 6F) exhibited significant surface area increment from 10.8 $m^2 \cdot g^{-1}$ of BHMI-6F to 1130 $m^2 \cdot g^{-1}$ of TR-BMI- 6F, an increment of about 105-times. An excellent gas transport performance was observed with CO_2 permeability of 5540 Barrer and CO_2/CH_4 selectivity of 37 for the membranes thermally treated at 450 °C. This is well above the latest trade-off curves for CO_2/CH_4 gas pairs.

The chemical structures of the *ortho*-hydroxyl-functionalized polyimide precursors have obvious influences on the gas transport properties of the resulting TR-PBOs. Ma et al. synthesized another *ortho*-hydroxyl-functionalized diamine using spir-

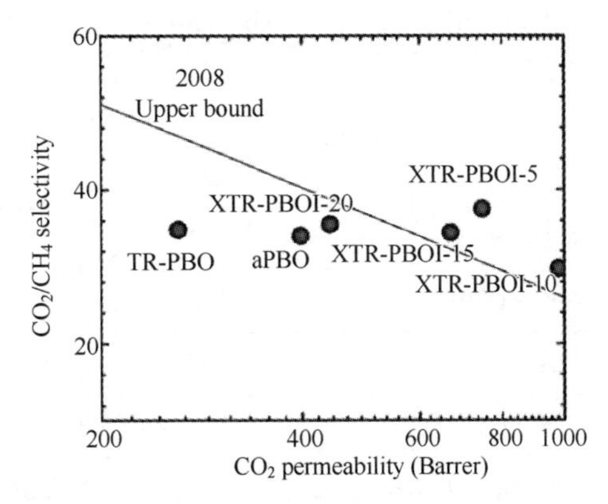

FIGURE 6.56 The permeability and selectivity for CO_2/CH_4 of the crosslinked PBO membranes on the 2008 trade-off curve.

obifuorene as site of contortion (HSBF) (Fig. 6.58), which was then reacted with 6FDA and SPDA as two distinct dianhydrides to obtain two *ortho*-hydroxyl-contained polyimides (6FDA-HSBF and SPDA-HSBF) [87]. After thermal rearrangement at 420 °C and 450 °C for 4 and 3 h, respectively, the corresponding PBO membranes were obtained. The TR process had a significant effect on the gas transport properties of the low-surface-area polyimides. For instance, after the TR process, sixfold enhancement in BET surface areas was observed in the relatively low-surface-area 6FDA-HSBF (from 70 to 417 $m^2 \cdot g^{-1}$). This is accordance with the trends of the permeability increment, the precursor with smaller surface area (6FDA-HSBF) showed ~10 fold increased permeability, whereas the relatively higher surface area SPDA-HSBF exhibited only ~2 times increased CO_2 permeability (Table 6.52).

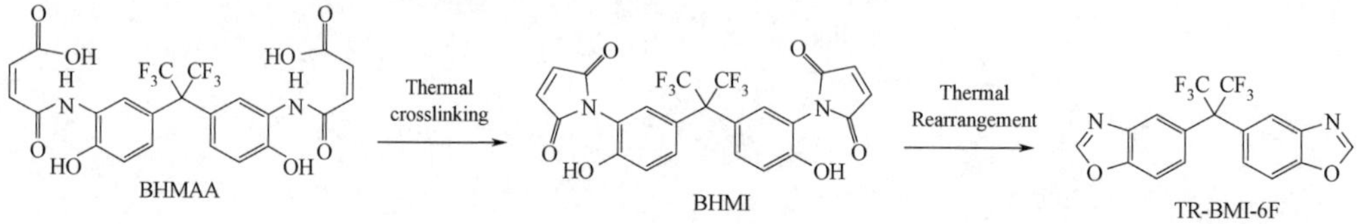

FIGURE 6.57 Crosslinked TR-PBO membranes derived from bismaleimides.

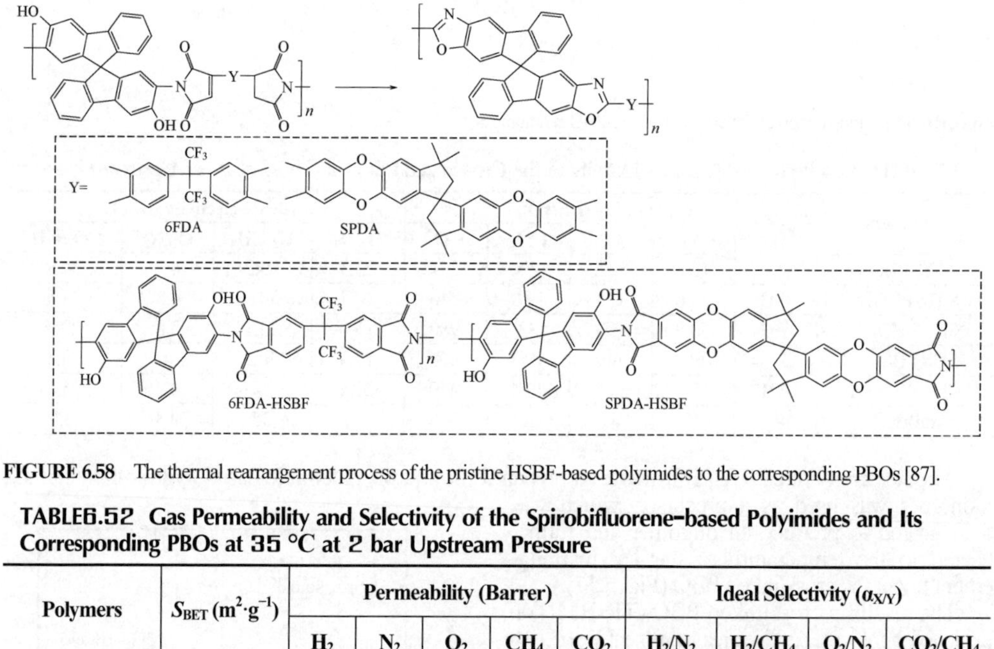

FIGURE 6.58 The thermal rearrangement process of the pristine HSBF-based polyimides to the corresponding PBOs [87].

TABLE6.52 Gas Permeability and Selectivity of the Spirobifluorene-based Polyimides and Its Corresponding PBOs at 35 °C at 2 bar Upstream Pressure

Polymers	S_{BET} ($m^2 \cdot g^{-1}$)	Permeability (Barrer)					Ideal Selectivity ($\alpha_{X/Y}$)			
		H_2	N_2	O_2	CH_4	CO_2	H_2/N_2	H_2/CH_4	O_2/N_2	CO_2/CH_4
6FDA-HSBF	70	162	3.8	19.3	2.4	100	42.5	67.5	5.1	41.7
SPDA-HSBF	460	600	27.7	114	34	657	22.3	17.6	4.3	19.2
6FDA-SP-PBO[a]	417	985	55	215	56	1158	17.9	17.6	3.9	20.7
SPDA-SP-PBO[b]	480	574	61.6	225	84.8	1279	9.3	6.8	3.7	15.1

[a] *6FDA-HSBF heated to 420 °C for 4 h to form 6FDA-SP-PBO.*
[b] *SPDA-HSBF heated to 450 °C for 2 h.*

6.5.3 Applications of TR-PBO Membranes

Apart from the gas separation applications, the TR-PBO membranes were also studied in the application of pervaporation for dehydration of biofuels in 2012. A TR-PBO membrane derived from 6FDA-HAB-based polyimide as precursor were been prepared [88]. The TR-PBO membranes showed a stable performance separation of water from isopropanol and *n*-butanol at 80 °C for 250 h, indicating the feasibility in purification of biofuels. The TR-PBO membrane has been used in the membrane distillation process, which showed very stable performance with an excellent water flux (80 kg $(m^2 \cdot h^{-1})^{-1}$) and salt rejection (99.99%) as well as convenient control of crystal nucleation mechanisms [89].

6.6 Polyimide-Derived Carbon Molecular Sieve Membranes

6.6.1 Formation of CMSMs

Carbon molecular sieve (CMS) membranes have been considered as a kind of high-performance gas separation membrane. A CMS membrane is a porous solid membrane with extreme rigidity and microporosity. In general, the CMSMs exhibited significantly higher gas transport properties than the corresponding polymeric membranes. In addition, CMSMs showed excellent solvent and pressure resistance (Table 6.53). The drawback of CMSMs is the brittleness, which restricts their practical applications.

TABLE6.53 The Difference Between CMSMs and Polymeric Membranes

	Polymeric Membrane	CMSM
Separation performance	Good to high	High
Separation mechanism	Solution-diffusion	Molecular sieve
Advantages	Low production cost, mass production	Excellent chemical stability, surpass the trade-off curves. Excellent thermal stability, Can be used under aggressive conditions
Disadvantages	Poor thermal and chemical stabilities, relatively low performance	Brittle, high cost

The gas transport in CMSMs is different from that in polymeric membranes. The difference is illustrated in Fig. 6.59. In a polymer system, the gases are separated by a solution-diffusion mechanism. In CMSMs, the gas molecules are separated by the size sieving effect. Gas molecules of certain sizes that are smaller than the critical distance (d_c) can pass, whereas those gas molecules larger than d_c will be rejected. Table 6.55 compares the advantages of CMSMs and polymeric membranes.

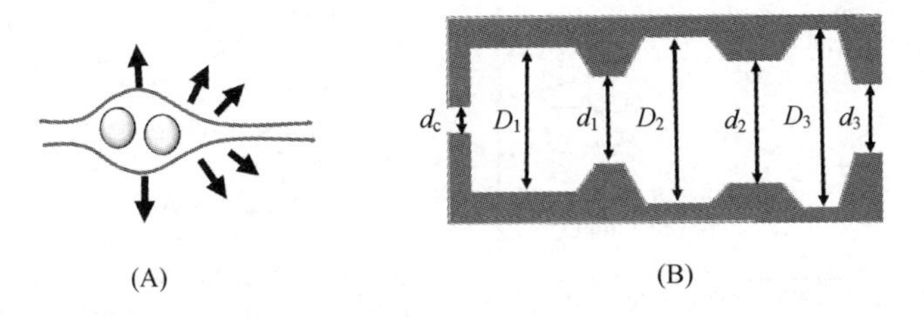

FIGURE 6.59 Schematic gas transport through polymeric membrane and carbon molecular sieve membranes.

CMSMs are usually produced by pyrolysis of polymeric membranes at high temperature. The polymer decomposed by losing certain atoms such as O, N, H, and other elements during pyrolysis, leaving most of the carbon atoms remaining in the membrane system. The procedure of making CMSMs generally includes four steps: (1) polymeric material selection; (2) polymeric membrane preparation; (3) posttreatment of the polymeric membrane; (4) pyrolysis at high temperature of the membrane to give the carbon molecule sieves membranes (Fig. 6.60).

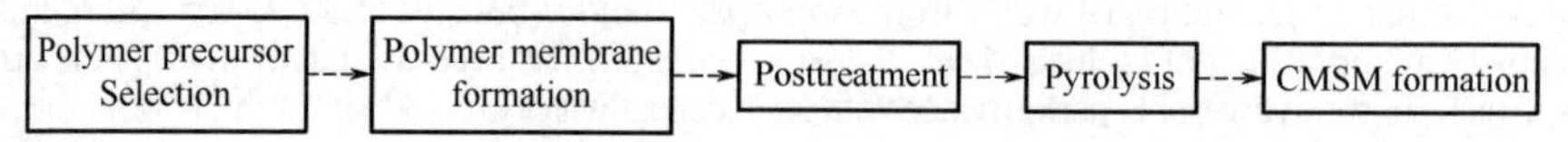

FIGURE 6.60 Procedure of formation carbon molecular sieve membranes.

The pyrolysis at high temperature has been considered the key step, in which the polymeric membrane was thermally treated at very high temperature and then soaked for some time in a particular atmosphere. The pyrolysis was generally performed in a tube furnace. Fig. 6.61 is a schematic set-up for a CMSM formation system.

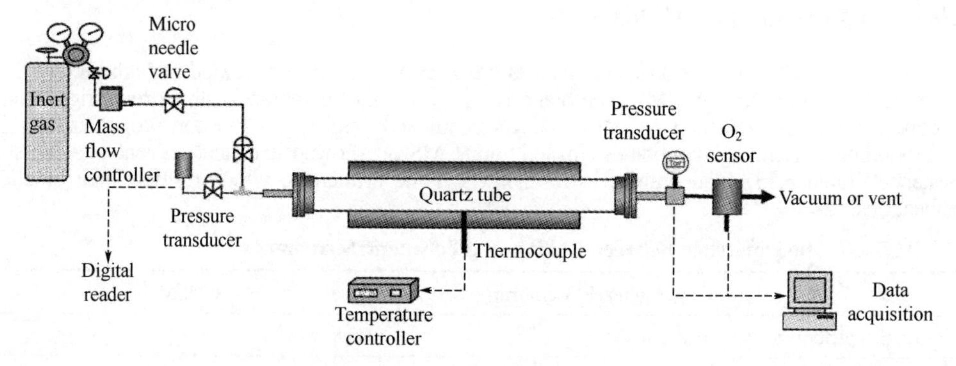

FIGURE 6.61 A schematic set-up for pyrolysis of polymeric membranes to carbon molecular sieve membranes [109].

6.6.2 Conversion From Polyimide to CMSMs

A typical TGA of the PI-(BTDA/ODA) membrane is shown in Fig. 6.62 [90]. The polyimide started to decompose at around 450 °C. From 500 °C to 650 °C, a rapid degradation speed indicated by rapid weight loss in a very short temperature range was observed, which is considered as the region I, corresponding to the high speed of creating pores. The region II started from 650 °C to 900 °C. There is relatively less weight loss during this temperature range, corresponding to the reduction or shrinking of the pore size.

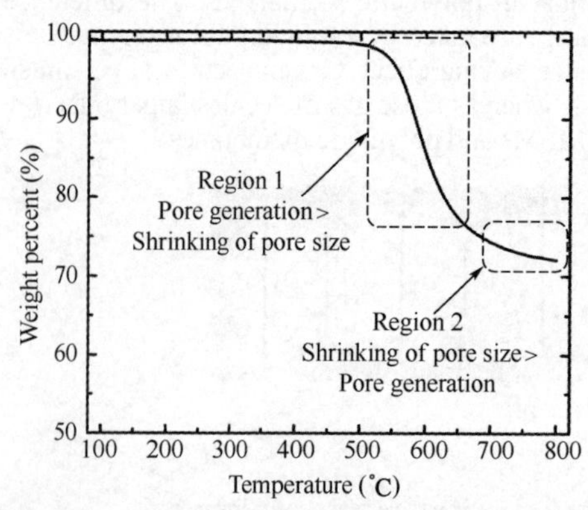

FIGURE 6.62 TGA-curve of polyimide membrane and their pores generation and shrinking upon heating.

Most of the polyimides are stable up to ~ 450 °C. For instance, the 6FDA/Durene showed almost the same FT-IR spectra as compared to the precursor when heated from room temperature to 250 °C, 350 °C, 400 °C, and 425 °C [91], indicating a minor change in the backbone structures of the polyimide owing to the excellent thermal stability of polyimide. When the treatment temperature increased to around 450 °C-550 °C, rapid weight loss (about 10%-20% of the total weight) in the polyimides was observed because most of the chemical bonds, including the imide bond, aliphatic carbon-hydrogen bond, CF_3 group, as well as some weak aromatic bonds, decomposed at such temperatures [91]. In region I, the FT-IR signal for the above-mentioned functional groups became very weak and almost completely disappeared. The polyimide membrane goes into an intermediate state.

In this intermediate state, graphite signals start to appear in the Raman spectrum with the wavelength of 1355 cm^{-1} and 1575 cm^{-1} (Fig. 6.63, [92]). Meanwhile, Ma et al. investigated the X-ray of PIM-6FDA-OH polyimide during this range [93]. The result indicated that during the above temperature range, a very broad peak in the X-ray of the membrane can be observed.

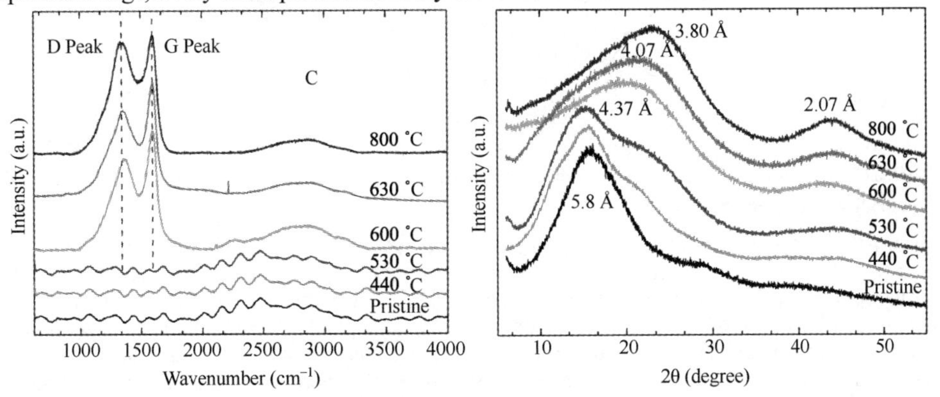

FIGURE 6.63 Raman spectra of PIM-6FDA-OH and wide angle X-ray scatterings of PIM-6FDA treated at different temperatures [92].

When the pyrolysis temperature of the polymeric membrane reaches above 550 °C-900 °C, the weight loss by TGA will be much smaller as compared to the temperature from 450 °C to 550 °C (Fig. 6.64). The new pore generation speed is much slower because the derivative of weight loss is much lower than the peak decomposition rate at around 500 °C. However, the pores in the carbon membrane gradually shrink to the ultramicropores, resulting in a significant molecule sieving effect. The ultramicroporosity was confirmed in the X-ray by an abrupt halo ~ 25 degrees corresponding to the *d*-spacing around 3.7 Å. The high temperature resulted in a characteristic graphite halo (100 phase) which can also become more clear as the gradually enhanced halo at ~45 degrees correlated to the *d*-spacing of 2.05 Å.

Fig. 6.64 depicts the CO_2 adsorption of the polymeric membrane during pyrolysis range. There is a very slight increase in CO_2 adsorptions correlated to smaller pore formation during the thermal treatment process at the temperature range of 600 °C-800 °C. However, an abrupt enhanced ultramicropore percent to 90% in the total incremental pore volume was observed with increased treatment temperature, as shown in the pore size distribution.

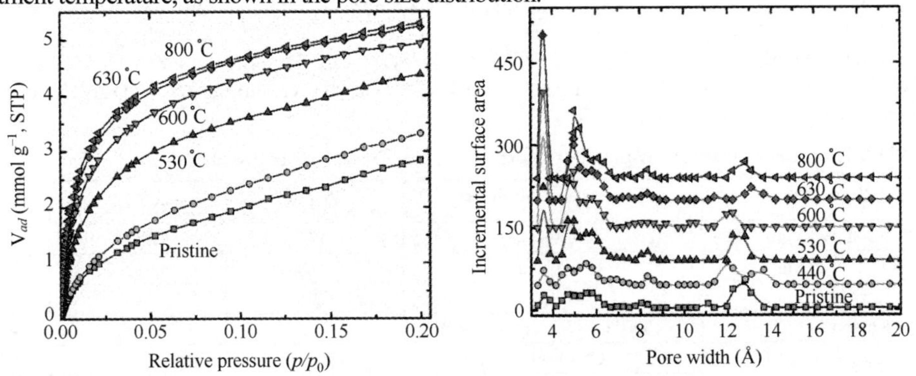

FIGURE 6.64 CO_2 adsorption of PIM-6FDA-OH and the pore size distributions at different treatment temperature [92].

As a result, the whole pyrolysis process of polyimide and the pore size distributions can be summarized in Fig. 6.65.

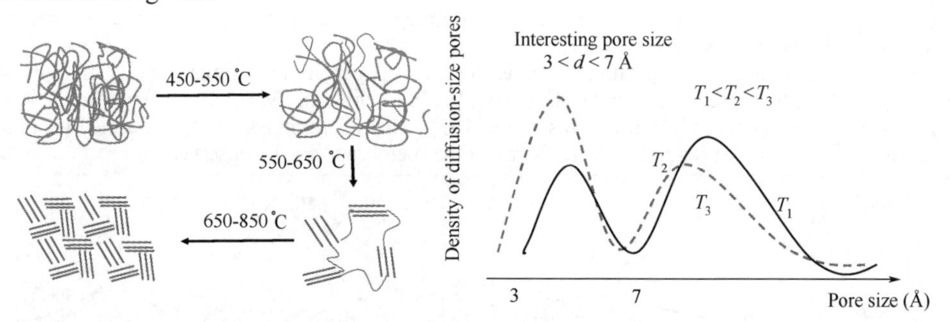

FIGURE 6.65 Structure changes of polyimide membranes during thermal treatment and the bi-model pore size distribution of the density of diffusion-size pores changes during the heating.

From the onset decomposition temperature to around 550 °C, some initial amorphous carbon structures appeared containing large amounts of irregularly distributed aromatic rings. The ultramicropores started to develop at this temperature. As the temperature continued to increase to around 650 °C, most of the polymer precursors changed to graphite structure, but they were organized in a very loosely packed structure. During this temperature range, a higher concentration of ultramicr-opores distribution can be observed and a decrease in density of pore size above 1 nm. The continued increase in temperature will result in a further closed packing of the graphite membrane with a sharp increase in the density of pores in the range of ultramicropore region.

6.6.3 Structures and Properties of CMSMs

The structure of CMSMs have obvious influence on their gas separation performance. There are several factors in the CMSM forming process which affects CMSM structures, including the chemical structures of the polyimide precursors, the casting solvents, the atmosphere conditions, the heating rates, and the aging, etc.

6.6.3.1 Structures of the Polyimide Precursors

Many polymer precursor candidates can be used to produce gas separation CMSMs such as PAN, phenolic resin, polyfurfuryl alcohol (PFA), polyphenylene oxide (PPO), poly(vinylidene chloride) (PVDC), phenol formaldehyde resin (PFR), polypyrrolone, poly(phthalazinone ether sulfone ketone) (PPESK) and polymer blends, etc. [94]. Aromatic polyimides are the most advanced polymer precursors and have been investigated by many researchers. Kapton H film can be converted to CMSMs of different thicknesses [95]. The properties of the resulting CMSMs are shown in Table 6.54. After being thermally treated at 600 °C, the CMSM showed CO_2 and permeability of 1820 Barrer with O_2/N_2 selectivity of 4.7. After thermal treatment to 1000 °C in an inert atmosphere, a substantial O_2/N_2 (36) selectivity can be obtained.

TABLE6.54 Gas Transport Properties of CMSM Based on Kapton as Precursor

Precursor	Treatment Conditions	P_{O_2}	α_{O_2/N_2}	P_{CO_2}
Kapton 113 μm	600 °C (vac)	383	4.7	1820
Kapton 113 μm	800 °C (vac)	34.8	11.5	128
Kapton 113 μm	1000 °C (vac)	0.96	23.4	4.15
Kapton 113 μm	1000 °C (13.3 K·min^{-1}, Ar)	0.92	21.6	3.54
Kapton 113 μm	1000 °C (4.5 K·min^{-1}, Ar)	0.66	23.0	2.51
Kapton 113 μm	1000 °C (1.33 K·min^{-1}, Ar)	0.15	36.0	0.5

The CMSMs have been prepared derived from Matrimid and PI-(6FDA-BPDA/DAM) (Fig. 6.66) using the following pyrolysis protocol: protocol A (start at 50 °C, heat to 250 °C at a rate of 13.3 °C·min^{-1}; followed by heat to 535 °C at a rate of 3.85 °C·min^{-1}; and then heat to 550 °C at a rate of 0.25 °C·min^{-1}; thereafter, soak the membrane at the 550 °C for 2 h) [96].

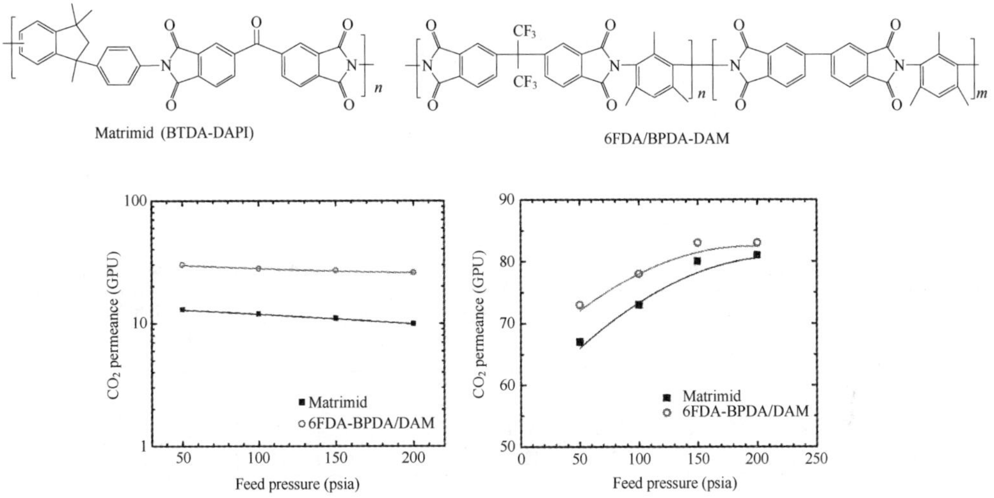

Matrimid (BTDA-DAPI)

6FDA/BPDA-DAM

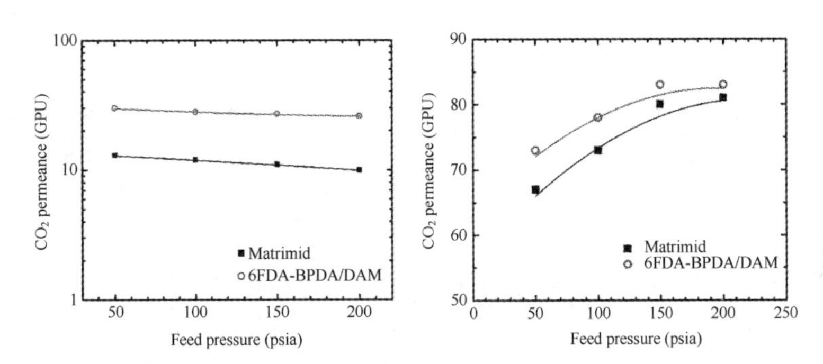

FIGURE 6.66 Structures of Matrimid 5218 (BTDA-DAPI) and PI-(6FDA-BPDA/DAM) (up) and the CO_2 permeability and CO_2/CH_4 selectivity with variable pressures of 10% CO_2/90% CH_4 (down, [96]).

The CMSMs derived from PI-(6FDA-BPDA/DAM) showed not only 130% higher CO_2 permeability than that of CMS fibers from Matrimid, but also a slightly enhanced CO_2/CH_4 selectivity (Fig. 6.67). These results indicated that CMSM with similar molecular-sieving and ultramicroporous morphologies but different ultramicropore volume can be obtained by identical pyrolysis protocols from two different polyimide precursors.

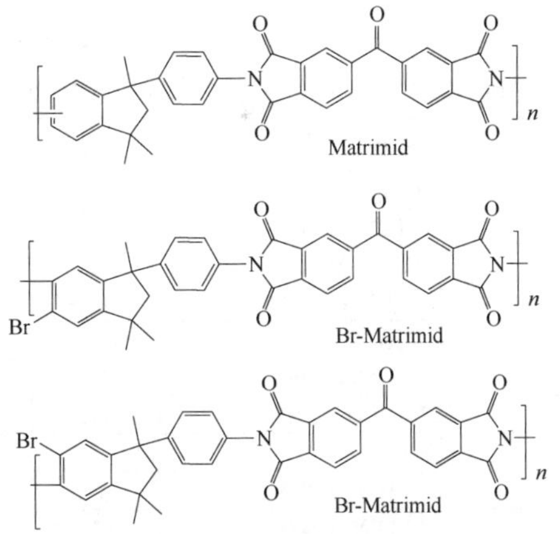

The structure effects of bromine-substituted Matrimid on the gas separation properties of the resulting CMSMs have been investigated [97]. Fig. 6.68 shows the chemical structures of the brominated Matrimid (Br-Matrimid). The CMSMs derived from Matrimid and Br-Matrimid are obtained by thermal pyrolysis of the polymer membranes at 550 °C, and show much higher gas permeability (Table 6.55). In comparison, Br-Matrimid-based CMSM has higher permeability than the Matrimid-based one, and the selectivity remained competitive to the original Matrimid precursor under the same conditions.

Matrimid

Br-Matrimid

Br-Matrimid

FIGURE 6.67 Structures of Matrimid and brominated Matrimid.

A novel diamine (DAI) was synthesized, and used to react with different aromatic dianhydrides including BTDA, OPDA, BPDA, and 6FDA to form a series of polyimides as precursors for carbon membranes (Fig. 6.69). The membranes were pyrolyzed at 550 °C-800 °C to produce a series of CMSMs [35]. The gas transport properties of the pristine polyimide membranes and their derived CMSMs are listed in Table 6.56. The CMSM pyrolyzed at 550 °C showed a huge increase in permeability, and the CMSMs pyrolyzed at 800 °C exhibited excellent gas separation performances with high selectivity. The larger free volume PI-(6FDA/DAI) derived CM-6FDA-550 has a CO_2 permeability of 4800 Barrer coupled with a

CO_2/CH_4 selectivity of 28. Further heating of the membranes to 800 °C, the CMSM derived from BTDA and BPDA demonstrated excellent results with O_2 permeability of 144 Barrer and 118 Barrer and O_2/N_2 selectivity of 8.8 and 8.5, respectively. These results are much better than those derived from 6FDA and ODPA.

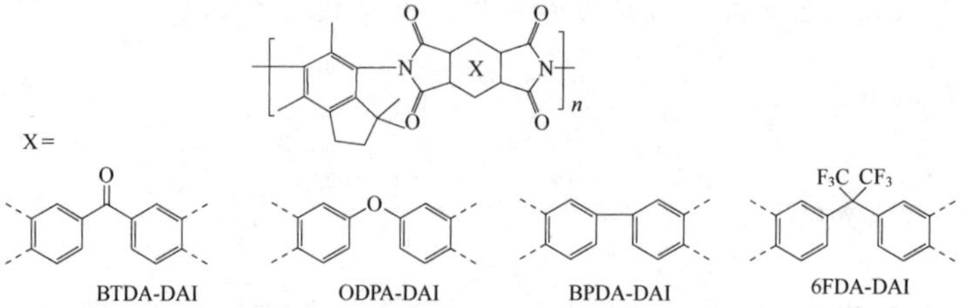

FIGURE 6.68 Structures of DAI-based polyimide precursors.

TABLE 6.55 Pure Gas Permeation Through Polyimides and Their Derived Carbon Membranes

Polymer	Permeability (Barrer)					Selectivity ($\alpha_{X/Y}$)	
	He	O_2	N_2	CH_4	CO_2	O_2/N_2	CO_2/CH_4
Br-Matrimid	38	3.6	0.55	0.44	13.3	6.5	30
Matrimid	21	1.7	0.25	0.19	6.5	6.6	34
CMSM from Br-Matrimid (550 °C)	1960	850	168	116	2900	5.1	25
CMSM from Matrimid (550 °C)	537	280	59	48	871	4.8	18

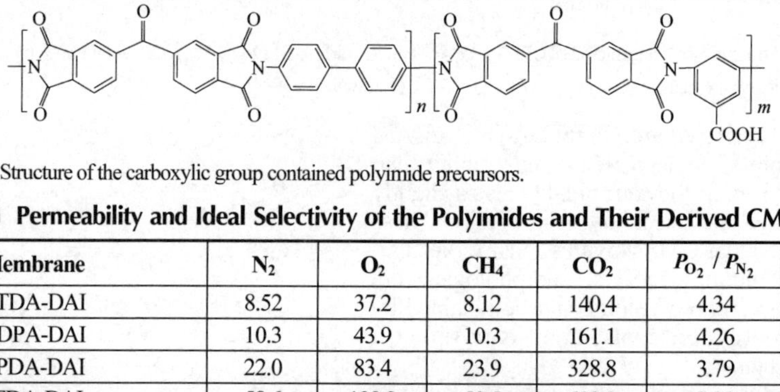

FIGURE 6.69 Structure of the carboxylic group contained polyimide precursors.

TABLE 6.56 Permeability and Ideal Selectivity of the Polyimides and Their Derived CMSMs

Membrane	N_2	O_2	CH_4	CO_2	P_{O_2}/P_{N_2}	P_{CO_2}/P_{CH_4}
BTDA-DAI	8.52	37.2	8.12	140.4	4.34	17.3
ODPA-DAI	10.3	43.9	10.3	161.1	4.26	15.6
BPDA-DAI	22.0	83.4	23.9	328.8	3.79	13.7
6FDA-DAI	58.6	198.2	50.5	692.3	3.38	13.7
CM-BTDA-550	128	578	94	1923	4.5	21
CM-ODPA-550	86	404	73	1321	4.7	18
CM-BPDA-550	112	509	82	1564	4.5	19
CM-6FDA-550	193	909	174	4800	4.7	28
CM-BTDA-800	16.3	144	4.75	500	8.8	105
CM-ODPA-800	14.9	116	6.4	344	7.8	54
CM-BPDA-800	13.8	118	4.3	353	8.5	82
CM-6FDA-800	23	160	8.2	580	6.9	71

The membranes were tested at 10 atm and 35 °C.

The pyrolysis of P84 at different temperatures to form CMSMs for CO_2/CH_4 separation has been investigated [98]. The CMSMs were obtained by thermally pyrolyzed at 550 °C, 650 °C, and 800 °C in vacuum, respectively. The gas permeability and gas pair selectivity of the CMSMs are shown in Table 6.57. The CO_2 permeability reduced while CO_2/CH_4 selectivity increased with rising in pyrolysis temperatures. The CMSM derived from P84 pyrolyzed at 800 °C showed the best CO_2/CH_4 separation performance with CO_2 permeability of 499 Barrer, and CO_2/CH_4 selectivity of 89, respectively. The CMSMs derived from PI-(BTDA/ODA) pyrolyzed at 700 °C/0 min. showed the highest O_2 and CO_2

permeability [99]. Further higher temperatures resulted in higher O_2/N_2 separation performance with reducing in O_2 permeability. The longer the soaking time (from 0, 30, to 60 min), the lower the permeability and the higher the selectivity for O_2/N_2.

TABLE6.57 Gas Separation Properties of the CMSMs Derived From P84 and PI-(BTDA/ODA)

Precursor	Treatment Conditions	P_{O_2}	α_{O_2/N_2}	P_{CO_2}	P_{CO_2}/P_{CH_4}
P84	550 °C	—	—	1808	22
	650 °C	—	—	738	37
	800 °C	—	—	499	89
BTDA-ODA	500 °C 30 min	509	9	1535	—
	700 °C 0 min	632	9	1989	—
	700 °C 30 min	204	13	505	—
	700 °C 60 min	136	15	350	—
	800 °C 30 min	61	15	176	—

1 Barrer=1×10^{-10} cm^3 (STP)·cm·cm^{-2}·cm Hg^{-1}.

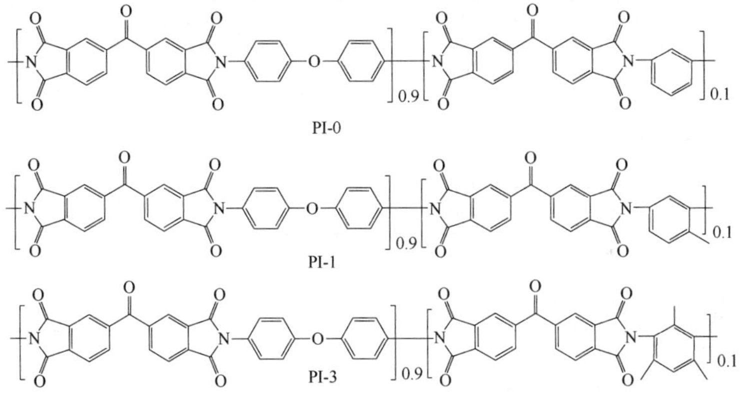

PI-0

PI-1

PI-3

FIGURE 6.70 Structure of copolyimides of BTDA-ODA and BTDA and different methyl substituted *meta*-phenylenediamine.

The carboxylic-containing polyimides have been used to produce the CMSMs by thermally pyrolyzing at 700 °C (Fig. 6.70) [100]. The CMSM membranes derived from polyimide PI-(BTDA/ODA/BTDA/*m*-PDA=80/20, PI-0), PI-(BTDA/ODA/BTDA/DBA=80/20, PI-2), and PI-(BTDA/ODA/BTDA/DBA=50/50, PI-5) were produced by pyrolyzing at 700 °C to give CMS700-0, CMS700-2, and CMS700-5, respectively. The gas transport properties were shown in Table 6.58. The higher the carboxylic percent, the higher the permeability of the resulting CMSM. For instance, the 50% DBA contained polyimide precursor showed an O_2 permeability of 2863 Barrar and a selectivity of O_2/N_2 of 9, which is more than threefold higher than CMSM derived from the pristine noncarboxylic-contained polyimide precursors.

Copolyimides have also been used as a precursor to produce CMSMs. A series of copolyimides PI-0, PI-1, and PI-3 have been prepared [100]. By changing the substitution in the *meta*-phenylenediamine (*m*PDA) from no-substitution (PI-0), one methyl (PI-1) to trimethyl (PI-3), the copolyimide membranes with 10 %(mol) of *m*PDA were thermally pyrolyzed at 600 °C and 800 °C, respectively to give the CMSMs. The gas transport properties are listed in Table 6.59. The permeability of the CMSM is closely related to the permeability of the polyimide precursor. The higher permeability of the polyimide precursor, the higher the permeability of its corresponding CMSM. Whereas the selectivity does not change obviously (Fig. 6.71).

TABLE 6.58 Gas Transport Properties of CMSM Derived From Carboxylic Contained Polyimide Precursors

Polymer	Permeability (Barrer)				Selectivity ($\alpha_{X/Y}$)		
	He	O_2	N_2	CO_2	He/N_2	CO_2/N_2	O_2/N_2
CMS 700-0	2763	829	256	24	106	35	11
CMS 700-2	3208	1674	501	49	64	34	10
CMS 700-5	4193	2863	707	83	51	34	9

1 Barrer=1×10^{-10} cm^3 (STP)·cm·cm^{-2}·$cmHg^{-1}$.

TABLE 6.59 Gas Separation Properties of the CMSMs at Different Pyrolizing Temperatures

Precursor	Treatment Conditions (°C)	P_{O_2}	α_{O_2/N_2}	P_{CO_2}
PI-0	—	0.21	10.5	0.81
PI-1	—	0.23	9.6	0.90
PI-3	—	0.38	8.3	0.38
C0-600	600	139	12	840
C1-600	600	188	11	925
C3-600	600	229	11	1017
C0-800	800	18	18	71
C1-800	800	22	17	80
C3-800	800	24	16	87

1 Barrer=1×10^{-10} cm^3 (STP)·cm·cm^{-2}·$cmHg^{-1}$.

(A) (B)

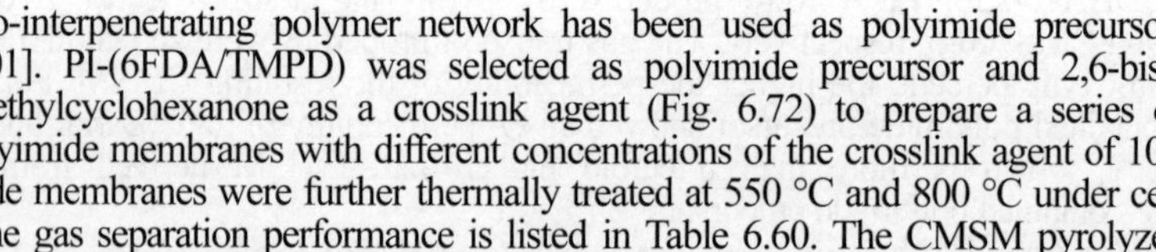

FIGURE 6.71 The structure of pristine polyimide (A), 6FDA-Durene and crosslink agent (B).

A pseudo-interpenetrating polymer network has been used as polyimide precursor to prepare CMSMs [101]. PI-(6FDA/TMPD) was selected as polyimide precursor and 2,6-bis(4-azidobenzylidene)-4-,ethylcyclohexanone as a crosslink agent (Fig. 6.72) to prepare a series of semiinterpenetrate polyimide membranes with different concentrations of the crosslink agent of 10% and 30%. The polyimide membranes were further thermally treated at 550 °C and 800 °C under certain heating protocols. The gas separation performance is listed in Table 6.60. The CMSM pyrolyzed at 550 °C exhibited high permeability. The 10% azide crosslinked 6FDA/TMPD showed higher permeability than the corresponding noncrosslinked version, the CO_2 permeability was measured at 9290±170 Barrer and the CO_2/CH_4 selectivity can reach as high as 39.6±2.5. The 30% azide crosslinked 6FDA/TMPD resulted in a reduced CO_2 permeability of 3640 Barrer but much higher CO_2/CH_4 selectivity of 66.4±1.0. The CMSMs pyrolyzed at 800 °C showed much lower CO_2 permeability while much higher CO_2/CH_4 selectivity, in which (6FDA/TMPDA)/azide (70-30) showed a significant high CO_2/CH_4 selectivity of 166.

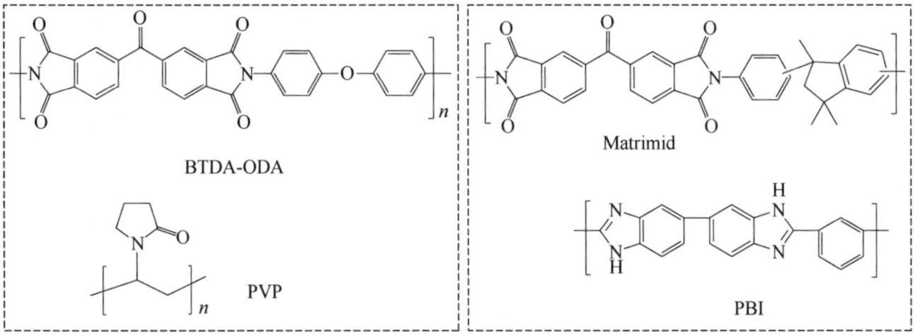

FIGURE 6.72 Structures of polyimide blend precursor membranes BTDA-ODA/PVP (left) and Matrimid/PBI (right).

A polymer blend BTDA-ODA/PVP was also used as precursor to prepare the CMSMs for gas separation applications (Fig. 6.73, [102]). First, PI/PVP blend membranes were prepared by casting PAA/PVP solutions onto a glass plate and then thermally imidizing in a vacuum oven to give a series of pristine precursor membranes with different PVP loadings, which were then pyrolyzed at 550 °C and 700 °C in a flowing of argon with a speed of 300 $mL \cdot min^{-1}$ to produce the CMSMs.

TABLE6.60 Gas Transport Properties of the Carbon Molecular Sieve Membranes Derived From the above Semiinterpenetrated Polyimides

Polyimide Precursor	Treatment Conditions (°C)	P_{O_2}	α_{CO_2/N_2}	P_{CO_2}/P_{CH_4}
6FDA/TMPDA	550	6810±125	24.6±0.4	59.4±0.1
6FDA/TMPDA/azide (90-10)	550	9290±170	26.0±0.8	39.6±2.5
6FDA/TMPDA/azide (70-30)	550	3640±17	24.2±0.1	66.4±1.0
6FDA/TMPDA	800	1460±17	31.3±0.3	62.2±1.3
6FDA/TMPDA/azide (90-10)	800	851±9.1	33.2±0.1	110±1.3
6FDA/TMPDA/azide (70-30)	800	280±7.0	31.7±0.1	166±6.0

All gases were tested at 10 bar upstream pressure

1 Barrer=1×10^{-10} cm^3 (STP)·cm·cm^{-2}·$cmHg^{-1}$.

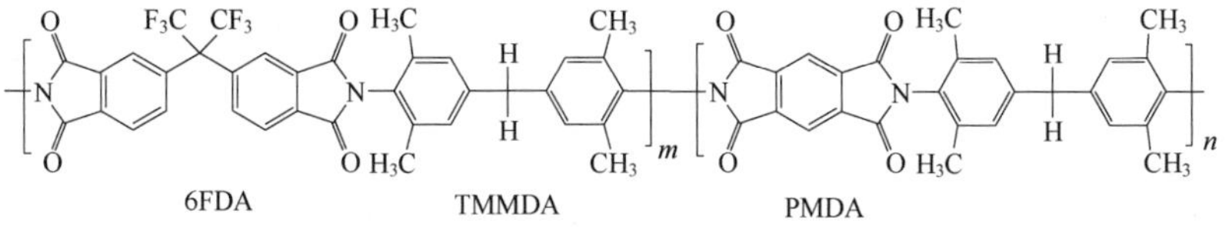

FIGURE 6.73 Structure of 6FDA/PMDA-TMMDA.

The gas permeability of the CMSM membranes were enhanced by increasing of PVP loadings. The PI/PVP-10 pyrolyzed at 550 °C showed O_2 and CO_2 permeability of 630 and 1900 Barrer, respectively, much higher than that pyrolyzed at 700 °C (Table 6.61). The CMSMs derived from the polyimide blend (PI/PBI) precursor showed better gas transport properties than pure P84 [103]. The Matrimid/PBI membranes with weight ratios of 25/75, 50/50 and 75/25 were prepared and pyrolyzed at 800 °C to give a series of CMSM membranes. The gas transport properties of the CMSMs are also listed in Table 6.61. The gas separation results clearly indicated that the higher temperature (from 600 °C, 700 °C to 800 °C) resulted in lower O_2 and CO_2 permeability. However, slight increases in O_2/N_2 and CO_2/CH_4 selectivity were observed. Compared with the pristine Matrimid, the higher PBI loadings result in an obvious reduction in permeability. The O_2/N_2 selectivity is slightly higher than pure Matrimid and the

CO_2/CH_4 selectivity is decreased with the increasing of PBI loadings in the blend polymers.

TABLE 6.61 Treatment Condition of the PI/PVP Blend Membrane and the Gas Separation Properties

Precursor	Treatment Conditions (°C)	P_{O_2}	α_{O_2/N_2}	P_{CO_2}	P_{CO_2}/P_{CH_4}
PI/PVP-0	550	490	—	1500	—
PI/PVP-0	700	130	—	300	—
PI/PVP-5	550	550	—	1750	—
PI/PVP-5	700	160	—	400	—
PI/PVP-10	550	630	10	1900	—
PI/PVP-10	700	230	14	480	—
Matrimid	800	79.4	6.14	300.9	67.1
PBI/Matrimid (50/50(*w*))	rt	0.58	8.05	2.16	48.0
PBI/Matrimid (50/50(*w*))	600	113.2	7.21	305.5	52.3
PBI/Matrimid (50/50(*w*))	700	54.2	8.60	169.6	121.1
PBI/Matrimid (50/50(*w*))	800	11.0	8.70	36.6	131.7
PBI/Matrimid (25/75(*w*))	800	30.4	8.04	96.5	204
PBI/Matrimid (75/25(*w*))	800	4.22	8.61	16.1	94.9

Multiwall carbon nanotubes as a filler was mixed with Ultem resin polyimide membrane, which was then thermally pyrolyzed at 750 °C to give CMSM membranes [104]. The composite membrane was obtained using mesoporous aluminum oxide as support. The resulting CMSM membranes with carbon nanotube showed both higher flux and selectivity for CO_2/N_2 and O_2/N_2 as compared with the pristine Ultem membranes.

CMSM as a filler was ball milled to small particles, which was mixed with pristine polyimide membranes, which was thermally pyrolyzed at high temperature to give a mixed CMSM membranes [105]. The milling process apparently alters the molecular scale structure and properties of the carbon material.

6.6.3.2 Effect of Pyrolysis Conditions on Gas Separation Properties

The solvent effect on the gas transport properties of the resulting polyimide membranes and their corresponding CMSM membranes have been investigated [106]. The precursor structures of PI-(6FDA/PMDA-TMMDA) polyimides is shown in Fig. 6.74. The polyimide membranes were obtained by casting its polyimide solutions in DMF or CH_2Cl_2, respectively. The polymer films were heated to 250 °C from room temperature at a rate of 13 °C·min^{-1}. Subsequently, the temperature was raised to different temperatures (550 °C, 650 °C, and 800 °C) by different protocols. The membrane prepared using DMF as a solvent (PI-DMF) showed a much lower permeability than that using CH_2Cl_2 as solvent (PI-CH_2Cl_2), because PI-DMF is a semicrystalline film whereas the PI-CH_2Cl_2 is totally amorphous. The resulting CMSM treated at 550 °C exhibited highest permeability for many gases but lowest selectivity for the interesting gas pairs (Table 6.62). The higher the carbonization temperature, the lower permeability and the higher selectivity for CO_2/CH_4 and O_2/N_2. The CMSM derived from

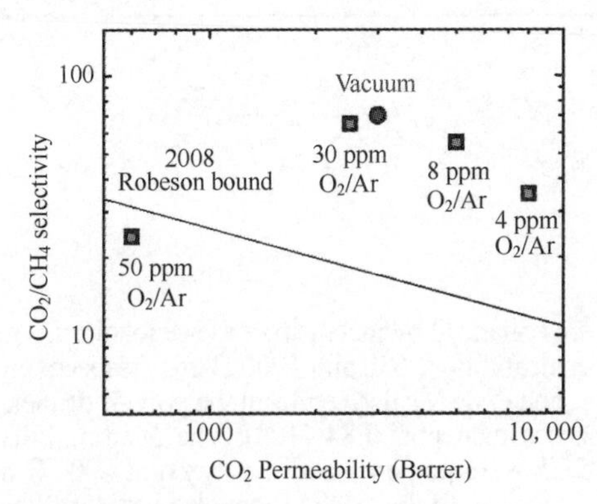

FIGURE 6.74 Gas separation performance of the 6FDA/BPDA- DAM and the resultingCMSM dense membranes in different treatment conditions.

PI-DCM at 800 °C showed the O_2 permeability of 166 Barrer with the O_2/N_2 selectivity of 9.9.

The nonsolvent pretreatment of the polyimide precursor membranes has also significant influence on the gas transport properties of the resulting CMSM membranes. Different solvents including methanol, ethanol, isopropanol and butanol have been used to treat the Matrimid and P84 precursor membranes, which was then converted into the CMSM membranes pyrolyzed at 800 °C [107]. The gas transport properties of the CMSMs depend on the pretreatment solvents (Table 6.63). The ethanol as a nonsolvent treatment gives the lowest permeability and highest selectivity of the corresponding CMSMs whereas the methanol, propanol and butanol treated membrane exhibited almost the same permeability and selectivity. Moreover, extension in soaking time in methanol also does not give prominent effect on the gas transport properties.

TABLE6.62 Gas Transport Performance of Polymeric Membrane and the CMSMs Derived From PI-CH_2Cl_2 and PI-DMF Membranes (10 atm, 35 °C)

Solvent	Treatment Conditions	P_{O_2}	P_{O_2}/P_{N_2}	P_{CO_2}	P_{CO_2}/P_{CH_4}
CH_2Cl_2	PI-CH_2Cl_2	45	3.8	187	14
	CMSM 550 °C	1173	4.3	3263	16.1
	CMSM 650 °C	344	7.5	1032	37.9
	CMSM 800 °C	168	9.9	465	77.5
DMF	PI-DMF	21	4.0	89	16.5
	CMSM 550 °C	418	5.4	1300	28.9
	CMSM 650 °C	259	6.6	751	35.8
	CMSM 800 °C	184	10.2	519	96.1

1 Barrer=1×10^{-10} cm^3 (STP)·cm·cm^{-2}·$cmHg^{-1}$.

TABLE6.63 Gas Transport Properties of the CMSMs Derived From Matrimid and P84 as Precursors Soaked in Different Solvents

Solvent	Treatment Conditions(°C)	P_{O_2}	P_{O_2}/P_{N_2}	P_{CO_2}	P_{CO_2}/P_{CH_4}
Matrimid	800	227	7.5	611	61
Matrimid-MeOH	800	138	8.8	423	88
Matrimid-EtOH	800	75.4	12	191	169
Matrimid-PrOH	800	204	8.5	565	84
Matrimid-BuOH	800	186	8.9	547	78
Matrimid- 2h-MeOH	800	227	7.5	611	61
Matrimid- 6h-MeOH	800	205	7.7	563	74
Matrimid-1d-MeOH	800	196	8.6	501	78
P84	800	158	8.9	499	89
P84-MeOH	800	132	9.7	402	109
P84-EtOH	800	101	11.2	278	139
P84-PrOH	800	144	9.9	428	110

1 Barrer=1×10^{-10} cm^3 (STP) ·cm·cm^{-2}·$cmHg^{-1}$.

TABLE6.64 O_2/N_2 and H_2/N_2 Separation Performance of the CMSMs Produced in Different Atmosphere

Atmosphere	O_2 flux	P_{O_2}/P_{N_2}	H_2 flux	P_{H_2}/P_{N_2}
Vacuum	25-50	7.4-9.0	372-473	64-110
Argon	71-284	2.8-6.1	451-731	6.8-31.2
Helium	73-140	4.7-6.1	428-676	15.2-35.7
Carbon dioxide	75-306	2.6-6.1	400-654	10.4-30.0

Flux, scale GPU. 1 GPU=10^{-6} cm^3 (STP)·cm^{-2}·s^{-1}·$cmHg^{-1}$ or 7.6×10^{-12} m^3 (STP)·m^{-2}·s^{-1}·Pa^{-1}.

TABLE6.65 Selectivity of the CMSM Membranes Pyrolyzed at Different Atmospheres With Different Gas Purge Rates

Atmosphere	Flow Rate (cm^3 (STP)·min^{-1})	O_2 Flux	P_{O_2} / P_{N_2}
Argon	200	71-284	2.8-6.1
Argon	20	0.050-0.54	4.7-6.1
Helium	200	73-140	2.4-7.0
Helium	20	0.05-0.11	4.0-5.2
Carbon dioxide	200	75-306	2.6-6.1
Carbon dioxide	20	0.05-15	2.0-7.5

Flux, scale GPU, 1 GPU=10^{-6} cm^3 (STP)·cm^{-2}·s^{-1}·$cmHg^{-1}$ or 7.6×10^{-12} m^3 (STP)·m^{-2}·s^{-1}·Pa^{-1}.

The pyrolysis of asymmetric PI-(6FDA/BPDA-DAM) hollow fiber into CMSM membranes in different gas atmospheres (N_2, He and CO_2) have been investigated [108]. The selectivity for O_2/N_2 and H_2/N_2 of the CMSMs pyrolyzed in helium, carbon dioxide, or argon was not changed obviously (Table 6.64). However, the CMSM pyrolyzed in vacuum showed much higher selectivity than those pyrolyzed in other gas atmospheres. In comparing with the inert gas flows, it indicated that high purge gas flow rates (i.e., 200 cm^3 (STP)·min^{-1}) resulted in CMSM with much higher permeability but lower selectivity (Table 6.65).

The gas separation properties of the CMSMs can deviate in atmospheres with different oxygen concentrations [109]. The pyrolysis of a fluorinated polyimide precursor into CMSM at 550 °C and 2 h soaking time have been investigated. The polymer was pyrolyzed in vacuum or inert atmosphere containing specific amount of oxygen (4×10^{-6}, 8×10^{-6}, 30×10^{-6}, or 50×10^{-6}). The vacuum pyrolized CMSMs resulted in the best performance with both relatively high permeability for CO_2 and highest CO_2/CH_4 selectivity. In 6FDA/DAM/BPDA, as the increase of the oxygen percentage from 4×10^{-6}, 8×10^{-6} to 30×10^{-6} in the carbonization atmosphere, a continuous decrease in permeability and enhanced selectivity was observed. Whereas the 50×10^{-6} oxygen will totally destroy the performance by both lower permeability and selectivity. In Matrimid, a continuous decrease in both permeability and selectivity was observed when the O_2 content gradually increased from 3×10^{-6} to 100×10^{-6}.

The carbonization of PI-(BPDA/ODA) membranes at 700 °C to form CMSMs, which was thereafter heated in air or pure O_2 atmosphere to 300 °C. The membrane was further oxidized with either an O_2/N_2 mixture or pure O_2 at 100-300 °C. Oxidation at 300 °C increased permeability without sacrificing the selectivity. The permeability of H_2 can reach 10^{-7} mol·m^{-2}·s^{-1}·Pa^{-1}. No difference in permeability between single and binary CO_2/N_2 (~ 50) systems was observed [110].

The heating speed of the polymer precursor membrane can also affect the gas transport properties of the resulting CMSMs. PEI/PVP was used as a polymer precursor, which was then thermally treated at 650 °C with different heating speeds from 1, 3, 5, 7, and 9 °C·min^{-1}, respectively. The resulting CMSM membranes showed excellent permeability and selectivity combinations for CO_2/N_2 and CO_2/CH_4, indicating that the heating rate of 3-5 °C·min^{-1} is the best condition to achieve the optimized permeability/selectivity performance (Fig. 6.75).

The CMSM membranes are very sensitive to environment conditions. The CMSM membranes in different environments (air, oxygen, nitrogen, propylene, etc.) were long-term aged for several months and the performance stability with time was analyzed periodically [111]. The N_2 and CO_2 permeability reduced sharply on exposure to oxygen or air atmosphere, whereas N_2 and propylene atmosphere only had slightly influence on the permeability. The CMSM membrane can adsorb oxygen from the air environment. At room temperature, exposure to oxygen resulted in a number of constrictions of the pore structure and hence added diffusional restrictions through the microporous space. The oxygen chemisorbed in the CMSM membrane can be removed partially from thermal regeneration after heating in N_2 at high temperatures (893 K). After regeneration, the O/C ratio in the membrane was decreased by 22%. This confirmed that the treatment was effective in removing oxygen surface groups [112].

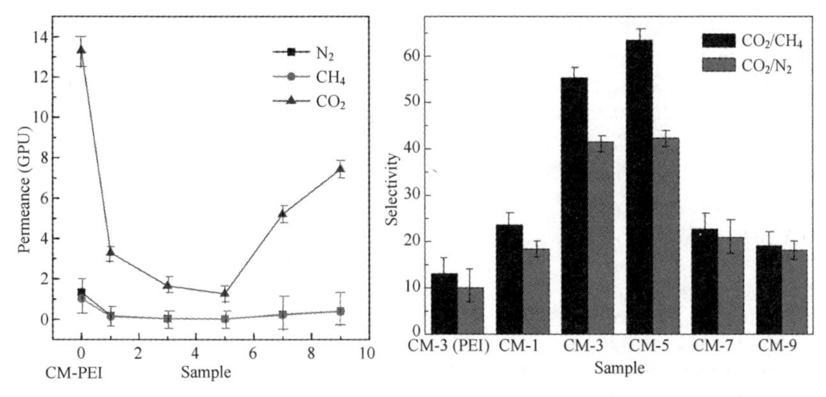

FIGURE 6.75 The permeability of N_2, CH_4 and CO_2 (left) as well as CO_2/CH_4 and CO_2/N_2 selectivity of the CMSMs heated at 650 °C with different heating rates [110].

Besides the O_2/N_2, CO_2/CH_4 separation, CMSM membranes have been investigated for ethylene/ethane, propylene/propane separations. Membrane technique is a very promising alternative for these separation applications. Matrimid as a precursor was used to produce the carbonized molecular sieve membrane and studied for the energy-intensive ethylene/ethane separation [113]. By heating the membrane gradually from 500 °C to 800 °C, especially at around 675 °C, a high C_2H_4 permeability of ~ 20 Barrer and C_2H_4/C_2H_6 selectivity over 10 was observed (Fig. 6.76).

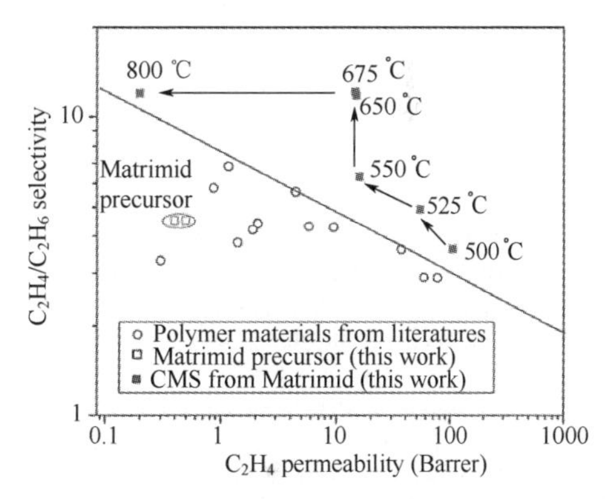

FIGURE 6.76 Comparison on Matrimid CMSM dense film membrane for C_2H_4/C_2H_6 separation [113].

A series of polyimides including PIM-6FDA-OH and PIM-6FDA for ethylene/ethane separation have been investigated [114,93]. Using the above two intrinsically microporous polyimide precursors, the membranes were heated to different temperatures of 500 °C, 600 °C, and 800 °C. The resulting CMSMs exhibited excellent gas transport properties for ethylene/ethane even at the mixed-gas pressure of 20 bar (Fig. 6.77). The PIM-6FDA demonstrated a selectivity for ethylene/ethane of 14 in the 50/50 ethylene/ethane mixed-gas under the upstream pressure of 20 Bar.

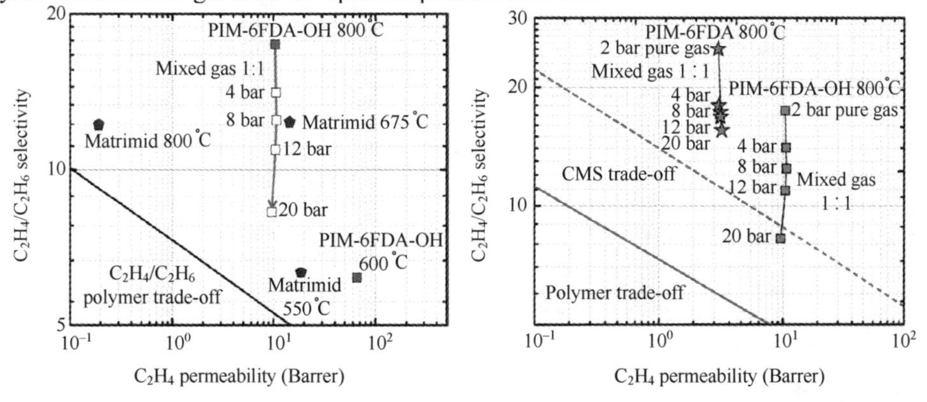

FIGURE 6.77 Pure-gas ethylene/ethane separation properties of the CMSMs derived from PIM-6FDA-OH (left) and PIM-6FDA (right) [93,114].

In summary, the CMSM membranes derived from polyimide precursors show significant permeability/selectivity performance that successfully outperform the trade-off curves. However, the performance of the CMSM membranes can be affected by many factors, including the chemical structures of the polyimide precursors, the casting solvents, the pretreatment method and posttreatment method. In addition, the CMSMs are sensitive to O_2 and moisture. Overall, the CMSM membranes show the great potential for applications in energy-intensive separations.

6.7 In Summary

Polyimide-based gas separation membranes have attracted great attention both in industrial and academic investigations. Since the 1970s, great progress has been achieved in searching for advanced gas separation polyimide membranes. The early development including the proposed design principles for the polyimide aromatic structures, which should include both rigidity and weaken the inter-chain interactions. An extensive study was devoted to the structure-properties relationship by changing the chemical structures of diamines or dianhydrides, including the introduction of bulky bridge structures in either diamine or dianhydrides, isolation groups for the inter-chain interactions, and isomers to adjust the gas separation properties during the 1980s. Later, some experienced criteria were established as 1991 or 2008 Robeson upper bond to evaluate the newly developed gas separation polyimide membranes. The ideal of introducing micropore into polyimide matrix was confirmed in 2007 by Weber et al. A series of high microporous polyimide membranes had been discovered by sophisticated structure design, and thereafter, some intrinsically microporous polyimides even demonstrated a performance much better than the 2008 Robeson upper bond or even redefined the 2015 upper bond. These microporous polyimides exhibited excellent gas separation performance for H_2/N_2, H_2/CH_4, and O_2/N_2, which had great potential in application of O_2/N_2 separations such as conventional nitrogen (99.9%) production or oxygen fluent gas generation. However, the introduction of micropores into the polyimides had very limited effect on the separation of CO_2/CH_4 and CO_2/N_2 after a long time aging.

The hydroxyl-containing polyimide membranes had very good separation performance for CO_2/CH_4 owing to the polar hydroxyl groups, which can create extra hydrogen bonding that further modifies the pore size of the polyimide membranes, resulting in higher activation energy for CO_2 and CH_4. The introduction of micropores into hydroxyl-containing polyimides have improved the permeability of CO_2 while retaining high selectivity for CO_2/CH_4. There are potential applications in acid gas removal fields such as natural gas sweetening, biogas upgrading, and so on. Moreover, the *ortho*-hydroxyl-containing polyimides could be thermally treated to produce polybenzoxazole membranes. The resulting polymer exhibited promising potential for gas separations, especially for CO_2/CH_4 separation. Different polyimide precursors, processing methods, and thermal treatment temperatures have significantly affected the properties of the resulting TR polymer performance.

The polyimide membranes are also good candidates to produce CMSMs. The CMSMs derived from polyimides not only demonstrate good separation performance for O_2/N_2, CO_2/N_2, CO_2/CH_4, H_2/N_2, H_2/CH_4, C_2H_4/C_2H_6, and C_3H_6/C_3H_8, etc., but also show good endurance to thermal, chemical, and environmental factors. The performance of the polyimide-derived CMSMs are affected by the structures of polyimide precursors, casting solvent, pretreatment method, posttreatment method, heating rate, the target temperature, the heating atmosphere, and soaking time, etc. The drawback of the CMSMs is their high cost and brittleness. Taking the high gas separation performance into account, the CMSMs are considered great potential candidates for the energy-intensive separation areas.

The real practical applications of polymeric membranes for gas separation still need further investigation. Two important challenges must be completely overcome before pilot-scale applications: the fast aging of ultrathin membranes and concentration polarization in the practical gas separations. Meanwhile, the long-term stability, environment tolerability, stability, and modulus manufacture are also required for further intensive investigation.

REFERENCES

[1] M. Mulder, J. Mulder, Basic Principles of Membrane Technology, Kluwer Academic Publishers, Dordrecht, 2003.

[2] J.A. Nollet, Lecons de physique experimentale. Hippolyte Louis Guerin, 1759.

[3] S. Alexander Stern, Polymers for gas separations: the next decade, J. Membrane Sci. 94 (1) (1994) 1-65.
[4] A.F. Ismail, K. Khulbe, T. Matsuura, Gas separation membranes polymeric and inorganic. Springer, 2015.
[5] W. Baldus, D. Tillman, Conditions which need to be fulfilled by membrane systems in order to compete with existing methods for gas separation, in: tr Barry (Ed.), Membranes in Gas Separation and Enrichment, Royal Society of Chemistry, 1986.
[6] D.S. Sholl, R.P. Lively, Seven chemical separations to change the world, Nature 532 (7600) (2016) 435-437.
[7] S. Loeb, S. Sourirajan, Sea water demineralization by means of an osmotic membrane, Adv. Chem. Ser. 38 (1963) 117-132.
[8] R.L. Riley, G.R. Hightower, C.R. Lyons, Thin Film Composite Membrane for Single-Stage Seawater Desalination by Reverse Osmosis, John Wiley & Sons, New York, NY, 1973.
[9] R.W. Baker, Membrane Technology and Applications, John Wiley & Sons, New York, NY, 2012.
[10] D.W. Breck, Zeolite Molecular Sieves, John Wiley & Sons, New York, NY, 1974.
[11] W.J. Koros, G.K. Fleming, Membrane-based gas separation, J. Membrane Sci. 83 (1) (1993) 1-80.
[12] M. Knudsen, The laws of molecular flow and the internal viscous streaming of gases through tubes, Ann. Phys. 28 (1908) 75.
[13] T. Graham, On the absorption and dialytic separation of gases by collid septa, Philos. Mag. J. Esc. 32 (1866) 401-420.
[14] Y. Mi, S.A. Stern, S. Trohalaki, Dependence of the gas-permeability of some polyimide isomers on their intrasegmental mobility, J. Membrane Sci. 77 (1) (1993) 41-48.
[15] I. Pinnau, B.D. Freeman, Polymeric materials for gas separations, ACS Symposium Series, American Chemical Society, Washington, DC, 1999. edn.
[16] A. Bondi, Physical Properties of Molecular Crystals, Liquids, and Gases, John Wiley & Sons, New York, 1968.
[17] L.M. Robeson, Correlation of separation factor versus permeability for polymeric membranes, J. Membrane Sci. 62 (2) (1991) 165-185.
[18] L.M. Robeson, The upper bound revisited, J. Membrane Sci. 320 (1-2) (2008) 390-400.
[19] B.D. Freeman, Basis of permeability/selectivity tradeoff relations in polymeric gas separation membranes, Macromolecules 32 (2) (1999) 375-380.
[20] D.W. Van Krevelen, Properties of Polymers: Their Correlation With Chemical Structure; Their Numerical Estimation and Prediction From Additive Group Contributions, Elsevier, Amsterdam, 1990.
[21] K.C. Obrien, W.J. Koros, T.A. Barbari, E.S. Sanders, A new technique for the measurement of multicomponent gas-transport through polymeric films, J. Membrane Sci. 29 (3) (1986) 229-238.
[22] S.W. Rutherford, D.D. Do, Review of time lag permeation technique as a method for characterisation of porous media and membranes, Adsorption—J. Int. Adsorption Soc. 3 (4) (1997) 283-312.
[23] H.H. Hoehn, Heat treatment of membranes of selected polyimides, polyesters and polyamides, 1972.
[24] M.R. Coleman, W.J. Koros, Isomeric polyimides based on fluorinated dianhydrides and diamines for gas separation applications, J. Membrane Sci. 50 (3) (1990) 285-297.
[25] N. Muruganandam, W.J. Koros, D.R. Paul, Gas sorption and transport in substituted polycarbonates, J. Polym. Sci. Part B: Polym. Phys. 25 (9) (1987) 1999-2026.
[26] K. Tanaka, M. Okano, H. Toshino, H. Kita, K.-I. Okamoto, Effect of methyl substituents on permeability and permselectivity of gases in polyimides prepared from methyl-substituted phenylenediamines, J. Polym. Sci. Part B: Polym. Phys. 30 (8) (1992) 907-914.
[27] M. Al-Masri, H.R. Kricheldorf, D. Fritsch, New polyimides for gas separation. 1. Polyimides derived from substituted terphenylenes and 4,4′-(hexafluoroisopropylidene)diphthalic anhydride, Macromolecules 32 (23) (1999) 7853-7858.
[28] M. Langsam, W.F. Burgoyne, Effects of diamine monomer structure on the gas-permeability of polyimides. 1. Bridged diamines, J.f Polym. Sci. Part A: Polym. Chem. 31 (4) (1993) 909-921.
[29] A.K. St. Clair, T.L. St. Clair, W.S. Slemp, Recent Advances in Polyimides: Synthesis, Characterization and Applications, Plenum Press, New York, 1987. W. Weber and M. Gupta, eds.
[30] D. Plaza-Lozano, B. Comesaña-Gándara, M. de la Viuda, J.G. Seong, L. Palacio, P. Prádanos, et al., New aromatic polyamides and polyimides having an adamantane bulky group, Mater. Today Commun. 5 (2015) 23-31.
[31] C. Álvarez, A.E. Lozano, J.G. de la Campa, High-productivity gas separation membranes derived from pyromellitic dianhydride and nonlinear diamines, J. Membrane Sci. 501 (2016) 191-198.
[32] J.W. Xu, M.L. Chng, T.S. Chung, C.B. He, R. Wang, Permeability of polyimides derived from non-coplanar diamines and 4,4′-(hexafluoroisopropylidene) diphthalic anhydride, Polymer 44 (16) (2003) 4715-4721.
[33] J.L. Santiago-García, C. Álvarez, F. Sánchez, J.G. de la Campa, Gas transport properties of new aromatic polyimides based on 3,8-diphenylpyrene-1,2,6,7-tetracarboxylic dianhydride, J. Membrane Sci. 476 (2015) 442-448.
[34] K. Tanaka, H. Kita, M. Okano, K.-i Okamoto, Permeability and permselectivity of gases in fluorinated and non-fluorinated polyimides, Polymer 33 (3) (1992) 585-592.
[35] Y.C. Xiao, T.S. Chung, M.L. Chng, S. Tamai, A. Yamaguchi, Structure and properties relationships for aromatic polyimides and their derived carbon membranes: experimental and simulation approaches, J. Phys. Chem. B 109 (40) (2005) 18741-18748.

[36] K. Tanaka, Y. Osada, H. Kita, K.I. Okamoto, Gas-permeability and permselectivity of polyimides with large aromatic rings, J. Polym. Sci. Part B: Polym. Phys. 33 (13) (1995) 1907-1915.

[37] Y.-H. Kim, S.-K. Ahn, H.S. Kim, S.-K. Kwon, Synthesis and characterization of new organosoluble and gas-permeable polyimides from bulky substituted pyromellitic dianhydrides, J. Polym. Sci. Part A: Polym. Chem. 40 (23) (2002) 4288-4296.

[38] D. Ayala, A.E. Lozano, J. de Abajo, C. Garcia-Perez, J.G. de la Campa, K.V. Peinemann, et al., Gas separation properties of aromatic polyimides, J. Membrane Sci. 215 (1-2) (2003) 61-73.

[39] H.-S. Kim, Y.-H. Kim, S.-K. Ahn, S.-K. Kwon, Synthesis and characterization of highly soluble and oxygen permeable new polyimides bearing a noncoplanar twisted biphenyl unit containing tert-butylphenyl or trimethylsilyl phenyl groups†, Macromolecules 36 (7) (2003) 2327-2332.

[40] Y.H. Kim, H.S. Kim, S.K. Kwon, Synthesis and characterization of highly soluble and oxygen permeable new polyimides based on twisted biphenyl dianhydride and spirobifluorene diamine, Macromolecules 38 (19) (2005) 7950-7956.

[41] Z. Qiu, G. Chen, Q. Zhang, S. Zhang, Synthesis and gas transport property of polyimide from 2,2′-disubstituted biphenyltetracarboxylic dianhydrides (bpda), Eur. Polym. J. 43 (1) (2007) 194-204.

[42] J. Weber, O. Su, M. Antonietti, A. Thomas, Exploring polymers of intrinsic microporosity-microporous, soluble polyamide and polyimide, Macromol. Rapid Commun. 28 (18-19) (2007) 1871-1876.

[43] B.S. Ghanem, N.B. McKeown, P.M. Budd, J.D. Selbie, D. Fritsch, High-performance membranes from polyimides with intrinsic microporosity, Adv. Mater. 20 (14) (2008) 2766-2771.

[44] B.S. Ghanem, N.B. McKeown, P.M. Budd, N.M. Al-Harbi, D. Fritsch, K. Heinrich, et al., Synthesis, characterization, and gas permeation properties of a novel group of polymers with intrinsic microporosity: pim-polyimides, Macromolecules 42 (20) (2009) 7881-7888.

[45] X.H. Ma, O. Salinas, E. Litwiller, I. Pinnau, Novel spirobifluorene- and dibromospirobifluorene-based polyimides of intrinsic microporosity for gas separation applications, Macromolecules 46 (24) (2013) 9618-9624.

[46] Y.J. Cho, H.B. Park, High performance polyimide with high internal free volume elements, Macromol. Rapid. Commun. 32 (7) (2011) 579-586.

[47] S.A. Sydlik, Z. Chen, T.M. Swager, Triptycene polyimides: soluble polymers with high thermal stability and low refractive indices, Macromolecules 44 (4) (2011) 976-980.

[48] J.R. Wiegand, Z.P. Smith, Q. Liu, C.T. Patterson, B.D. Freeman, R. Guo, Synthesis and characterization of triptycene-based polyimides with tunable high fractional free volume for gas separation membranes, J. Mater. Chem. A 2 (33) (2014) 13309.

[49] F. Alghunaimi, B. Ghanem, N. Alaslai, R. Swaidan, E. Litwiller, I. Pinnau, Gas permeation and physical aging properties of iptycene diaminebased microporous polyimides, J. Membrane Sci. 490 (2015) 321-327.

[50] S. Luo, Q. Liu, B. Zhang, J.R. Wiegand, B.D. Freeman, R. Guo, Pentiptycene-based polyimides with hierarchically controlled molecular cavity architecture for efficient membrane gas separation, J. Membrane Sci. 480 (2015) 20-30.

[51] Z. Wang, D. Wang, J. Jin, Microporous polyimides with rationally designed chain structure achieving high performance for gas separation, Macromolecules 47 (21) (2014) 7477-7483.

[52] Z.G. Wang, D. Wang, F. Zhang, J. Jin, Troger's base-based microporous polyimide membranes for high-performance gas separation, Acs Macro Lett. 3 (7) (2014) 597-601.

[53] Y. Zhuang, J.G. Seong, Y.S. Do, H.J. Jo, Z. Cui, J. Lee, et al., Intrinsically microporous soluble polyimides incorporating Tröger's base for membrane gas separation, Macromolecules 47 (10) (2014) 3254-3262.

[54] B. Ghanem, N. Alaslai, X.H. Miao, I. Pinnau, Novel 6FDA-based polyimides derived from sterically hindered Tröger's base diamines: synthesis and gas permeation properties, Polymer 96 (2016) 13-19.

[55] M. Lee, C.G. Bezzu, M. Carta, P. Bernardo, G. Clarizia, J.C. Jansen, et al., Enhancing the gas permeability of Tröger's base derived polyimides of intrinsic microporosity, Macromolecules 49 (11) (2016) 4147-4154.

[56] Y. Zhuang, J.G. Seong, Y.S. Do, W.H. Lee, M.J. Lee, M.D. Guiver, et al., High-strength, soluble polyimide membranes incorporating Tröger's base for gas separation, J. Membrane Sci. 504 (2016) 55-65.

[57] N. Alaslai, B. Ghanem, F. Alghunaimi, I. Pinnau, High-performance intrinsically microporous dihydroxyl-functionalized triptycene-based polyimide for natural gas separation, Polymer 91 (2016) 128-135.

[58] Y. Rogan, L. Starannikova, V. Ryzhikh, Y. Yampolskii, P. Bernardo, F. Bazzarelli, et al., Synthesis and gas permeation properties of novel spirobisindane-based polyimides of intrinsic microporosity, Polym. Chem. 4 (13) (2013) 3813-3820.

[59] Y. Rogan, R. Malpass-Evans, M. Carta, M. Lee, J.C. Jansen, P. Bernardo, et al., A highly permeable polyimide with enhanced selectivity for membrane gas separations, J. Mater. Chem. A 2 (14) (2014) 4874.

[60] X.H. Ma, B. Ghanem, O. Salines, E. Litwiller, I. Pinnau, Synthesis and effect of physical aging on gas transport properties of a microporous polyimide derived from a novel spirobifluorene-based dianhydride, Acs Macro Lett. 4 (2) (2015) 231-235.

[61] B.S. Ghanem, R. Swaidan, E. Litwiller, I. Pinnau, Ultra-microporous triptycene-based polyimide membranes for high-performance gas separation, Adv. Mater. 26 (22) (2014) 3688-3692.

[62] R. Swaidan, M. Al-Saeedi, B. Ghanem, E. Litwiller, I. Pinnau, Rational design of intrinsically ultramicroporous polyimides

containing bridgehead-substituted triptycene for highly selective and permeable gas separation membranes, Macromolecules 47 (15) (2014) 5104-5114.

[63] B. Ghanem, F. Alghunaimi, X. Ma, N. Alaslai, I. Pinnau, Synthesis and characterization of novel triptycene dianhydrides and polyimides of intrinsic microporosity based on 3,3′-dimethylnaphthidine, Polymer 101 (2016) 225-232.

[64] S. Alexander Stern, H.K. Ajay, Y. Houde, G. Zhou, Material and process for separating carbon dioxide from methane, 1997.

[65] N. Alaslai, B. Ghanem, F. Alghunaimi, E. Litwiller, I. Pinnau, Pure- and mixed-gas permeation properties of highly selective and plasticization resistant hydroxyl-diamine-based 6FDA polyimides for CO_2/CH_4 separation, J. Membrane Sci. 505 (2016) 100-107.

[66] B. Comesaña-Gándara, A. Hernández, J.G. de la Campa, J. de Abajo, A.E. Lozano, Y.M. Lee, Thermally rearranged polybenzoxazoles and poly (benzoxazole-co-imide)s from ortho-hydroxyamine monomers for high performance gas separation membranes, J. Membrane Sci. 493 (2015) 329-339.

[67] W. Qiu, K. Zhang, F.S. Li, K. Zhang, W.J. Koros, Gas separation performance of carbon molecular sieve membranes based on 6FDA-MPDA/DABA (3:2) polyimide, ChemSusChem 7 (4) (2014) 1186-1194.

[68] A.C. Puleo, D.R. Paul, S.S. Kelley, The effect of degree of acetylation on gas sorption and transport behavior in cellulose acetate, J. Membrane Sci. 47 (3) (1989) 301-332.

[69] C.H. Jung, Y.M. Lee, Gas permeation properties of hydroxyl-group containing polyimide membranes, Macromol. Res. 16 (6) (2008) 555-560.

[70] R. Swaidan, B. Ghanem, E. Litwiller, I. Pinnau, Effects of hydroxyl-functionalization and sub-tg thermal annealing on high pressure pure- and mixed-gas CO_2/CH_4 separation by polyimide membranes based on 6FDA and triptycene-containing dianhydrides, J. Membrane Sci. 475 (2015) 571-581.

[71] S.L. Yi, X.H. Ma, I. Pinnau, W.J. Koros, A high-performance hydroxyl-functionalized polymer of intrinsic microporosity for an environmentally attractive membrane-based approach to decontamination of sour natural gas, J. Mater. Chem. A 3 (45) (2015) 22794-22806.

[72] B. Kraftschik, W.J. Koros, J.R. Johnson, O. Karvan, Dense film polyimide membranes for aggressive sour gas feed separations, J. Membrane Sci. 428 (2013) 608-619.

[73] C.S.K. Achoundong, N. Bhuwania, S.K. Burgess, O. Karvan, J.R. Johnson, W.J. Koros, Silane modification of cellulose acetate dense films as materials for acid gas removal, Macromolecules 46 (14) (2013) 5584-5594.

[74] G. Chatterjee, A.A. Houde, S.A. Stern, Poly(ether urethane) and poly(ether urethane urea) membranes with high H_2S/CH_4 selectivity, J. Membrane Sci. 135 (1) (1997) 99-106.

[75] J.T. Vaughn, W.J. Koros, Analysis of feed stream acid gas concentration effects on the transport properties and separation performance of polymeric membranes for natural gas sweetening: a comparison between a glassy and rubbery polymer, J. Membrane Sci. 465 (2014) 107-116.

[76] X. Ma, O. Salinas, E. Litwiller, I. Pinnau, Pristine and thermally-rearranged gas separation membranes from novel o-hydroxyl-functionalized spirobifluorene-based polyimides, Polym. Chem. 5 (24) (2014) 6914-6922.

[77] F. Alghunaimi, B. Ghanem, N. Alaslai, M. Mukaddam, I. Pinnau, Triptycene dimethyl-bridgehead dianhydride-based intrinsically microporous hydroxyl-functionalized polyimide for natural gas upgrading, J. Membrane Sci. 520 (2016) 240-246.

[78] G. Tullos, L. Mathias, Unexpected thermal conversion of hydroxy-containing polyimides to polybenzoxazoles, Polymer 40 (12) (1999) 3463-3468.

[79] H.B. Park, C.H. Jung, Y.M. Lee, A.J. Hill, S.J. Pas, S.T. Mudie, et al., Polymers with cavities tuned for fast selective transport of small molecules and ions, Science 318 (5848) (2007) 254-258.

[80] S.H. Han, N. Misdan, S. Kim, C.M. Doherty, A.J. Hill, Y.M. Lee, Thermally rearranged (tr) polybenzoxazole: effects of diverse imidization routes on physical properties and gas transport behaviors, Macromolecules 43 (18) (2010) 7657-7667.

[81] M. Calle, Y.M. Lee, Thermally rearranged (tr) poly(ether-benzoxazole) membranes for gas separation, Macromolecules 44 (5) (2011) 1156-1165.

[82] H. Wang, T.-S. Chung, The evolution of physicochemical and gas transport properties of thermally rearranged polyhydroxyamide (pha), J. Membrane Sci. 385-386 (2011) 86-95.

[83] B. Comesaña-Gándara, M. Calle, H.J. Jo, A. Hernández, J.G. de la Campa, J. de Abajo, et al., Thermally rearranged polybenzoxazoles membranes with biphenyl moieties: monomer isomeric effect, J. Membrane Sci. 450 (2014) 369-379.

[84] R. Guo, D.F. Sanders, Z.P. Smith, B.D. Freeman, D.R. Paul, J.E. McGrath, Synthesis and characterization of thermally rearranged (tr)polymers: effect of glass transition temperature of aromatic poly(hydroxyimide) precursors on tr process and gas permeation properties, J. Mater. Chem. A 1 (19) (2013) 6063.

[85] M. Calle, C.M. Doherty, A.J. Hill, Y.M. Lee, Cross-linked thermally rearranged poly(benzoxazole-co-imide) membranes for gas separation, Macromolecules 46 (20) (2013) 8179-8189.

[86] Y.S. Do, W.H. Lee, J.G. Seong, J.S. Kim, H.H. Wang, C.M. Doherty, et al., Thermally rearranged (tr) bismaleimide-based network polymers for gas separation membranes, Chem. Commun. (Camb) 52 (93) (2016) 13556-13559.

[87] S. Li, H.J. Jo, S.H. Han, C.H. Park, S. Kim, P.M. Budd, et al., Mechanically robust thermally rearranged (tr) polymer

membranes with spirobisindane for gas separation, J. Membrane Sci. 434 (2013) 137-147.
[88] Y.K. Ong, H. Wang, T.-S. Chung, A prospective study on the application of pervaporation, Chem. Eng. Sci. 79 (2012) 41-53.
[89] J.H. Kim, S.H. Park, M.J. Lee, S.M. Lee, W.H. Lee, K.H. Lee, et al., Thermally rearranged polymer membranes for desalination, Energy Environ. Sci. 9 (3) (2016) 878-884.
[90] Y. Kim, H. Park, Y. Lee, Preparation and characterization of carbon molecular sieve membranes derived from btda-oda polyimide and their gas separation properties, J. Membrane Sci. 255 (1-2) (2005) 265-273.
[91] L. Shao, T.S. Chung, K.P. Pramoda, The evolution of physicochemical and transport properties of 6FDA-durene toward carbon membranes; from polymer, intermediate to carbon, Microp. Mesop. Mater. 84 (1-3) (2005) 59-68.
[92] X.H. Ma, R. Swaidan, B.Y. Teng, H. Tan, O. Salinas, E. Litwiller, et al., Carbon molecular sieve gas separation membranes based on an intrinsically microporous polyimide precursor, Carbon 62 (2013) 88-96.
[93] O. Salinas, X. Ma, Y. Wang, Y. Han, I. Pinnau, Carbon molecular sieve membrane from a microporous spirobisindane-based polyimide precursor with enhanced ethylene/ethane mixed-gas selectivity, RSC Adv. 7 (6) (2017) 3265-3272.
[94] W.N.W. Salleh, A.F. Ismail, T. Matsuura, M.S. Abdullah, Precursor selection and process conditions in the preparation of carbon membrane for gas separation: a review, Sep. Purif. Rev. 40 (4) (2011) 261-311.
[95] H. Suda, K. Haraya, Gas permeation through micropores of carbon molecular sieve membranes derived from kapton polyimide, J. Phys. Chem. B 101 (20) (1997) 3988-3994.
[96] D.Q. Vu, W.J. Koros, S.J. Miller, High pressure CO_2/CH_4 separation using carbon molecular sieve hollow fiber membranes, Ind. Eng. Chem. Res. 41 (3) (2002) 367-380.
[97] Y.C. Xiao, Y. Dai, T.S. Chung, M.D. Guiver, Effects of brominating matrimid polyimide on the physical and gas transport properties of derived carbon membranes, Macromolecules 38 (24) (2005) 10042-10049.
[98] P.S. Tin, T.S. Chung, Y. Liu, R. Wang, Separation of CO_2/CH_4 through carbon molecular sieve membranes derived from p84 polyimide, Carbon 42 (15) (2004) 3123-3131.
[99] M. Yan, T.W. Kim, A.G. Erlat, M. Pellow, D.F. Foust, H. Liu, et al., A transparent, high barrier, and high heat substrate for organic electronics, Proc. IEEE 93 (8) (2005) 1468-1477.
[100] H.B. Park, Y.K. Kim, J.M. Lee, S.Y. Lee, Y.M. Lee, Relationship between chemical structure of aromatic polyimides and gas permeation properties of their carbon molecular sieve membranes, J. Membrane Sci. 229 (1-2) (2004) 117-127.
[101] L. Bee Ting, C. Tai Shung, Carbon molecular sieve membranes derived from pseudo-interpenetrating polymer networks for gas separation and carbon capture, Carbon 49 (2011) 6.
[102] Y.K. Kim, H.B. Park, Y.M. Lee, Carbon molecular sieve membranes derived from thermally labile polymer containing blend polymers and their gas separation properties, J. Membrane Sci. 243 (1-2) (2004) 9-17.
[103] S.S. Hosseini, T.S. Chung, Carbon membranes from blends of pbi and polyimides for n2/ch4 and co2/ch4 separation and hydrogen purification, J. Membrane Sci. 328 (1-2) (2009) 174-185.
[104] H.-H. Tseng, I.A. Kumar, T.-H. Weng, C.-Y. Lu, M.-Y. Wey, Preparation and characterization of carbon molecular sieve membranes for gas separation-the effect of incorporated multi-wall carbon nanotubes, Desalination 240 (1-3) (2009) 40-45.
[105] M. Das, J.D. Perry, W.J. Koros, Effect of processing on carbon molecular sieve structure and performance, Carbon 48 (13) (2010) 3737-3749.
[106] L. Shao, T.S. Chung, G. Wensley, S.H. Goh, K.P. Pramoda, Casting solvent effects on morphologies, gas transport properties of a novel 6FDA/PMDA-TMMDA copolyimide membrane and its derived carbon membranes, J. Membrane Sci. 244 (1-2) (2004) 77-87.
[107] P.S. Tin, T.S. Chung, A.J. Hill, Advanced fabrication of carbon molecular sieve membranes by nonsolvent pretreatment of precursor polymers, Ind. Eng. Chem. Res. 43 (20) (2004) 6476-6483.
[108] V.C. Geiszler, W.J. Koros, Effects of polyimide pyrolysis conditions on carbon molecular sieve membrane properties, Ind. Eng. Chem. Res. 35 (9) (1996) 2999-3003.
[109] M. Kiyono, P.J. Williams, W.J. Koros, Effect of pyrolysis atmosphere on separation performance of carbon molecular sieve membranes, J. Membrane Sci. 359 (1-2) (2010) 2-10.
[110] K. Kusakabe, M. Yamamoto, S. Morooka, Gas permeation and micropore structure of carbon molecular sieving membranes modified by oxidation, J. Membrane Sci. 149 (1) (1998) 59-67.
[111] S. Lagorsse, F.D. Magalhães, A. Mendes, Aging study of carbon molecular sieve membranes, J. Membrane Sci. 310 (1-2) (2008) 494-502.
[112] L. Xu, M. Rungta, J. Hessler, W.L. Qiu, M. Brayden, M. Martinez, et al., Physical aging in carbon molecular sieve membranes, Carbon 80 (2014) 155-166.
[113] L. Xu, M. Rungta, W.J. Koros, Matrimid (r) derived carbon molecular sieve hollow fiber membranes for ethylene/ethane separation, J. Membrane Sci. 380 (1-2) (2011) 138-147.
[114] O. Salinas, X.H. Ma, E. Litwiller, I. Pinnau, Ethylene/ethane permeation, diffusion and gas sorption properties of carbon molecular sieve membranes derived from the prototype ladder polymer of intrinsic microporosity (pim-1), J. Membrane Sci. 504 (2016) 133-140.

Chapter 7

Polyimide Proton Exchange Membranes

Jian-Hua Fang
Shanghai Jiao Tong University, Shanghai, China

7.1 Introduction

A fuel cell is a device which directly converts chemical energy into electricity through electrochemical oxidation of a fuel (hydrogen, methanol, etc.). Basically, a fuel cell is composed of an electrolyte layer which is sandwiched between two porous electrodes (anode and cathode). At the anode a fuel such as hydrogen is oxidized into protons with releasing electrons, while at the cathode oxygen is reduced and combined with protons to form water (Fig. 7.1). The electrons flow through an outside circuit to generate current. The neat product of a FC reaction is water when hydrogen is used as fuel, which is very clean and environmentally friendly. In terms of the electrolyte, fuel cells can be classified into many types such as alkaline (normally KOH) fuel cells (AFCs), phosphoric acid fuel cells (PAFCs), molten carbonate (mixture of Li_2CO_3 and K_2CO_3) fuel cells (MCFCs), solid oxide (e.g., yttria-stabilized zirconia) fuel cells (SOFCs), and polymer electrolyte membrane fuel cells (PEMFCs), and among them, PEMFCs have been identified as one of the most promising power sources for vehicular transportation and for other applications requiring clean, quiet, and portable power.

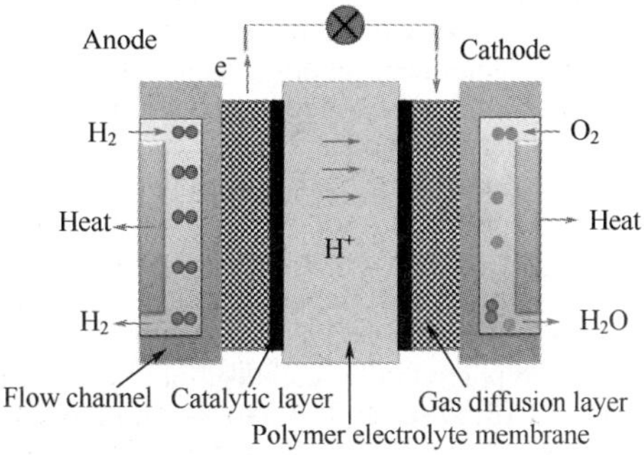

FIGURE 7.1 Schematic diagram of a PEMFC.

A proton exchange membrane (PEM) is one of the key components of a PEMFC system. From the viewpoint of practical use, an ideal PEM must meet the following requirements: (1) low cost, (2) low ionic resistance (i.e., high proton conductivity) under fuel cell operating conditions, (3) good mechanical strength, preferably with resistance to swelling, (4) long-term chemical and mechanical stability at elevated temperatures in oxidizing and reducing environments, (5) low, ideally no, fuel/oxygen crossover, (6) good interfacial compatibility with catalyst layers, and (7) electron insulation. Current state-of-the-art PEMs used in practical systems are sulfonated perfluoropolymers, typically, DuPont Nafion, which have the merits of high proton conductivity and excellent chemical and electrochemical stability. However, some drawbacks, such as high cost, low working temperature (< 90 °C) and high fuel permeability, seriously hinder their industrial applications. In the past two decades, a large number of sulfonated hydrocarbon polymers have been developed as alternatives such as sulfonated polystyrene and its derivatives, sulfonated poly(ether sulfone)s (SPESs), sulfonated poly(ether ether ketone)s (SPEEKs), sulfonated poly(aryl ether)s, sulfonated polyphosphazene, sulfonated poly(phenylene sulfide), sulfonated poly(sulfide

Advanced Polyimide Materials. DOI: https://doi.org/10.1016/B978-0-12-812640-0.00007-X

sulfone)s, sulfonated polybenzimidazoles, sulfonated polyphenylenes, and sulfonated polyimides (SPIs). Among them, SPIs have been extensively studied and some of them have been identified to be very promising as cost-effective and high-performance PEMs for fuel cells [1-107].

Aromatic polyimides, known for their excellent thermal stability, high mechanical strength and modulus, superior electric properties, and good chemical resistance and film-forming ability, have found wide applications in industry. Such merits of PIs are just required for the PEMs used in fuel cell systems. Mercier and coworkers first reported the fuel cell performance of six-membered ring SPI membranes [1,2]. They found that six-membered ring SPI membranes were much more stable to hydrolysis than five-membered ones. They also reported that a fuel cell assembled with a six-membered ring SPI membrane exhibited comparable fuel cell performance to the one assembled with Nafion, and meanwhile this fuel cell showed fairly high long-term durability (3000 h) at 60 °C. Thereafter, six-membered ring SPI membranes have attracted much attention and a large number of new six-membered ring SPI membranes have been developed as potential PEMs for fuel cells. In this chapter, the fundamentals of SPI membranes including the synthesis of various monomers and SPIs, the structure-property relationship, and fuel cell performances will be introduced. Recent progress on the development of high-performance SPI membranes will also be introduced.

7.2 Monomer Synthesis

7.2.1 Synthesis of Sulfonated Diamines

Because only a few sulfonated diamine monomers are commercially available, in the past two decades many novel sulfonated diamine monomers have been synthesized for the studies of membrane structure-property relationships [3-50]. The chemical structures and abbreviations of various sulfonated diamine monomers reported in literature are illustrated in Table 7.1. For convenience, the chemical structures and abbreviations of various nonsulfonated diamine monomers which have been used as comonomers for preparation of various SPIs are also shown in this table. The sulfonated diamine monomers can be classified into two types: main-chain-type and side-chain-type. Fig. 7.2 shows the chemical structures of some typical main-chain-type sulfonated diamines and side-chain-type sulfonated diamines. For main-chain-type sulfonated diamines, some of them can be synthesized by direct sulfonation of the commercial diamines using fuming sulfuric acid or concentrated sulfuric acid as the sulfonating reagent or via multistep reactions. For example, as shown in Fig. 7.3, 4,4′-diaminodiphenyl ether-2,2′-disulfonic acid (ODADS), 9,9-bis(4-aminophenyl)fluorene-2,7- disulfonic acid (BAPFDS), and 4,4′-bis(4-aminophenoxy)biphenyl-3,3′-disulfonic acid (BAPBDS) can be synthesized by direct sulfonation of the corresponding commercial diamines, 4,4′-diaminodiphenyl ether (ODA), 9,9-bis(4-aminophenyl)fluorene (BAPF), and 4,4′-bis(4-aminophenoxy)biphenyl (BAPB) using fuming sulfuric acid as the sulfonating reagent at appropriate temperatures (50-80 °C) [3,6,7]. Prior to sulfonation, the commercial diamines are allowed to carefully dissolve in minimal amounts of cool (below 0 °C) concentrated sulfuric acid to protect the amino groups via protonation. Because the protonated amino groups are strong electron-withdrawing groups, the sulfonation occurred at the *meta*-positions of the amino groups for ODA, 2,7-positions of fluorene ring for BAPF, and *ortho*-positions of ether bond of the central biphenyl groups for BAPB. However, in some cases there is no such sulfonation site selectivity. As a result, the relevant sulfonated diamines cannot be synthesized by direct sulfonation of the corresponding commercial diamines. For example, direct sulfonation of 4,4′-bis(4-aminophenoxy)benzophenone and 4,4′-bis(4-aminophenylthio)benzophenone failed to give the corresponding sulfonated diamines, 4,4′-bis(4-aminophenoxy)benzophenone-3,3′-disulfonic acid (BAPBPDS) and 4,4′-bis(4-aminophenylthio)benzophenone-3,3′-disulfonic acid (BAPTBPDS) because the electron-withdrawing carbonyl groups deactivated the central benzene rings, leading to loss of sulfonation site selection, as in the case of BAPF and BAPB. The synthesis of BAPBPDS and BAPTBPDS can be achieved via multiple-step reactions (Fig. 7.4).

TABLE7.1 Chemical Structures and Abbreviations of Various Sulfonated Diamine Monomers and Nonsulfonated Diamine Monomers

Diamine	Abbreviation	References
HO_3S H_2N NH_2 SO_3H	BDSA	1,2
SO_3H H_2N O NH_2 HO_3S	ODADS	3
H_2N NH_2 HO_3S SO_3H	BAPFDS	6
HO_3S SO_3H H_2N O O NH_2	BAPBDS	7
SO_3H HO_3S H_2N O O NH_2	*i*BAPBDS	5,47
H_2N NH_2 O O HO_3S SO_3H	*m*BAPBDS	5,47
HO_3S H_2N O O NH_2 SO_3H	*o*BAPBDS	10
HO_3S SO_3H CF_3 H_2N O O NH_2 CF_3	BAPHFDS	41
SO_3H O NH_2 H_2N O	BAPNS	42
HO_3S SO_3H H_2N O O NH_2	BAPNDS	27
SO_3H O C=O H_2N O S O NH_2 O O=C HO_3S	BAPSBPS	22
HO_3S O SO_3H O C O H_2N NH_2	BANBPDS	40
HO_3S O SO_3H H_2N O S O NH_2 O	BAPPSDS	43

TABLE7.1 (Continued)

Diamine	Abbreviation	References
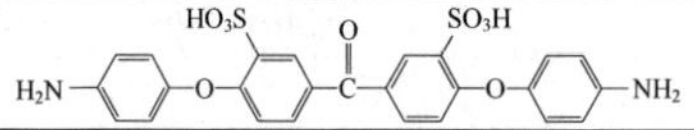 HO_3S O SO_3H H_2N O C O NH_2	BAPBPDS	39,40
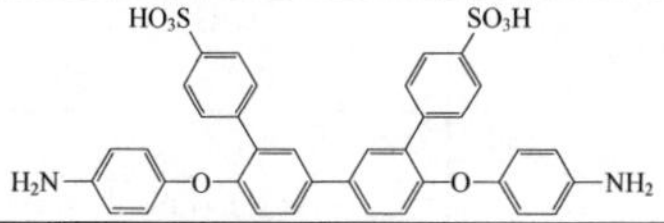 HO_3S SO_3H H_2N O O NH_2	BAPSPB	32,44
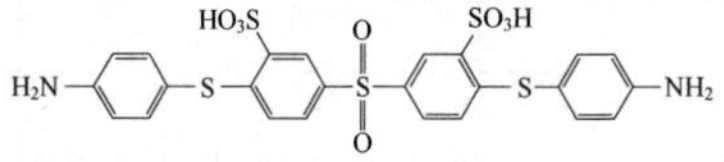 HO_3S O SO_3H H_2N S S S NH_2 O	BAPTPSDS	43
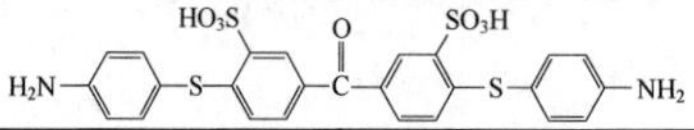 HO_3S O SO_3H H_2N S C S NH_2	BAPTBPDS	39
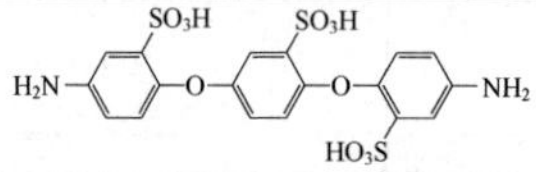 SO_3H SO_3H H_2N O O NH_2 HO_3S	*p*BABTS	34
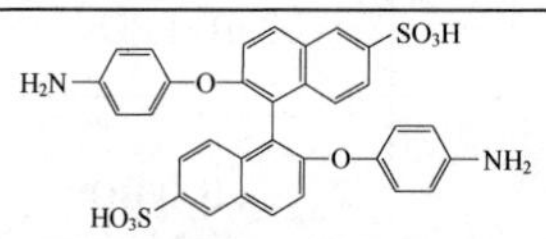 SO_3H H_2N O O NH_2 HO_3S	BNDADS	26
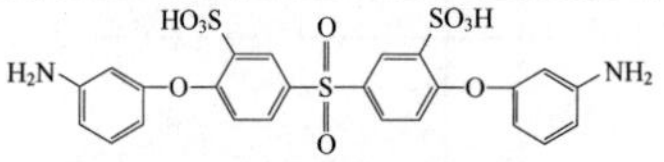 HO_3S O SO_3H H_2N O S O NH_2 O	SADADPS	11
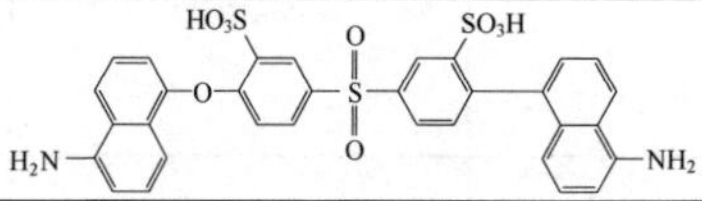 HO_3S O SO_3H O S O H_2N NH_2	DANPS	48
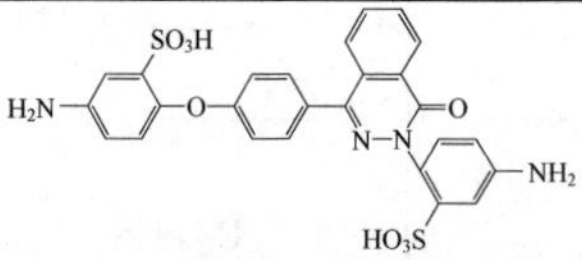 SO_3H H_2N O O N–N NH_2 HO_3S	DHPZDA	28
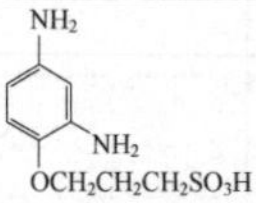 NH_2 NH_2 $OCH_2CH_2CH_2SO_3H$	DAPPS	8
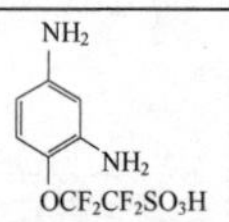 NH_2 NH_2 $OCF_2CF_2SO_3H$	DAPTFSA	36
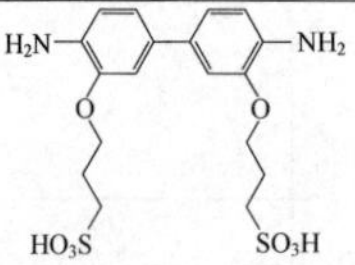 H_2N NH_2 O O HO_3S SO_3H	3,3′-BSPB	4,35
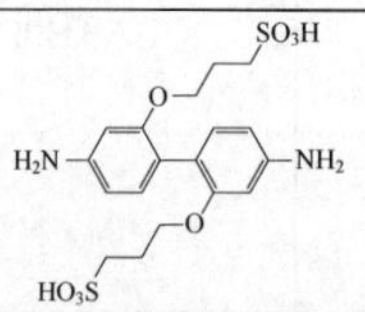 SO_3H O H_2N NH_2 O HO_3S	2,2′-BSPB	35,45

TABLE7.1 (Continued)

Diamine	Abbreviation	References
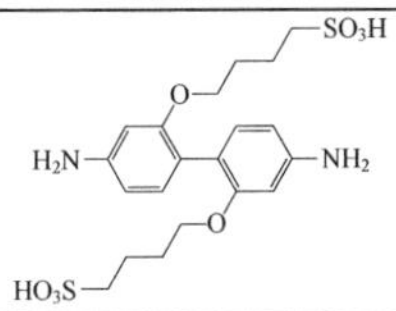 H_2N–biphenyl–NH_2 with O–$(CH_2)_4$–SO_3H / HO_3S–$(CH_2)_4$–O substituents	2,2′-BSBB	37
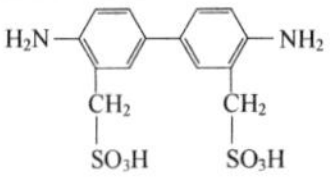 H_2N–biphenyl–NH_2 with CH_2–SO_3H substituents	3,3′-BSMB	24
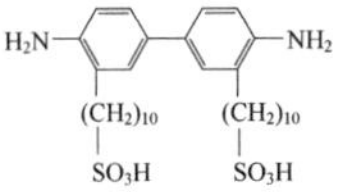 H_2N–biphenyl–NH_2 with $(CH_2)_{10}$–SO_3H substituents	3,3′-BSDcB	24
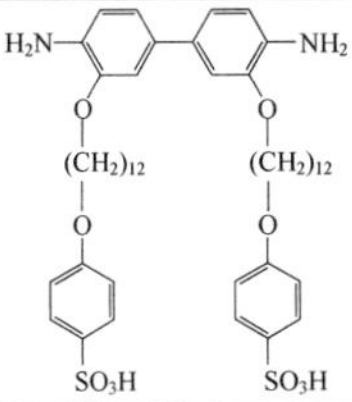 H_2N–biphenyl–NH_2 with O–$(CH_2)_{12}$–O–phenyl–SO_3H substituents	3,3′-BSDdB	24
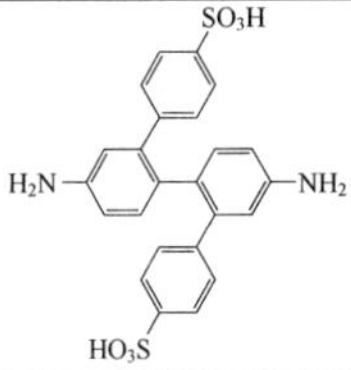 H_2N–biphenyl–NH_2 with phenyl–SO_3H / phenyl–SO_3H (HO_3S) substituents	BSPhB	46
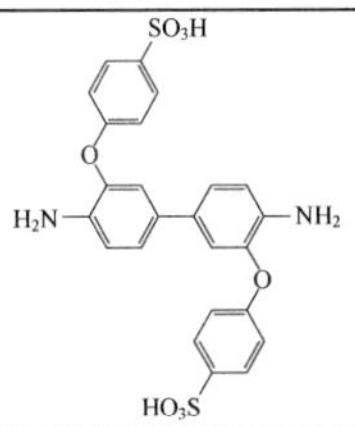 H_2N–biphenyl–NH_2 with O–phenyl–SO_3H / O–phenyl–SO_3H (HO_3S) substituents	3,3′-BSPOB	38
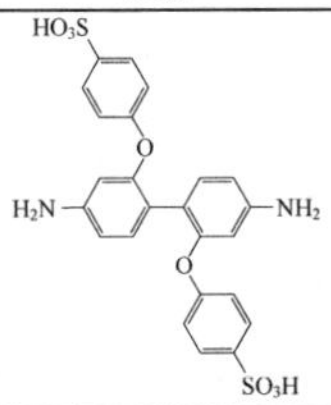 H_2N–biphenyl–NH_2 with O–phenyl–SO_3H (HO_3S) / O–phenyl–SO_3H substituents	2,2′-BSPOB	9,12-15
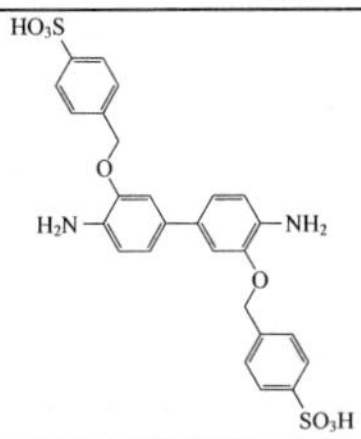 H_2N–biphenyl–NH_2 with O–CH_2–phenyl–SO_3H (HO_3S) / O–CH_2–phenyl–SO_3H substituents	3,3′-BSBOB	31

TABLE7.1 (Continued)

Diamine	Abbreviation	References
	S-BAPPO	29
	DASDSPB	23
	DASSPB	23
	SDAM	33
	DMBDSA	49
	DMMDADS	50
	DSDSA	51,52
	2,4-DABS	53,54
	2,5-DABS	53,54
	ODA	3,5,7
	BAHF	3
	BAPB	7
	*o*BAPB	1,10
	BAPBz	12-15
	BAPHF	1,16
	BAPP	1,55,56

TABLE7.1 (Continued)

Diamine	Abbreviation	References
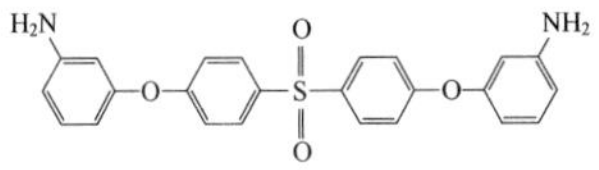	*m*BAPPS	6,15,23,25
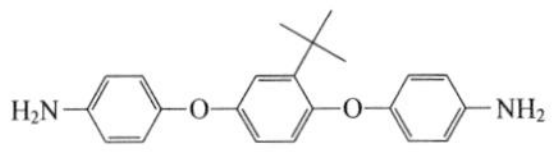	BAPBB	1
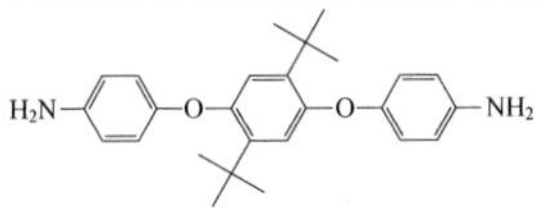	BAPDBB	1
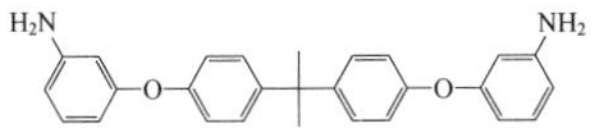	*m*BAPP	1
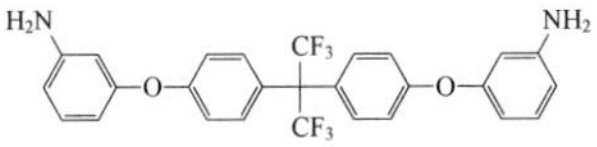	*m*BAPHF	1
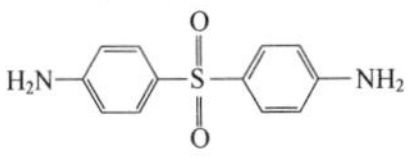	DADPS	57
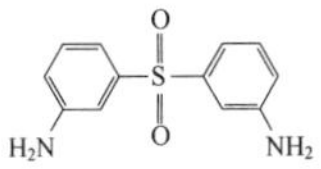	*m*DADPS	57
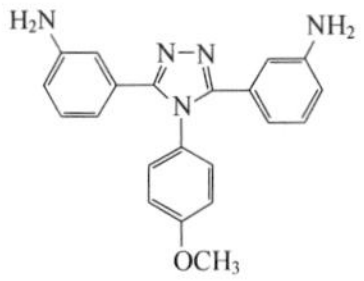	APT	58
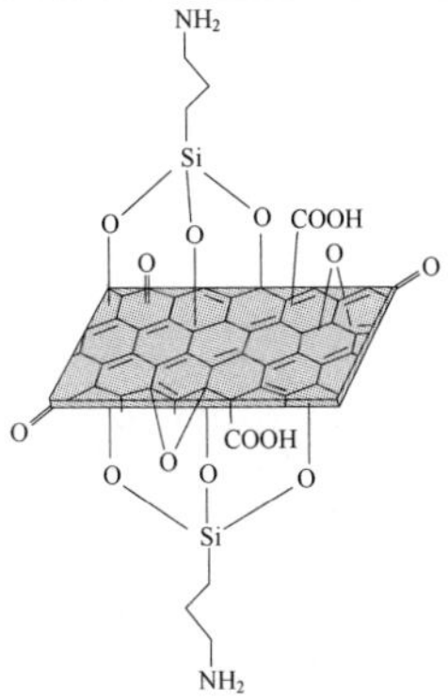	APSGO	59
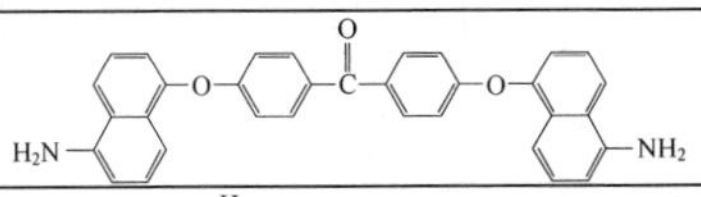	BANBP	40
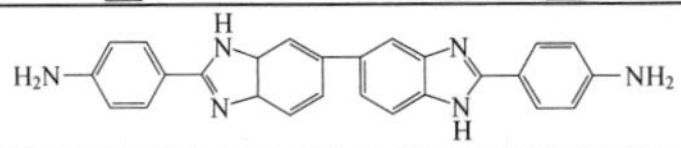	BAPBI	60,61

TABLE7.1 (Continued)

Diamine	Abbreviation	References
	BIPOB	62
	DAA	63
	DABA	64,66
	3DAC	65
	DAPBI	67
	DAN	68
	DATPA	69,70
	DMA	71
$NH_2 \left(CH_2 \right)_y NH_2$	DDA (y=12) DMDA (y=10) HMDA (y=6)	4
	DMMDA	50
	6FATFVP	72

TABLE7.1 (Continued)

Diamine	Abbreviation	References
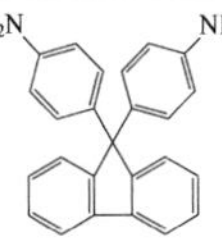	BAPF	6,20,24,73,74
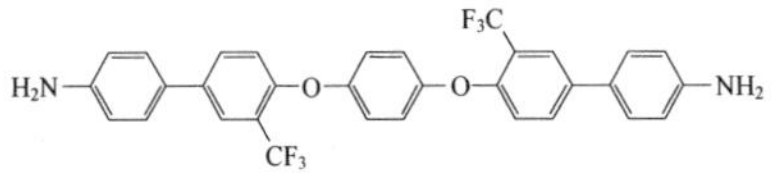	HQA	52
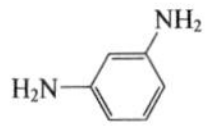	*m*-PDA	48
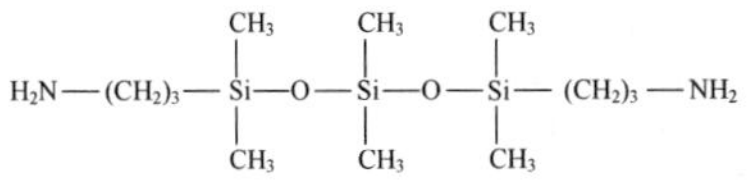	PMDS	75
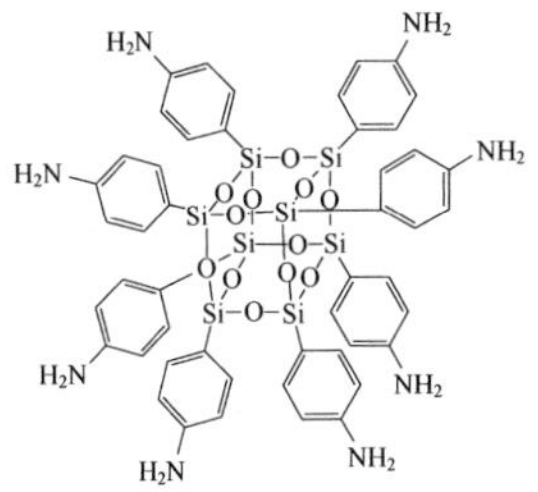	OPS	76
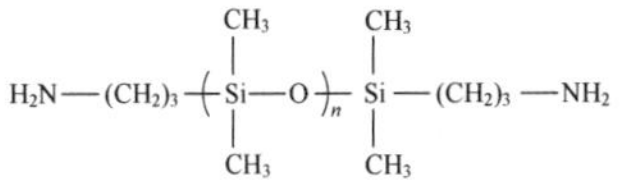	PSX	77
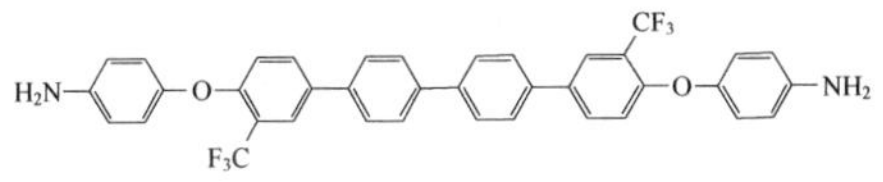	QA	78,79
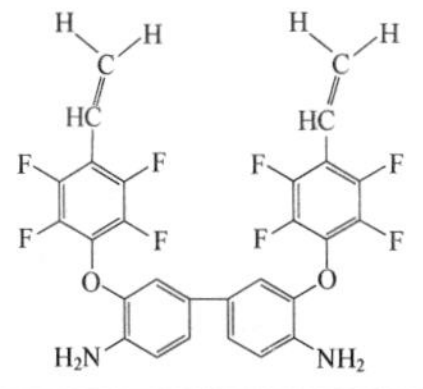	TFVBPA	79
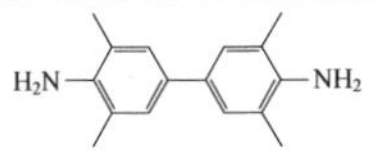	TMB	51
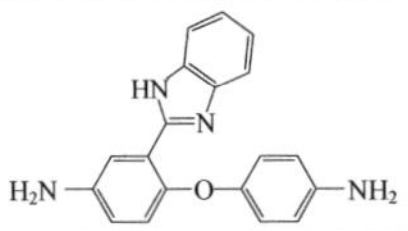	2-BIODA	81
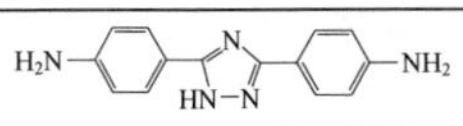	p-APTAz	36,82,83

TABLE7.1 (Continued)

Diamine	Abbreviation	References
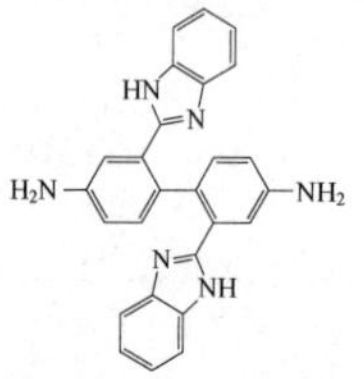	BBIBz	84
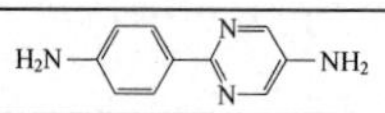	PAPRM	85
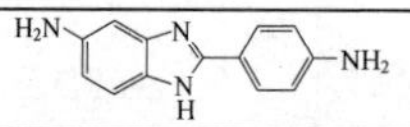	APABI	73
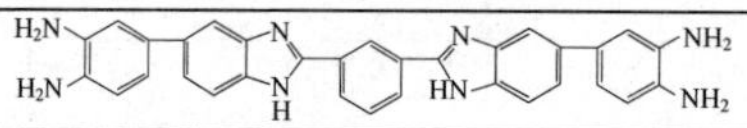	PBI-NH_2	86

Most side-chain-type sulfonated diamines are synthesized via multistep reactions [4,9,12-15,24, 32,35,37,38,45]. For example, as shown in Fig. 7.5, 2,2′-bis(3-sulfopropoxy)benzidine (2,2′-BSPB) is synthesized via four-step reactions [35,45].

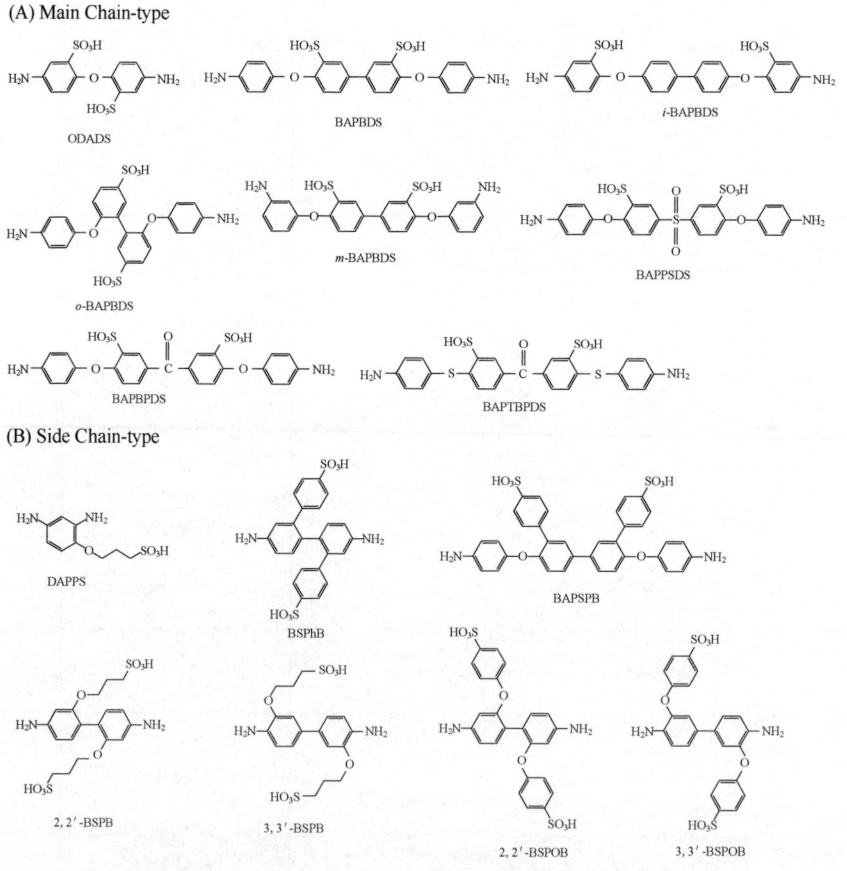

(A) Main Chain-type

ODADS, BAPBDS, *i*-BAPBDS, *o*-BAPBDS, *m*-BAPBDS, BAPPSDS, BAPBPDS, BAPTBPDS

(B) Side Chain-type

DAPPS, BSPhB, BAPSPB, 2, 2′-BSPB, 3, 3′-BSPB, 2, 2′-BSPOB, 3, 3′-BSPOB

FIGURE 7.2 Typical chemical structures of sulfonated diamines: (A) main-chain-type, (B) side-chain-type.

7.2.2 Synthesis of Six-membered Ring Dianhydrides

1,4,5,8-Naphthalenetetracarboxylic dianhydride (NTDA) and 3,4,9,10-perylenetetracarboxylic dianhydride (PTDA) are two commercial dianhydride monomers. Besides NTDA and PTDA, some other six-membered ring dianhydride monomers have also been reported [16,50,67,69,87-93]. Table 7.2 shows the chemical structures and abbreviations of various six-membered ring dianhydride monomers reported in the literature. The synthesis of 4,4′-binaphthyl-1,1′,8,8′-tetracarboxylic dianhydride (BNTDA) was first reported by Wang and coworkers [90]. As shown in Fig. 7.6, the anhydride group of 4-chloro-1,8-naphthalic anhydride was converted to di-*n*-butyl ester groups followed by the Ni(0)-catalyzed C—C coupling reaction to yield the tetraester intermediate product. Hydrolysis of the tetraester resulted in 4,4′-binaphthyl-1,1′,8,8′-tetracarboxylic acid which was dehydrated at elevated temperatures in a vacuum to give the objective product BNTDA. The total yield of this synthetic route was fairly high (66%). BNTDA could also be synthesized by direct coupling 4-chloro-1,8-naphthalic anhydride, but a low yield (44%) was obtained.

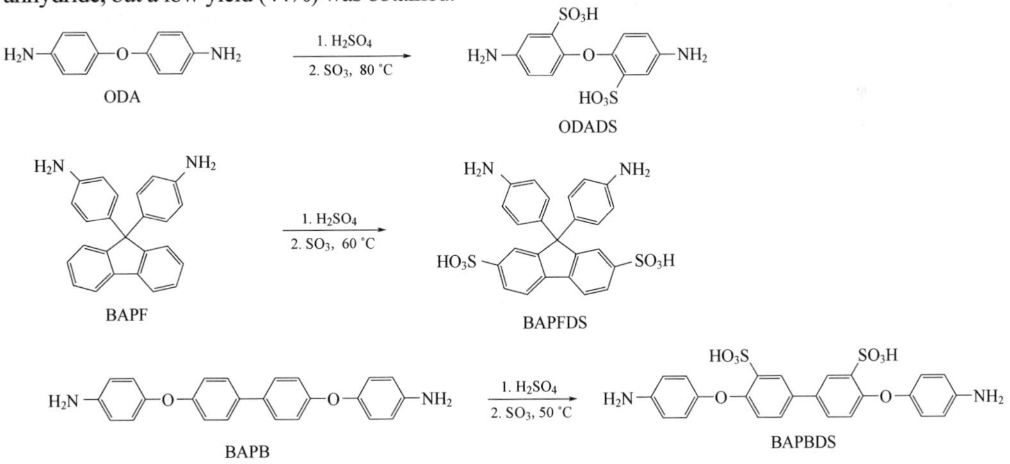

ODA → (1. H_2SO_4; 2. SO_3, 80 °C) → ODADS

BAPF → (1. H_2SO_4; 2. SO_3, 60 °C) → BAPFDS

BAPB → (1. H_2SO_4; 2. SO_3, 50 °C) → BAPBDS

FIGURE 7.3 Synthesis of sulfonated diamine monomers via direct sulfonation of the corresponding non-sulfonated diamines.

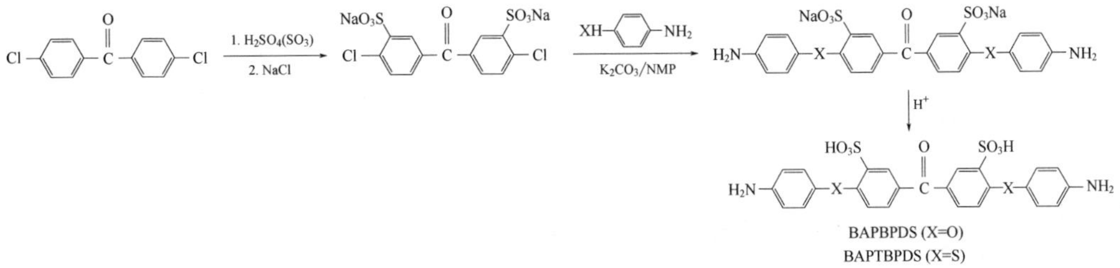

1. $H_2SO_4(SO_3)$; 2. NaCl; K_2CO_3/NMP; H^+

BAPBPDS (X=O)
BAPTBPDS (X=S)

FIGURE 7.4 Synthesis of two sulfonated diamine monomers via multiple step reactions.

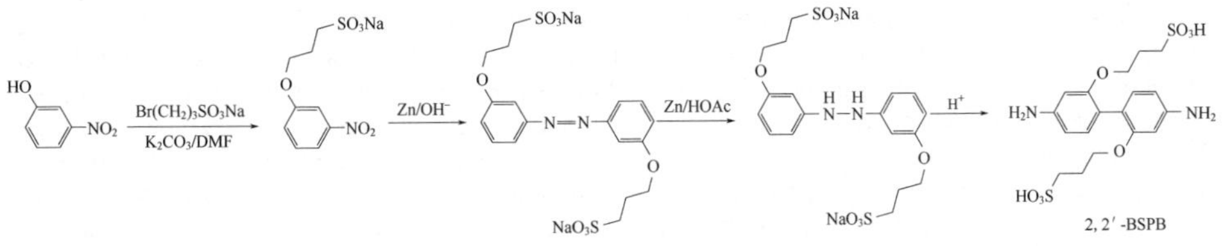

$Br(CH_2)_3SO_3Na$, K_2CO_3/DMF; Zn/OH^-; Zn/HOAc; H^+

2, 2′-BSPB

FIGURE 7.5 Synthesis of 2,2′-BSPB.

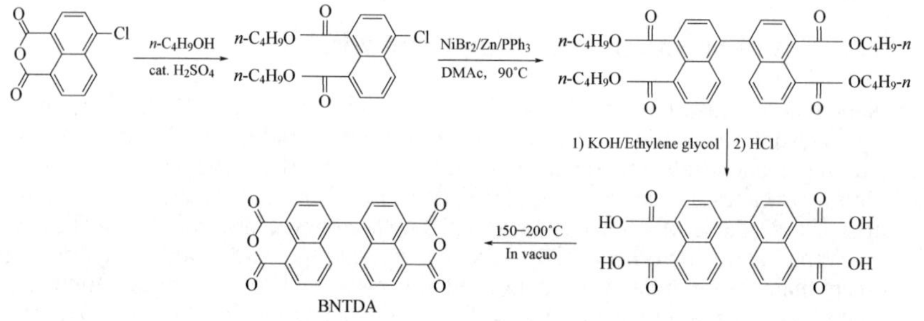

FIGURE 7.6 Synthesis of BNTDA.

One-step reaction of 4-bromo-1,8-naphthalic anhydride and 4,4′-biphenol gave 4,4′-(biphenyl-4,4′-diyldi(oxo))bis(1,8-naphthalic anhydride) (BPNDA) in high yield (75%, Fig. 7.7) [89]. 4,4′-Ketone-dinaphthalene-1,1′,8,8′-tetracarboxylic dianhydride (KDNTDA) was synthesized by Okamoto and coworkers [92] via six-step reactions using acenaphthene as the starting material (Fig. 7.8).

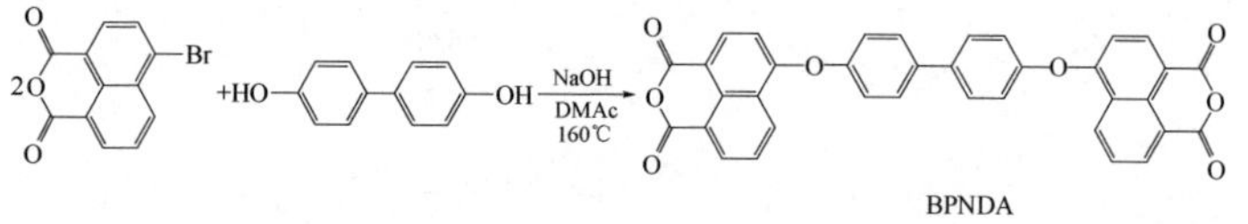

FIGURE 7.7 Synthesis of BPNDA.

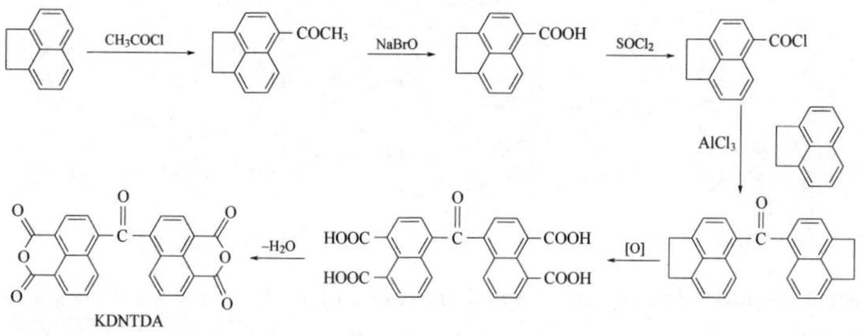

FIGURE 7.8 Synthesis of KDNTDA.

7.3 Polyimide Preparations

7.3.1 Preparation From Sulfonated Diamines

Six-membered ring SPIs are usually synthesized by one-step condensation polymerization of a naphthalenic (six-membered ring) dianhydride, a sulfonated diamine and a nonsulfonated diamine in *m*-cresol in the presence of triethylamine, benzoic acid and isoquinoline at 180 °C (see Fig. 7.9 using the synthesis of NTDA-based SPIs as an example). Because the amino groups of a sulfonated diamine are protonated resulting from the acid (sulfonic acid groups)-base (amino groups) interaction, a sulfonated diamine monomer does not directly react with a dianhydride monomer. To commence the polymerization a stronger base, typically triethylamine (Et_3N), is employed to liberate the protonated amino groups of a sulfonated diamine monomer. The resulting triethylammonium salt of the sulfonated diamine is generally much more soluble in *m*-cresol than the pristine sulfonated diamine. Because of the low reactivity of naphthalenic dianhydrides resulting from their low electron affinity (in comparison with that of phthalic dianhydrides), their polymerization with diamines needs to be conducted at elevated temperatures. Moreover, unlike the case of conventional polymerization of pththalic (five-membered ring) dianhydrides and diamines, no poly(amic acid) precursors are formed during the polymerization process of six-membered ring dianhydrides and diamines [94]. Benzoic acid acts as a catalyst to promote the formation of *trans*-form polyisoimide, while isoquinoline is used as a catalyst to convert the polymers from their isoimide form into their imide form [93,94]. By controlling the molar ratio between the sulfonated diamine and the nonsulfonated diamine, the ion exchange capacity (IEC) of the SPIs can be exactly controlled.

TABLE 7.2 Chemical Structures and Abbreviations of Various Six-Membered Ring Dianhydride Monomers

Dianhydride	Abbreviation	References
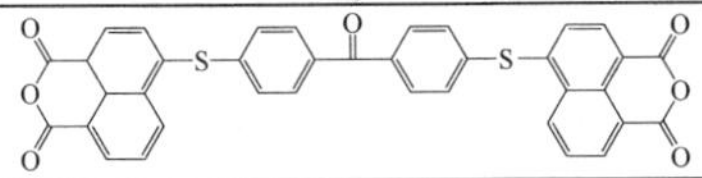	BPBTNA	87,88
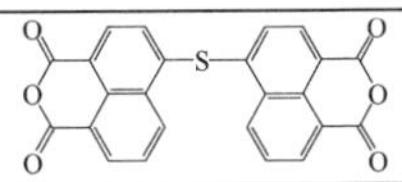	SBNA	88
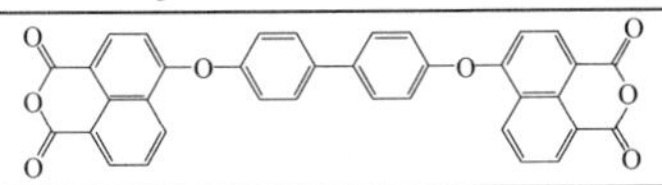	BPNDA	89
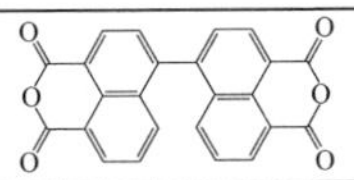	BTDA	50,67,69,90
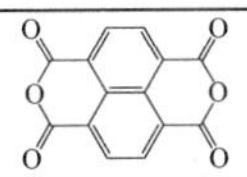	NTDA	1-86
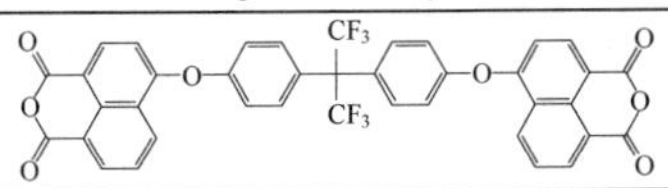 CF_3 / CF_3	HFBNA	16
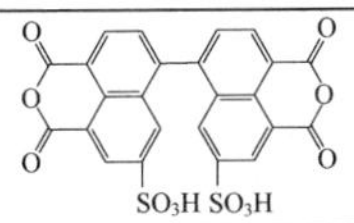 SO_3H SO_3H	SBTDA	50
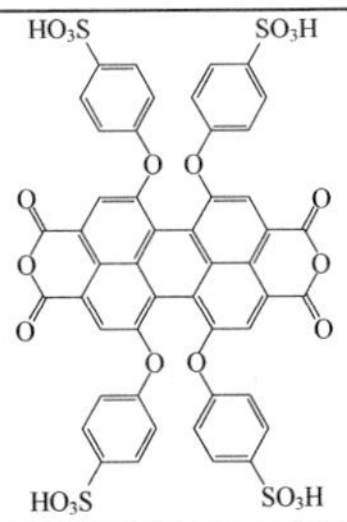 HO_3S SO_3H / HO_3S SO_3H	SPTDA	91
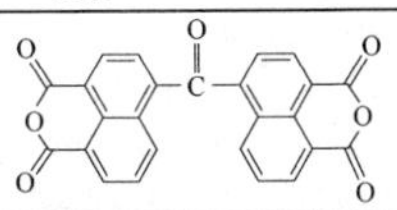	KDNTDA	92
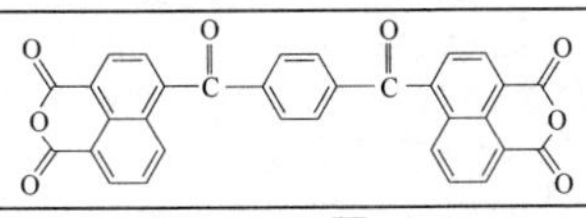	TBNDA	93
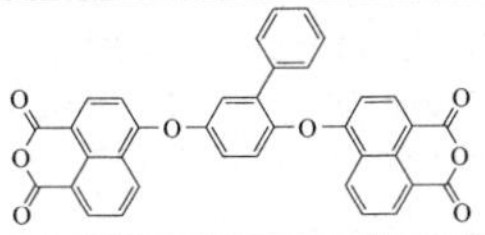	PPBNDA	93

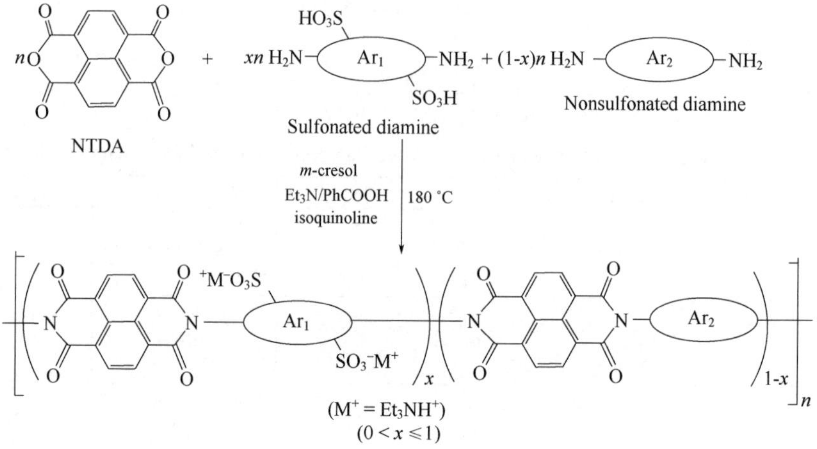

FIGURE 7.9 Synthesis of NTDA-based sulfonated polyimides.

The formation of six-membered ring SPIs is usually characterized by FT-IR spectroscopy. Many researchers reported that the characteristic absorption bands of six-membered SPIs were around 1712 cm^{-1} (C=O stretch, asymmetric), 1672 cm^{-1} (C=O stretch, symmetric), and 1375 cm (C–N stretch). However, Sek and coworkers reported that on the basis of model compound studies the absorption bands around 1712 and 1672 cm^{-1} should be ascribed to polyisoimides (C=O and C=N stretches), while the characteristic absorption band of polyimides should be around 1637 cm^{-1} (C=O stretch) [94]. Because quite few publications on spectral characterizations of six-membered ring polyimides are available, it is hard to make a clear judgement on this issue; to make it clear, many more fundamental works are needed. In the following discussions of this chapter, we employ the widely accepted concept "polyimide," while the factor "polyisoimide" is not involved.

Sulfonate groups generally exhibit multiple absorption bands in the region of 1400-1000 cm^{-1} and the most frequently appeared absorption bands around 1090 and 1020 cm^{-1} can be assigned to the asymmetric stretch and symmetric stretch of $-SO_2-$, respectively. They are fairly useful for characterization of sulfonic acid groups.

7.3.2 Preparation From Sulfonated Dianhydrides

Although quite few reports are available, SPIs can also be synthesized via condensation polymerization of sulfonated dianhydrides and common nonsulfonated diamines. Zhang and coworkers reported on the synthesis of a sulfonated dianhydride monomer, 1,6,7,12-tetra(4-sulfophenoxy) perylene-3,4,9,10-tetracarboxylic dianhydride (SPTDA), and a series of SPTDA-based SPIs [91]. The synthesis of the SPTDA is achieved via three-step reactions (Fig. 7.10). In the first step reaction, the anhydride groups are protected by reacting 1,6,7,12-tetrachloro perylene-3,4:9,10-tetracarboxylic dianhydride with *n*-butylamine to yield N,N′-di-*n*-butyl-1,6,7,12-tetrachloro-3, 4:9,10-perylenetetracar boxydiimide (**1**). In the second step reaction, the diimide intermediate product **1** is reacted with phenol in the presence of potassium carbonate to yield the intermediate product N,N′-di-*n*-butyl-1,6,7,12-tetraphenoxy-3,4:9,10-perylenetetracarboxydiimide (**2**). Finally, the intermediate product **2** is sulfonated in concentrated sulfuric acid at room temperature for 24 h followed by hydrolysis and thermal cycling to yield the sulfonated dianhydride product SPTDA.

The experimental procedures for synthesis of the SPTDA-based SPIs are similar to those performed with sulfonated diamine monomers, i.e., SPTDA, BNTDA and a nonsulfonated diamines (ODA, 1,12-diaminododecane) are copolymerized in *m*-cresol in the presence of benzoic acid and Et_3N at elevated temperatures (Fig. 7.11). The function of Et_3N is to prevent protonation of the amino groups of the diamine monomers, while benzoic acid acts as a catalyst.

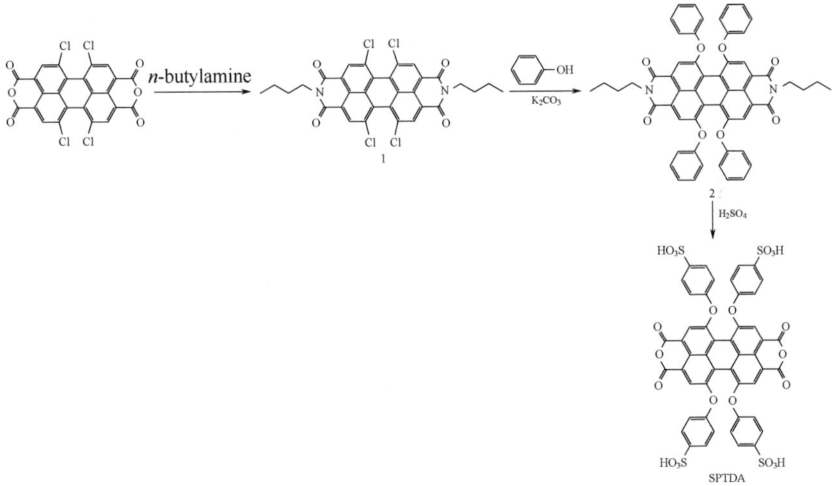

FIGURE 7.10 Synthesis of the sulfonated dianhydride monomer SPTDA.

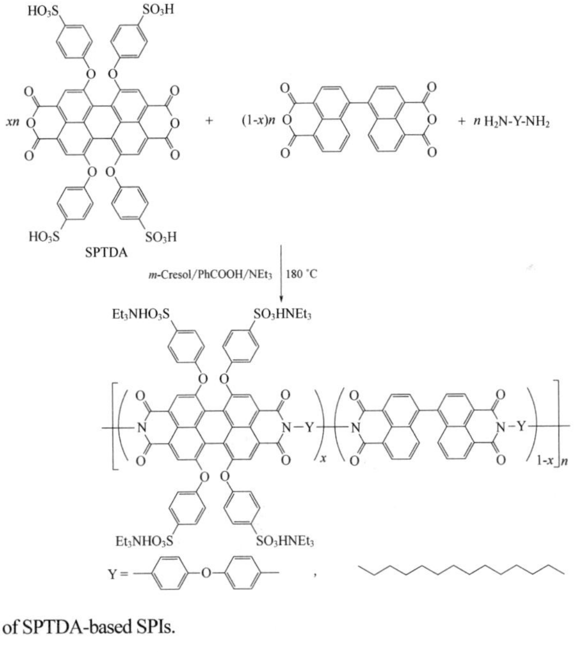

FIGURE 7.11 Synthesis of SPTDA-based SPIs.

7.3.3 Synthesis via Postsulfonation

Postsulfonation of non-SPIs in concentrated sulfuric acid (sulfonating reagent and reaction medium) at moderate temperatures (50-80 °C) is a rarely used but useful method for the preparation of some special SPIs [89,95,96]. For example, our research group previously found that it was difficult to get high molecular weight SPIs by polymerization of BPNDA with common sulfonated diamines such as BAPBDS and ODADS because of the insufficiently high reactivity of BPNDA as compared with that of NTDA [89]. However, the polymerization of BPNDA with common nonsulfonated diamines such as 1,3-bis(4-aminophenoxy)benzene (BAPBz) and 2-(4-aminophenyl)-5-aminobenzimidazole (APABI) could yield high-molecular-weight polyimides because nonsulfonated diamines generally have higher reactivity than the corresponding sulfonated ones. Postsulfonation of the BPNDA-based polyimides in concentrated sulfuric acid at 50 °C for 24 h resulted in the corresponding highly SPIs (Fig. 7.12) with good mechanical properties (tensile strength: 50-83 MPa, elongation at break (EB): 14.8%-44.8%) [89]. Our research group also successfully synthesized a series of highly sulfonated hyperbranched polyimides (SHBPI) via post-sulfonation of an amine-terminated hyperbranched polyimide (HBPI) which was prepared by polymerization of a difunctional monomer, 9,9-fluorenylidenebis (4,1-phenylene) bis(oxy)-4,4'-bis(1,8-naphthalic anhydride) (FBPNA) and a trifunctional monomer, tris(4-aminophenyl) amine (TAPA), at the molar ratio of FBPNA/TAPA = 1:1 in *m*-cresol at 180 °C for 20 h (Fig. 7.13) [95]. However, it should be noted that, despite these successful examples, the postsulfonated method is only suitable for those with excellent chemical stability (this issue will be discussed in detail in Section 7.4.5 on water-resistance below).

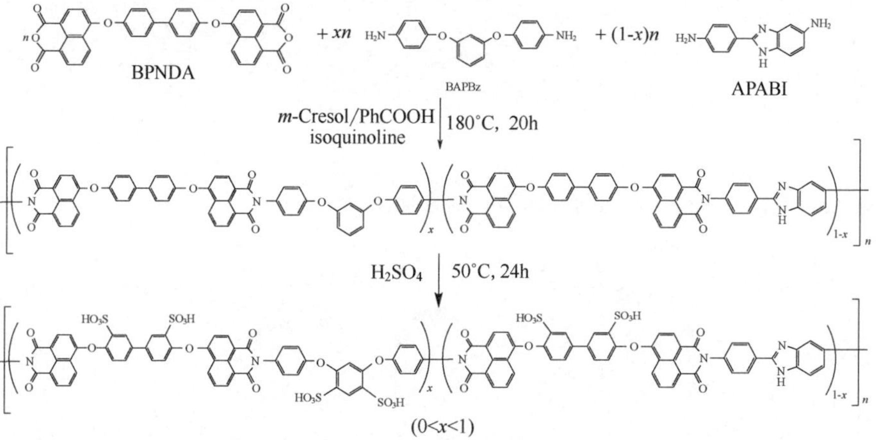

BPNDA + xn BAPBz + $(1-x)n$ APABI

m-Cresol/PhCOOH isoquinoline, 180°C, 20h

H_2SO_4, 50°C, 24h

$(0<x<1)$

FIGURE 7.12 Synthesis of various sulfonated copolyimides via post-sulfonation.

7.3.4 Block Copolymerization

As will be discussed in Section 7.4 on membrane properties, the morphology of a PEM has large effects on its properties (proton conductivity, swelling ratio, etc.). Block copolymerization is a powerful technique to prepare the PEMs with microphase-separated morphologies and thus superior membrane performances such as higher proton conductivities and lower in-plane swelling ratios (SR) than their random analogs may be obtained. Many SPI block copolymers have been developed by several research groups [15,16,44,50,64,97-104]. The synthesis of SPI block copolymers is usually performed by a two-pot method. In one pot, a sulfonated diamine is reacted with a dianhydride to yield an amine-terminated (excess sulfonated diamine is charged) or an anhydride-terminated (excess dianhydride is charged) hydrophilic SPI oligomer. In another pot, a nonsulfonated diamine is reacted with a dianhydride to yield an anhydride-terminated (excess dianhydride is charged) or an amine-terminated (excess nonsulfonated diamine is charged) hydrophobic PI oligomer. By controlling monomer molar ratio, the degrees of polymerization of both oligomers can be regulated. The resulting hydrophilic SPI oligomer can further react with the hydrophobic PI oligomer to give the multiblock copolyimide. Fig. 7.14 shows the synthesis of multiblock copolyimides consisting of benzimidazole- groups containing sulfonated polyimide hydrophilic blocks (averaged block length=20) and nonsulfonated polyimide hydrophobic blocks (averaged block length=5 or 10). A typical experimental procedure is described as follows using the multiblock copolyimide SPIBI20-*b*-NS05 as an example. Here, the "SPIBI" refers to the hydrophilic blocks derived from NTDA, BAPBDS and an amine-terminated polybenzimidazole oligomer PBI-NH_2 (degree of polymerization=1), while the "NS" refers to the hydrophobic blocks derived from NTDA and BAPPS. The numerals "20" and "05" refer to the averaged block lengths of the hydrophilic blocks and hydrophobic blocks, respectively.

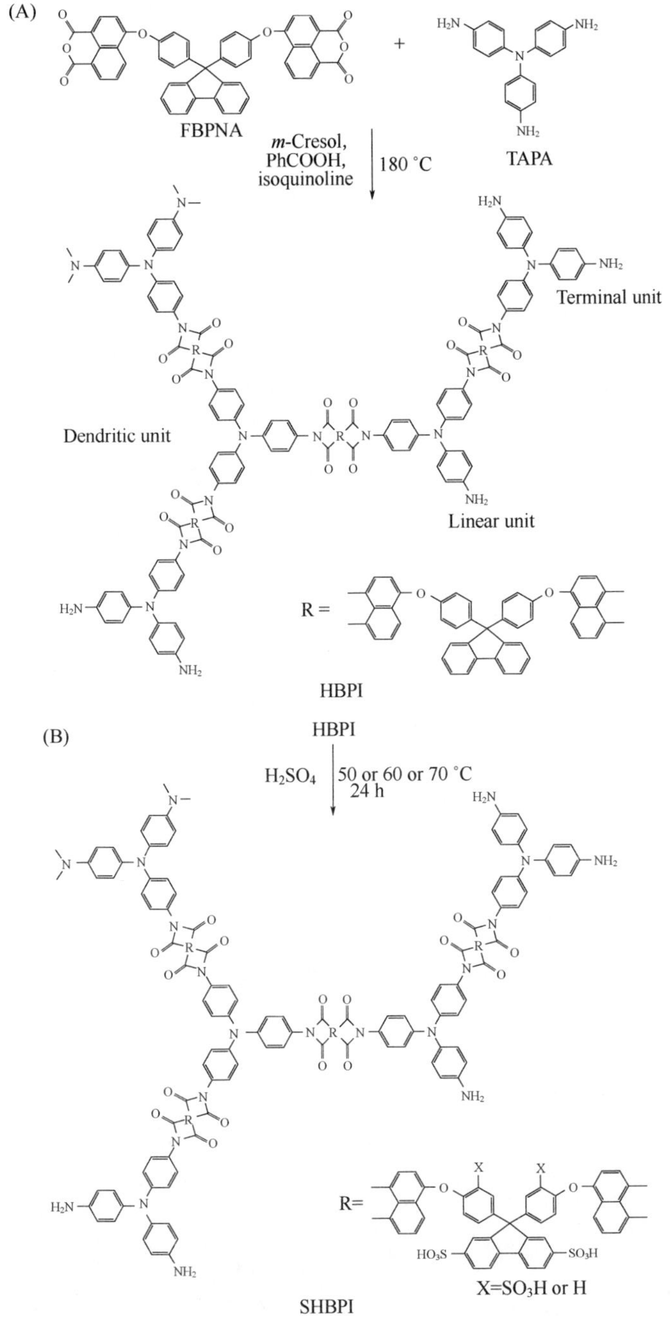

FIGURE 7.13 (A) Synthesis of an amine-terminated hyperbranched polyimide (HBPI). (B) Post-sulfonation of the HBPI.

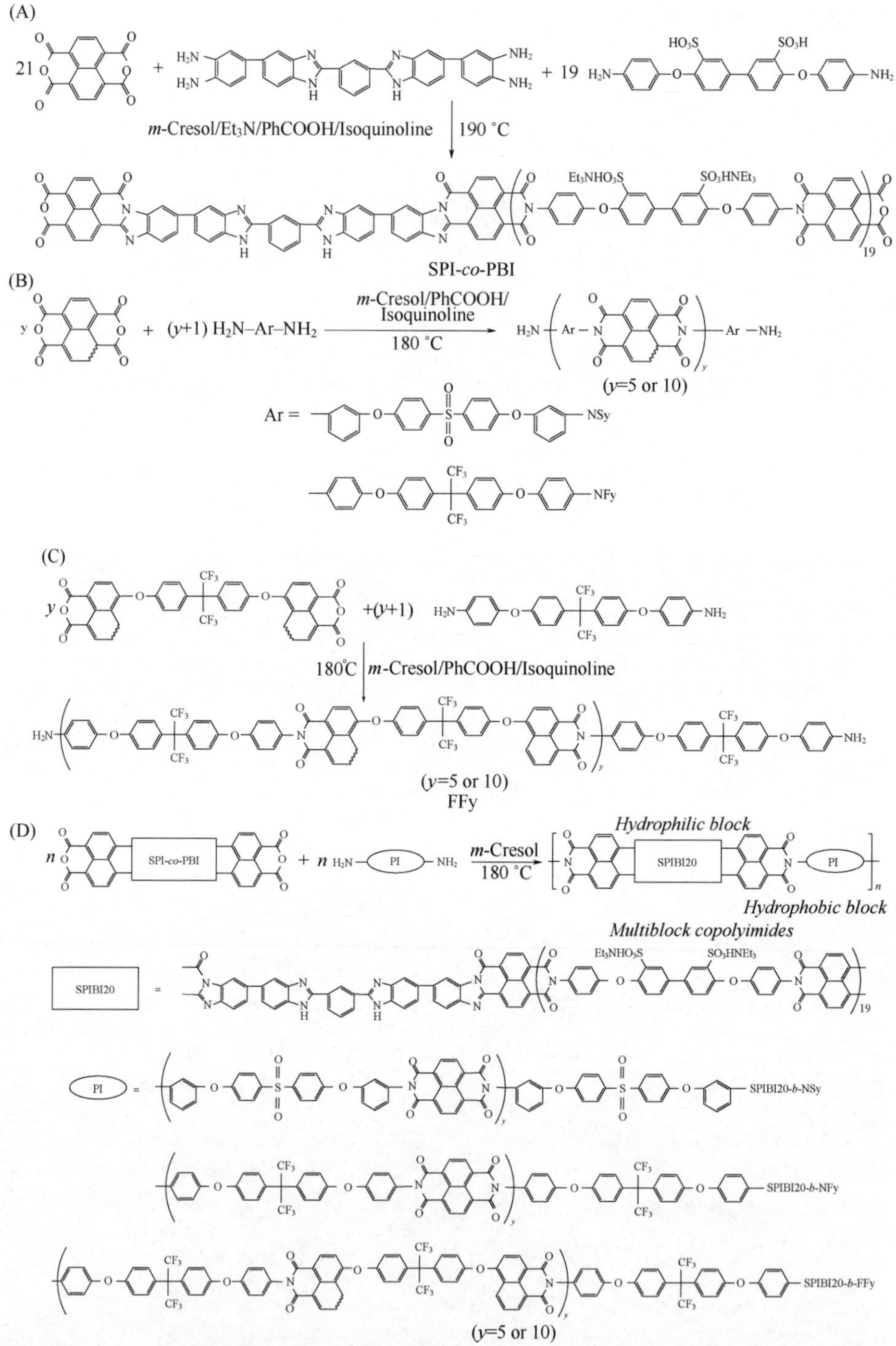

(A)

m-Cresol/Et_3N/PhCOOH/Isoquinoline 190 °C

SPI-*co*-PBI

(B)

m-Cresol/PhCOOH/Isoquinoline 180 °C

(y=5 or 10)

NSy

NFy

(C)

180°C *m*-Cresol/PhCOOH/Isoquinoline

(y=5 or 10)

FFy

(D)

m-Cresol 180 °C

Hydrophilic block

Hydrophobic block

Multiblock copolyimides

SPIBI20-*b*-NSy

SPIBI20-*b*-NFy

SPIBI20-*b*-FFy

(y=5 or 10)

FIGURE 7.14 (A) Synthesis of anhydride-terminated hydrophilic oligomer, (B) synthesis of amine-terminated hydrophobic oligomers NSy and NFy, (C) synthesis of amine-terminated hydrophobic oligomers FFy, (D) block copolymerization.

To a 100 mL dry three-necked flask equipped with a magnetic stirring device, a nitrogen inlet and a nitrogen outlet, and a condenser were added 2.0064 g (3.8 mmol) of BAPBDS, 0.1044 g (0.2 mmol) of H_2N-PBI, 15 mL of *m*-cresol, 1.2 mL of Et_3N and 1.2 mL of isoquinoline. The mixture was stirred at room temperature under nitrogen flow to form a clear solution. Then 1.1256 g (4.2 mmol) of NTDA and 0.9768 g (8.0 mmol) of benzoic acid were added and the reaction mixture was stirred at room temperature for 0.5 h followed by heating at 80 °C for 12 h and 190 °C for another 8 h yielding anhydride-terminated SPIBI oligomer (hydrophilic block). The reaction mixture was cooled to room temperature and was set aside for the next step reaction.

To another 100mL dry three-neck flask equipped with a magnetic stirring device, a nitrogen inlet and a nitrogen outlet, and a condenser were added 0.5190 g (1.2 mmol) of BAPPS, 0.2680 g (1.0 mmol) of NTDA, 0.2442 g (2.0 mmol) of benzoic acid, 7 mL of *m*-cresol and 0.3 mL of isoquinoline. The mixture was magnetically stirred at room temperature under nitrogen flow for 0.5 h and then heated at 80 °C for 12 h and 180 °C for another 8 h yielding the amine-terminated polyimide oligomer (hydrophobic block). After cooling to room temperature, the solution mixture was completely transferred to the flask where the foregoing anhydride-terminated SPI oligomer solution was produced. Additional 5 mL of *m*-cresol was added to the flask and the reaction mixture was magnetically stirred at room temperature under nitrogen flow. Subsequently the reaction mixture was heated at 80 °C for 4 h and 180 °C for another 20 h to fulfill the block copolymerization. After cooling to room temperature, the viscous solution mixture was poured into 150 mL of methanol. The resulting fiber-like precipitate was collected by filtration, thoroughly washed by methanol, and dried at 80 °C for 10 h in a vacuum.

Sequenced copolymers generally have less perfect microphase-separated morphologies than block copolymers. Sequential copolymerization is usually performed by a two-stage one-pot method. Typically, a dianhydride monomer (e.g., NTDA) is allowed to react with a sulfonated diamine monomer in the first step to yield an amine-terminated (excess sulfonated diamine is used) or anhydride-terminated (excess dianhydride is used) SPI oligomer. Subsequently, the oligomer is further randomly copolymerized with a nonsulfonated diamine and the same or another dianhydride to yield the sequenced SPI copolymer. By controlling the molar ratio between the dianhydride monomer and the sulfonated diamine monomer in the first step reaction, the average block length of the hydrophilic block can be controlled. Various sequenced SPIs have been reported [1,6]. One example is the BDSA-based SPI sequenced copolymers with average length of hydrophilic block of five (Fig. 7.15) reported by Mercier and coworkers [1]. Another example is the BAPFDS-based sequenced copolyimides (Fig. 7.16) reported by Okamoto and coworkers [6].

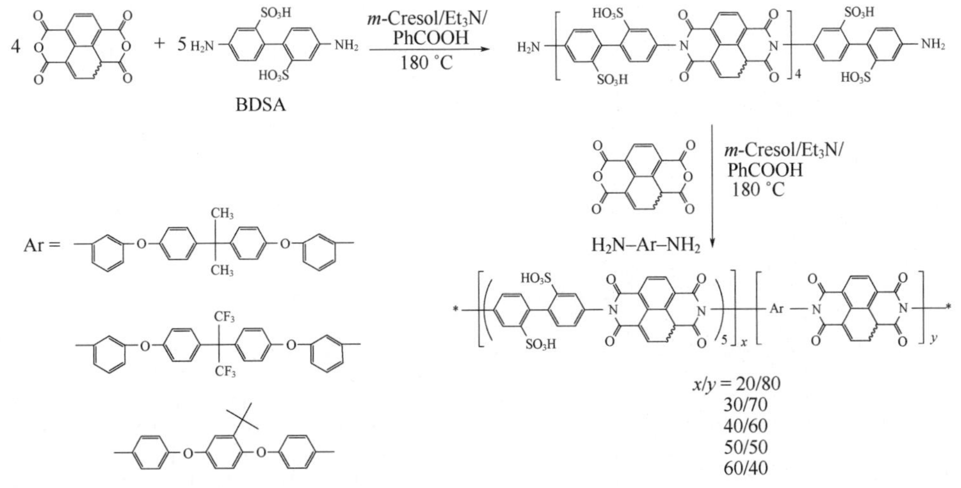

FIGURE 7.15 Synthesis of BDSA-based sequenced copolyimides.

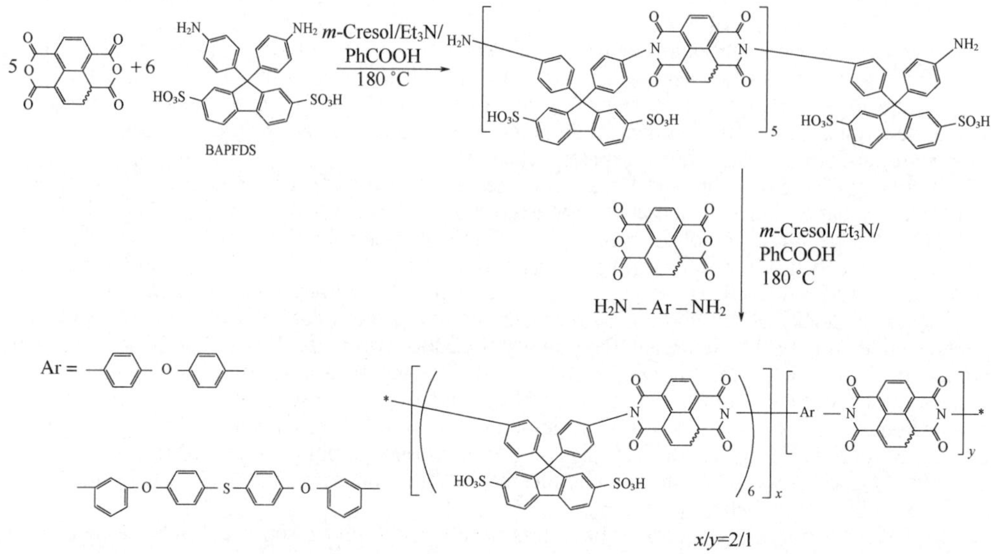

FIGURE 7.16 Synthesis of BAPFDS-based sequenced copolyimides.

7.4 Ion Exchange Membrane Properties

7.4.1 Solubility

Table 7.3 shows the solubility of some SPIs in common organic solvents such as *m*-cresol, dimethyl sulfoxide (DMSO), 1-methyl-2-pyrrolidinone (NMP), *N,N*-dimethylacetamide (DMAc), and methanol. These SPIs are prepared from different sulfonated diamines including BDSA, ODADS, BAPFDS, BAPBDS, 2,2′-bis(4-aminophenoxy)biphenyl-5,5′-disulfonic acid (*o*BAPBDS), 4,4′-bis(3-aminophenoxy) biphenyl-3,3′-disulfonic acid (*m*BAPBDS), 4,4′-bis(4-amino-2-sulfophenoxy)biphenyl (*i*BAPBDS), 2,2′-BSPB, 3,3′-BSPB, 3-(2′,4′-diaminophenoxy)propane sulfonic acid (DAPPS) and 3,3′-bis(4-sulfophenoxy)benzidine. Among the sulfonated diamines, BAPBDS, *o*BAPBDS, *m*BAPBDS, *i*BAPBDS, and 3,3′-BSPOB are isomers of each other. From this table, it can be seen that all the SPIs, except the NTDA-BDSA/BAPB(1/1), are well soluble in *m*-cresol, the polymerization medium for synthesis of the SPIs. NMP and DMAc are two poor solvents for most of the SPIs. In addition, except for NTDA-*o*BAPBDS, none of the SPIs are soluble in methanol. The solubility is closely related to the chemical structures of the SPIs. *o*BAPBDS, *m*BAPBDS and DAPPS-based SPIs exhibit the best solubility among the SPIs listed in this table. The homopolymers, NTDA-*o*BAPBDS, NTDA-*m*BAPBDS, and NTDA-DAPPS, are well soluble in *m*-cresol, DMSO, NMP, and DMAc. The NTDA-*o*BAPBDS is even partially soluble in methanol indicating excellent solubility of this polymer. This is attributed to the nonlinear (*meta*- or *ortho*-orientated) configurations of these sulfonated diamine moieties [5,47]. The BAPBDS and *i*BAPBDS-based SPIs show the poorest solubility due to the linear (*para*-orientated) configuration of the sulfonated diamine moieties. 2,2′-BSPB, 3,3′-BSPB and 3,3′-BSPOB are also *para*-orientated sulfonated diamines, however, the SPIs derived from these sulfonates display better solubility than those derived from BAPBDS and *i*BAPBDS. This is probably because the side-chain-type SPIs have less regular chain packing than the main-chain-type SPIs.

TABLE7.3 Organo–solubility of Some SPIs in Their Triethylammonium Salt Form

Polyimide	Solubility					References
	m-cresol	DMSO	NMP	DMAc	MeOH	
NTDA- ODADS	+	+	–	–	–	3
NTDA- ODADS/ODA(1/1)	+	–	–	–	–	3
NTDA- ODADS/BAPB(1/1)	+	–	–	–	–	3
NTDA- ODADS/BAPF(1/1)	+	+	–	–	–	3
NTDA- ODADS/BAPHF(1/1)	+	+	–	–	–	3
NTDA- BDSA	+	+	–	–	–	3
NTDA- BDSA/ODA(1/1)	+	–	–	–	–	3
NTDA- BDSA/BAPB(1/1)	+–	–	–	–	–	3
NTDA- BDSA/BAPF(1/1)	+	+	–	–	–	3
NTDA- BDSA/BAPHF(1/1)	+	+	–	–	–	3
NTDA- BAPFDS	+	+	–	–	–	6
NTDA-BAPFDS/ODA(4/1)	+	+	–	–	–	6
NTDA-BAPFDS/ODA(2/1)	+	+	–	–	–	6
NTDA-BAPFDS/ODA(1/1)	+	+	–	–	–	6
NTDA-BAPFDS/BAPB(4/3)	+	+	–	–	–	6
NTDA-BAPFDS/BAPPS(2/1)	+	+	–	–	–	6
NTDA-BAPFDS/ODA(2/1)-s[a]	+	+	–	–	–	6
NTDA-BAPFDS/*m*BAPPS(2/1)-s[a]	+	+	–	–	–	6
6NTDA-BAPBDS	+(–)	–(–)	–(–)	–(–)	–(–)	7
NTDA-BAPBDS/TFMB(4/1)	+	–	–	–	–	7
NTDA-BAPBDS/TFMB(1/1)	+	–	–	–	–	7
NTDA-BAPBDS/BAPB(4/1)	+	–	–	–	–	7
NTDA-BAPBDS/BAPB(2/1)	+	–	–	–	–	7
NTDA-*o*BAPBDS	+	+	+	+	+–	10
NTDA-*o*BAPBDS/BAPB(2/1)	+	+	+	–	–	10
NTDA-*o*BAPBDS/ODA(4/3)	+	+	+	–	–	10
NTDA-*o*BAPBDS/*m*BAPPS(2/1)	+	+	+	–	–	10
NTDA-*o*BAPBDS/*o*BAPB(2/1)	+	+	+	–	–	10
NTDA-*i*BAPBDS	+	–	–	–	–	47
NTDA-*i*BAPBDS/BAPB(2/1)	+	–	–	–	–	47
NTDA-*m*BAPBDS	+	+	+	+	–	47
NTDA-*m*BAPBDS(3/2)	+	+	+–	+	–	47
NTDA-2,2′-BSPB	+	+	–	–	–	45
NTDA-3,3′-BSPB	+	+	–	–	–	45
NTDA-DAPPS	+	+	+	+	–	8
NTDA-3,3′-BSPOB/TrMPD(9/1)	+	+	–	–	–	110
NTDA-3,3′-BSPOB/TrMPD(2/1)	+	+	–	–	–	110

The data in parenthesis refer to the membrane in its proton form.
[a]*Sequenced copolyimides. Other copolyimides in this table are random copolymers.*

The solubility of SPIs is also dependent on the chemical structures of nonsulfonated diamine moieties. For example, the sulfonated copolyimides derived from the nonsulfonated diamines, BAPF and 2,2-bis(4-aminophenoxy) hexafluoropropane (BAPHF), display better solubility than those derived from ODA and BAPB. This is likely because of the presence of the bulky fluorenylidene group of BAPF and the hexafluoropropylidene group of BAPHF which cause less dense packing of polymer chains.

It should be noted that since the sulfonic acid group itself is a bulky and polar group, SPIs generally have much better solubility than their nonsulfonated analogs. For example, NTDA-ODADS is well soluble in *m*-cresol and DMSO, whereas NTDA-ODA is insoluble in any organic solvents.

7.4.2 Thermal Stability

The thermal stability of a polymer is usually evaluated by thermogravimetric analysis (TGA) in nitrogen or air atmosphere. Thermogravimetry-mass spectroscopy (TG-MS), a technique combining TGA and mass spectroscopy, is very useful for studying the thermal degradation mechanism of a polymer. Fig. 7.17 shows the TG-MS curves of the SPI homopolymer NTDA-ODADS in its proton form [3]. The first-stage weight loss occurring in the range from room temperature to ~ 120 °C is due to evaporation of the absorbed moisture in the membrane. The second-stage weight loss starting from ~280 °C is due to the decomposition of sulfonic acid groups judging by the evolution of sulfur monoxide and sulfur dioxide. The third-stage weight loss occurring above ~500 °C is attributed to the decomposition of the polyimide backbone judging by the evolution of carbon monoxide and carbon dioxide. For the SPI derived from 2,2′-BSPB, the decomposition temperature of the sulfonic acid groups starts from ~ 240 °C [45] which is about 40 °C lower than that of the wholly aromatic polyimide NTDA-ODADS. This indicates that the sulfonic acid groups attached to aliphatic groups are less stable than those attached to benzene rings. Nevertheless, because the working temperature of a PEMFC is far below 200 °C, the thermal stability of all the SPIs is high enough for practical applications.

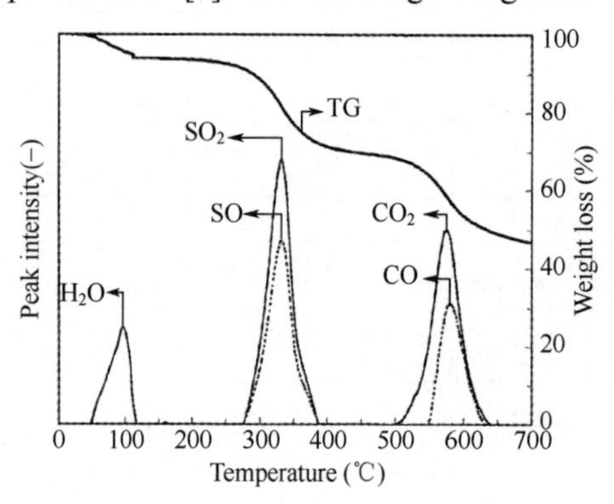

FIGURE 7.17 TG-MS curves of NTDA- ODADS in its proton form in nitrogen atmosphere.

7.4.3 Water Uptake and Swelling Ratios

Water uptakes (WU) and SR are two important properties of a PEM and are affected by many factors, such as IEC, polymer chemical structure and morphology, and temperature. They are generally measured by soaking a PEM sample in deionized water at a given temperature for a period of time when sorption equilibrium has been achieved. WU is calculated from the following equation:

$$\mathrm{WU} = (W - W_0) / W_0 \tag{7.1}$$

where W_0 and W refer to the weight of dry and wet membrane sample, respectively.

Through-plane swelling ratio (Δt) and in-plane swelling ratio (Δl) are frequently employed to describe the swelling behavior of a PEM. They are calculated from the following equations:

$$\Delta t = (t - t_0) / t_0 \tag{7.2}$$

$$\Delta l = (l - l_0) / l_0 \tag{7.3}$$

where t_0 and t refer to the thickness of the dry and wet membrane, respectively, and l_0 and l refer to the length of the dry and wet membrane, respectively.

In some papers, area swelling ratio (ΔS) and volume swelling ratio (ΔV) are used to characterize the swelling behavior of a PEM. They are calculated from the following equations:

$$\Delta S = (S - S_0) / S_0 \tag{7.4}$$

$$\Delta V = (V - V_0) / V_0 \tag{7.5}$$

where S_0 and S refer to the area of the dry and wet membrane, respectively, while V_0 and V refer to the volume of the dry and wet membrane, respectively.

Table 7.4 shows the IEC, WU, and SR values of some SPI membranes reported in the literature. IEC is an intrinsic factor affecting WU and SR and the membrane with higher IEC tends to have higher WU and higher volume SR. A membrane may even completely dissolve in water as its IEC is too high. For example, the SPI membranes, NTDA-BDSA and NTDA-ODADS, are soluble in hot water because of their extremely high IECs (3.47 mol·g^{-1} for NTDA-BDSA, 3.37 mol·g^{-1} for NTDA-ODADS) [3]. Copolymerization with a nonsulfonated diamine is a common method to decrease membrane IEC, WU and SR.

TABLE7.4 Ion Exchange Capacity (IEC), Water Uptake (WU) and Swelling Ratio (SR) in In-plane Direction (Δl) and Through-plane Direction (Δt) of Some SPI Membranes

Membrane	*IEC*[a] (mol·g^{-1})	*T* (°C)	W_a (%)	*SR* (%)		References
				Δt	Δl	
NTDA-BDSA	3.47	80	Dissolved			3
NTDA-ODADS	3.37	80	Dissolved			3
NTDA-BDSA/ODA(1/1)	1.98	80	79	NA	NA	3
NTDA-BDSA/BAPF(1/1)	1.73	80	63	NA	NA	3
NTDA-ODADS/ODA(1/1)	1.95	80	87	NA	NA	3
NTDA-ODADS/BAPB(1/1)	1.68	80	57	NA	NA	3
NTDA-BAPBDS	2.63	80	107	24	24	7
		80	133	28	52	7
		20	75	16	15	7
NTDA-BAPBDS/BAPB(4/1)	2.20	80	73	20	14	7
NTDA-BAPBDS/TFMB(4/1)	2.23	80	82	18	16	7
NTDA-*o*BAPBDS	2.63	80	Dissolved			10
NTDA-*o*BAPBDS/ODA(4/3)	1.85	80	153	NA	NA	10
NTDA-*o*BAPBDS/*o*BAPB(2/1)	1.89	50	730	NA	NA	10
NTDA-*o*BAPBDS/BAPB(2/1)	1.89	80	152	NA	NA	10
NTDA-*m*BAPBDS	2.63	50	184	NA	NA	10
		80	Dissolved			10
NTDA-*m*BAPBDS/BAPB (3/2)	1.73	30	47	NA	NA	10
NTDA-2,2′-BSPB	2.89	50	222[b]	230[c]	1.0[c]	45
NTDA-3,3′-BSPB	2.89	50	250[b]	180[c]	11[c]	45
NTDA-DAPPS	2.09	50	105	12[c]	15[c]	8
SPI-*co*-PBI-19/1	2.24	80	95	18	35	91
SPI-*co*-PBI-29/1	2.37	80	104	21	40	91
SPIBI20-*b*-NS05	1.94	80	79	26	14	16
SPIBI20-*b*-NS10	1.68	80	68	22	10	16
SPIBI20-*b*-NF05	1.85	80	69	21	12	16
SPIBI20-*b*-NF10	1.54	80	53	17	10	16
SPIBI20-*b*-FF05	1.64	80	44	18	9	16
SPIBI20-*b*-FF10	1.28	80	36	12	7	16

[a]*Theoretical values.*
[b]*50 °C.*
[c]*Room temperature.*

Besides IEC, the chemical structures of the sulfonated diamine moieties and the nonsulfonated diamine moieties also have large effects on WU and SR. The SPIs derived from the isomeric sulfonated diamines, BAPBDS, *o*BAPBDS and *m*BAPBDS are good examples for studying the relationship between the chemical structures of sulfonated diamine moieties on WU and SR. As shown in Table 7.4, the WU, Δt and Δl values at 80 °C of NTDA-BAPBDS are 107 %(w), 24%(w) and 24%(w), respectively, while NTDA-*o*BAPBDS and NTDA-*m*BAPBDS are completely dissolved at 80 °C despite exactly the same IEC of these membranes. A similar trend is also observed with the SPI copolymer membranes, i.e., NTDA-BAPBDS/BAPB(4/1) and NTDA-BAPBDS/TFMB(4/1) show significantly lower WU than NTDA-*o*BAPBDS/ODA(4/3) and NTDA-*o*BAPBDS/ODA(4/3), despite the higher IECs of the former. This should be ascribed to the different configurations of the BAPBDS, *o*BAPBDS, and *m*BAPBDS. As discussed in Section 7.4.1, BAPBDS is a *para*-orientated linear diamine, whereas *o*BAPBDS and *m*BAPBDS are *ortho*- and *meta*-orientated nonlinear diamines. As a result, the chain packing is much more regular for BAPBDS-based SPI membranes than for *o*BAPBDS and *m*BAPBDS-based ones. The huge W_a of NTDA-*o*BAPBDS/*o*BAPB(2/1) is related to the *ortho*-orientated nonlinear configurations of both the sulfonated diamine and the nonsulfonated diamine.

NTDA-2,2′-BSPB and NTDA-3,3′-BSPB are two side-chain-type SPIs. They have fairly high WU because of their very high IEC (2.89 mol·g^{-1}). However, their swelling behaviors are quite different from that of most main-chain-type SPI membranes, i.e., NTDA-2,2′-BSPB and NTDA-3,3′-BSPB exhibit extremely strong anisotropic swelling behaviors (little Δl but very large Δt) [45], whereas most main-chain-type SPI membranes (e.g., NTDA-BAPBDS) show almost isotropic or only moderate anisotropic [7,10,21] swelling behaviors. The extremely strong anisotropic swelling behavior of NTDA-2,2′-BSPB and NTDA-3,3′-BSPB should be ascribed to their unique laminate morphology resulting from its rod-like stiff backbone as well as the microphase-separated morphology. The in-plane swelling ratio of NTDA-2,2′-BSPB approaches to almost zero which is very favorable for enhancing fuel cell durability.

Unlike NTDA-2,2′-BSPB and NTDA-3,3′-BSPB, NTDA-DAPPS shows almost isotropic swelling behavior despite the fact that it is also a side-chain-type SPI membrane. This should be due to the *meta*-orientated nonlinear configuration of the DAPPS moiety.

7.4.4 Proton Conductivity

Proton conductivity (σ) is one of the most important properties of a PEM. It is usually measured by using a four-point-probe electrochemical impedance spectroscopy technique over a wide range of frequencies (e.g., 10 Hz to 100 KHz). To measure *in-plane* proton conductivity ($\sigma_{\parallel}$), a sheet of membrane sample and two platinum plate electrodes are mounted on a Teflon cell and the two parallel platinum electrodes are set on one side of the membrane. The cell is placed in either a thermo-controlled humid chamber for measurement at relative humidity (RH) lower than 100% or distilled deionized water for measurement in liquid water. To measure *through-plane* proton conductivity ($\sigma_{\perp}$), a sheet of membrane sample is sandwiched between two platinum electrodes. The resistance value is determined from high-frequency intercept of the impedance with the real axis. *In-plane* proton conductivity is calculated from the following equation:

$$\sigma_{\parallel}=d/(tlR) \tag{7.6}$$

where d is the distance between the two electrodes, t and l are the thickness and width of the membrane, respectively, and R is membrane resistance measured. For measurement in liquid water, the dimensions in fully hydrated state are used.

Through-plane proton conductivity is calculated from the Eq. 7.7:

$$\sigma_{\perp} = t/(AR) \tag{7.7}$$

where t, A, and R refer to membrane thickness, effective area of electrode, and membrane resistance measured.

Because *through-plane* conductivity is more difficult to measure than *in-plane* conductivity, *in-plane* conductivity has been widely used for characterization of membrane proton conducting performance and the proton conductivities reported in most literature refer to *in-plane* conductivity.

Proton conductivity is affected by many factors such as IEC, polymer structure and morphology, RH, and temperature. Among them, IEC is an intrinsic factor determining proton conductivity of a PEM. To achieve high proton conductivity, high IEC (e.g., IEC > 2.0 mol·g^{-1}) is generally required.

However, too-high IEC often causes excess swelling or even dissolution of the membranes. As a result, IEC must be controlled at an appropriate level. Although the optimal IEC values may differ case by case depending on individual membranes and applications, it seems that the IECs of the SPI membranes usually should be controlled in the range of 1.7-2.3 mol·g^{-1}.

Proton conductivity is strongly dependent on RH and for a given membrane the higher RH tends to give higher proton conductivity. This is because water molecules absorbed in the membranes play an important role in proton conduction. First, the process of dissociation of sulfonic acid groups requires water molecules. Secondly, the transport of protons is facilitated with the aid of water. Fig. 7.18 shows the variation of *in-plane* proton conductivity of some SPI membranes as a function of RH at 50 °C. For comparison purposes, the data of Nafion117 are also shown in this figure. The theoretical values of IEC of the SPI membranes are in the range of 2.63-2.89 mol·g^{-1} which are much larger than that of Nafion 117 (0.91 mol·g^{-1}). It can be seen that for all the membranes the proton conductivity increases rapidly as the RH increases. At high RH (> 80%), most SPI membranes show comparable or even higher proton conductivities than Nafion 117. However, at low RH (<40%) except BAPFDS and NTDA-3,3′-BSPOB/ *m*BAPPS the SPI membranes exhibit significantly lower proton conductivities than Nafion 117 despite the much higher IECs of the former. This is likely because, unlike Nafion 117, the SPI membranes do not possess ionic channels to promote proton transport. Another possible reason is that the sulfonic acid groups of the SPI membranes are less acidic than that of Nafion 117. NTDA-BAPFDS displays the highest proton conductivities at low RH range (close to those of Nafion 117) among the SPI membranes. This is probably related to the bulky fluorenylidene groups which produce high fractional free volumes to allow the sorption of more water molecules. The side-chain-type SPI membrane NTDA-2,2′-BSPB shows significantly lower proton conductivities than that of Nafion 117 in low RH range (< 40%) despite its high IEC and the microphase-separated morphology resulting from the hydrophobic polyimide backbone and the hydrophilic ionic domains (average size: about 5 nm) [45]. This is because the hydrophilic ionic domains of NTDA-2,2′-BSPB are not well connected making it difficult to form ionic channels. The low acidity of the alkyl sulfonic acid groups may be another possible reason for the low conductivities of this SPI membrane in low RH range.

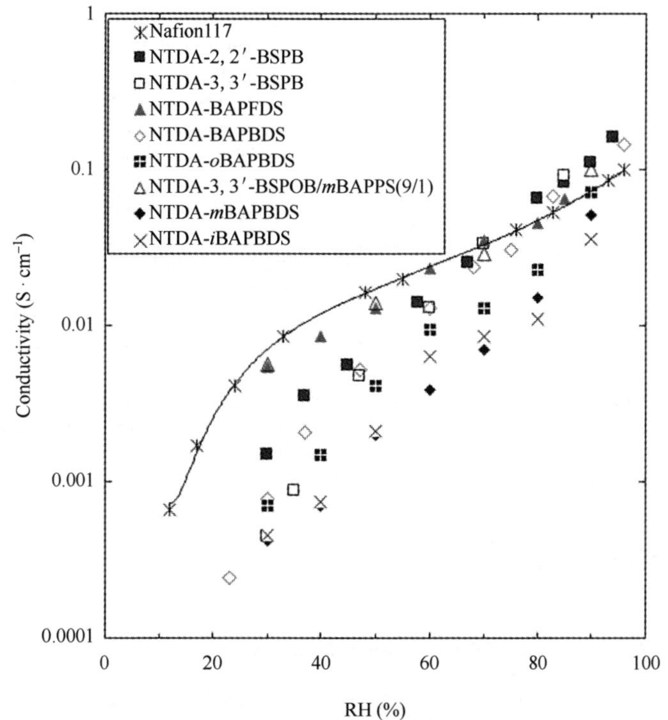

FIGURE 7.18 *In-plane* proton conductivities of some SPI membranes and Nafion 117 as a function of relative humidity (RH) at 50 °C.

Proton conductivity is also temperature-dependent and usually, proton conductivity increases with increasing temperature. In normal cases, proton conduction can be elucidated in terms of the Arrhenius equation:

$$\sigma=\sigma_0\exp(-E_a/RT) \tag{7.8}$$

where E_a is the activation energy of proton conduction, and σ_0, R and T are the preexponential factor, universal gas constant and absolute temperature, respectively.

From Eq. 7.8, it can be deduced that the natural logarithm of proton conductivity value is proportional to the reciprocal of temperature. Thus, plotting natural logarithm of proton conductivity versus reciprocal of temperature gives a straight line and the slope is equal to $-E_a/R$. This indicates that by measuring the temperature dependence of proton conductivity, the activation energy for proton conduction can be obtained.

Fig. 7.19 shows the variation of *in-plane* proton conductivity of some SPI membranes derived from the sulfonated diamines, 4,4′-bis(4-aminophenoxy)-3,3′-bis(4-sulfophenyl)biphenyl (BAPSPB) in deionized water as a function of temperature. It can be seen that for all the SPI membranes the conductivity increases with increasing temperature. The activation energies of the SPIs calculated from the Arrhenius equation are in the range 11-13 kJ • mol^{-1} which is similar to that of Nafion 117 (12 kJ • mol^{-1}) [45].

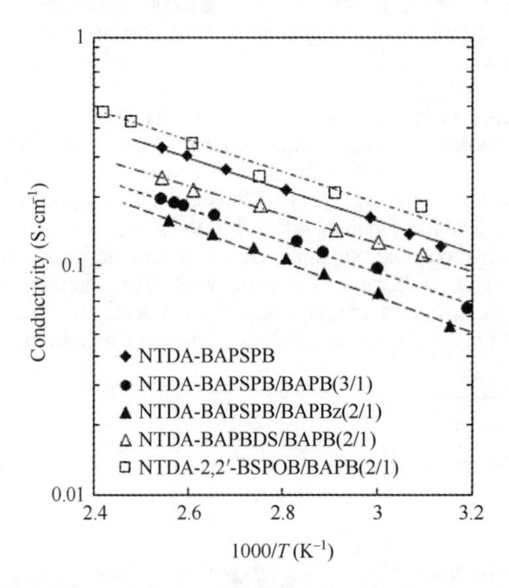

FIGURE 7.19 Variation of *in-plane* proton conductivities in water of BAPSPB, 2,2′-BSPOB and BAPBDS-based SPI copolymer membranes as a function of temperature.

Proton conductivity is closely related to membrane morphologies and many block copolymer membranes have been reported to show higher *in-plane* proton conductivity than their random analogs [15,16,44,50,64]. However, Okamoto and coworkers reported an important phenomenon that SPI membranes generally exhibited anisotropic proton conductivities, i.e., the through-plane conduct- ivity was significantly lower than that of *in-plane* conductivity and such a difference was larger for multiblock copolymer membranes than for random copolymer membranes [15,44]. The ratios of *through-plane* conductivity to *in-plane* conductivity are around 0.61-0.77 for BAPBDS and 2,2′-BSPOB-based random copolyimide membranes and 0.11-0.68 for the multiblock copolyimide membranes in their fully hydrated state [15]. In contrast, Nafion 112 exhibits isotropic proton conductivity, i.e., the *in-plane* conductivity (0.139 S • cm^{-1}, 60 °C, in water) and *through-plane* conductivity (0.136 S •cm^{-1}, 60 °C, in water) are almost identical [15]. The anisotropic proton conductivities of the random copolyimide membranes are likely related to the polymer chain alignment in the membrane plane direction due to the highly stiff polyimide backbones, while the microphase-separated structure with layer-like domains orientated in the plane direction of the multiblock copolyimide membranes should be responsible for the larger difference in their *in-plane* conductivities and *through-plane* conductivities.

7.4.5 Water Resistance

7.4.5.1 Methods for Evaluation of Water Resistance

Water resistance, or water stability of PEMs, is an extremely important performance which is closely related to fuel cell durability. The water resistance of SPI membranes is usually tested by soaking them in deionized water at elevated temperatures and evaluated either by the elapsed time when they started to lose mechanical properties or by the loss in sample weight, polymer molecular weight, viscosity, and tensile strength for a given soaking time. In early studies, Mercier and coworkers reported a criterion for the judgement of the loss of mechanical properties that the SPI membranes became extremely brittle at 80 °C [1]. Later, Okamoto and coworkers reported a much stricter criterion that the membranes were broken when lightly bent [3,5-10,45]. The water resistance of the SPI membrane NTDA-BDSA/ODA(1/1) at 80 °C, e.g., is around 1000 h by Mercier's criterion but only 5.5 h by Okamoto's criterion. Viscosity loss and tensile strength loss are more precise and effective methods. The viscosity loss method is suitable for organo-soluble SPI samples, while the mechanical strength loss method is applicable to all kinds of samples. Generally, the viscosity loss method is more accurate than the weight loss method for detecting polymer degradation because the products due to degradation may be insoluble in water. The molecular weight loss method has hardly been used because most SPIs are insoluble in tetrahydrofuran (THF) or *N*,*N*-dimethylformamide (DMF). In addition, for highly water-stable SPI membranes, the aging test is often performed at high temperatures such as 130 °C or 140 °C in an autoclave to save test time [12,25,86].

7.4.5.2 Effect of IEC

In many publications, for SPI membranes the concept of water resistance is considered to be identical to the concept of hydrolytic stability. However, this is not always true. In fact, one of the following situations may occur during the process of soaking an SPI membrane in deionized water at elevated temperatures: (1) they are hydrolyzed into low-molecular-weight species and completely dissolved; (2) they are hydrolyzed into low-molecular-weight species leading to a dramatic loss in mechanical properties, but membrane integrity was maintained; (3) they are hardly hydrolyzed but completely dissolved due to good solubility in deionized water; (4) they are hardly hydrolyzed and meanwhile membrane integrity was maintained. Only the last situation, i.e., the membranes having both excellent hydrolytic stability and excellent resistance to swelling/dissolution in water can possibly meet the requirements for long-term durability PEMFC applications. The behaviors of SPI membranes upon soaking in deionized water are closely dependent on their IEC, test temperatures, and polymer chemical structures, in particular the chemical structures of the sulfonated diamine moieties and the dianhydride moieties. Covalent crosslinking has also been identified to be a very effective method to improve SPI membrane water resistance [13,38,72,80]. In the following, the effects of various factors on water resistance of SPI membranes will be discussed in detail.

Because sulfonic acid groups are highly hydrophilic, it is easy to understand that the SPIs with high IECs generally exhibit poor water resistance and some of them are even water soluble. NTDA-BDSA, NTDA-ODADS, and NTDA-BAPFDS, e.g., are soluble in hot water because of their extremely high IECs (3.47-2.70 mol·g^{-1}) [3,5]. However, there are some exceptions. NTDA-2,2′-BSPB, e.g., could maintain reasonable mechanical properties (Okamoto's criterion) after being soaked in deionized water at 100 °C for 2500 h despite its very high IEC (2.89 mol·g^{-1}). This is mainly attributed to its excellent hydrolytic stability resulting from the very high basicity of the amino groups of the 2,2′-BSPB moiety as well as the nanophase-separated structure (sulfonic acid groups at the end of flexible side chains aggregate to form hydrophilic domains, while the aggregation of polymer backbones yields hydrophobic domains) [45]. Another possible and important reason is that the SPI membrane has unique laminate morphology resulting from its rod-like stiff backbone. The strong interchain and intrachain charge transfer complex interactions result in these rod-like chains being tightly packed which prevents the SPI from dissolving in water.

7.4.5.3 Hydrolytic Stability

Hydrolysis-induced polymer chain degradation is an extremely important issue for SPI membranes because imide rings generally have a tendency to hydrolyze. A possible hydrolytic reaction mechanism of SPI membranes (Fig. 7.20) has been proposed by Okamoto and coworkers [47]. Such a mechanism involves two key-step reactions: (1) water molecules attack the carbonyl carbons of the imide rings leading to ring disclosure and formation of amide linkages; (2) amide groups further hydrolyze leading to chain scission of the SPI membranes. Mercier and coworkers first reported that generally six-membered ring SPIs are much more stable to hydrolysis than the five-membered ring SPIs [1,2]. This can be explained by the above hydrolytic reaction mechanism. The electron-donating effect of a naphthalene ring is much stronger than that of a benzene ring because the former contains more π electrons than the latter. As a result, the carbonyl carbons of naphthalenic (six-membered ring) SPIs are less positive than that of the phthalic (five-membered ring) SPIs, i.e., the electrophilicity of naphthalenic carbonyl carbons is lower than that of phthalic carbonyl carbons. This makes it more difficult for water molecules to attack the naphthalenic carbonyl carbons than to attack phthalic carbonyl carbons (Fig. 7.21), and therefore naphthalenic SPIs generally exhibit much better hydrolytic stability than phthalic ones.

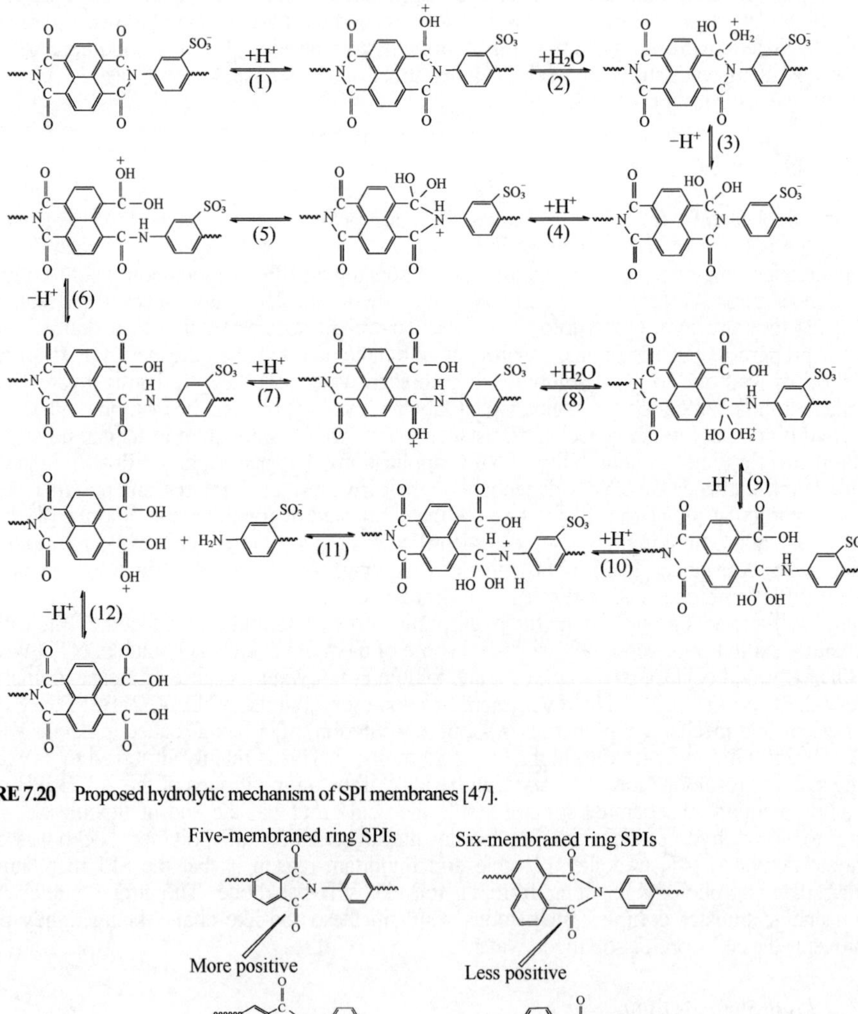

FIGURE 7.20 Proposed hydrolytic mechanism of SPI membranes [47].

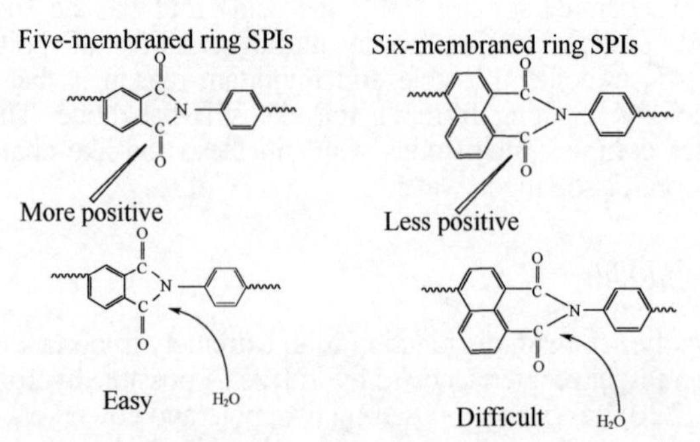

FIGURE 7.21 Difference in hydrolytic rate between the five-membered ring SPIs and the six-membered ring SPIs.

According to the above hydrolytic mechanism, the electron affinity of the positive carbonyl carbon atoms of imido rings is the crucial factor determining the hydrolytic stability of SPI membranes. The lower electron affinity (less positive) of the positive carbonyl carbon atoms of imido rings results in higher hydrolytic stability and vice versa. There are two ways to decrease the electron affinity of the positive carbonyl carbon atoms of imido rings: (1) to increase the basicity of the amino groups of the diamine moieties especially the sulfonated diamine moieties, and (2) to decrease the carbonyl group density of the dianhydride moieties. Okamoto and coworkers have synthesized many sulfonated diamine monomers and systematically studied the effects of the chemical structures of sulfonated diamines on water resistance of SPI membranes [3,5-10,45,47]. They found that the SPI membranes derived from the sulfonated diamines with high basicity indeed exhibited much better hydrolytic stability than those derived from the sulfonated diamines with low basicity [7,8,10,45,47]. A good example is the SPIs synthesized from NTDA and four isomeric sulfonated diamines: BAPBDS, *i*BAPBDS, *m*BAPBDS, and *o*BAPBDS (for structures, see Fig. 7.22). First, let us look at the water resistance of NTDA-BAPBDS and NTDA-*i*BAPBDS. The NTDA-BAPBDS displayed a water stability of 1000 h at 100 °C which is much longer than that of NTDA-*i*BAPBDS at 80 °C (80 h). Please note that BAPBDS and *i*BAPBDS have exactly the same chemical structure except for the sulfonic acid group bond position. For BAPBDS, the sulfonic acid groups are attached to the central biphenyl unit, while for *i*BAPBDS the sulfonic acid groups are attached to the two end benzene rings where the amino groups are attached. Because sulfonic acid itself is a strong electron-withdrawing group, the amino groups of BAPBDS are much more basic than that of *i*BAPBDS. Therefore, it is reasonable to ascribe the higher water stability of NTDA-BAPBDS to the higher basicity of the amino groups of the BAPBDS moiety. Unlike NTDA-BAPBDS, NTDA-*m*BAPBDS and NTDA-*o*BAPBDS exhibit extremely poor water resistance, i.e., NTDA-*m*BAPBDS is soluble in hot water (80 °C), while NTDA-*o*BAPBDS is even soluble in water at room temperature [10]. This does not mean that NTDA-*m*BAPBDS and NTDA-*o*BAPBDS have poor hydrolytic stability. In fact, after they were recovered from their water solutions via vacuum distillation, tough films were obtained again with both samples via solution cast, indicating that no serious hydrolysis-induced polymer degradation occurred. The poor water resistance of these two SPI membranes is due to their good solubility in water resulting from the nonlinear (twisty) configuration of *m*BAPBDS and *o*BAPBDS and high IEC [47]. Numerous studies have identified that the polyimides derived from the diamines with nonlinear configuration (*meta*-and *ortho*-substituted) exhibited much better solubility in organic solvents than the corresponding analogs derived from the one with linear configuration structure (*para*-substituted). This is likely because solvent molecules can more readily access the free volume spaces of the polyimides derived from the diamines with nonlinear configuration resulting from the less regular packing of polymer chains. Since *o*BAPBDS has a higher twisty degree than *m*BAPBDS, NTDA-*o*BAPBDS is easier to dissolve in water. Decreasing the IEC of *m*BAPBDS- and *o*BAPBDS-based SPI membranes by copolymerization with a nonsulfonated diamine cause a great improvement in water resistance. For example, NTDA-*o*BAPBDS/*m*BAPPS(2/1) having a theoretical IEC of 1.83 mol·g^{-1} (NTDA-*o*BAPBDS: 2.63 mol·g^{-1}) exhibited a fairly good water resistance of 1000 h at 80 °C [10], while NTDA-*m*BAPBDS/BAPB(3/2) could maintain fairly high mechanical properties even after being soaked in pressurized water vapor at 130 °C for 96 h [10]. These results further confirmed the good hydrolytic stability of the *m*BAPBDS- and *o*BAPBDS-based SPI membranes resulting from the high basicity of the amino groups of the *m*BAPBDS and *o*BAPBDS moieties. By far, BAPBDS-based SPI membranes represent the most water-stable PEMs among the main-chain-type SPI membranes. However, it should be noted that the above discussed improvement in hydrolytic stability of the SPI membranes derived from the sulfonated diamines with high basicity of their amino groups does *not* mean any occurrence of hydrolysis-induced polymer degradation, but such a degradation rate is greatly decreased. Under harsh aging conditions (high temperatures and long aging times), the effect of hydrolysis-induced polymer degradation is still significant even for BAPBDS-based SPI membranes. For example, Okamoto and coworkers found that for NTDA-BAPBDS/BAPB (2/1) after being soaked in deionized water at 100 °C for 300 h the tensile strength, EB and tensile modulus decreased from 120 to 49 MPa, 120% to 8%, and 1.8 to 1.6 GPa, respectively. The significant loss in mechanical properties should be ascribed to the hydrolysis-induced polymer degradation, though the degradation rate is rather slow. A similar phenomenon has been observed with NTDA-*m*BAPBDS/BAPB(3/2). The stress-strain curves of NTDA-BAPBDS/BAPB(2/1) and NTDA-*m*BAPBDS BAPB(3/2) before and after the aging test are shown in Fig. 7.23 [47].

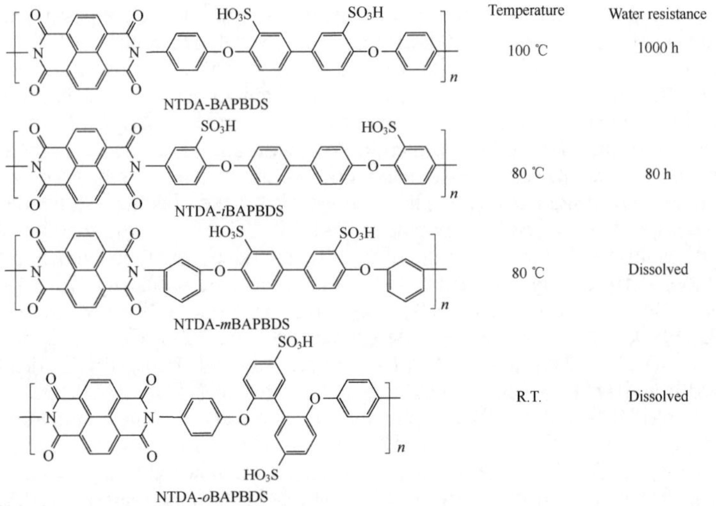

	Temperature	Water resistance
NTDA-BAPBDS	100 ℃	1000 h
NTDA-*i*BAPBDS	80 ℃	80 h
NTDA-*m*BAPBDS	80 ℃	Dissolved
NTDA-*o*BAPBDS	R.T.	Dissolved

FIGURE 7.22 Chemical structures of the SPIs derived from four sulfonated isomeric diamine monomers.

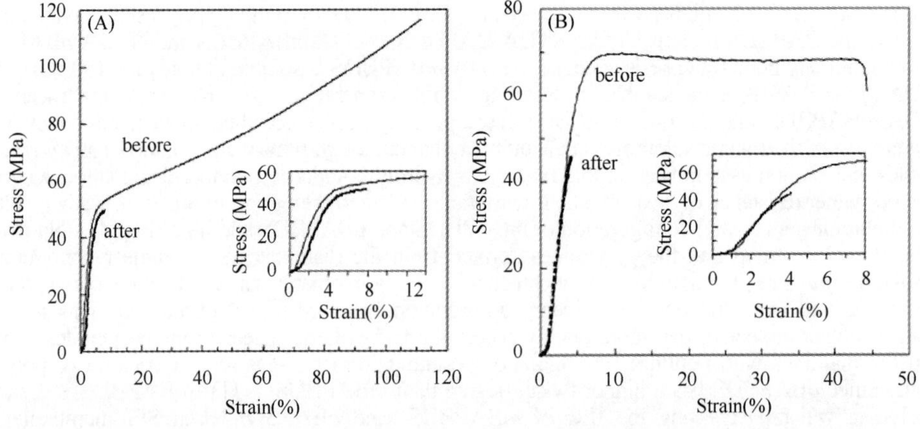

FIGURE 7.23 Stress-strain curves before and after aging test: (A) NTDA-BAPBDS/BAPB(2/1) in water at 100 °C for 300 h, (B) NTDA-*m*BAPBDS/BAPB(3/2) in water vapor at 130 °C for 96 h.

Side-chain-type SPI membranes generally exhibit better water resistance than main-chain-type ones because the former readily form microphase-separated structures and water molecules mainly surround the sulfonic acid groups which are relatively far away from the imido rings in polymer backbones. 2,2′-BSPB and 2,2′-BSPOB are two typical side-chain-type sulfonated diamines and the relevant SPI membranes exhibit the best water resistance by far. As mentioned above, NTDA-2,2′-BSPB exhibits an extremely high water stability of 2500 h at 100 °C. NTDA-2,2′-BSPOB/BAPB(2/1) could also maintain reasonable mechanical properties after being soaked in deionized water at 130 °C for 500 h [12]. These sulfonated diamines have the same rigid benzidine backbone and the highly basic amino groups due to the presence of electron-donating propoxy groups and phenoxy groups. Theoretically, the electron-donating effect of a propoxy group is stronger than that of a phenoxy group. As a result, 2,2′-BSPB-based SPI membranes are expected to have even better hydrolytic stability than 2,2′-BSPOB-based ones. However, it is well known that aliphatic ether bonds are much less stable to hydrolysis at elevated temperatures than aromatic ether bonds. Okamoto and coworkers indeed found that the cleavage of ether bond of sulfopropoxy group took place fairly easily at 130 °C for the 2,2′-BSPB-based SPIs, whereas the cleavage of the wholly aromatic ether bond hardly occurred at the same temperature for the BAPBDS-based SPIs [106]. And therefore, 2,2′-BSPOB should be an even better sulfonated diamine monomer than 2,2′-BSPB. It should be noted that the molecular configuration of sulfonated diamines also has large effects on the water resistance of side-chain-type SPI membranes. The membranes derived from the sulfonated diamines with linear configuration (*para*-substitution) tend to have high water resistance, whereas the ones derived from the sulfonated diamines with nonlinear configuration (*meta*-and *ortho*-substitution) tend to have low water resistance. For example, the SPI membrane prepared from NTDA and 3-(2′,4′-diaminophenoxy)propane (DAPPS) (Fig. 7.24) exhibits a water resistance of only 200 h at 80 °C despite the high basicity of the amino groups resulting from the strong electron-donating effects of propoxy groups [8]. This should be related to the nonlinear configuration structure of the DAPPS moiety. Among the BSPB and BSPOB-based SPIs, 2,2′-BSPB and 2,2′-BSPOB give better water resistance than 3,3′-BSPB and 3,3′-BSPOB. As mentioned above, NTDA-2,2′-BSPB exhibits a very high water resistance of 2500 h at 100 °C which is significantly longer than that of NTDA-3,3′-BSPB under the same aging conditions (700 h) [45]. A similar phenomenon has been observed with 2,2′-BSPOB and 3,3′-BSPOB-based SPI membranes [9,12-15,105]. This is probably due to the fact that 3,3′-BSPB and 3,3′-BSPOB-based SPIs have the substituents in the *ortho*-position of amino groups, which causes less dense packing of polymer chains and thus larger free volume leading to relatively weaker interchain interaction. Okamoto and coworkers reported that the density of NTDA-3,3′-BSPB membrane was 1.402 $g \cdot cm^{-3}$ at 70% RH or 1.130 $g \cdot cm^{-3}$ in water which is slightly lower than that of NTDA-2,2′-BSPB in water (1.438 $g \cdot cm^{-3}$ at 70% RH or 1.142 $g \cdot cm^{-3}$) [45]. This is a solid evidence of less dense packing of NTDA-3,3′-BSPB chains than NTDA-2,2′-BSPB.

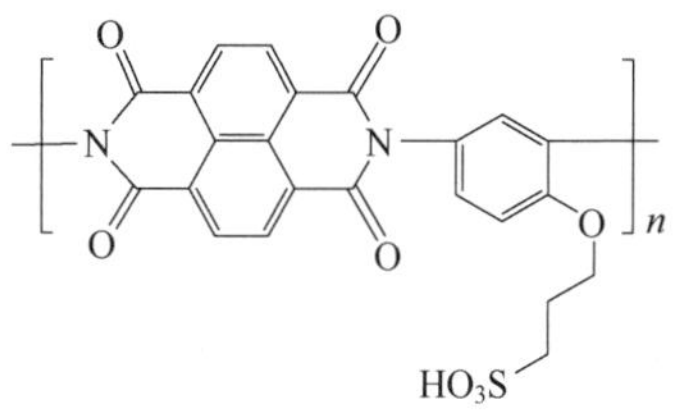

FIGURE 7.24 Chemical structure of NTDA- DAPPS.

Decreasing the carbonyl group density of the dianhydride moieties is another effective method of enhancing SPI hydrolytic stability. Ding and coworkers first reported that the SPIs derived from 4,4′-binaphthyl- 1,1′,8,8′-tetracarboxylic dianhydride (BNTDA) showed much better hydrolytic stability than the corresponding NTDA-based ones because of the lower electron affinity of the BNTDA moiety [107]. For example, for the copolyimide membrane BNTDA-ODADS/ODA (3/1) (for the structure, see Fig. 7.25) the tensile strength only slightly decreased from 101 to 94 MPa, EB from 35% to 16%, and tensile modulus from 1.46 to 1.26 GPa after the aging test in deionized water at 100 °C for 800 h, indicating excellent hydrolytic stability of this SPI membrane [107]. However, as mentioned above, the corresponding NTDA-based analog, NTDA-ODADS/ODA (1/1), exhibits a fairly poor hydrolytic stability (25 h, 80 °C) [3]. Fang and coworkers reported the synthesis of a dianhydride monomer, benzophenone-4,4′-bis(4-thio-1,8-naphthalic anhydride) (BPBTNA) and a series of SPIs derived from this dianhydride monomer, ODADS and ODA (for structures, see Fig. 7.26) [87,88]. Comparing with NTDA, BPBTNA also shows lower electron affinity due to the lower carbonyl group density as well as the electron-donating effect of the sulfide linkages. As expected, the BPBTNA-based SPI membranes displayed excellent hydrolytic stability. For example, for the BPBTNA-ODADS/ODA(6/4), BPBTNA-ODADS/ODA(8/2), and BPBTNA-ODADS/ODA(9/1) no reduction in tensile strength and only slight reduction in EB and tensile modulus were observed after the aging test in deionized water at 140 °C for 24 h [87,88]. Our research group synthesized a series of ether-type dianhydride monomers from 4-bromo-1,8-naphthalic anhydride and various dibasic phenols (Fig. 7.27). Polymerization of these dianhydride monomers with common diamines such as 4,4′-diaminodiphenyl ether (ODA), 1,3-bis(4-aminophenoxy)benzene (BAPBz), and 9,9-bis(4-aminophenyl)fluorene (BAPF) yielded various high-molecular-weight non-SPIs. The electron-donating effects of ether bonds are expected to be very favorable for enhancing the hydrolytic stability of the polyimides. Indeed, an aging test in 10% NaOH aqueous solution in an autoclave at 140 °C for 50 h revealed that these polyimides displayed only moderate loss in viscosity (1.7%-16%) and tensile strength (2.3%-29%), indicating excellent stability under such harsh conditions (unpublished data). It is well known that conventional polyimides are very unstable in alkaline media. For example, Kapton completely dissolved in 10% NaOH aqueous solution in an autoclave at 140 °C within a few minutes.

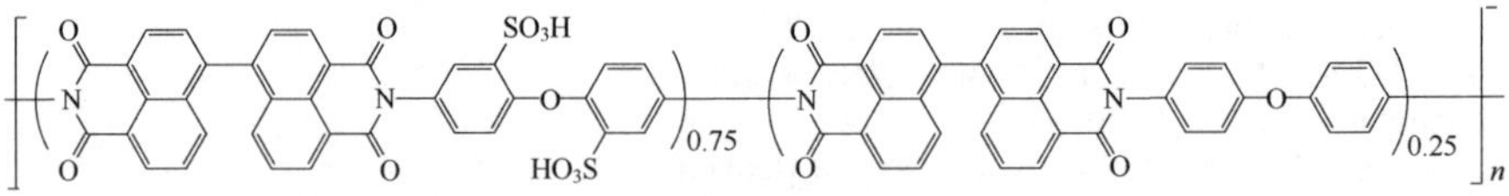
SO3H
HO3S
0.75
0.25
n
BNTDA-ODADS/ODA(3/1)

FIGURE 7.25 Chemical structure of a sulfonated copolyimide derived from BNTDA.

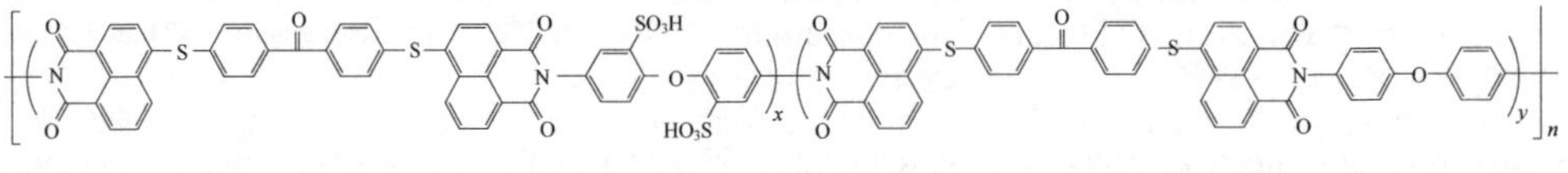
SO3H
HO3S
x
y
n
BPBTNA-ODADS/ODA(x/y)
(x/y = 6/4, 8/2, 9/1)

FIGURE 7.26 Chemical structures of the SPIs derived from BPBTNA.

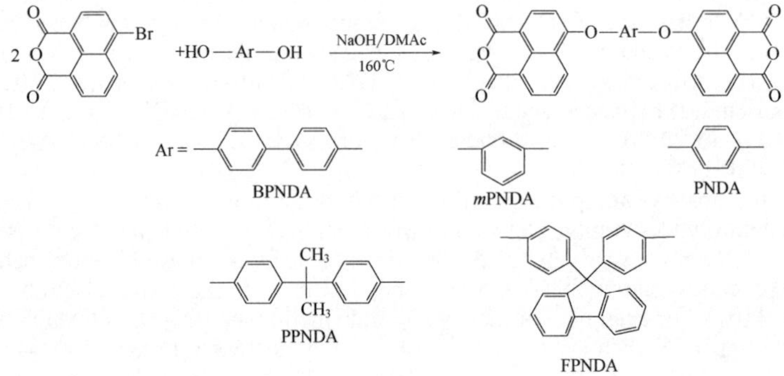
2
Br
+HO—Ar—OH
NaOH/DMAc
160℃
O—Ar—O
Ar =
BPNDA
mPNDA
PNDA
CH3
CH3
PPNDA
FPNDA

FIGURE 7.27 Synthesis of ether-type naphthalenic dianhydride monomers.

7.4.5.4 Effect of Crosslinking in SPI

Besides structural optimization, covalent crosslinking is another important method of enhancing water stability of SPI membranes. It is also a common method to suppress membrane swelling degree and to improve membrane mechanical properties. In the past decades, a large number of crosslinked sulfonated polymer membranes have been developed by many research groups. Kerres and coworkers extensively studied the preparation of various ionically crosslinked acid-base blends and ionomermembranes (acid: SPES, SPEEK, sulfonated poly(phenylene oxide); base: polybenzimidazole), covalently crosslinked membranes through the reaction between sulfinate groups of sulfonated polymers and alkyl or aryl dihalides (crosslinkers), and covalent-ionically crosslinked membranes. However, just as Kerres pointed out in the review article [108], the ionically crosslinked acid-base (blend) membranes showed unacceptable swelling in water above 70-90 °C leading to possible destruction of the membranes in fuel cell operation. The covalently crosslinked (blend) membranes are fairly stable but have the disadvantage of much effort involved in materials synthesis. Pintauro et al.reported on a photocrosslinking method on the basis of the photochemical reaction between the α-methyl groups of the sulfonated polyphosphazene and benzophenone (photoinitiator) under the exposure of UV [109]. However, it is well known that α-methyl-group-containing polymers such as sulfonated polystyrene are unstable toward radical oxidation. Guiver and coworkers developed a facile crosslinking approach based on the esterification reaction between the sulfonic acid groups of SPEEK and the hydroxyl groups of polyatomic alcohols [110]. This method seems, in principle, applicable to all kinds of sulfonated polymers, but no information on the water stability of the resulting crosslinked membranes is available. Theoretically, the resulting sulfonic acid ester linkages can still undergo hydrolysis and therefore they are not expected to be very stable. Han and coworkers prepared a series of maleimido-end-capped SPI oligomers which could undergo self-crosslinking reaction in the presence of an initiator, 4,4′-azobis(4-cyanovaleric acid), and a diacrylate compound (another crosslinker). However, these crosslinked membranes displayed rather poor water resistance (around 200 h at 80 °C) probably because of the poor hydrolytic stability of the maleimido rings (the electron affinity of maleimido carbonyl carbon is much higher than that of naphthalenic imido carbonyl carbon) [111]. Okamoto's research group prepared a series of interesting branched/crosslinked SPI membranes on the basis of the reaction between the anhydride end groups of SPI oligomers and the amino groups of a triamine monomer. These SPI membranes showed good water stability as well as high proton conductivity [19]. Lee and coworkers prepared a series of crosslinked SPI membranes based on the chemical reaction between the carboxyl groups from a nonsulfonated diamine moiety and the hydroxyl groups of *N,N*-bis(2-hydroxyethyl)-2-aminothanesulfonic acid (crosslinker) [112]. However, the demerit of this crosslinking method is that the ester linkages of the crosslinked SPI membranes are not very stable (hydrolyze at elevated temperatures). Our groups developed a facile crosslinking technique without the need for the introduction of special crosslinking groups to polymer structure. The chemical structures of the SPIs used for the crosslinking treatment are shown in Fig. 7.28. Two typical experimental procedures described in the following have been developed for preparation of the crosslinked SPI membranes [38].

Method 1: Dry SPI membranes in their proton form were immersed into Eaton's reagent (phosphorus pentoxide/methanesulfonic acid, 1/10 by weight, PPMA) in a glass vessel at 80 °C for 3-48 h under nitrogen atmosphere. The membranes were taken out, thoroughly rinsed with deionized water until the water was neutral and dried in vacuo at 100 °C for 20 h.

Method 2: Dry SPIs in their proton form were dissolved in DMSO containing 5%(w) phosphorus pentoxide at room temperature. The weight ratio between SPI and phosphorus pentoxide was controlled at 1:1. The solution mixture was cast onto glass plates and dried at 80 °C for 10 h in an air oven. The glass plates were moved to a vacuum oven and dried at 80 °C for 1 h, 120 °C for 2 h, and 170 °C for 10 h, successively. The membranes were thoroughly rinsed with deionized water until the water was neutral and then dried in vacuo at 100 °C for 20 h.

For both methods, the crosslinking reaction is based on the chemical reaction between the sulfonic acid groups of sulfonated diamine moiety and the activated hydrogen atoms of nonsulfonated diamine moiety (BAPF) in the presence of a condensation agent (PPMA in method 1 or phosphorus pentoxide in method 2), and the resulting crosslinking bonds are the very stable sulfonyl groups (Fig. 7.29). The crosslinking sites are 2,7-positions of BAPF because these two positions are highly reactive. Such a crosslinking mechanism has been confirmed by FT-IR spectra and model compound synthesis [38].

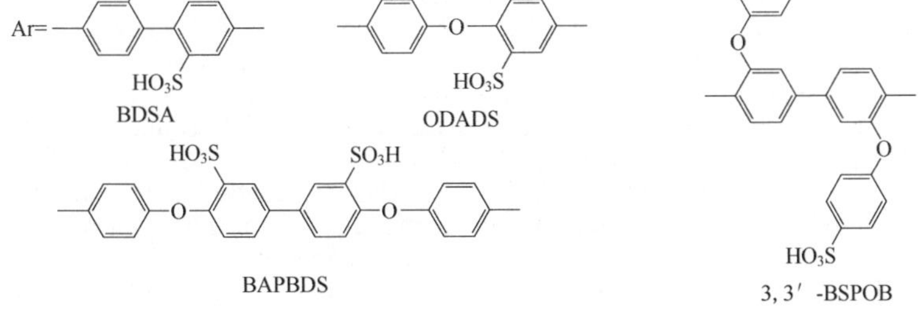

(x/y =3/1 or 5/1 or 9/1)

Ar=

BDSA

ODADS

BAPBDS

3,3′-BSPOB

FIGURE 7.28 Chemical structures of the SPIs used for the covalent crosslinking treatment.

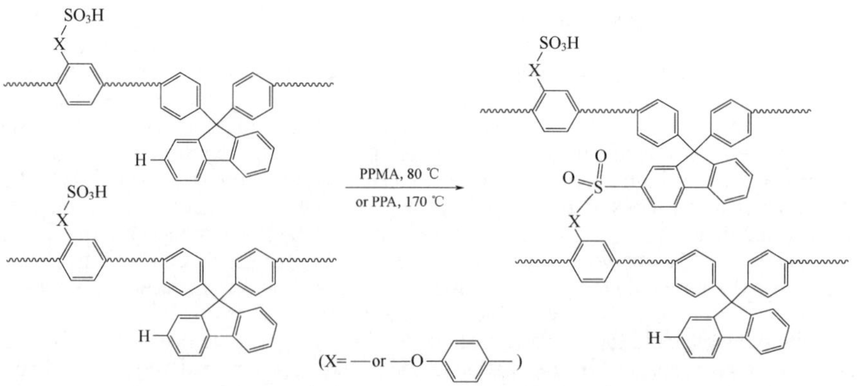

FIGURE 7.29 PPMA and PPA-catalyzed covalent crosslinking reaction.

Method 1 is suitable for the crosslinking of the SPIs which are insoluble in PPMA but can absorb this medium rapidly, while method 2 is applicable to the SPIs in their proton form which are soluble in common organic solvents (inert to phosphorus pentoxide) such as DMSO. The occurrence of crosslinking is readily judged from the insolubility of the membranes in organic solvents (DMSO and *m*-cresol) in which the membranes are well soluble before the crosslinking. For method 1, because the absorption rate of PPMA in SPI membranes is far faster than the crosslinking rate, membrane thickness (< 100 μm) shows little effect on crosslinking rate. Table 7.5 shows the crosslinking conditions for the SPI membranes. It can be seen that, for method 1, the crosslinking rate is in the order: NTDA-3,3′-BSPOB/BAPF(9/1) >> NTDA-BAPBDS/BAPF(3/1) > NTDA-BDSA/BAPF(3/1), NTDA- ODADS/ BAPF(3/1), and NTDA-ODADS/BAPF(5/1). This indicates that the crosslinking rate is mainly dependent on the chemical structure of the sulfonated diamine moiety. 3,3′-BSPOB is a side-chain-type sulfonated diamine and its sulfonic acid groups are at the end of the side chains without any adjacent (*ortho-*) substituent groups leading to little steric effect to crosslinking reaction. This is the main reason that NTDA-3,3′-BSPOB/BAPF(9/1) has a much faster crosslinking rate than the others. The mobility of the side chains resulting from the flexible ether bonds might also contribute to the fast crosslinking rate of this SPI membrane. Unlike 3,3′-BSPOB, BAPBDS, ODADS, and BDSA are main-chain-type sulfonated diamines and significant steric effects exist when the sulfonic acid groups approach the reactive sites (2 or 7 positions) of BAPF moieties (crosslinking reaction), and therefore the reactivity of the SPIs derived from these sulfonated diamines is significantly lower than that of the SPI derived from 3,3′-BSPOB. The higher crosslinking rate of NTDA-BAPBDS/BAPF (3/1) than that of NTDA-BDSA/BAPF(3/1) and NTDA-ODADS/BAPF(3/1) may also be explained by the slight difference in the reactivity of sulfonic acid groups of different sulfonated diamine moieties. According to Ueda's discovery, for the reactions of aryl sulfonic acid compounds with aromatic hydrocarbons electron-rich phenyl rings to which sulfonic acid groups are attached are favorable for the formation of diaryl sulfone products, i.e., electron-donating effect leads to an enhancement in reactivity of sulfonic acid groups. In the present case, the electron-donating effect of the phenyl rings to which sulfonic acid groups are attached is stronger for BAPBDS than for ODADS and BDSA leading to higher reactivity of the former. In addition, crosslinking rate hardly depends on the molar fraction of BAPF because the same crosslinking rate was observed for NTDA-ODADS/BAPF (3/1) and NTDA-ODADS/BAPF (5/1). For method 2, all the SPIs except NTDA-BAPBDS/BAPF (3/1) could be readily crosslinked under the same conditions. The inapplicability of method 2 to NTDA-BAPBDS/BAPF (3/1) is because of the insolubility of this polymer in DMSO. The crosslinking reaction mainly occurred during the process of thermal treatment at 170 °C for 10 h. In comparison with method 1, method 2 needs shorter crosslinking time because of the higher crosslinking temperature (170 °C vs 80 °C). It should be noted that for method 2 the effect of thermal treatment time on crosslinking formation has not been investigated in detail. NTDA-BSPOB/BAPF (9/1) might be readily crosslinked in a short period of time (<< 10 h) via the technique of method 2.

TABLE 7.5 Solubility Changes of SPI Membranes Before and After Immersing Into PPMA at 80 °C

Membrane	Method	Crosslinking Treatment	Solubility[a]	
			m-Cresol	DMSO
NTDA-3,3′-BSPOB/BAPF(9/1)		No	+	+
	1	3 h	–	–
	2	10 h	–	–
NTDA-BAPBDS/BAPF(9/1)		No	+	–
	1	10 h	+–	+–
	1	24 h	–	–
NTDA-ODADS/BAPF(5/1)		No	+	+
	1	24 h	+–	+–
	1	48 h	–	–
	2	10 h	–	–
NTDA-ODADS/BAPF(3/1)		No	+	+
	1	24 h	+–	+–
	1	48 h	–	–
	2	10 h	–	–
NTDA-BDSA/BAPF(3/1)		No	+	+
	1	24 h	+–	+–
	1	48 h	–	–
	2	10 h	–	–

[a]*Key: "+", soluble; "+–," partially soluble; "–," insoluble on heating.*

Table 7.6 shows the IEC, WU at 100 °C, tensile stress (TS), EB, proton conductivity (σ) measured in water at 60 °C, and water stability of the crosslinked and the noncrosslinked SPI membranes. Obviously the crosslinking treatment caused great improvements in water stability of the SPI membranes. The noncrosslinked NTDA-3,3′-BSPOB/BAPF (9/1) and NTDA-BAPBDS/BAPF (3/1) exhibit a similar water stability of 150 h at 100 °C. However, after crosslinking, both membranes exhibit a greatly improved water stability which is more than 1 month (the aging test was ceased after 720 h) at 100 °C. In fact, after this aging test the crosslinked membranes were still tough and had a tensile strength of about 10 MPa in a fully hydrated state, which should be high enough for fuel cell use. For NTDA-ODADS/BAPF (5/1) and NTDA-BDSA/BAPF (3/1), their water stability dramatically increased from 0.5 h and 10 min to >720 h and 400 h at 100 °C, respectively, due to the crosslinking treatment. Moreover, it is worth noting that the crosslinking can not only improve the water stability, but can also reduce the swelling ratio and enhance the TS of the hydrate membranes which are favorable for improving membrane durability. However, this crosslinking technique has a demerit of reduction in proton conductivity because part of the sulfonic acid groups had been consumed due to the crosslinking. This negative effect can be minimized by controlling the crosslinking density at a reasonable level.

TABLE 7.6 IEC, Water Uptake (WU) at 100 °C, Tensile Stress (TS), Elongation at Break (EB), Proton Conductivity (σ) Measured in Water at 60 °C and Water Stability of the Crosslinked and the Noncrosslinked SPI Membranes

Membrane	Crosslinking Treatment	*IEC* ($mol \cdot g^{-1}$)	W_a(%)	*TS*[c] (MPa)	EB[c] (%)	σ ($S \cdot cm^{-1}$)	Water Stability (h)
NTDA-	No	2.02	130	114 (44)	5.5 (10)	0.21	150
3,3′-BSPOB/BAPF (9/1)	Yes (Method 1)	1.96	60	128 (62)	13 (13)	0.16	>720
NTDA-	No	2.16	110	59 (19)	14 (5.0)	0.16	150
BAPBDS/BAPF (3/1)	Yes (Method 1)	1.80	90	44 (24)	21 (9.5)	0.12	>720
NTDA-	No	2.82[a]	120[b]	70 (8.0)	15 (1.8)	0.22	0.5
ODADS/BAPF (5/1)	Yes (Method 1)	2.08	85	33 (11)	22 (5.2)	0.13	>720
NTDA-	No	2.55	200[b]	97 (18)	11 (8.7)	NM	10 min
BDSA/BAPF (3/1)	Yes (Method 2)	2.11	120	77 (27)	8.3 (14)	0.15	400

[a]*Theoretical value.*
[b]*Measured at room temperature.*
[c]*The data in parenthesis refer to wet membranes.*

Besides the above-mentioned crosslinking media (PPMA and phosphorus pentoxide), polyphosphoric acid (PPA) is also a good crosslinking medium and various crosslinked SPI membranes have been successfully prepared by immersing the SPI membranes in their proton form in PPA at elevated temperatures (e.g., 180 °C) in our lab. The crosslinking mechanism is also based on the condensation reaction of sulfonic acid groups of SPIs with the activated phenyl rings of the nonsulfonated diamine moieties yielding the highly stable sulfonyl groups as discussed previously. SPIs are generally insoluble in PPA but can absorb considerable amounts of PPA at elevated temperatures. Fig. 7.30 shows the chemical structures of some SPIs which were crosslinked via this method in our lab. They were synthesized by random copolymerization of NTDA with the sulfonated diamine BAPBDS, two nonsulfonated diamine BAPF and 1,3-bis (4-aminophenoxy) benzene (BAPBz), and two benzimidazole ring-containing diamines, 2,2′-bis (4-aminophenyl)-5, 5′-bibenzimidazole (BAPBI) and 2,2′-bis(*1H*-benzimidazol-2-yl) benzidine (BBIBz). BAPBI was synthesized by one-step reaction of 4-aminobenzoic acid and 3,3′- diaminobenzidine (DABz) in PPA at 190 °C [113], while BBIBz was synthesized by multiple-step reactions using 2,2′-biphenyldicarboxylic acid as the starting material (Fig. 7.31). The purpose of incorporating benzimidazole ring-containing diamines into polymer structure is to improve the radical oxidative stability of the SPI membranes which will be discussed in the following section. The crosslinking reaction was performed at 180 °C for 10 h and the crosslinking reaction occurred at 2 or 7 position of BAPF and *ortho*-positions of ether bonds of the central benzene ring of BAPBz moiety.

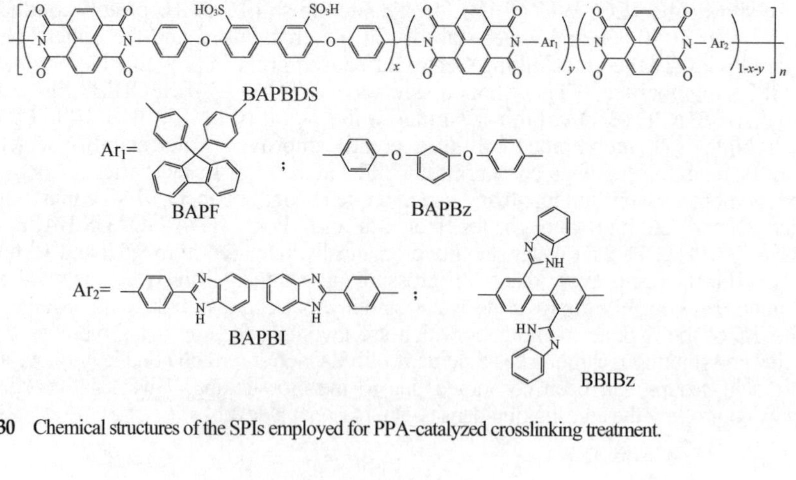

BAPBDS

Ar_1= BAPF ; BAPBz

Ar_2= BAPBI ; BBIBz

FIGURE 7.30 Chemical structures of the SPIs employed for PPA-catalyzed crosslinking treatment.

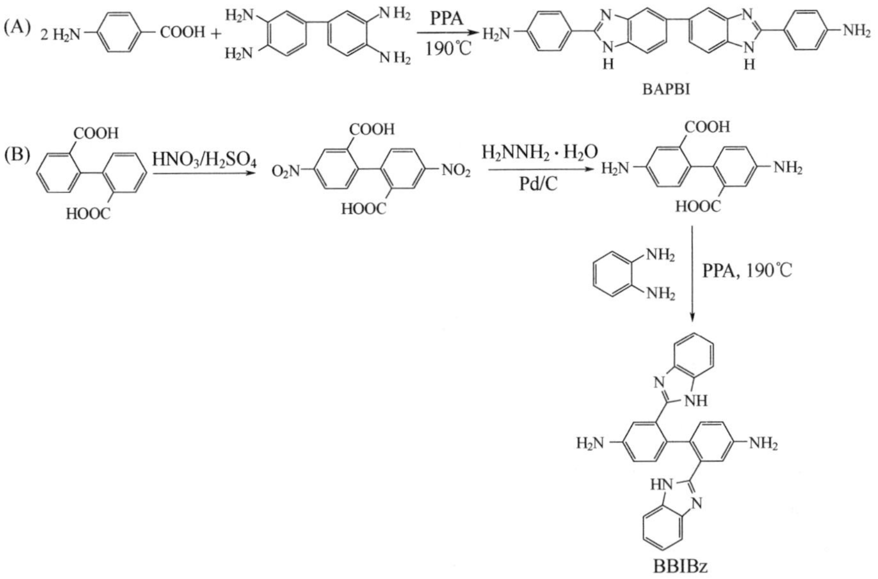

(A) 2 H2N–C6H4–COOH + (H2N)2C6H3–C6H3(NH2)2 —PPA, 190℃→ BAPBI

(B) HNO3/H2SO4; H2NNH2·H2O, Pd/C; PPA, 190℃ → BBIBz

FIGURE 7.31 Synthesis of two benzimidazole groups-containing diamines, BAPBI, and BBIBz.

Table 7.7 shows the PPA uptake and dimensional changes of the SPI membranes after they were immersed in PPA at 180 °C for 10 h. All the membranes showed exceptionally high PPA uptake (800%-1200 %(w)) due to their highly hydrophilic feature as well as the presence of benzimidazole groups. It is interesting that despite the extremely high PPA uptake these SPI membranes could maintain good mechanical properties with little dimensional changes in the inplane direction. The dimensional changes were observed in the through-plane direction (100%-200%). Such a sorption phenomenon is quite different from that of polybenzimidazole membranes. These PPA-doped SPI membranes might be applicable to high-temperature PEMFCs.

The water stability of the crosslinked and the noncrosslinked SPI membranes was examined by immersing them in deionized water at 100 °C for 30 h and evaluated by the changes of tensile properties of the wet membranes. As shown in Table 7.8, for all the membranes the water stability was improved due to the crosslinking treatment. However, it should be noted that even for the crosslinked membranes, the water stabilities are not very good. This is because the protonated benzimidazole (benzimidazolium) groups are extremely strong electron-withdrawing groups which drastically deteriorate the hydrolytic stability of the SPI membranes. To further improve the hydrolytic stability, replacing the benzimidazole-containing diamine with an amine-terminated polybenzimidazole oligomer is an effective method which will be discussed in the following section.

TABLE7.7 PPA Uptake and Dimensional Change of Various SPI Membranes in PPA at 180 °C for 10 h

SPI	PPA Uptake(%(w))	Dimensional Change (%)	
		In-plane	Through-plane
NTDA-BAPBDS/BAPBz/BAPBI(5/1/1)	1100	0	160
NTDA-BAPBDS/BAPF/BAPBI(5/1/1)	900	0	185
NTDA-BAPBDS/BAPF/BAPBI(3/1/1)	1100	0	200
NTDA-BAPBDS/BAPF/BBIBz(9/1/1)	800	0	133
NTDA-BAPBDS/BAPF/BBIBz(5/1/1)	1100	0	100
NTDA-BAPBDS/BAPBz/BBIBz(3/1/1)	1200	0	200

TABLE7.8 Tensile Stress (*TS*) and Elongation at Break (*EB*) of the Crosslinked and Noncrosslinked SPI Membranes in Their Fully Hydrate State Before and After the Aging Test (100 °C/30 h)

Membrane	Crosslinking Treatment	*TS* (MPa)		*EB* (%)	
		Before	After	Before	After
NTDA-BAPBDS/BAPBz/BAPBI (5/1/1)	No	56	4.6	78	1.0
	Yes	48	18	51	3.3
NTDA-BAPBDS/BAPBz/BBIBz (3/1/1)	No	58	30	51	5.0
	Yes	50	36	11	5.0
NTDA-BAPBDS/BAPF/BAPBI (5/1/1)	No	42	13	30	2.8
	Yes	21	18	11	3.4
NTDA-BAPBDS/BAPF/BAPBI (3/1/1)	No	37	4.2	15	0.67
	Yes	29	9.2	22	1.1
NTDA-BAPBDS/BAPF/BBIBz (9/1/1)	No	45	NM	32	NM
	Yes	28	10.4	42	3.3
NTDA-BAPBDS/BAPF/BBIBz (5/1/1)	No	38	10.7	37	2.3
	Yes	29	14.0	30	2.3

NM, not measured because the sample was too brittle to be measured.

7.4.6 Radical Oxidative Stability

Besides hydrolytic stability, radical oxidative stability is another important property of PEMs. It has been identified that radical oxidation-induced degradation of polymer main chains is one of the common reasons that cause deterioration of fuel cells. Peroxide radicals (hydroxyl and hydroperoxy radicals, etc.) are formed in fuel cells due to oxygen permeation through the PEM from the cathode side and incomplete reduction at the anode side [114]. It is greatly desirable to develop PEMs with high radical oxidative stability. The ex situ evaluation of radical oxidative stability of a PEM is usually performed by Fenton's test, i.e., the membrane samples are immersed in Fenton's reagent at room temperature or elevated temperatures (e.g., 80 °C) and characterized by the elapsed time when the samples start to break into pieces (τ_1), start to dissolve (τ_2), and are completely dissolved (τ_3), or by the weight loss after a certain period of soaking time.

One of the frequently used experimental procedures for measurement of radical oxidative stability is that a SPI membrane is immersed in a 30 %(w). hydrogen peroxide solution containing 30×10^{-6} ferrous sulfate at room temperature. The results obtained by this method of some SPI membranes are summarized in Table 7.9. It can be seen that for most of the membranes, the τ_1 values are around 20-30 h, while the τ_2 values are slightly longer than the τ_1, and the τ_3 values are a few hours longer than the τ_2. IEC seems to be one of the factors affecting radical oxidative stability, i.e., for the membranes derived from the same sulfonated diamine higher IECs tend to give lower radical oxidative stability. This is reasonable because acidic environment is favorable for formation of hydroxide radicals in the presence of ferrous cation catalyst. However, in some cases such an effect is not very significant. Membrane thickness has little effect on radical oxidative stability and thus such a factor can be neglected as long as the thickness is in a normal range (e.g., 10-100 μm). It is interesting to compare the Fenton's test results of NTDA-2,2′-BSPB and NTDA-3,3′-BSPB. The former exhibits significantly better radical oxidative stability than the latter despite the fact that 2,2′-BSPB and 3,3′-BSPB are isomers to each other. As discussed above, NTDA-2,2′-BSPB shows much better water resistance than NTDA-3,3′-BSPB. Therefore, we can conclude that 2,2′-BSPB is superior to 3,3′-BSPB and the fuel cells assembled with 2,2′-BSPB-based SPI membranes are expected to have better durability than those with 3,3′-BSPB-based SPI membranes under the same conditions. From this table it can be seen that the sequenced copolyimides,

NTDA-BAPFDS/ODA (2/1)-s and NTDA-BAPFDS/BAPPS (2/1)-s show significantly shorter τ_1 and τ_2 than various random copolyimides. This indicates that copolymerization manner also has a large effect on radical oxidative stability and random copolymerization is superior to sequential copolymerization.

TABLE7.9 IEC, Thickness and Radical Oxidative Stability of Various SPI Membranes Evaluated by Fenton's Test (30×10^{-6} $FeSO_4$ in 30% H_2O_2) at Room Temperature

Membrane	*IEC* (mol·g^{-1})	Thickness (μm)	τ_1 (h)	τ_2 (h)	τ_3 (h)	References
NTDA-ODADS/ODA(1/1)	1.95	29	20	24	—	3
NTDA-ODADS/BAPB(1/1)	1.68	37	29	32	—	3
NTDA-ODADS/BAPF(1/1)	1.71	40	29	32	—	3
NTDA-BDSA/ODA(1/1)	1.98	21	13	20	—	3
NTDA-BDSA/BAPF(1/1)	1.73	34	23	26	—	3
NTDA-BAPFDS/ODA(4/1)	2.36	58	17	21	—	6
NTDA-BAPFDS/ODA(1/1)	1.71	23	18	22	—	6
NTDA-BAPFDS/BAPB(4/3)	1.68	31	22	26	—	6
NTDA-BAPFDS/ODA(2/1)-s[a]	2.09	30	12	16	—	6
NTDA-BAPFDS/BAPPS(2/1)-s[a]	1.87	21	7	10	—	6
NTDA-BAPBDS	2.63	—	—	22	25	7
NTDA-BAPBDS/TFMB(4/1)	2.23	—	—	22	25	7
NTDA-BAPBDS/TFMB(1/1)	1.52	—	—	23	38	7
NTDA-BAPBDS/BAPB(2/1)	1.89	—	—	20	29	7
NTDA-2,2′-BSPB	2.89	16	—	18	22	45
NTDA-3,3′-BSPB	2.89	22	—	13	16	45

—, not recorded.
[a]*Here "s" refers to sequenced copolymers, and BAPPS denotes 4,4'-bis(3-aminophenoxy)diphenyl sulfone.*

In principle, polymer molecular weight should have a large effect on the radical oxidative stability and for the same SPI the membrane with higher molecular weight is expected to have better radical oxidative stability. However, little information about this issue is available from the literature, and thus it needs to be investigated in the future. For this reason, the effect of polymer molecular weight on radical oxidative stability is not discussed in this book.

Fenton's test can be shortened by raising the test temperature. The widely used experimental procedure is that a membrane sample is immersed in a 3 %(w). hydrogen peroxide solution containing 2×10^{-6} or 3×10^{-6}ferrous sulfate at 80 °C and the elapsed time when the sample starts to dissolve or break into pieces (τ_1) and completely dissolve (τ_2) are recorded, and/or the weight loss after a certain period of test time is measured. Table 7.10 shows the radical oxidative stability data of some SPI membranes measured by this method. The τ_1 values of the BAPBPDS and BAPTBPDS-based SPI membranes are in the range of 34-85 min (3×10^{-6} $FeSO_4$ in Fenton's reagent), and the τ_2 values are about one-and-a-half- to twofold that of τ_1. Banerjee and coworkers reported that the random copolyimide membranes DHNH60 and DHNH70 which were synthesized from NTDA, 4,4′-diaminostilbene-2, 2′-disulfonic acid (DSDSA), and 1,4-bis(2′-trifluoromethyl-4′-(4″-aminophenyl)phenoxy) benzene (HQA) exhibited τ_1 values of 204 and 174 min (2×10^{-6} $FeSO_4$ in Fenton's reagent), respectively. They ascribed the better radical oxidative stability of these SPIs to the presence of hydrophobic trifluoromethyl groups which could protect the polymer main chain from being attacked by water molecules containing highly oxidizing radical species [52]. However, as already discussed (see Table 7.9), the two copolyimides derived from 2,2′-bis(trifluoromethyl)benzidine, NTDA-BAPBDS/TFMB (4/1) and NTDA-BAPBDS/TFMB (1/1), do not exhibit significant better radical oxidative stability than those containing no trifluoromethyl groups in their structures. Theoretically, the possibility of radical attack on sulfonated (hydrophilic) units should be much higher than on nonsulfonated (hydrophobic) units because water molecules containing radical species are mainly concentrated in sulfonated units. Thus, it seems that, though the mechanism is not clear, the DSDSA moiety may be related to the radical oxidative stability. In addition, despite the relatively good radical oxidative stability, the DHNH60 and DHNH70 membranes displayed only moderate water resistance (450 h at 80 °C) resulting from the low basicity of the amino groups of the DSDSA moiety [52].

TABLE7.10 IEC and Radical Oxidative Stability of Some SPI Membranes Evaluated by Fenton's Test (2×10^{-6} or 3×10^{-6} $FeSO_4$ in 3%(w) H_2O_2) at 80 °C

Membrane	*IEC* ($mol\cdot g^{-1}$)	τ_1 (min)	τ_2 (min)	τ_3 (min)	Weight Loss (%)	References
NTDA-BAPBPDS/ODA(2/1)	1.99	57	81	—	—	39
NTDA-BAPBPDS/BAPB(3/1)	2.02	85	80	—	—	39
NTDA-BAPBPDS/BAPF(3/1)	2.04	61	101	—	—	39
NTDA-BAPTBPDS/ODA(2/1)	1.93	36	48	—	—	39
NTDA-BAPTBPDS/BAPB(3/1)	1.96	34	34	—	—	39
NTDA-BAPTBPDS/BAPF(3/1)	1.97	47	89	—	—	39
DHNH60[a]	1.75	204[b]	—	672[b]		54
DHNH70[a]	2.10	174[b]	—	378[b]		54
1a(50)	1.92	—	—	—	6	4
NTDA-2,2'-BSBB	2.78	—	—	—	10	45
NTDA-2,2'-BSBB/ODA(2/1)	2.14	—	—	—	6	37

[a]*Random copolyimide derived from NTDA, 4,4'-diaminostilbene-2,2'-disulfonic acid (DSDSA) and 1,4-bis(2'-trifluoromethyl-4'-(4"-aminophenyl)phenoxy benzene (HQA).*

[b]*Fenton's reagent: 3%(w) H_2O_2 + 2×10^{-6} $FeSO_4$.*

It has been identified that the incorporation of benzimidazole groups into polymer structures is a very effective approach for improving radical oxidative stability of PEMs [16,62,73,86]. This is based on the discovery that the crosslinked sulfonated polybenzimidazole membranes (SOPBI) exhibit unusually high radical oxidative stability (weight loss: 8.5%-16% after the membranes were soaked in 3 %(w). H_2O_2 containing 3×10^{-6} $FeSO_4$ at 80 °C for 12 h) [115]. Although the exact mechanism is not clear yet, such an excellent radical oxidative stability should be attributed to the presence of benzimidazole groups. Because of the ionic crosslinking resulting from the interaction between the acidic sulfonate groups and the basic imidazole groups, these SOPBI membranes displayed either rather low proton conductivities at low degrees of sulfonation or unacceptable SR at too-high degrees of sulfonation, making them unsuitable as PEMs for practical applications [115,116]. The incorporation of small amounts of benzimidazole groups into SPI structures causes significant improvement in radical oxidative stability, while high proton conductivity and low swelling ratio can be maintained. One example is the sulfonated copolyimides synthesized by random copolymerization of NTDA, 2-(4-aminophenyl)-5-aminobenzimidazole (APABI), ODADS, and BAPF (Fig. 7.32). The SPI membranes were further treated in PPA at 180 °C for 6 h to form covalent crosslinking. Fenton's test (3% hydrogen peroxide containing 3×10^{-6} ferrous sulfate, 80 °C) results of the covalently crosslinked and noncrosslinked SPI membranes are shown in Table 7.11. For the membranes without covalent crosslinking, the τ_1 values are in the order: $\mathbf{5}<\mathbf{1}<\mathbf{2}\approx\mathbf{6}<\mathbf{3}<\mathbf{4}\approx\mathbf{7}$, which follows the decreasing order of IECs. The membrane with the lowest IEC (**4**) displayed three-times longer τ_1 values than the one with the highest IEC (**5**). This is a common phenomenon which has been observed with many other SPI membranes as previously discussed. Moreover, all the benzimidazole groups-containing SPI membranes (**1-4**, **6**, **7**) exhibit significantly better radical oxidative stability than the one without benzimidazole groups (**5**). From this table it can also be seen that the covalent crosslinking treatment caused further improvements in radical oxidative stability and such an effect is much more significant for the benzimidazole groups-containing SPI membranes than for the one without benzimidazole groups. The membrane **4**, e.g., shows a τ_1 and τ_2 of 480 and 500 min, respectively, after covalent crosslinking, which are about three times longer than those (180 and 140 min) of the corresponding noncrosslinked one, whereas for the membrane without benzimidazole groups (**5**), crosslinking causes only slight increases in τ_1 (from 60 to 85 min) and τ_2 (from 70 to 120 min). This indicates that the synergetic effect of benzimidazole groups and covalent crosslinking is much more effective for enhancing the radical oxidative stability than each factor alone. Fig. 7.33 shows the weight residue of the covalently crosslinked and noncrosslinked membranes (**4** and **5**) as a function of test time in Fenton's reagent. For the covalently crosslinked SPI membrane **4** the weight residue is about 85% and the mechanical strength is still maintained (not broken) after a 4h test, indicating excellent radical oxidative stability of this membrane. For the noncrosslinked membrane **4** the weight residue in the initial 2 h is also fairly high, but thereafter rapidly decreased with test time and the membrane finally became very brittle. The noncrosslinked membrane **5** shows the worst radical oxidative stability of which weight residual is below 70% after only a 1h test. For the covalently crosslinked membrane **5**, however, the weight residue ~90% under the same conditions decreased to nearly 60% after a 2h test. These results clearly indicate that the synergetic effect of the covalent crosslinking and the incorporation of benzimidazole groups into polymer structures is very effective for the improvement of radical oxidative stability of the SPI membranes.

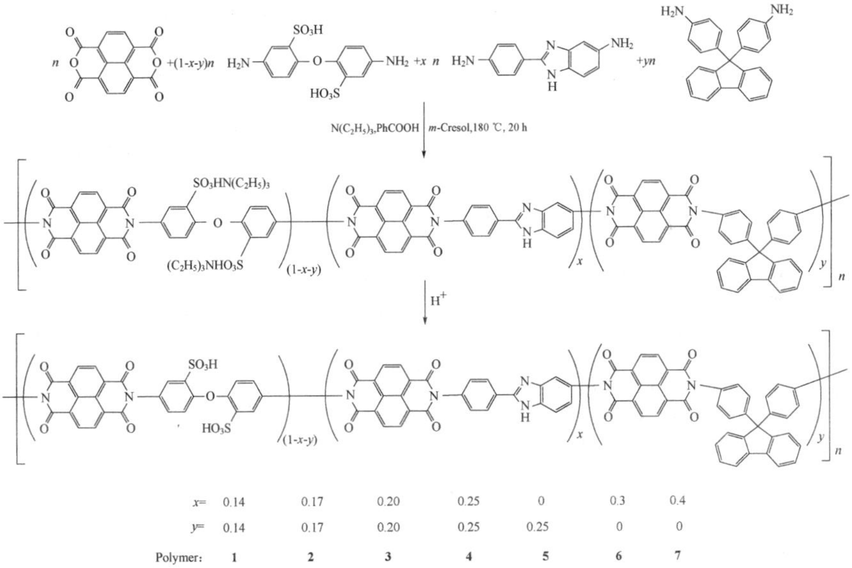

x=	0.14	0.17	0.20	0.25	0	0.3	0.4
y=	0.14	0.17	0.20	0.25	0.25	0	0
Polymer:	1	2	3	4	5	6	7

FIGURE 7.32 Chemical structures of the benzimidazole groups containing SPI copolymers.

TABLE7.11 Radical Oxidative Stability of Various SPI Membranes by Fenton's Test in 3% H_2O_2 Containing 3×10^{-6} $FeSO_4$ at 80 °C

Membrane	Crosslinking Treatment[a]	*IEC*[a] ($mmol\cdot g^{-1}$)	τ_1^b (min)	τ_2^c (min)
1	No	2.13 (2.25)	80	70
	Yes	2.04	270	240
2	No	1.99 (2.06)	110	120
	Yes	1.74	330	300

TABLE7.11 (Continued)

Membrane	Crosslinking Treatment[a]	*IEC*[a] (mmol·g^{-1})	τ_1^b (min)	τ_2^c (min)
3	No	1.72 (1.78)	150	150
	Yes	1.59	450	360
4	No	1.32 (1.35)	180	140
	Yes	1.22	480	500
5	No	2.36 (2.55)	60	70
	Yes	2.28	85	120
6	No	1.85 (1.99)	120	130
7	No	1.39 (1.49)	190	180

[a]*In PPA at 180 °C for 6 h.*
[b]*Elapsed time when the membranes started to break into pieces.*
[c]*Elapsed time when the membranes started to dissolve.*

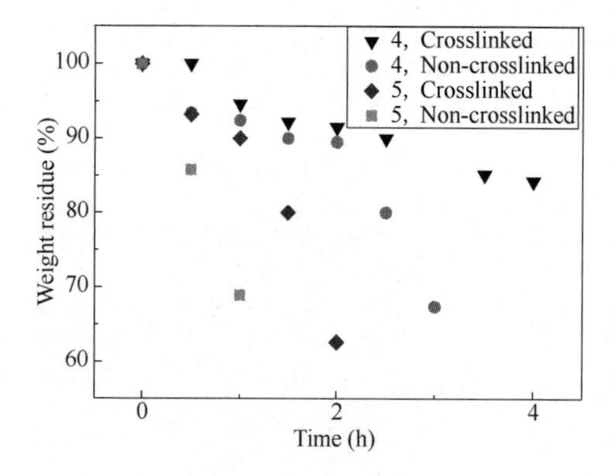

FIGURE 7.33 Variation of weight residue of the SPI membranes as a function of aging time in Fenton's reagent (3%(*w*) H_2O_2 + 3×10^{-6} $FeSO_4$, 80 °C).

It should be noted that, as mentioned in Section 7.4.4.4, the benzimidazole groups-containing SPI membranes have the drawback that they are less stable to hydrolysis than those without benzimidazole groups. The deterioration in hydrolytic stability of the benzimidazole groups containing SPI membranes is due to the strong electron-withdrawing effect of the protonated benzimidazole groups, and such a drawback can be overcome by replacing the weak imido rings adjacent to benzimidazole groups with the highly stable benzimidazobenzisoquinolinone groups (Fig. 7.34). To synthesize the SPIs containing benzimid- azobenzisoquinolinone groups, an amine-terminated polybenzimidazole oligomer (PBI-NH_2) is needed which can be synthesized by condensation polymerization of isophthalic acid with excess 3,3′-diaminobenzidine in PPA at 190 °C (Fig. 7.35). It is found that as the degree of polymerization is below five, the resulting PBI-NH_2 oligomers are soluble in *m*-cresol and thus can be utilized as a "diamine" monomer for preparation of various SPIs. To avoid too much decrease in IEC due to the acid-base interactions (ionic crosslinking) between sulfonic acid groups and benzimidazole groups as well as benzimid-azobenzisoquinolinone groups, the PBI-NH_2 oligomer with minimal degree of polymerization (*m*=1) is employed to copolymerize with NTDA and BAPBDS at the molar ratio of PBI-NH_2/BAPBDS= 1:(19-49) to yield a series of sulfonated polyimide-polybenzimidazole random copolymers (Fig. 7.36). The resulting copolymers are denoted as SPI-*co*-PBI-*x*/*y*, where *x*/*y* refers to the molar ratio between PBI-NH_2 and BAPBDS.

Hydrolytic stability

Highly positive
(easier to be attacked by water molecules)

Strong linkage
(Nonhydrolyticbond)

FIGURE 7.34 Comparison of hydrolytic stability of the SPIs with different benzimidazole groups containing moieties.

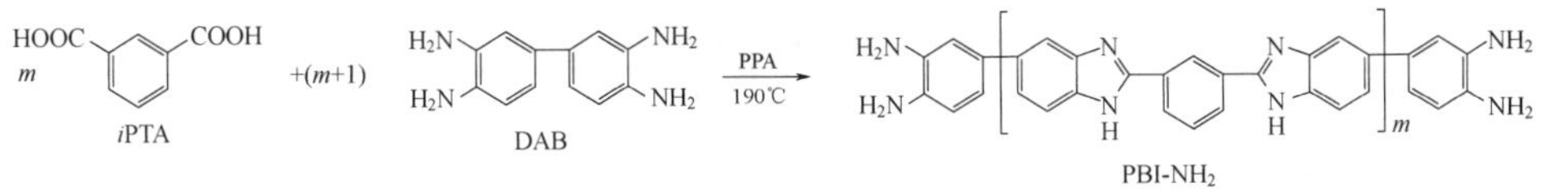

PPA 190℃

*i*PTA DAB $PBI-NH_2$

FIGURE 7.35 Synthesis of the amine-terminated polybenzimidazole oligomer ($PBI-NH_2$).

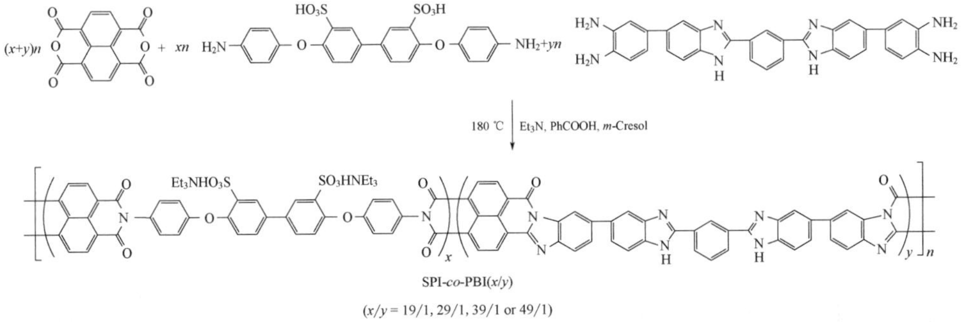

180 ℃ Et_3N, PhCOOH, *m*-Cresol

SPI-*co*-PBI(*x*/*y*)

(*x*/*y* = 19/1, 29/1, 39/1 or 49/1)

FIGURE 7.36 Synthesis of the sulfonated polyimide-polybenzimidazole copolymers (SPI-*co*-PBI).

Table 7.12 shows Fenton's test results (3% H_2O_2 + 3×10^{-6} $FeSO_4$, 80 °C) of the SPI-*co*-PBI membranes [86]. For comparison purposes, the data of the homopolymer NTDA-BAPBDS are also shown in this table. It is clear that the radical oxidative stability is in the order: SPI-*co*-PBI-19/1 > SPI-*co*-PBI-29/1 > SPI-*co*-PBI-39/1 > SPI-*co*- PBI-49/1 > NTDA-BAPBDS, which follows the decreasing trend of the content of $PBI-NH_2$ moiety. The SPI-*co*-PBI-19/1 membranes showed a τ_1 of 200 min and a τ_3 of 270 min which are much longer than those of the NTDA-BAPBDS membrane (τ_1=90 min, τ_3=140 min). Moreover, the weight residue of SPI-*co*-PBI-19/1 reaches 99.4% after being soaked in Fenton' reagent for 150 min, whereas the NTDA-BAPBDS membrane exhibits complete dissolution under the same conditions, indicating much higher oxidative stability of the former. Furthermore, although the maximum molar fraction of the PBI oligomer is only 5% (relative to the sum of the BAPBDS and the PBI oligomer), the improvement in radical oxidative stability is very significant.

Table 7.13 shows the hydrolytic stability of the SPI-*co*-PBI-19/1 membrane evaluated by the weight loss or viscosity loss after it was soaked in the deionized water at 140 °C in an autoclave for 24-168 h. For comparison purposes, the hydrolytic stability of another type of SPI membrane NTDA-BAPBDS/APABI (5/1) under the same conditions is also listed in this table. NTDA-BAPBDS/APABI (5/1) is synthesized by random copolymerization of the NTDA, BAPBDS, and APABI at the molar ratio of BAPBDS/APABI=5/1. The two kinds of polymers have quite similar chemical structure except that different benzimidazole moieties ($PBI-NH_2$ and APABI) are used. As shown in this table, the weight loss of SPI-*co*-PBI-19/1 membrane is only 1.0% after aging for 24 h which is one order lower than that of the NTDA-BAPBDS/APABI (5/1), indicating much better hydrolytic stability of the former. The difference in viscosity loss between these two types of membranes is even more significant, i.e., the viscosity loss after aging for 24 h is only 2.0% for the SPI-*co*-PBI-19/1 but 42% for the NTDA-BAPBDS/APABI (5/1) under the same conditions. This indicates that the employment of the $PBI-NH_2$ oligomer instead of the APABI is very effective for improvement of membrane hydrolytic stability. From this table, it can also be seen that the weight loss of the SPI-*co*-PBI-19/1 membrane increases with aging time. However, the weight loss is still at a moderate level (15%) even after an aging test of 1 week indicating excellent hydrolytic stability of this membrane. The high hydrolytic stability of the SPI-*co*-PBI-19/1 membrane should be attributed to the high basicity of the amino groups of the BAPBDS as well as the nonhydrolytic benzimidazobenzisoquinolinone unit, whereas the relatively poor hydrolytic stability of the NTDA-BAPBDS/APABI (5/1) membrane is due to the lower basicity of the amino groups of the APABI moiety resulting from the strong electron-withdrawing effect of the protonated imidazole rings.

TABLE7.12 Radical Oxidative Stability of the SPI-*co*-PBI and the Pure SPI (Homopolymer) Membranes Measured by the Fenton's Test in 3% Hydrogen Peroxide Containing 3×10^{-6} Ferrous Sulfate at 80 °C

Ionomer	Oxidative Stability		
	τ_1^b (min)	τ_3^b (min)	Weight Residue[c] (%)
SPI-*co*-PBI-19/1	200	270	99.4
SPI-*co*-PBI-29/1	175	240	97.0
SPI-*co*-PBI-39/1	155	220	93.6
SPI-*co*-PBI-49/1	140	200	72
SPI[a]	90	140	0

[a]*The homopolymer synthesized from the NTDA and the BAPBDS.*

[b]*τ_1 and τ_3 refer to the elapsed time when the membrane samples started to break into pieces and completely dissolved, respectively.*

[c]*Soaking time: 150 min.*

TABLE7.13 Comparison of the Weight Loss and Viscosity Loss of the SPI-*co*-PBI-19/1 and the NTDA-BAPBDS/APABI(5/1) Membranes After Being Soaked at 140 °C for a Certain Period of Time

Membrane	Soaking Time (h)	Weight Loss (%)	Viscosity Loss (%)
SPI-*co*-PBI-19/1	24	1.0	2.0
	72	7.0	15
	168	15	32
NTDA-BAPBDS/APABI(5/1)	24	10	42

7.4.7 Methanol Permeability

Methanol permeability (P_M) of a PEM is an important parameter to a direct methanol fuel cell (DMFC). The lower methanol permeability of a PEM, the higher the fuel utilization efficiency of the DMFC. From the viewpoint of practical applications, it is greatly desirable to develop PEMs with high proton conductivity but low methanol permeability. Methanol permeability is usually measured with a liquid permeation cell which is composed of two compartments separated by a vertical hydrate membrane. One compartment of the cell is filled with methanol feed solution, and the other compartment is filled with distilled water. Methanol gradually permeates through the membrane from the feed side into the permeate side and the compositions of feed and permeate solutions can be analyzed with gas chromatography apparatus. Methanol permeability coefficient is calculated from the following equation:

$$P=C_b V_b L/(A C_a t) \tag{7.9}$$

where C_a and C_b refer to the methanol concentration in feed and permeate, respectively. V_b is the solution volume of permeate. L, A, and t refer to membrane thickness, membrane effective area, and permeation time, respectively.

Methanol permeability is affected by many factors such as IEC, polymer chemical structure, methanol concentration, and temperature. Table 7.14 shows the IEC, *in-plane* proton conductivity, methanol permeability at 30 °C at an initial methanol concentration of 8.6 %(w) in feed and the ratio (Φ) of proton conductivity to methanol permeability of some SPI membranes. The chemical structures of the SPI membranes are shown in Fig. 7.37. The theoretical IEC values of most SPI membranes in this table are around 2.0 $mol\cdot g^{-1}$. It is clear that both BAPBPDS and BAPTBPDS-based SPI membranes [39] exhibit significantly lower methanol permeability ($(2.4\text{-}4.5)\times10^{-7}$ $cm^2\cdot s^{-1}$) than the ones derived from other sulfonated diamines (BAPBDS [117], 3,3′-BSPB [117], 4,4′-bis(3-aminophenoxy)diphenyl sulfone-3,3′-disulfonic acid [118]). This is probably due to the highly rigid benzophenone moiety which causes high stiffness and dense packing of polymer chains. NTDA-BAPBPDS/BAPF(3/1) displays the lowest methanol permeability (2.4×10^{-7} $cm^2\cdot s^{-1}$) which is one order lower than that of Nafion 112.

Since the membranes ODA-70 and ODA-60 have quite similar chemical structures, the higher methanol permeability of ODA-70 should be ascribed to its higher IEC.

TABLE7.14 *IEC*, Proton Conductivity (σ), P_M at 30 °C at an Initial Methanol Concentration of 8.6%(w) in Feed and Φ of Some SPI Membranes

Membrane	*IEC*[a] (mmol·g^{-1})	σ (S·cm^{-1})	P_M (10^{-7} cm^2·s^{-1})	Φ (10^4 S·cm^{-3}·s^{-1})	References
NTDA-BAPBPDS/ODA(2/1)	1.99	0.084	3.7	23	39
NTDA-BAPBPDS/BAPB(3/1)	2.02	0.066	4.5	15	39
NTDA-BAPBPDS/BAPF(3/1)	2.04	0.096	2.4	40	39
NTDA-BAPTBPDS/ODA(2/1)	1.93	0.084	3.2	26	39
NTDA-BAPTBPDS/BAPB(3/1)	1.97	0.090	4.1	22	39
NTDA-BAPTBPDS/BAPF(3/1)	1.96	0.085	2.7	32	39
BAPS-80	2.03	0.12	8.0	15	118
ODA-70	1.98	0.11	6.9	16	118
ODA-60	1.80	0.07	5.1	14	118
NTDA-BAPBDS/BAPB(2/1)	1.89	0.091	11.5	7.9	117
NTDA-3,3′-BSPB/BAPF(2/1)	2.04	0.092	5.0	18	117
Nafion 112	0.91	0.10	24[b]	4.2[b]	117

[a] *Theoretical values.*
[b] *Initial methanol concentration in feed: 10%(w).*

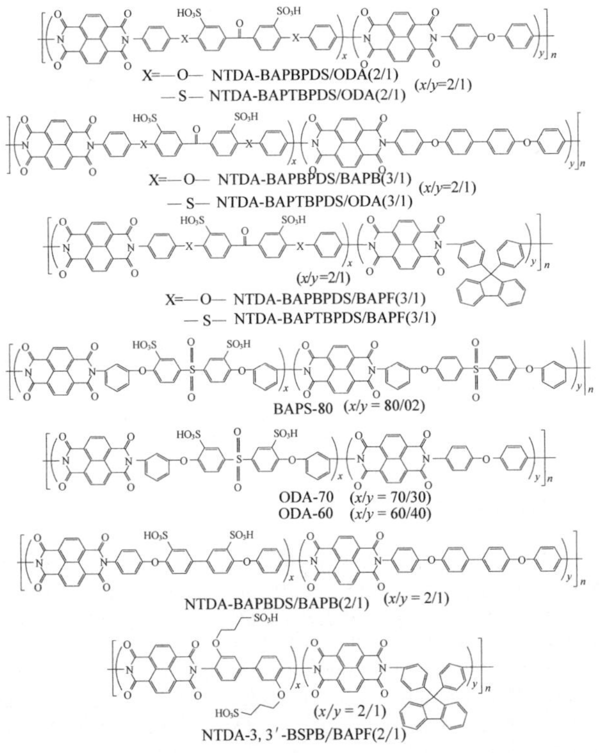

FIGURE 7.37 Chemical structures of the SPIs for methanol permeability test.

Φ is often used to evaluate the membrane performance in a DMFC system, and the higher Φ, the better performance of the membrane. As shown in this table, except NTDA- BAPBPDS/BAPB(3/1) the BAPBPDS- and BAPTBPDS-based SPI membranes show significantly larger Φ values (> 20) than other SPI membranes. NTDA-BAPBPDS/BAPF(3/1) displayed the highest Φ (40), while the proton conductivity is close to that of Nafion 112, indicating fairly high performance of this membrane.

Okamoto and coworkers studied the effects of methanol concentration and test temperature on methanol permeability of some SPI membranes [117]. The methanol permeability of three SPI membranes, NTDA-BAPBDS/BAPB(2/1), NTDA-2,2′-BSPB/BAPB(2/1), and NTDA-3,3′-BSPB/ODA(2/1), and Nafion112 are shown in Fig. 7.38. The initial methanol concentrations in feed are at 10 or 30 %(w) and the test temperatures are 30 °C or 50 °C. It can be seen that for all the membranes methanol permeability increases with increasing temperature. This is likely because of the increased membrane swelling ratio and methanol mobility at higher temperatures. The effect of methanol concentration on methanol permeability is somewhat complex. As shown in Fig. 7.38, except for NTDA-BAPBDS/BAPB(2/1) other membranes exhibit increased methanol permeability with increasing methanol concentration in feed. Because there is little information on the difference in membrane swelling ratio at different concentration methanol solutions, it is hard to give a proper explanation for the results of methanol permeability of these SPI membranes.

The Φ values of the above SPI membranes and Nafion112 at different temperatures (30 °C or 50 °C) and different initial methanol concentrations in feed (10% and 30%(w)) are shown in Fig. 7.39. All the SPI membranes exhibit significantly higher Φ values than Nafion112. Among the SPIs, NTDA-2,2′-BSPB/BAPB (2/1) displays the highest Φ value. Increasing temperature tends to give lower Φ values. However, the effects of temperature and methanol concentration in feed on Φ are not clear.

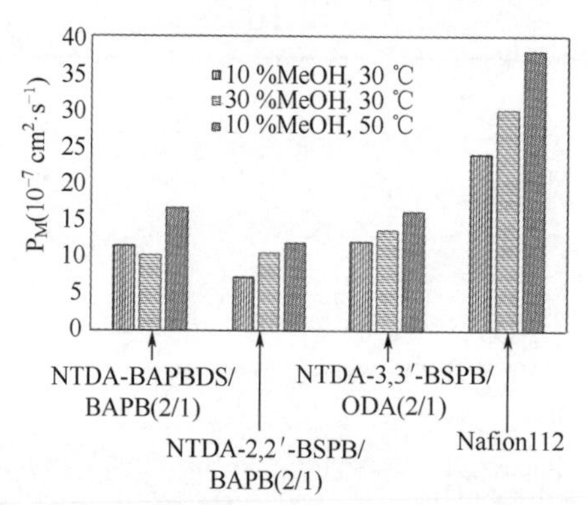

FIGURE 7.38 Methanol permeability of some SPI membranes and Nafion112 at different temperatures and different initial methanol concentrations in feed.

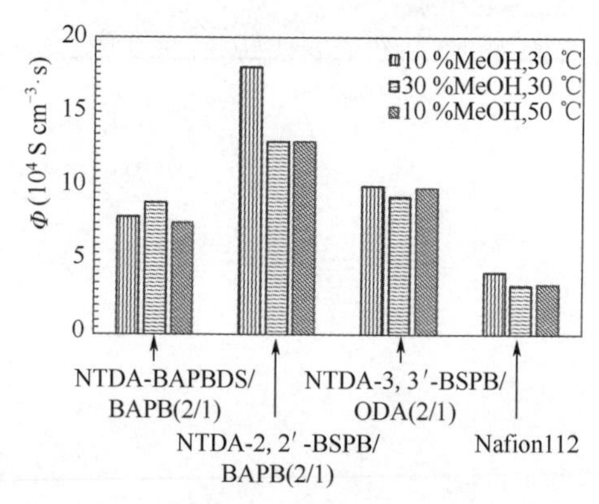

FIGURE 7.39 The ratios (Φ) of proton conductivity to methanol permeability of some SPI membranes and Nafion112 at different temperatures and initial methanol concentrations in feed.

7.5 Fuel Cell Performance

Fig. 7.40 shows the schematic diagram of a single cell. Prior to installing a single cell, a membrane electrode assembly (MEA) needs to be prepared. Generally, an MEA can be prepared via two methods: catalyst-coated membrane (CCM) method and catalyst-coated substrate (CCS) method. For the CCM method, a catalyst ink which is a mixture of the catalyst (platinum supported on carbon) suspended in an ionomer solution (e.g., Nafion solution in ethanol/isopropanol mixture) is sprayed onto both sides of a PEM and is subsequently dried at ~ 60 °C to remove the residual solvent. The ionomer of ink functions as the binder to fix the Pt/C catalyst nanoparticles as well as the proton conductor. For the CCS method, the same catalyst ink is employed to spray onto one side of a gas diffusion layer (GDL, carbon paper or carbon cloth) and dried at ~ 60 °C to remove the residual solvent. The resulting catalyst-coated GDL is called gas diffusion electrode (GDE). Then, a pair of the prepared GDE and a PEM is sandwiched (the catalyst layer directly contacts with the PEM) and hot-pressed to form an MEA. Because SPI membranes are generally very stiff, prior to hot-press both surfaces of a SPI membrane need to be coated with an extremely thin (nanometer-scale) soft electrolyte layer (typically Nafion) as a binder to strengthen the adhesion between the SPI membrane and the catalyst layers. Since GDE is commercially available for laboratory use, the CCS method is more frequently used than the CCM method. The key issues associated with MEA fabrication are: (1) keeping ohmic contact between the catalyst layer and the membrane, (2) keeping high porosity of the catalyst layer to allow fast mass (hydrogen, oxygen, etc.) transfer.

The performance of a single PEMFC is affected by many factors such as PEM *through-plane* proton conductivity, catalyst (Pt) loading level and efficiency, MEA fabrication techniques, fuel (hydrogen, methanol, etc.) and oxygen/air feeding conditions (pressure, flow rate, humidification, etc.), and operation temperatures. Fuel cell performance is usually characterized by recording the variation of cell voltage (V) as a function of current density (I) and the I-V curve is called the polarization curve. Fig. 7.41 shows the polarization curves of two single cells using the SPI random copolymer membrane NTDA-2,2′-BSPOB/BAPB(4/1) and Nafion212, respectively, as the electrolyte membranes at 80 °C under the external humidification of 100% for anode and 82% for cathode (unpublished data of our group). The chemical structure of NTDA-2,2′-BSPOB/BAPB (4/1) is illustrated in Fig. 7.42. The thickness values are 46 μm for NTDA-2,2′-BSPOB/BAPB (4/1) and 52 μm for Nafion 212. The Pt-loading amounts are 0.3 $g\cdot cm^{-2}$ for anode and 0.7 $mg\cdot cm^{-2}$ for cathode, and hydrogen (0.2 MPa) and oxygen (0.2 MPa) are used as fuel and oxidant, respectively. A very fast voltage decay in the initial stage (0 - 200 $mA\cdot cm^{-2}$) is observed for both fuel cells, which is mainly due to the slow kinetics of the oxygen reduction reaction (ORR). The decay rate in this stage reflects the efficiency of the catalyst and it is still a big challenge to develop a catalyst with faster ORR kinetics. The approximate linear voltage decay in the second stage is ascribed to the ohmic resistance including the electrolyte membrane resistance, catalyst layer resistance, and contact resistance. The last stage voltage decay is due to the mass (hydrogen and oxygen) transfer limit (local starvation of hydrogen and/or oxygen due to their too-rapid consumptions at high current densities).

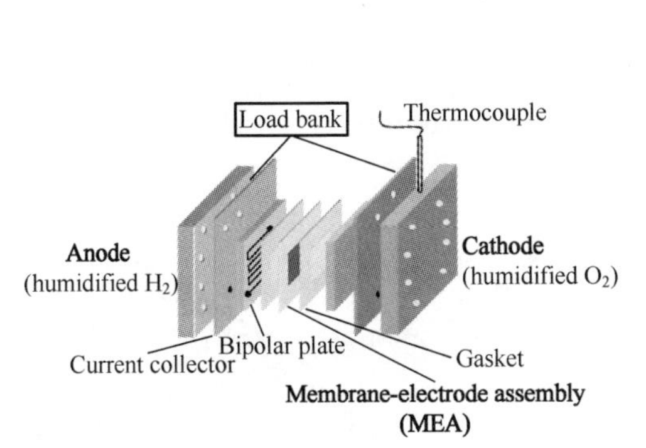

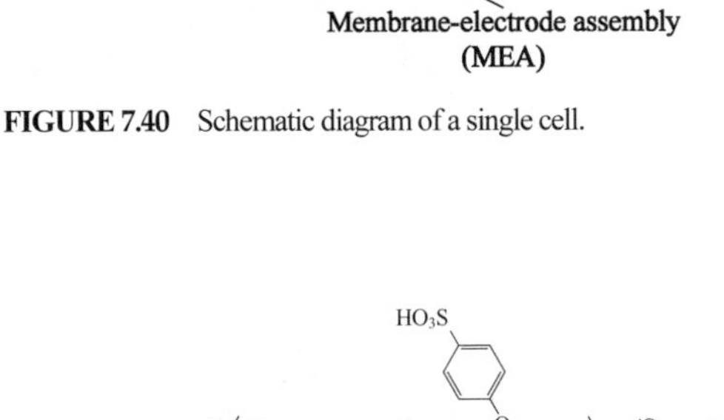

FIGURE 7.40 Schematic diagram of a single cell.

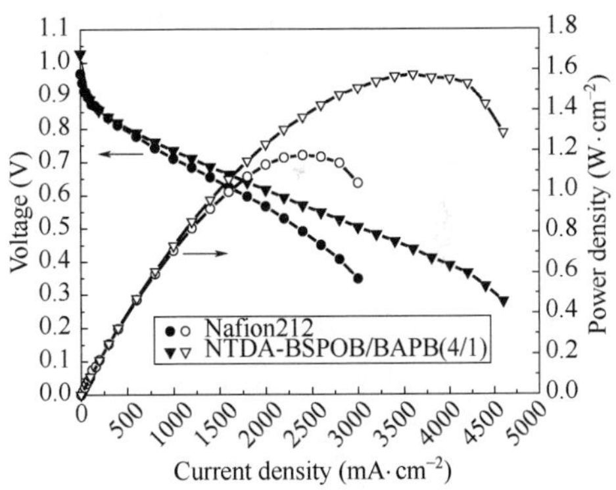

FIGURE 7.41 H_2-O_2 fuel cell performances of the single cells assembled with NTDA-2,2′-BSPOB/BAPB(4/1) and Nafion212 at 80 °C under the external humidification of 100% anode and 82% for cathode.

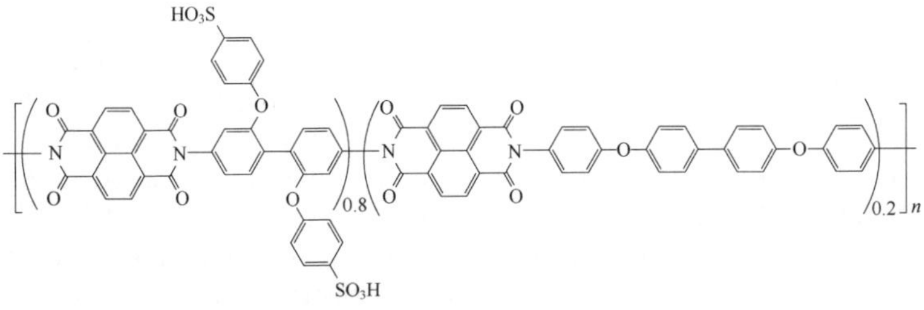

FIGURE 7.42 Chemical structure of the SPI random copolymer NTDA-2,2′-BSPOB/BAPB (4/1).

As shown in Fig. 7.41, the peak power density reaches 1.58 W·cm^{-2} for the cell assembled with NTDA-2,2′-BSPOB/BAPB (4/1) which is 0.40 W·cm^{-2} higher than that of the one with Nafion 212 (1.18 W·cm^{-2}), indicating much better performance of the former. This is mainly due to the higher proton conductivity of NTDA-2,2′-BSPOB/BAPB (4/1) (*in-plane* direction, 0.30 S·cm^{-1}, 80 °C, in water) than that of Nafion212 (*in-plane* direction, 0.17 S cm^{-1}, 80 °C, in water). In addition, the open circuit voltage (OCV) values are 1.027 V for the cell assembled with NTDA-2,2′-BSPOB/BAPB (4/1) and 0.967 V for the one with Nafion 212. Because NTDA-2,2′-BSPOB/BAPB (4/1) is even slightly thinner than Nafion 212, the higher OCV of the former indicates its significantly lower gas permeability than the latter.

Fig. 7.43 shows the H_2-air fuel cell performances of four single cells assembled with three SPI membranes SPIBI20-*b*-NF05, SPIBI20-*b*-FF05, and SPI-*co*-PBI (19/1) (for the chemical structures, see Fig. 7.14) and Nafion112 [16]. The fuel cell tests are carried out at 90 °C under the external humidification of 92% RH for the anode side and 82% for the cathode side. The Pt-loading amounts are 0.5 g·cm^{-2} for both electrodes and the gas pressure is set at 0.2 MPa for both hydrogen and air. These membranes have approximately the same thickness (52-55 μm). The OCV values of these fuel cells are quite similar to each other (0.96-0.97 V). The single cell equipped with the SPIBI20-*b*-NF05 membrane shows fairly good fuel cell performance with a peak power density of 0.70 W·cm^{-2} which is comparable to those of the SPI-*co*-PBI(19/1) and Nafion112. However, the single cell equipped with the SPIBI20-*b*-FF05 membrane exhibits a lower peak power density of 0.60 W·cm^{-2} because of its lower proton conductivity resulting from its lower IEC. Nevertheless, at the current density, below 0.9 A cm^{-2} this fuel cell exhibits almost the same performance as the others.

The fuel cell performances of the single cells have been systematically studied using 2,2′-BSPOB-based SPIs as the electrolyte membranes under various operation conditions [12-14]. Fig. 7.44 shows the H_2-O_2 polarization curves of the single cell using the crosslinked SPI random copolymer membrane NTDA-2,2′-BSPOB/BAPBz (2/1)-CL ('CL' refers to crosslinking, membrane thickness: 35 μm) as the PEM under the external humidification of 84% for anode and 68% for cathode at 90 °C and different gas pressures (Pt-loading amounts: 0.5 mg·cm^{-2} for each electrode). The covalent crosslinking was performed by immersing the SPI membrane into Eaton's reagent (phosphorous pentoxide in methanesulfonic acid at the weight ratio of 1:10) at 80 °C (Fig. 7.45) as discussed in Section 7.4.5.4. It is clear that the fuel cell performance increases with increasing gas pressure and a big increase in fuel cell performance is observed as the gas press increased from 0.1 to 0.15 MPa. This is likely because at higher pressures (> 0.15 MPa) the water molecules in the cathode side may more readily diffuse back to the SPI membrane leading to higher proton conductivity.

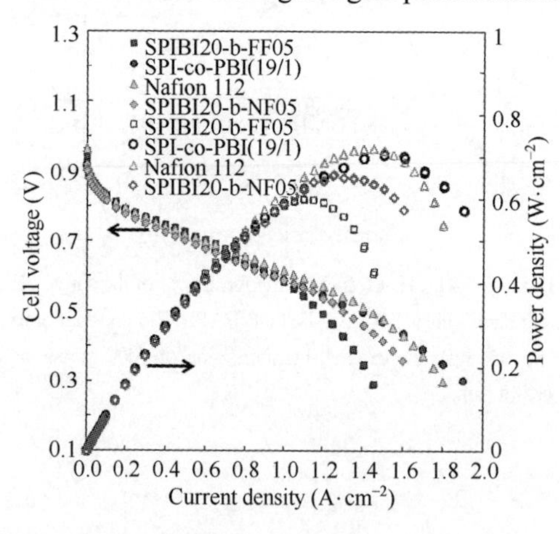

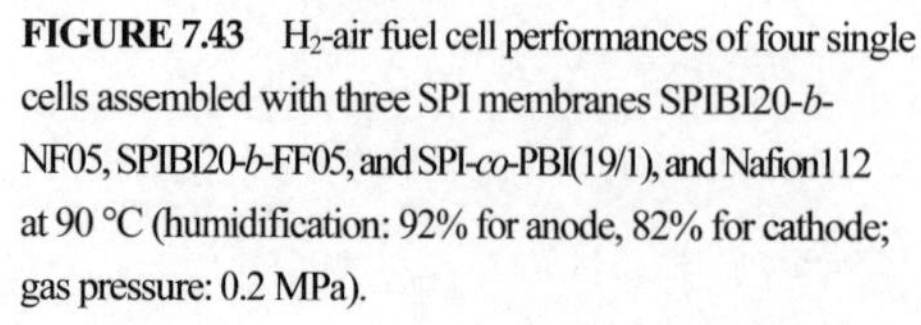

FIGURE 7.43 H_2-air fuel cell performances of four single cells assembled with three SPI membranes SPIBI20-*b*-NF05, SPIBI20-*b*-FF05, and SPI-*co*-PBI(19/1), and Nafion112 at 90 °C (humidification: 92% for anode, 82% for cathode; gas pressure: 0.2 MPa).

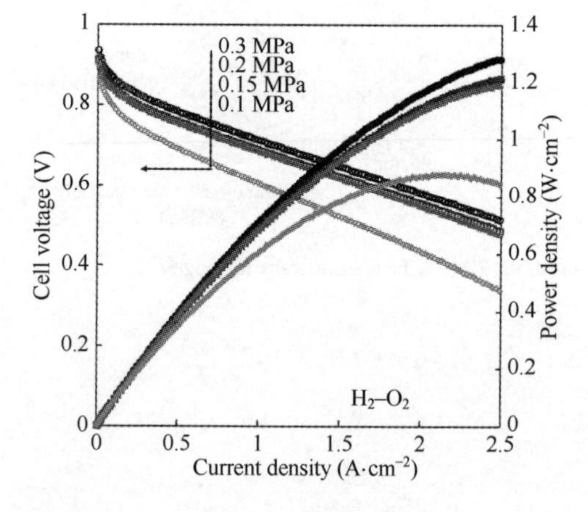

FIGURE 7.44 H_2-O_2 fuel cell performance of the single cell using the crosslinked SPI membrane NTDA-2,2′-BSPOB/BAPBz(2/1)-CL as the PEM under the external humidification of 84% for anode and 68% for cathode at 90 °C and different gas pressures (Pt-loading amounts: 0.5 mg·cm^{-2} for each electrode).

The effect of humidification conditions on fuel cell performance is shown in Fig. 7.46 using the above crosslinked SPI membrane as an example. The tests were carried out at 90 °C and 0.2 MPa for both hydrogen and air. It can be seen that the fuel cell performance is strongly dependent on humidification conditions. The peak power density reaches 0.74 $W \cdot cm^{-2}$ under the humidification of 84% for anode and 68% for cathode but decreases to 0.46 and 0.28 $W \cdot cm^{-2}$ as the RH decreases to 50% for both anode and cathode and 30% for both anode and cathode, respectively. This is because, as discussed in Section 7.4.4, the membrane proton conductivity is strongly dependent on RH and the conductivity rapidly decreases with decreasing RH.

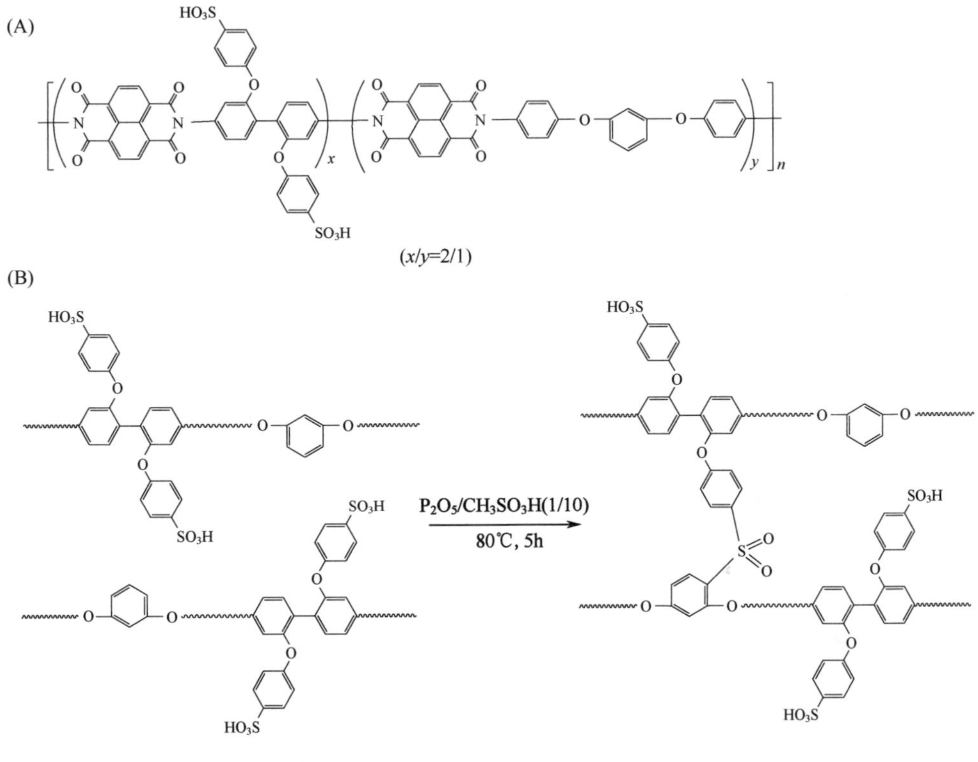

FIGURE 7.45 (A) Chemical structure of NTDA-2,2′-BSPOB/BAPBz (2/1), (B) covalent crosslinking.

Enhancing membrane IEC is an effective method to improve fuel cell performance especially under weak humidification conditions. Fig. 7.47 shows the H_2-air single cell performances using two crosslinked SPI copolymer membranes, NTDA-2,2′-BSPOB/BAPBz(3/1)-CL (thickness: 36 μm) and NTDA-2,2′-BSPOB/BAPBz(2/1)-CL (thickness: 35 μm), and Nafion112 (55 μm) as the PEMs under rather weak humidification conditions (27%RH) at 90 °C and 0.2 MPa. The IEC values determined by titration are 1.95 $mol \cdot g^{-1}$ for NTDA-2,2′-BSPOB/BAPBz(3/1)-CL and 1.73 $mol \cdot g^{-1}$ for NTDA-2,2′-BSPOB/BAPBz(2/1)-CL. The single cell assembled with NTDA-2,2′-BSPOB/BAPBz (3/1)-CL exhibits fairly good performance of which peak power density reaches 0.59 $W \cdot cm^{-2}$, which is even slightly larger than that of the one assembled with Nation112 (0.57 $W \cdot cm^{-2}$), whereas the single cell assembled with NTDA-2,2′-BSPOB/BAPBz(2/1)-CL displays rather poor performance (peak power density: 0.27 $W \cdot cm^{-2}$). The significantly better performance of the single cell assembled with NTDA-2,2′-BSPOB/BAPBz(3/1)-CL should be due to its higher proton conductivity resulting from its higher IEC.

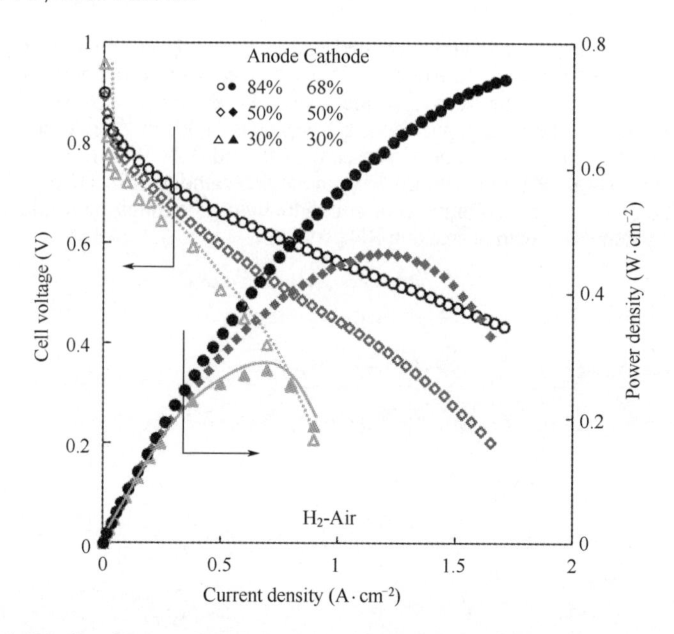

FIGURE 7.46 H_2-air fuel cell performance of the single cell using the crosslinked SPI membrane NTDA-2,2′-BSPOB/BAPBz(2/1)-CL as the PEM under different external humidification conditions (84%, 50%, and 30% RH) for both anode and cathode at 90 °C and 0.2 MPa gas pressure (Pt-loading amounts: 0.5 mg • cm^{-2} for each electrode).

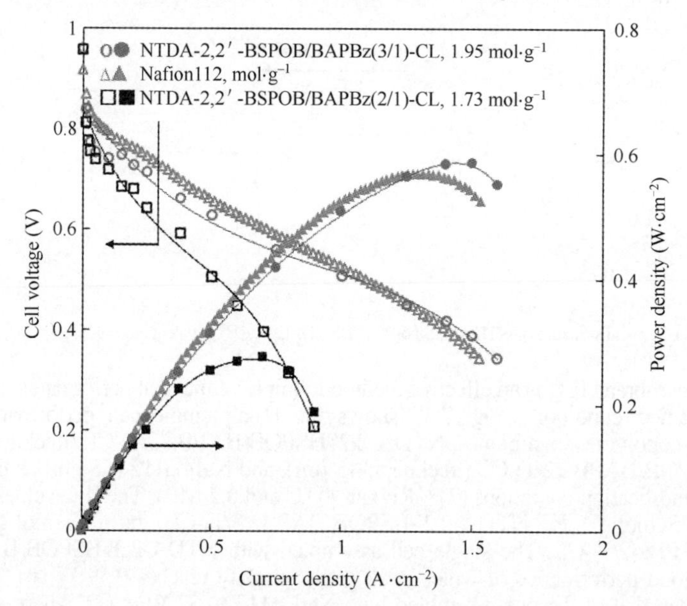

FIGURE 7.47 H_2-air single cell performances using the crosslinked SPI membranes, NTDA-2,2′-BSPOB/BAPBz(3/1)-CL (thickness: 36 μm) and NTDA-2,2′-BSPOB/BAPBz(2/1)-CL (thickness: 35 μm), and Nafion112 (55 μm) as the PEMs under rather weak humidification conditions (27% for both anode and cathode) at 90 °C and 0.2 MPa.

The effects of temperature on fuel cell performance is complex. On the one hand, increasing temperature has the benefit of accelerating electrochemical reaction rate; on the other hand, at higher temperatures membrane dehydration may become a big problem which would cause deterioration of fuel cell performance and durability. Figs. 7.48 and 7.49 show the temperature dependence of single cell performance using Nafion112 as the PEM under 30% and 70% humidification, respectively. The gas pressure is set at 0.1 MPa and the Pt-loading amounts are 0.5 mg·cm^{-2} for both anode and cathode. It is clear that under 30% humidification, the fuel cell performance deteriorates significantly with increasing temperature. The peak power density decreased from 1.01 W·cm^{-2} at 70 °C to 0.796 W·cm^{-2} at 80 °C and 0.387 W·cm^{-2} at 90 °C. This is because under weak (30%) humidification membrane dehydration is a serious problem leading to significant reduction in proton conductivity and such a negative effect is predominant as compared with the positive effect of increasing electrochemical reaction rate. However, under 70% humidification the best fuel cell performance is achieved at 80 °C and the performance at 90 °C is also significantly improved. This is likely because under 70% humidification membrane dehydration is less serious than under 30% humidification, which weakened the negative effect.

Dehydration at elevated temperatures is also observed with SPI membrane. Fig. 7.50 shows the performances of the single cell using the already-mentioned crosslinked SPI membrane NTDA-2,2′-BSPOB/BAPBz (2/1)-CL as the PEM at 90 °C and 110 °C, respectively. At 110 °C and 0.2 MPa, the fuel cell exhibits a peak power density of 0.146 W·cm^{-2} which is less than one-third of that (0.462 W·cm^{-2}) at 90 °C and 0.2 MPa. The fuel cell performance at 110 °C can be significantly improved by raising gas pressure from 0.2 to 0.3 MPa and by using the SPI membranes with higher IEC. Fig. 7.51 shows the fuel cell performance of the single cells using NTDA-2,2′-BSPOB/BAPBz (2/1)-CL, NTDA-2,2′-BSPOB/BAPBz (3/1)-CL, and Nafion212 as the PEMs at 110 °C and 0.3 MPa under 49% humidification. The fuel cell using NTDA-2,2′-BSPOB/BAPBz (2/1)-CL as the PEM exhibits a peak power density of 0.322 W·cm^{-2} which is more than double that measured at 0.2 MPa. The fuel cell using NTDA-2,2′-BSPOB/BAPBz (3/1)-CL as the PEM reaches a peak power density of 0.561 W·cm^{-2} which is 1.7-times as large as that of the fuel cell using 2,2′-BSPOB/BAPBz(2/1)-CL as the PEM because of the higher proton conductivity resulting from the higher IEC of the former.

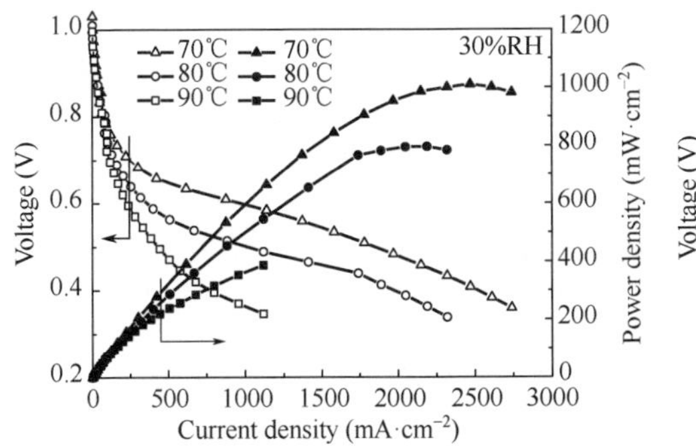

FIGURE 7.48 Performance of the single cell using Nafion112 as the PEM under 30% humidification for both anode and cathode at gas pressure of 0.1 MPa at different temperatures.

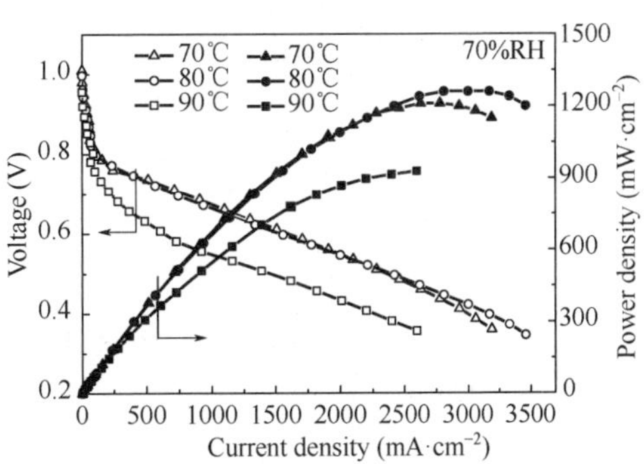

FIGURE 7.49 Performance of the single cell using Nafion112 as the PEM under 70% humidification for both anode and cathode at gas pressure of 0.1 MPa at different temperatures.

The OCV decay of the single cell using the crosslinked SPI membrane NTDA-2,2′-BSPOB/BAPBz (3/1)-CL as the PEM is fairly slow (from 0.96 to 0.85 V) after being tested at 110 °C and 0.2 MPa for 1000 h indicating fairly high durability of this membrane. In contrast, the single cell using Nafion112 is completely damaged within 3 days under the same conditions (Fig. 7.52) [13]. The high durability of the cell using NTDA-2,2′-BSPOB/BAPBz (2/1)-CL should be attributed to the good chemical stability and high thermal resistance of this SPI membrane, whereas the poor durability of the cell using Nafion212 is mainly due to the low glass transition temperature (~105 °C) of the membrane.

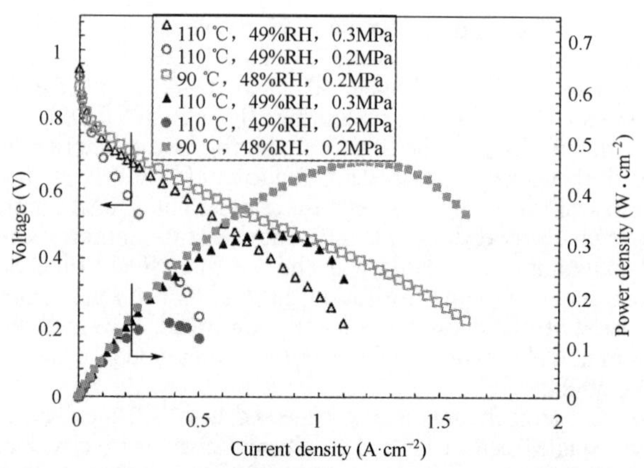

FIGURE 7.50 Fuel cell performance of the crosslinked SPI membrane NTDA-2,2′-BSPOB/BAPBz (2/1)-CL at 90 and 110 °C.

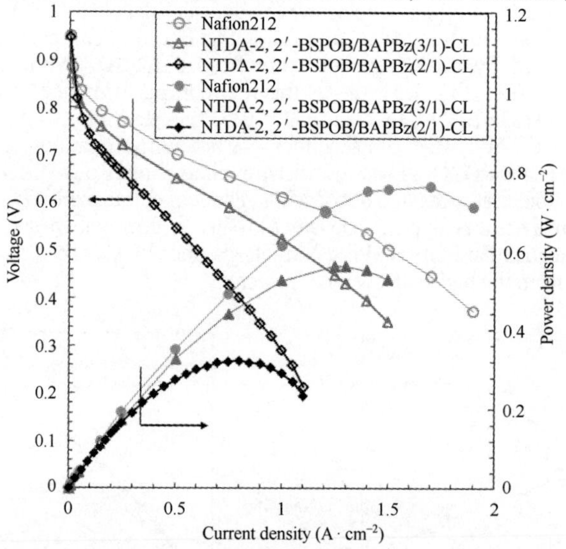

FIGURE 7.51 Performance of the single cells using NTDA-2,2′-BSPOB/BAPBz (3/1)-CL, NTDA-2,2′-BSPOB/BAPBz(2/1)-CL and Nafion212 as the PEMs at 110 °C and 0.3 MPa under 49% humidification.

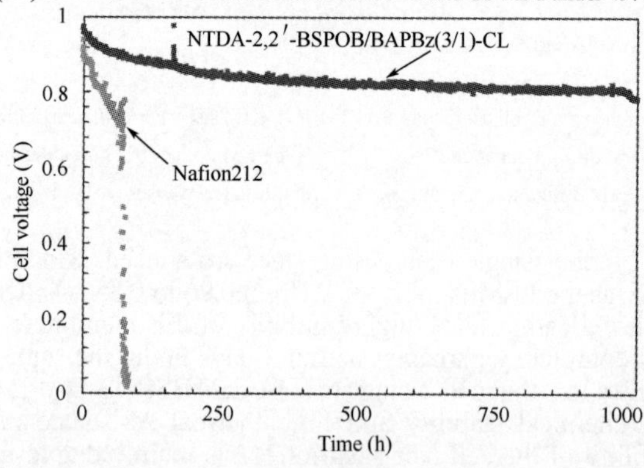

FIGURE 7.52 Durability of the single cell using the crosslinked SPI membrane NTDA-2,2′-BSPOB/BAPBz (3/1)-CL as the PEM at 110 °C and 0.2 MPa.

7.6 Summary

In the past two decades, the SPIs have been extensively studied as PEMs for fuel cell applications owing to their structural variety, excellent thermal stability, high mechanical strength and modulus, superior electric properties, and good film-forming ability. Hydrolysis-induced polymer degradation is the major drawback associated with sulfonated polyimide membranes. Structural modification using the sulfonated diamines with high basicity and linear (*para*-oriented) configuration and/or the six-membered ring dianhydrides with low electron affinity have greatly enhanced the water resistance of the membranes. Covalent crosslinking, especially the phosphorous pentoxide-catalyzed crosslinking, is also very effective for improving the water resistance of the SPI membranes. Radical oxidation-induced degradation of polymer main chains has been identified to be one of the common reasons for the deterioration C19 : D19 of fuel cells. Incorporation of benzimidazole groups into the sulfonated polyimide structures leads to significant improvements in the radical oxidative stability of the sulfonated membranes. The synergetic effect of benzimidazole groups and covalent crosslinking is much more effective than each factor alone. The SPIs derived from 2,2′-BSPB and 2,2′-BSPOB display almost zero swelling ratio in the in-plane direction despite their high IECs, which is very favorable for achieving long-term durable fuel cells. Because the sulfonic acid groups of 2,2′-BSPB are less stable to hydrolysis than those of 2,2′-BSPOB, the 2,2′-BSPOB-based SPIs are superior to the 2,2′-BSPB-based ones.

Proton conductivity is one of the most important properties of PEM. As in the case of many other sulfonated polymer membranes, the proton conductivity of the SPIs exhibited strong humidity dependence. In a low RH environment, the sulfonated polyimide membranes generally show rather low proton conductivities, making their fuel cell performance strongly dependent on external humidification conditions. Moreover, the sulfonated polyimide membranes generally exhibit anisotropic proton conductivity, i.e., the in-plane conductivity is significantly lower than the through-plane conductivity. This is unfavorable for achieving high fuel cell performance which needs to be paid more attention and solved in the future.

Currently, the covalently crosslinked SPI membranes derived from 2,2′-BSPOB such as NTDA-2,2′-BSPOB/BAPBz (3/1)-CL represent the best PEMs with the highest fuel cell performance and fairly good long-term durability at elevated temperatures (e.g., 110 °C). The radical oxidative stability of the 2,2′-BSPOB-based SPI membranes still needs to be enhanced which can be achieved by incorporating benzimidazole groups into polymer structures. The crosslinked benzimidazole groups-containing 2,2′-BSPOB-based SPI membranes are expected to have much improved long-term durability at elevated temperatures.

REFERENCES

[1] C. Genies, R. Mercier, B. Sillion, R. Petiaud, N. Cornet, G. Gebel, et al., Soluble sulfonated naphthalenic polyimides as materials for proton exchange membranes, Polymer 42 (2001) 359-373.

[2] C. Genies, R. Mercier, B. Sillion, R. Petiaud, N. Cornet, G. Gebel, et al., Stability study of sulfonated phthalic and naphthalenic polyimide structures in aqueous medium, Polymer 42 (2001) 5097-5105.

[3] J. Fang, X. Guo, S. Harada, T. Watari, K. Tanaka, H. Kita, et al., Novel sulfonated polyimides as polyelectrolytes for fuel cell application. 1. Synthesis, proton conductivity, and water stability of polyimides from 4,4′-diaminodiphenyl ether-2,2′-disulfonic acid, Macromolecules 35 (2002) 9022-9028.

[4] N. Asano, M. Aoki, S. Suzuki, K. Miyatake, H. Uchida, M. Watanabe, Aliphatic/aromatic polyimide ionomers as a proton conductive membrane for fuel cell applications, J. Am. Chem. Soc. 128 (2006) 1762-1769.

[5] J. Fang, X. Guo, H. Xu, K.I. Okamoto, Sulfonated polyimides: synthesis, proton conductivity and water stability, J. Power Sources 159 (2006) 4-11.

[6] X. Guo, J. Fang, T. Watari, K. Tanaka, H. Kita, K.I. Okamoto, Novel sulfonated polyimides as polyelectrolytes for fuel cell application. 2. Synthesis and proton conductivity of polyimides from 9,9-bis (4-aminophenyl) fluorene-2,7-disulfonic acid, Macromolecules 35 (2002) 6707-6713.

[7] T. Watari, J. Fang, K. Tanaka, H. Kita, K.I. Okamoto, T. Hirano, Synthesis, water stability and proton conductivity of novel sulfonated polyimides from 4,4′-bis(4-aminophenoxy)biphenyl-3,3′-disulfonic acid, J. Membr. Sci. 230 (2004) 111-120.

[8] Y. Yin, J. Fang, Y. Cui, K. Tanaka, H. Kita, K.I. Okamoto, Synthesis, proton conductivity and methanol permeability of a novel sulfonated polyimide from 3-(2′, 4′-diaminophenoxy) propane sulfonic acid, Polymer 44 (2003) 4509-4518.

[9] Y. Sutou, Y. Yin, Z. Hu, S. Chen, H. Kita, K.I. Okamoto, et al., Synthesis and properties of sulfonated polyimides derived from

bis(sulfophenoxy) benzidines, J. Polym. Sci., Part A: Polym. Chem. 47 (2009) 1463-1477.
[10] X. Guo, J. Fang, K. Tanaka, H. Kita, K.I. Okamoto, Synthesis and properties of novel sulfonated polyimides from 2,2′-bis(4-aminophenoxy) biphenyl-5,5′-disulfonic acid, J. Polym. Sci., Part A: Polym. Chem. 42 (2004) 1432-1440.
[11] B.R. Einsla, Y.T. Hong, Y. Seung Kim, F. Wang, N. Gunduz, J.E. McGrath, Sulfonated naphthalene dianhydride based polyimide copolymers for proton-exchange-membrane fuel cells. I. Monomer and copolymer synthesis, J. Polym. Sci., Part A: Polym. Chem. 42 (2004) 862-874.
[12] N. Endo, K. Matsuda, K. Yaguchi, Z. Hu, K. Chen, M. Higa, et al., Cross-linked sulfonated polyimide membranes for polymer electrolyte fuel cells, J. Electrochem. Soc. 156 (2009) B628-B633.
[13] K. Yaguchi, K. Chen, N. Endo, M. Higa, K.I. Okamoto, Crosslinked membranes of sulfonated polyimides for polymer electrolyte fuel cell applications, J. Power Sources 195 (2010) 4676-4684.
[14] K.I. Okamoto, K. Yaguchi, H. Yamamoto, K. Chen, N. Endo, M. Higa, et al., Sulfonated polyimide hybrid membranes for polymer electrolyte fuel cell applications, J. Power Sources 195 (2010) 5856-5861.
[15] Z. Hu, Y. Yin, K. Yaguchi, N. Endo, M. Higa, K.I. Okamoto, Synthesis and properties of sulfonated multiblock copolynaphthalimides, Polymer 50 (2009) 2933-2943.
[16] X. Guo, W. Li, J. Fang, Y. Yin, Synthesis and properties of novel multiblock copolyimides consisting of benzimidazole-groups-containing sulfonated polyimide hydrophilic blocks and non-sulfonated polyimide hydrophobic blocks as proton exchange membranes, Electrochim. Acta 177 (2015) 151-160.
[17] N. Asano, K. Miyatake, M. Watanabe, Hydrolytically stable polyimide ionomer for fuel cell applications, Chem. Mater. 16 (2004) 2841-2843.
[18] Y. Yin, O. Yamada, Y. Suto, T. Mishima, K. Tanaka, H. Kita, et al., Synthesis and characterization of proton-conducting copolyimides bearing pendant sulfonic acid groups, J. Polym. Sci., Part A: Polym. Chem. 43 (2005) 1545-1553.
[19] Y. Yin, S. Hayashi, O. Yamada, H. Kita, K.I. Okamoto, Branched/crosslinked sulfonated polyimide membranes for polymer electrolyte fuel cells, Macromol. Rapid Commun. 26 (2005) 696-700.
[20] H. Zhou, K. Miyatake, M. Watanabe, Polyimide electrolyte membranes having fluorenyl and sulfopropoxy groups for high temperature PEFCs, Fuel Cells 5 (2005) 296-301.
[21] Y. Yin, O. Yamada, K. Tanaka, K.I. Okamoto, On the development of naphthalene-based sulfonated polyimide membranes for fuel cell applications, Polym. J. 38 (2006) 197-219.
[22] S. Chen, Y. Yin, K. Tanaka, H. Kita, K.I. Okamoto, Synthesis and properties of novel side-chain-sulfonated polyimides from bis[4-(4-aminophenoxy)-2-(3-sulfobenzoyl)] phenyl sulfone, Polymer 47 (2006) 2660-2669.
[23] Z. Hu, Y. Yin, S. Chen, O. Yamada, K. Tanaka, H. Kita, et al., Synthesis and properties of novel sulfonated (co) polyimides bearing sulfonated aromatic pendant groups for PEFC applications, J. Polym. Sci., Part A: Polym. Chem. 44 (2006) 2862-2872.
[24] T. Yasuda, Y. Li, K. Miyatake, M. Hirai, M. Nanasawa, M. Watanabe, Synthesis and properties of polyimides bearing acid groups on long pendant aliphatic chains, J. Polym. Sci., Part A: Polym. Chem. 44 (2006) 3995-4005.
[25] Y. Yin, O. Yamada, S. Hayashi, K. Tanaka, H. Kita, K.I. Okamoto, Chemically modified proton-conducting membranes based on sulfonated polyimides: Improved water stability and fuel-cell performance, J. Polym. Sci., Part A: Polym. Chem. 44 (2006) 3751-3762.
[26] Y. Li, R. Jin, Z. Wang, Z. Cui, W. Xing, L. Gao, Synthesis and properties of novel sulfonated polyimides containing binaphthyl groups as proton-exchange membranes for fuel cells, J. Polym. Sci., Part A: Polym. Chem. 45 (2007) 222-231.
[27] Y. Li, R. Jin, Z. Cui, Z. Wang, W. Xing, X. Qiu, et al., Synthesis and characterization of novel sulfonated polyimides from 1, 4-bis (4-aminophenoxy)- naphthyl-2,7-disulfonic acid, Polymer 48 (2007) 2280-2287.
[28] H.Y. Pan, Y.F. Liang, X.L. Zhu, X.G. Jian, Synthesis and characterization of novel sulfonated polyimide containing phthalazinone moieties as PEM for PEMFC, Chin. Chem. Lett. 18 (2007) 1148-1150.
[29] M. Cakir, S. Karata¸s, Y. Menceloğlu, N. Kayaman-Apohan, A. Güngör, Phosphorus-containing sulfonated polyimides for proton exchange membranes, Macromol. Chem. Phys. 209 (2008) 919-929.
[30] K. Miyatake, T. Yasuda, M. Watanabe, Substituents effect on the properties of sulfonated polyimide copolymers, J. Polym. Sci., Part A: Polym. Chem. 46 (2008) 4469-4478.
[31] O. Savard, T.J. Peckham, Y. Yang, S. Holdcroft, Structure-property relationships for a series of polyimide copolymers with sulfonated pendant groups, Polymer 49 (2008) 4949-4959.
[32] K. Chen, X. Chen, K. Yaguchi, N. Endo, M. Higa, K.I. Okamoto, Synthesis and properties of novel sulfonated polyimides bearing sulfophenyl pendant groups for fuel cell application, Polymer 50 (2009) 510-518.
[33] R. Lei, C. Kang, Y. Huang, Y. Li, X. Wang, R. Jin, et al., Novel sulfonated polyimide ionomers by incorporating pyridine functional group in the polymer backbone, J. Appl. Polym. Sci. 114 (2009) 3190-3197.
[34] B.K. Chen, T.Y. Wu, J.M. Wong, Y.M. Chang, H.F. Lee, W.Y. Huang, et al., Highly sulfonated diamine synthesized polyimides and protic ionic liquid composite membranes improve PEM conductivity, Polymers 7 (2015) 1046-1065.
[35] Y. Yin, J. Fang, H. Kita, K.I. Okamoto, Novel sulfoalkoxylated polyimide membrane for polymer electrolyte fuel cells, Chem. Lett. 32 (4) (2003) 328-329.
[36] J. Saito, M. Tanaka, M. Hirai, M. Nanasawa, K. Miyatake, M. Watanabe, Polyimide ionomer containing superacid groups,

Polym. Adv. Technol. 22 (8) (2011) 1305-1310.
[37] Y. Yin, Q. Du, Y. Qin, Y. Zhou, K.I. Okamoto, Sulfonated polyimides with flexible aliphatic side chains for polymer electrolyte fuel cells, J. Membr. Sci. 367 (2011) 211-219.
[38] J. Fang, F. Zhai, X. Guo, H. Xu, K.I. Okamoto, A facile approach for the preparation of cross-linked sulfonated polyimide membranes for fuel cell application, J. Mater. Chem. 17 (11) (2007) 1102-1108.
[39] F. Zhai, X. Guo, J. Fang, H. Xu, "Synthesis and properties of novel sulfonated polyimide membranes for direct methanol fuel cell application," J. Membr. Sci. 296 (2007) 102-109.
[40] L. Akbarian-Feizi, S. Mehdipour-Ataei, H. Yeganeh, Synthesis of new sulfonated copolyimides in organic and ionic liquid media for fuel cell application, J. Appl. Polym. Sci. 124 (2012) 1981-1992.
[41] K. Tanaka, M.N. Islam, M. Kido, H. Kita, K.I. Okamoto, Gas permeation and separation properties of sulfonated polyimide membranes, Polymer 47 (2006) 4370-4377.
[42] Y.H. Li, W.J. Wang, R.Z. Jin, L.X. Gao, "Synthesis and characterization of novel sulfonated polyimides from 4,6-bis(4-aminophenoxy)-naphthalene-2-sulfonic acid,", Chem. Res. Chin. Univ. 29 (2013) 1225-1228.
[43] X. Guo, F. Zhai, J. Fang, M.F. Laguna, M. López-González, E. Riande, Permselectivity and conductivity of membranes based on sulfonated naphthalenic copolyimides, J. Phys. Chem. B 111 (2007) 13694-13702.
[44] K. Chen, Z. Hu, N. Endo, M. Higa, K.I. Okamoto, Sulfonated multiblock copolynaphthalimides for polymer electrolyte fuel cell application, Polymer 52 (2011) 2255-2262.
[45] Y. Yin, J. Fang, T. Watari, K. Tanaka, H. Kita, K.I. Okamoto, Synthesis and properties of highly sulfonated proton conducting polyimides from bis(3-sulfopropoxy)benzidine diamines, J. Mater. Chem. 14 (2004) 1062-1070.
[46] Z. Hu, Y. Yin, H. Kita, K.I. Okamoto, Y. Suto, H. Wang, et al., Synthesis and properties of novel sulfonated polyimides bearing sulfophenyl pendant groups for polymer electrolyte fuel cell application, Polymer 48 (2007) 1962-1971.
[47] Y. Yin, S. Chen, X. Guo, J. Fang, K. Tanaka, H. Kita, et al., Structure-property relationship of polyimides derived from sulfonated diamine isomers, High Perform. Polym. 18 (2006) 617-635.
[48] A. Rabiee, S. Mehdipour-Ataei, Physical and mechanical properties of sulfonated aromatic copolyimide membranes, e-Polymers 9 (2009) 1314-1323.
[49] T. Watari, H. Wang, K. Kuwahara, K. Tanaka, H. Kita, K.I. Okamoto, Water vapor sorption and diffusion properties of sulfonated polyimide membranes, J. Membr. Sci. 219 (2003) 137-147.
[50] N. Li, J. Liu, Z. Cui, S. Zhang, W. Xing, Novel hydrophilic-hydrophobic multiblock copolyimides as proton exchange membranes: enhancing the proton conductivity, Polymer 50 (2009) 4505-4511.
[51] F.M. Abu-Orabi, M.H. Kailani, B.A. Sweileh, M.Y. Mustafa, M. Al-Hussein, Sulfonated polyimide copolymers based on 4,4′-diaminostilbene-2,2′-disulfonic acid and 3,5,3′,5′-tetramethylbenzidine with enhanced solubility, Polym. Bull. 74 (2017) 895-909.
[52] P. Sarkar, A.K. Mohanty, P. Bandyopadhyay, S. Chattopadhyay, S. Banerjee, Proton exchange properties of flexible diamine-based new fluorinated sulfonated polyimides, RSC Adv. 4 (2014) 11848-11858.
[53] H. Deligöz, S. Vatansever, F. Ö ksüzömer, S.N. Koc,, S. Ö zgümü ,s, M.A. Gürkaynak, Synthesis and characterization of sulfonated homo-and copolyimides based on 2,4 and 2,5-diaminobenzenesulfonic acid for proton exchange membranes, Polym. Adv. Technol. 19 (2008) 1792-1802.
[54] H. Deligöz, S. Vantansever, S.N. Koc,, F. Ö ksüzömer, S. Ö zgümü ,s, M.A. Gürkaynak, Preparation of sulfonated copolyimides containing aliphatic linkages as proton-exchange membranes for fuel cell applications, J. Appl. Polym. Sci. 110 (2008) 1216-1224.
[55] B.K. Chen, T.Y. Wu, C.W. Kuo, Y.C. Peng, I.C. Shih, L. Hao, et al., 4,4′-Oxydianiline (ODA) containing sulfonated polyimide/protic ionic liquid composite membranes for anhydrous proton conduction, Int. J. Hydrogen Energy 38 (2013) 11321-11330.
[56] B.K. Chen, J.M. Wong, T.Y. Wu, L.C. Chen, I. Shih, Improving the conductivity of sulfonated polyimides as proton exchange membranes by doping of a protic ionic liquid, Polymers 6 (2014) 2720-2736.
[57] M. Rodgers, Y. Yang, S. Holdcroft, A study of linear versus angled rigid rod polymers for proton conducting membranes using sulfonated polyimides, Eur. Polym. J. 42 (2006) 1075-1085.
[58] R. Lei, L.X. Gao, R.Z. Jin, X.P. Qiu, Sulfonated polyimides containing 1, 2, 4-triazole groups for proton exchange membranes, Chinese J. Polym. Sci. 32 (2013) 941-952.
[59] R.P. Pandey, V.K. Shahi, Sulphonated imidized graphene oxide (SIGO) based polymer electrolyte membrane for improved water retention, stability and proton conductivity, J. Power Sources 299 (2015) 104-113.
[60] X. Gu, N. Xu, X. Guo, J. Fang, Synthesis, proton conductivity and chemical stability of novel sulfonated copolyimides-containing benzimidazole groups for fuel cell applications, High Perform. Polym. 25 (2013) 508-517.
[61] Z. Yue, Y.B. Cai, S. Xu, Phosphoric acid-doped organic-inorganic cross-linked sulfonated poly (imide-benzimidazole) for high temperature proton exchange membrane fuel cells, Int. J. Hydrogen Energy 41 (2016) 10421-10429.
[62] W. Li, X. Guo, D. Aili, S. Martin, Q. Li, J. Fang, Sulfonated copolyimide membranes derived from a novel diamine monomer with pendant benzimidazole groups for fuel cells, J. Membr. Sci. 481 (2015) 44-53.
[63] S. Sundar, W. Jang, C. Lee, Y. Shul, H. Han, Crosslinked sulfonated polyimide networks as polymer electrolyte membranes in

fuel cells, J. Polym. Sci., Part B: Polym. Phys. 43 (2005) 2370-2379.

[64] G. Wang, K. Yamazaki, M. Tanaka, H. Kawakami, Polymer electrolyte characteristics of sulfonated block-graft polyimide membranes: Influence of block ratio, J. Photopolym. Sci. Technol. 29 (2016) 259-263.

[65] H. Liu, M.H. Lee, J. Lee, Synthesis of new sulfonated polyimide and its photo-crosslinking for polymer electrolyte membrane fuel cells, Macromol. Res. 17 (2009) 725-728.

[66] K. Krishnan, H. Iwatsuki, M. Hara, S. Nagano, Y. Nagao, Influence of molecular weight on molecular ordering and proton transport in organized sulfonated polyimide thin films, J. Phys. Chem. C 119 (2015) 21767-21774.

[67] N. Li, Z. Cui, S. Zhang, W. Xing, Sulfonated polyimides bearing benzimidazole groups for proton exchange membranes, Polymer 48 (2007) 7255-7263.

[68] K. Miyatake, N. Asano, M. Watanabe, Synthesis and properties of novel sulfonated polyimides containing 1,5-naphthylene moieties, J. Polym. Sci., Part A: Polym. Chem. 41 (2003) 3901-3907.

[69] F. Zhang, N. Li, Z. Cui, S. Zhang, S. Li, Novel acid-base polyimides synthesized from binaphthalene dianhydride and triphenylaminecontaining diamine as proton exchange membranes, J. Membr. Sci. 314 (2008) 24-32.

[70] A. Ganeshkumar, D. Bera, E.A. Mistri, S. Banerjee, Triphenyl amine containing sulfonated aromatic polyimide proton exchange membranes, Eur. Polym. J. 60 (2014) 235-246.

[71] R. Lei, C.Q. Kang, Y.J. Huang, X.P. Qiu, X.L. Ji, W. Xing, et al., Sulfonated polyimides containing pyridine groups as proton exchange membrane materials, Chinese J. Polym. Sci. 29 (2011) 532-539.

[72] H. Yao, N. Song, K. Shi, S. Feng, S. Zhu, Y. Zhang, et al., Highly sulfonated co-polyimides containing hydrophobic cross-linked networks as proton exchange membranes, Polym. Chem. 7 (2016) 4728-4735.

[73] G. Zhang, X. Guo, J. Fang, K. Chen, K.I. Okamoto, Preparation and properties of covalently cross-linked sulfonated copolyimide membranes containing benzimidazole groups, J. Membr. Sci. 326 (2009) 708-713.

[74] S. Yuan, C. del Rio, M. López-González, X. Guo, J. Fang, E. Riande, Impedance spectroscopy and performance of cross-linked new naphthalenic polyimide acid membranes, J. Phys. Chem. C 114 (2010) 22773-22782.

[75] C.H. Lee, S.H. Chen, Y.Z. Wang, C.C. Lin, C.K. Huang, C.N. Chuang, et al., Preparation and characterization of proton exchange membranes based on semi-interpenetrating sulfonated poly (imide-siloxane)/epoxy polymer networks, Energy 55 (2013) 905-915.

[76] C. Gong, Y. Liang, Z. Qi, H. Li, Z. Wu, Z. Zhang, et al., Solution processable octa (aminophenyl) silsesquioxane covalently cross-linked sulfonated polyimides for proton exchange membranes, J. Membr. Sci. 476 (2015) 364-372.

[77] L. Zou, M. Anthamatten, Synthesis and characterization of polyimide-polysiloxane segmented copolymers for fuel cell applications, J. Polym. Sci., Part A: Polym. Chem. 45 (16) (2007) 3747-3758.

[78] E.A. Mistri, A.K. Mohanty, S. Banerjee, H. Komber, B. Voit, Naphthalene dianhydride based semifluorinated sulfonated copoly (ether imide) s: Synthesis, characterization and proton exchange properties, J. Membr. Sci. 441 (2013) 168-177.

[79] E.A. Mistri, A.K. Mohanty, S. Banerjee, Synthesis and characterization of new fluorinated poly (ether imide) copolymers with controlled degree of sulfonation for proton exchange membranes, J. Membr. Sci. 411 (2012) 117-129.

[80] H. Yao, K. Shi, N. Song, N. Zhang, P. Huo, S. Zhu, et al., Polymer electrolyte membranes based on cross-linked highly sulfonated copolyimides, Polymer 103 (2016) 171-179.

[81] J.C. Chen, J.A. Wu, C.Y. Lee, M.C. Tsai, K.H. Chen, Novel polyimides containing benzimidazole for temperature proton exchange membrane fuel, J. Membr. Sci. 483 (2015) 144-154.

[82] K. Miyatake, H. Furuya, M. Tanaka, M. Watanabe, Durability of sulfonated polyimide membrane in humidity cycling for fuel cell applications, J. Power Sources 204 (2012) 74-78.

[83] T. Okanishi, Y. Tsuji, Y. Sakiyama, S. Matsuno, B. Bae, K. Miyatake, et al., Effect of PEFC operating conditions on the durability of sulfonated polyimide membranes, Electrochim. Acta 58 (2011) 589-598.

[84] N. Xu, PhD thesis, Shanghai Jiao Tong University, 2009.

[85] R. Jin, Y. Li, W. Xing, X. Qiu, X. Ji, L. Gao, Preparation and properties of ionic cross-linked sulfonated copolyimide membranes containing pyrimidine groups, Polym. Adv. Technol. 23 (2012) 31-37.

[86] W. Li, X. Guo, J. Fang, Synthesis and properties of sulfonated polyimide-polybenzimidazole copolymers as proton exchange membranes, J. Mater. Sci. 49 (2014) 2745-2753.

[87] H. Wei, G. Chen, L. Cao, Q. Zhang, Q. Yan, X. Fang, Enhanced hydrolytic stability of sulfonated polyimide ionomers using bis(naphthalic anhydrides) with low electron affinity, J. Mater. Chem. A 1 (2013) 10412-10421.

[88] H. Wei, X. Fang, Novel aromatic polyimide ionomers for proton exchange membranes: Enhancing the hydrolytic stability, Polymer 52 (2011) 2735-2739.

[89] X. Guo, S. Yuan, J. Fang, Synthesis and properties of novel sulfonated polyimides from 4, 4′-(biphenyl-4, 4′-diyldi (oxo))bis (1, 8-naphthalic anhydride), Polymer 59 (2015) 207-214.

[90] J.P. Gao, Z.Y. Wang, Synthesis and properties of polyimides from 4,40-binaphthyl-l,1′,8,8′-tetracarboxylic dianhydride, J. Polym. Sci., Part A: Polym. Chem. 33 (1995) 1627-1635.

[91] F. Zhang, N. Li, S. Zhang, S. Li, Ionomers based on multisulfonated perylene dianhydride: Synthesis and properties of water resistant sulfonated polyimides, J. Power Sources 195 (2010) 2159-2165.

[92] X. Chen, Y. Yin, K. Tanaka, H. Kita, K.I. Okamoto, Synthesis and characterization of novel sulfonated polyimides derived from naphthalenic dianhydride, High Perform. Polym. 18 (2006) 637-654.

[93] D. Sek, A. Wanic, E. Schab-Balcerzak, Investigation of polyamides containing naphthalene units. III. Influence of monomers structure on polymers properties, J. Polym. Sci., Part A: Polym. Chem. 35 (1997) 539-545.

[94] D. Sek, A. Wanic, E. Schab-Balcerzak, Investigation of polyamides containing naphthalene units. II. Model compound synthesis, J. Polym. Sci., Part A: Polym. Chem. 33 (1995) 547-554.

[95] L. Ying, X. Guo, J. Fang, Synthesis, freestanding membrane, formation, and properties of novel sulfonated hyperbranched polyimides, High Performance Polymers 30 (2018) 3-15.

[96] H. Deligöz, S. Vatansever, F. Ö ksüzömer, S.N. Koc,, S. Ö zgümü ,s, M.A. Gürkaynak, Preparation and characterization of sulfonated polyimide ionomers via post-sulfonation method for fuel cell applications, Polym. Adv. Technol. 19 (2008) 1126-1132.

[97] H. Bi, S. Chen, X. Chen, K. Chen, N. Endo, M. Higa, et al., Poly(sulfonated phenylene)-block-polyimide copolymers for fuel cell applications, Macromol. Rapid Commun. 30 (2009) 1852-1856.

[98] K. Yamazaki, M. Tanaka, H. Kawakami, Preparation and characterization of sulfonated block-graft copolyimide/sulfonated polybenzimidazole blend membranes for fuel cell application, Polym. Int. 64 (2015) 1079-1085.

[99] M.R. Hibbs, C.J. Cornelius, Ion transport within random-sulfonated and block-sulfonated copolyimides, J. Mater. Sci. 48 (2013) 1303-1309.

[100] C.F. Kins, E. Sengupta, A. Kaltbeitzel, M. Wagner, I. Lieberwirth, H.W. Spiess, et al., Morphological anisotropy and proton conduction in multiblock copolyimide electrolyte membranes, Macromolecules 47 (2014) 2645-2658.

[101] C.H. Park, C.H. Lee, J.Y. Sohn, H.B. Park, M.D. Guiver, Y.M. Lee, Phase separation and water channel formation in sulfonated block copolyimide, J. Phys. Chem. B 114 (2010) 12036-12045.

[102] Y. Iizuka, M. Tanaka, H. Kawakami, Preparation and proton conductivity of phosphoric acid-doped blend membranes composed of sulfonated block copolyimides and polybenzimidazole, Polym. Int. 62 (2013) 703-708.

[103] K. Yamazaki, Y. Tang, H. Kawakami, Proton conductivity and stability of low-IEC sulfonated block copolyimide membrane, J. Membr. Sci. 362 (2010) 234-240.

[104] D. Jamróz, Y. Maréchal, Hydration of sulfonated polyimide membranes, II. Water uptake and hydration mechanisms of protonated homopolymer and block copolymers, J. Phys. Chem. B 109 (2005) 19664-19675.

[105] J. Fang, X. Guo, M. Litt, Synthesis and properties of novel sulfonated polyimides for fuel cell application, Trans. Mater. Soc. Japan 29 (6) (2004) 2541-2546.

[106] Y. Yin, Y. Suto, T. Sakabe, S. Chen, S. Hayashi, T. Mishima, et al., Water stability of sulfonated polyimide membranes, Macromolecules 39 (2006) 1189-1198.

[107] J. Yan, C. Liu, Z. Wang, W. Xing, M. Ding, Water resistant sulfonated polyimides based on 4,4′-binaphthyl-1, 1′,8,8′-tetracarboxylic dianhydride (BNTDA) for proton exchange membranes, Polymer 48 (2007) 6210-6214.

[108] J. Kerres, Blended and cross-linked ionomer membranes for application in membrane fuel cells, Fuel Cells 5 (2005) 230-247.

[109] Q. Guo, P.N. Pintauro, H. Tang, S. O'Connor, Sulfonated and crosslinked polyphosphazene-based proton-exchange membranes, J. Membr. Sci. 154 (1999) 175-181.

[110] S.D. Mikhailenko, G.P. Robertson, M.D. Guiver, S. Kaliaguine, Properties of PEMs based on cross-linked sulfonated poly(ether ether ketone), J. Membr. Sci. 285 (2006) 306-316.

[111] S.-J. Yang, W. Jang, C. Lee, Y.G. Shul, H. Han, The Effect of crosslinked networks with poly(ethylene glycol) on sulfonated polyimide for polymer electrolyte membrane fuel cell, J. Polym. Sci. Part B: Polym. Phys. 43 (2005) 1455-1464.

[112] C.H. Lee, H.B. Park, Y.S. Chung, Y.M. Lee, B.D. Freeman, Water sorption, proton conduction, and methanol permeation properties of sulfonated polyimide membranes cross-linked with N,N-bis(2-hydroxyethyl)-2-aminoethanesulfonic acid (BES), Macromolecules 39 (2006) 755-764.

[113] V. Ayala, D. Munoz, A.E. Lozano, J.G.D.L. Campa, J.D. Abajo, Synthesis, characterization, and properties of new sequenced poly(ether amides)s based on 2-(4-aminophenyl)-5-aminobenzimidazole and 2-(3-aminophenyl)-5-aminobenzimidazole, J. Polym. Sci., Part A: Polym.Chem. 44 (2006) 1414-1423.

[114] R. Borup, J. Meyers, B. Pivovar, Y.S. Kim, R. Mukundan, N. Garland, et al., Scientific aspects of polymer electrolyte fuel cell durability and degradation, Chem. Rev. 107 (2007) 3904-3951.

[115] H. Xu, K. Chen, X. Guo, J. Fang, J. Yin, Synthesis of novel sulfonated polybenzimidazole and preparation of cross-linked membranes for fuel cell application, Polymer 48 (2007) 5556-5564.

[116] J. Jouanneau, R. Mercier, L. Gonon, G. Gebel, Synthesis of sulfonated polybenzimidazoles from functionalized monomers: preparation of ionic conducting membranes, Macromolecules 40 (2007) 983-990.

[117] K.I. Okamoto, Y. Yin, O. Yamada, M.N. Islam, T. Honda, T. Mishima, et al., Methanol permeability and proton conductivity of sulfonated co-polyimide membranes, J. Membr. Sci. 258 (2005) 115-122.

[118] B.R. Einsla, Y.S. Kim, M.A. Hickner, Y.-T. Hong, M.L. Hill, B.S. Pivovar, et al., Sulfonated naphthalene dianhydride based polyimide copolymers for proton-exchange-membrane fuel cells II. Membrane properties and fuel cell performance, J. Membr. Sci. 255 (2005) 141-148.

Chapter 8

Soluble and Low-k Polyimide Materials

Yi Zhang[1] and Wei Huang[2]
[1]*Sun Yat-sen University, Guangzhou, China*
[2]*Shanghai Jiao Tong University, Shanghai, China*

8.1 Introduction

Aromatic polyimides (PIs) are considered as a class of high-performance polymers and are widely applied in many high-tech fields including aerospace, microelectronics, engineering, and so on [1,2]. In the 1960s, aromatic PIs were first developed and soon became very important materials because of their high thermal stability, excellent mechanical properties, outstanding dielectric properties, and wonderful chemical resistance, etc. However, their applications are limited by the difficulties in synthesis and processing. On the one hand, traditional aromatic PIs do not show any suitable flow properties and are unable to process in the molten state (i.e., injection or extrusion molding). Therefore, some special methods, such as compression or sintering molding, must be adopted to process them. On the other hand, traditional aromatic PIs possess an extremely rigid molecular structure (i.e., heterocyclic imide rings and aromatic rings in the backbone) which makes them insoluble in any organic media such that their solution-process is also found to be impracticable. Fortunately, they can be processed through their soluble precursors (polyamic acids, PAAs) and then converted into PIs through the thermal imidization at high temperatures (up to 300 °C). However, such high temperatures damage the substrate dramatically and thus harm the final material properties. For example, when aromatic PIs are applied in the fields of microelectronics or opto-electric displays by introducing some functional groups (e.g., long-chain alkyl, nonlinear optical chromophore, etc.), most of them cannot endure a high-temperature process. The thermal decomposition of these functional groups occurs in the thermal imidization procedure at high temperature and destroys the optoelectronic properties of the resulting PIs [1,2]. Accordingly the chemical imidization at relatively low temperature (below 180 °C) is often used to prepare these functional PIs, but this method usually requires that the PI has a high solubility in organic solvents. Otherwise, it would be precipitated from the reaction mixture and cannot be further processed. In addition, other inherent problems affect the final properties of PIs, e.g., the precursor PAA is usually unstable with a decrease in molecular weight in storage and the byproduct (H_2O) is released during the imidization process, etc.

To overcome the above problems of aromatic PIs, many efforts have been made to synthesize soluble aromatic PIs by designing novel diamine or dianhydride monomers, which can decrease molecular order, torsional mobility, and intermolecular bonding. The detailed approaches have been described as follows: (1) introduction of flexible linkages to reduce chain stiffness; (2) introduction of side groups to hinder molecular chain packing and crystallization; (3) introduction of asymmetric units to suppress coplanar structures; (4) introduction of alicyclic units to lower regularity and molecular ordering. It is noted that the factors that are beneficial to improving the solubility of aromatic PIs often conflict with other important properties, such as mechanical properties, thermal resistance, or chemical resistance. Therefore, the degree of modification should be adjusted to optimize the balance of various properties of aromatic PIs. Furthermore, the practical or potential

Advanced Polyimide Materials. DOI: https://doi.org/10.1016/B978-0-12-812640-0.00008-1

applications of soluble aromatic PIs are also discussed here.

8.2 Structures and Properties of Soluble Aromatic Polyimides

8.2.1 Soluble Polyimides With Flexible Backbones

The introduction of flexible linkages into the backbones of aromatic PIs is an effective strategy to improve their solubility. When flexible linkages appear in PI main chains, the chain rigidity can be decreased while the chain mobility can be enhanced, which results in an increase in their solubility. In addition, some flexible linkages, including —C═O—, —SO_2—, and —CHOH—, etc., have a strong interaction with the polar organic solvents and are also helpful in improving the solubility of corresponding PIs.

The commercial soluble aromatic polyimide P84 is manufactured by Evonik Fibres and its chemical structure is shown in Fig. 8.1 [3]. As can be seen, P84 contains the flexible linkage of —C═O— in its backbone and can be dissolved in common polar solvents, such as DMF and NMP. The good solubility of P84 is attributed to the increased flexibility of its backbone.

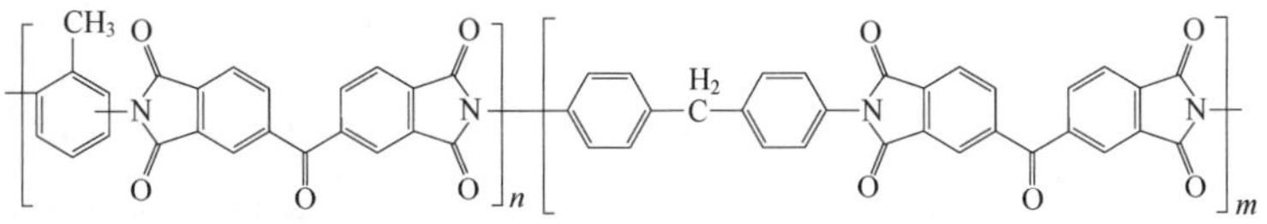

FIGURE 8.1 The structure of the commercial soluble and fusible polyimide P84.

Ghatge et al. [4] synthesized a series of PIs containing the flexible linkage —C═O— through the polycondensation of BTDA and a series of aromatic diamines with different flexible side substituents (Fig. 8.2A). All of them could be dissolved in polar solvents such as DMSO and NMP. On the contrary, when PMDA instead of BTDA was used to react with the same diamines, the resulting PIs (Fig. 8.2B) were insoluble in any organic solvents. This also confirmed the function of flexible linkage —C═O— to enhance the solubility of PIs. However, the T_{10} (the temperature for 10% weight loss at TGA measurement) in the air of the PIs from BTDA (from 425 °C to 470 °C) were relatively lower than that of the PIs from PMDA (from 460 °C to 490 °C). This indicates that the introduction of flexible linkages will decrease the thermal stability of aromatic PIs to a certain extent.

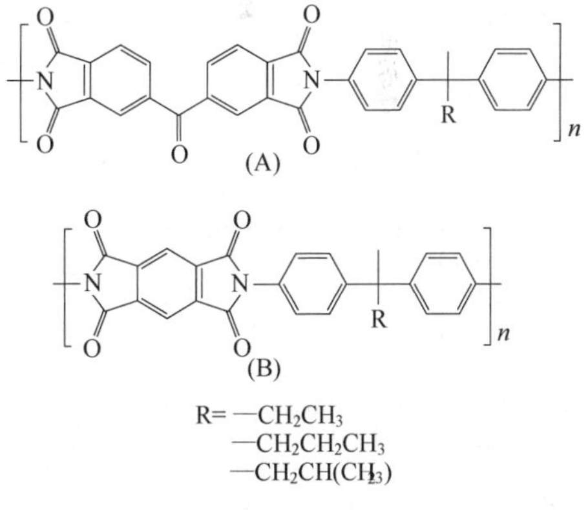

FIGURE 8.2 The polyimides with different flexible side groups.

An aromatic diamine containing two kinds of flexible linkages (—O— and —C═O—) at the same time was synthesized by Zhang et al. [5] through the Friedel-Crafts acylation. The corresponding PI (Fig. 8.3) prepared from this diamine and the dianhydride OPDA by the chemical imidization dissolved easily in common organic solvents such as DMAc and THF. The excellent solubility of this PI was first attributed to the flexibility of the linkages (—O— and —C═O—), and then to the polarizability of the nitrogen atom in the pyridine ring. The glass transition temperature (T_g) of this PI was only 209 °C and its T_{10} is 531 °C. Its relatively lower T_g was ascribed to the decrease in the chain rigidity by introducing flexible linkages.

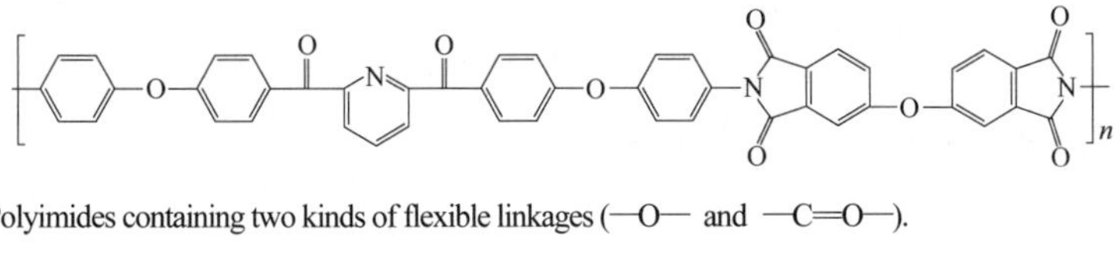

FIGURE 8.3 Polyimides containing two kinds of flexible linkages (—O— and —C═O—).

—CHOH— was another kind of flexible linkage and was also usually introduced into PI backbones to enhance their solubility. For example, Connell et al. synthesized two PIs containing the —CHOH— linkage and their chemical structures are shown in Fig. 8.4. Both of them could also easily dissolve in polar organic solvents such as DMAc and *m*-cresol [6].

Malinge et al. [7] also reported some PIs containing the linkage —CHOH— in Fig. 8.5, and a similar result was obtained, i.e., these PIs displayed the high solubility in polar solvents. They argued that the improved solubility of these PIs could be attributed to the strong interaction of the —OH in the linkage of —CHOH— with the polar solvents.

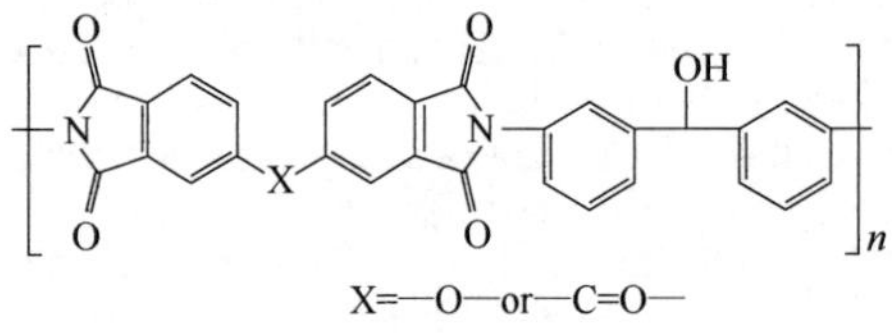

X=—O—or—C=O—

FIGURE 8.4 Polyimides containing the flexible linkage —CHOH—.

X=—CH$_2$—or—O—

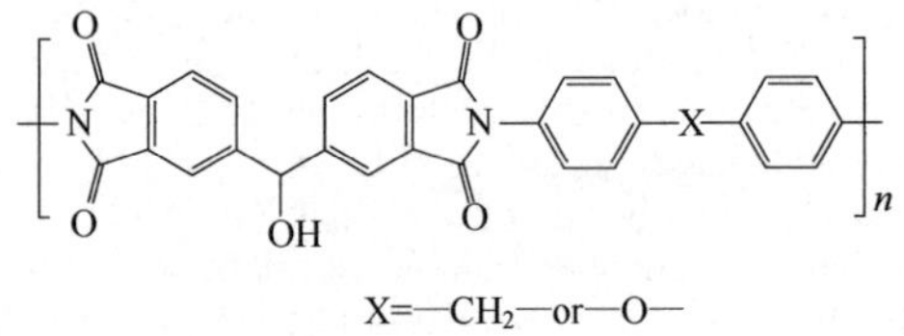

FIGURE 8.5 Polyimides containing the flexible linkage —CHOH—.

The linkage —S=O— was often introduced into the backbones of PIs to increase their solubility. 3,3′,4,4′-Diphenylsulfonetetracarboxylic dianhydride (DSDA, Fig. 8.6) was a commercial aromatic dianhydride that was often used to polymerize with various aromatic diamines to synthesize soluble PIs. For example, Kawashima [8] prepared a series of PIs from DSDA with diamines, and the resulting PIs showed better solubility in aprotic polar solvents than that of the conventional PI. They even displayed the increased thermoplasticity and their T_{10}s ranged from 475 °C to 515 °C.

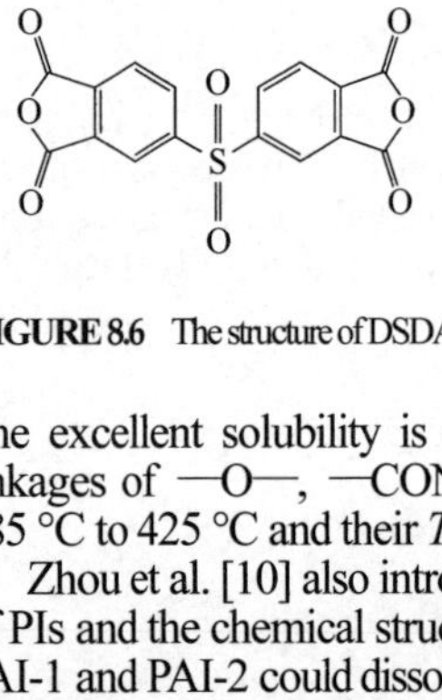

FIGURE 8.6 The structure of DSDA.

Thiruvasagam et al. [9] also synthesized a series of PIs containing the linkage —S=O— and their chemical structures are shown in Fig. 8.7. These PIs could even dissolve in the low-boiling-point solvent THF. The excellent solubility is not only attributed to —S=O— linkage, but also to the other flexible linkages of —O—, —CONH—, and —CH$_3$CCH$_3$—. The T_{10} in nitrogen of these PIs ranged from 385 °C to 425 °C and their T_g ranged from 174 °C to 185 °C.

Zhou et al. [10] also introduced the flexible linkages of —S— and —CONH— into the backbone of PIs and the chemical structure of the resulting poly(amide-imide)s (PAIs) are shown in Fig. 8.8. Both PAI-1 and PAI-2 could dissolve in the organic solvent NMP. The T_gs of PAI-1 and PAI-2 are 191 °C and 193 °C, while the T_{10}s of PAI-1 and PAI-2 are 507 °C and 498 °C, respectively. In addition, the PAI films show good optical transparency in the visible light region with the cutoff wavelengths around 400 nm and optical transmittances higher than 90% at 500 nm. The high sulfur contents and the flexible thioether linkages endow the PAI films with refractive indices higher than 1.73 and birefringences lower than 0.04 at 632.8 nm.

Although introducing flexible linkage could evidently enhance the solubility of PIs, the flexible linkage would decrease their thermal stability at the same time. For example, the T_{10}s of most of the soluble PIs mentioned above containing flexible linkages are below 500 °C and the T_gs of most of them are below 300 °C which are lower than those of the traditional aromatic PIs.

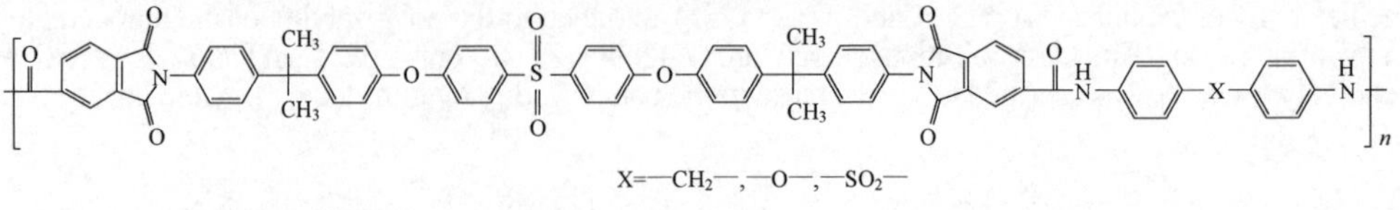

X=—CH$_2$—,—O—,—SO$_2$—

FIGURE 8.7 The polyimides containing the flexible linkage —S=O—.

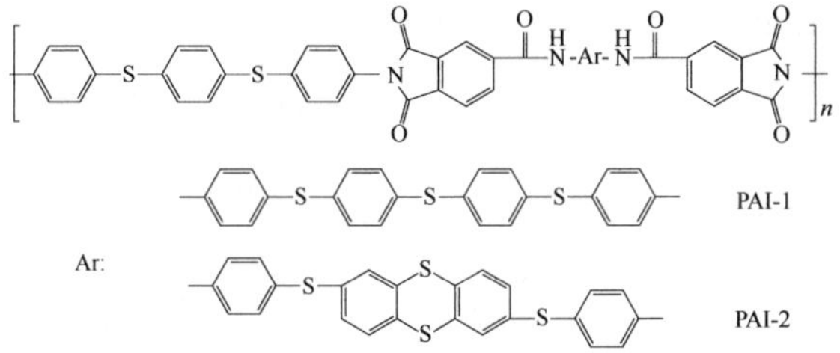

FIGURE 8.8 The polyimides containing the flexible linkage –S–.

8.2.2 Soluble Polyimides With Asymmetric Structures

For traditional aromatic PIs, most of them are prepared from dianhydride and diamine monomers possessing symmetric units, resulting in the regular packing of PI main chains with poor solubility in organic solvents. Thus, the introduction of asymmetric units into the PIs through dianhydride and diamine monomers is an effective approach to destroying the regular packing of main chains and increasing their solubility in turn.

Rusanov et al. [11] first synthesized two asymmetric diamines with the chemical structures shown in Fig. 8.9 and then prepared a series of PIs from them with dianhydrides OPDA and BTDA. The resulting PIs could dissolve in polar solvents such as NMP and DMF, and some of them could even dissolve in the low-boiling solvent $CHCl_3$. The T_{10} of these of PIs ranged from 430 °C to 465 °C and the soften temperature ranged from 260 °C to 305 °C. The authors considered that the relatively lower thermal stability of these PIs was attributed to the presence of methyl and methoxy in them.

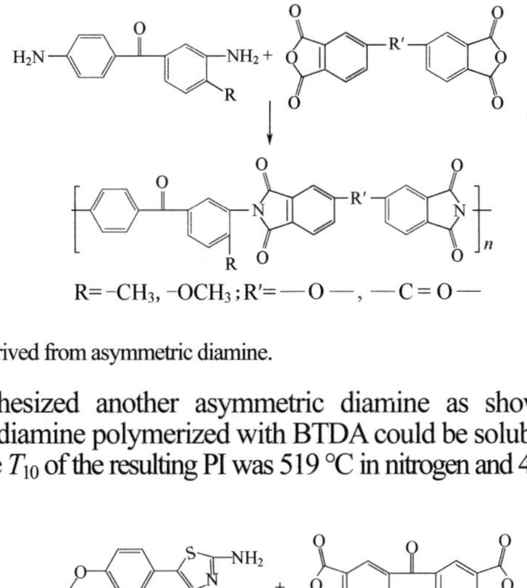

FIGURE 8.9 The polyimides derived from asymmetric diamine.

Zhao et al. [12] synthesized another asymmetric diamine as shown in Fig. 8.10, and the corresponding PI from this diamine polymerized with BTDA could be soluble in polar solvents, such as NMP, DMAc, and so on. The T_{10} of the resulting PI was 519 °C in nitrogen and 470 °C in air, respectively, and its T_g was only 216 °C.

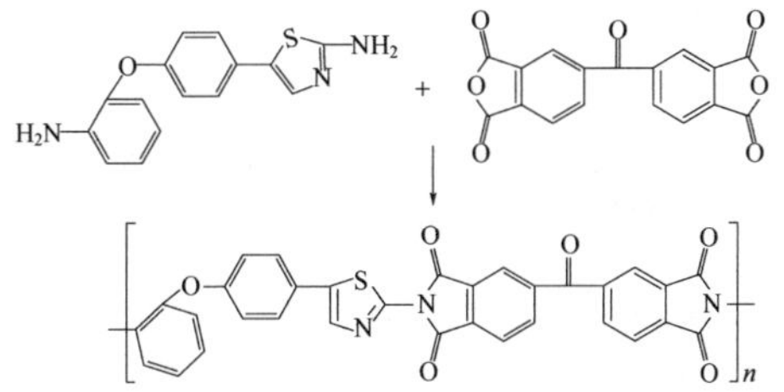

FIGURE 8.10 The polyimide derived from asymmetric diamine.

Mushtaq et al. [13] reported a asymmetric diamine (Fig. 8.11) containing naphthalimide moieties, and the PIs made from this diamine with commercial dianhydrides (BPDA, BTDA, OPDA, and BPADA) could dissolve in NMP and *m*-cresol. In addition, all of them displayed high T_g above 300 °C, which was comparable to that of traditional aromatic PIs. But their 5% weight loss temperature (T_5) in nitrogen was relatively low (ranging from 435 °C to 481 °C).

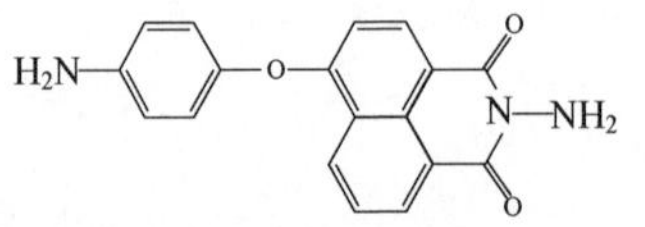

FIGURE 8.11 The molecular structure of the asymmetric diamine synthesized.

Hsiao et al. [14] prepared three PIs from dianhydride OPDA with an asymmetric diamine and two symmetric diamines, respectively, and further compared their solubilities. The chemical structures of diamines and the resulting PIs are shown in Fig. 8.12. Only the PI from the asymmetric diamine could dissolve in polar solvents, which indicated that the asymmetric unit could indeed enhance the solubility of PI. The T_g and T_{10} of the PI containing asymmetric units are 307 °C and 592 °C, respectively.

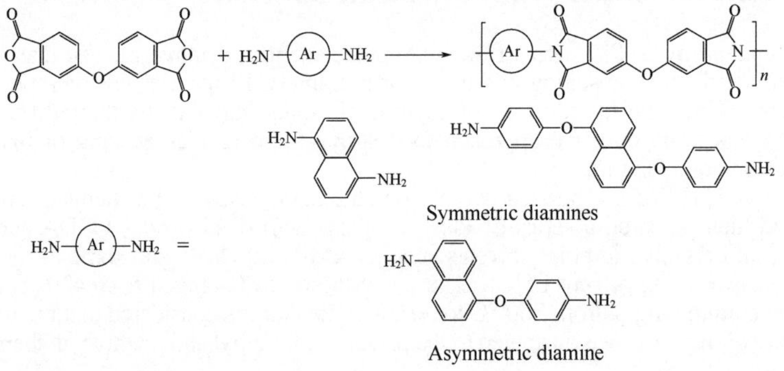

FIGURE 8.12 The polyimides from symmetric diamines and asymmetric diamine.

Overall, introduction of asymmetric units into PI can improve their solubility but not destroy their thermal stability simultaneously. This is an effective approach to increasing the solubility of PIs, but there are still some difficulties in the design and synthesis of the dianhydride and diamine monomers containing asymmetric units.

8.2.3 Soluble Polyimides With Alicyclic Structures

When aromatic units in traditional PIs were replaced partially or entirely by alicyclic units, the rigidity and intermolecular interaction of PI main chains could be decreased evidently and then their solubility was enhanced.

Oishi et al. [15] synthesized a series of PIs containing various alicyclic moieties by the silylation method and their chemical structures are shown in Fig. 8.13. All of them were soluble in the polar solvent NMP. The T_gs of these alicyclic PIs ranged from 223 °C to 295 °C. They also had lower refractive index and dielectric constant. However, the T_{10}s in air of them were in the range from 340 °C to 405 °C.

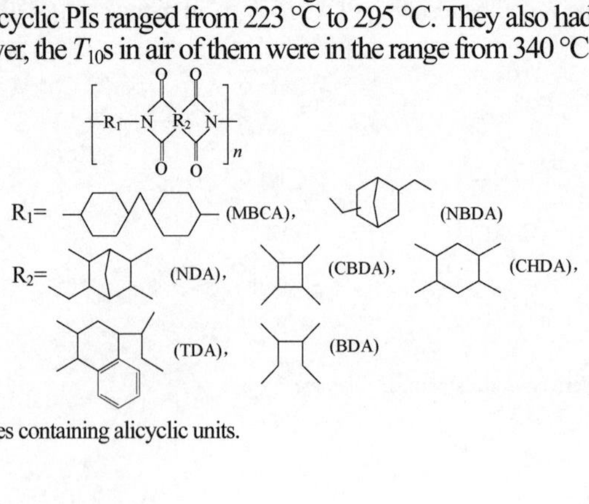

FIGURE 8.13 The polyimides containing alicyclic units.

Yamada et al. [16] reported a polyalicyclic dianhydride and its corresponding PIs with commercial diamines (such as DDE, DDS, and DDM, etc.) through the two-step method shown in Fig. 8.14. All of them could dissolve in polar organic solvents, including DMAc, DMF, and so on; their T_5s in air ranged from 396 to 455 °C and the T_gs ranged from 178 °C to 292 °C.

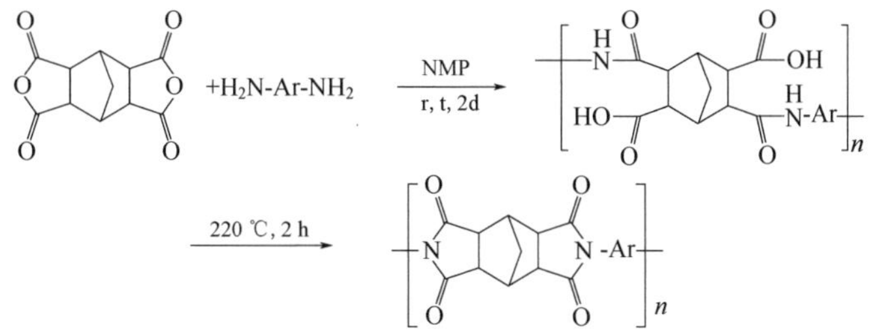

+H₂N-Ar-NH₂ NMP r, t, 2d
220 ℃, 2 h

FIGURE 8.14 The polyimides containing alicyclic units.

Yang et al. [17,18] synthesized another two alicyclic dianhydrides DMBD and TTDA. Their chemical structures are shown in Figs. 8.15 and 8.16, respectively. The PIs prepared from DMBD or TTDA with some commercial diamines could dissolve in many organic solvents such as DMAc and DMF, etc. These PIs could be used to prepare colorless and transparent films with low absorption cutoff wavelengths. They were promising candidates as the alignment layers for advanced nematic LCDs such as TFT-AM-LCDs. The T_{10}s of the PIs from DMBD were in the range from 426 °C to 598 °C, while the T_{10}s of those from TTDA ranged from 447 °C to 455 °C. Similarly, the T_gs of the PIs from DMBD were in the range from 238 °C to 325 °C, while T_gs of those from TTDA ranged from 210 °C to 278 °C.

Furthermore, Yang et al. also investigated the thermal degradation behavior of the PI made from TTDA with DDE by TG-FTIR, RT-FTI and TG-MS techniques. The results revealed that the aliphatic single bond in the TTDA moiety might be the weak bond in the PI main chains, which was prone to be cleaved in the elevated-temperature process. This indicated that the thermal stabilities of alicyclic moieties were lower than those of the aromatic units. Although the T_{10}s and T_gs of the PIs containing alicyclic moieties were lower than those containing aromatic moieties, their color was decreased and transparency was improved. At the same time, their solubility was also increased, and they could be processed into colorless and transparent films conveniently.

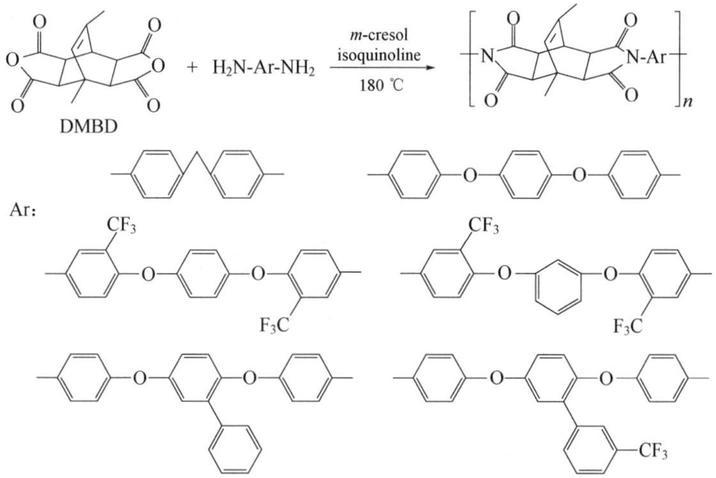

DMBD + H₂N-Ar-NH₂ m-cresol isoquinoline 180 ℃
Ar:

FIGURE 8.15 The polyimides derived from DMBD.

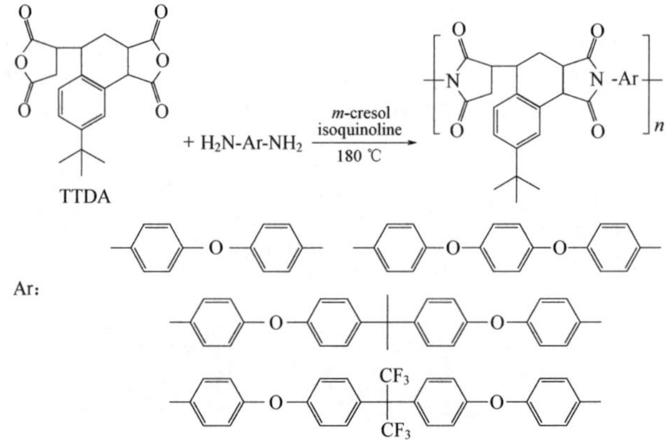

FIGURE 8.16 The polyimides derived from TTDA.

8.2.4 Soluble Polyimides With Side Groups

The introduction of side groups into PI backbone is a beneficial approach to improve their solubility, because the presence of side groups can effectively prevent the coplanarity of aromatic rings and reduce the packing efficiency of their main chains. Compared with the introduction of flexible linkage and alicyclic moieties into PIs, the introduction of most side groups would not dramatically decrease the inherent thermal stability of PIs.

8.2.4.1 Soluble Polyimides With Halogens or Halogenated Side Groups

Harris et al. [19] synthesized PIs containing bromine or iodine atoms as side groups and their chemical structures are shown in Fig. 8.17. These PIs could dissolve in THF and NMP solvents, etc., at a low concentration of 2-3 mg·mL^{-1}. To some extent, the introduction of halogen atoms as side groups could increase the solubility of PIs. The T_5 in air of the PIs with X as Cl and Br were 542 °C and 522 °C, respectively, but that of the PI with X as I was only 448 °C. This indicated that the atoms Cl or Br would be suitable as side groups and not destroy the thermal stability of PIs. On the contrary, the atom I would not fit as a side group and dramatically decrease the thermal stability. The T_g of them ranged from 329 °C to 335 °C, which could be attributed to their rigid-rod backbones.

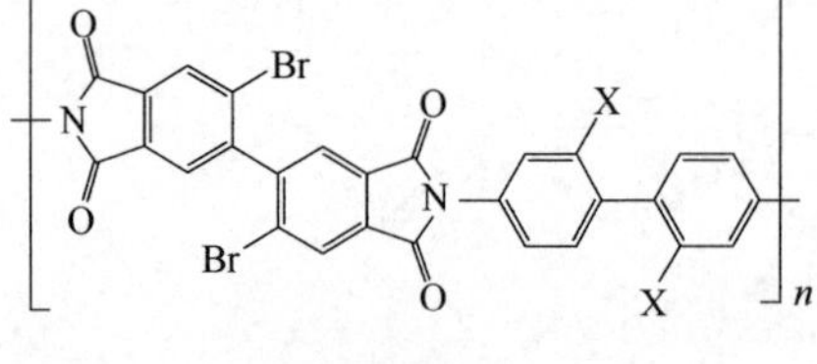

FIGURE 8.17 The polyimides containing halogen side groups.

Liu et al. [20] synthesized a novel aromatic diamine 4FMA containing fluorine atoms and a phenyl unit as a side group, as shown in Fig. 8.18. The PIs made from 4FMA with four dianhydrides including BPDA, OPDA, BTDA, and 6FDA showed good solubility not only in polar aprotic solvents, but also in low-boiling-point solvents, such as THF. The excellent solubility was attributed to the synergetic effects of the bulky lateral phenyl unit and fluorine groups in 4FMA. This indicated that the introduction of fluorine atoms as side groups could also enhance the solubility of PIs in the same way as Cl, Br, and I atoms. The T_{10}s and T_gs of these PIs were high, up to 578-628 °C and 292-338 °C, respectively, which indicated that the fluorine atoms as side groups did not decrease the thermal stability of PIs.

Besides, the halogenated groups were often introduced into PIs as side groups to enhance their solubility. Among of them, trifluoromethyl is the typical one and most reported, because it has great effects on the solubility and optical transparency of the resulting PIs. For example, the commercial dianhydride 6FDA containing trifluoromethyl side groups is often used to synthesize soluble and transparent PIs [21-23].

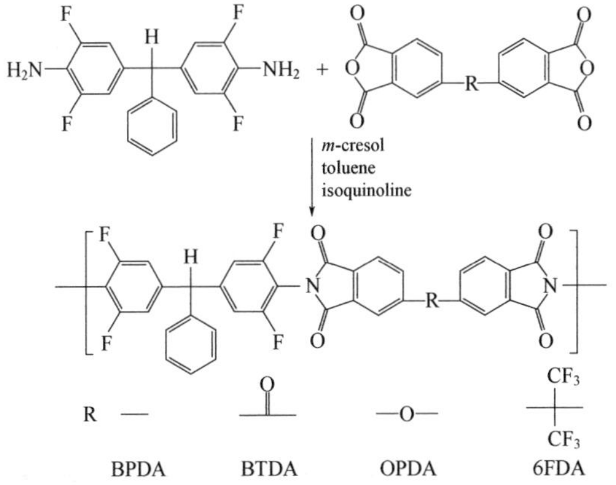

FIGURE 8.18 The polyimides containing fluorine side groups.

Various aromatic diamines containing trifluoromethyl could be easily prepared from the 4-nitro-2-(trifluoromethyl)-chlorobenzene shown in Fig. 8.19. They could be used to prepare soluble PIs with different aromatic dianhydrides through the one-pot polycondensation. Some of them could dissolve not only in polar solvents, such as DMAc, bus also in the low-boiling-point solvents, such as $CHCl_3$ or THF [24-46].

A series of diamine and dianhydride monomers containing trifluoromethyl as side groups were synthesized by Yang and colleagues [26,38,47-51] and their chemical structures was shown in Fig. 8.20. The resulting PIs prepared from these monomers were soluble in many polar organic solvents such as NMP, DMAc, DMF, and *m*-cresol, as well as in some low-boiling-point organic solvents such as THF. The excellent solubility could be attributed to the synergistic effect of the steric hindrance and electron-withdrawing effect of the side group trifluoromethyl. In addition, the T_{10}s of these PIs were above 500 °C, which indicates that the introduction of trifluoromethyl would not destroy the thermal stability of PIs. Most of these PIs possessed T_gs lower than 300 °C, which was due to the flexible linkages in these monomers. The films prepared from these PIs are transparent and colorless. The coPI from the diamine **4** could be processed into porous membranes and applied in lithium-ion batteries [52].

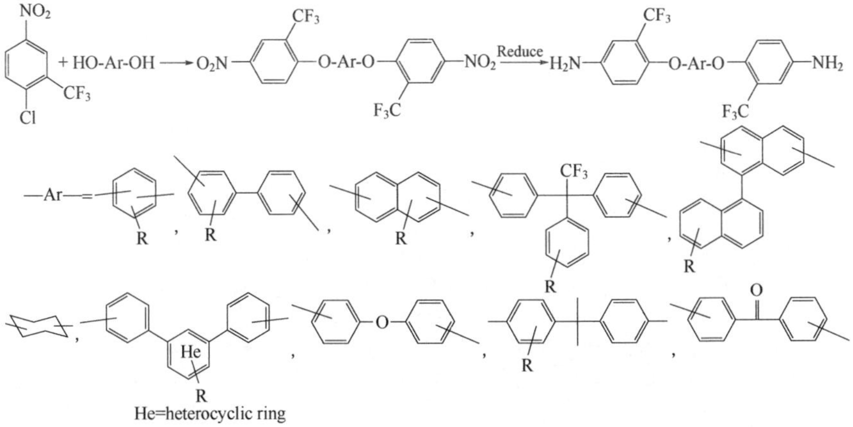

FIGURE 8.19 Preparation of diamines containing trifluoromethyl side groups.

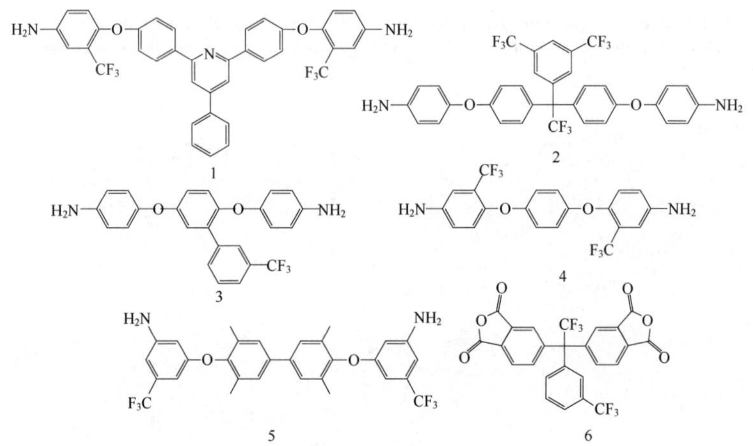

FIGURE 8.20 Diamines and dianhydrides containing trifluoromethyl groups.

In addition, Yang and colleagues [53] also synthesized a novel side-chain-type sulfonated aromatic diamine BABSA containing trifluoromethyl, and its chemical structure is shown in Fig. 8.21. The sulfonated PIs (SPI) from NTDA or BNTDA with BABSA and other diamines displayed good solubility in many common organic solvents such as NMP and DMAc. The SPI membranes exhibited good dimensional stability with isotropic swelling less than 22% and high thermal stability with the

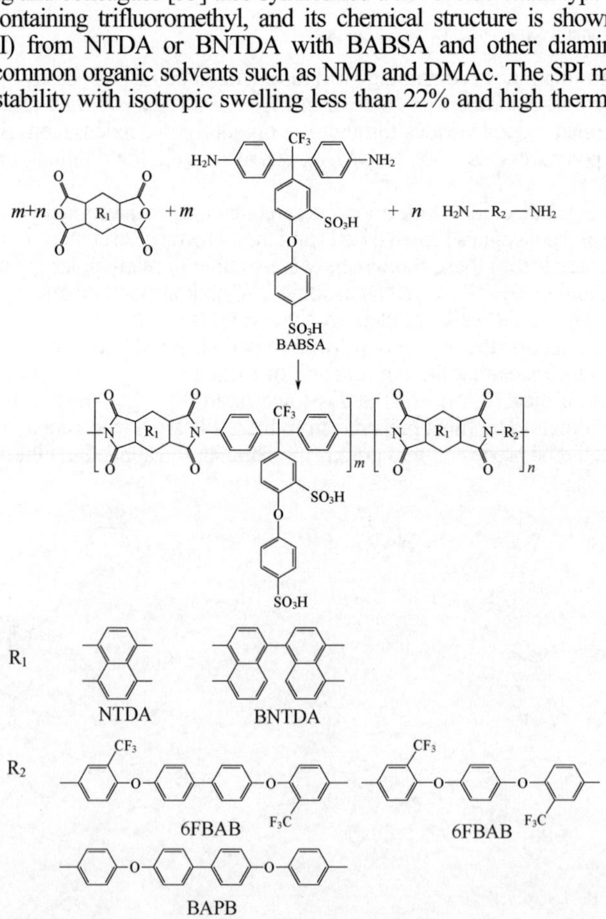

FIGURE 8.21 Sulfonated polyimides containing trifluoromethyl groups.

desulfonation temperature of 283-330 °C. These membranes also displayed excellent oxidation stability, which was improved by the introduction of trifluoromethyl. All the SPI membranes had the better permselectivity with the ratios of proton conductivity to methanol permeability (V) of nearly two- to three-times that for Nafion 115. All of these excellent properties make them promising candidates for direct methanol fuel cell applications.

Yang and coworkers also introduced the trifluoromethyl into the diamine containing lateral phenylphosphine oxide (PPO) and the chemical structure of BATFDPO is shown in Fig. 8.22 [54]. The PIs from BATFDPO could dissolve in common solvents such as DMAc and NMP with T_5 ranging from 472 °C to 477 °C and T_g from 244 to 266 °C. The PI films were transparent and their refractive indexes were as low as 1.5511 at 1310 nm. They were promising candidates for optical waveguides applications.

Besides, other side groups containing fluorines were also introduced into PIs to enhance their solubility. For example, an aromatic diamine as shown in Fig. 8.23, 5-[1*H*,1*H*-2-bis(trifluoromethyl) heptafluoropentyl]-1,3-phenylenediamine(RfbMPD) containing 1*H*,1*H*-2-bis(trifluoromethyl) heptafluoropentyl as a side group, was synthesized by Auman et al [55]. The PIs based on RfbMPD could be easily dissolved in the common organic solvents and their T_gs were relatively low (257 °C).

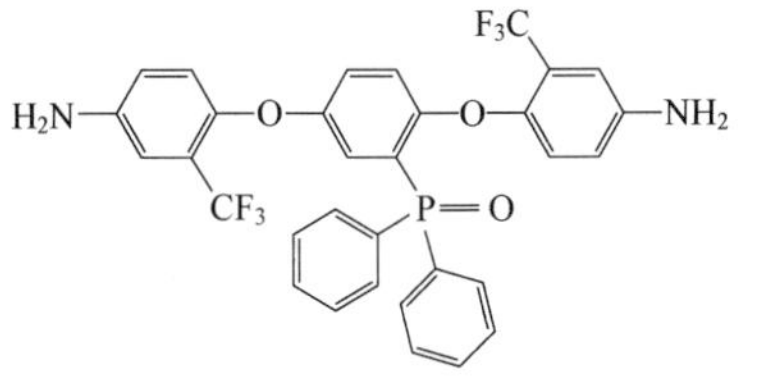

FIGURE 8.22 Diamine BATFDPO containing lateral phenylphosphine oxide (PPO) and trifluoromethyl groups.

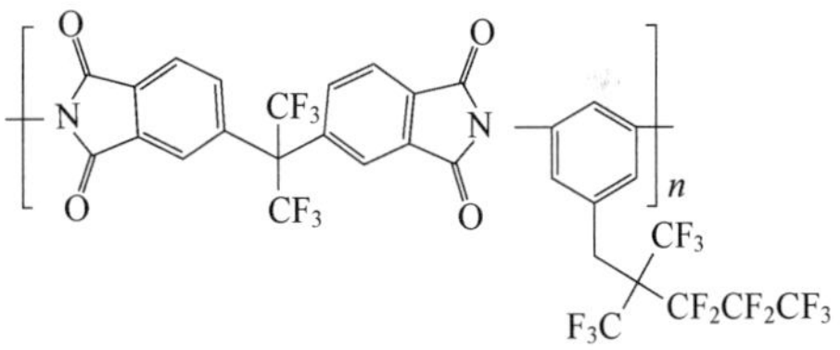

FIGURE 8.23 The polyimides containing 1H,1H-2-bis(trifluoromethyl)heptafluoropentyl as side group.

Overall, the introduction of trifluoromethyl as side groups could evidently increase the solubility of PIs without decreasing their thermal performance. However, the cost of these PIs containing trifluoromethyl was ordinarily increased and limited their wide applications.

8.2.4.2 Soluble Polyimides With Aliphatic Side Groups

The introduction of aliphatic side groups into PIs could also enhance their solubility. Long-chain alkyls were often introduced into PIs and the solubility of the resulting PIs was increased with the length of long chain alkyl. Kim and Jung [56] synthesized a series of PIs with long-chain alkyl side groups and their chemical structure is shown in Fig. 8.24. When *R* was butyl, the PI could only dissolve in DMAc and NMP under the heating. When R was C_8H_{17}, the PI could dissolve in NMP at room temperature, but dissolve in DMAc still under heating. When R is $C_{12}H_{25}$, the PIs could dissolve in DMAc and NMP at room temperature. The solubility of the PIs increased with the length of alkyl chain side group. However, with the increase in the length of the alkyl chain, the thermal stability of the PIs would dramatically decrease. When R is C_4H_9, C_8H_{17}, and $C_{12}H_{25}$, the T_{10}s values were 459 °C, 440 °C, and 418 °C, respectively, and the T_gs were 292 °C, 176 °C, and 138 °C.

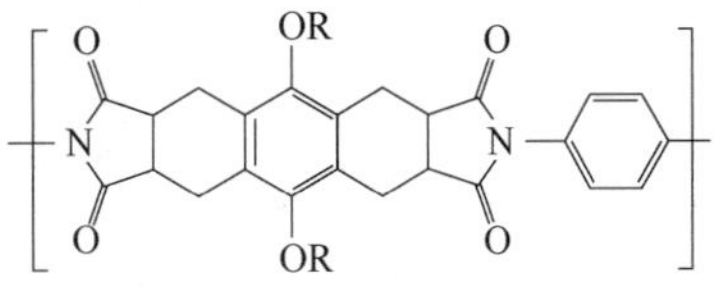

FIGURE 8.24 The polyimides containing long-chain alkyls as side groups.

As an aliphatic group, *tert*-butyl was usually introduced into PIs as a side group to increase their solubility. Interestingly, the unique structure of *tert*-butyl displayed different effects on the properties of PIs compared to those of the long-chain alkyl and alicyclic moieties. Our group synthesized a series of PIs containing *tert*-butyl as side groups and their chemical structures are shown in Fig. 8.25 [57-61]. We found that the introduction of *tert*-butyl as side groups could dramatically increase the solubility of PIs because the large pendant *tert*-butyl could increase the distance between PI chains. All resulting PIs could dissolve in polar solvents such as DMAc, and some of them could even dissolve in the low-boiling-point solvents, such as THF and $CHCl_3$. At the same time, *tert*-butyl containing four aliphatic carbon atoms in branched structure would not dramatically decrease the thermal stability of PIs. All of these PIs were stable up to 450 °C.

In addition, the PI prepared from aromatic diamine **7** with PMDA could not dissolve in *m*-cresol under heating, while the PI made from aromatic diamine **8** and PMDA was soluble in *m*-cresol, NMP, and even in THF at room temperature. The difference is that diamine **8** contains one more *tert*-butyl than diamine **7**. This result clearly proved that the number of *tert*-butyls introduced also affected the solubility of PIs. At the same time, the T_{10} of the PI from aromatic diamine **8** was a little lower than that of the PI from aromatic diamine **7**, which indicated the number of *tert*-butyl side groups did not dramatically decrease the thermal stability of PIs.

Commercial dianhydrides + *tert*-butylated diamines

m-cresol isoquinoline, Heating

PMDA BPDA OPDA BTDA

7 8

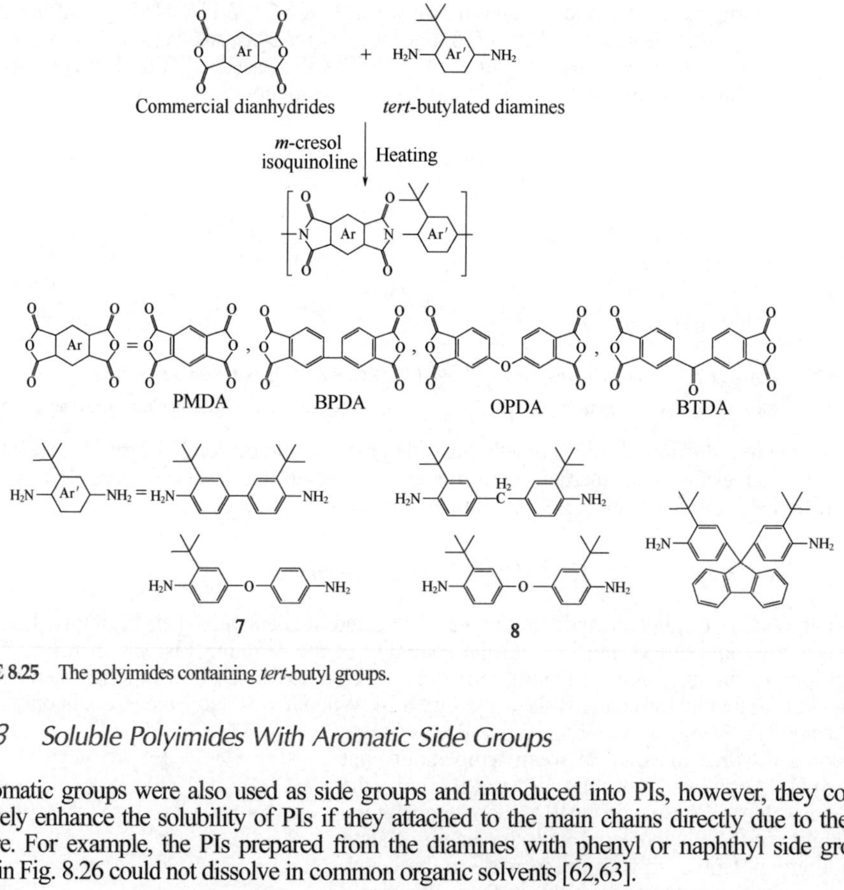

FIGURE 8.25 The polyimides containing *tert*-butyl groups.

8.2.4.3 Soluble Polyimides With Aromatic Side Groups

Aromatic groups were also used as side groups and introduced into PIs, however, they could not effectively enhance the solubility of PIs if they attached to the main chains directly due to their rigid structure. For example, the PIs prepared from the diamines with phenyl or naphthyl side groups as shown in Fig. 8.26 could not dissolve in common organic solvents [62,63].

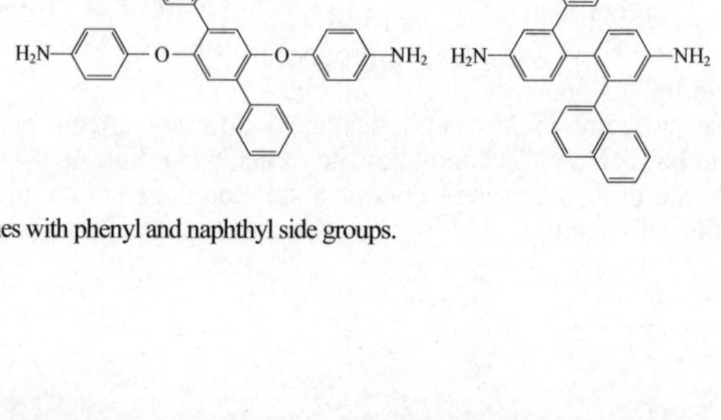

FIGURE 8.26 The diamines with phenyl and naphthyl side groups.

But when the aromatic side groups were introduced into PIs through a flexible linkage, such as the most commonly used linkage, —O—, they could evidently enhance the solubility of PIs due to the increased activity of side groups. Qui and Zhang [64] and Zeng et al. [65] synthesized a series of aromatic dianhydrides with phenyloxy side groups (Fig. 8.27) and the PIs from these dianhydrides could easily dissolve in common organic solvents. The T_5s in N_2 of these PIs ranged from 468 °C to 537 °C.

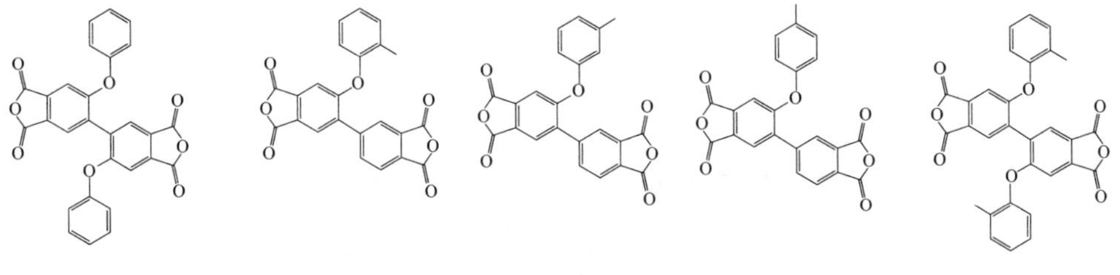

FIGURE 8.27 The dianhydrides with phenyloxy side groups.

The chemical structure and solubility of PI (DAB-BPDA) and its corresponding derivatives with various side groups are outlined in Fig. 8.28 [59,64,66-68]. Obviously, the introduction of trifluoromethyl, *tert*-butyl, and phenyloxy into PIs as side groups could increase the solubility of PIs evidently; however, the introduction of phenyl into PI as the side group was not useful to improving the solubility of the PI. The effect of these side groups on the solubility of PIs decreased in the order of trifluoromethyl > phenyloxy ≈ *tert*-butyl > phenyl.

The chemical structures and the effects of various side groups on the thermal properties of PI 4,4′-(1,4-phenylenebis(oxy))dianiline-PMDA (1,4-BAPBz- PMDA) and its derivatives are shown in Fig. 8.29 [62,69-74].

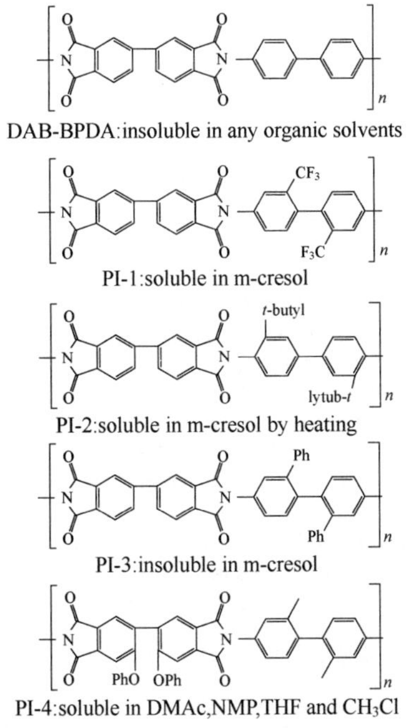

DAB-BPDA:insoluble in any organic solvents

PI-1:soluble in m-cresol

PI-2:soluble in m-cresol by heating

PI-3:insoluble in m-cresol

PI-4:soluble in DMAc,NMP,THF and CH_3Cl

FIGURE 8.28 The solubility of DAB-BPDA and its derivatives.

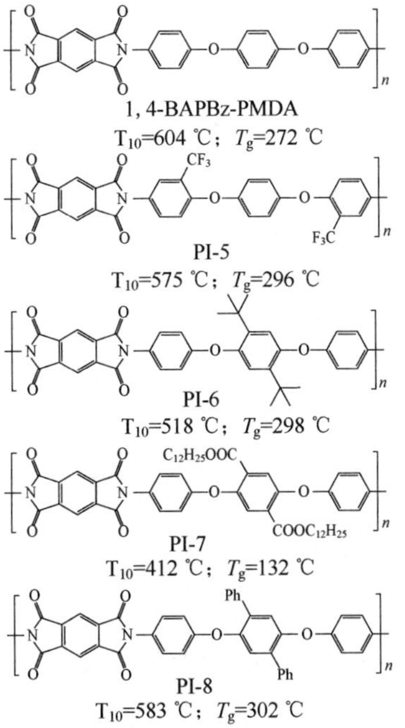

1, 4-BAPBz-PMDA

T_{10}=604 ℃; T_g=272 ℃

PI-5

T_{10}=575 ℃; T_g=296 ℃

PI-6

T_{10}=518 ℃; T_g=298 ℃

PI-7

T_{10}=412 ℃; T_g=132 ℃

PI-8

T_{10}=583 ℃; T_g=302 ℃

FIGURE 8.29 The thermal properties of 1,4-BAPBz-PMDA and its derivatives.

Overall, the long-chain alkyls as side groups would dramatically decrease the thermal stability and T_g of PIs. The *tert*-butyls as side groups decreased the thermal stability of PIs slightly, while the phenyls and trifluoromethyls as side groups hardly affected the thermal stability of PIs. In addition, trifluoromethyl, *tert*-butyl and phenyl enhanced the T_g of PIs because these pendant side groups could restrict the free-activity of PI main chains.

8.3 Applications of Soluble Aromatic PIs

8.3.1 Second-order Nonlinear Optical (NLO) Materials

NLO materials have been paid much attention over the past decades because of their various potential applications, such as optical communication, optical data storage, optical switching, frequency modulation, and so on [75-77]. Up until now, inorganic NLO materials, especially lithium niobate, have been extensively studied for these applications, but they have some unavoidable disadvantages, such as slow response time, high absorption, and degradative photorefractive effects, etc. On the other hand, polymeric NLO materials exhibit great advantages over inorganic ones, such as large EO coefficients, ultrafast response times, and easy processability [75,76], which makes them promising candidates for future NLO applications. For polymeric NLO materials, considerable interest was focused on the second-order optical nonlinearity and its corresponding applications. Here, symbols $x^{(2)}$ and β stand for the macroscopic and microscopic second-order nonlinearity, respectively. In addition, second-harmonic generation (SHG) and linear EO effect are the most common second-order nonlinear effects and the second-harmonic coefficients d_{33} of SHG and EO coefficient r_{33} of the linear EO effect are often used to characterize the second-order optical nonlinearity of second-order polymeric NLO materials.

Actually, there are still some crucial issues that need to be addressed thoroughly before second-order NLO polymers have any commercial value, including the poor stability of dipole orientation, relatively low optical nonlinearity, and large optical loss. Generally, the optical nonlinearity of the polymers is endowed by the chromophores blended in them or embedded into their chemical structure. The chromophores can align in noncentrosymmetrical order by corona poling, which generates optical nonlinearity. The poor orientational stability of second-order polymeric NLO materials usually rises from the relaxation of the chromophores in them after corona poling [78]. The high T_gs of polymers can be utilized to restrain the relaxation of the noncentrosymmetric chromophores [79-82]. The second-order polymeric NLO materials with low T_gs, such as polyacrylate [83], polystyrene [84], polyester [85], polyurethane [86], and so on, display rather poor orientational stability relative to the inorganic NLO materials. Due to possessing high T_gs, PIs are considered one of the most promising candidates for preparing high-performance second-order polymeric NLO materials. In addition, most NLO active chromophores could not endure the process of the high temperature above 300 °C. This means that thermal imidization should be avoided in the preparation of NLO PIs. Therefore, soluble PIs synthesized through chemical imidization are the best candidates for preparing polymeric NLO materials.

The NLO PIs can be divided into four categories according to their chemical structures, including guest-host NLO PIs, main-chain NLO PIs, side-chain NLO PIs, and crosslinked NLO PIs.

8.3.1.1 Guest-host NLO Polyimides

The guest-host NLO PIs are prepared by blending the guest NLO active chromophores into the host-matrix PIs. They can display second-order nonlinearity after corona-poling. For example, as shown in Fig. 8.30, Jeng et al [87]. blended an NLO active alkoxysilane chromophore ASD into the organosoluble PI BPPI containing trifluoromethyl side group. After corona-poling, this material could display second-order optical nonlinearity with an E-O coefficient r_{33} of 2.2-17.0 pm·V^{-1}.

The guest-host NLO PIs are convenient to prepare compared with other types of NLO PIs, in which NLO active chromophores are attached to the PI backbone through covalent bonds. However, the concentration of NLO active chromophores should be limited at a low level in order to avoid the phase-separation between guest NLO active chromophores and host-matrix PI, which results in a low nonlinearity generally [88]. On the other hand, the mobility of the chromophores in guest-host NLO PIs is much higher than that of those attached to the PI backbone by covalent bonds. Thus they cause the relaxation of the poled order more easily, i.e., the orientational stability of guest-host NLO PIs is relatively lower than that of other NLO PIs.

8.3.1.2 Main-chain NLO Polyimides

The main-chain NLO PIs are defined as the PIs whose active chromophores are incorporated directly into the repeating units of PI main-chains. Due to the chromophores located in polymer backbones through head-to-tail linkage, their poled-order relaxation can be restricted evidently. In other words, the orientational stability of main-chain NLO PIs can be obviously enhanced. For example, as shown in Fig. 8.31, the main-chain NLO polyimide **PI-M1** prepared by Verbiest et al. [89] exhibited much higher orientational stability than that of the side-chain NLO polyimide **PI-S1** containing the same chromophore. The **PI-M1** could retain 93% SHG coefficient at 225 °C for 1000 h, while the **PI-S1** was only stable at 100 °C. The improvement in orientational stability could be attributed to the lower mobility of the chromophore in **PI-M1**. However, the optical nonlinearity of **PI-M1** is a little lower than that of **PI-S1** because of the relatively rigid structure of the main-chain NLO PI decreasing the poling efficiency of the chromophores. Therefore, the trade-off between optical nonlinearity and orientational stability should be considered.

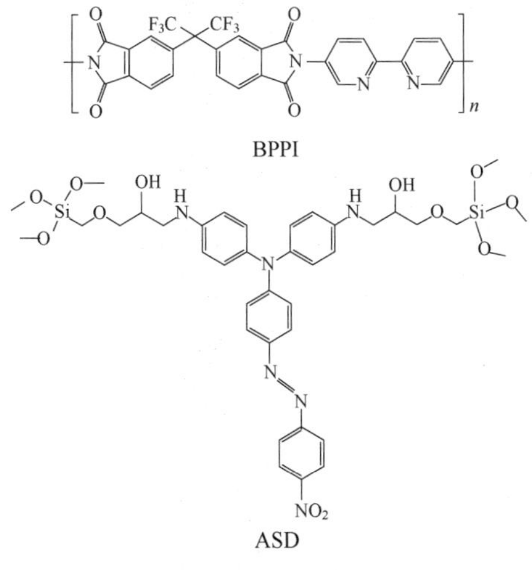

FIGURE 8.30 Guest-host NLO polyimides.

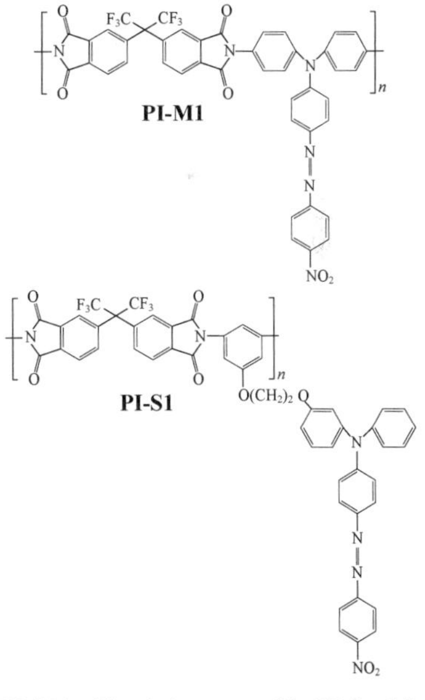

FIGURE 8.31 Chemical structure of the NLO polyimides.

The X-shaped and Λ-shaped chromophores can increase the optical nonlinearity of second-order polymeric NLO materials. Such chromophores were also embedded into PIs to prepare main-chain NLO PIs. For example, as shown Fig. 8.32, Qin et al. [75] reported a main chain nonlinear PI containing X-shaped chromophore and its second-harmonic coefficients d_{33} of 16.5 pm·V^{-1} after poling and onset depoling temperature up to 200 °C, which is much higher than that of guest-host and side-chain NLO PIs. Another main-chain NLO PI synthesized by Tsai et al. [90] containing a two-dimensional Λ-shaped chromophore was shown in Fig. 8.33. Benefiting from the unique Λ-shaped molecular structure, the dipole of the chromophore traverses the PI backbone and makes the chromophore orientate easily under corona poling. This PI exhibits second-harmonic coefficients d_{33} of 17.2 pm·V^{-1} at 1064 nm. On the other hand, its onset depoling temperature is up to 240 °C and it can maintain a large SH signal at 100 °C beyond 600 hours. It can also exhibit excellent orientational stability.

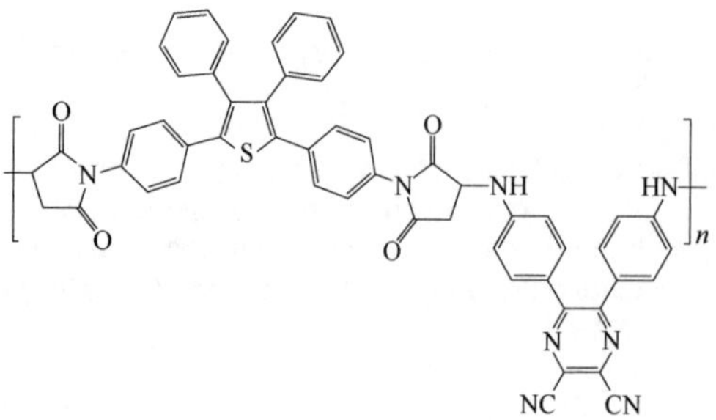

FIGURE 8.32 X-shaped NLO polyimide.

8.3.1.3 Side-chain NLO Polyimides

For side-chain NLO PIs, the active chromophores are introduced into PIs as side-chains by a flexible tether group. Due to one end of the chromophores being attached to the PI backbone through a covalent flexible linkage, they display much higher orientational stability than that of guest-host NLO PIs. On the other hand, the mobility of the chromophores located in side-chains is much higher than that of the chromophores embedded into the main chains and results in a higher poling efficiency [88]. Therefore, side-chain NLO PIs often display higher optical nonlinearity than that of main-chain NLO PIs. Due to the above advantages, side-chain NLO PIs are often reported and many efforts have been made to improve their combination properties. For example, Saadeh et al. [91] synthesized a series of side-chain NLO PIs as shown in Fig. 8.34 and all of them display high NLO performance. In particular, the d_{33} and r_{33} of the NLO PI **PI-S2** could reach up to 169 pm·V^{-1} (at 532 nm) and 25 pm·V^{-1} (at 1300 nm), respectively. Eighty percent of the SHG signal of these NLO PIs could be retained for more than 1000 h at 100 °C in air.

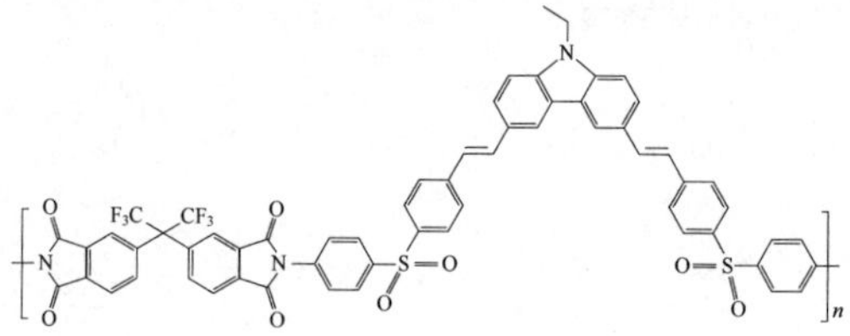

FIGURE 8.33 Λ-shaped NLO polyimide.

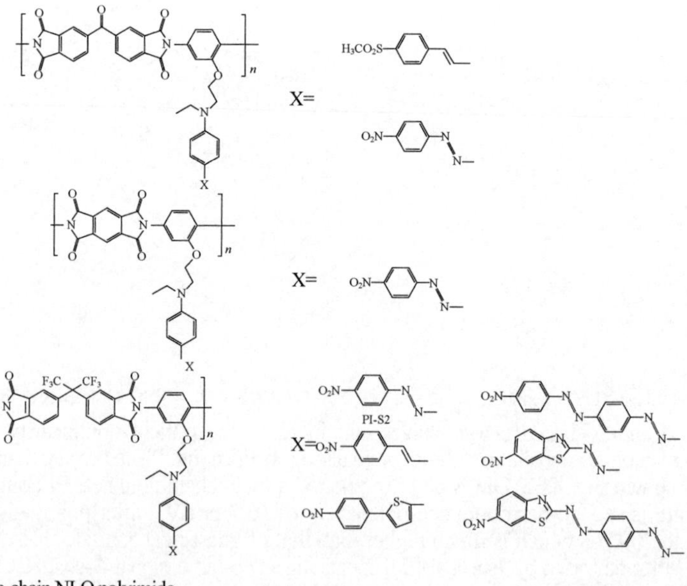

FIGURE 8.34 Side-chain NLO polyimide.

For side-chain NLO PIs, the high concentration of chromophores often caused the heterogeneous molecular aggregation due to the large dipole-dipole interactions, and result in the increases of optical loss. Moreover, the large dipole-dipole interactions also complicated their noncentrosymmetric orientation and decreased the poling efficiency during the poling process [92]. Then their optical nonlinearity was also decreased in turn. To overcome the above problems, the dendronized chromophores were introduced into side-chain NLO PIs. The pendant dendronized chromophores offer larger internal free-volume and can enhance the poling efficiency. In addition, the dendronized chromophores possess a three-dimensional orientation and can reduce the dipole-dipole interaction. For example, He et al. [92] synthesized an NLO polyimide **FPI-DEHNT** (Fig. 8.35) with the dendronized chromophores DEHNT as side-chains. This PI displayed a larger order parameter r_{33} (34 pm·V^{-1}) than that of the polyimide **FPI-EHNT** (r_{33}: 20 pm·V^{-1}) containing normal chromophore **EHNT** as side-chain at the similar chromophore concentration. This can be attributed to the higher poling efficiency of the dendronized **DEHNT.**

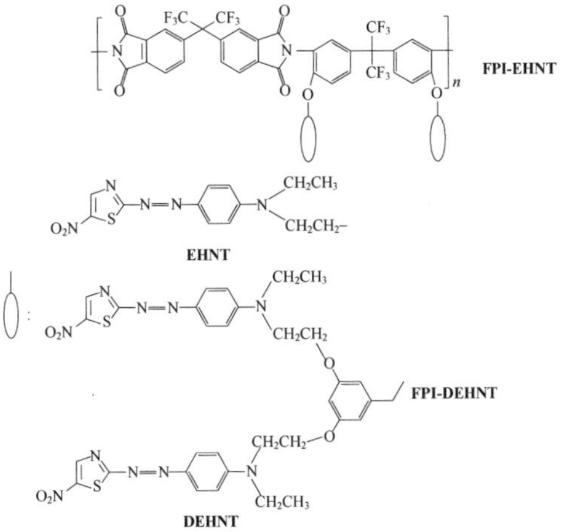

FPI-EHNT

EHNT

FPI-DEHNT

DEHNT

FIGURE 8.35 Side-chain NLO polyimide containing dendronized chromophores.

Do et al. [93] synthesized an NLO polyimide **PI-Drdas** with a dendronized chromophore Drdas as side-chain and its chemical structure is shown in Fig. 8.36. The chromophore concentration in **PI-Drdas** is higher than that in **PI-Rdas** due to the dendronized molecular structure. The d_{33} and r_{33} of PI-Drdas are 23.3 and 12.9 pm·V^{-1}, respectively, while those of PI-Rdas are 19.7 and 10.9 pm·V^{-1}. The higher optical nonlinearity of PI-Drdas could be attributed to the higher chromophore concentration and the enhanced poling efficiency. On the other hand, the optical loss of PI-Drdas is equal to that of PI-Rdas, which indicates the dendronized chromophores can serve to resolve the trade-off between the optical loss and nonliearity [93]. However, the introduction of dendronized chromophores often decrease the T_g of the PI dramatically and further decrease its orientational stability. For example, the T_g of PI-Rdas is 208 °C, while that of PI-Drdas is only 155 °C. When the poling PIs were treated at 80 °C for 40 h, the nonlinearity reduction of the PI-Rdas was less than 10%, but that of PI-Drdas reached up to 28%.

8.3.1.4 Crosslinked NLO Polyimides

As mentioned above, the orientational stability of most NLO PIs is relatively low due to the relaxation of the chromophores after poling. The crosslinked framework could evidently inhibit the chromophores' relaxation and then enhance the orientational stability. For example, as shown in Fig. 8.37, Sui et al. [94] prepared a side-chain NLO polyimide **PUI** and blended with a photocrosslinkable polyimide **PSPI** to afford a complex **PUI/PSPI-20** (the molar ratio of PSPI/PUI is 20%). The **PSPI** could be crosslinked under the irradiation of near UV light and the resulting crosslinked structure could increase the orientational stability of the poled **PUI/PSPI-20**. The decay of order parameters of **PUI** and **PUI/PSPI-20** at 150 °C was shown in Fig. 8.38. The **PUI/PSPI-20** displayed similar orientational stability to that of the PUI before the photocrosslinkig. After the photocrosslinking, its orientational stability was dramatically enhanced. This result confirmed that crosslinked NLO PIs possess higher orientational stability than that of the side-chain ones.

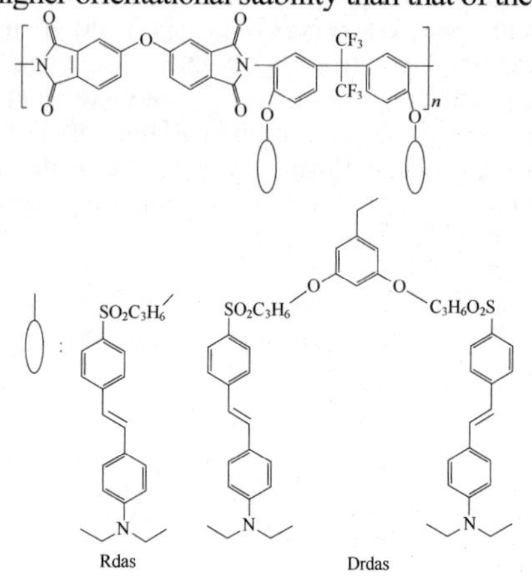

Rdas Drdas

FIGURE 8.36 Side-chain NLO polyimide.

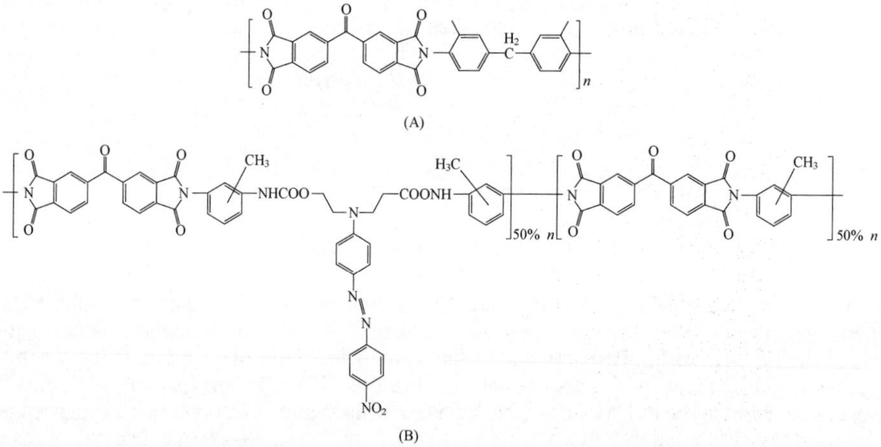

(A)

(B)

FIGURe 8.37 The structure of PSPI (A) and PUI (B).

Although NLO PIs have attracted much more attention from scientists in academia and industry due to the higher orientational stability than that of the other NLO polymers, their optical nonlinearity and orientational stability still need to be enhanced further. In addition, the optimal trade-off between optical nonlinearity and optical loss of chromophores should be studied further.

8.3.2 Memory Device Materials

During the past several decades, the capacity of memory devices based on inorganic materials has drastically increased while their size has dramatically decreased by traditional lithography technologies. However, these technologies are now facing many inherent problems, such as the physical limitation of the lithography pattern resolution, high process costs, and so on [95]. Therefore, organic and polymeric memory materials have been paid more attention because of their advantages, including good scalability, low process cost, three-dimensional stacking capability, and excellent flexibility [96]. In recent years, many polymeric materials have been reported to be used in resistive-type memory devices [97,98].

In order to store data, a memory device based on polymeric materials also needs two distinct electronic states at the same applied voltage, which can be assigned as "0" and "1" or "on" and "off." [99] In the past few years, various polymeric materials, such as conjugated polymers, vinyl polymers with specific pendant groups, PIs etc., were reported to have bistable electrical switching properties. Among them, PIs have attracted significant interest because of their excellent thermal stability, chemical resistance, and mechanical properties. However, traditional PIs usually exhibit poor solubility or are insoluble in organic solvents. Therefore, the precursor PAA film needs to be heated through thermal imidization to produce a PI film. In this process, the intermolecular imidization is promoted and the packing behavior highly depends on the film-forming method, which might significantly change the electrical performance of the PI [100]. In contrast, soluble PIs can be easily processed into complex structures like the multistack layer structures without destroying their electrical performance. Hence, the design of a novel donor-acceptor PI, which is not only soluble in organic solvents but also can maintain the high thermal stability, mechanical properties, and has good memory properties, is imminently required.

8.3.2.1 Soluble Polyimides With Triphenylamine (TPA) Groups

The first memory PI containing triphenylamine units was reported in 2009. Liu et al. [101] synthesized a soluble functional PI (PYTPA-PI) containing triphenylamine-substituted diphenyl-pyridine moieties (PYTPA) as electron donors and phthalimide moieties as electron acceptors, and its chemical structure is shown in Fig. 8.39. A simple metal(Al)/polymer/metal (indium-tin oxide) sandwich structure switching device was fabricated and exhibited two accessible conductivity states, the low-conductivity (OFF) state and the high-conductivity (ON) state, and the ON/OFF current ratio was more than 10^3. Under the refreshing voltage pulse of 2 V, the ON state can be sustained, while under the reverse bias of -0.9 V it can be reset to the OFF state. Thus, the device, which can be written, erased, read, and the electrical states refreshed, is determined to be a dynamic random access memory (DRAM).

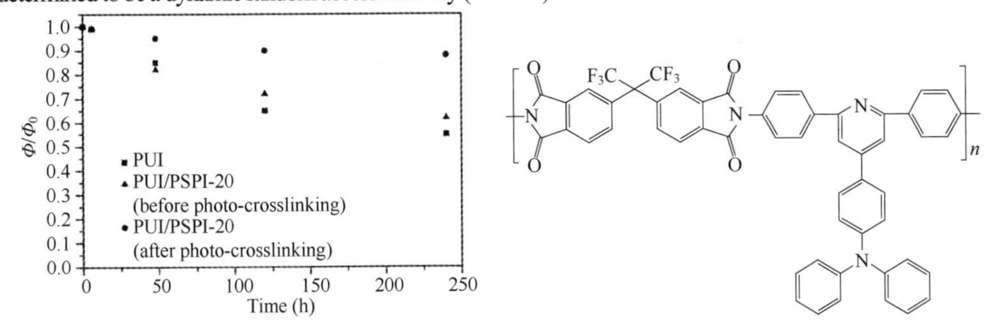

FIGURE 8.38 Decay of the order parameter of poled PUI and PUI/PSPI-20 films before and after photo-crosslinking at 150 °C.

FIGURE 8.39 Soluble functional polyimide containing triphenylamine.

Kim et al. [102,103] also reported another two novel PIs in 2009. They used poly(*N*-(*N*′,*N*′-diphenyl-*N*′-1,4-phenyl)-*N*,*N*-4,4′-diphenylene hexafluoro-isopropylidene-diphthalimide) (6F-2TPA PI) and poly(4,4′-amino(4-hydroxyphenyl)-diphenylene hex afluoroisopropylidene-diphthalimide) (6F-HTPA PI) (Fig. 8.40) to fabricate some high-performance programable memory devices. The former PI films with the thicknesses in the range 34-74 nm exhibited excellent write-once-read-many-times (WORM) memory characteristics while the 100 nm-thick PI films showed DRAM memory behaviors. Both of them had high ON/OFF current ratio, up to 10^{10}. The latter PI films with a thickness of less than 77 nm were found to exhibit excellent unipolar WORM memory behavior with a high ON/OFF current ratio up to 10^6. And the memory devices were all electrically stable at high temperatures up to 150 °C.

In 2012, they further synthesized three functional PIs, poly(*N*,*N*-bis(4-aminophenyl)-*N*′,*N*′-diphenyl-1,4-phenylene-3,3′,4,4′-diphenyl-sulfonyltetracarboximide) (DSDA-2TPA PI), poly

(N,N-bis(4-aminophenyl)-N',N'-di(4-methoxyphenyl)-1,4-phenylene-3,3′,4,4′-diphenylsulfonyl-tetrac - arboximide) (DSDA-2TPA-OMe_2 PI), and poly(N,N-bis(4-aminophenyl)-N',N'-di(4-cyanophenyl)-1,4-phenylene-3,3′,4,4′-diphenylsulfonyltetracarboximide) (DSDA-2TPA-CN_2 PI) and their chemical structures are shown in Fig. 8.41 [104]. The memory devices based on these PI films showed good retention abilities in both the ON and OFF states, and had high ON/OFF current ratios (10^6-10^{10}). PIs with different substituents revealed different memory behaviors: DSDA-2TPA PI and DSDA-2TPA-OMe_2 PI exhibited WORM memory behavior, while DSDA-2TPA-CN_2 PI showed DRAM behavior.

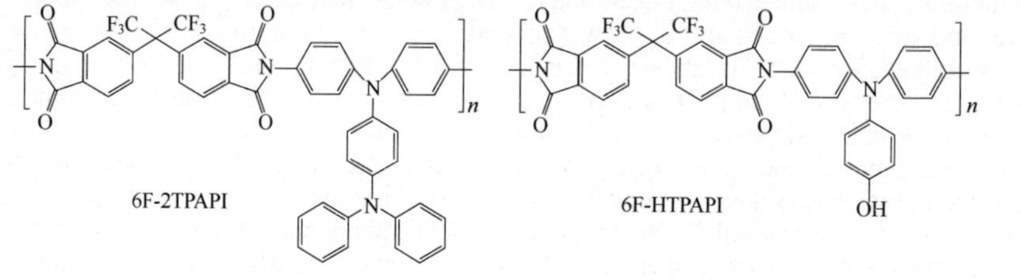

6F-2TPAPI 6F-HTPAPI

FIGURE 8.40 Soluble functional polyimide containing triphenylamine.

DSDA-2TPA DSDA-2TPA-OMe_2 DSDA-2TPA-CN_2

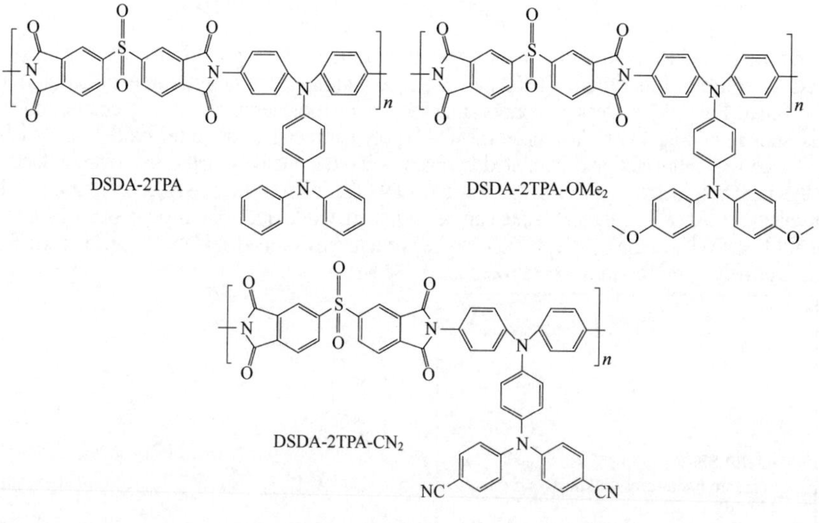

FIGURE 8.41 Soluble functional polyimide containing triphenylamine.

Liu et al. [105] synthesized a PI poly(4,4′-bis(4-methloxytriphenylamine) -3,3′-biphenylenehexafluoroisopropylidenediphthalimide) (MTPA-PI) and its chemical structure was shown in Fig. 8.42. Then it was used as the based thin film sandwiched between different bottom electrode materials such as Al or indium tin oxide (ITO). Different switching behaviors were investigated in the different devices which used different bottom electrodes. Compared with the ITO/MTPA-PI/Al device, the Al/MTPA-PI/Al device presented higher ON state current and lower OFF state current, which resulted in the higher ON/OFF current ratios (up to 10^9 compared with 10^4).

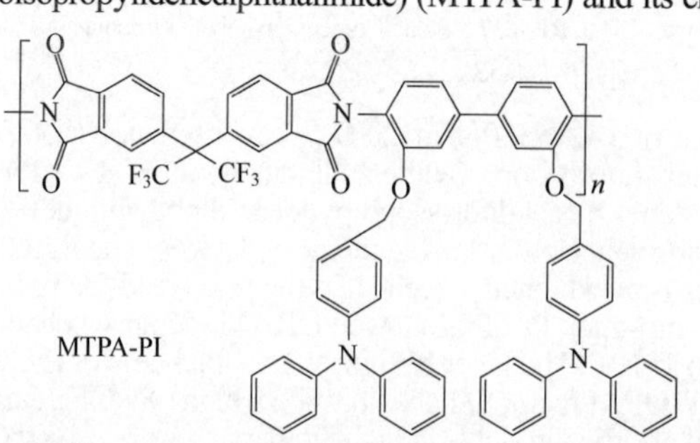

MTPA-PI

FIGURE 8.42 Soluble functional polyimide containing triphenylamine.

Kurosawa et al. [106] synthesized two

PIs, poly(*N*-(2,4-diaminophenyl)-*N*, *N*-diphenylamine-hexafluoroisopropylidenediphthalimide) (PI(DAT-6FDA)) and poly(*N*-(4-(2′,4′-diaminophenoxy)phenyl)-*N*,*N*-diphenylaminehexafluoroisopropylidenediphthalimide) (PI(DAPT- 6FDA)) and their chemical structures are shown in Fig. 8.43. The PI(DAT-6FDA)-based memory device exhibited an unstable volatile characteristic, while the device that used PI(DAPT-6FDA) which had a very bulky donor showed a stable nonvolatile flash-type memory behavior and a retention time over 10^4 s.

Chen et al. [107] prepared a novel functional PI $(OMe)_2$TPPA-6FPI, as shown in Fig. 8.44. When it was used in memory devices, the PI performed DRAM behavior with high ON/OFF current ratios of 10^7.

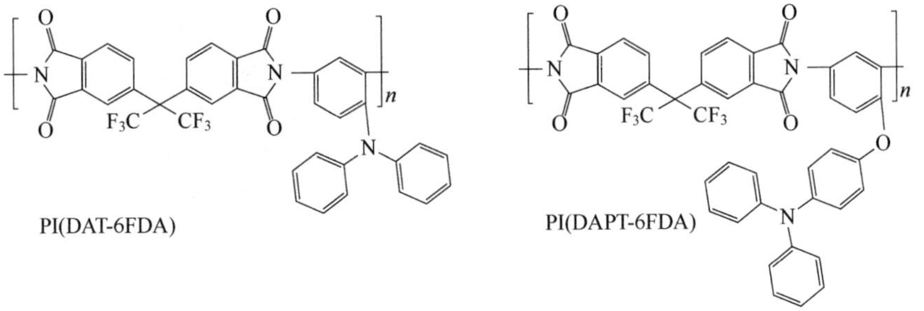

PI(DAT-6FDA) PI(DAPT-6FDA)

FIGURE 8.43 Soluble functional polyimide containing triphenylamine.

$(OMe)_2$TPPA-6FPI

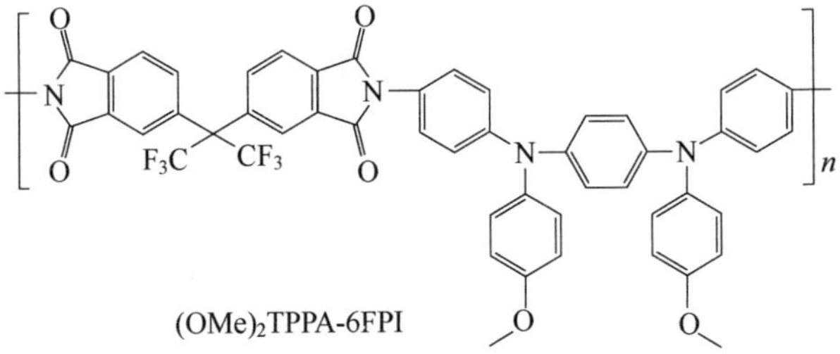

FIGURE 8.44 Soluble functional polyimide containing triphenylamine.

Song et al. [108] synthesized three aromatics-endcapped hyperbranched PIs (HBPIs), PA-HBPI, NA-HBPI, and PDA-HBPI, as shown in Fig. 8.45. When they were used as polymer memory materials, NA-HBPI and PDA-HBPI exhibited enhanced nonvolatile WORM memory behaviors and outstanding thermal stability.

8.3.2.2 Soluble Polyimide with Carbazole Moieties

In 2010, Hahm et al. [109] reported a soluble PI poly(3,3′-bis(*N*-ethylenyloxycarbazole)-4,4′-biphenylene hexafluoro-isopropylidenediphthalimide) (6F-HAB-CBZ PI) and its chemical structure is shown in Fig. 8.46. The memory devices based on this PI exhibited write-read-erase memory characteristics with a high ON/OFF current ratio of 10^{11}.

In 2014, Shi et al. [110,111] synthesized three functional PIs (6F/CzTPA PI, 6F-CzTZ PI and 6F-OCzTZ PI) for memory device applications and their chemical structures are shown in Fig. 8.47. 6F/CzTPA PI exhibited a volatile static random access memory (SRAM) characteristic and its ON/OFF current ratio was up to 10^5. The 6F-CzTZ PI and 6F-OCzTZ PI were different in the incorporation of phenoxy linkages in the main chain, thus they exhibited different memory behaviors. The 6F-CzTZ PI possessed a WORM memory effect with no polarity, while the 6F-OCzTZ PI displayed volatile SRAM behaviors and had polarity.

8.3.2.3 Other Soluble Polyimides

Li et al. [112] synthesized two novel fluorinated PIs (BTFBPD-DPBPDA and BTFBPD-

BTFBPDA) with noncoplanar twisted biphenyl units and their chemical structures are shown in Fig. 8.48. Both of them exhibited high solubility and thermal stability. Furthermore, the PI (BTFBPD- DPBPDA) showed flash-type memory behaviors, while the PI (BTFBPD-BTFBPDA) presented WORM memory characteristics. The memory devices based on them had high ON/OFF current ratios of 10^3-10^4.

Tian et al. [113] synthesized another functional PI (ferrocene-*g*-6FDA/DHTM) by grafting ferrocene units to its backbone and its chemical structure is shown in Fig. 8.49. The PI possessed nonvolatile flash memory behaviors and its ON/OFF current ratio was up to 10^3.

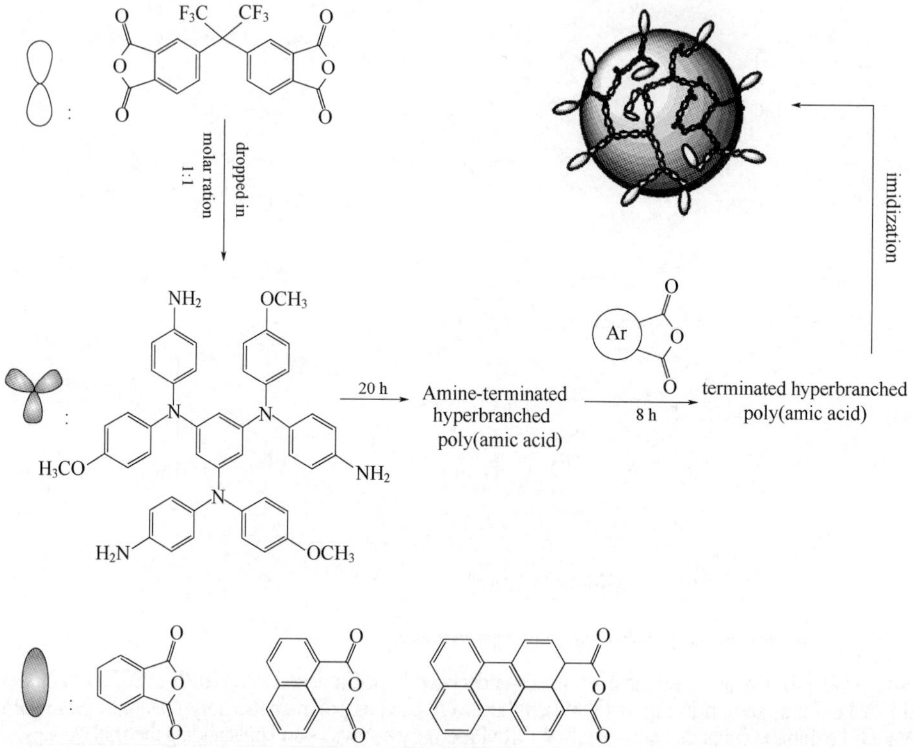

FIGURE 8.45 Hyperbranched polyimide containing triphenylamine.

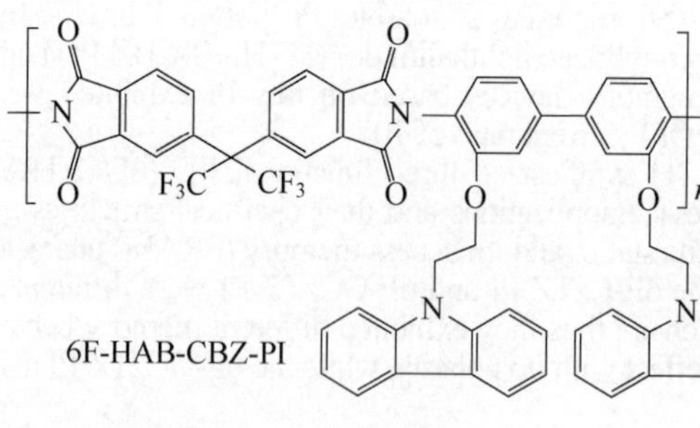

FIGURE 8.46 Soluble functional polyimide containing carbazole moieties.

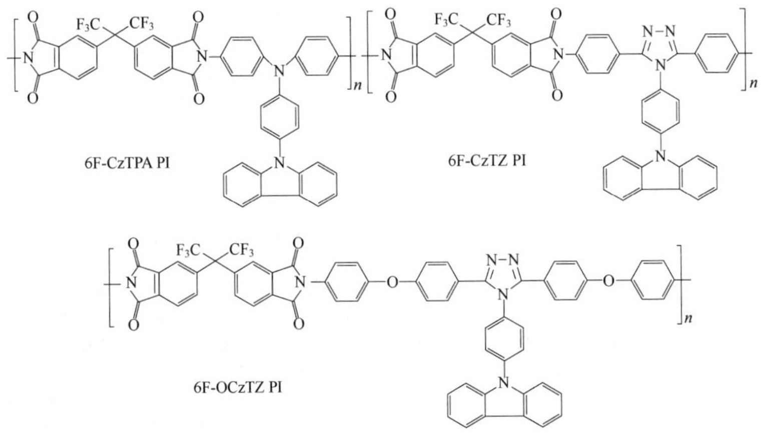

6F-CzTPA PI

6F-CzTZ PI

6F-OCzTZ PI

FIGURE 8.47 Soluble functional polyimide containing carbazole moieties.

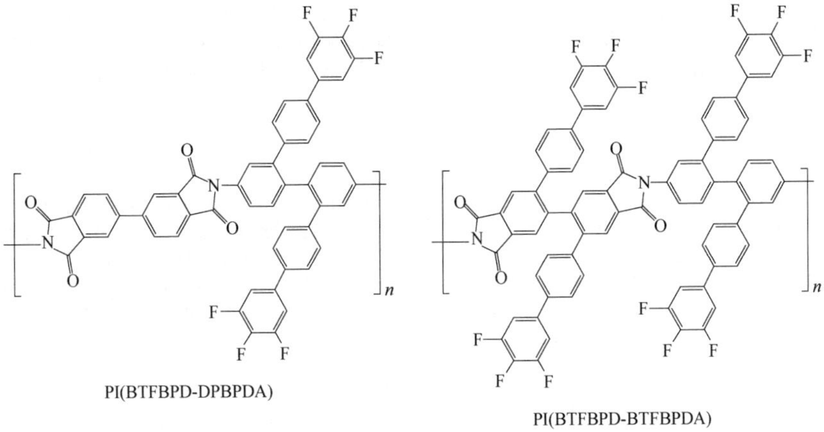

PI(BTFBPD-DPBPDA)

PI(BTFBPD-BTFBPDA)

FIGURE 8.48 Soluble functional polyimide.

Ferrocene-g-6FDA/DHTM PI

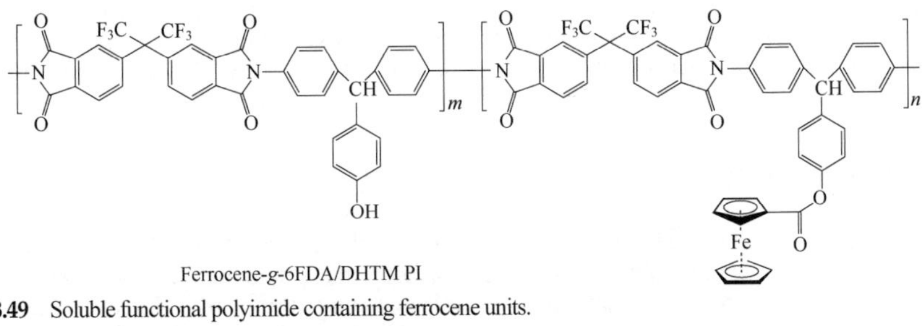

FIGURE 8.49 Soluble functional polyimide containing ferrocene units.

8.3.3 Compensator Materials for Liquid Crystal Displays

Thin-film-transistor active matrix twisted nematic (TN) liquid crystal displays (LCDs) are widely used in televisions, lap-top computers, and information-related equipments because they can offer many advantages, including high information content, color capability, gray scale, and fast response. However, some problems continue to exist in TN-LCDs. The major problem with TN-LCDs is asymmetric

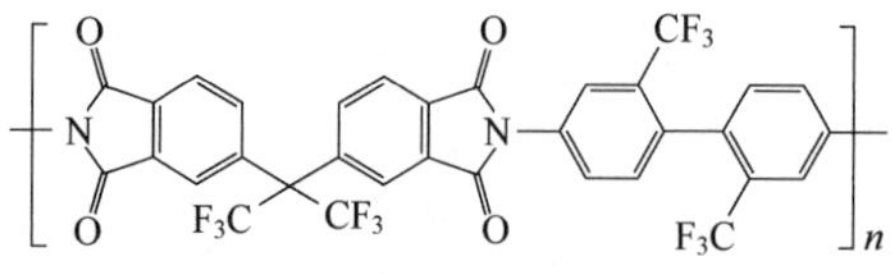

6FDA-TFMB

FIGURE 8.50 6FDA-TFMB used in uniaxial negative birefringence com pensators for LCD.

viewing angle characteristics along the horizontal and vertical directions, as well as narrow viewing angles [114]. Luckily, uniaxial negative birefringence compensator materials can be used to improve the viewing angles and the contrast ratio of TN-LCDs. Aromatic PIs are considered as prom- ising candidates for uniaxial negative birefringence compensator materials due to their anisotropic optical properties on the base of their anisotropic orientation. However, most aromatic PIs are insoluble and must be processed through their soluble intermediate PAAs and then thermally imidized at high temperature. Such high-temperature imidization history often affects the morphology and structure of the resulting films, and finally affects their optical properties [115,116]. Therefore, soluble PIs draw much attention from researchers in this field because they can be processed into films without the imidization procedure. As shown in Fig. 8.50, Li et al. [117] synthesized a soluble PI 6FDA-TFMB containing trifluoromethyl side groups to prepare the anisotropic optical films. Then these films were used as the uniaxial negative birefringence compensators. The introduction of trifluoromethyls could improve the solubility of the PIs, while the rigid-rod structure of main chains can retain their anisotropic optical properties. When these films were assembled into liquid crystal displays, the isocontrast map of the displays has shown that the viewing angle has been extended 10 degrees or more on each side along the horizontal direction compared to that of the uncompensated displays.

8.3.4 Gas Separation Materials

Soluble aromatic PIs are convenient to be processed into gas-separation membranes through the solution-casting, which generally exhibited relatively high gas permselectivity. For example, Tong et al. [118] synthesized a series of PIs containing flexible aliphatic linkages and bulky phenyl side groups, as shown in Fig. 8.51. These PIs could dissolve in common organic solvents such as NMP and THF. The membranes with thicknesses of 40-50 μm could be prepared by casting the DMAc solution containing these PIs onto a clean glass plate. These PI membranes displayed high gas permeability and permselectivity. In particular, the membranes provided good selectivity for CO_2/CH_4 and CO_2/N_2 gas pairs because their appropriate cavity size is favorable to separate the CO_2 from the other gases.

In addition, soluble PIs could be complexed with inorganic nanoparticles to afford the inorganic-organic hybrid membranes (mixed-matrix membranes, MMMs) which combined the processability of polymers and the superior gas separation properties of inorganic materials. For example, the commercial soluble PI Matrimid, as shown in Fig. 8.52, was often used to prepare various MMMs for gas separation by complexing with metal organic framework (MOF) [119-121], fullerene (C_{60}) [122], ordered mesoporous silica spheres (MCM-41) [123]. The resulting MMMs display much higher gas permselectivities than that of traditional polymeric gas separation membranes.

8.3.5 Other Applications

Soluble PIs could also be used to prepare photoreactive PI films applied as the LC alignment layers due to their excellent thermal stability and electrical properties. The PI alignment layer should be pretreated with a rubbing process to make the LC molecules align uniformly. However, there are some drawbacks of the rubbing process, such as dust generation, electrostatic problems, and poor control of rubbing strength and uniformity. Fortunately, some photoreactive soluble PI films could align the LC molecules uniformly after the treatment based on irradiation of the polymer with linearly polarized ultraviolet light (LPUVL) without rubbing. Lee et al. [124] synthesized a photoreactive soluble PI 6F-HAB-CI PSPI by incorporation of cinnamate (CI) side groups into the PIs and the synthesis route was shown in Fig. 8.53. Under the irradiation of LPUVL, the CI side groups would undergo [2+2] photodimerization and the CI side groups parallel to the polarization direction of LPUVL react more rapidly than those that are perpendicular to the polarization direction of the LPUVL. After the irradiation, most of the residual CI side groups are perpendicular to the polarization of the LPUVL and the regular order of these CI side groups could induce the LC molecules to align uniformly.

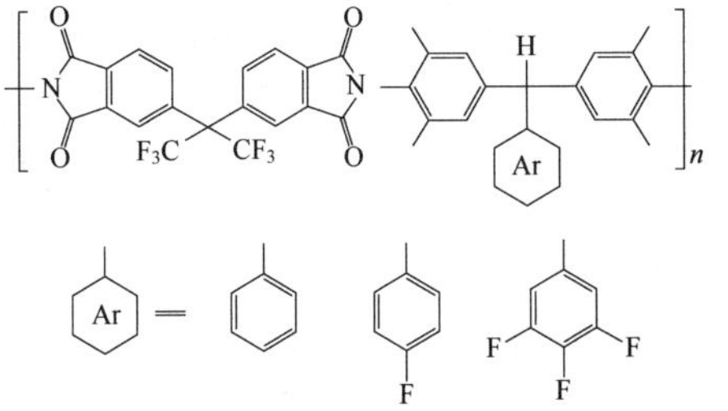

FIGURE 8.51 Chemical structures of the gas separation membranes.

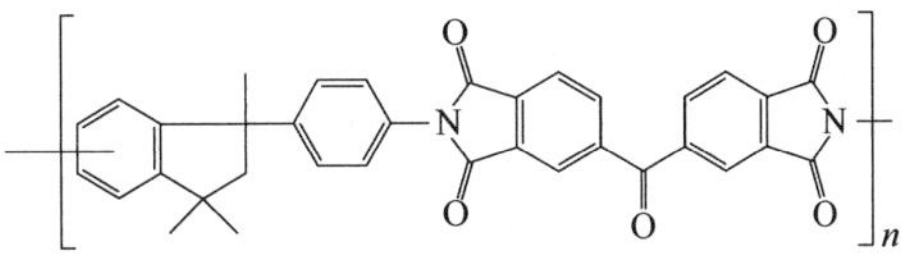

Matrimid 5218/Matrimid 9725

FIGURE 8.52 Chemical structure of Matrimid.

Soluble PIs could also be used to prepare fibers. For example, the commercial PI Tetrimide could be use as optical fiber coatings for avionics with high thermal performance and excellent self-repair properties [125]. They could also be used as adhesives due to the low dielectric constant, low dielectric tangent, and excellent thermal properties in the high-frequency printed circuit board [126]. The commercial soluble PI p84 could be used as an asymmetric membrane with superior selectivity and relatively high flux for pervaporative dehydration of isopropanol [127].

In summary, the solubility of PIs could be enhanced by introduction of flexible linkages, asymmetric units, alicyclic moieties, and side groups. The introduction of appropriate side groups is the most promising method because most of the side groups could not only increase their solubility but also retain their inherent thermal stability. The soluble PIs can broaden the application scope of PIs extensively and are widely used in TN-LCD, second-order NLO materials, memory materials, compensator materials for liquid crystal displays, gas separation materials, and so on.

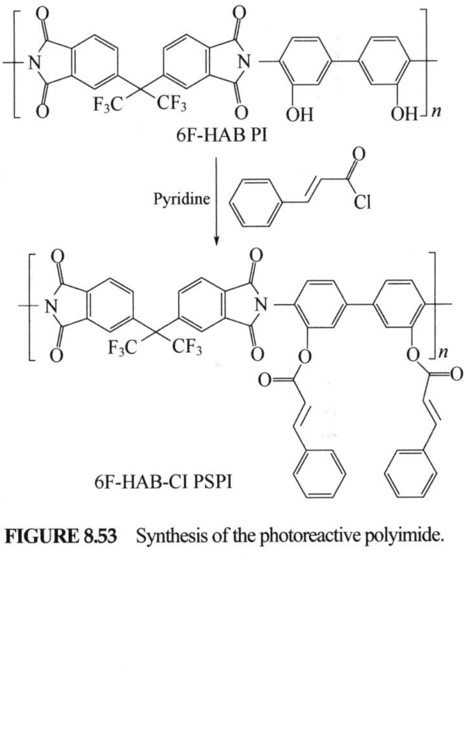

6F-HAB PI

Pyridine

6F-HAB-CI PSPI

FIGURE 8.53 Synthesis of the photoreactive polyimide.

8.4 Low-*k* Polyimide Materials

8.4.1 Introduction

With the development of ULSI to high-speed transmission and high integration in the semiconductor industry, and with the continuing miniaturization in the dimensions of electronic devices utilized in ULSI circuits, an urgent need for high-performance low- and ultralow-*k* dielectric materials (low-*k*: $k \leqslant 2.5$; ultralow-*k*: $k \leqslant 2.0$) has arisen [128-130]. Such dielectric materials would reduce the capacitance between the metal interconnects, the resistance-capacitance delay, the line-to-line crosstalk noise, and the power dissipation [131-133], and have important application prospects in the fields of interlayer dielectric, semiconductor packaging (chips modules, etc.), and high-frequency, low loss boards, etc. So far, research into low dielectric materials as an alternative to the workhorse dielectric silicon dioxide (k=3.9-4.3) is continually being pursued today, which mainly includes organosilicates and organic polymers [134-136].

Compared with inorganic dielectric materials, organic polymer materials often have a lower dielectric constant due to the lower materials density and lower individual bond polarizability. Moreover, they show distinct advantages in terms of easy chemical and geometric structural design [137-139]. Thus, they have attracted much interest. Among them, PIs have been widely used as materials for electronic packaging and electrical insulation in microelectronics industries, due to their excellent thermal, mechanical, and dielectric properties [140,141]. PIs are some of the most promising polymer candidates for the next-generation high-performance interlayer dielectric. However, the k-value of the traditional PIs, like commercial DuPont Kapton PI films, is typically 3.1-3.5, which makes it difficult to meet the requirements of the ultralow-k of less than 2.0 for the technology nodes below 130 nm [142]. Hence, research on PIs with ultralow dielectric constants is of great significance. In this review, we would like to introduce the recent research progress on the design, preparation, and properties of low-k dielectric constant PI materials, especially focusing on the methods to achieve the low-k properties.

8.4.2 Impact Factors on Dielectric Properties [143]

Dielectric constant k (also called relative permittivity ε_r) is the ratio of the permittivity of a substance to that of free space. A material containing polar components, such as polar chemical bonds, which are presented as electric dipoles in Fig. 8.54, has an elevated dielectric constant, in which the electrical dipoles align under an external electric field. This alignment of dipoles adds to the electric field. As a result, a capacitor with a dielectric medium of higher k will hold more electric charge at the same applied voltage or, in other words, its capacitance will be higher. The dipole formation is a result of electronic polarization (displacement of electrons), distortion polarization (displacement of ions), or orientation polarization (displacement of molecules) in an alternating electric field. These phenomena have characteristic dependencies on the frequency of the alternating electric field, giving rise to a change in the real and imaginary part of the dielectric constant between the microwave, ultraviolet, and optical frequency ranges [144,145].

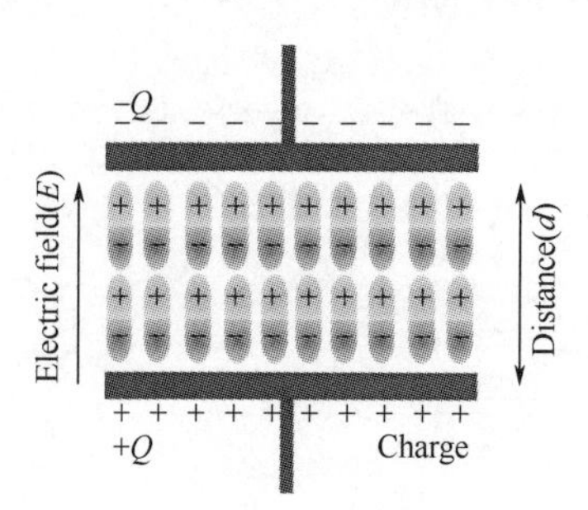

FIGURE 8.54 Schematic illustration of a capacitor.

8.4.2.1 Capacitance and Relative Permittivity

The relative permittivity ε_r of a medium is defined as the ratio of the capacitances of a capacitor with and without the dielectric in place. The capacitance is described by its charge density σ and plate area A (Fig. 8.54).

The charge density σ and the magnitude of the applied field E are related as

$$E=\sigma/\varepsilon \tag{8.1}$$

The charge on the plates arises from the polarizing medium which induces a net charge density p. The electric field between the plates can be written as

$$E=(\sigma-p)/\varepsilon_0 \tag{8.2}$$

Since the electric field in both these equations is the same, one obtains

$$p=\left(\frac{\varepsilon-\varepsilon_0}{\varepsilon}\right)\sigma=(\varepsilon_r-1)\varepsilon_0 E \tag{8.3}$$

The electric susceptibility χ_e is defined as

$$\chi_e=\varepsilon_r-1 \tag{8.4}$$

and writing the polarization and electric field as vectors, one obtains

$$\boldsymbol{P}=\chi_e\varepsilon_0\boldsymbol{E} \tag{8.5}$$

The next stage is relating the polarization of the medium, ***P***, to the polarizability of its molecules. The polarization is the dipole-moment density and equals the mean dipole moment of a molecule, ***p***, in the medium, multiplied by the number density of molecules, *N*.

The induced dipole moment is proportional to the local electric field $\boldsymbol{E}^*$. The local electric field is the total field arising from the applied field and the electric dipoles which that field stimulates in the medium. The polarization of the medium becomes

$$\boldsymbol{P}=\boldsymbol{N}_{\mathbf{p}}=\alpha \boldsymbol{N}\boldsymbol{E}^* \tag{8.6}$$

where α is the polarization constant. For a continuous dielectric, the Lorentz local field can be derived from electrostatics and is given by

$$\boldsymbol{E}^*=\boldsymbol{E}+\boldsymbol{P}/3\varepsilon_0 \tag{8.7}$$

8.4.2.2 Polarization Phenomena

The polarizability of a molecule is a measure of its ability to respond to an electric field and acquire an electric dipole moment ***p***. There are several microscopic mechanisms of polarization in a dielectric material [146-148]. Electric dipole moments can be permanent or can be induced by the electric field.

The *induced electric dipole moment* can be a result of two polarization phenomena, i.e., electronic polarization and distortion polarization.

Electronic polarization, α_e, describes the displacement of the cloud of bound electrons with reference to the nucleus under an applied electric field. The atom distorts, and the center of the atom's negative charge no longer coincides with the position of the nucleus, resulting in an electric dipole moment. The electric dipole moment of each atom is described by

$$\boldsymbol{P}=\alpha_e\boldsymbol{E}^* \tag{8.8}$$

Distortion polarization α_d (also often referred to as ionic polarization) relates to the distortion of the position of the nuclei by the applied field, thereby stretching or compressing the bond length, depending on the relative orientation between the ionic bond and the electric field. The molecule is bent and stretched by the applied field and its dipole moment changes accordingly. Nonpolar molecules may acquire an induced dipole moment in an electric field on account of the distortion the field causes in their electronic distributions and nuclear positions. A polar molecule is a molecule with a permanent electric dipole moment. The permanent dipole moment is a result of the partial charges on the atoms in the molecule that arise from the different electronegativity or other features of bonding. Polar molecules may have their existing dipole moments modified by the applied field. Orientation polarization relates to the phenomenon of a permanent dipole moment as a result of polar molecules. The total polarization of a medium, composed of polarizable polar molecules, is therefore

$$\boldsymbol{P}=N\left(\alpha_e+\alpha_d+\frac{\mu^2}{3kT}\right)\boldsymbol{E}^* \tag{8.9}$$

where N is the number of molecules per m^3, μ is the orientation polarizability, k is the Boltzmann constant, and T is the temperature in K.

The terms α_e and α_d represent the electronic and distortion polarization in the molecule, while the term $\mu^2/3kT$ stems from the thermal averaging of permanent electric dipole moments in the presence of an applied field. The quantitative relation between the relative permittivity and properties of the molecules is described by the Debye equation:

$$\frac{\varepsilon_r-1}{\varepsilon_r+2}=\frac{N}{3\varepsilon_0}\left(\alpha_e+\alpha_d+\frac{\mu^2}{3kT}\right) \tag{8.10}$$

The relative permittivity ε_r of materials is high if its molecules are polar and highly polarizable. This equation shows that the permittivity is smaller if materials do not contain polar molecules. The reduction of density polarizabilities α_e and α_d and N are also possible ways of decreasing the dielectric constant. Reducing the number of ionic bonds in the material minimizes distortion polarization. The electronic polarization is minimized by lowering the electron density in the material, i.e., introducing smaller elements. The same expression, but without the permanent dipole moment contribution, is called the Clausius-Mossotti equation:

$$\frac{\varepsilon_r - 1}{\varepsilon_r + 2} = \frac{N}{3\varepsilon_0}(\alpha_e + \alpha_d) \tag{8.11}$$

Although the value of the electric dipole moments is extremely important for predicting properties of dielectric materials, their calculation is difficult. In the case of simple molecules, the Stark effect [149] is used to measure the electric dipole moment of molecules for which rotational spectra can be observed. In the case of more complicated systems, the polarizability and permanent dipole moment of molecules can be determined by measuring ε_r at a series of temperatures. These measurements facilitate the determination of molar polarizations and their slope and intersect versus $1/T$ give values of dipole moment and polarizability.

8.4.2.3 Film Density and Relative Permittivity

The possibility to lower the molecular polarizability is limited. Eqs. (8.9)-(8.11) show that the number of molecules per unit of volume (film density) plays an important role in the reduction of the film permittivity. The effect of the density on the film permittivity is stronger than the effect of molecular polarizability because decreasing density allows decreasing of the dielectric constant to the extreme value close to unity.

Technologically, a preferred way to reduce the film density is the introduction of pores. Generally, porous films can be considered as two-component materials where the solid skeleton has a dielectric constant close to the dense prototype and the second component (pores) has dielectric constant equal to 1. The relative permittivity of porous film ε_r directly depends on porosity

$$\frac{\varepsilon_r - 1}{\varepsilon_r + 2} = \boldsymbol{P} \cdot \frac{\varepsilon_1 - 1}{\varepsilon_1 + 2} + (1 - \boldsymbol{P}) \cdot \frac{\varepsilon_2 - 1}{\varepsilon_2 + 2} = (1 - \boldsymbol{P}) \cdot \frac{\varepsilon_2 - 1}{\varepsilon_2 + 2} \tag{8.12}$$

In this equation ε_1 is permittivity of material inside the pores, ε_2 is permittivity of the film skeleton, and P is the film porosity. The term $P\cdot(\varepsilon_1-1)/(\varepsilon_1+2)$ is equal to 0 if the pores are empty, as experimentally demonstrated by Hrubesh and coworkers.

8.4.2.4 Relative Permittivity and Frequency

When the applied electric field is an AC field, the frequency of the signal comes into play. The polarization phenomena described above are very different for applied electric fields with different frequencies. For an applied field

$$\boldsymbol{E} = \boldsymbol{E}_0 \boldsymbol{e}^{-i\omega t} \tag{8.13}$$

The polarization vector is of the form

$$\boldsymbol{P} = \varepsilon_0(\varepsilon_r - 1)\boldsymbol{E}_0 \boldsymbol{e}^{-i\omega t} \tag{8.14}$$

and gives rise to a current density

$$\boldsymbol{J} = -i\omega\varepsilon_0(\varepsilon_r - 1)\boldsymbol{E}_0 \boldsymbol{e}^{-iwt} \tag{8.15}$$

which is imaginary as long as ε_r is real. This will not always be the case as will be discussed later. Let us first consider the consequences for the polarization phenomena. Electronic polarization follows the electric field almost instantaneously as only the displacement of bound electrons is involved. The distortion polarization cannot respond as rapidly to fast changing fields because it involves the displacement of entire ions. Both electronic and distortion polarization are subject to a counteractive restoring force, which gives rise to a resonant frequency. In contrast, orientation polarization requires the motion of complete molecules. For orientation polarization there is no counteractive restoring force. Hence, it does not give rise to a resonance at a critical frequency, as distortion and electronic polarization do. The orientation polarization is, however, opposed by thermal disorder. At low frequencies the three polarization phenomena contribute to the real part of the dielectric constant. The maximum frequency for orientation polarization is of the order of 10^9 Hz. Above this frequency, distortion and electronic polarization contribute to the dielectric constant up to the resonance frequency for distortion polarization, which is typically of the order of $\sim 10^{13}$ Hz, and beyond that only electronic polarization defines the dielectric constant. The resonant frequency of the electronic polarization is typically beyond the frequency of visible light at ~10^{15} Hz. It follows from Maxwell's equations that the refractive index relates to the relative permittivity, beyond optical frequencies, as

$$n_r = \varepsilon_r^{1/2} \tag{8.16}$$

Because of the relation between the relative permittivity and the molecular polarizability, one can relate n_r to the molecular properties as follows:

$$n_r = \left(\frac{1 + \left(\frac{2\alpha N}{3\varepsilon_0} \right)}{1 - \left(\frac{\alpha N}{3\varepsilon_0} \right)} \right)^{1/2} \approx 1 + \left(\frac{\alpha N}{2\varepsilon_0} \right) \tag{8.17}$$

The expression leads to the Lorenz-Lorentz formula

$$\frac{n_r^2 - 1}{n_r^2 + 2} = \frac{N\alpha(\omega)}{3\varepsilon_0} \tag{8.18}$$

The characteristic response of the different polarization phenomena to the electric field results in a frequency dependence which has not only an impact on the real part of the dielectric constant, but the imaginary part is related to the counteractive restoring forces in case of electronic and distortion polarization, and to the thermal disorder in case of orientation polarization. The imaginary component of the dielectric constant corresponds to a current density within the dielectric that is no longer exactly $\pi/2$ out of phase with the electric field. From the frequency dependence, one should note that orientation polarization in low-*k* dielectrics should be prevented as much as possible, not only for its contribution to the dielectric constant but more importantly for the imaginary part of the dielectric constant. Indeed, orientation polarization leads to heat dissipation in a frequency range that is envisioned within the application frequencies of microelectronics systems.

TABLE8.1 The Electronic Polarizability and Bond Enthalpies

Bond	Polarizability (Å^3)	Average Bond Energy (kCal·mol^{-1})
C—C	0.531	83
C—F	0.555	116
C—O	0.584	84
C—H	0.652	99
O—H	0.706	102
C=O	1.020	176
C=C	1.643	146
C≡C	2.036	200
C≡N	2.239	213

8.4.3 Structures and Dielectric Properties of Polyimides

According to the above discussion, it can be concluded that there are mainly two strategies for designing low dielectric materials, including decreasing dipole strength or the number of dipoles or a combination of both. In the first strategy, materials with chemical bonds of lower polarizability than Si—O or lower density would be used. Today, the microelectronics industry has already moved to certain low-*k* materials, where some silica Si—O bonds have been replaced with less polar Si—F or Si—C bonds. A more elementary reduction of the polarizability can be attained by utilizing all nonpolar bonds, such as C—C or C—H, as in the case of organic polymers. Table 8.1 shows the electronic polarizability and bond enthalpies.

The second strategy involves decreasing the number of dipoles within the ILD material by effectively decreasing the density of a material. This can be achieved by increasing the free volume through rearranging the material structure or introducing porosity. Porosity can be constitutive or subtractive. Constitutive porosity refers to the self-organization of a material. After manufacturing, such a material is porous without any additional treatment. Constitutive porosity is relatively low (usually less than 15%) and pore sizes are ~ 1 nm in diameter. According to International Union of Pure and Applied Chemistry (IUPAC) classification, pores of less than 2 nm are denoted "micropores." Subtractive porosity involves selective removal of part of the material. This can be achieved via an artificially added ingredient (e.g., a thermally degradable substance called a "porogen," which is removed by annealing to leave behind pores) or by selective etching (e.g., Si—O bonds in SiOCH materials removed by HF).

8.4.4 Low-*k* Polyimides With Porous Structures

The introduction of air gaps into interconnected structures [150,151] and nanopores/mesopores into polymers [152,153] to reduce their dielectric constants has become an attractive approach. The incorporation of air, which has a dielectric constant of about 1, can greatly reduce the dielectric constant of the resulting porous structure. The probability of this technique is readily apparent via the prediction by Maxwell-Garnett modeling of porous structures based on a matrix polymer [154]. Supported porous films with low-*k*-value are considered to be the most promising materials for the sub-100 nm technology node and future integrated circuit (IC) process technologies. The nano/mesopores can be introduced into the polymers by more than five methods such as reprecipitation, sol-gel, thermolysis, supercritical foaming, and electrochemical etch. The pore size, morphologies, porosities, and film thicknesses of the products can be controlled by raw materials and processing conditions in these methods. The low-*k* dielectric materials based on nano/mesoporous polymers have great potential applications in ULSI in light of their special dielectric, optical, thermal, and mechanical properties.

8.4.4.1 Porous Fillers—Added Polyimides

Recently, many new nanomaterials (porous nanoparticles, nanotubes, etc.), such as zeolite [155], SiO_2 nanotubes [156,157], phyllosilicates [158], etc., have been reported. These nanomaterials have large specific surface areas or hollow structures, which make them potentially suitable for applications as catalysts, storage devices, and templates to grow other nanosized materials. One of the new potential applications is to incorporate the porous particles or hollow tubes into a polymer matrix to reduce the total dielectric constant of the composite. Meanwhile, the inorganic/organic hybrids with inorganic particles are also quite helpful in improving the thermal stability of the materials, which is very important in the microelectronics industry.

Zhang et al. [156] reported a series of novel silica tube/PI composite films with variable low dielectric constant. Silica tubes (STs) were synthesized by self-assembly, and the ST/PI composite films were then prepared by mixing the STs and PI precursors in a solution of *N,N*-dimethylacetamide (DMAc) under ultrasonic agitation. The ST contents in the composites ranged from 1 to 20%(*w*). The silica tubes possessed lengths of several micrometers and diagonal lengths of tens to hundreds of nanometers (Fig. 8.55A and B). The dielectric constant of the composite was reduced from about 3.3 for pure PI to 2.9 for the 3%(*w*) ST/PI composite (Fig. 8.55C). The authors concluded that the low dielectric constant was ascribed to the effect of air stored in the silica tubes, pores, and interfaces between the ST and the PI matrix in the composites. They also predicted that it might be possible to further reduce the dielectric constant of the composite by choosing a matrix with a smaller dielectric constant.

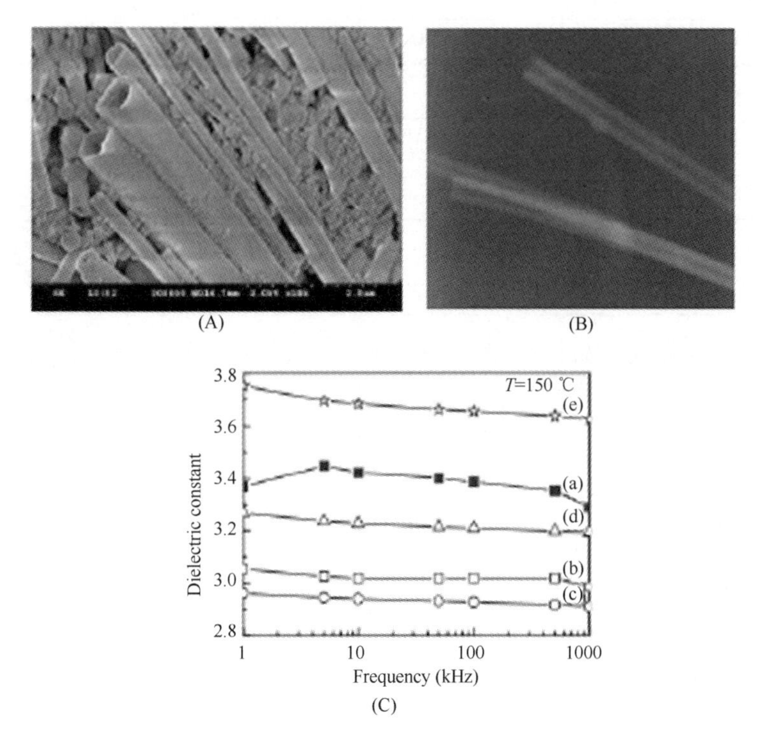

FIGURE 8.55 SEM (A) and TEM (B) images of silica tubes; frequency dependence of the dielectric constant of pure PI and ST/PI composites with frequency ranging from 1 kHz to 1 MHz at 150 °C (C). The silica tube content: (a) 0, (b) 1, (c) 3, (d) 10, and (e) 20%(w).

Yudin et al. [159] synthesized novel magnesium hydrosilicate [$Mg_3Si_2O_5(OH)_4$] nanotubes (SNTs). New polymer nanocomposites containing SNTs and PI matrices such as PMDA-ODA (denoted by $PI_{PAA\text{-}PM}$) and Ultem-1000 were prepared with SNT concentrations ≤ 12 %(vol) (or ~20%(w)) to yield new materials with improvements in stiffness, strength, and barrier properties. The dielectric and gas barrier measurements of the PI nanocomposites revealed a reduction in the dielectric constant and improvements in the gas barrier properties as a function of the SNT concentration (Table 8.2). The k-value was about 2.2 when the addition of SNT was 10%. Because of their facile synthesis and desirable properties for a number of applications in protective coatings and films for microelectronic applications and flammability reduction, these PI nanocomposites are expected to be excellent model systems for exploring mechanisms of reduction in dielectric constant of PIs filled with hollow nanoscale inorganic tubes, a very important and desirable characteristic in microelectronic applications.

Lin and Wang [160] successfully prepared a series of novel low-k PI composite films containing the SBA-15 or the SBA-16-type mesoporous silica via in situ polymerization and following thermal imidization (Fig. 8.56). They investigated their morphologies, dielectric constants, thermal, and dynamic mechanical properties, and found that the k-value of the composite films could be reduced from 3.34 of the pure PI to 2.73 and 2.61 by incorporating 3%(w) SBA-15 and 7%(w) SBA-16, respectively. The reduction in the dielectric constant is attributed to the incorporation of the air voids (k=1) stored within the mesoporous silica materials, the air volume existing in the gaps on the interfaces between the mesoporous silica and the PI matrix, and the free volume created by introducing large-sized domains. The as-prepared PI/mesoporous silica composite films also present stable dielectric constants across the wide frequency range and a good phase interconnection. The improvements in dynamic mechanical properties and the thermal stability of the PI film are achieved by incorporation of the mesoporous silica materials. The enhanced interfacial interaction between the PI matrix and the surface-treated mesoporous silica has led to the minimization of the deterioration of the mechanical properties. The incorporation of the mesoporous silica materials is a promising approach to prepare the low-k PI films.

TABLE8.2 Dielectric Constants of $PI_{(PAA-PM)}$/SNT Nanocomposite Films

SNT (%(w))	SNT (%(vol))	ε_1' Initial Film	ε_2' After Drying	ε_3' After Steam Storage
0	0	2.9	2.9	2.9
10	5.9	2.2	2.2	2.7
15	9.1	2.4	2.3	2.8
20	12.4	2.5	2.3	3.1
30	19.6	2.8	2.5	3.3

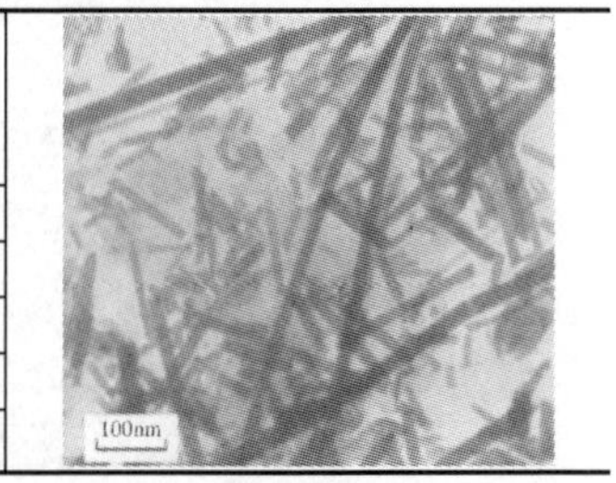

A series of fluorinated random copolyimide/amine-functionalized zeolite MEL50 hybrid films (*co*-FPI/MEL50-NH_2) was successfully prepared with the addition of MEL50-NH_2 (e.g. 0%, 1%, 3%, 5%, 7% %(*w*). of the *co*-FPI) by Li et al. [161] The zeolite MEL50 was synthesized by a hydrothermal method and *co*-FPI was obtained by random copolycondensation from 4,4′-(hexafluoroisopropylidene) diphthalic anhydride (6FDA), 2,2-bis[4-(3,4-dicarboxyphenoxy)phenyl] propane dianhydride (BPADA) and 4,4′-diaminodiphenyl ether (ODA) (Fig. 8.57). The results showed that low dielectric properties and the thermal properties of the hybrid films were greatly enhanced to achieve the optimum with the addition of 3 %(*w*). MEL50-NH_2. The T_g increased up to 266 °C from 243 °C of the pristine *co*-FPI film and 5% weight loss temperature increased up to 527 °C from 517 °C of the pristine *co*-FPI film in nitrogen. The dielectric constant decreased down to 2.68 from 3.21 of the pristine *co*-FPI film. This convenient approach provides a possibility of making PI possess both low dielectric constant and enhanced thermal properties, which is more applicable to microelectronics.

Park and Ha [162] reported a series of PI/hollow silica (HS) sphere hybrid films with low-*k* values via thermal imidization processes using pyromellitic dianhydride (PMDA)/4,4′-oxydianiline (ODA) as the polymer matrix and HS spheres as inorganic particles with the closed air voids (Fig. 8.58). The monodispersed HS spheres were synthesized via a one-step process, which means that the formation of silica shells and dissolution of the polystyrene particles occurs in the same medium. The HS particles have uniform size of ca. 1.5 μm in diameter and ca. 100 nm in shell thickness. PI/HS sphere hybrid films were synthesized using a mixture of PAA and HS spheres prepared via one-pot process. HS spheres of two different kinds (pristine HS spheres (PHS spheres) and amine-modified HS spheres (AHS spheres)) were used for the preparation of the hybrid films. With the varying contents of AHS spheres in the range of 1%-10%(*w*), the dielectric constants of the PI/AHS sphere hybrid films were decreased from 3.1 of pure PI to 1.81 by incorporating 5%(*w*) AHS. The dielectric constants of the PI/PHS sphere hybrid films were decreased to 1.86 by incorporating 5%(*w*) PHS. Organic-inorganic hybrid porous PIs may be expected as prime candidates for polymeric insulators due to their high thermal stability, good mechanical properties, solvent resistance, and low-*k*.

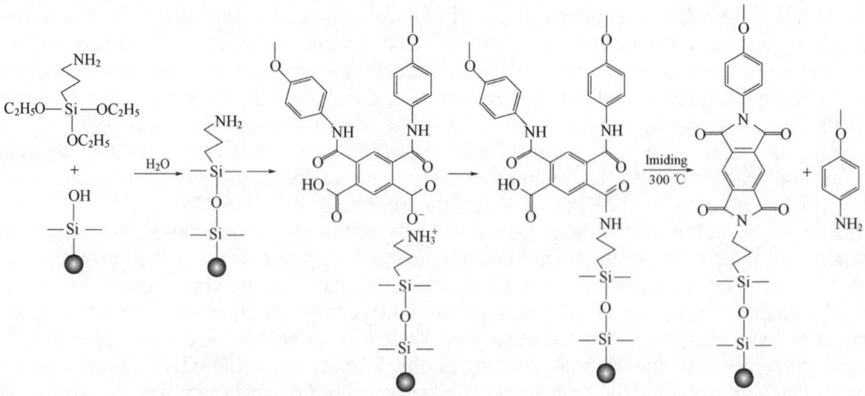

FIGURE 8.56 Scheme of possible reaction between the mesoporous silica's surface bound APTS and PAA.

Pure silica zeolite nanocrystal particles functionalized with aminopropyl groups (APSZN) were introduced into a 2,2-bis-(3,4-dicarboxyphenyl) hexafluoropropane dianhydride-2,2′-bis(trifluoromethyl)-4,4′-diaminobiphenyl polyimide (FPI) matrix to obtain FPI/APSZN hybrid films [163]. The effect of APSZN content on the hybrid films was evaluated in terms of UV-visible transmittance, k-value, coefficient of thermal expansion (*CTE*), mechanical properties, dynamic mechanical properties, and thermal degradation behavior. The k-value of the FPI/APSZN hybrid film decreases to 2.56 with 7%(w) APSZN (Fig. 8.59). With increasing content of APSZN, the tensile strength and the Young's modulus of FPI/APSZN films increase, while the *CTE* of FPI/APSZN hybrid films decreases. This study provides a novel approach for preparing low-k FPI hybrids with enhanced properties.

Kurinchyselvan et al. [164] reported the concept of reduction of polarization by introducing the porous material (amine-functionalized mobil composition of matter no. 41 (FMCM-41)) and long-chain aliphatic diamine in the skeleton of PI, which results in the reduction of the dielectric constant and dielectric loss of a material (Fig. 8.60). The mechanical (tensile strength) properties of the nanocomposites were lowered when pores are introduced into the film. However, the lowest k-value (2.21) and dielectric loss (0.0082) were achieved for 7%(w) FMCM-41/PI, and beyond this concentration (10%(w) FMCM-41/PI) the reverse trend was noticed, due to the agglomeration of nanoparticles in the PI matrix. The particular concentration of FMCM-41 reinforced with PI has the lowest value of dielectric constant with significantly improved thermal stability, thus this kind of material can be used for high-performance interlayer dielectrics.

Ye et al. [165] reported a new and simple method for preparing PI nanocomposites that have very low dielectric constants and good thermal properties: simply through blending the PI precursor with two multifunctional POSS derivatives, one containing eight fluorinated ether groups (octakis (dimethylsiloxyhexafluoropropyl) silsesquioxane (OF)) and the other eight ether groups (octakis (dimethylsiloxypropyl) silsesquioxane (OP)) (Fig. 8.61). Results showed that the existence of effective intermolecular interactions between PAA and OF increased their compatibility, resulting in better dispersion of these POSS cages within the PI matrix and improved thermal and dielectric properties for the PI-OF nanocomposites relative to the PI-OP system. The incorporation of OF into the PI resulted in a gradual decrease of the dielectric constant from 3.19 (OF=0%(w)) to 2.12 (OF=15%(w)). They observed a similar trend for the PI-OP composites, except that the value of k decreased relatively slower than that of the PI-OF composites at the same loading of the POSS derivative. This simple method for enhancing the properties of PIs might have potential applicability in the electronics industry.

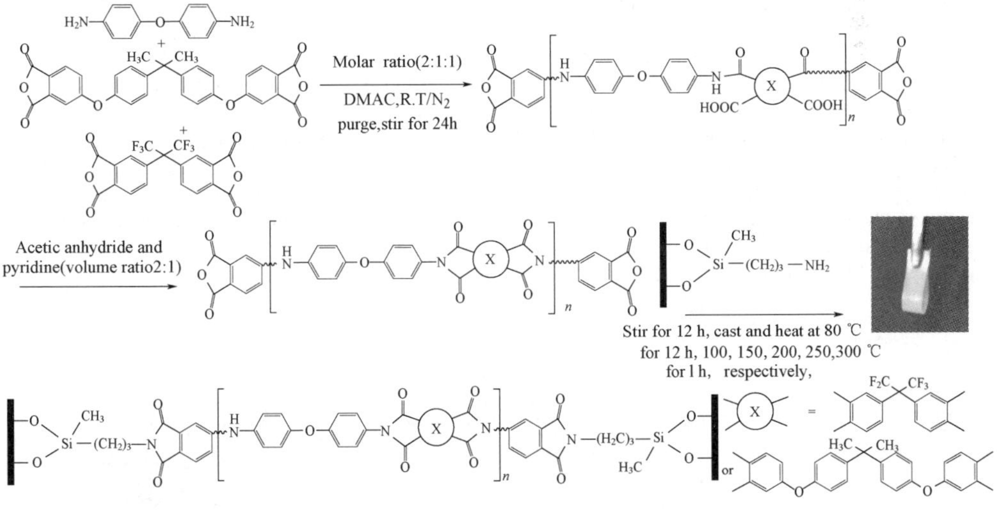

Molar ratio(2:1:1)
DMAC,R.T/N_2 purge,stir for 24h
Acetic anhydride and pyridine(volume ratio2:1)
Stir for 12 h, cast and heat at 80 ℃ for 12 h, 100, 150, 200, 250,300 ℃ for 1 h, respectively,

FIGURE 8.57 Synthesis process of co-FPI/MEL50 mixed matrix membranes. *Reproduced with permission from Li, Q., Wang, Y., Zhang, S., Pang, L., Tong, H., Li, J. et al. Novel fluorinated random co-polyimide/amine-functionalized zeolite MEL50 hybrid films with enhanced thermal and low dielectric properties. J. Mater. Sci., 52 (9) (2017), 5283-5296. Copyright 2017, Springer.*

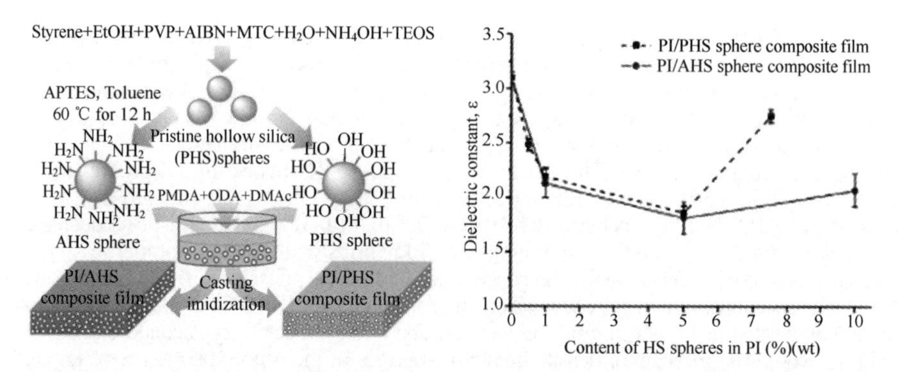

FIGURE 8.58 Illustration of the synthesis of the polyimide hybrid films containing pristine hollow silica (PHS) spheres (PI/PHS sphere hybrid film) and amine-modified hollow silica (AHS) spheres (PI/AHS sphere hybrid film); the dielectric constant of the composites.*Reproduced with permission from Park, S.S., Ha, C.-S. Polyimide/hollow silica sphere hybrid films with low dielectric constant. Composite Interfaces, 23 (8) (2016),831-846. Copyright 2017, Taylor & Francis.*

Kim et al. [166] used a PI aerogel (PIA) to synthesize low-dielectric-constant PI composite films. The resulting films maintained the original thermal and chemical properties of PI, which contribute to their homogeneity. The dielectric constant of the aerogel composite film was decreased to ~ 2.4, a value similar to that of a silica aerogel (SIA) composite film. The increase in dielectric constant as a result of water adsorption exhibited by SIA films was counteracted by the hydrophobicity of the PI in the PIA film. The PIA composite film exhibited a 3% water uptake, compared with a 29% water uptake in the SIA film, indicating that the PI film is the preferred choice in wet conditions. The PI-PIA films exhibited excellent thermal properties with a T_g of 298 °C and a 1% thermal decomposition temperature of 589 °C.

8.4.4.2 Nanoporous Polyimides

Carter et al. [167] prepared porous, low-*k* PI films by a "nanofoam" approach. The nanoporous foams are generated by preparing triblock copolymers with the major phase comprising PI and the minor phase consisting of a thermally labile block (Fig. 8.62). Films of the copolymers are cast and then heated to effect solvent removal and annealing, resulting in microphase separation of the two dissimilar blocks. The labile blocks are selectively removed via thermal treatments, leaving pores the size and shape of the original copolymer morphology. The PI derived from 2,2-bis(4-aminophenyl) hexafluoropropane (6FDAm) and 9,9-bis(trifluoromethyl) xanthene-tetracarboxylic dianhydride (6FXDA) was used as the matrix material for the generation of nanofoams, and specially functionalized poly(propylene oxide) oligomers were used as the thermally labile constituents. The synthesis and characterization of the copolymers were performed and the process for obtaining nanofoams was optimized. Thin-film, high-modulus nanoporous films with good mechanical properties and dielectric constants ~2.3 have been synthesized by the copolymer/nanofoam approach.

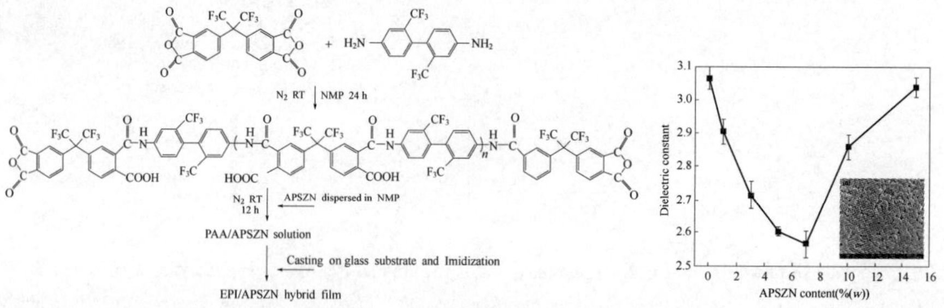

FIGURE 8.59 Preparation procedure of FPI/APSZN hybrid films and dielectric constant of FPI/APSZN hybrid films as a function of APSZN content at 1 MHz. Insert: SEM images of APSZN.*Reproduced with permission from reference [163]. Copyright 2015, Royal Society of Chemistry.*

Wang et al. [168] reported on a simple method for preparing ultralow-k nanoporous films based on a soluble fluorinated PI. Initially, molecular modification of the ozone-pretreated FPI via thermally induced graft copolymerization with acrylic acid (AAc) was carried out in NMP solution. Films of the copolymers were subjected to thermal treatment to decompose the AAc polymer (PAAc) side-chains, leaving behind nanosized pores and gaps in a matrix of preserved FPI backbones. The processes are shown schematically in Fig. 8.63. The k-value of the resulting nanoporous FPI film is governed by the intrinsic k-value of the FPI matrix and the morphology (porosity) of the porous structure. The dielectric constant of the pristine FPI film is about 3.1 under ambient conditions. As anticipated, all the nanoporous FPI films exhibit considerably lower dielectric constants and the dielectric constant decreases with the increase in porosity. A dielectric constant of about 1.9 is obtained for the nanoporous FPI film prepared from the PAAc-g-FPI copolymer with an initial bulk graft concentration of about 1.67 and a final porosity of about 8%. This dielectric constant is even lower than that of the poly(tetrafluoroethylene) (PTFE) film (k~ 2.1). Graft copolymerization with AAc reduces the structural rigidity and the intermolecular packing density of the FPI chains and, thus, increases the molar free volume of the polymer. As a result, the dielectric constant decreases significantly, especially, after the decomposition of the grafted AAc side-chains to form the nanovoids.

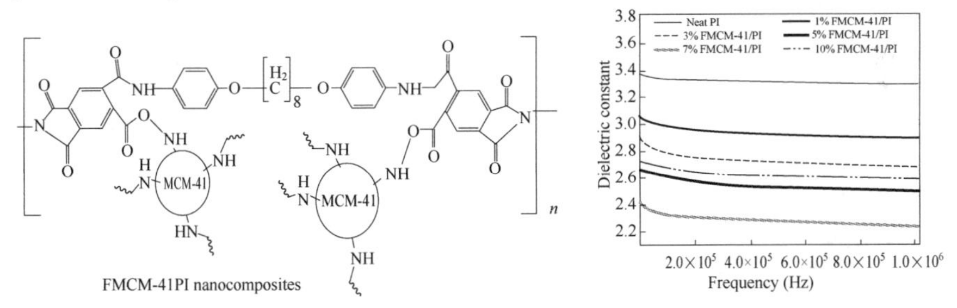

FIGURE 8.60 Structure of FMCM-41/PI hybrid nanocomposite (left) and dielectric constant of neat PI and FMCM-41/PI nanocomposites (right).

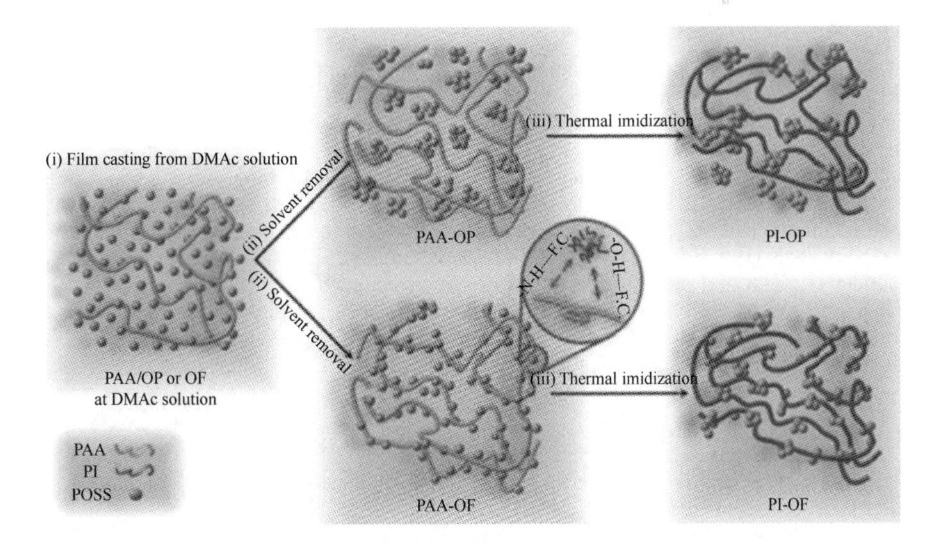

FIGURE 8.61 The deformation processes occurring during the imidization processes in PI-OF and PI-OP systems. *Reproduced with permission from reference [165]. Copyright 2008, John Wiley & Sons..*

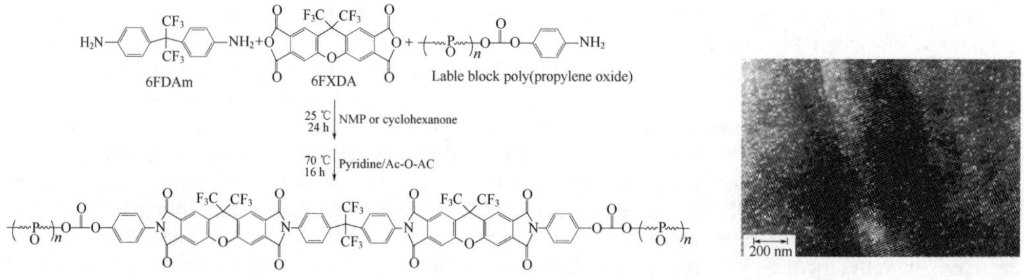

FIGURE 8.62 Synthesis of polyimide copolymers (left) and TEM micrograph of a 6FXDA/ 6FDAm polyimide nanofoam (right). *Reproduced with permission from reference [167]. Copyright 2001, American Chemical Society.*

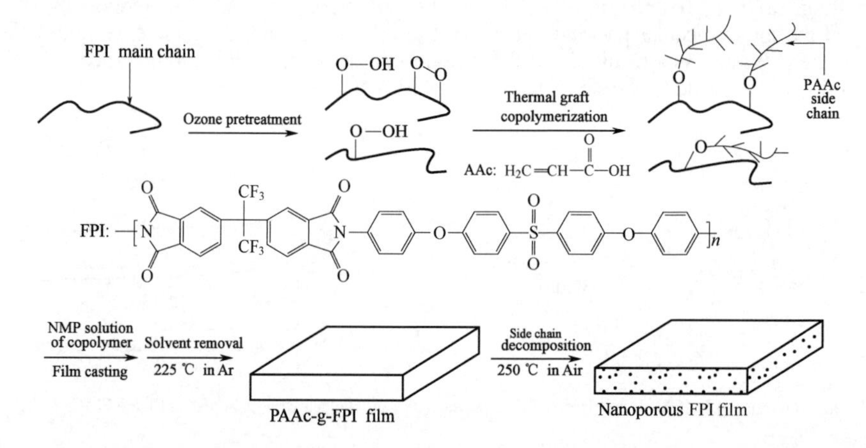

FIGURE 8.63 The process of thermally induced graft copolymerization of AAc with ozone-preactivated FPI backbones and the preparation of a nanoporous FPI film.*Reproduced with permission from reference [168]. Copyright 2004, John Wiley & Sons.*

Lee et al. [169] described a novel method for preparing nanoporous PI films through the use of a hybrid PEO-POSS template. They successfully synthesized a PEO-functionalized POSS (PEO-POSS) possessing a silsesquioxane unit at its core. Incorporation of oligomeric PEO chains onto the silsesquioxane core significantly reduced PEO-POSS selfaggregation and enhances its solubility in the polyamic acid matrix prior to imidization. To obtain nanoporous PI films, they coated the soluble precursor polymers onto a glass plate to yield uniform films, converted them into phase-separated hybrids through imidization, and then foamed them upon decomposition of the thermally labile PEO-POSS moieties (Fig. 8.64). The PMDA-ODA polyimide/PEO-POSS hybrids were prepared at 300 °C under nitrogen and transformed into a porous structure by annealing at 280 °C in air. The PEO-POSS units were microphase-separated in the PI matrix with domain sizes of 20 nm, and the final nanoporous with sizes of 10-40 nm were generated from these films through a foaming process. In so doing, the bulk dielectric constant of the thin film was lowered from 3.25 to 2.25.

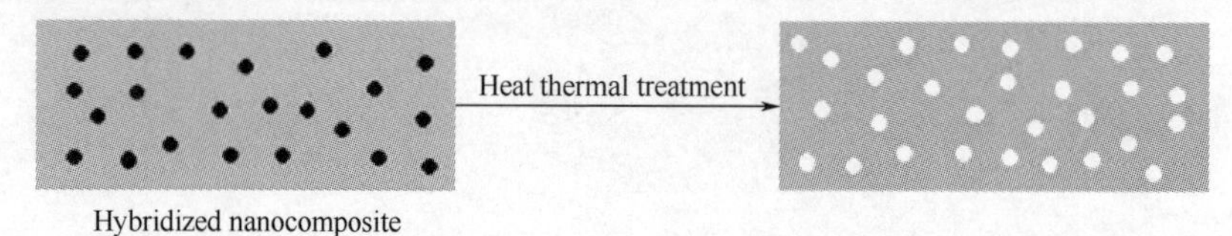

FIGURE 8.64 Illustration of the generation of nanoporous polyimide films.

Zha et al. [170] reported a convenient and simple method to fabricate the nanoporous fluorinated PI films with low relative dielectric permittivity. The SiO_2/F–PI nanohybrid films with 2,2-bis[4(4-

aminophenoxy) phenyl] hexafluoropropane (3FEDAM) and 6FDA were successfully synthesized by a wet phase inversion process. Then, the removal of the SiO_2 nanoparticles from the SiO_2/F-PI nanohybrid films by HF acid etching gave rise to an amount of nanopores with a diameter of about 40 nm in F-PI films. The relative dielectric permittivity of nanoporous F-PI films was decreased to 2.45 while the films still displayed good mechanical properties and thermal stability (Fig. 8.65).

Using the template method, i.e., thermolysis of polystyrene nanospheres in the matrix, Ma et al. [171] prepared a series of PI membranes with nanopores following the synthesis of the template. The typical diameter of the nanopores in the membrane was 200 nm corresponding to the template proportion of 5%, and the diameter was decreased with the increasing of the template proportion while the pore proportion was increased (Fig. 8.66). As 10% template was applied in the membrane to prepare the nanoporous PI, the *k*-value of the membrane was decreased to the minimum of 2.08 from 3.34 (i.e., the dielectric constant of PI membrane without nanopore). The trends of pore diameter and dielectric constant might be due to the collapse of the nanopores in the membrane.

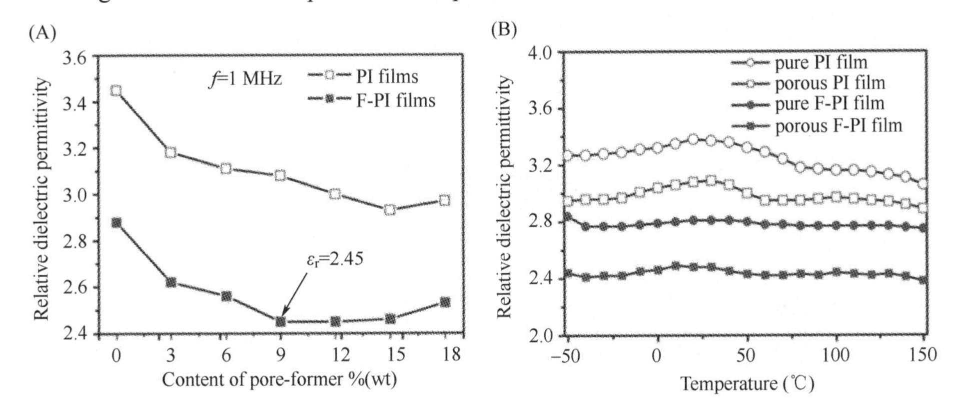

FIGURE 8.65 Relative dielectric permittivity dependences on (A) content of the pore-former and (B) temperature in the PI and F–PI films.*Reproduced with permission from reference [170]. Copyright 2012, American Chemical Society.*

Lv et al [172]. prepared low-dielectric PI nanofoams by introducing nanopores into the PI matrix containing fluorine groups. The nanopores were formed by thermolysis of the thermally labile content, namely polyethylene glycol (PEG) oligomers, in air (Fig. 8.67). Results indicated that the PI nanofoams showed nanosized closed pores, excellent thermal stability, and a low dielectric constant of 2.12. The dielectric constant of the as-prepared nanofoams was stable within –150 °C to 150 °C. The thermal decomposition process of PEG in the PI matrix was designed and optimized to control the decomposition rate of PEG and the diffusion rate of the decomposition products of PEG. The dielectric constant of the nanofoams significantly decreased from 2.45 to 2.12 as the heating rate decreased from 5 to 1 $°C·min^{-1}$. The as-prepared PI nanofoams exhibited excellent properties and thus could be used in the microelectronics industry as a dielectric layer, multichip modules, or integrated circuit chips.

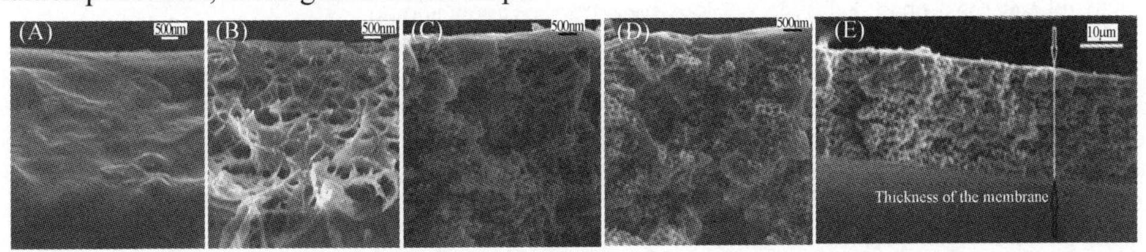

FIGURE 8.66 Fracture surface scanning electron microscopy photographs of the polyimide membranes with different proportions of template. (A), (B), (C), and (D) correspond to the samples with 0%, 5%, 10%, and 15% template, respectively; and the thickness measurement (E).*Reproduced with permission from reference [171]. Copyright 2016, John Wiley & Sons.*

8.4.4.3 Polyimide Aerogels

Very recently, aerogel dielectrics have become one of the most important research topics for high-performance ILDs using ULSI due to their extremely high porosity, low density, low thermal conductivity, and ultralow dielectric constants [173]. The porosity of an aerogel material can usually reach over 80% (volume ratio). The trapped air efficiently decreases the *k*-values of the aerogels to a low level extremely close to 1.0 for air. Among various aerogels, organic polymeric aerogels are more suitable to be used as ILDs for ULSI due to their intrinsically flexible and tough nature compared with their fragile inorganic counterparts, such as silica, alumina aerogels, etc [174]. Various polymer aerogels, such as polyurethane, polyurea, polystyrene, and polydicyclopentadiene aerogels have been widely investigated in the literature and have found a variety of applications in high-tech fields [175]. However, common organic aerogels usually suffer from their low thermal and dimensional stability at elevated temperatures; thus they cannot meet the severe demands of the interlayer electrical insulation process for ULSI fabrication. Thus, as representative high-temperature-resistant organic aerogels, PI aerogels have developed rapidly in recent years [176-180]. The dielectric constants and dissipation factors for PI aerogels have also been investigated in detail.

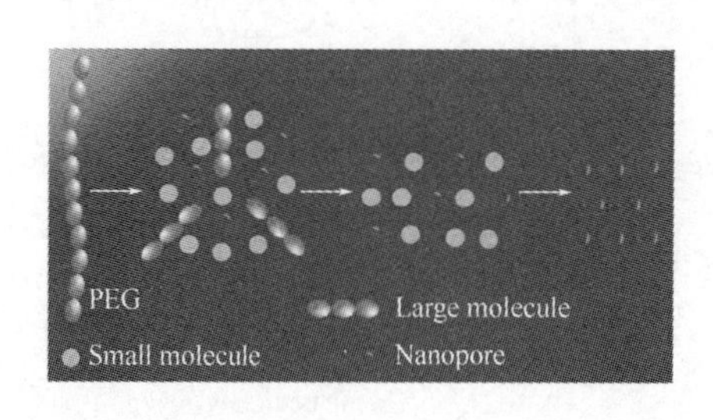

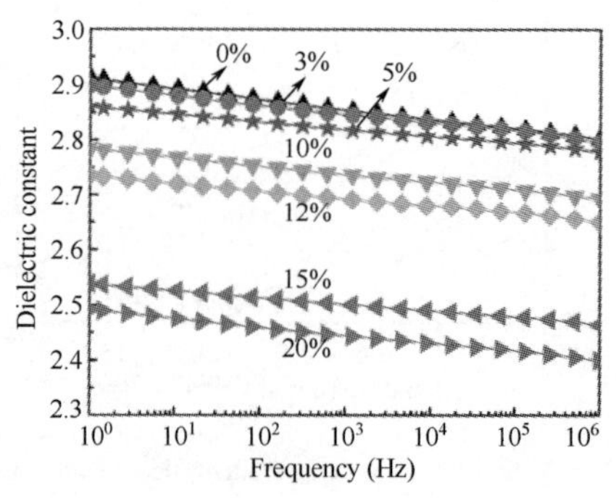

FIGURE 8.67 The schematic of the process of PEG thermal decomposition. *Reproduced with permission from reference [172]. Copyright 2017, Royal Society of Chemistry.*

Meador et al. [181] reported PI aerogels made from 2,2′-dimethylbenzidine (DMBZ) and biphenyl 3,3′,4,4′-tetracarbozylic dianhydride (BPDA) cross-linked with 1,3,5-triaminophenoxybenzene (TAB) (Fig. 8.68). It was found that relative dielectric constant varied linearly with density similar to silica aerogels. Relative dielectric constants as low as 1.16 at a frequency of X-band (~11-12 GHz) were measured for PI aerogels made from DMBZ and BPDA crosslinked with TAB, with densities of the order of 0.11 g cm^{-3}. This formulation was used as the substrate to fabricate and test prototype microstrip antennas and benchmark against state of practice commercial antenna substrates. The PI aerogel antennas exhibited significantly broader bandwidth, higher gain, and lower mass than the antennas made using commercial substrates, hence demonstrating the feasibility of aerogel-based antennas as building blocks for high-gain, broadband, low mass phased array technology for aerospace applications.

They further examined the effect of increasing the amount of 2,2-bis(3,4-dicarboxyphenyl) hexafluoropropane dianhydride (6FDA) in the backbone of the PI aerogels on the dielectric properties (Fig. 8.69) [182]. Two different anhydride-capped PI oligomers were synthesized: one from 6FDA and 4,4′-oxidianiline (ODA) and the other from biphenyl-3,3′,4,4′-tetracarboxylic dianhydride and ODA. The oligomers were combined with 1,3,5-triaminophenoxy- benzene to form a block copolymer networked structure that gelled in under 1 h. The PI gels were supercritically dried to give aerogels with relative dielectric constants as low as 1.08. Increasing the amount of 6FDA blocks by up to 50% of the total dianhydride decreased the density of the aerogels, presumably by increasing the free volume and also by decreasing the amount of shrinkage seen upon processing, resulting in a concomitant decrease in the dielectric properties. They have also changed the density independent of fluorine substitution by changing the polymer concentration in the gelation reactions and showed that the change in dielectric due to density is the same with and without fluorine substitution. The aerogels with the lowest dielectric properties (1.084) and lowest densities (0.078 g • cm^{-3}) still had compressive moduli of 4–8 MPa (40 times higher than silica aerogels at the same density), making them suitable as low dielectric substrates for lightweight antennas for aeronautic and space applications.

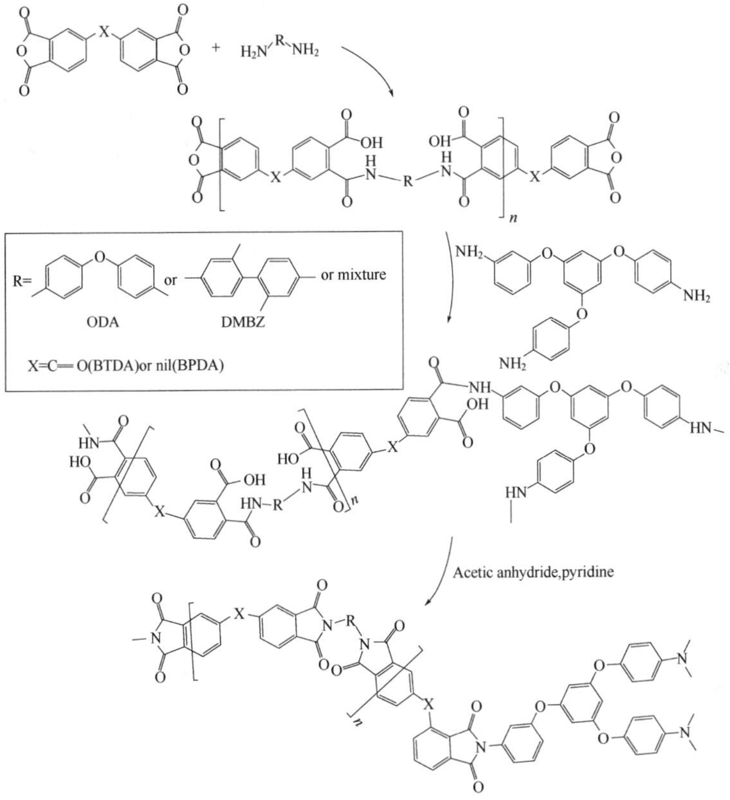

FIGURE 8.68 Synthesis of polyimide aerogels with TAB crosslinks. *Reproduced with permission from reference [181]. Copyright 2012, American Chemical Society.*

Shen et al. [183] reported intrinsically highly hydrophobic semialicyclic fluorinated PI aerogel with ultralow-*k*-values of 1.17-1.19 in the frequency range of 2-12 GHz by the polycondensation of an alicyclic dianhydride, an aromatic fluorinated diamine, and an amino-substituted polyhedral oligomeric silsesquioxane (POSS) end-capper, followed by a supercritical carbon dioxide drying procedure (Fig. 8.70).

Zhang et al. [184] successfully prepared a PI aerogel with excellent combined thermal and dielectric properties by the polycondensation of 3,3′,4,4′-biphenyltetracarboxylic dianhydride (BPDA), 5-amino-2-(4-aminophenyl) benzoxazole (APBO) and octa(amino-phenyl)silsesquioxane (OAPS) crosslinker, followed by a supercritical carbon dioxide ($scCO_2$) drying treatment (Fig. 8.71). The developed PI aerogel exhibited an ultralow-*k*-value of 1.15 at a frequency of 2.75 GHz, a volume resistivity of 5.45×10^{14} Ω·cm, and a dielectric strength of 132 kV cm^{-1}. The flexible PI aerogel exhibited an open-pore microstructure consisting of a three-dimensional network with tangled nanofibers morphology with a porosity of 85.6% (volume ratio), an average pore diameter of 19.2 nm, and a Brunauer-Emmet-Teller (BET) surface area of 428.6 $m^2 \cdot g^{-1}$. In addition, the PI aerogel showed excellent thermal stability with a T_g of 358.3 °C, a 5% weight loss temperature over 500 °C, and a residual weight ratio of 66.7% at 750 °C in nitrogen.

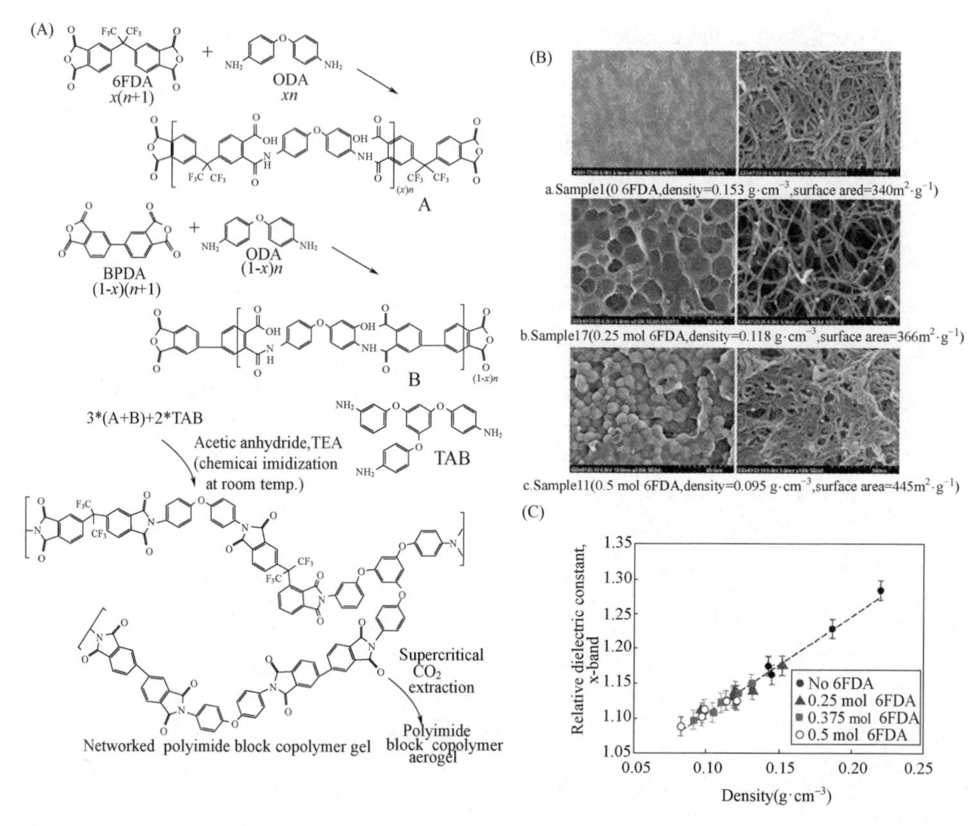

FIGURE 8.69 (A) Synthesis of polyimide block copolymer aerogels with TAB crosslinks. (B) SEM images of the fracture surfaces of polyimide aerogels at low and high magnification for (a) 6 FDA fraction=0, (b) 6FDA fraction=0.25 mol; (c) 6FDA fraction=0.5 mol. (C) The relative dielectric constant versus density for various formulations of aerogels showing linear correlation regardless of the 6FDA concentration.*Reproduced with permission from reference [182]. Copyright 2014, American Chemical Society.*

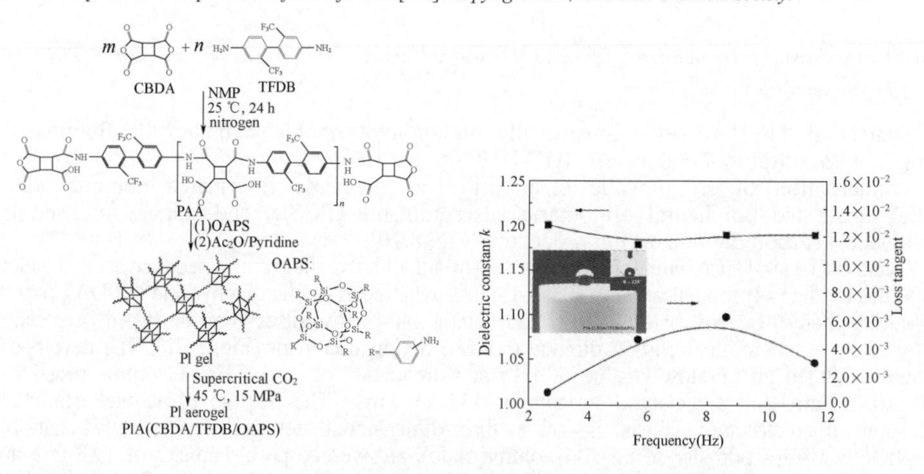

FIGURE 8.70 Polyimide aerogel with intrinsically hydrophobic nature and low dielectric properties.

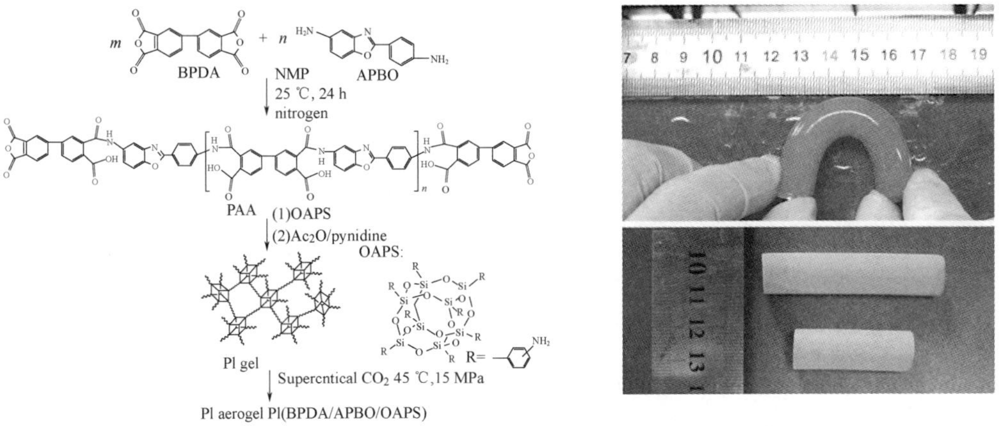

FIGURE 8.71 Synthesis of PI (BPDA/APBO/OAPS) aerogel (left), and appearance of PI wet gel (above) and PI aerogel (below) (right).

8.4.4.4 POSS-containing Polyimides

Compared with other porous materials, POSS moieties possess nanometer-sized structures, high thermal stabilities, monodispersity, controlled porosity, and adjustable chemical functionality. They are important starting materials for the design of several hybrid nanostructure materials [185-189]. The use of functionalized POSS moieties allows well-distributed nanosized POSS domains to be generated within polymer matrices; they can also function as physical crosslinking sites. Because of their porous nature, POSS units possess low dielectric constants; [190-192] therefore, POSS-containing PI composites will have low dielectric constants and good mechanical and thermal properties if their POSS units are distributed evenly and adhered well to the PI matrix.

Lee et al. [192] prepared a novel PI hybrid nanocomposite containing POSS with well-defined architecture by copolymerization of 4,4′-carbonyldiphthalic anhydride (BTDA), 4,4′-oxydianiline diamine (ODA), and octakis(glycidyldimethylsiloxy) octasilsesquioxane (Epoxy-POSS) (Fig. 8.72). In these nanocomposite materials, the equivalent ratio of the Epoxy-POSS and ODA are adjustable, and the resulting PI-POSS nanocomposites give variable mechanical and thermal properties. More importantly, they explored the possibility of incorporating POSS moiety through the Epoxy-POSS into the PI network to achieve the PI hybrid with lower dielectric constant (low-*k*) and thermal expansion. The lowest dielectric constant achieved of the POSS/PI material (PI-10P) is 2.65 by incorporating 10 %(*w*). Epoxy-POSS (pure PI, *k*=3.22). In addition, when contents of the POSS in the hybrids are 0, 3, 10 %(*w*). (PI-0P, PI-3P, PI-10P), and the resulting coefficients of thermal expansion (*CTE*) are 66.23, 63.28, and 58.25×10^{-6} $°C^{-1}$, respectively. The reduction in the dielectric constants and the resulting thermal expansion coefficients of the PI-POSS hybrids can be explained in terms of creating silsesquioxane cores of the POSS and the free volume increase by the presence of the POSS-tethers network resulting in a loose PI structure.

Wang et al. [193] prepared a series of fluorinated PI/POSS hybrid polymers (FPI-4−FPI-16) via a facile synthetic route using 2,2′-bis(trifluoromethyl) benzidine, 4,4′-oxydiphthalic dianhydride, and monofunctional POSS as starting materials (Fig. 8.73). The hybrid polymers showed excellent solubility and film formation ability. Flexible and robust hybrid films could be conveniently obtained via solution-casting. The hybrid films demonstrated low dielectric constants and high thermal stability. Their *k*-values were in the range of 2.47-2.92 at 1 MHz measured for their capacitance, and were tunable and decreased with an increase of POSS content. Their 10% weight loss temperatures were in the range of 539−591 °C and the weight residual at 800 °C ranged from 48% to 53% in nitrogen atmosphere. These hybrid films also possessed hydrophobic characteristics and good mechanical properties.

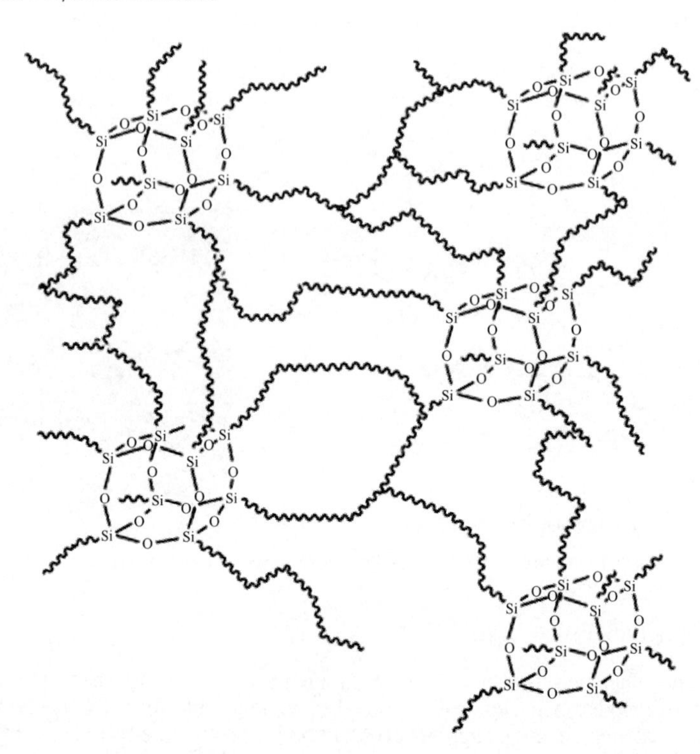

FIGURE 8.72 Network structure of PI-POSS nanocomposite. The mole ratio of the terminal amine (–NH_2) to the epoxy group is 1:1.*Reproduced with permission from reference [192]. Copyright 2005, Elsevier.*

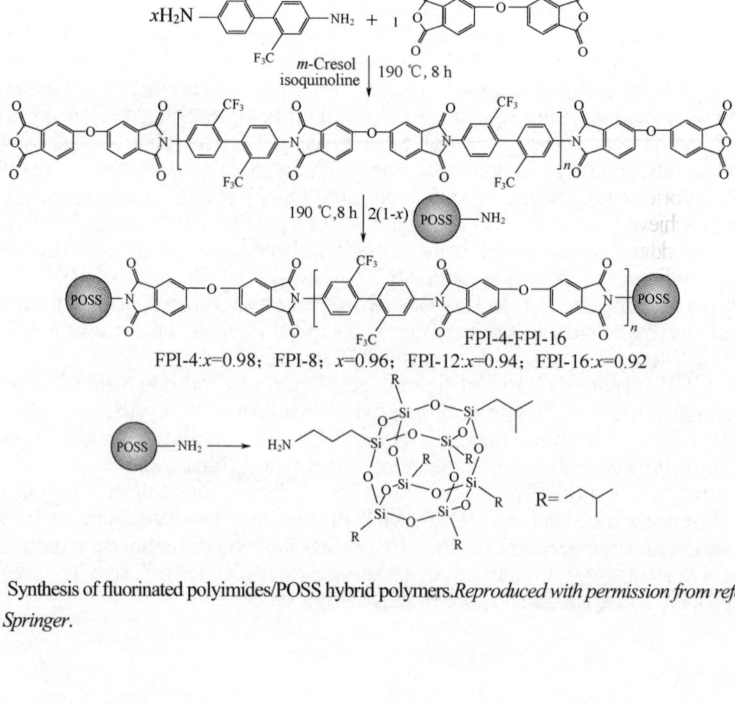

CF_3
xH_2N — — NH_2 + 1 O O O O O O O
F_3C
m-Cresol
isoquinoline
190 ℃, 8 h

O O O CF_3 O O CF_3 O O
O O N N O N N O
O O F_3C O O F_3C *n* O O

190 ℃,8 h 2(1-*x*) POSS –NH_2

O O CF_3 O O
POSS N O N N O N POSS
O O F_3C O O *n*

FPI-4-FPI-16

FPI-4:*x*=0.98; FPI-8; *x*=0.96; FPI-12:*x*=0.94; FPI-16:*x*=0.92

POSS –NH_2 → H_2N
R O Si O Si O Si O Si O R R O O Si O Si R O Si O Si R R

R=

FIGURE 8.73 Synthesis of fluorinated polyimides/POSS hybrid polymers.*Reproduced with permission from reference [193]. Copyright 2016, Springer.*

Leu et al. [194] reported porous POSS/PI nanocomposites by the reaction of nanoporous POSS containing amine groups (NH_2-POSS) and poly(amic acid) having anhydride end groups. The van der Waals interaction between tethered POSS molecules is incompatible with the polar-polar interaction of imide segments, resulting in a self-assembled system of zigzag-shaped cylinders or lamellas about 60 nm long and 5 nm wide, as determined by transmission electron microscopy (Fig. 8.74). The incorporation of 2.5%(mol) nanoporous POSS results in a reduction of the dielectric constant of the PI nanocomposite from 3.40 for the pure PI to 3.09 without any degradation in mechanical strength.

Also by the method of self-assembly, they reported another nanocomposite with well-defined architecture (Fig. 8.75) [190]. PI-tethered polyhedral oligomeric silsesquioxane ($R_7R'Si_8O_{12}$) (POSS), is prepared by the copolymerization reaction of a new type of diamine monomer: POSS-diamine, 4,4′-oxydianiline (ODA), and pyromellitic dianhydride (PMDA). This type of PI-side-chain-tethered POSS nanocomposite presents self-assembly characteristics when the amount of POSS exceeds10%(mol) Furthermore, POSS/PI nanocomposites have both lower and tunable dielectric constants, with the lowest value of 2.3, and controllable mechanical properties, as compared to that of pure PI.

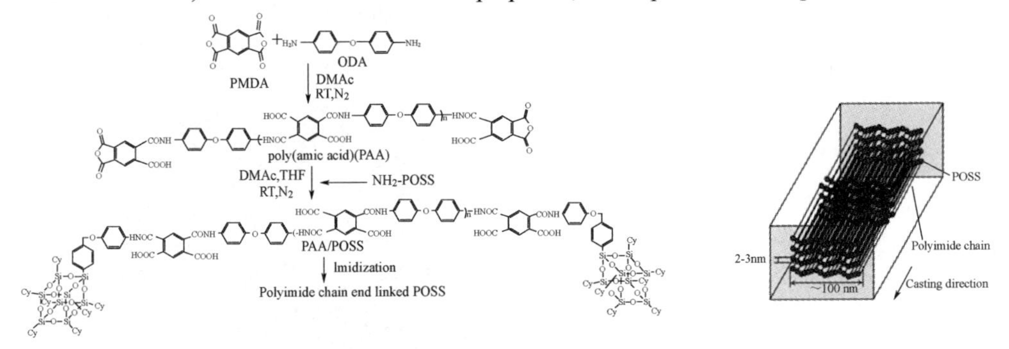

FIGURE 8.74 Process for preparing the POSS/polyimide nanocomposite and the illustration of self-assembly in polyimide covalently bonded polyhedral oligomeric silsesquioxane nanocomposites. *Reproduced with permission from reference [194]. Copyright 2003, American Chemical Society.*

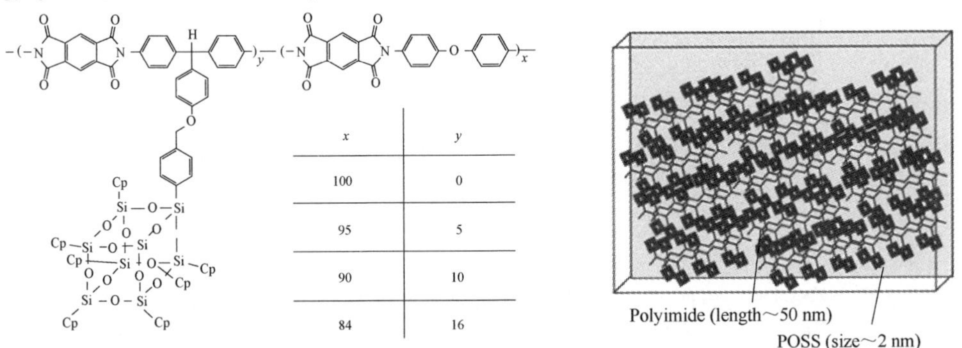

x	y
100	0
95	5
90	10
84	16

FIGURE 8.75 Schematic drawing of the self-assembled architecture of polyimide-tethered POSS. *Reproduced with permission from reference [190]. Copyright 2003, American Chemical Society.*

8.4.4.5 Controlling Methods of the Porous Structures in Polyimides

The growing relevance of high-performance PI nanocomposites for electrical and dielectric applications underscores the importance of understanding intrinsic dielectric characteristics such as the temperature dependence and dynamics of primary and secondary relaxations. Consequentially, Jacobs et al. [195] chose CP2, a widely used high-performance aerospace-grade PI derived from 1,3-bis(3-5 aminophenoxy) benzene (APB) and 2,2-bis(phthalic anhydride)-1,1,1,3,3,3-hexafluoroisopropane (6FDA), as the subject to be investigated, using impedance spectroscopy across a wide range of temperatures (10^{-2}-10^{6} Hz, 40-225 °C) with emphasis near the T_g (199 °C), to gain insight into the molecular mechanisms underpinning its dielectric properties and kinetic characteristics. Particular attention was placed on elucidating the electric and dielectric properties in the neighborhood of CP2's T_g. Modeling the impedance spectra revealed the classic primary (α) and subglass secondary relaxation processes (β [phenyl ring rotations] and β'[CF_3 rotation]) common to many types of amorphous polymer systems. In addition, an extrinsic low-frequency interfacial polarization linked to residual DMAc was identified. Temperature-variable impedance analysis near T_g enabled the temperature dependencies of the DC conductivity and each dielectric event to be characterized. The presence of DMAc solvent remaining after the imidization process significantly increased the film's charge transport and molecular mobility properties. Solvent-free CP2 films maintain low-loss dielectric characteristics ($\varepsilon'' < 10^{-2}$) up to considerably high temperatures (170 °C). The results of this study provideed a detailed account of the electrical and dielectric properties, relaxation dynamics, and temperature dependent characteristics of PI CP2. The goal has been to enumerate these basic properties to reconcile a thorough understanding of CP2 as a low dielectric loss matrix for emerging composite materials.

Krause et al. [196] reported an entirely different route by introducing porosity in a physical and cleaner way. Their batch process invokes the foaming created by the evaporation of dissolved supercritical CO_2 (Fig. 8.76). This strategy has the following attractive features. Physical foaming by the escape of

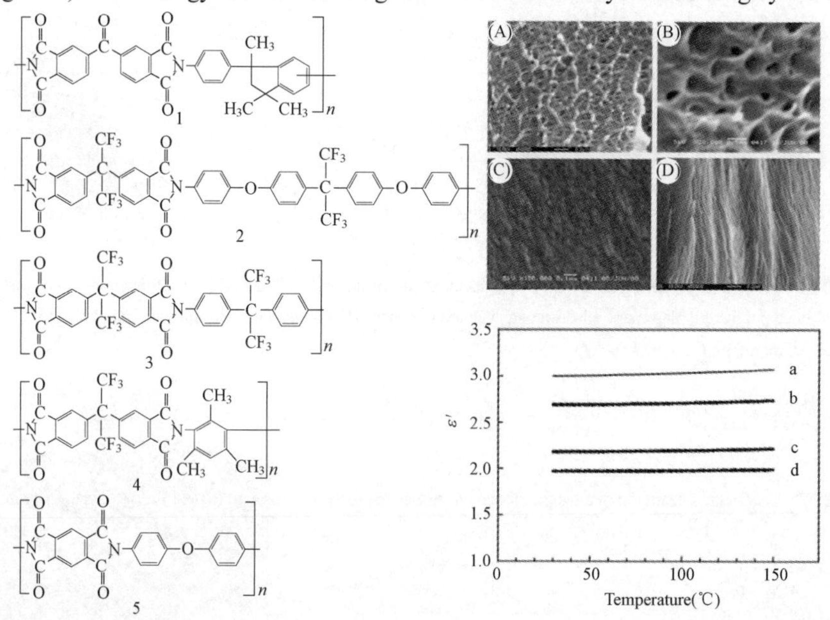

FIGURE 8.76 Chemical structure, morphology, and dielectric properties of the polyimides. SEM images: (A) Matrimid (1) foamed at 270 °C. Magnification 50,000, white horizontal bar equals 200 nm. (B) 6-FDA-4,4′-6F (3) saturated at 25 °C and foamed at 250 °C. Magnification 30,000, white horizontal bar equals 500 nm. (C) 6-FDA-4,4′-6F (3) saturated at 0 °C and foamed at 250 °C. Magnification 100,000, white horizontal bar equals 100 nm. (D) Kapton HN (5) foamed at 300 °C. Magnification 50,000, white horizontal bar equals 200 nm.*Reproduced with permission from reference [196]. Copyright 2002, John Wiley & Sons.*

dissolved gases from a polymer is not only nonpolluting; it also produces no impurities. It further enables precise tailoring of the morphology from microcellular to bicontinuous, nanoporous structures. It is feasible, starting from amorphous low-*k* rigid PIs and by introducing porosities of up to 40% to get well into the ultralow-*k* regime (*k*- or ε-values below 1.8). In addition, by applying a mesoscopically ordered polymer one can create specially structured foams with disk-like pores, thus reducing the *k*-value even further into hitherto unknown areas of mechanically stable ultralow-*k* materials.

Zhao et al. [197] employed the reprecipitation method and subsequent imidization to prepare unique hollow PI nanoparticles (NPs) by blending a suitable polymer porogen with the PI precursor in the form of individual NPs (Fig. 8.77). The hollow structures were induced through phase separation between the PI and the porogen, a process that was influenced by the compatibility between the two polymers. They then assembled multilayered films by depositing the hollow PI NPs electrophoretically onto a substrate. After spin-coating a solution of the PI precursor and then subjecting the system to thermal treatment, the individual PI NPs were bound to their adjacent neighbors, thereby improving the film strength to some extent. They obtained dense, uniformly packed films having controlled thicknesses in the range from 500 nm to 10 μm. This strategy provided films in which air voids existed between and within the composite PI NPs; as a result, the dielectric constant reached as low as 1.9.

Porous PI synthesized from 4,4′-oxydiphthalic anhydride (ODPA) and 4,4′-diaminodiphenyl ether (ODA) monomers is a promising material with an ultralow-dielectric constant. Li et al. [198] reported a strategy toward engineering PI films of various porous textures using a small molecular phase dispersion agent, dibutyl phthalate (DBP), as porogen (Fig. 8.78). In the presence of DBP, the ODPA-ODA poly(amic acid) solution, the precursor to PI, undergoes phase separation as *N,N*-dimethylacetamide (DMAc) solvent is slowly evaporated, forming spherical domains of DBP phase uniformly dispersed in the polyamidic acid matrix. Upon thermal imidization, a PI film with high porosity is attained after acetone extraction of DBP. It is demonstrated in this study that the porous texture of PI films can be readily engineered by tailoring the initial DBP content. The average pore size increases with increasing concentration of DBP, but was no larger than 6 mm. The PI film achieves a dielectric constant of 1.7 at an optimal porosity of 72%, This study examined the pore formation mechanism, the imidization chemistry, the surface morphology, the density, the thermal stability and mechanical properties of the formed porous PI films. Thermo-gravimetric analysis indicated that porous films retain the inherent exceptional thermal stability of PIs, with thermal decomposition onset above 500 °C in nitrogen atmosphere.

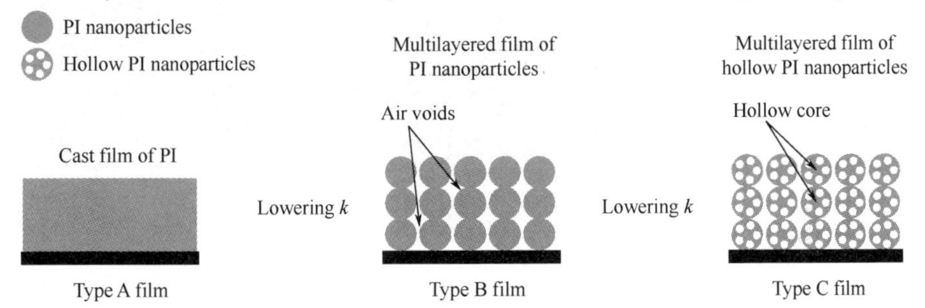

FIGURE 8.77 Schematic illustration of a simple strategy to reduce the dielectric constant (k) of PI films (Type A) by introducing air voids between PI NPs (Type B) and hollow cores into separate PI NPs (Type C).*Reproduced with permission from reference [197]. Copyright 2009, American Chemical Society.*

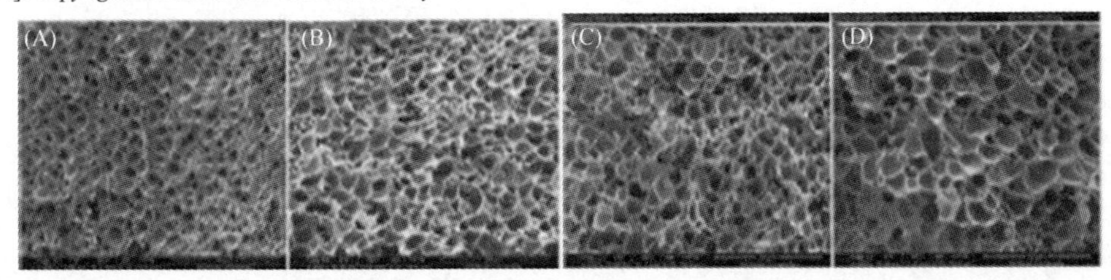

FIGURE 8.78 SEM images of cross-section of porous PI films prepared at: (A) 5 %(w). DBP, (B) 10 %(w). DBP, (C) 15 %(w). DBP, and (D) 20 %(w). DBP.*Reproduced with permission from reference [198]. Copyright 2015, Royal Society of Chemistry.*

The proper combination of material (i.e., fluorinated PIs) and processing technique (electrospinning) could lead to the formation of PIs with low dielectric constant, high thermo-oxidative stability and T_g, and high hydrophobicity.

With the aim of providing low-dielectric-constant PI membranes with good processability, high thermal stability, and low water absorption, Chen et al. [199] chose fluorinated PIs and used electrospinning for processing into nanomats. The PIs in this work were based on 4,4-bis [3′-trifluoromethyl-4′(4′-amino benzoxy) benzyl] biphenyl (*Q*) and various fluorinated and nonfluorinated dianhydrides namely 3,3′,4,4′-biphenyl-tetracarboxylic dianhydride, benzophenone-3, 3′,4,4′-tetracarboxylic dianhydride, benzene-1,2,4,5-tetracarboxylic dianhydride, and 6FDA. Processing of the PIs was carried out in the PAA stage by two different methods-electrospinning and solution casting for comparison purposes. The processing of PIs by electrospinning led to enhancement in mechanical properties (dianhydride-structure dependent) and hydrophobicity without sacrificing thermo-oxidative stability and T_gs significantly. Also, low dielectric constants (as low as 1.43) could be attained by suitable combination of dianhydride (6FDA) with 4,4-bis [3′-trifluoromethyl-4′(4′-amino benzoxy) benzyl] biphenyl diamine.

Li et al. [200] fabricated three kinds of PI fiber membranes by electrospinning of PAA solutions which are from polycondensation of 4,4′-oxidianiline (ODA) and three dianhydrides, pyromellitic dianhydride (PMDA), 2,2′-bis(3,4-dicarboxyphenyl) hexafluropropane dianhydride (6FDA) and 1,2,4,5-cyclohexanetetracarboxylic dianhydride (HPMDA), followed by imidization at higher temperature (Fig. 8.79). The relationship of the fiber morphology, thermostability, and dielectric properties of the membranes with the polymer structure were discussed. Under the same conditions, PAAs with more flexible structure are easier to form low-viscosity solution and fabricate high pore fraction membranes which are low-*k* materials. Under the coupling effect of fluorine-containing groups and the contribution of pores, the dielectric constant of 6FDA-containing PI is lowered to 1.21 at 1 kHz with lower dielectric loss, which accords with the calculated value. Also, the 5% weight loss temperature of the three kinds of PIs are all higher than 400 °C. The formed electrospun membranes are thermostable low-*k* materials.

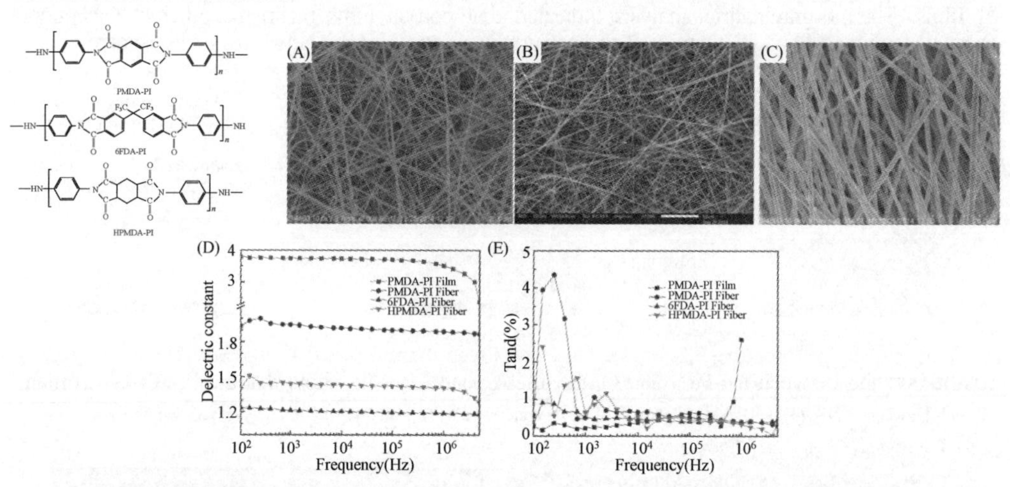

FIGURE 8.79 Morphology of PMDA-PI (A), 6FDA-PI (B), and HPMDA-PI (C) fibers from 25 %(*w*). of PAA solutions; dielectric constant (D) and dielectric loss (E) of PMDA-PI, 6FDA-PI, and HPMDA-PI fibers from 25 %(*w*). PAA solutions.*Reproduced with permission from reference [200]. Copyright 2016, John Wiley & Sons.*

Liu et al. [201] prepared a series of ultralow-*k* silica/polyimide (SiO_2/PI) composite nanofiber membranes by the combined sol-gel and electrospinning techniques. The emulsion composed of partially hydrolyzed tetraethoxysilane (TEOS) and PAA is spun to yield the precursor of the SiO_2/PI fibers with a core-shell structure due to phase separation (Fig. 8.80). The *k*-value of the composite membranes varies from 1.78 to 1.32 with increasing content of SiO_2. The fibers accumulate and form the film with a large number of pores leading to a lower k. In addition, the interfacial reaction between SiO_2 and the PI matrix reduces the value of k as the SiO_2 concentration is increased. The thermal stability of PIs increase after mixing with SiO_2 and the SiO_2/PI composite fibers have large commercial potential in the electronics industry.

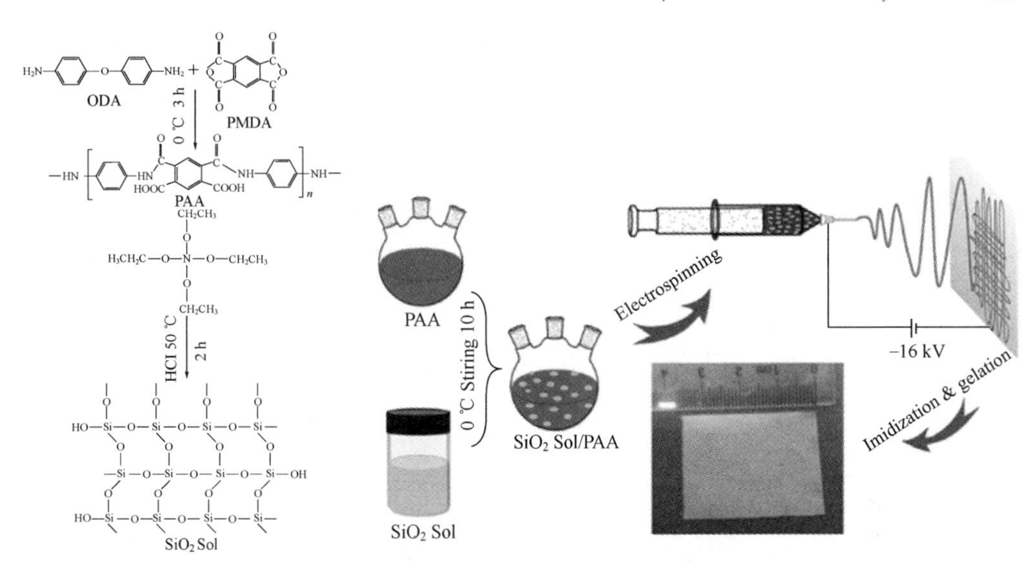

FIGURE 8.80 Flow chart illustrating the preparation of the SiO_2/PI composite nanofibers. *Reproduced with permission from reference [201]. Copyright 2016, Elsevier.*

8.4.5 Organic–Inorganic Hybrid Polyimide Materials

In recent years, research has been carried out on the organic-inorganic or other composite materials [202]. The fabricated PI composites with well-designed organic-inorganic structures can reduce the *k*-value by utilizing the effects of air volume, air gap or pore voids, low polarization, and increased free volume. Meanwhile, the processability and mechanical properties of the composites can also be improved.

Tsai and Whang [203] successfully prepared a new type of low-dielectric polyimide/poly(silsesquioxane)-like (PI/PSSQ-like) hybrid nanocomposite material from the PI (ODA-ODPA) precursor containing phenyltrialkoxysilane (PTS) at two chain ends and monoaryltrialkoxysilane with a self-catalyzed sol-gel process (Fig. 8.81). They employed *p*-aminophenyl trimethoxysilane (APTS) to provide bonding between the PTS and ODPA-ODA phase. Results show that the PSSQ-like domain sizes with uniform size are fairly well separated in the hybrid films. The dielectric constant can be 2.79 for 5000-PIS-140-PTS with fairly good mechanical properties. The PI/PSSQ-like hybrid films have higher onset decomposition temperature and char yield in TGA and higher T_g in DSC than the pure PI. Moreover, the PI/PSSQ-like hybrid films have excellent transparency even under high PTS content. In the series of X-PIS hybrid films, the coefficient of thermal expansion (*CTE*) below T_g increases with the PI block chain length, but in the series of X-PIS-y-PTS films, it slightly increases with the PTS content. However, above T_g the *CTE* of X-PIS and X-PIS-24-PTS is much lower than that of the pure PI. The dielectric constant and water absorption of X-PIS-y-PTS films decrease with the PTS content because of the higher free volume and hydrophobicity.

Wang et al. [204] proposed an effective approach using in situ polymerization, to fabricate large-area graphene oxide (GO)/PI composite films with outstanding mechanical properties (Fig. 8.82). The GO/PI composite films provide ultrahigh tensile strength (up to 844 MPa) and Young's modulus (20.5 GPa). The NH_2-functionalized GO (ODA-GO) is a versatile starting platform for polymer grafting, promoting excellent dispersion of GO within the polymer matrix, and forming strong links with the polymer to facilitate load transfer. The Young's modulus of the integrated GO-PI composite films with 3.0 %(*w*). ODA-GO loading is 15-times greater, and the tensile strength is nine-times greater than comparable properties of pure PI film. The dielectric constant decreases with increasing GO content and a dielectric constant of 2.0 was achieved.

To explore the effects of fluorographene on the dielectric properties of hybrid materials, Wang et al.

[205] synthesized sufficient amounts of fluorographene sheets with different sheet-sizes and fluorine/carbon ratios for preparation of fluorographene/PI hybrids. It is found that the fluorine/carbon ratio, width of band gap, and sheet-size of fluorographene play the important roles in determining the final dielectric properties of hybrids (Fig. 8.83). The fluorographene with high fluorine/carbon ratio ($F/C \approx 1$) presents broadened band gap, enhanced hydrophobicity, good dispersity, and thermal stability, etc. Even at a very low filling, only 1%(w), its PI hybrids exhibit drastically reduced dielectric constants as low as 2.1 without sacrificing thermal stability, improved mechanical properties obviously and decreased water absorption by about 120% to 1.0 %(w).. This provides a novel route for improving the dielectric properties of materials and a new thought to carry out the application of fluorographene as an advanced material.

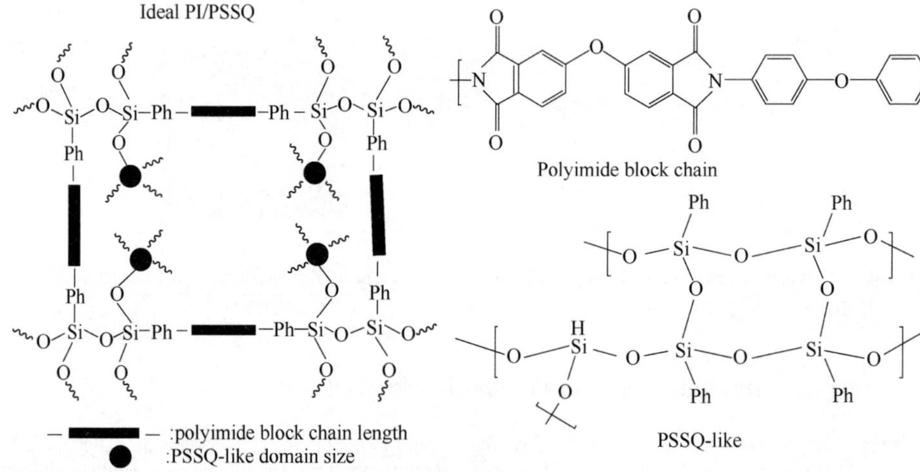

Ideal PI/PSSQ

Polyimide block chain

PSSQ-like

— ▬ — :polyimide block chain length

● :PSSQ-like domain size

FIGURE 8.81 Structure of PI/PSSQ-like hybrid films.

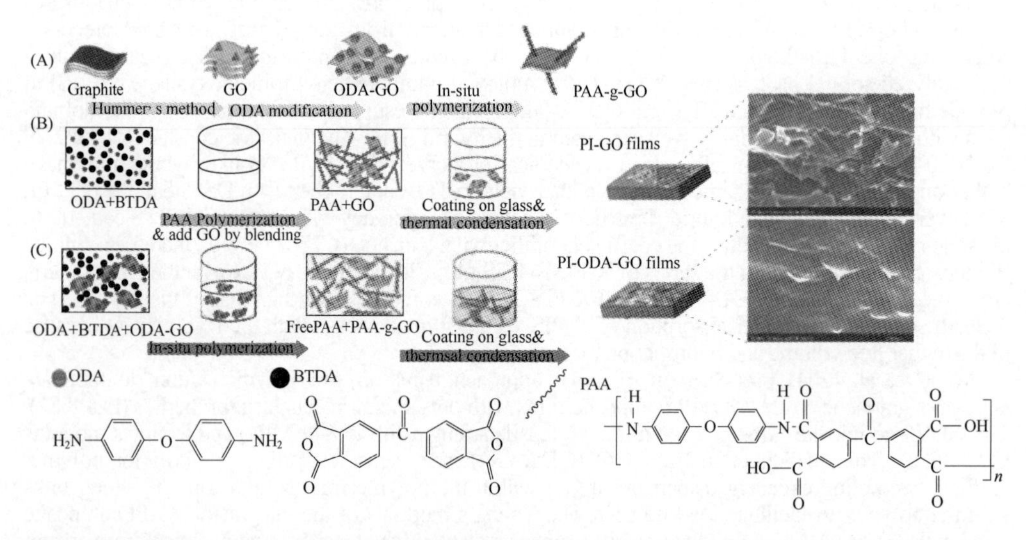

FIGURE 8.82 The procedure for preparation of (A) ODA-GO, (B) PI-GO films, and (C) PI-ODA-GO films. *Reproduced with permission from reference [204]. Copyright 2011, Royal Society of Chemistry.*

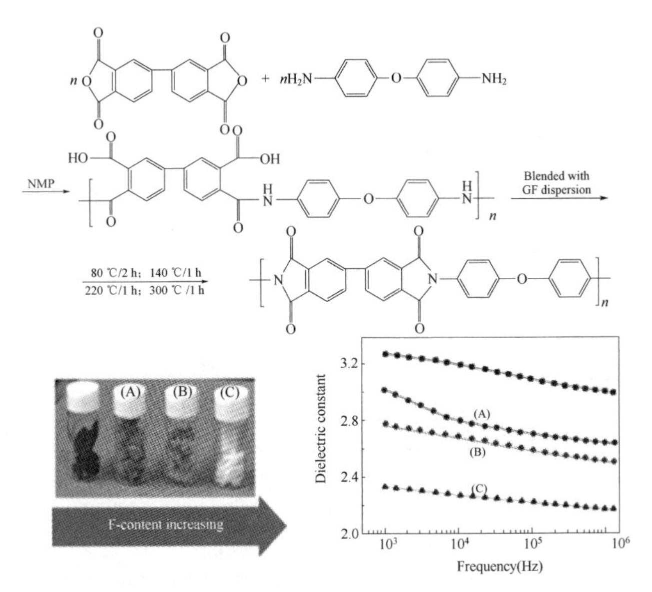

FIGURE 8.83 Photographs of nonfluorinated and fluorinated graphene: neat SG, FSG-1, FSG-2, and FSG-3 (left to right); and the dielectric constant of the neat polyimide and fluorographene hybrid polyimide films at room temperature.

Liao et al. [206] proposed an effective and simple approach for fabricating high-performance graphene oxide (GO)/soluble polyimide (SPI) composite films through a novel and effective process (Fig. 8.84). In this method, GO is dispersed in a dissolved SPI (R-SPI) polymeric matrix with curing state, preventing the reduction of crosslinking reactions of the polymeric matrix, and resulting in substantial improvements in the mechanical and dielectric properties of the composite. The GO/R-SPI composite film containing 1.0%(*w*) GO possesses high tensile strength (up to 288.6 MPa) and Young's modulus (7.58 GPa), which represent an increase of 260% in tensile strength and 402% in Young's modulus, compared with the neat SPI film (80.3 MPa and 1.51 GPa, respectively). The *k*-symbol decreases with an increase in the GO content; the *k* of the GO/R-SPI composite film can be as low as 2.1 (compared with 2.8 for the neat SPI film). This novel fabricating method provides a path for developing high-performance GO/R-SPI composite materials as next-generation low-*k* dielectric materials.

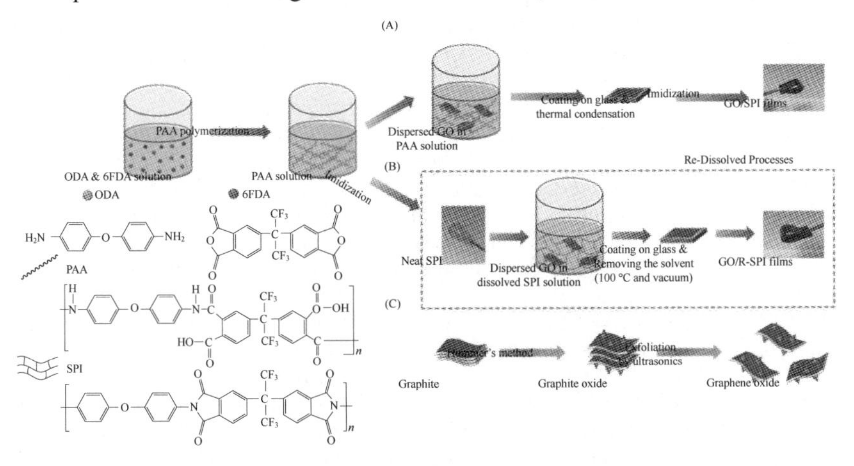

FIGURE 8.84 Scheme of the procedure for preparation of (A) GO/SPI films, (B) GO/R-SPI films, and (C) GO sheets.*Reproduced with permission from reference [206]. Copyright 2014, Royal Society of Chemistry.*

They further proposed another effective method to prepare octa(aminophenyl) silsesquioxane (OAPS) functionalized graphene oxide (GO) reinforced PI composites with a low dielectric constant and ultrastrong mechanical properties (Fig. 8.85) [207]. The amine-functionalized surface of OAPS-GO was a versatile starting platform for in situ polymerization, which promoted the uniform dispersion of OAPS-GO in the PI matrix. Compared with GO/PI composites, the strong interfacial interaction between OAPS-GO and the PI matrix through covalent bonds facilitated a load transfer from the PI matrix to the OAPS-GO. The OAPS-GO/PI composite film with 3.0 %(*w*). OAPS-GO exhibited an 11.2-fold increase in tensile strength, and a 10.4-fold enhancement in tensile modulus compared with neat PI. The *k*-value decreased with the increasing content of 2D porous OAPS-GO, and a *k*-value of 1.9 was achieved.

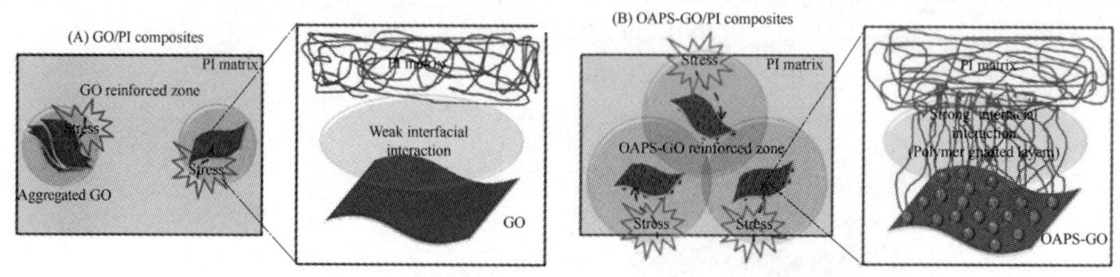

FIGURE 8.85 Proposed model of the reinforcing (A) GO and (B) OAPS-GO in PI polymeric matrix.*Reproduced with permission from reference [207]. Copyright 2014, American Chemical Society.*

Kim et al. [208] reported the synthesis of a hydroxyl-terminated hyperbranched PI via the A2 +B3 reaction between dianhydride and triamine monomers (Fig. 8.86). The hydroxyl groups at the peripheral positions were then introduced by modification of the anhydride end groups via a reaction with 4-aminophenol. They successfully fabricated hybrid ternary composites, which were comprised of a linear PI (PI6FDA-APB), HBPIBPADA-TAP(OH), and an inorganic SiO_2 component. Because of the appropriate choice of the hybrid ternary composite systems with HBPIBPADA-TAP(OH) and inorganic silica, it is sensible to improve the dielectric properties and thermal resistant properties of unary systems or improve the disadvantages of the dielectric and optical properties of binary systems. For an optimized composition, the *k*-value of the PI6FDA-APB-HBPIBPADA-TAP(OH)-30%-SiO_2-20% composite reaches the lowest value of 2.24 at 100 kHz. Research also showed that the optical transparency was significantly improved with the increase of the HBPIBPADA-TAP(OH) content in the composite. Compared with the binary linear PI 6FDA-APB-SiO_2 composite, the transmittance increases from 1% to 75% at the wavelength of 450 nm. The incorporation of SiO_2 can preserve the good thermal properties of the hybrid composites containing HBPIBPADA-TAP(OH). By adding 10% of HBPIBPADA-TAP(OH) to the PI6FDA-APB-SiO_2 20% system, the *CTE* of the hybrid ternary composite is 20.9×10^{-6} $°C^{-1}$ in the temperature range from 100 °C to 150 °C, which is significantly lower than that of the linear PI (37.1×10^{-6} $°C^{-1}$ for PI6FDA-APB). Because of these optimized properties, hybrid ternary composites have the potential for use in applications in the microelectronic insulator fields, such as interlayer dielectrics of advanced electronic devices.

Also based on the mentioned hydroxyl-terminated hyperbranched PI (HBPIBPADA-TAP(OH)), Kim et al. [209] fabricated two series of hybrid ternary composites composed of a linear PI (PI6FDA-APB), HBPIBPADA-TAP(OH) (10% and 20%), and various concentrations of SiO_2 by sol-gel method (Fig. 8.87). Research showed that properties of the composites were closely related with the ratio of HBPIBPADA-TAP(OH) to SiO_2, which could be optimized by proper choice of the compositions. Under optimized conditions, the hybrid ternary composites show properties complementary to the drawbacks of unary or binary systems. The dielectric constant of the PI6FDA-APB_HBPIBPADA-TAP(OH)-10%_SiO_2-10% composite reaches the lowest value of 2.26. The silica loading and reinforcement binding with PI6FDA-APB_HBPIBPADA-TAP(OH) matrix are confirmed by the SEM morphology of the hybrid ternary composite films. When a suitable amount (20%) of HBPIBPADA-TAP(OH) is added into the PI6FDA-APB_SiO_2-10% composite, the optical transmittance is significantly improved to be 63% at the wavelength of 450 nm. With increase in the content of SiO_2, the thermal expansion is obviously reduced. For the PI6FDA-APB HBPIBPADA-TAP(OH)-10% SiO_2-30% composite, the coefficient of thermal expansion (*CTE*) of the film is reduced to 15.9×10^{-6} $°C^{-1}$ compared to 37.1×10^{-6} $°C^{-1}$ for the PI6FDA-APB film (reducing about 59%).

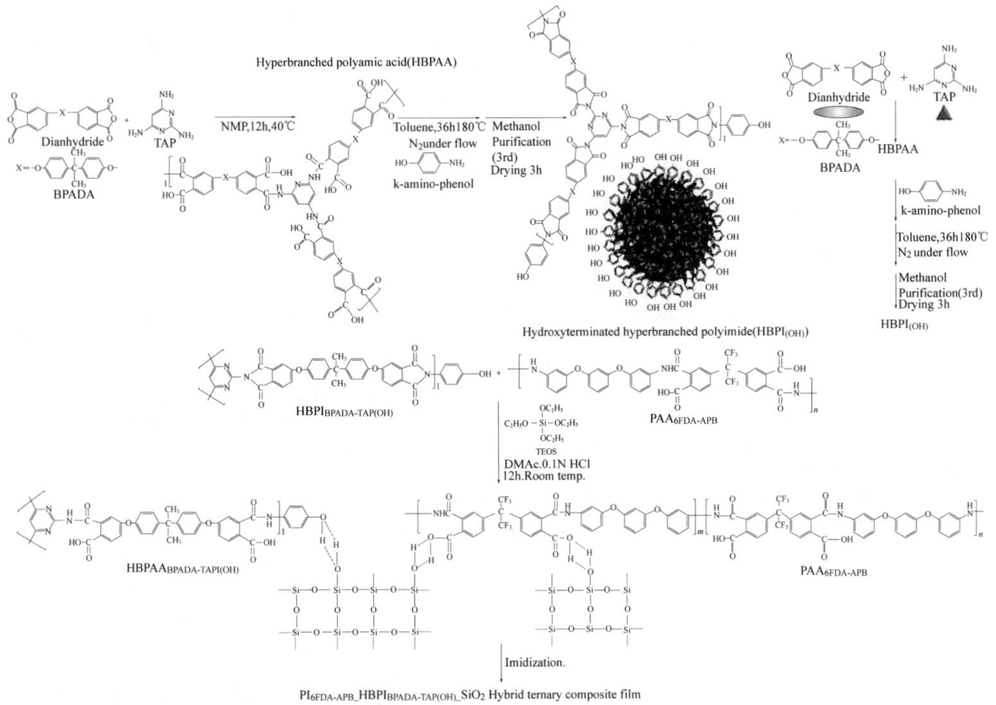

FIGURE 8.86 Synthetic route for the hydroxyl terminated hyperbranched $HBPI_{BPADA-TAP(OH)}$ and the $PI_{6FDA-APB}$-$HBPI_{BPADA-TAP(OH)}$-SiO_2 hybrid ternary composites. *Reproduced with permission from reference [208]. Copyright 2014, Royal Society of Chemistry.*

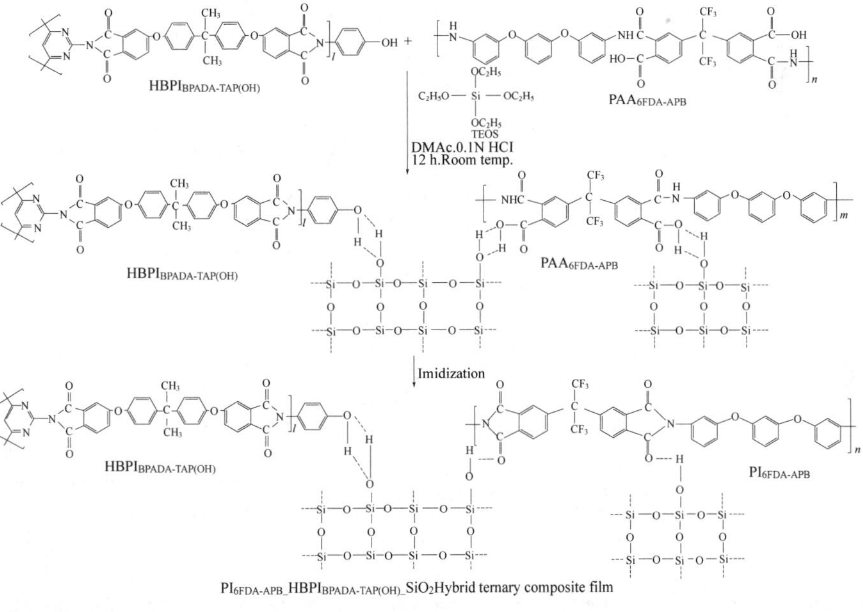

FIGURE 8.87 Synthetic route for the $PI_{6FDA-APB}$_$HBPI_{BPADA-TAP(OH)}$_SiO_2 hybrid ternary composites. *Reproduced with permission from reference [209]. Copyright 2014, Royal Society of Chemistry.*

The strong interchain packing and polarity of PI compromise its outstanding thermal and mechanical properties. Crown ethers are theoretically expected to form host–guest inclusion complex with PI and improve its disadvantages. On the basis of this hypothesis, Li et al. [210] synthesized a series of PI/crown ether composite films and investigated their synthesis mechanism, structures and properties. Results suggested that the introduction of crown ethers increased the free volume of PI matrix and generated a special necklacelike supramolecular structure, which simultaneously and greatly improved PI's mechanical, dielectric and hydrophobic properties (Fig. 8.88). The Young modulus, elongation and tensile energy at break of PI composite films were maximally increased by 73.0%, 135.5%, and 190.0%, respectively. The dielectric constant of PI composite film decreased obviously in the frequency range of 10^5–10^6 Hz with the addition of crown ether, and the maximal decreases were from ~3.59 of neat PI to 3.08 of PI-15-0.4 and to 2.99 of PI-180.4 at 10^6 Hz, respectively, i.e., the several maximal decrease amplitude was 14.2% for 15-crown ether-5 and 16.6% for 18-crown ether-6. Crown ethers with different molecular sizes demonstrated different improvement effects on PI's properties. Their inclusion rates stabilized at ~50%, which were related to the equal reaction probability between anhydride and amino groups.

To improve the mechanical and hydrophobic properties of dielectric materials, Zhang et al. [211] prepared a series of low dielectric constant fluorographene/polyimide (FG/PI) composite films by a facile solution blending method, suggesting that the mechanical, electrical, hydrophobic and thermal properties were significantly enhanced in the presence of FG (Fig. 8.89). With addition of 1%(w) FG, the tensile strength, Young's modulus and elongation at break were dramatically increased by 139%, 33%, and 18%, respectively, when compared with pure PI film. Furthermore, composite films exhibit superior hydrophobic and thermal stability performance. Especially, the FG/PI film with 0.5 %(w). of FG possessing a low dielectric constant of 2.48 and a good electrical insulativity that is lower than 10^{-14} $S\cdot m^{-1}$.

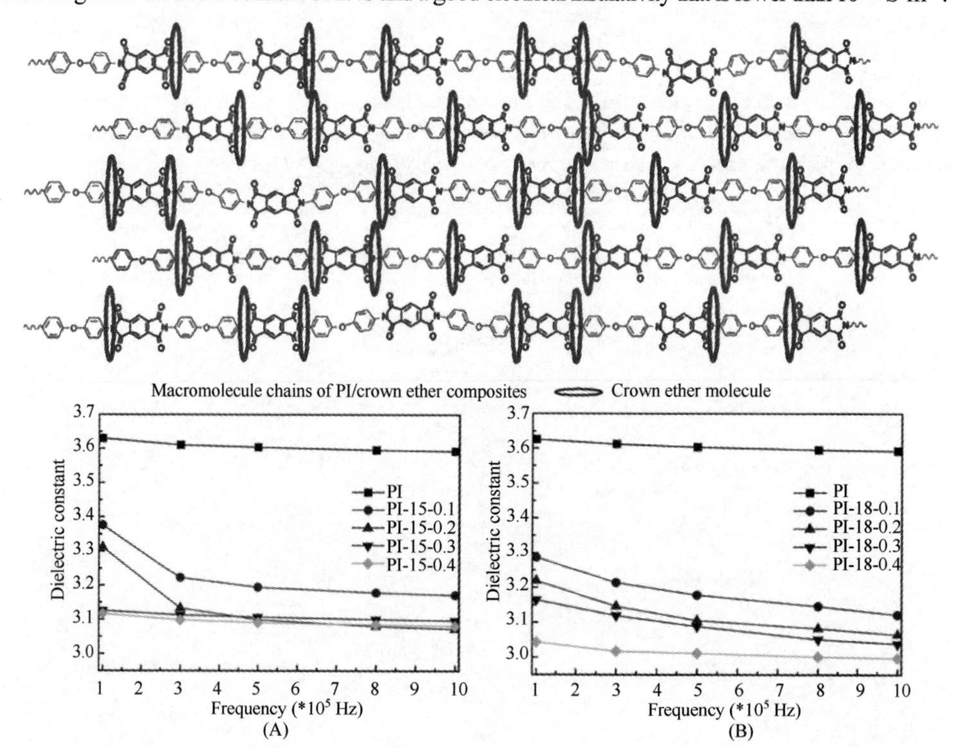

FIGURE 8.88 Structure of PI/Crown Ether Composite and Dielectric constants of (A) PI/15-crown ether-5 composite films and (B) PI/18-crown ether-6 composite films. *Reproduced with permission from reference [210]. Copyright 2015, American Chemical Society.*

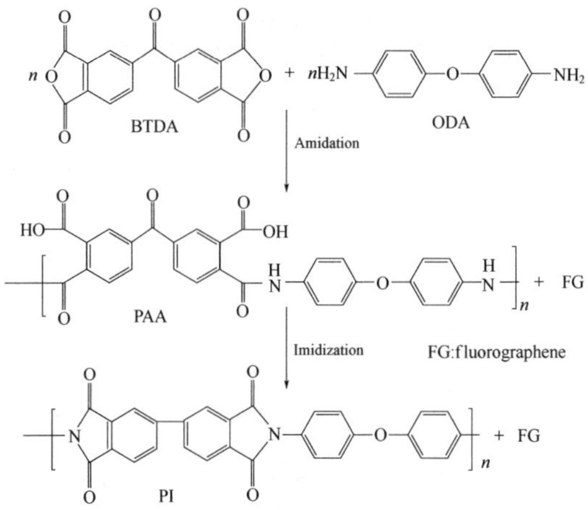

FIGURE 8.89 Synthesis procedure for FG/PI composite films.

Chen et al. [212] introduced an effective approach to fabricating novel graphene/PI nanocomposite films with low dielectric constants. Graphene fluoroxide (GFO) nanosheets were produced through the exfoliation of graphite fluoroxide (GiFO) in *N*-methyl-2-pyrrolidone (NMP). GFO nanosheets reacted with 4,4″-diaminodiphenyl ether (ODA) to become ODA-functioned GFO (GFO-ODA) with amine end groups, which provided reactive sites for covalent bonding with PI chains. GFO-ODA/PI composite films were prepared via the thermal imidization of a GFO-ODA/PAA solution (Fig. 8.90). In this composite, GFO-ODA nanosheets were chemically bonded to the PI matrix. The dielectric constant of the films depended on the loading of GFO-ODA and its minimum value reached 2.75 (at 10^6 Hz) when the content of GFO-ODA was 1.0 %(w). Strong GFO-ODA nanosheets facilitate the load transfer and enhance the mechanical properties of PI films. GFO-ODA/PI film with 1.0%(w) loading had a Young's modulus ~40% larger than pure PI film. In addition, the incorporation of GFO-ODA increased the T_g of the films and had no significant effect on the thermal stability.

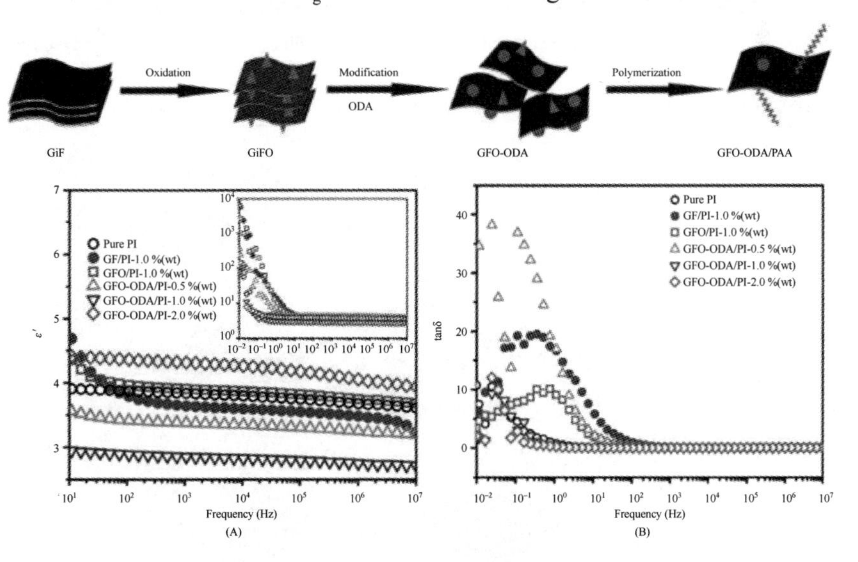

FIGURE 8.90 Preparation procedure of GFO-ODA/PAA and dependence of dielectric constant (A) and loss tangent (B) on the frequency for pure PI and PI composites. The inset shows the frequency dependence of dielectric constants in the range from 10^{-2} to 10^7 Hz. *Reproduced with permission from reference [212]. Copyright 2017, Royal Society of Chemistry.*

Xu et al. [213] designed an inexpensive, simple, and facile approach to the preparation of low dielectric constant PI. A series of low dielectric constant PI-based films hybridized with polytetrafluoroethylene (PTFE) are made by an aqueous solution blending route, where a synthesized water soluble poly(amic acid) ammonium salt was blended with a PTFE aqueous emulsion, followed by spin-coating and thermal imidization (Fig. 8.91). The PI hybrid film (40 %(*w*). PTFE) showed a lowest dielectric constant of 2.25 (at 1 kHz). Meanwhile, the PI/PTFE hybrid films exhibited good thermal stability of the 5% weight loss temperatures ($T_{5\%}$) higher than 520 °C, and T_gs higher than 285 °C; as well as excellent mechanical properties of the tensile stress, modulus, and elongation at break being 84-127 MPa, 0.94-1.86 GPa, and 56%-118%, respectively.

To decrease the *k*-value and simultaneously maintain its mechanical and thermal properties of dielectric materials, Lei et al. [214] fabricated a series of robust PI films by copolymerizing amine-functionalized hyperbranched polysiloxane (HBPSi) with pyromellitic dianhydride (PMDA) and 4,4′-oxydianiline (ODA) (Fig. 8.92). The outstanding dielectric properties were achieved in a 35%(*w*) HBPSi PI film, which exhibited a *k*-value as low as 2.24 (1 MHz), mainly owing to the enhanced free volume and dielectric confinement effect afforded by the bulky HBPSi. Meanwhile, 35%(*w*) HBPSi PI demonstrates remarkable thermal stability and admirable mechanical properties, with the T_g of 388 °C, 5% weight loss temperature in argon flow up to 554 °C, a tensile strength of 80.6 MPa, elongation at break of 13.7%, and a tensile modulus of 1.36 GPa. It also demonstrates conspicuous film homogeneity and planarity with the surface roughness as low as 0.42 nm and good moisture resistance with water uptake less than 1.5%.

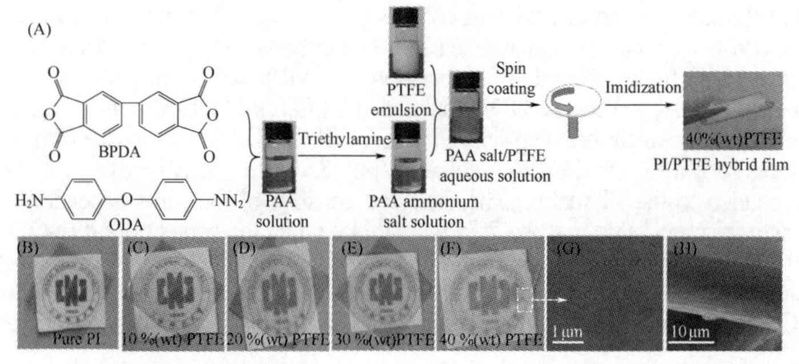

FIGURE 8.91 (A) Schematic illustration of preparation of the PI/PTFE hybrid films; (B-F) optical images of the prepared pure PI film (B), and PI-based hybrid films with the PTFE contents of 10%(*w*) (C), 20%(*w*) (D), 30%(*w*) (E), and 40%(*w*) (F); (G, H) SEM images of the surface (G) and cross-section (H) of the PI/PTFE hybrid film (40 %(*w*). PTFE).*Reproduced with permission from reference [213]. Copyright 2017, Springer.*

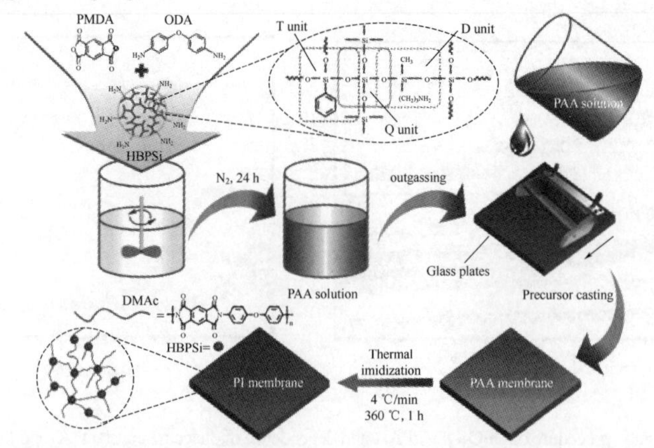

FIGURE 8.92 Reaction scheme for the fabrication of HBPSi PI films. Note: the molecular structure herein presented represents one of the possible molecular structures.*Reproduced with permission from reference [214]. Copyright 2016, Royal Society of Chemistry.*

Using the same materials system, Dong et al. [215] proposed an effective method to prepare PI composite fibers with exceptional dielectric behavior, mechanical and thermal properties (Fig. 8.93). The composite fiber containing 10%(w) NH_2-HBPSi exhibited a 10% and a 26% increase in tensile strength and modulus compared to the pure PI fiber. Owing to the dielectric confinement effect from NH_2-HBPSi, the dielectric constants of the NH_2-HBPSi/PI composites were reduced drastically, and the value could reach as low as 2.2 at 10^8 Hz. The composite fiber demonstrated better specific strength, specific modulus and lower dielectric constant than the commonly used E-glass, S-glass, and quartz fibers.

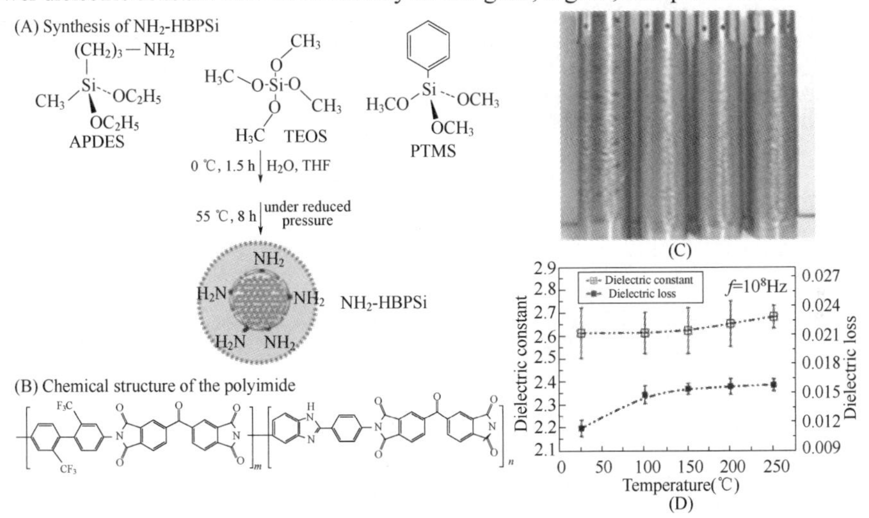

FIGURE 8.93 (A) Synthesis of the NH_2-HBPSi; (B) chemical structure of the organo-soluble polyimide (m/n=1/1); (C) Photographs of NH_2-HBPSi/PI composite fibers with NH_2-HBPSi loadings of 0, 5, 10, and 16%(w), and (D) the dielectric properties of the 10%(w) -NH_2-HBPSi/PI composite at various temperatures.*Reproduced with permission from reference [215]. Copyright 2017, Royal Society of Chemistry.*

8.4.6 Intrinsic Low-k Polyimide Materials

The intrinsic low-k PI films refer to a class of functional polymer materials, of which the k-value can be decreased only by changing the chemical structure of the monomers through introducing some special chemical groups, without the aid of other means such as copolymerization, pore-forming or film forming processes. These specific chemical groups may have a lower molar polarizability, or can occupy a larger volume of space. Compared with other methods, to lower the dielectric constant of materials by this means is beneficial to maintaining the mechanical properties and thermal stability of PIs, as well as the stability of performance during operating process.

Simpson and Clair [216] carried out fundamental research to examine the effect of polarizability, free volume, and fluorine content on the dielectric properties of PIs. Results showed that minimizing polarizability, maximizing free volume and fluorination all lowered dielectric constants in the PIs studied. Polarizability is the primary variable influencing dielectric constants whereas free volume and fluorine content are secondary variables which can alter a polymer's polarizability. Enhanced free volume lowers polarization by decreasing the number of polarizable groups per unit volume. Fluorination increases free volume, lowers electronic polarization and can either increase or have no effect on dipole polarization depending on whether the fluorination is asymmetric or symmetric. Several types of PIs have been developed which show lower dielectric constants, including incorporating diamine and dianhydride reactants which minimize polarizability; incorporating diamine and dianhydride reactants which impart a high degree of free volume; and incorporating fluorine atoms into the molecular structure of the PI.

8.4.6.1 Multimethyl-substituted Copolyimides

Poly(2,6-dimethyl-1,4-phenylene ether) (PPE), a typical engineering plastic, is well known as a low-ε material (ε=2.5) [129,217]. These findings prompted the development of new nonfluorinated PIs with the very low ε by introducing phenylene ether units in the main chains. Watanabe et al [218]. developed a series of new aromatic nonfluorinated PIs with low dielectric constant from aromatic dianhydrides and an aromatic diamine containing phenylene ether units (Fig. 8.94). The diamine monomer was prepared from 4-bromophenyl ether in five steps. Polycondensations were performed in 1-methyl-2-pyrrolidinone at room temperature for 18 h, giving PAAs with inherent viscosities up to 0.53 dL • g^{-1}. PAAs were converted to corresponding PIs by thermal treatment at 300 °C. New PIs showed good thermal stability (5% weight loss around 450 °C) and the low dielectric constant (2.74).

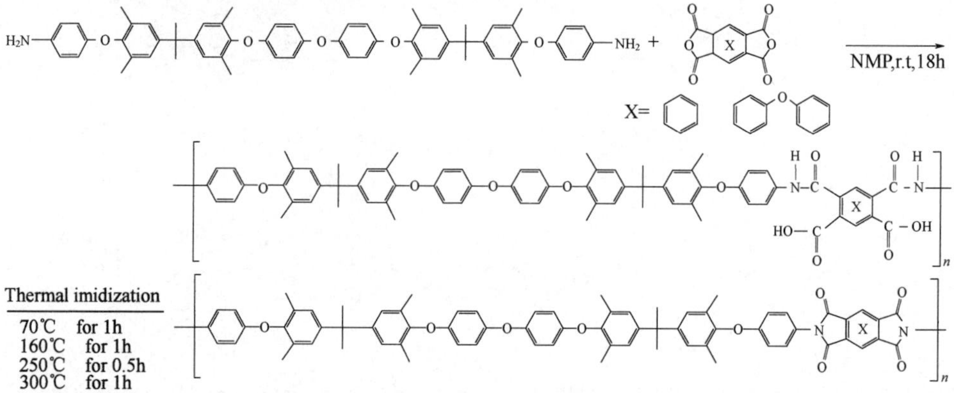

FIGURE 8.94 Synthesis of new aromatic nonfluorinated polyimides.*Reproduced with permission from reference [218]. Copyright 2005, Elsevier.*

Polyoxometalates (POMs) are also porous inorganic nanoparticles or molecular clusters and have found wide applications in many fields [219]. As nanobuilding blocks, they have been embedded into some polymers and resulted in pronounced property improvements of the resulting materials [220,221]. Chen et al. [202] reported novel PI hybrid films in which nanoscale POM clusters covalently linked with the polymer chains exhibit not only drastically reduced dielectric constants relative to neat PI film, as low as 1.22, but also improved thermal stability and mechanical properties, without sacrificing the transparency of the polymer films (Fig. 8.95).

Yuan et al. [222] designed and synthesized a series of copolymerized PI thin films with low dielectric constants with different molars ratio of bis[3,5-dimethyl-4-(4aminophenoxy) phenyl]methane and 9,9-bis[4-(4-aminophenoxy) phenyl]fluorene as diamines and 4,4′-(4,4′isopropylidenediphenoxy)bis(phthalic anhydride) as a dianhydride (Fig. 8.96). Some films possessed good dielectric properties with an ultralow dielectric constant of 2.3 at 1 MHz. The structures and properties of the thin films were measured with Fourier transform infrared and NMR spectroscopy, thermogravimetric analysis, and dynamic mechanical analysis. The PI films exhibited glass-transition temperatures in the range 223-243 °C and possessed initial thermal decomposition temperatures reaching up to 475-486 °C in air and 464-477 °C in nitrogen. All of the PI films exhibited excellent solubility in organic solvents. The mechanical properties of these films were also examined.

8.4.6.2 Hyperbranched and Crosslinked Polyimides

Deligöz et al. [223] prepared a novel crosslinked polyimide film (CPI) and conventional PI by thermal imidization of cross-linked PAA and conventional PAA (Fig. 8.97). The dielectric constant and dielectric loss of the conventional and novel crosslinked PI films were found to be frequency and temperature dependent. It was found that this type of crosslinked PI was very promising for electrical applications due to its good thermal stability, excellent solvent resistance, low dielectric constant, low moisture absorption and stability in various regions of frequencies. For CPI the activation energy was calculated for 300-350 K temperature interval and β-relaxation was observed above room temperature. They conclude by saying that the dielectric constant of the novel crosslinked PI film has lower values compared to dielectric constant of conventional PI film. Thus, this type of novel PI film can be used as alternative dielectric layers in the microelectronics industry.

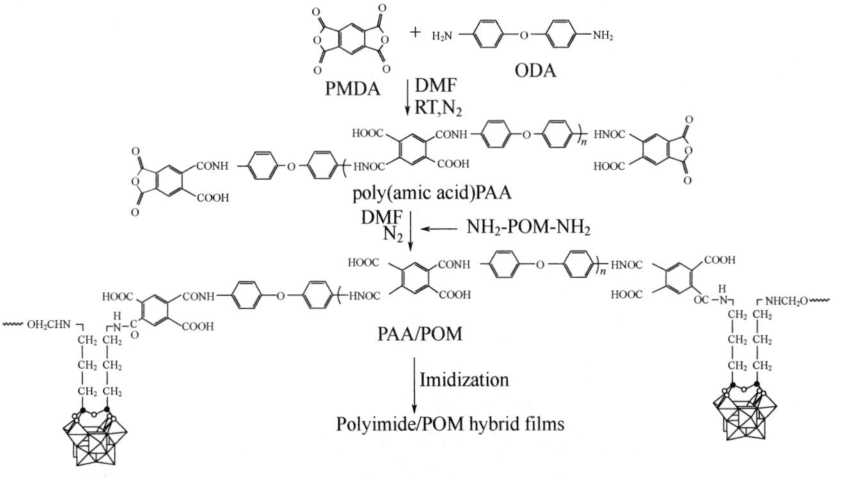

PMDA + ODA → (DMF, RT, N_2) poly(amic acid)PAA → (DMF, N_2; NH_2-POM-NH_2) PAA/POM → (Imidization) Polyimide/POM hybrid films

FIGURE 8.95 Preparation of polyimide/POM hybrid films.*Reproduced with permission from reference [202]. Copyright 2007, Royal Society of Chemistry.*

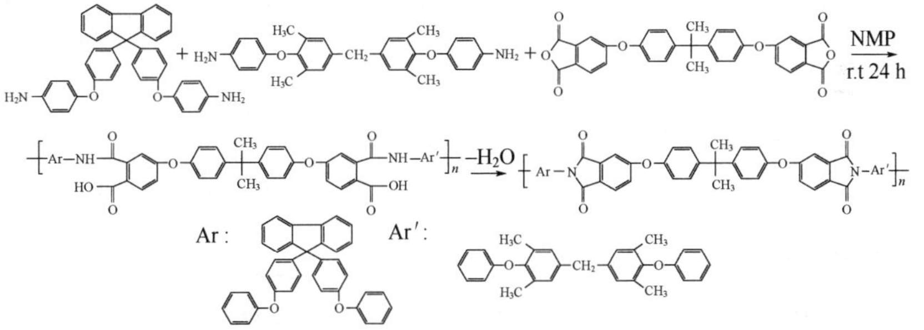

NMP, r.t 24 h; $-H_2O$; Ar :; Ar′ :

FIGURE 8.96 Synthesis schemes of the PIs.*Reproduced with permission from reference [222]. Copyright 2008, John Wiley & Sons.*

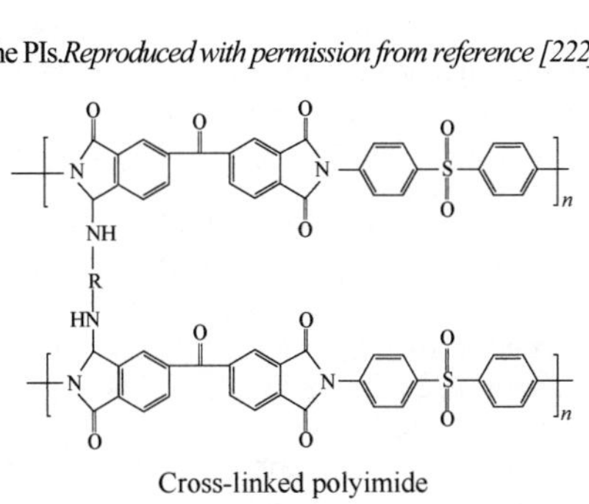

Cross-linked polyimide

FIGURE 8.97 Chemical structure of CPI.

Currently, low-dielectric PIs such as fluorinated or porous PIs exhibit a low-dielectric property but have undesirable mechanical and/or thermal properties. Therefore, it is crucial to find a more considerate method that could lower the permittivity, while maintaining or improving the mechanical and thermal properties. Lei et al. [224] synthesized a series of hyperbranched (HB) PI films by adjusting the content of the rigid diamine, 2,2′-dimethylbenzidine (DMBZ) (Fig. 8.98). The dielectric properties of the HBPIs were accordingly tuned, i.e., the permittivity of the resulting HBPIs decreased with increasing the DMBZ fraction owing to the enlarged free volume and the hindered dipole orientations afforded by the rigid DMBZ. The maximum mechanical strength of the resulting HBPIs located at the formulation were made using 50% DMBZ and 50% ODA. At this formulation, the optimal comprehensive performances were achieved, i.e., excellent tensile strength (124.1 MPa), desirable thermal stability (5% weight loss temperature up to 505 °C with weight residual of 56.7% at 800 °C under argon), high glass-transition temperature (324 °C), low relative permittivity (2.69, 1 MHz), reduced water absorption (~1.86%), and good solubility.

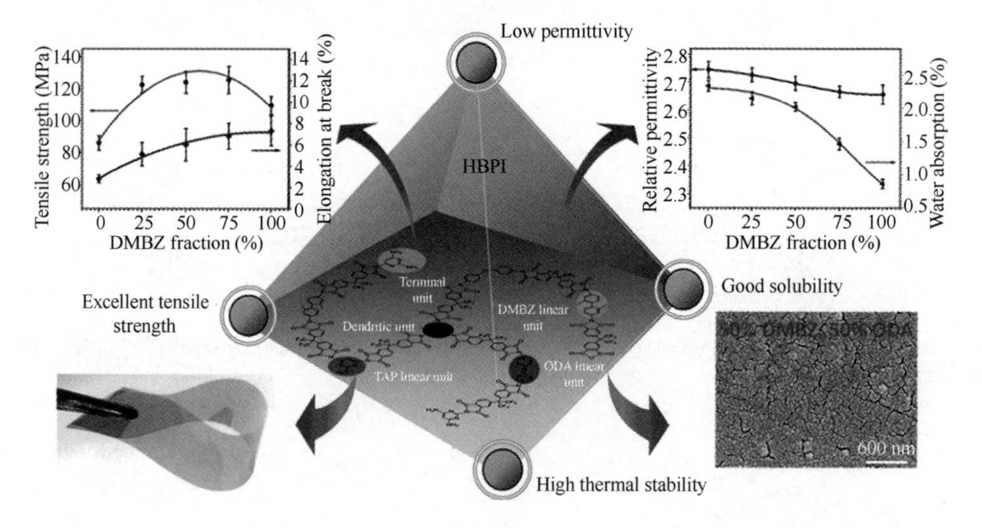

FIGURE 8.98 Tunable permittivity in high-performance hyperbranched polyimide films by adjusting backbone rigidity.*Reproduced with permission from reference [224]. Copyright 2016, American Chemical Society.*

8.4.6.3 Bulky Structure-contained Polyimides

Introducing bulky structure into the polymer leads to decreased crystallinity, increased solubility, and enhanced T_g and thermal stability of the polymer materials, and thus may help to decrease the dielectric properties of materials. Chern and Shiue [225] synthesized new adamantane-based PIs by reacting 1,3-bis[4-(4-aminophenoxy)phenyl]adamantane with various aromatic tetracarboxylic dianhydrides (Fig. 8.99). The poly(amic acid)s have number-average molecular weights (Mn) of 39,000-139,000. These films have low dielectric constants ranging from 2.77 to 2.91 and low moisture absorptions of less than 0.46%. All seven PIs formed tough and transparent films. These films have tensile strengths of 88.2-113.5 MPa, elongations to break of 5.6%-12.5%, and initial moduli of 2.0- 2.2 GPa. Dynamic mechanical analysis (DMA) reveals that adamantane-based PIs have two relaxations on the temperature scale between 0 °C and 350 °C. The subglass relaxations of the seven polyimides do not occur at the same temperature; instead, they range from 100 °C to 175 °C. Differential scanning calorimetry (DSC) and DMA reveal their T_gs were found to be 232-330 and 262-343 °C, respectively. These T_gs lie in a manageable temperature range for processing in melt.

They also used diamantane, which is a cycloaliphatic-cage hydrocarbon containing an "extended-cage" adamantane structure, to synthesize new diamantane-based PIs by reacting of 4,9-bis[4-(4-aminophenoxy)phenyl] diamantane with various aromatic tetracarboxylic dianhydrides (Fig. 8.100) [226]. Due to the low hydrophobicity and polarity of the rigid and bulky diamantane, films of these PIs have low dielectric constants, ranging from 2.58 to 2.74, and low moisture absorptions of less than 0.3%. Four of these films have good solubilities. Three nonfluorinated PIs were soluble in *o*-chlorophenol, *m*-cresol, *N*-methyl-2-pyrrolidone (NMP), and chloroform. The hexafluoroisopropylidene-containing PI was soluble in *m*-cresol, chloroform, and tetrahydrofuran (THF). The soluble PIs have molecular weights (Mn) ranging from 33000 to 96000. All PIs formed tough transparent films, with tensile strengths of 55.3-101.4 MPa, elongations to break of 6.1%-22.3%, and initial moduli of 1.9-2.2 GPa. Dynamic mechanical analysis (DMA) reveals that diamantane-based PIs have two transitions on the temperature scale between 0 °C and 400 °C. Their glass relaxations, characterized by DMA, occur at high temperatures, ranging from 296 °C to 413 °C.

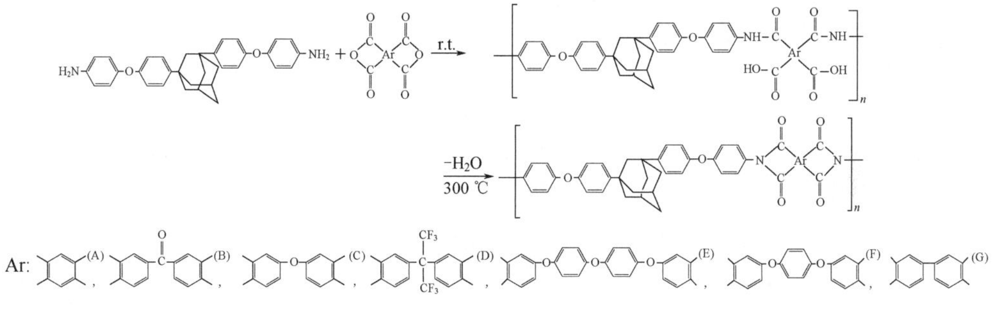

H_2N O O NH_2 + O C Ar C O r.t. O NH C C NH Ar HO C C OH O O n
$-H_2O$ 300 ℃ O O N C Ar C N O O n
Ar: (A), C O (B), O (C), CF_3 C CF_3 (D), O O O O (E), O O (F), (G)

FIGURE 8.99 Synthesis of adamantane-based polyimides. *Reproduced with permission from reference [225]. Copyright 1997, American Chemical Society.*

H_2N O O NH_2 + O C Ar C O 50 ℃ O NH C C NH Ar HO C C OH O O n
$-H_2O$ 320 ℃ O O N C Ar C N O O n
Ar: (A), C O (B), O (C), CF_3 C CF_3 (D), O O O O (E), O O (F), (G)

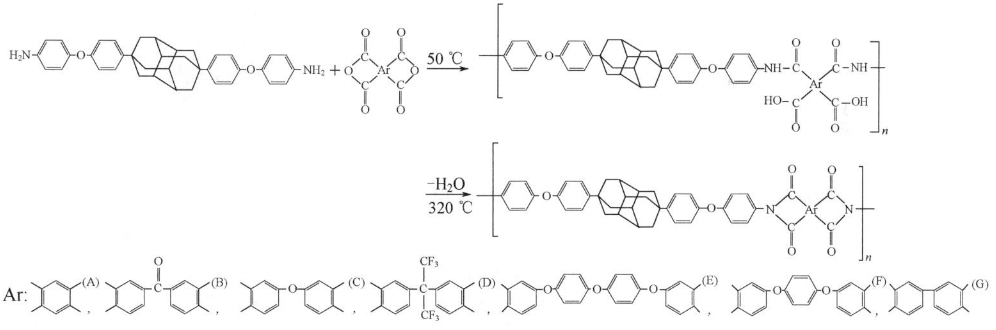

FIGURE 8.100 Synthesis of diamantane-based polyimides. *Reproduced with permission from reference [226]. Copyright 1997, American Chemical Society.*

Chern [227] further synthesized another diamantane-based PI from 1,6-bis[4-(4-aminophenoxy) phenyl] diamantane and various aromatic tetracarboxylic dianhydrides (Fig. 8.101). They also exhibit a low dielectric constant (2.56-2.78), low moisture absorption (<0.292%), good solubility, and high number-average molecular weights (Mn: 3.7×10^4 to 12.8×10^4). This kind of diamantane-based PI can form a tough and transparent film after cyclodehydration. The films have tensile strengths of 74.4-118.2 MPa, elongations to break of 5.2%-24.3%, and initial moduli of 1.9-2.2 GPa.

Long and Swager [228] described the use of the triptycene moiety as a rigid and shape-persistent component as a method to introduce molecular-scale free volume into a polymer film (Fig. 8.102). Triptycenes having restricted rotation by multiple point attachment to the polymer backbone are shown to introduce free volume into the films, thereby lowering their dielectric constants. The triptycene-containing polymers exhibit a number of desirable properties including low dielectric constant (in the range of 3.35-2.41), low-water absorption (in the range of 5.1% to 0.3%) and high thermal stability. Systematic studies wherein comparisons are made between two separate classes of triptycene polymers and their nontriptycene-containing analogs demonstrate that proper insertion of triptycenes into a polymer backbone can give rise to a reduction in the material's dielectric constant while also improving its mechanical properties.

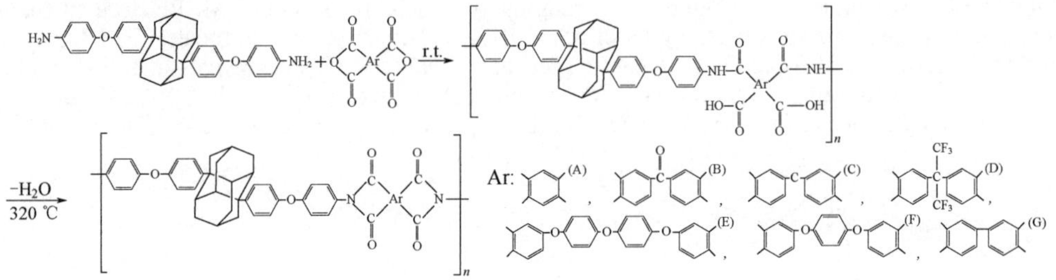

FIGURE 8.101 Synthesis of polyimides from 1,6-bis[4-(4-aminophenoxy)phenyl] diamantine.*Reproduced with permission from reference [227]. Copyright 1998, American Chemical Society.*

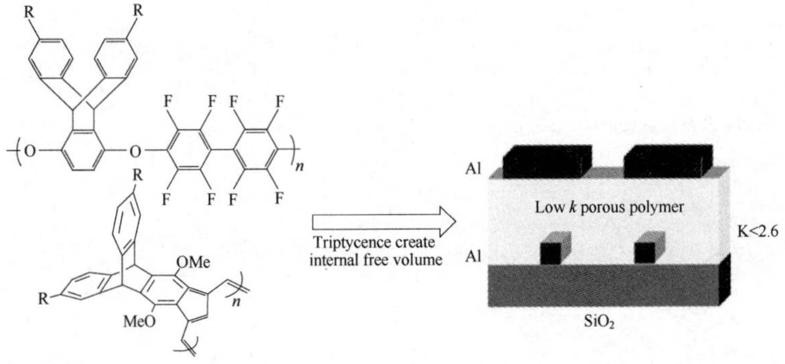

FIGURE 8.102 Low-*k*-polyimide-containing triptycene subunit.*Reproduced with permission from reference [228]. Copyright 2003, American Chemical Society.*

Sydlik et al. [229] synthesized a series of novel triptycene PIs that exhibited high thermal stability and solubility (Fig. 8.103). All of these triptycene polyimides (TPIs) were soluble in common organic solvents despite their completely aromatic structure due to the three-dimensional triptycene structure that prevented strong interchain interactions. Low solution viscosities (0.07-0.47 $dL \cdot g^{-1}$) and versatile solubilities allow for easy solution processing of these polymers. Nanoporosity in the solid state gives rise to high surface areas (up to 430 $m^2 \cdot g^{-1}$), low refractive indices (1.19-1.79 at 633 nm), and low dielectric constants (1.42-3.20). Polymer films were found to be amorphous. The decomposition temperature (T_d) for all of the polymers is above 500 °C, and no T_gs can be found below 450 °C by differential scanning calorimetry (DSC), indicating excellent prospects for high-temperature applications. This combination of properties makes these polymers candidates for spin-on dielectric materials.

8.4.6.4 Fluorinated Polyimides

The incorporation of fluorine atoms provides a means of obtaining optical transparency and lowering the dielectric constants of PIs. Seino [145] developed a positive-working photosensitive PI precursor based on fluorinated polyisoimide (FPII) and 2,3,4-tris(1-oxo-2-diazo-naphthoquinone-5-sulfonyloxy)benzophenone (DSSB) as a photosensitive compound (Fig. 8.104). The FPII film showed a good solubility in a wide range of organic solvents. The dissolution behavior of FPII containing

30 %(w). of DSSB after exposure was studied and it was found that the difference of dissolution rate between the exposed and the unexposed parts was enough to get a positive pattern due to photochemical reaction of DSSB in the polymer film. The photosensitive fluorinated polyimide (FPI) precursor containing 30 %(w). of DSSB showed a sensitivity of 250 $mJ \cdot cm^{-1}$ and a contrast of 1.5 with 436 nm light, when it was developed with a mixture of 2.38% aqueous tetramethylammonium thermal expansion of 10.8×10^{-6} $°C^{-1}$ and a low dielectric constant of 2.89.

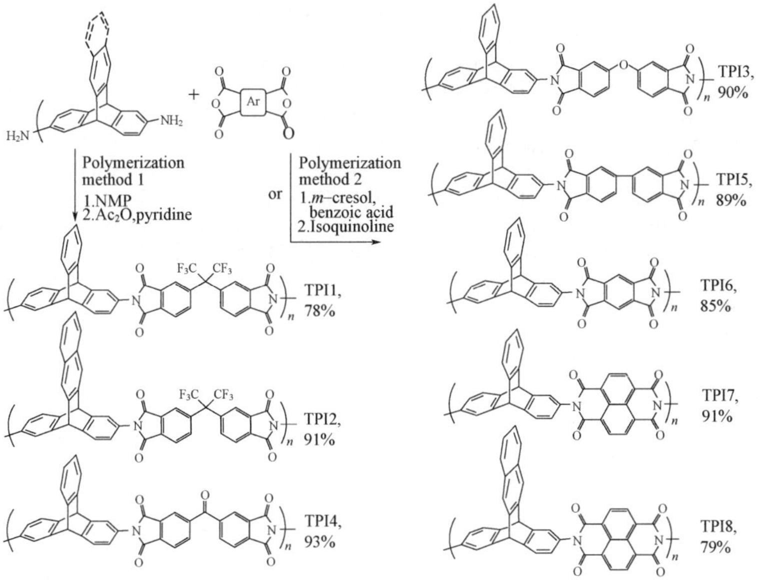

FIGURE 8.103 Low-*k*-polyimide-containing triptycene subunit.*Reproduced with permission from reference [229]. Copyright 2011, American Chemical Society.*

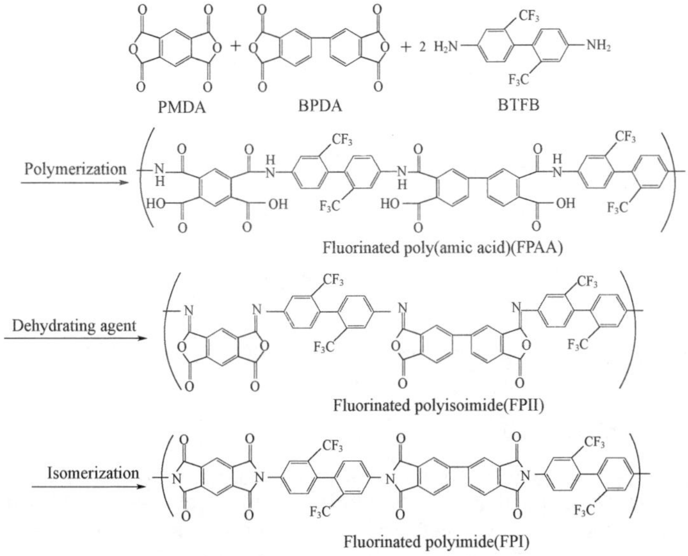

FIGURE 8.104 Synthesis of fluorinated polyimide (FPI).*Reproduced with permission from reference [145]. Copyright 1999, Elsevier.*

Stoakley et al. [230] prepared a series of copolymers of FPIs and 3,3′,4,4′- biphenyltetracarboxylic dianhydride (BPDA) as films and composite laminates (Fig. 8.105). The addition of BPDA was used as a means to achieve insolubility, making the polymers suitable as aircraft matrix resins. T_gs, thermooxidative stabilities, and tensile strengths were increased with increasing BPDA content in the copolymers. Although the addition of BPDA did increase the UV cutoff and decrease the percent transmission slightly, the optical transparency of the polymers was still excellent. Dielectric constants of the copolyimide films ranged from 2.6 to 2.9. Flexural strengths on unidirectional specimens were in the 1.24-1.41 GPa range and flexural moduli were 41 GPa.

Yang et al. [231] reported a series of fluorine-containing copolyimides from 6FDA dianhydride and different ratios of BisAAF and PPD diamines (Fig. 8.106). The inherent viscosity increased with increasing PPD mole fraction, from 0.40 dL·g^{-1} of pure 6FDA-BisAAF to 0.84 dL·g^{-1} of pure 6FDA-PPD. The dielectric constant decreased with increasing fluorine content. The glass-transition temperature increased with an increasing PPD mole fraction; the values increased from 317 °C with pure 6FDA-BisAAF PI to 364 °C with pure 6FDA-PPD PI. The 5% weight loss temperature (T_d) of the copolyimides was around 530 °C in air and 540 °C in a nitrogen atmosphere. The tensile modulus and tensile strength gradually increased with an increasing PPD molar fraction. The transmittance of 6FDA-BisAAF-PPD copolyimides was greater than 90% at wavelengths above 500 nm.

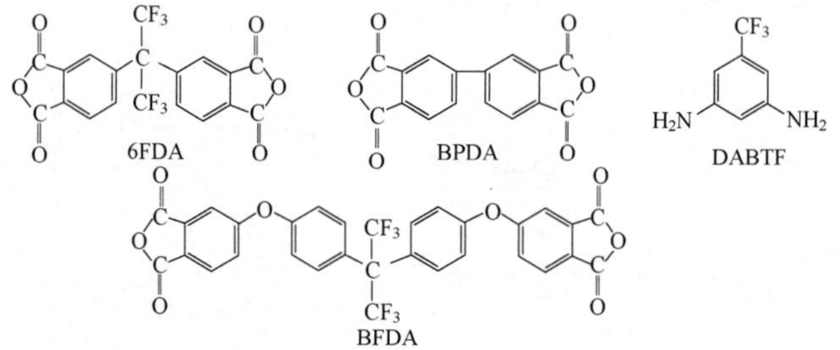

FIGURE 8.105 Dianhydride and diamine monomers.

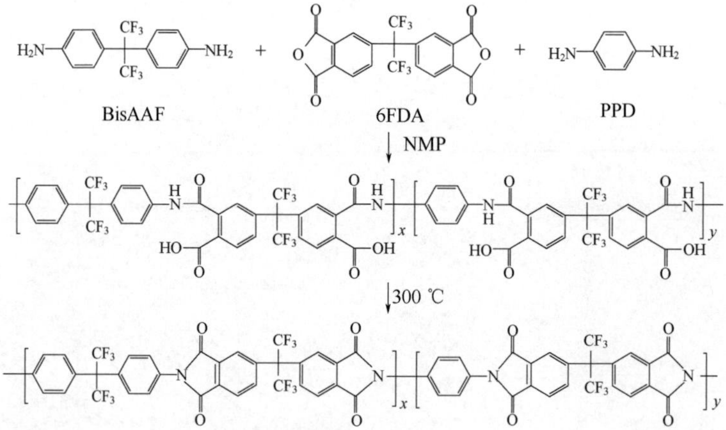

FIGURE 8.106 Synthesis of fluorine-containing copolyimides.

To acquire low-k PI films with good mechanical and thermal properties and low coefficient of thermal expansion (CTE) applied in microelectronic fields, Dong et al. [232] developed three novel diamines containing pyridine and —C(CF_3)$_2$— groups to employ polymerization with 2,2′-bis(3,4-dicarboxyphenyl) hexafluoropropanedianhydride (6FDA) via a two-stage process with a heating imidization method (Fig. 8.107). Three diamine monomers included one unsubstituted pyridine ring, and another two methyl-substituent groups on two pyridine rings at the 6-and

4-positions. The structure-property relationships between the different pyridine rings of the fluorinated PI films, including dielectric constant, thermal stability, mechanical strength, optical transparency, and solubility, were systematically investigated. The fluorinated PI films exhibit low dielectric constant in the range of 2.36-2.52 at 1 MHz, while they still display excellent mechanical properties with tensile strengths as high as 114 MPa. Meanwhile, the PI films show good thermal stability with T_gs in the range of 262-275 °C, low coefficients of thermal expansion (*CTE*s) ranging from 64×10^{-6} to 68×10^{-6} °C^{-1} and $T_{5\%}$ located between 468 °C and 499 °C. Further, PI films possess outstanding solubility for easy fabrication.

Aiming to develop a dielectric polymer with superior durability of low-*k* value and high thermal stability, Jia et al. [233] reported a perfluorocyclobutyl (PFCB) biphenyl ether-based PI, PFCBBPPI (Fig. 8.108). This polymer possesses a T_g of 310.3 °C and a $T_{5\%}$ of 510.5 °C. PFCBBPPI exhibited an extremely low water uptake of 0.065%±0.018%, representing the best water resistance in PIs. The increasing percentage in *k*-value was below 2% for PFCBBPPI film exposed to moisture under various humidity conditions for 6 h. PFCBBPPI film equilibrated at 75% R.H. for 2 weeks still kept its *k*-value below 2.50, remarkably outperforming the Kapton film. The remarkable water resistance and resulting high durability of low-*k* property displayed by PFCBBPPI originate from the hydrophobic nature and small free volume fraction of the polymer, as confirmed by contact angle test and positron annihilation lifetime spectroscopy results.

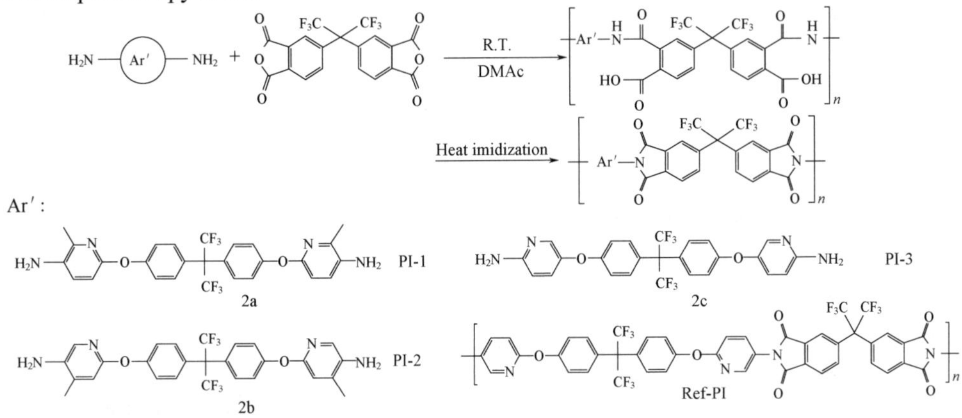

FIGURE 8.107 Preparation of fluorinated polyimides.

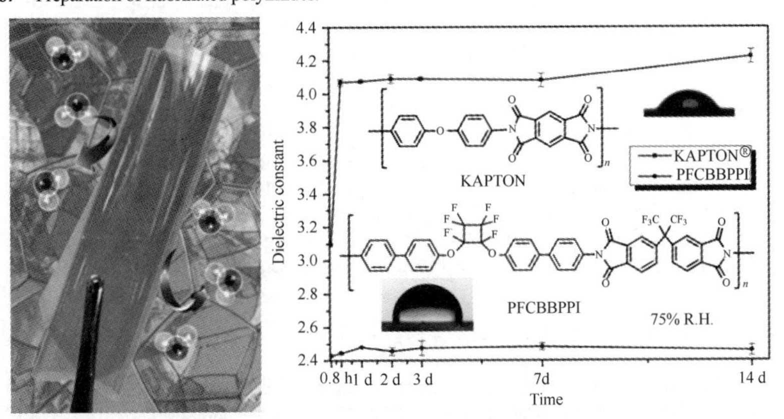

FIGURE 8.108 Perfluorocyclobutyl biphenyl ether-based low-*k* polyimide. *Reproduced with permission from reference [233]. Copyright 2016, American Chemical Society.*

Nawaz et al. [234] synthesized two novel isomeric fluorinated diamines via Williamson etherification reaction. Their polymerization was carried out with commercially available anhydrides, i.e., pyromellitic dianhydride (PMDA), 3,3′,4,4′-benzophenone tetracarboxylic dianhydride (BTDA), 2,2-bis(3,4-dicarboxyphenyl) hexafluoropropanedianhydride (6FDA) and oxydipthalic anhydride (ODPA), respectively (Fig. 8.109). A series of meta substituted fluorinated PIs PD-11, PD-13, PD-14, PD-15 and *ortho*-substituted fluorinated PIs PD-41, PD-43, PD-44, and PD-45 were synthesized from each respective anhydride and diamine. Solubility studies revealed comparable solubility attitude for meta and *ortho*-substituted fluorinated PIs, only PD-14 and PD-44 were soluble at room temperature. The other PIs showed solubility at elevated temperatures in all solvents tested. The inherent viscosities of the PIs also exhibited a descending order PD-43 > PD-11 > PD-45 > PD-44 > PD-14 > PD-41 > PD13 > PD-15 that lies in the range of 0.372-0.875 $dL \cdot g^{-1}$, respectively. Dielectric properties studies showed that *meta*-substituted fluorinated PIs had higher dielectric constant values ranging from 3.26 to 4.53 and dielectric tangent loss in the range of 0.0023-0.010. Whilst, the *ortho*-substituted fluorinated PIs showed dielectric constant values in the range of 1.86-3.46 with a dielectric tangent loss of 0.0034-0.0092, respectively.

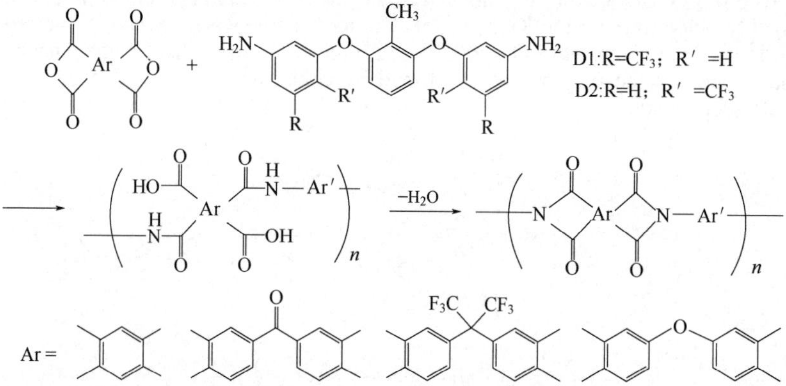

D1:R=CF_3; R′ =H

D2:R=H; R′ =CF_3

FIGURE 8.109 Synthesis of polyimides from respective diamines species and dianhydrides.

Wang et al. [235] successfully synthesized a new aromatic diamine monomer, containing four pendant trifluoromethylphenyl group substituents via a three-step reaction from readily available reagents (Fig. 8.110). A new series of fluorinated PIs with multibulky pendant groups was prepared from the diamine monomer with three commercially available aromatic tetracarboxylic dianhydrides using one-step high-temperature polycondensation. The incorporation of multibulky pendant fluorinated groups and large noncoplanar structures brought about a great improvement in their properties. The resulting PIs had excellent solubility and the maximum solubility even reached 20%(*w*) in NMP, $CHCl_3$, and THF at room temperature. Their film displayed a light color and high optical transparency with the cut-off wavelengths ranging from 327 to 343 nm. They also possessed intrinsically low dielectric constant values of 2.69-2.85 (at 1 MHz) and low water absorption (0.59%-0.68%). Moreover, these PIs showed high T_gs (259-281 °C) and excellent thermal stability with 5% weight loss at temperatures over the range of 551-561 °C and 515-520 °C under nitrogen and air atmospheres, respectively. Compared to some trifluoromethyl-substituted PIs, these PIs with multibulky pendant groups possessed better solubility and lower cut-off wavelength and dielectric constants.

8.4.6.5 Alicyclic and Steric-substituted Polyimides

Tang et al. [236] prepared a spirobichroman dianhydride (SBCDA) through oxidation of an octamethyl spirobichroman (OMSBC), which was synthesized from acid-fragmentation of bisphenol A by 3,4-dimethylphenol, followed by Diels-Alder reaction (Fig. 8.111). The reaction mechanism was proposed, and the optimal reaction conditions were discussed. Based on a high-temperature solution polymerization of SBCDA and 4,4′-diaminodiphenylmethane (DDM), a spirobichroman-containing PI, SBC-DDM, was successfully prepared. Because of the contorted spiro-structure and rigid polymer backbone, SBC-DDM exhibits a large free volume, leading to outstanding organo-solubility and a low dielectric constant (2.34 at 1 GHz). In addition, the resulting film of SBC-DDM shows foldability, a high T_g, and good thermal stability.

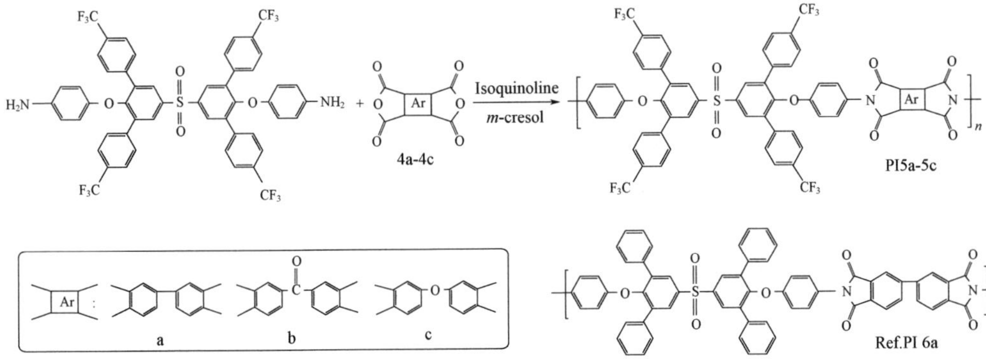

FIGURE 8.110 Synthesis of polyimides containing trifluoromethylphenyl group substituents.

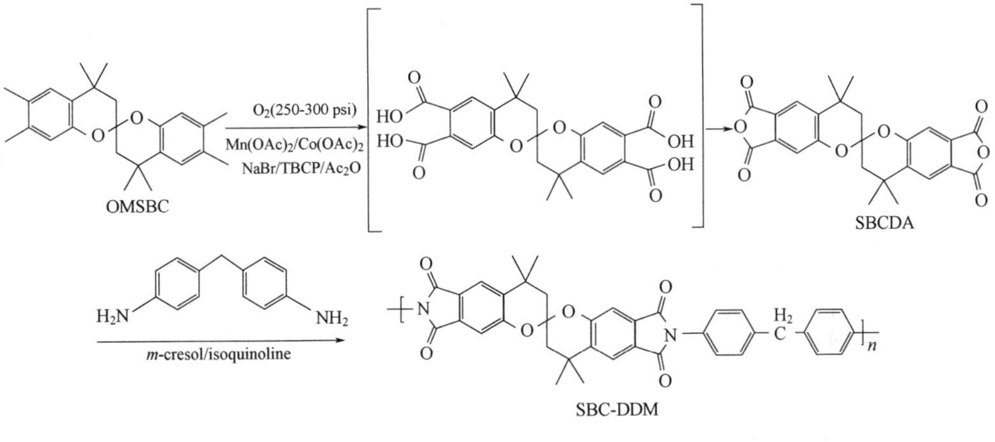

FIGURE 8.111 Synthesis of SBCDA and its derived polyimide SBC-DDM.

Liaw and Tseng [237] reported a new kink diamine with trifluoromethyl group on either side, bis[4-(2-trifluoromethyl-4-aminophenoxy)phenyl]diphenylmethane (BTFAPDM). It was reacted with various aromatic dianhydrides to prepare PIs via poly(amic acid) precursors followed by thermal imidization. PIs were prepared using 3,3′,4,4′-biphenyltetracarboxylic dianhydride(1), 4,4′-oxydiphthalic anhydride(2), 3,3′,4,4′-benzophenonetetracarboxylic dianhydride (3), 4,4′-sulfonyldiphthalic anhydride(4), and 4,4′-hexafluoroisopropylidene diphathalic anhydride(5) (Fig. 8.112). The fluoro-PIs exhibited low dielectric constants between 2.46 and 2.98, light color, and excellent high solubility. They exhibited T_gs of between 227 °C and 253 °C, and possessed a coefficient of thermal expansion (*CTE*) of $(60\text{-}88)\times10^{-6}$ $°C^{-1}$. Polymers PI-2, PI-3, PI-4, PI-5 showed excellent solubility in the organic solvents. The inherent viscosities of the PIs were found to range between 0.58 and 0.72 $dL\cdot g^{-1}$. Thermogravimetric analysis of the PIs revealed a high thermal stability decomposition temperature in excess of 500 °C in nitrogen. Temperature at 10% weight loss was found to be in the range 506-563 °C and 498-557 °C in nitrogen and air, respectively. The PI films had a tensile strength in the range of 75-87 MPa; tensile modulus, 1.5-2.2 GPa; and elongation at break, 6%-7%.

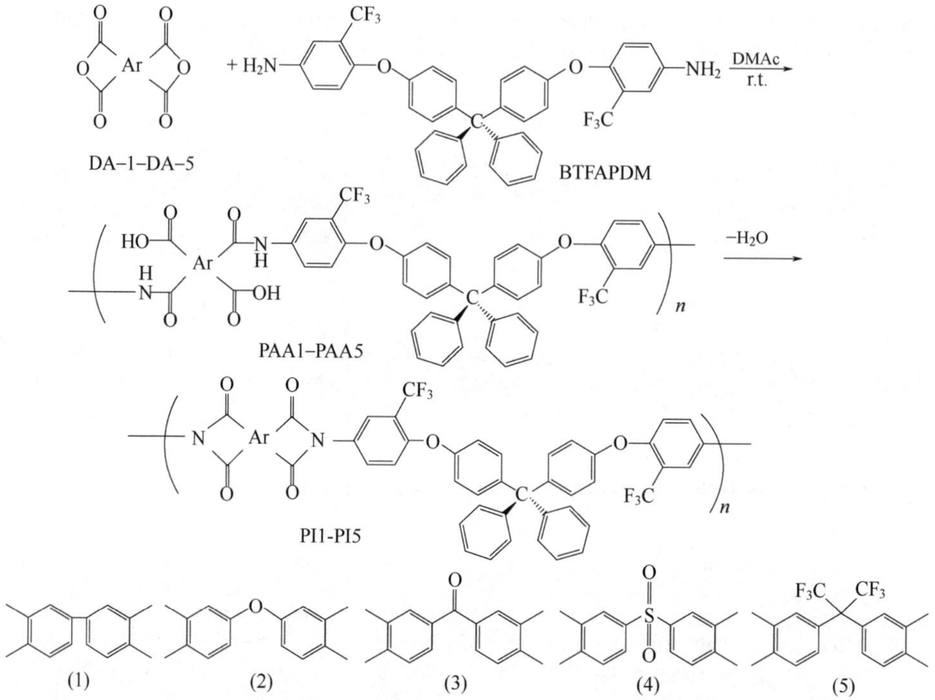

FIGURE 8.112 Synthesis of trifluoromethyl-containing polyimides.

8.4.6.6 Asymmetric Structure-Contained Polyimides

One attractive method for improved processability has been to incorporate geometrically asymmetric diamine or dianhydride, which may have positive effect to lower the dielectric properties of materials

8.4.6.7 Polyimides With Rigid and Nonplanar Large Conjugated Structures

Recently, a highly-efficient design strategy for high-performance intrinsic low-*k* PIs has been developed by Zhang and Xu's groups [238-242], using some propeller-like structures such as triphenylamine, triphenylmethane, and trisstyrene units, etc., to form a series of novel functional diamine monomers with rigid nonplanar large conjugated structures. By regulating the chemical structures and aggregation structures of the polymers at the molecular level, low dielectric and even ultralow dielectric constant PI films with excellent comprehensive properties can be achieved.

Liu et al. [238,239] designed and synthesized novel, simple but efficient diamine (TriPEDA and TetraPEDA) containing rigid nonplanar conjugated triphenylethylene and tetraphenylethylene moieties through Wittig-Horner and Suzuki coupling reactions. A series of high-performance functional PIs were thus prepared by the dipolymerization of TriPEDA/TetraPEDA and four dianhydrides, respectively (Fig. 8.113). Because of the introduction of the aromatic rigid nonplanar triphenylethylene/tetraphenylethylene structure, the PIs exhibited special fluorescent and resistive switching (ONOFF) characteristics, as the maximum fluorescence emission of the four PIs was observed at 425-505 nm in NMP solution and at 470-541 nm in film state. Also, these organo-soluble PIs showed outstanding properties, such as low dielectric constant (even without fluorinated substituents), light color, high T_gs (359-443 °C), thermal stability, and excellent mechanical properties.

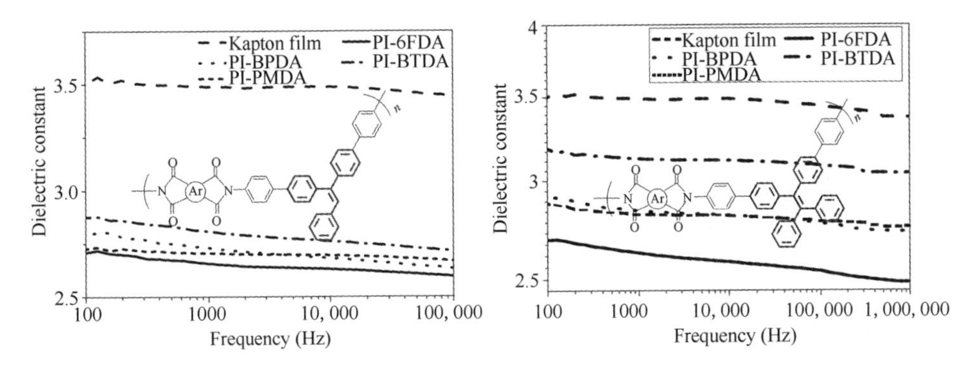

FIGURE 8.113 Low-*k* polyimide containing triphenylethylene and tetraphenylethylene moieties.

Chen et al. [240] prepared novel fluorinated aromatic PIs by the conventional two-step imidization of 4.4′(hexafluoroisopropylidene) diphthalic anhydride (6FDA) and diamines (TriPMPDA and TriPMMDA) bearing triphenyl methane moiety (Fig. 8.114). Both of the flexible and tough PI films exhibited an intrinsic low dielectric constant (low-*k*) values of 2.56 and 2.33 at the frequency of 10 kHz, respectively, due to the introduction of the bulky triphenyl methane side groups and the tortuous backbone structures. In addition, they showed light color, high thermal stability, moderate mechanical property, and more importantly, excellent solubility in common organic solvents (even completely dissolved in dichloromethane and chloroform). Thus, both of the functional PIs possessed attractive potential applications in the field of high performance flexible polymer interlayer materials.

A series of high-performance multifunctional PIs with exceptional thermostability and solubility (T_g as high as 494 °C and, at the same time, well soluble in common organic solvents) were successfully designed and synthesized by introducing a typical aromatic rigid trifluoromethyl-containing moiety with special nonplanar and conjugated characteristics into the polymer backbone (Fig. 8.115). Additionally, these PIs show light color (one even colorlessness) and transparency, intrinsic ultralow dielectric constant (k, k ≈ 1.93), and electrical bistability characteristics (ON/OFF ratio as high as 10^7,working voltage as low as 1.5 V) simultaneously. The excellent thermal stability and solubility allow them to undergo the high-temperature process (over 400 °C) in the preparation of photoelectric devices (like PVD or PECVD), or the highly efficient, continuous roll-to-roll process. The as-synthesized polymers are ideal potential candidates for practical applications in the fields of ULSI, high-performance polymer memory devices, flexible displays, thin-film PV industries, and wearable electronics [242].

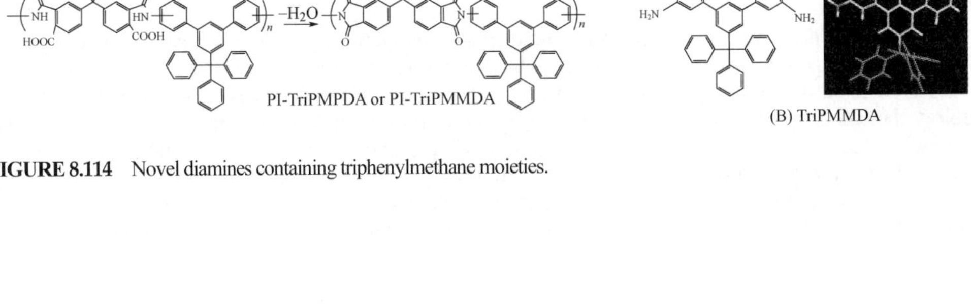

FIGURE 8.114 Novel diamines containing triphenylmethane moieties.

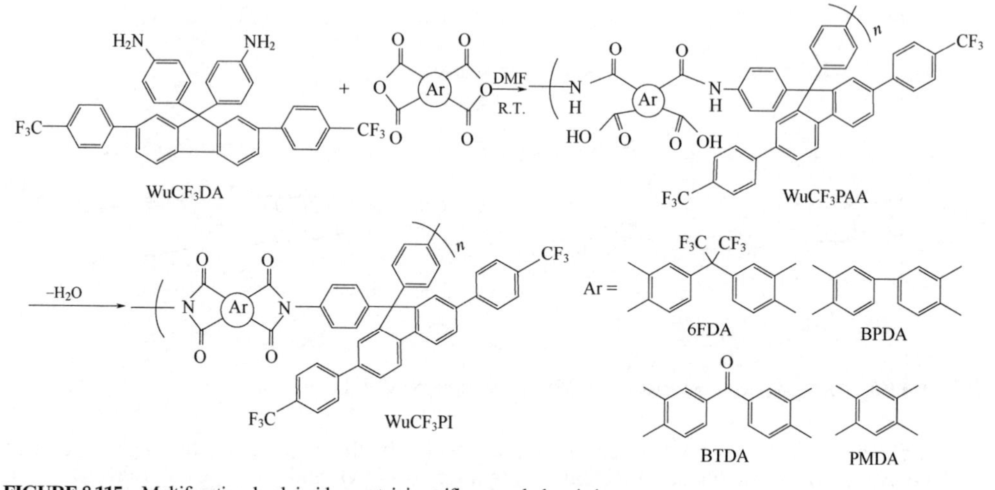

FIGURE 8.115 Multifunctional polyimides containing trifluoromethyl moieties.

A bulk dielectric polymer film with an intrinsic ultralow-*k*-value of 1.52 at 10 kHz has been successfully synthesized based on a novel polyimide FPTTPI [241] (Fig. 8.116). More importantly, such outstanding dielectric properties remain stable up to 280 °C. The excellent ultralow dielectric properties are mainly because of the larger free volume (subnanoscale), which intrinsically exists in the amorphous region of polymeric materials. Meanwhile, FPTTPI also shows excellent thermal stability and mechanical properties, with a T_g of 280 °C, 5%(*w*) loss temperature of 530 °C, and a residual of 63% at 800 °C under N_2. It was soluble in common solvents, which made it possible to undergo simple spin-on or efficient, low-cost and continuous roll-to-roll processes.

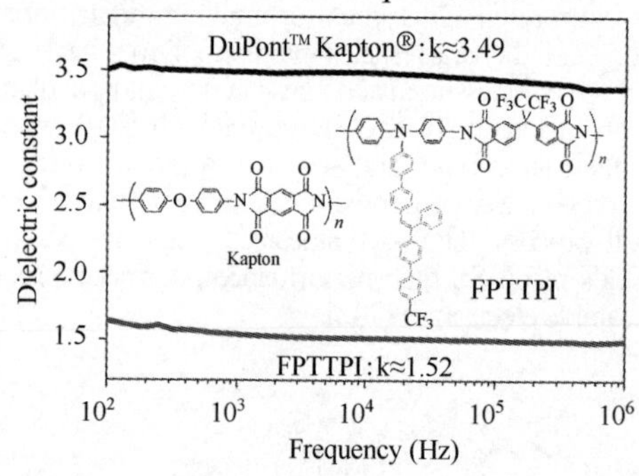

FIGURE 8.116 Ultralow-*k* polyimide containing triphenylamine and triphenylethylene moieties. *Reproduced with permission from reference [241]. Copyright 2015, American Chemical Society.*

8.4.7 Summary

A series of fruitful research work in the field of low dielectric PI materials have been performed, which effectively promoted the development of the semiconductor industry, and laid a good foundation for research for a new generation of ultralow dielectric constant PI materials. Low-dielectric or even ultralow dielectric PI materials can be obtained by the introduction of pore structures. However, these processes are relatively complex and difficult to control, and the production cost is high. Moreover, no matter what the method, pore structure, pore size, and pore distribution will have a significant impact on the uniformity of material properties, which is not conducive to large-scale preparation processes. In addition, the introduction of the whole structure will greatly reduce the mechanical properties and air impermeability of materials. The mechanical properties of the obtained materials are low and the barrier properties are poor, which cannot meet the needs of practical applications.

Compared with the porous PI materials, the intrinsic low-dielectric-constant PI materials have practical application values, both from the preparation process and the comprehensive performance. However, the dielectric constant of most of the developed low-k PI materials can only reach 2.3 or so, which still cannot meet the requirements of ultralarge-scale integrated circuits. A rigid ultralow dielectric constant PI with dielectric constant as low as 1.52 and excellent comprehensive performance has been developed by introducing a rigid nonplanar large conjugate structure into PI main chain in our group. On this basis, we hope to further develop a much simpler and more efficient design strategy to guide the preparation of high-performance ultralow-k dielectric PI. The research and development of the new generation of high-performance intrinsic ultralow dielectric constant PI materials is imminent, and its successful development will greatly promote the leap-forward development of the semiconductor industry in China and even the whole world, which has important theoretical and practical significance.

REFERENCES

[1] M. Ghosh, Polyimides: Fundamentals and Applications, CRC Press, Boca Raton, FL, 1996.

[2] K.L. Mittal, Polyimides: Synthesis, Characterization, and Applications, Springer Science & Business Media, Netherlands, 2013.

[3] J. Ren, Z. Li, R. Wang, Effects of the thermodynamics and rheology of BTDA-TDI/MDI co-polyimide (P84) dope solutions on the performance and morphology of hollow fiber UF membranes, J. Membr. Sci. 309 (1) (2008) 196-208.

[4] N.D. Ghatge, B.M. Shinde, U.P. Mulik, Polyimides from dianhydrides and bis(p-aminophenyl)alkane diamines. I, J. Polym. Sci., Polym. Chem. Ed. 22 (11) (1984) 3359-3365.

[5] S. Zhang, Y. Li, X. Wang, X. Zhao, S. Yu, D. Yin, et al., Synthesis and properties of novel polyimides derived from 2,6-bis(4-aminophenoxy-4′-benzoyl)pyridine with some of dianhydride monomers, Polymer 46 (25) (2005) 11986-11993.

[6] J. Connell, C. Croall, P. Hergenrother, Chemistry and properties of polyimides containing benzhydrol groups, Polym. Preprints(USA) 33 (1) (1992) 1101-1102.

[7] J. Malinge, J. Garapon, B. Sillion, New developments in polybenzhydrolimide resins: application in the field of heat resistant coatings, adhesives and laminates, Br. Polym. J. 20 (5) (1988) 431-439.

[8] Y. Kawashima, 3,3',4,4'-Diphenylsulfonetetracarboxylic dianhydride (DSDA)-based polyimides and their properties, Polyimides: Mater., Chem. Charact., Proc. Int. Conf. Polyimides, 3rd (1989) 123-137.

[9] P. Thiruvasagam, M. Vijayan, Synthesis of new diacid monomers and poly(amide-imide)s: study of structure-property relationship and applications, J. Polym. Res. 19 (3) (2012) 9845.

[10] L.I. Zhuo, J. Liu, Z. Gao, J. Chen, S. Yang, Synthesis and properties of organo-soluble poly(amide imide)s with high refractive indices, Acta Polym. Sinica 009 (1) (2009) 17-22.

[11] A.L. Rusanov, L.G. Komarova, M.P. Prigozhina, R.S. Begunov, O.I. Nozdracheva, Asymmetric benzophenone-series aromatic diamines and related soluble polyimides, Polym. Sci. Ser. B 50 (1) (2008) 6-10.

[12] X. Zhao, Q.-F. Geng, T.-H. Zhou, X.-H. Gao, G. Liu, Synthesis and characterization of novel polyimides derived from unsymmetric aldiamine: 2-Amino-5-[4-(2′-aminophenoxy)phenyl]-thiazole, Chin. Chem. Lett. 24 (1) (2013) 31-33.

[13] N. Mushtaq, G. Chen, L.R. Sidra, X. Fang, Organosoluble and high T_g polyimides from asymmetric diamines containing N-amino and N-aminophenyl naphthalimide moieties, RSC Adv. 6 (30) (2016) 25302-25310.

[14] S.H. Hsiao, K.H. Lin, Polyimides derived from novel asymmetric ether diamine, J. Polym. Sci. Part A: Polym. Chem. 43 (2) (2005) 331-341.

[15] Y. Oishi, K. Ogasawara, H. Hirahara, Synthesis of alicyclic polyimides by the silylation method, J. Photopolym. Sci. Technol. 14 (1) (2001) 37-40.

[16] M. Yamada, M. Kusama, T. Matsumoto, T. Kurosaki, Soluble polyimides with polyalicyclic structure. 2. Polyimides from bicyclo[2.2.1]heptane-2-exo-3-exo-5-exo-6-exo-tetracarboxylic 2,3:5,6-dianhydride, Macromolecules 26 (18) (1993) 4961-4963.

[17] J.G. Liu, M.H. He, H.W. Zhou, Z.G. Qian, F.S. Wang, S.Y. Yang, Organosoluble and transparent polyimides derived from alicyclic dianhydride and aromatic diamines, J. Polym. Sci., Part A: Polym. Chem. 40 (1) (2002) 110-119.

[18] Y.Z. Guo, D.X. Shen, H.J. Ni, J.G. Liu, S.Y. Yang, Organosoluble semi-alicyclic polyimides derived from 3,4-dicarboxy-1,2,3,4-tetrahydro-6-tert-butyl-1-naphthalene succinic dianhydride and aromatic diamines: synthesis, characterization and thermal degradation investigation, Prog. Org. Coat. 76 (4) (2013) 768-777.

[19] F.W. Harris, S.-H. Lin, F. Li, S.Z.D. Cheng, Organo-soluble polyimides: synthesis and polymerization of 2,2′-disubstituted-4,4′,5,5′-biphenyltetracarboxylic dianhydrides, Polymer 37 (22) (1996) 5049-5057.

[20] X. Zhao, J. Liu, Novel polyfluorinated polyimides derived from α,α-bis(4-amino-3,5- difluorophenyl)phenylmethane and aromatic dianhydrides: synthesis and characterization, Eur. Polym. J. 44 (3) (2008) 808-820.

[21] S.H. Park, K.J. Kim, W.W. So, S.J. Moon, S.B. Lee, Gas separation properties of 6FDA-based polyimide membranes with a polar group, Macromol. Res. 11 (3) (2003) 157-162.

[22] Y.S. Hsieh, C.R. Yang, G.Y. Hwang, Y.D. Lee, Preparation of organic soluble polyimides and their applications in KrF excimer laser LIGA process, Macromol. Chem. Phys. 202 (11) (2001) 2394-2401.

[23] G.H. Liu, X.P. Qiu, S.Q. Bo, X.L. Ji, Chain conformation and local rigidity of soluble polyimides(II): isomerized polyimides in THF, Chem. Res. Chin. Univ. 28 (2) (2012) 329-333.

[24] K. Han, K. You, E. Kim, J. Kim, W.-H. Jang, T.H. Rhee, Polymeric optical waveguides using fluorinated polyimides, Mol. Cryst. Liq. Cryst. 349 (1) (2000) 71-74.

[25] K. Xie, S. Zhang, J. Liu, M. He, S. Yang, Synthesis and characterization of soluble fluorine-containing polyimides based on 1,4-bis (4-amino-2-trifluoromethylphenoxy) benzene, J. Polym. Sci., Part A: Polym. Chem. 39 (15) (2001) 2581-2590.

[26] H. Zhou, J. Liu, Z. Qian, S. Zhang, S. Yang, Soluble fluorinated polyimides derived from 1,4-(4′-aminophenoxy)-2-(3′-trifluoromethylphenyl) benzene and aromatic dianhydrides, J. Polym. Sci., Part A: Polym. Chem. 39 (14) (2001) 2404-2413.

[27] C.-P. Yang, R.-S. Chen, H.-C. Chiang, Organosoluble and light-colored fluorinated polyimides based on 1, 2-bis (4-amino-2-trifluoromethylphenoxy) benzene and aromatic dianhydrides, Polym. J. 35 (8) (2003) 662-670.

[28] C.P. Yang, F.Z. Hsiao, Synthesis and properties of fluorinated polyimides based on 1, 4-bis (4-amino-2-trifluoromethylphenoxy)-2, 5-di-tertbutylbenzene and various aromatic dianhydrides, J. Polym. Sci., Part A: Polym. Chem. 42 (9) (2004) 2272-2284.

[29] C.P. Yang, Y.Y. Su, K.L. Wu, Synthesis and properties of new aromatic polyimides based on 2, 5-bis (4-amino-2-trifluoromethylphenoxy)-tertbutylbenzene and various aromatic dianhydrides, J. Polym. Sci., Part A: Polym. Chem. 42 (21) (2004) 5424-5438.

[30] Z. Qiu, J. Wang, Q. Zhang, S. Zhang, M. Ding, L. Gao, Synthesis and properties of soluble polyimides based on isomeric ditrifluoromethyl substituted 1, 4-bis (4-aminophenoxy) benzene, Polymer 47 (26) (2006) 8444-8452.

[31] Y. Shao, Y.F. Li, X. Zhao, X.L. Wang, T. Ma, F.C. Yang, Synthesis and properties of fluorinated polyimides from a new unsymmetrical diamine: 1,4-(2′-Trifluoromethyl-4′, 4′-diaminodiphenoxy) benzene, J. Polym. Sci., Part A: Polym. Chem. 44 (23) (2006) 6836-6846.

[32] C.P. Yang, Y.Y. Su, Y.C. Chen, Light-colored fluorinated polyimides based on 2, 5-bis (4-amino-2-trifluoromethylphenoxy) biphenyl and various aromatic dianhydrides, J. Appl. Polym. Sci. 102 (5) (2006) 4101-4110.

[33] C.-P. Yang, Y.-Y. Su, H.-C. Chiang, Organosoluble and light-colored fluorinated polyimides from 4-tert-butyl-[1, 2-bis (4-amino-2-trifluoromethylphenoxy) phenyl] benzene and aromatic dianhydrides, Reactive Funct. Polym. 66 (7) (2006) 689-701.

[34] Q. Mi, Y. Ma, L. Gao, M. Ding, Synthesis and characterization of optically active aromatic polyimides derived from 2,2′-bis(2-trifluoro-4-aminophenoxy)-1,1′-binaphthyl and aromatic tetracarboxylic dianhydrides, J. Polym. Sci., Part A: Polym. Chem. 37 (24) (1999) 4536-4540.

[35] C.P. Yang, S.H. Hsiao, M.F. Hsu, Organosoluble and light-colored fluorinated polyimides from 4,4′-bis (4-amino-2-trifluoromethylphenoxy) biphenyl and aromatic dianhydrides, J. Polym. Sci., Part A: Polym. Chem. 40 (4) (2002) 524-534.

[36] C.-P. Yang, S.-H. Hsiao, K.-H. Chen, Organosoluble and optically transparent fluorine-containing polyimides based on 4,4′-bis(4-amino-2-trifluoromethylphenoxy)-3,3′,5, 5′-tetramethylbiphenyl, Polymer 43 (19) (2002) 5095-5104.

[37] D.-J. Liaw, C.-C. Huang, W.-H. Chen, Color lightness and highly organosoluble fluorinated polyamides, polyimides and poly (amide-imide) s based on noncoplanar 2, 2′-dimethyl-4, 4′-biphenylene units, Polymer 47 (7) (2006) 2337-2348.

[38] J.-G. Liu, X.-J. Zhao, L. Fan, S.-Y. Yang, G.-L. Wu, F.-Q. Zhang, et al., Organo-soluble fluorinated polyimides derived from 4,4″-bis (3′amino-5-triuoromethylphenoxy)-3,3′,5,5′-tetramethyl biphenyl (TFMDA) and aromatic dianhydrides, High Perform. Polym. 18 (2) (2006) 145-161.

[39] S.H. Hsiao, C.P. Yang, C.L. Chung, Synthesis and characterization of novel fluorinated polyimides based on 2,7-bis (4-amino-2-trifluoromethylphenoxy) naphthalene, J. Polym. Sci., Part A: Polym. Chem. 41 (13) (2003) 2001-2018.

[40] C.-P. Yang, S.-H. Hsiao, K.-L. Wu, Organosoluble and light-colored fluorinated polyimides derived from 2, 3-bis (4-amino-2-trifluoromethylphenoxy) naphthalene and aromatic dianhydrides, Polymer 44 (23) (2003) 7067-7078.

[41] S.H. Hsiao, C.P. Yang, S.C. Huang, Preparation and properties of new polyimides and polyamides based on 1,4-bis (4-amino-2-trifluoromethylphenoxy) naphthalene, J. Polym. Sci., Part A: Polym. Chem. 42 (10) (2004) 2377-2394.

[42] S.-H. Hsiao, C.-P. Yang, S.-C. Huang, Polyimides from 1,5-bis (4-amino-2-trifluoromethyl- phenoxy) naphthalene and aromatic tetracarboxylic dianhydrides, Eur. Polym. J. 40 (6) (2004) 1063-1074.

[43] C.P. Yang, S.H. Hsiao, C.L. Chung, Organosoluble, low-dielectric-constant fluorinated polyimides based on 2,6-

bis(4-amino-2-trifluoromethylphenoxy) naphthalene, Polym. Int. 54 (4) (2005) 716-724.

[44] C.-L. Chung, S.-H. Hsiao, Novel organosoluble fluorinated polyimides derived from 1, 6-bis (4-amino-2-trifluoromethylphenoxy) naphthalene and aromatic dianhydrides, Polymer 49 (10) (2008) 2476-2485.

[45] C.L. Chung, W.F. Lee, C.H. Lin, S.H. Hsiao, Highly soluble fluorinated polyimides based on an asymmetric bis (ether amine): 1,7-bis(4-amino-2-rifluoromethylphenoxy) naphthalene, J. Polym. Sci., Part A: Polym. Chem. 47 (7) (2009) 1756-1770.

[46] S.-H. Hsiao, W. Guo, C.-L. Chung, W.-T. Chen, Synthesis and characterization of novel fluorinated polyimides derived from 1, 3-bis (4-amino-2-trifluoromethylphenoxy) naphthalene and aromatic dianhydrides, Eur. Polym. J. 46 (9) (2010) 1878-1890.

[47] Y. Shang, L. Fan, S. Yang, X. Xie, Synthesis and characterization of novel fluorinated polyimides derived from 4-phenyl-2,6-bis[4-(4′-amino-2′-trifluoromethyl-phenoxy)phenyl]pyridine and dianhydrides, Eur. Polym. J. 42 (5) (2006) 981-989

[48] D. Yin, Y. Li, H. Yang, S. Yang, L. Fan, J. Liu, Synthesis and characterization of novel polyimides derived from 1,1-bis[4-(4′-aminophenoxy) phenyl]-1-[3″,5″-bis(trifluoromethyl)phenyl] -2,2,2-trifluoroethane, Polymer 46 (9) (2005) 3119-3127.

[49] S.Y. Yang, Z.Y. Ge, D.X. Yin, J.G. Liu, Y.F. Li, L. Fan, Synthesis and characterization of novel fluorinated polyimides derived from 4,4′-[2,2,2-trifluoro-1-(3-trifluoromethylphenyl)ethylidene]diphthalic anhydride and aromatic diamines, J. Polym. Sci., Part A: Polym. Chem. 42 (17) (2004) 4143-4152.

[50] D. Yin, Y. Li, Y. Shao, X. Zhao, S. Yang, L. Fan, Synthesis and characterization of soluble polyimides based on trifluoromethylated aromatic dianhydride and substitutional diaminetriphenylmethanes, J. Fluorine Chem. 126 (5) (2005) 819-823.

[51] K. Xie, S.Y. Zhang, J.G. Liu, H.E. Min Hui, S.Y. Yang, New soluble fluorinated polyimides, Chin. Chem. Lett. 11 (12) (2000) 1049-1052.

[52] H. Wang, T. Wang, S. Yang, F. Lin, Preparation of thermal stable porous polyimide membranes by phase inversion process for lithium-ion battery, Polymer 54 (23) (2013) 6339-6348.

[53] F. Sun, T. Wang, S. Yang, F. Lin, Synthesis and characterization of sulfonated polyimides bearing sulfonated aromatic pendant group for DMFC applications, Polymer 51 (17) (2010) 3887-3898.

[54] Z. Li, J. Liu, Z. Gao, Z. Yin, L. Fan, S. Yang, Organo-soluble and transparent polyimides containing phenylphosphine oxide and trifluoromethyl moiety: synthesis and characterization, Eur. Polym. J. 45 (4) (2009) 1139-1148.

[55] B.C. Auman, D.P. Higley Jr, K.V. Scherer, E.F. Mccord, W.H. Shaw Jr, Synthesis of a new fluoroalkylated diamine, 5-[1 H, 1 H-2-bis(trifluoromethyl)-heptafluoropentyl]-1,3-phenylenediamine, and polyimides prepared therefrom, Polymer 36 (3) (1995) 651-656.

[56] Y.S. Kim, J.C. Jung, Synthesis and properties of polyimides derived from 9,10-dialkyloxy-1,2,3,4,5,6,7,8-octahydro-2, 3,6,7-anthracenetetracarboxylic-2,3:6,7-dianhydrides, J. Polym. Sci., Part A: Polym. Chem. 40 (11) (2002) 1764-1774.

[57] W. Huang, D. Yan, Q. Lu, Synthesis and characterization of a highly soluble aromatic polyimide from 4,4′-methylenebis(2-tert-butylaniline), Macromol. Rapid Commun. 22 (18) (2015) 1481-1484.

[58] W. Huang, D. Yan, Q. Lu, P. Tao, Preparation of aromatic polyimides highly soluble in conventional solvents, J. Polym. Sci., Part A: Polym. Chem. 40 (2) (2002) 229-234.

[59] L. Yi, C. Li, W. Huang, D. Yan, Soluble aromatic polyimides with high glass transition temperature from benzidine containing tert-butyl groups, J. Polym. Res. 21 (11) (2014) 572.

[60] L. Yi, C. Li, W. Huang, D. Yan, Soluble polyimides from 4, 4′-diaminodiphenyl ether with one or two tert-butyl pedant groups, Polymer 80 (2015) 67-75.

[61] L. Yi, C. Li, W. Huang, D. Yan, Soluble and transparent polyimides with high T_g from a new diamine containing tert-butyl and fluorene units, J. Polym. Sci., Part A: Polym. Chem. 54 (7) (2016) 976-984.

[62] S.H. Hsiao, Y.H. Chang, Synthesis and properties of soluble trifluoromethyl-substituted polyimides containing laterally attached p -Terphenyls, J. Polym. Sci., Part A: Polym. Chem. 42 (5) (2004) 1255-1271.

[63] D.J. Liaw, F.C. Chang, M.K. Leung, M.Y. Chou, K. Muellen, High thermal stability and rigid rod of novel organosoluble polyimides and polyamides based on bulky and noncoplanar naphthalene-biphenyldiamine, Macromolecules 38 (9) (2005) 4024-4029.

[64] Z. Qiu, S. Zhang, Synthesis and properties of organosoluble polyimides based on 2, 2′-diphenoxy-4, 4′, 5, 5′-biphenyltetracarboxylic dianhydride, Polymer 46 (5) (2005) 1693-1700.

[65] K. Zeng, Q. Guo, S. Gao, D. Wu, H. Fan, G. Yang, Studies on organosoluble polyimides based on a series of new asymmetric and symmetric dianhydrides: structure/solubility and thermal property relationships, Macromol. Res. 20 (1) (2012) 10-20.

[66] M. Koton, Y.N. Sazanov, Thermogravimetric study of the effect of the chemical structure of polyimides on their thermal stability, J. Therm. Anal. Calorim. 7 (1) (1975) 165-171.

[67] F. Li, J.G. Jason, P.S. Honigfort, S. Fang, J.-C. Chen, F.W. Harris, et al., Dianhydride architectural effects on the relaxation

behaviors and thermal and optical properties of organo-soluble aromatic polyimide films, Polymer 40 (18) (1999) 4987-5002.
[68] F.W. Harris, Y. Sakaguchi, M. Shibata, S.Z. Cheng, Organo-soluble polyimides: synthesis and characterization of polyimides containing phenylated p-biphenyl and p-terphenyl units, High Perform. Polym. 9 (3) (1997) 251-261.
[69] S.-H. Hsiao, Y.-J. Chen, Structure-property study of polyimides derived from PMDA and BPDA dianhydrides with structurally different diamines, Eur. Polym. J. 38 (4) (2002) 815-828.
[70] D.J. Liaw, B.Y. Liaw, Synthesis and properties of polyimides derived from 1,4-bis(4-aminophenoxy)2,5-di-tert-butylbenzene, J. Polym. Sci., Part A: Polym. Chem. 35 (8) (1997) 1527-1534.
[71] D.H. Kim, J.C. Jung, Synthesis and properties of polyimides having alkyloxycarbonyl side chain, Polym. Bull. 50 (5) (2003) 311-318.
[72] Y. Lu, G. Xiao, H. Chi, Y. Dong, Z. Hu, Effects of tert-butyl substitutes of fluorinated diamine on the properties of polyimides, High Perform. Polym. 25 (8) (2013) 894-900.
[73] X. Fang, Z. Yang, S. Zhang, L. Gao, M. Ding, Polyimides derived from mellophanic dianhydride, Macromolecules 35 (23) (2002) 8708-8717.
[74] Y. Lu, G. Xiao, Z. Hu, Y. Wang, Q. Fang, Synthesis and properties of soluble and transparent fluorine-containing polyimide thin films, Polym. Mater. Sci. Eng. 29 (2) (2013) 13-16.
[75] A. Qin, Z. Yang, F. Bai, C. Ye, Design and synthesis of a thermally stable second-order nonlinear optical chromophore and its poled polymers, J. Polym. Sci., Part A: Polym. Chem. 41 (41) (2003) 2846-2853.
[76] L. Dalton, Nonlinear Optical Polymeric Materials: From Chromophore Design to Commercial Applications, Springer Berlin Heidelberg, Heidelberg, 2002.
[77] J. Luo, Y.J. Cheng, T.D. Kim, S. Hau, S.H. Jang, Z. Shi, et al., Facile synthesis of highly efficient phenyltetraene-based nonlinear optical chromophores for electrooptics, Org. Lett. 8 (7) (2006) 1387-1390.
[78] M.A. Mortazavi, B.G. Higgins, A. Dienes, A. Knoesen, S.T. Kowel, Second-harmonic generation and absorption studies of polymer-dye films oriented by corona-onset poling at elevated temperatures. J. Opt. Soc. Am. B: Optical Phys., 6 (4) (1989), 733-741.
[79] R.J. Jeng, C.C. Chang, C.P. Chen, C.T. Chen, W.C. Su, Thermally stable crosslinked NLO materials based on maleimides, Polymer 44 (1) (2003) 143-155.
[80] R.J. Jeng, W.Y. Hung, C.P. Chen, G.H. Hsiue, Organic/Inorganic NLO materials based on reactive polyimides and a bulky alkoxysilane dye via sol/Gel process, Polym. Adv. Technol. 14 (1) (2003) 66-75.
[81] J. Lu, J. Yin, Synthesis and characterization of photocrosslinkable, side-chain, second-order nonlinear optical poly(ester imide)s with great film-forming ability and long-term dipole orientation stability, J. Polym. Sci., Part A: Polym. Chem. 41 (2) (2003) 303-312.
[82] J.G. Liu, Y. Nakamura, Y. Shibasaki, S. Ando, M. Ueda, Synthesis and characterization of highly refractive polyimides from 4,4′-thiobis[(pphenylenesulfanyl) aniline] and various aromatic tetracarboxylic dianhydrides, J. Polym. Sci., Part A: Polym. Chem. 45 (23) (2010) 5606-5617.
[83] D.R. Robello, Linear polymers for nonlinear optics. I. Polyacrylates bearing aminonitro-stilbene and -azobenzene dyes, J. Polym. Sci., Part A: Polym. Chem. 28 (1) (1990) 1-13.
[84] C. Ye, J. Wang, Z. Feng, Synthesis of NLO active polystyrene derivatives with p -(40-nitro-4-alkoxyl)stilbene, Synth. Met. 57 (1) (1993) 3951-3954.
[85] G. Lee, D. Won, J. Lee, Synthesis and electro-optic properties of novel polyester containing dioxybenzylidenecyanoacetate as a NLOchromophore, Mol. Cryst. Liq. Cryst. 504 (1) (2009) 189-195.
[86] N. Tsutsumi, O. Matsumoto, A. Wataru Sakai, T. Kiyotsukuri, Nonlinear optical polymers. 2. Novel NLO linear polyurethane with dipole moments aligned transverse to the main backbone, Macromolecules 29 (2) (1996) 592-597.
[87] W.J. Kuo, M.C. Chang, T.Y. Juang, C.P. Chen, C.T. Chen, H.L. Chang, et al., Stable second-order NLO semi-IPN system based on bipyridine-containing polyimide and alkoxysilane dye, Polym. Adv. Technol. 16 (7) (2005) 515-523.
[88] J.C. Min, H.C. Dong, P.A. Sullivan, A.J.P. Akelaitis, L.R. Dalton, Recent progress in second-order nonlinear optical polymers and dendrimers, Prog. Polym. Sci. 33 (11) (2008) 1013-1058.
[89] T. Verbiest, D.M. Burland, M.C. Jurich, V.Y. Lee, R.D. Miller, W. Volksen, Exceptionally thermally stable polyimides for second-order nonlinear optical applications, Science 268 (5217) (1995) 1604-1606.
[90] H.C. Tsai, W.J. Kuo, G.H. Hsiue, Highly thermal stable main-chain nonlinear optical polyimide based on two-dimensional carbazole chromophores, Macromol. Rapid Commun. 26 (12) (2005) 986-991.
[91] H. Saadeh, D. Yu, L.M. Wang, L.P. Yu, Highly stable, functionalized polyimides for second order nonlinear optics, J. Mater. Chem. 9 (9) (1999) 1865-1873.
[92] M. He, Y. Zhou, Y. Gao, Y. Wang, X. Bu, T. Zhang, et al., Molecular design and synthesis of branched bichromophore-attached linear fluorinated polyimides for nonlinear optical applications, J. Mater. Sci. 48 (9) (2013) 3370-3377.
[93] J.Y. Do, S.K. Park, J.J. Ju, M.S. Kim, S. Park, M.H. Lee, et al., Alkyl sulfone-containing optical polyimide for an efficient

wavelength conversion, Opt. Mater. 29 (12) (2007) 1563-1570.
[94] Y. Sui, D. Wang, J. Yin, G.Z. Tan, Z.K. Zhu, Z.G. Wang, Side-chain second-order nonlinear optical poly(urethane-imide)/ photosensitive polyimide blends with the improved dipole orientation stability by photo-crosslinking, Mater. Lett. 52 (1-2) (2002) 53-56.
[95] R. Compañó, Trends in nanoelectronics, Nanotechnology 12 (2) (2001) 85.
[96] Y. Yang, J. Ouyang, L. Ma, R.J.H. Tseng, C.W. Chu, Electrical switching and bistability in organic/polymeric thin films and memory devices, Adv. Funct. Mater. 16 (8) (2010) 1001-1014.
[97] Q.D. Ling, D.J. Liaw, C. Zhu, S.H. Chan, E.T. Kang, K.G. Neoh, Polymer electronic memories: materials, devices and mechanisms, Prog. Polym. Sci. 33 (10) (2008) 917-978.
[98] Q.D. Ling, D.J. Liaw, Y.H. Teo, C. Zhu, S.H. Chan, E.T. Kang, et al., Polymer memories: bistable electrical switching and device performance, Polymer 48 (18) (2007) 5182-5201.
[99] L.P. Ma, J. Liu, Y. Yang, Organic electrical bistable devices and rewritable memory cells, Appl. Phys. Lett. 80 (16) (2002) 2997-2999.
[100] C.H. Tu, D.L. Kwong, Y.S. Lai, Negative differential resistance and electrical bistability in nanocrystal organic memory devices, Appl. Phys. Lett. 89 (25) (2006) 062105.
[101] Y.L. Liu, Q.D. Ling, E.T. Kang, K.G. Neoh, D.J. Liaw, K.L. Wang, et al., Volatile electrical switching in a functional polyimide containing electron-donor and -acceptor moieties, J. Appl. Phys. 105 (4) (2009) 31.
[102] T.J. Lee, C.W. Chang, S.G. Hahm, K. Kim, S. Park, D.M. Kim, et al., Programmable digital memory devices based on nanoscale thin films of a thermally dimensionally stable polyimide, Nanotechnology 20 (13) (2009) 135204.
[103] D.M. Kim, S. Park, T.J. Lee, S.G. Hahm, K. Kim, J.C. Kim, et al., Programmable permanent data storage characteristics of nanoscale thin films of a thermally stable aromatic polyimide, Langm Acs J. Surf. Colloids 25 (19) (2009) 11713-11719.
[104] K. Kim, H.J. Yen, Y.G. Ko, C.W. Chang, W. Kwon, G.S. Liou, et al., Electrically bistable digital memory behaviors of thin films of polyimides based on conjugated bis(triphenylamine) derivatives, Polymer 53 (19) (2012) 4135-4144.
[105] Q. Liu, K. Jiang, L. Wang, Y. Wen, J. Wang, Y. Ma, et al., Distinct electronic switching behaviors of triphenylamine-containing polyimide memories with different bottom electrodes, Appl. Phys. Lett. 96 (21) (2010) 106.
[106] T. Kurosawa, A.D. Yu, T. Higashihara, W.C. Chen, M. Ueda, Inducing a high twisted conformation in the polyimide structure by bulky donor moieties for the development of non-volatile memory, Eur. Polym. J. 49 (10) (2013) 3377-3386.
[107] C.J. Chen, H.J. Yen, W.C. Chen, G.S. Liou, Novel high-performance polymer memory devices containing (OMe) 2 tetraphenyl-pphenylenediamine moieties, J. Polym. Sci., Part A: Polym. Chem. 49 (17) (2011) 3709-3718.
[108] Y. Song, H. Yao, H. Tan, S. Zhu, B. Dong, S. Guan, Changing the memory behaviors from volatile to nonvolatile via end-capping of hyperbranched polyimides with polycyclic arenes, Dyes Pigments 139 (2017) 730-736.
[109] S.G. Hahm, S. Choi, S.H. Hong, T.J. Lee, S. Park, M.K. Dong, et al., Novel rewritable, non-volatile memory devices based on thermally and dimensionally stable polyimide thin films, Adv. Funct. Mater. 18 (20) (2010) 3276-3282.
[110] L. Shi, G. Tian, H. Ye, S. Qi, D. Wu, Volatile static random access memory behavior of an aromatic polyimide bearing carbazole-tethered triphenylamine moieties, Polymer 55 (5) (2014) 1150-1159.
[111] L. Shi, N. Jia, L. Kong, S. Qi, D. Wu, Tuning resistive switching memory behavior from non-volatile to volatile by phenoxy linkages in soluble polyimides containing carbazole-tethered triazole groups, Macromol. Chem. Phys. 215 (23) (2014) 2374-2388.
[112] Y. Li, H. Xu, X. Tao, K. Qian, S. Fu, Y. Shen, et al., Synthesis and memory characteristics of highly organo-soluble polyimides bearing a noncoplanar twisted biphenyl unit containing aromatic side-chain groups, J. Mater. Chem. 21 (6) (2011) 1810-1821.
[113] G. Tian, S. Qi, F. Chen, L. Shi, W. Hu, D. Wu, Nonvolatile memory effect of a functional polyimide containing ferrocene as the electroactive moiety, Appl. Phys. Lett. 98 (20) (2011) 92.
[114] A. Lien, H. Takano, S. Suzuki, H. Uchlda, The symmetry property of a 90° twisted nematic liquid crystal cell, Mol. Cryst. Liq. Cryst. 198 (1) (1991) 37-49.
[115] D.T. Grubb, I.Mita, D.Y. Yoon, Materials Science of High Temperature Polymers for Microelectronics, 1991.
[116] L.A. Hornak, Polymers for Lightwave and Integrated Optics: Technology and Applications, 1992.
[117] F. Li, F.W. Harris, S.Z.D. Cheng, Polyimide films as negative birefringent compensators for normally white twisted nematic liquid crystal displays, Polymer 37 (23) (1996) 5321-5325.
[118] H. Tong, C. Hu, S. Yang, Y. Ma, H. Guo, L. Fan, Preparation of fluorinated polyimides with bulky structure and their gas separation performance correlated with microstructure, Polymer 69 (2015) 138-147.
[119] S. Basu, A. Cano, Asymmetric Matrimid® /[Cu 3(BTC) 2] mixed-matrix membranes for gas separations, J. Membr. Sci. 362 (1) (2010) 478-487.
[120] M.J.C. Ordoñez Jr, K.J. Balkus, J.P. Ferraris, I.H. Musselman, Molecular sieving realized with ZIF-8/Matrimid®; mixed-matrix membranes, J. Membr. Sci. 361 (1-2) (2010) 28-37.

[121] Y. Zhang, I.H. Musselman, J.P. Ferraris, K.J. Balkus Jr, Gas permeability properties of Matrimid®; membranes containing the metal-organic framework Cu-BPY-HFS, J. Membr. Sci. 313 (1-2) (2008) 170-181.

[122] T.S. Chung, S.S. Chan, R. Wang, Z. Lu, C. He, Characterization of permeability and sorption in Matrimid/C60 mixed matrix membranes, J. Membr. Sci. 211 (1) (2003) 91-99.

[123] A.L. Khan, C. Klaysom, A. Gahlaut, A.U. Khan, I.F.J. Vankelecom, Mixed matrix membranes comprising of Matrimid and -SO 3 H functionalized mesoporous MCM-41 for gas separation, J. Membr. Sci. 447 (447) (2013) 73-79.

[124] S.W. Lee, I.K. Sang, B. Lee, W. Choi, B. Chae, S.B. Kim, et al., Photoreactions and photoinduced molecular orientations of films of a photoreactive polyimide and their alignment of liquid crystals, Macromolecules 36 (17) (2003) 6527-6536.

[125] A.T. Haldeman, J.R. Dimaio, M. Shaughnessy, T. Duniho, B. Sawders, Tetrimide™: soluble polyimide opitcal fiber coatings for avionics; proceedings of the Avionics. Avionics, Fiber-Optics and Photonics Technology Conference(AVFOP), Atlanta, GA, USA, 2014.

[126] T. Tasaki, A. Shiotani, M. Tsuji, T. Nakamura, T. Yamaguchi, The low Dk / Df adhesives for high frequency printed circuit board using the novel solvent soluble polyimide; proceedings of the Microsystems, Packaging. Microsystems, Packaging, Assembly and Circuits Technology Conference (IMPACT), Taipei, Taiwan, 2015.

[127] X. Qiao, T.S. Chung, K.P. Pramoda, Fabrication and characterization of BTDA-TDI/MDI (P84) co-polyimide membranes for the pervaporation dehydration of isopropanol, J. Membr. Sci. 264 (1-2) (2005) 176-189.

[128] R.D. Miller, Device physics: in search of low-*k* dielectrics, Science 286 (5439) (1999) 421-423.

[129] G. Maijer, Low dielectric constant polymers for microelectronics, Prog. Polym. Sci. 26 (2001) 3-65.

[130] A. Grill, S.M. Gates, T.E. Ryan, S.V. Nguyen, D. Priyadarshini, Progress in the development and understanding of advanced low-*k* and ultralow-*k* dielectrics for very large-scale integrated interconnects—state of the art, Appl. Phys. Rev. 1 (2014) 011306.

[131] W. Volksen, R.D. Miller, D. Geraud, Low dielectric constant materials, Chem. Rev. 110 (2010) 56-110.

[132] D. Shamiryan, T. Abell, F. Iacopi, K. Maex, Low-*k* dielectric materials, Mater. Today 7 (2004) 34-39.

[133] M. Morgen, E.T. Ryan, J.H. Zhao, C. Hu, T.H. Cho, P.S. Ho, Low dielectric constant materials For ULSI interconnects, Annu. Rev. Mater. Sci. 30 (2000) 645-680.

[134] F. Hoffmann, M. Cornelius, J. Morell, M. Froba, Silica-based mesoporous organic-inorganic hybrid materials, Angewan. Chem. Int. Ed. 45 (20) (2006) 3216-3251.

[135] Z. Li, M.C. Johnson, M. Sun, E.T. Ryan, D.J. Earl, W. Maichen, et al., Mechanical and dielectric properties of pure-silica-zeolite low-*k* materials, Angewan. Chem. Int. Ed. 45 (38) (2006) 6329-6332.

[136] M. Seino, W. Wang, J.E. Lofgreen, D.P. Puzzo, T. Manabe, G.A. Ozin, Low-*k* periodic mesoporous organosilica with air walls: POSS-PMO, J. Am. Chem. Soc. 133 (45) (2011) 18082-18085.

[137] H.J. Lee, E.K. Lin, H. Wang, W.L. Wu, W. Chen, E.S. Moyer, Structural comparison of hydrogen silsesquioxane based porous low-*k* thin films prepared with varying process conditions, Chem. Mater. 14 (2002) 1845-1852.

[138] Z.B. Wang, H.T. Wang, A.P. Mitra, L.M. Huang, Y.S. Yan, Pure-silica zeolite low-*k* dielectric thin films, Adv. Mater. 13 (2001) 746-749.

[139] Y.Z. Zhu, T.E. Müller, J.A. Lercher, Single step preparation of novel hydrophobic composite films for low-*k* applications, Adv. Funct. Mater. 18 (21) (2008) 3427-3433.

[140] J.J. Ge, C.Y. Li, G. Xue, I.K. Mann, D. Zhang, F.W. Harris, et al., Rubbing-induced molecular reorientation on an alignment surface of an aromatic polyimide containing cyanobiphenyl side chains, J. Am. Chem. Soc. 123 (2001) 5768-5776.

[141] H. Lim, W.J. Cho, C.S. Ha, S. Ando, Y.K. Kim, C.H. Park, et al., Flexible organic electroluminescent devices based on fluo-rine-containing colorless polyimide substrates, Adv. Mater. 14 (2002) 1275-1279.

[142] Z. Wang, A. Mitra, H. Wang, L. Huang, Y. Yan, Pure silica zeolite films as low-*k* dielectrics by spin-on of nanoparticle suspensions, Adv. Mater. 13 (2001) 1463-1466.

[143] K. Maex, M.R. Baklanov, D. Shamiryan, F. lacopi, S.H. Brongersma, Z.S. Yanovitskaya, Low dielectric constant materials for microelectronics, J. Appl. Phys. 93 (11) (2003) 8793-8841.

[144] H.S. Lee, A.S. Lee, K.-Y. Baek, S.S. Hwang, Low Dielectric Materials for Microelectronics, 2012.

[145] H. Seino, O. Haba, M. Ueda, A. Mochizukib, Photosensitive polyimide-precursor based on polyisoimide: dimensionally stable polyimide with a low dielectric constant, Polymer 40 (3) (1999) 551-558.

[146] P.W. Atkins, Physical Chemistry, fifth ed, Oxford University Press, New York, 1994.

[147] J.D. Livingston, Electronic Properties of Engineering Materials, Willey, New York, 1999.

[148] R.P. Feynman, R.B. Leighton, M. Sands, The Feynman Lectures on Physics., Adison Wesley, Reading, PA, 1966.

[149] P.W. Atkins, R.S. Friedman, Molecular Quantum Mechanics, 3rd ed., Oxford University Press, New York, 1997.

[150] L.S. Loo, K.K. Gleason, Hot filament chemical vapor deposition of polyoxymethylene as a sacrificial layer for fabricating air gaps, Electrochem. Solid-State Lett. 4 (11) (2001) G81-G84.

[151] P.A. Kohl, D.M. Bhusari, M. Wedlake, C. Case, F.P. Klemens, J. Miner, et al., Air-gaps in 0.3 mu m electrical

interconnections, IEEE Electron Device Lett. 21 (12) (2000) 557-559.

[152] A.M. Padovani, L. Rhodes, L. Riester, G. Lohman, B. Tsuie, J. Conner, et al., Porous methylsilsesquioxane for low-*k* dielectric applications, Electrochem. Solid-State Lett. 4 (11) (2001) F25-F28.

[153] C.V. Nguyen, K.R. Carter, C.J. Hawker, J.L. Hedrick, R.L. Jaffe, R.D. Miller, et al., Low-dielectric, nanoporous organosilicate films prepared via inorganic/organic polymer hybrid templates, Chem. Mater. 11 (11) (1999) 3080-3085.

[154] J.C. Maxwell Garnett B.A., Colours in metal glasses and in metallic films, Philos. Trans. Royal Soc. Ser. A 203 (1904) 385-420.

[155] Z.-M. Dang, L.-J. Ma, J.-W. Zha, S.-H. Yao, D. Xie, Q. Chen, et al., Origin of ultralow permittivity in polyimide/mesoporous silicate nanohybrid films with high resistivity and high breakdown strength, J. Appl. Phys. 105 (4) (2009) 044104.

[156] Y.H. Zhang, S.G. Lu, Y.Q. Li, Z.M. Dang, J.H. Xin, S.Y. Fu, et al., Novel silica tube/polyimide composite films with variable low dielectric Constant, Adv. Mater. 17 (8) (2005) 1056-1059.

[157] F. Miyaji, Y. Watanabe, Y. Suyama, Morphology of silica derived from various ammonium carboxylate templates, Mater. Res. Bull. 38 (13) (2003) 1669-1680.

[158] L.-Y. Jiang, C.-M. Leu, K.-H. Wei, Layered silicates/fluorinated polyimide nanocomposites for advanced dielectric materials applications, Adv. Mater. 14 (6) (2002) 426-429.

[159] V.E. Yudin, J.U. Otaigbe, S. Gladchenko, B.G. Olson, S. Nazarenko, E.N. Korytkova, et al., New polyimide nanocomposites based on silicate type nanotubes: dispersion, processing and properties, Polymer 48 (5) (2007) 1306-1315.

[160] J. Lin, X. Wang, Novel low-*k* polyimide/mesoporous silica composite films: preparation, microstructure, and properties, Polymer 48 (1) (2007) 318-329.

[161] Q. Li, Y. Wang, S. Zhang, L. Pang, H. Tong, J. Li, et al., Novel fluorinated random co-polyimide/amine-functionalized zeolite MEL50 hybrid films with enhanced thermal and low dielectric properties, J. Mater. Sci. 52 (9) (2017) 5283-5296.

[162] S.S. Park, C.-S. Ha, Polyimide/hollow silica sphere hybrid films with low dielectric constant, Compos. Interfaces 23 (8) (2016) 831-846.

[163] Z. Huang, S. Liu, Y. Yuan, J. Zhao, High-performance fluorinated polyimide/pure silica zeolite nanocrystal hybrid films with a low dielectric constant, RSC Adv. 5 (93) (2015) 76476-76482.

[164] S. Kurinchyselvan, R. Sasikumar, M. Ariraman, P. Gomathipriya, M. Alagar, Low dielectric behavior of amine functionalized MCM-41 reinforced polyimide nanocomposites, High Perform. Polym. 28 (7) (2016) 842-853.

[165] Y.-S. Ye, Y.-C. Yen, W.-Y. Chen, C.-C. Cheng, F.-C. Chang, A simple approach toward low-dielectric polyimide nanocomposites: blending the polyimide precursor with a fluorinated polyhedral oligomeric silsesquioxane, J. Polym. Sci., Part A: Polym. Chem. 46 (18) (2008) 6296-6304.

[166] J. Kim, J. Kwon, M. Kim, J. Do, D. Lee, H. Han, Low-dielectric-constant polyimide aerogel composite films with low water uptake, Polym. J. 48 (7) (2016) 829-834.

[167] K.R. Carter, R.A. DiPietro, M. Sanchez, A. Sally Swanson, Nanoporous polyimides derived from highly fluorinated polyimide/poly(propylene oxide) copolymers, Chem. Mater. 13 (1) (2001) 213-221.

[168] W.C. Wang, R.H. Vora, E.T. Kang, K.G. Neoh, C.K. Ong, L.F. Chen, Nanoporous ultralow-*k* films prepared from fluorinated polyimide with grafted poly(acrylic acid) side chains, Adv. Mater. 16 (1) (2004) 54-57.

[169] Y.-J. Lee, J.-M. Huang, S.-W. Kuo, F.-C. Chang, Low-dielectric, nanoporous polyimide films prepared from PEO-POSS nanoparticles, Polymer 46 (23) (2005) 10056-10065.

[170] J.-W. Zha, H.-J. Jia, H.-Y. Wang, Z.-M. Dang, Tailored ultralow dielectric permittivity in high-performance fluorinated polyimide films by adjusting nanoporous characterisitics, J. Phys. Chem. C 116 (44) (2012) 23676-23681.

[171] S. Ma, Y. Wang, C. Liu, Q. Xu, Z. Min, Preparation and characterization of nanoporous polyimide membrane by the template method as low-*k* dielectric material, Polym. Adv. Technol. 27 (3) (2016) 414-418.

[172] P. Lv, Z. Dong, X. Dai, Y. Zhao, X. Qiu, Low-dielectric polyimide nanofoams derived from 4,4'-(hexafluoroisopropylidene) diphthalic anhydride and 2,2'-bis(trifluoromethyl)benzidine, RSC Adv. 7 (8) (2017) 4848-4854.

[173] S.V. Nitta, V. Pisupatti, A. Jain, J.P.C. Wayner, W.N. Gill, J.L. Plawsky, Surface modified spin-on xerogel films as interlayer dielectrics, J. Vacuum Sci. Technol. B 17 (1999) 205-212.

[174] P. Yan, B. Zhou, A. Du, Synthesis of polyimide crosslinked silica aerogels with good acoustic performance, RSC Adv. 4 (2014) 58252-58259.

[175] N. Leventis, C. Sotiriou-Leventis, N. Chandrasekaran, S. Mulik, Z.-J. Larimore, H.-B. Lu, et al., Multifunctional polyurea aerogels from isocyanates and water. A structure-property case study, Chem. Mater. 22 (2010) 6692-6710.

[176] A. Bang, C. Buback, C. Sotiriou-Leventis, N. Leventis, Flexible aerogels from hyperbranched polyurethanes: probing the role of molecular rigidity with poly(urethane acrylates) versus poly(urethane norbornenes), Chem. Mater. 26 (2014) 6979-6993.

[177] M.A.B. Meador, E.J. Malow, R. Silva, S. Wright, D. Quade, S.L. Vivod, et al., Mechanically strong, flexible polyimide aerogels crosslinked with aromatic triamine, ACS Appl. Mater. Interfaces 4 (2012) 536-544.

[178] H.Q. Guo, M.A.B. Meador, L. McCorkle, D.J. Quade, J. Guo, B. Hamilton, et al., Polyimide aerogels cross-linked through amine functionalized polyoligomeric silsesquioxane, ACS Appl. Mater. Interfaces 3 (2011) 546-552.

[179] M.A.B. Meador, C.R. Alemán, K. Hanson, N. Ramirez, S.L. Vivod, N. Wilmoth, et al., Polyimide aerogels with amide cross-liznks: a low cost alternative for mechanically strong polymer aerogels, ACS Appl. Mater. Interfaces 7 (2015) 1240-1249.

[180] P. Liu, T.-Q. Tran, Z. Fan, H.M. Duong, Formation mechanisms and morphological effects on multi-properties of carbon nanotube fibers and their polyimide aerogel-coated composites, Compos. Sci. Technol. 117 (2015) 114-120.

[181] M.A. Meador, S. Wright, A. Sandberg, B.N. Nguyen, F.W. Van Keuls, C.H. Mueller, et al., Low dielectric polyimide aerogels as substrates for lightweight patch antennas, ACS Appl. Mater. Interfaces 4 (11) (2012) 6346-6353.

[182] M.A. Meador, E. McMillon, A. Sandberg, E. Barrios, N.G. Wilmoth, C.H. Mueller, et al., Dielectric and other properties of polyimide aerogels containing fluorinated blocks, ACS Appl. Mater. Interfaces 6 (9) (2014) 6062-6068.

[183] D.X. Shen, J.G. Liu, H.X. Yang, S.Y. Yang, Intrinsically highly hydrophobic semi-alicyclic fluorinated polyimide aerogel with ultralow dielectric constants, Chem. Lett. 42 (10) (2013) 1230-1232.

[184] X.M. Zhang, J.G. Liu, S.Y. Yang, Synthesis and characterization of flexible and high-temperature resistant polyimide aerogel with ultra-low dielectric constant, eXPRESS Polym. Lett. 10 (10) (2016) 789-798.

[185] M. Seino, T. Hayakawa, Y. Ishida, M. Kakimoto, Hydrosilylation polymerization of double-decker-shaped silsesquioxane having hydrosilane with diynes, Macromolecules 39 (10) (2006) 3473-3475.

[186] R.H. Baney, M. Itoh, A. Sakakibara, T. Suzuki, Silsesquioxanes, Chem. Rev. 95 (1995) 1409-1430.

[187] R.M. Laine, Nanobuilding blocks based on the $[OSiO1.5]_x$ (x=6, 8, 10) octasilsesquioxanes, J. Mater. Chem. 15 (2005) 3725-3744.

[188] H.Y. Xu, B.H. Yang, J.F. Wang, S.Y. Guang, L. Cun, You have full text access to this content preparation, T_g improvement, and thermal stability enhancement mechanism of soluble poly(methyl methacrylate) nanocomposites by incorporating octavinyl polyhedral oligomeric silsesquioxanes, J. Polym. Sci., Part A: Polym. Chem. 44 (22) (2007) 5308-5317.

[189] A. Mariani, V. Alzari, O. Monticelli, J.A. Pojman, G. Caria, Polymeric nanocomposites containing polyhedral oligomeric silsesquioxanes prepared via frontal polymerization, J. Polym. Sci., Part A: Polym. Chem. 45 (19) (2007) 4514-4521.

[190] C.-M. Leu, Y.-T. Chang, K.-H. Wei, Synthesis and dielectric properties of polyimide -tethered polyhedral oligomeric silsesquioxane (POSS) nanocomposites via POSS-diamine, Macromolecules 36 (24) (2003) 9122-9127.

[191] C.M. Leu, Y.T. Chang, K.H. Wei, Polyimide-side-chain tethered polyhedral oligomeric silsesquioxane nanocomposites for low-dielectric film applications, Chem. Mater. 15 (2003) 3721-3727.

[192] Y.-J. Lee, J.-M. Huang, S.-W. Kuo, J.-S. Lu, F.-C. Chang, Polyimide and polyhedral oligomeric silsesquioxane nanocomposites for low-dielectric applications, Polymer 46 (1) (2005) 173-181.

[193] C.-y Wang, W.-t Chen, C. Xu, X.-y Zhao, J. Li, Fluorinated polyimide/POSS hybrid polymers with high solubility and low dielectric constant, Chinese J. Polym. Sci. 34 (11) (2016) 1363-1372.

[194] C.-M. Leu, G.M. Reddy, K.-H. Wei, S. Ching-Fong, Synthesis and dielectric properties of polyimide -chain-end tethered polyhedral oligomeric silsesquioxane nanocomposites, Chem. Mater. 15 (11) (2003) 2261-2265.

[195] J.D. Jacobs, M.J. Arlen, D.H. Wang, Z. Ounaies, R. Berry, L.-S. Tan, et al., Dielectric characteristics of polyimide CP2, Polymer 51 (14) (2010) 3139-3146.

[196] B. Krause, G.-H. Koops, N.F.A. van der Vegt, M. Wessling, M. Wübbenhorst, J. v Turnhout, Ultralow-k dielectrics made by supercritical foaming of thin polymer films, Adv. Mater. 14 (15) (2002) 1041-1046.

[197] G. Zhao, T. Ishizaka, H. Kasai, M. Hasegawa, T. Furukawa, H. Nakanishi, et al., Ultralow-delectric-constant films prepared from hollow polyimide nanoparticles possessing controllable core sizes, Chem. Mater. 21 (2) (2009) 419-424.

[198] Z. Li, H. Zou, P. Liu, Morphology and properties of porous polyimide films prepared through thermally induced phase separation, RSC Adv. 5 (47) (2015) 37837-37842.

[199] F. Chen, D. Bera, S. Banerjee, S. Agarwal, Low dielectric constant polyimide nanomats by electrospinning, Polym. Adv. Technol. 23 (6)(2012)951-957.

[200] P. Li, L. Liu, L. Ding, F. Lv, Y. Zhang, Thermal and dielectric properties of electrospun fiber membranes from polyimides with different structural units, J. Appl. Polym. Sci. 133 (9) (2016) 43081.

[201] L. Liu, F. Lv, P. Li, L. Ding, W. Tong, P.K. Chu, et al., Preparation of ultra-low dielectric constant silica/polyimide nanofiber membranes by electrospinning, Composites, Part A:84 (2016) 292-298.

[202] H. Chen, L. Xie, H. Lu, Y. Yang, Ultra-low-k polyimide hybrid films via copolymerization of polyimide and polyoxometalates, J. Mater. Chem. 17 (13) (2007) 1258-1261.

[203] M.-H. Tsai, W.-T. Whang, Low dielectric polyimide-poly(silsesquioxane)-like nanocomposite material, Polymer 42 (2001) 4197-4207.

[204] J.-Y. Wang, S.-Y. Yang, Y.-L. Huang, H.-W. Tien, W.-K. Chin, C.-C.M. Ma, Preparation and properties of graphene oxide/polyimide composite films with low dielectric constant and ultrahigh strength via in situ polymerization, J. Mater.

Chem. 21 (35) (2011) 13569-13575.

[205] X. Wang, Y. Dai, W. Wang, M. Ren, B. Li, C. Fan, et al., Fluorographene with high fluorine/carbon ratio: a nanofiller for preparing low-*k* polyimide hybrid films, ACS Appl. Mater. Interfaces 6 (18) (2014) 16182-16188.

[206] W.-H. Liao, S.-Y. Yang, S.-T. Hsiao, Y.-S. Wang, S.-M. Li, H.-W. Tien, et al., A novel approach to prepare graphene oxide/soluble polyimide composite films with a low dielectric constant and high mechanical properties, RSC Adv. 4 (93) (2014) 51117-51125.

[207] W.H. Liao, S.Y. Yang, S.T. Hsiao, Y.S. Wang, S.M. Li, C.C. Ma, et al., Effect of octa(aminophenyl) polyhedral oligomeric silsesquioxane functionalized graphene oxide on the mechanical and dielectric properties of polyimide composites, ACS Appl. Mater. Interfaces 6 (18) (2014) 15802-15812.

[208] S. Kim, X. Wang, S. Ando, X. Wang, Low dielectric and thermally stable hybrid ternary composites of hyperbranched and linear polyimides with SiO_2, RSC Adv. 4 (52) (2014) 27267-27276.

[209] S. Kim, X. Wang, S. Ando, X. Wang, Hybrid ternary composites of hyperbranched and linear polyimides with SiO_2: a research for low dielectric constant and optimized properties, RSC Adv. 4 (80) (2014) 42737-42746.

[210] Y. Li, J.Q. Zhao, Y.C. Yuan, C.Q. Shi, S.M. Liu, S.J. Yan, et al., Polyimide/crown ether composite films with necklace-like supramolecular structure and improved mechanical, dielectric, and hydrophobic properties, Macromolecules 48 (7) (2015) 2173-2183.

[211] P. Zhang, J. Zhao, K. Zhang, R. Bai, Y. Wang, C. Hua, et al., Fluorographene/polyimide composite films: mechanical, electrical, hydrophobic, thermal and low dielectric properties, Composites, Part A 84 (2016) 428-434.

[212] X. Chen, H. Huang, X. Shu, S. Liu, J. Zhao, Preparation and properties of a novel graphene fluoroxide/polyimide nanocomposite film with a low dielectric constant, RSC Adv. 7 (4) (2017) 1956-1965.

[213] X. Xu, T. Yang, Y. Yu, W. Xu, Y. Ding, H. Hou, Aqueous solution blending route for preparing low dielectric constant films of polyimide hybridized with polytetrafluoroethylene, J. Mater. Sci.: Mater. Electron. (2017). Available from: https://doi.org/10.1007/s10854-017-7093-1.

[214] X. Lei, Y. Chen, M. Qiao, L. Tian, Q. Zhang, Hyperbranched polysiloxane (HBPSi)-based polyimide films with ultralow dielectric permittivity, desirable mechanical and thermal properties, J. Mater. Chem. C 4 (11) (2016) 2134-2146.

[215] J. Dong, C. Yang, Y. Cheng, T. Wu, X. Zhao, Q. Zhang, Facile method for fabricating low dielectric constant polyimide fibers with hyperbranched polysiloxane, J. Mater. Chem. C 5 (11) (2017) 2818-2825.

[216] J.O. Simpson, A.K.S. Clair, Fundamental insight on developing low dielectric constant polyimides, Thin Solid Films 308-309 (1997) 480-485.

[217] R.H. Vora, P.S.G. Krishnan, S.H. Goh, T.-S. Chung, Synthesis and properties of designed low-*k* fluoro-copolyetherimides, Part 1. Adv. Funct. Mater. 11 (5) (2001) 361-373.

[218] Y. Watanabe, Y. Shibasaki, S. Ando, M. Ueda, Synthesis and characterization of polyimides with low dielectric constants from aromatic dianhydrides and aromatic diamine containing phenylene ether unit, Polymer 46 (16) (2005) 5903-5908.

[219] T.J. Barton, L.M. Bull, W.G. Klemperer, D.A. Loy, B. McEnaney, M. Misono, et al., Tailored porous materials, Chem. Mater. 11 (10) (1999) 2633-2656.

[220] U. Schubert, Polymers reinforced by covalently bonded inorganic clusters, Chem. Mater. 13 (10) (2001) 3487-3494.

[221] C. Sanchez, B. Julian, P. Belleville, M. Popall, Applications of hybrid organic-inorganic nanocomposites, J. Mater. Chem. 15 (35-36) (2005) 3559-3592.

[222] Y. Yuan, B.-P. Lin, X.-Q. Zhang, L.-W. Wu, Y. Zhan, Nonfluorinated copolymerized polyimide thin films with ultralow dielectric constants, J. Appl. Polym. Sci. 110 (3) (2008) 1515-1519.

[223] H. Deligöz, S. Ö zgümü¸s, T. Yalc¸inyuva, S. Yildirim, D. Dğer, K. Uluta¸s, A novel cross-linked polyimide film: synthesis and dielectric properties, Polymer 46 (11) (2005) 3720-3729.

[224] X. Lei, M. Qiao, L. Tian, Y. Chen, Q. Zhang, Tunable permittivity in high-performance hyperbranched polyimide films by adjusting backbone rigidity, J. Phys. Chem. C 120 (5) (2016) 2548-2561.

[225] Y.-T. Chern, H.-C. Shiue, Low dielectric constants of soluble polyimides based on adamantane, Macromolecules 30 (16) (1997) 4646-4651.

[226] Y.-T. Chern, H.-C. Shiue, Low dielectric constants of soluble polyimides derived from the novel 4,9-bis[4-(4-aminophenoxy) phenyl]diamantane, Macromolecules 30 (19) (1997) 5766-5772.

[227] Y.-T. Chern, Low dielectric constant polyimides derived from novel 1,6-bis[4-(4-aminophenoxy)phenyl]diamantane, Macromolecules 31 (17) (1998) 5837-5844.

[228] T.M. Long, T.M. Swager, Molecular design of free volume as a route to low-*k* dielectric materials, J. Am. Chem. Soc. 125 (46) (2003) 14113-14119.

[229] S.A. Sydlik, Z. Chen, T.M. Swager, Triptycene polyimides: soluble polymers with high thermal stability and low refractive indices, Macromolecules 44 (4) (2011) 976-980.

[230] D.M. Stoakley, A.K.S. Clair, C.I. Croall, low dielectric, fluorinated polyimide copolymers, J. Appl. Polym. Sci. 51 (1994)

1479-1483.

[231] C.-Y. Yang, S.L.-C. Hsu, J.S. Chen, Synthesis and properties of 6FDA-BisAAF-PPD copolyimides for microelectronic applications, J. Appl. Polym. Sci. 98 (5) (2005) 2064-2069.

[232] W. Dong, Y. Guan, D. Shang, Novel soluble polyimides containing pyridine and fluorinated units: preparation, characterization, and optical and dielectric properties, RSC Adv. 6 (26) (2016) 21662-21671.

[233] M. Jia, Y. Li, C. He, X. Huang, Soluble perfluorocyclobutyl aryl ether-based polyimide for high-performance dielectric material, ACS Appl. Mater. Interfaces 8 (39) (2016) 26352-26358.

[234] H. Nawaz, Z. Akhter, N. Iqbal, Study of physicochemical properties of meta and ortho trifluoromethyl substituted isomeric aromatic polyimides, Polym. Bull. 74 (9) (2017) 3889-3906.

[235] C. Wang, S. Cao, W. Chen, C. Xu, X. Zhao, J. Li, et al., Synthesis and properties of fluorinated polyimides with multi-bulky pendant groups, RSC Adv. 7 (42) (2017) 26420-26427.

[236] I.C. Tang, M.W. Wang, C.H. Wu, S.A. Dai, R.J. Jeng, C.H. Lin, A strategy for preparing spirobichroman dianhydride from bisphenol A and its resulting polyimide with low dielectric characteristic, RSC Adv. 7 (2) (2017) 1101-1109.

[237] D.-J. Liaw, W.-T. Tseng, New organosoluble polyimides with low dielectric constants derived from bis[4-(2-trifluoromethyl-4-aminophenoxy) phenyl] diphenylmethylene, Macromol. Symp. 199 (1) (2003) 351-362.

[238] Y. Liu, Y. Zhang, Q. Lan, S. Liu, Z. Qin, L. Chen, et al., High-performance functional polyimides containing rigid nonplanar conjugated triphenylethylene moieties, Chem. Mater. 24 (6) (2012) 1212-1222.

[239] Y. Liu, Y. Zhang, Q. Lan, Z. Qin, S. Liu, C. Zhao, et al., Synthesis and properties of high-performance functional polyimides containing rigid nonplanar conjugated tetraphenylethylene moieties, J. Polym. Sci., Part A: Polym. Chem. 51 (6) (2013) 1302-1314.

[240] W. Chen, Z. Zhou, T. Yang, R. Bei, Y. Zhang, S. Liu, et al., Synthesis and properties of highly organosoluble and low dielectric constant polyimides containing non-polar bulky triphenyl methane moiety, Reactive Funct. Polym. 108 (2016) 71-77.

[241] Y. Liu, C. Qian, L. Qu, Y. Wu, Y. Zhang, X. Wu, et al., A bulk dielectric polymer film with intrinsic ultralow dielectric constant and outstanding comprehensive properties, Chem. Mater. 27 (19) (2015) 6543-6549.

[242] Y. Liu, Z. Zhou, L. Qu, B. Zou, Z. Chen, Y. Zhang, et al., Exceptionally thermostable and soluble aromatic polyimides with special characteristics: intrinsic ultralow dielectric constant, static random access memory behaviors, transparency and fluorescence, Mater. Chem. Front. 1 (2) (2017) 326-337.

Appendix

Unit of Measurement Conversion Table

SI	CGS
Length	
$1m=10^{10}Å$	$1Å=10^{-10}m$
$1m=10^{9}nm$	$1nm=10^{-9}m$
$1m=10^{6}\mu m$	$1\mu m=10^{-6}m$
$1m=10^{3}mm$	$1mm=10^{-3}m$
$1m=10^{2}cm$	$1cm=10^{-2}m$
1mm=0.0394in	1in=25.4mm
1cm=0.394in	1in=2.54cm
1m=39.37in=3.28ft	1ft=12in=0.3048m
1mm=39.37mil	$1mil=10^{-3}in=0.0254mm$
$1\mu m=39.37\mu in$	$1\mu in=0.0254\mu m$
Area	
$1m^2=10^4cm^2$	$1cm^2=10^{-4}m^2$
$1cm^2=10^2mm^2$	$1mm^2=10^{-2}cm^2$
$1m^2=10.76ft^2$	$1ft^2=0.093m^2$
$1cm^2=0.1550in^2$	$1in^2=6.452cm^2$
Volume	
$1m^3=10^6cm^3$	$1cm^3=10^{-6}m^3$
$1cm^3=10^3mm^3$	$1mm^3=10^{-3}cm^3$
$1m^3=35.32ft^3$	$1ft^3=0.0283m^3$
$1cm^3=0.0610in^3$	$1in^3=16.39cm^3$
Mass	
$1t=10^3kg$	$1kg=10^{-3}t$
$1kg=10^3g$	$1g=10^{-3}kg$
$1kg=2.205lb_m$	$1lb_m=0.4536kg$
$1g=2.205\times10^{-3}lbm$	$1lb_m=453.6g$
1g=0.035oz	1oz=28.35g
Density	
$1kg/m^3=10^{-3}g/cm^3$	$1g/cm^3=10^3kg/m^3$
$1Mg/m^3=1g/cm^3$	$1g/cm^3=1Mg/m^3$
$lkg/m^3=0.0624lb_m/ft^3$	$1lb_m/ft^3=16.02kg/m^3$
$lg/cm^3=62.4lbm/ft^3$	$1lb_m/ft^3=1.602\times10^{-2}g/cm^3$
$lg/cm^3=0.0361lb_m/in^3$	$1lb_m/in^3=27.7g/cm^3$

(Continued)

(Continued)

Force	
1N=10^5dyn	1dyn=10^{-5}N
1N=0.2248lb_f	1lb_f=4.448N
Stress, Pressure	
1Pa=lN/m^2=10dyn/cm^2	1dyn/cm^2=0.10Pa
lMPa=145psi	1psi=1lb/in^2=6.90×10^{-3}MPa
1MPa=0.102kg/mm^2	1kg/mm^2=9.806MPa
lkg/mm^2=1422psi	1psi=7.03×10^{-4}kg/mm^2
lkPa=0.00987atm	1atm=101.325kPa
lkPa=0.01bar	1bar=100kPa
1Pa=0.0075torr=0.0075mmHg	1torr=1mmHg=133.322Pa
Fracture toughness	
1MPa$(m)^{1/2}$=910psi$(in)^{1/2}$	1psi$(in)^{1/2}$=1.099×10^{-3} MPa$(m)^{1/2}$
Energy,Work,Heat	
1J=10^7erg	1erg=10^{-7}J
1J=6.24×10^{18}eV	1eV=1.602×10^{-19}J
1J=0.239cal	1cal=4.184J
1J=9.48×10^{-4}Btu	1Btu=1054J
1J=0.738ft·lbf	1ft·1bf=1.356J
1eV=3.83×10^{-20}cal	lcal=2.61×10^{19}eV
1cal=3. 97×10^{-3}Btu	1Btu=252.0cal
Power	
1W=0.239cal/s	1cal/s=4.184W
1W=3.414Btu/h	1Btu/h=0.293W
1cal/s=14.29Btu/h	1Btu/h=0.070cal/s
Viscosity	
1Pa·s=10P	1P=0.1Pa·s
1mPa·s=1cP	1cP=10^{-3}Pa·s
Temperature	
T(K)=273+T(℃)	T(℃)=T(K)−273
T(K)=5/9[T(℉)−32]+273	T(℉)=9/5[T(K)−273]+32
T(℃)=5/9[T(℉)−32]	T(℉)=9/5T(℃)+32
Specific heat capacity	
1J/(kg·K)=2. 39×10^{-4}cal/(g·K)	1cal/(g·K)=4184J/(kg·K)
1J/(kg·K)=2.39×10^{-4}Btu/(lbm·℉)	1Btu/(lbm·℉)=4184J/(kg·K)
1cal/(g·℃)=1.0Btu/(lbm·℉)	1Btu/(lbm·℉)=1.0cal/(g·K)
Thermal conductivity	
1W/(m·K)=2.39×10^{-3}cal/(cm·s·K)	1cal/(cm·s·K)=418.4W/(m·K)
1W/(m·K)=0.578Btu/(ft·h·℉)	1Btu/(ft·h·℉)=1.730W/(m·K)
1cal/(cm·s·K)=241.8Btu/(ft·h·℉)	1Btu/(ft·h·℉)=4.136×10^{-3}cal/(cm·s .K)
Electromagnetism	
1A=4π/10 Gb	1Gb=(Gilbert)=10/4π A
1A/m=4π/10^3 Oe	1Oe(Oesterd)=10^3/4π A/m
1Wb(Weber)=108Mx	1Mx(Maxwell)=10^{-8}Wb
1T(Tesla)=1Wb/m^2=10^4G	1G(Gauss)=10^{-4}Wb/m^2=10^{-4}T
Other	
1Gy=1J/kg=10^2rad	1rad=10^2erg/g=10^{-2}Gy

(Continued)

(Continued)

Unit Symbol		
A=Ampere	Gb=Gibb	mm=millimeter
Å=Ångstrom	Gy=Gray	N=Newton
bar=bar	h=hour	nm=nanometer
Btu=British Thermal Unit	H=Henry	Oe=Ørsted
C=Coulomb	Hz=Hertz	psi=pounds per square inch
℃=degree Celsius	in=inch	P=Poise
cal=calorie	J=Joule	Pa=Pascal
cm=centimeter	K=Kelvins	rad= radiation absorbed dose
cP=centipoise	kg=kilogram	s=second
dB=decibel	ksi= kilopounds per square inch	S=Siemens
dyn=dyne	L=Liter	T=Tesla
erg=erg	lbf=lbf	Torr=Torr
eV=electron volt	lbm=lbm	μm=micron
F=farad	m=meter	V=Volt
℉=Fahrenheit	Mg=Megagram	W=Watt
ft=foot	mil=mil	Wb=Web
g=gram	min=minut	Ω=Ohm

Prefix Symbols in SI		
Factor	Prefix	Symbol
10^{9}	giga	G
10^{6}	mega	M
10^{3}	kilo	k
10^{-2}	centi	c
10^{-3}	milli	m
10^{-6}	mirco	μ
10^{-9}	nano	n
10^{-12}	pico	p